Applications of Mathematics

1 Fleming/Rishel, **Deterministic and Stochastic Optimal Control** (1975)
2 Marchuk, **Methods of Numerical Mathematics,** Second Ed. (1982)
3 Balakrishnan, **Applied Functional Analysis,** Second Ed. (1981)
4 Borovkov, **Stochastic Processes in Queueing Theory** (1976)
5 Liptser/Shiryayev, **Statistics of Random Processes I: General Theory,** Second Ed. (1977)
6 Liptser/Shiryayev, **Statistics of Random Processes II: Applications,** Second Ed. (1978)
7 Vorob'ev, **Game Theory: Lectures for Economists and Systems Scientists** (1977)
8 Shiryayev, **Optimal Stopping Rules** (1978)
9 Ibragimov/Rozanov, **Gaussian Random Processes** (1978)
10 Wonham, **Linear Multivariable Control: A Geometric Approach,** Third Ed. (1985)
11 Hida, **Brownian Motion** (1980)
12 Hestenes, **Conjugate Direction Methods in Optimization** (1980)
13 Kallianpur, **Stochastic Filtering Theory** (1980)
14 Krylov, **Controlled Diffusion Processes** (1980)
15 Prabhu, **Stochastic Storage Processes: Queues, Insurance Risk, Dams, and Data Communication,** Second Ed. (1998)
16 Ibragimov/Has'minskii, **Statistical Estimation: Asymptotic Theory** (1981)
17 Cesari, **Optimization: Theory and Applications** (1982)
18 Elliott, **Stochastic Calculus and Applications** (1982)
19 Marchuk/Shaidourov, **Difference Methods and Their Extrapolations** (1983)
20 Hijab, **Stabilization of Control Systems** (1986)
21 Protter, **Stochastic Integration and Differential Equations** (1990)
22 Benveniste/Métivier/Priouret, **Adaptive Algorithms and Stochastic Approximations** (1990)
23 Kloeden/Platen, **Numerical Solution of Stochastic Differential Equations** (1992)
24 Kushner/Dupuis, **Numerical Methods for Stochastic Control Problems in Continuous Time,** Second Ed. (2001)
25 Fleming/Soner, **Controlled Markov Processes and Viscosity Solutions** (1993)
26 Baccelli/Brémaud, **Elements of Queueing Theory** (1994)
27 Winkler, **Image Analysis, Random Fields, and Dynamic Monte Carlo Methods: An Introduction to Mathematical Aspects** (1994)
28 Kalpazidou, **Cycle Representations of Markov Processes** (1995)
29 Elliott/Aggoun/Moore, **Hidden Markov Models: Estimation and Control** (1995)
30 Hernández-Lerma/Lasserre, **Discrete-Time Markov Control Processes: Basic Optimality Criteria** (1996)
31 Devroye/Györfi/Lugosi, **A Probabilistic Theory of Pattern Recognition** (1996)
32 Maitra/Sudderth, **Discrete Gambling and Stochastic Games** (1996)
33 Embrechts/Klüppelberg/Mikosch, **Modelling Extremal Events** (1997)
34 Duflo, **Random Iterative Models** (1997)

(continued after index)

Paul Glasserman

Monte Carlo Methods in Financial Engineering

With 99 Figures

 Springer

Paul Glasserman
403 Uris Hall
Graduate School of Business
Columbia University
New York, NY 10027, USA
pg20@columbia.edu

Managing Editors

B. Rozovskii
Denney Research Building 308
Center for Applied Mathematical
  Sciences
University of Southern California
1042 West Thirty-Sixth Place
Los Angeles, CA 90089, USA
rozovski@math.usc.edu

M. Yor
Laboratoire de Probabilités
  et Modèles Aléatoires
Université de Paris VI
175, rue du Chevaleret
75013 Paris, France

Cover illustration: Cover pattern by courtesy of Rick Durrett, Cornell University, Ithaca, New York.

Mathematics Subject Classification (2000): 65C05, 91Bxx

Library of Congress Cataloging-in-Publication Data
Glasserman, Paul, 1962–
    Monte Carlo methods in financial engineering / Paul Glasserman.
       p. cm. — (Applications of mathematics ; 53)
    Includes bibliographical references and index.
    ISBN 0-387-00451-3 (alk. paper)
    1.   Financial engineering. 2. Derivative securities. 3. Monte Carlo method. I. Title.
    II. Series.
    HG176.7.G57   2003
    658.15`5`01519282—dc21   2003050499

ISBN 0-387-00451-3   Printed on acid-free paper.

Printed in the United States of America.   (EB)

9 8 7 6 5

springer.com

Preface

This is a book about Monte Carlo methods from the perspective of financial engineering. Monte Carlo simulation has become an essential tool in the pricing of derivative securities and in risk management; these applications have, in turn, stimulated research into new Monte Carlo techniques and renewed interest in some old techniques. This is also a book about financial engineering from the perspective of Monte Carlo methods. One of the best ways to develop an understanding of a model of, say, the term structure of interest rates is to implement a simulation of the model; and finding ways to improve the efficiency of a simulation motivates a deeper investigation into properties of a model.

My intended audience is a mix of graduate students in financial engineering, researchers interested in the application of Monte Carlo methods in finance, and practitioners implementing models in industry. This book has grown out of lecture notes I have used over several years at Columbia, for a semester at Princeton, and for a short course at Aarhus University. These classes have been attended by masters and doctoral students in engineering, the mathematical and physical sciences, and finance. The selection of topics has also been influenced by my experiences in developing and delivering professional training courses with Mark Broadie, often in collaboration with Leif Andersen and Phelim Boyle. The opportunity to discuss the use of Monte Carlo methods in the derivatives industry with practitioners and colleagues has helped shaped my thinking about the methods and their application.

Students and practitioners come to the area of financial engineering from diverse academic fields and with widely ranging levels of training in mathematics, statistics, finance, and computing. This presents a challenge in setting the appropriate level for discourse. The most important prerequisite for reading this book is familiarity with the mathematical tools routinely used to specify and analyze continuous-time models in finance. Prior exposure to the basic principles of option pricing is useful but less essential. The tools of mathematical finance include Itô calculus, stochastic differential equations, and martingales. Perhaps the most advanced idea used in many places in this book is the concept of a change of measure. This idea is so central both to derivatives pricing and to Monte Carlo methods that there is simply no avoiding it. The prerequisites to understanding the statement of the Girsanov theorem should suffice for reading this book.

Whereas the language of mathematical finance is essential to our topic, its technical subtleties are less so for purposes of computational work. My use of mathematical tools is often informal: I may assume that a local martingale is a martingale or that a stochastic differential equation has a solution, for example, without calling attention to these assumptions. Where convenient, I take derivatives without first assuming differentiability and I take expectations without verifying integrability. My intent is to focus on the issues most important to Monte Carlo methods and to avoid diverting the discussion to spell out technical conditions. Where these conditions are not evident and where they are essential to understanding the scope of a technique, I discuss them explicitly. In addition, an appendix gives precise statements of the most important tools from stochastic calculus.

This book divides roughly into three parts. The first part, Chapters 1–3, develops fundamentals of Monte Carlo methods. Chapter 1 summarizes the theoretical foundations of derivatives pricing and Monte Carlo. It explains the principles by which a pricing problem can be formulated as an integration problem to which Monte Carlo is then applicable. Chapter 2 discusses random number generation and methods for sampling from nonuniform distributions, tools fundamental to every application of Monte Carlo. Chapter 3 provides an overview of some of the most important models used in financial engineering and discusses their implementation by simulation. I have included more discussion of the models in Chapter 3 and the financial underpinnings in Chapter 1 than is strictly necessary to run a simulation. Students often come to a course in Monte Carlo with limited exposure to this material, and the implementation of a simulation becomes more meaningful if accompanied by an understanding of a model and its context. Moreover, it is precisely in model details that many of the most interesting simulation issues arise.

If the first three chapters deal with running a simulation, the next three deal with ways of running it better. Chapter 4 presents methods for increasing precision by reducing the variance of Monte Carlo estimates. Chapter 5 discusses the application of deterministic *quasi*-Monte Carlo methods for numerical integration. Chapter 6 addresses the problem of discretization error that results from simulating discrete-time approximations to continuous-time models.

The last three chapters address topics specific to the application of Monte Carlo methods in finance. Chapter 7 covers methods for estimating price sensitivities or "Greeks." Chapter 8 deals with the pricing of American options, which entails solving an optimal stopping problem within a simulation. Chapter 9 is an introduction to the use of Monte Carlo methods in risk management. It discusses the measurement of market risk and credit risk in financial portfolios. The models and methods of this final chapter are rather different from those in the other chapters, which deal primarily with the pricing of derivative securities.

Several people have influenced this book in various ways and it is my pleasure to express my thanks to them here. I owe a particular debt to my frequent collaborators and co-authors Mark Broadie, Phil Heidelberger, and Perwez Shahabuddin. Working with them has influenced my thinking as well as the book's contents. With Mark Broadie I have had several occasions to collaborate on teaching as well as research, and I have benefited from our many discussions on most of the topics in this book. Mark, Phil Heidelberger, Steve Kou, Pierre L'Ecuyer, Barry Nelson, Art Owen, Philip Protter, and Jeremy Staum each commented on one or more draft chapters; I thank them for their comments and apologize for the many good suggestions I was unable to incorporate fully. I have also benefited from working with current and former Columbia students Jingyi Li, Nicolas Merener, Jeremy Staum, Hui Wang, Bin Yu, and Xiaoliang Zhao on some of the topics in this book. Several classes of students helped uncover errors in the lecture notes from which this book evolved.

Paul Glasserman
New York, 2003

Contents

1 Foundations 1
  1.1 Principles of Monte Carlo 1
    1.1.1 Introduction 1
    1.1.2 First Examples 3
    1.1.3 Efficiency of Simulation Estimators 9
  1.2 Principles of Derivatives Pricing 19
    1.2.1 Pricing and Replication 21
    1.2.2 Arbitrage and Risk-Neutral Pricing 25
    1.2.3 Change of Numeraire 32
    1.2.4 The Market Price of Risk 36

2 Generating Random Numbers and Random Variables 39
  2.1 Random Number Generation 39
    2.1.1 General Considerations 39
    2.1.2 Linear Congruential Generators 43
    2.1.3 Implementation of Linear Congruential Generators ... 44
    2.1.4 Lattice Structure 47
    2.1.5 Combined Generators and Other Methods 49
  2.2 General Sampling Methods 53
    2.2.1 Inverse Transform Method 54
    2.2.2 Acceptance-Rejection Method 58
  2.3 Normal Random Variables and Vectors 63
    2.3.1 Basic Properties 63
    2.3.2 Generating Univariate Normals 65
    2.3.3 Generating Multivariate Normals 71

3 Generating Sample Paths 79
  3.1 Brownian Motion 79
    3.1.1 One Dimension 79
    3.1.2 Multiple Dimensions 90
  3.2 Geometric Brownian Motion 93

| | 3.2.1 | Basic Properties | 93 |

3.2.1   Basic Properties 93
3.2.2   Path-Dependent Options 96
3.2.3   Multiple Dimensions 104
3.3   Gaussian Short Rate Models 108
3.3.1   Basic Models and Simulation 108
3.3.2   Bond Prices 111
3.3.3   Multifactor Models 118
3.4   Square-Root Diffusions 120
3.4.1   Transition Density 121
3.4.2   Sampling Gamma and Poisson 125
3.4.3   Bond Prices 128
3.4.4   Extensions 131
3.5   Processes with Jumps 134
3.5.1   A Jump-Diffusion Model 134
3.5.2   Pure-Jump Processes 142
3.6   Forward Rate Models: Continuous Rates 149
3.6.1   The HJM Framework 150
3.6.2   The Discrete Drift 155
3.6.3   Implementation 160
3.7   Forward Rate Models: Simple Rates 165
3.7.1   LIBOR Market Model Dynamics 166
3.7.2   Pricing Derivatives 172
3.7.3   Simulation 174
3.7.4   Volatility Structure and Calibration 180

4   Variance Reduction Techniques 185
4.1   Control Variates 185
4.1.1   Method and Examples 185
4.1.2   Multiple Controls 196
4.1.3   Small-Sample Issues 200
4.1.4   Nonlinear Controls 202
4.2   Antithetic Variates 205
4.3   Stratified Sampling 209
4.3.1   Method and Examples 209
4.3.2   Applications 220
4.3.3   Poststratification 232
4.4   Latin Hypercube Sampling 236
4.5   Matching Underlying Assets 243
4.5.1   Moment Matching Through Path Adjustments 244
4.5.2   Weighted Monte Carlo 251
4.6   Importance Sampling 255
4.6.1   Principles and First Examples 255
4.6.2   Path-Dependent Options 267
4.7   Concluding Remarks 276

5 Quasi-Monte Carlo 281
  5.1 General Principles 281
    5.1.1 Discrepancy 283
    5.1.2 Van der Corput Sequences 285
    5.1.3 The Koksma-Hlawka Bound 287
    5.1.4 Nets and Sequences 290
  5.2 Low-Discrepancy Sequences 293
    5.2.1 Halton and Hammersley 293
    5.2.2 Faure 297
    5.2.3 Sobol' 303
    5.2.4 Further Constructions 314
  5.3 Lattice Rules 316
  5.4 Randomized QMC 320
  5.5 The Finance Setting 323
    5.5.1 Numerical Examples 323
    5.5.2 Strategic Implementation 331
  5.6 Concluding Remarks 335

6 Discretization Methods 339
  6.1 Introduction 339
    6.1.1 The Euler Scheme and a First Refinement 339
    6.1.2 Convergence Order 344
  6.2 Second-Order Methods 348
    6.2.1 The Scalar Case 348
    6.2.2 The Vector Case 351
    6.2.3 Incorporating Path-Dependence 357
    6.2.4 Extrapolation 360
  6.3 Extensions 362
    6.3.1 General Expansions 362
    6.3.2 Jump-Diffusion Processes 363
    6.3.3 Convergence of Mean Square Error 365
  6.4 Extremes and Barrier Crossings: Brownian Interpolation 366
  6.5 Changing Variables 371
  6.6 Concluding Remarks 375

7 Estimating Sensitivities 377
  7.1 Finite-Difference Approximations 378
    7.1.1 Bias and Variance 378
    7.1.2 Optimal Mean Square Error 381
  7.2 Pathwise Derivative Estimates 386
    7.2.1 Method and Examples 386
    7.2.2 Conditions for Unbiasedness 393
    7.2.3 Approximations and Related Methods 396
  7.3 The Likelihood Ratio Method 401
    7.3.1 Method and Examples 401

      7.3.2   Bias and Variance Properties 407
      7.3.3   Gamma 411
      7.3.4   Approximations and Related Methods 413
   7.4 Concluding Remarks 418

8 **Pricing American Options** 421
   8.1 Problem Formulation 421
   8.2 Parametric Approximations 426
   8.3 Random Tree Methods 430
      8.3.1   High Estimator 432
      8.3.2   Low Estimator 434
      8.3.3   Implementation 437
   8.4 State-Space Partitioning 441
   8.5 Stochastic Mesh Methods 443
      8.5.1   General Framework 443
      8.5.2   Likelihood Ratio Weights 450
   8.6 Regression-Based Methods and Weights 459
      8.6.1   Approximate Continuation Values 459
      8.6.2   Regression and Mesh Weights 465
   8.7 Duality 470
   8.8 Concluding Remarks 478

9 **Applications in Risk Management** 481
   9.1 Loss Probabilities and Value-at-Risk 481
      9.1.1   Background 481
      9.1.2   Calculating VAR 484
   9.2 Variance Reduction Using the Delta-Gamma Approximation .. 492
      9.2.1   Control Variate 493
      9.2.2   Importance Sampling 495
      9.2.3   Stratified Sampling 500
   9.3 A Heavy-Tailed Setting 506
      9.3.1   Modeling Heavy Tails 506
      9.3.2   Delta-Gamma Approximation 512
      9.3.3   Variance Reduction 514
   9.4 Credit Risk 520
      9.4.1   Default Times and Valuation 520
      9.4.2   Dependent Defaults 525
      9.4.3   Portfolio Credit Risk 529
   9.5 Concluding Remarks 535

A **Appendix: Convergence and Confidence Intervals** 539
   A.1 Convergence Concepts 539
   A.2 Central Limit Theorem and Confidence Intervals 541

B   Appendix: Results from Stochastic Calculus 545
    B.1 Itô's Formula 545
    B.2 Stochastic Differential Equations 548
    B.3 Martingales 550
    B.4 Change of Measure 553

C   Appendix: The Term Structure of Interest Rates 559
    C.1 Term Structure Terminology 559
    C.2 Interest Rate Derivatives 564

References 569

Index 587

1

Foundations

This chapter's two parts develop key ideas from two fields, the intersection of which is the topic of this book. Section 1.1 develops principles underlying the use and analysis of Monte Carlo methods. It begins with a general description and simple examples of Monte Carlo, and then develops a framework for measuring the efficiency of Monte Carlo estimators. Section 1.2 reviews concepts from the theory of derivatives pricing, including pricing by replication, the absence of arbitrage, risk-neutral probabilities, and market completeness. The most important idea for our purposes is the representation of derivative prices as expectations, because this representation underlies the application of Monte Carlo.

1.1 Principles of Monte Carlo

1.1.1 Introduction

Monte Carlo methods are based on the analogy between probability and volume. The mathematics of measure formalizes the intuitive notion of probability, associating an event with a set of outcomes and defining the probability of the event to be its volume or measure relative to that of a universe of possible outcomes. Monte Carlo uses this identity in reverse, calculating the volume of a set by interpreting the volume as a probability. In the simplest case, this means sampling randomly from a universe of possible outcomes and taking the fraction of random draws that fall in a given set as an estimate of the set's volume. The law of large numbers ensures that this estimate converges to the correct value as the number of draws increases. The central limit theorem provides information about the likely magnitude of the error in the estimate after a finite number of draws.

A small step takes us from volumes to integrals. Consider, for example, the problem of estimating the integral of a function f over the unit interval. We may represent the integral

$$\alpha = \int_0^1 f(x)\, dx$$

as an expectation $\mathsf{E}[f(U)]$, with U uniformly distributed between 0 and 1. Suppose we have a mechanism for drawing points $U_1, U_2, \ldots$ independently and uniformly from $[0, 1]$. Evaluating the function f at n of these random points and averaging the results produces the Monte Carlo estimate

$$\hat{\alpha}_n = \frac{1}{n} \sum_{i=1}^{n} f(U_i).$$

If f is indeed integrable over $[0, 1]$ then, by the strong law of large numbers,

$$\hat{\alpha}_n \to \alpha \quad \text{with probability 1 as } n \to \infty.$$

If f is in fact square integrable and we set

$$\sigma_f^2 = \int_0^1 (f(x) - \alpha)^2\, dx,$$

then the error $\hat{\alpha}_n - \alpha$ in the Monte Carlo estimate is approximately normally distributed with mean 0 and standard deviation $\sigma_f / \sqrt{n}$, the quality of this approximation improving with increasing n. The parameter σ_f would typically be unknown in a setting in which α is unknown, but it can be estimated using the sample standard deviation

$$s_f = \sqrt{\frac{1}{n-1} \sum_{i=1}^{n} (f(U_i) - \hat{\alpha}_n)^2}.$$

Thus, from the function values $f(U_1), \ldots, f(U_n)$ we obtain not only an estimate of the integral α but also a measure of the error in this estimate.

The form of the standard error $\sigma_f / \sqrt{n}$ is a central feature of the Monte Carlo method. Cutting this error in half requires increasing the number of points by a factor of four; adding one decimal place of precision requires 100 times as many points. These are tangible expressions of the square-root convergence rate implied by the $\sqrt{n}$ in the denominator of the standard error. In contrast, the error in the simple trapezoidal rule

$$\alpha \approx \frac{f(0) + f(1)}{2n} + \frac{1}{n} \sum_{i=1}^{n-1} f(i/n)$$

is $O(n^{-2})$, at least for twice continuously differentiable f. Monte Carlo is generally not a competitive method for calculating one-dimensional integrals.

The value of Monte Carlo as a computational tool lies in the fact that its $O(n^{-1/2})$ convergence rate is not restricted to integrals over the unit interval.

Indeed, the steps outlined above extend to estimating an integral over $[0, 1]^d$ (and even $\Re^d$) for all dimensions d. Of course, when we change dimensions we change f and when we change f we change σ_f, but the standard error will still have the form $\sigma_f/\sqrt{n}$ for a Monte Carlo estimate based on n draws from the domain $[0, 1]^d$. In particular, the $O(n^{-1/2})$ convergence rate holds for all d. In contrast, the error in a product trapezoidal rule in d dimensions is $O(n^{-2/d})$ for twice continuously differentiable integrands; this degradation in convergence rate with increasing dimension is characteristic of all deterministic integration methods. Thus, Monte Carlo methods are attractive in evaluating integrals in high dimensions.

What does this have to do with financial engineering? A fundamental implication of asset pricing theory is that under certain circumstances (reviewed in Section 1.2.1), the price of a derivative security can be usefully represented as an expected value. Valuing derivatives thus reduces to computing expectations. In many cases, if we were to write the relevant expectation as an integral, we would find that its dimension is large or even infinite. This is precisely the sort of setting in which Monte Carlo methods become attractive.

Valuing a derivative security by Monte Carlo typically involves simulating paths of stochastic processes used to describe the evolution of underlying asset prices, interest rates, model parameters, and other factors relevant to the security in question. Rather than simply drawing points randomly from $[0, 1]$ or $[0, 1]^d$, we seek to sample from a space of paths. Depending on how the problem and model are formulated, the dimension of the relevant space may be large or even infinite. The dimension will ordinarily be at least as large as the number of time steps in the simulation, and this could easily be large enough to make the square-root convergence rate for Monte Carlo competitive with alternative methods.

For the most part, there is nothing we can do to overcome the rather slow rate of convergence characteristic of Monte Carlo. (The quasi-Monte Carlo methods discussed in Chapter 5 are an exception — under appropriate conditions they provide a faster convergence rate.) We can, however, look for superior sampling methods that reduce the implicit constant in the convergence rate. Much of this book is devoted to examples and general principles for doing this.

The rest of this section further develops some essential ideas underlying Monte Carlo methods and their application to financial engineering. Section 1.1.2 illustrates the use of Monte Carlo with two simple types of option contracts. Section 1.1.3 develops a framework for evaluating the efficiency of simulation estimators.

1.1.2 First Examples

In discussing general principles of Monte Carlo, it is useful to have some simple specific examples to which to refer. As a first illustration of a Monte Carlo method, we consider the calculation of the expected present value of the payoff of a call option on a stock. We do not yet refer to this as the option *price*; the connection between a price and an expected discounted payoff is developed in Section 1.2.1.

Let $S(t)$ denote the price of the stock at time t. Consider a call option granting the holder the right to buy the stock at a fixed price K at a fixed time T in the future; the current time is $t = 0$. If at time T the stock price $S(T)$ exceeds the strike price K, the holder exercises the option for a profit of $S(T) - K$; if, on the other hand, $S(T) \leq K$, the option expires worthless. (This is a *European* option, meaning that it can be exercised only at the fixed date T; an *American* option allows the holder to choose the time of exercise.) The payoff to the option holder at time T is thus

$$(S(T) - K)^+ = \max\{0, S(T) - K\}.$$

To get the present value of this payoff we multiply by a discount factor e^{-rT}, with r a continuously compounded interest rate. We denote the expected present value by $\mathsf{E}[e^{-rT}(S(T) - K)^+]$.

For this expectation to be meaningful, we need to specify the distribution of the random variable $S(T)$, the terminal stock price. In fact, rather than simply specifying the distribution at a fixed time, we introduce a model for the dynamics of the stock price. The Black-Scholes model describes the evolution of the stock price through the stochastic differential equation (SDE)

$$\frac{dS(t)}{S(t)} = r\,dt + \sigma\,dW(t), \tag{1.1}$$

with W a standard Brownian motion. (For a brief review of stochastic calculus, see Appendix B.) This equation may be interpreted as modeling the percentage changes dS/S in the stock price as the increments of a Brownian motion. The parameter σ is the volatility of the stock price and the coefficient on dt in (1.1) is the mean rate of return. In taking the rate of return to be the same as the interest rate r, we are implicitly describing the *risk-neutral* dynamics of the stock price, an idea reviewed in Section 1.2.1.

The solution of the stochastic differential equation (1.1) is

$$S(T) = S(0)\exp\left([r - \tfrac{1}{2}\sigma^2]T + \sigma W(T)\right). \tag{1.2}$$

As $S(0)$ is the current price of the stock, we may assume it is known. The random variable $W(T)$ is normally distributed with mean 0 and variance T; this is also the distribution of $\sqrt{T}Z$ if Z is a standard normal random variable (mean 0, variance 1). We may therefore represent the terminal stock price as

$$S(T) = S(0)\exp\left([r - \tfrac{1}{2}\sigma^2]T + \sigma\sqrt{T}Z\right). \tag{1.3}$$

The logarithm of the stock price is thus normally distributed, and the stock price itself has a lognormal distribution.

The expectation $\mathsf{E}[e^{-rT}(S(T) - K)^+]$ is an integral with respect to the lognormal density of $S(T)$. This integral can be evaluated in terms of the standard normal cumulative distribution function Φ as $\mathrm{BS}(S(0), \sigma, T, r, K)$ with

$$\mathrm{BS}(S, \sigma, T, r, K) =$$
$$S\Phi\left(\frac{\log(S/K) + (r + \frac{1}{2}\sigma^2)T}{\sigma\sqrt{T}}\right) - e^{-rT}K\Phi\left(\frac{\log(S/K) + (r - \frac{1}{2}\sigma^2)T}{\sigma\sqrt{T}}\right). \quad (1.4)$$

This is the Black-Scholes [50] formula for a call option.

In light of the availability of this formula, there is no need to use Monte Carlo to compute $\mathsf{E}[e^{-rT}(S(T) - K)^+]$. Moreover, we noted earlier that Monte Carlo is not a competitive method for computing one-dimensional integrals. Nevertheless, we now use this example to illustrate the key steps in Monte Carlo. From (1.3) we see that to draw samples of the terminal stock price $S(T)$ it suffices to have a mechanism for drawing samples from the standard normal distribution. Methods for doing this are discussed in Section 2.3; for now we simply assume the ability to produce a sequence $Z_1, Z_2, \ldots$ of independent standard normal random variables. Given a mechanism for generating the Z_i, we can estimate $\mathsf{E}[e^{-rT}(S(T) - K)^+]$ using the following algorithm:

> for $i = 1, \ldots, n$
>   generate Z_i
>   set $S_i(T) = S(0) \exp\left([r - \frac{1}{2}\sigma^2]T + \sigma\sqrt{T}Z_i\right)$
>   set $C_i = e^{-rT}(S(T) - K)^+$
> set $\hat{C}_n = (C_1 + \cdots + C_n)/n$

For any $n \geq 1$, the estimator $\hat{C}_n$ is *unbiased*, in the sense that its expectation is the target quantity:

$$\mathsf{E}[\hat{C}_n] = C \equiv \mathsf{E}[e^{-rT}(S(T) - K)^+].$$

The estimator is *strongly consistent*, meaning that as $n \to \infty$,

$$\hat{C}_n \to C \quad \text{with probability 1.}$$

For finite but at least moderately large n, we can supplement the point estimate $\hat{C}_n$ with a confidence interval. Let

$$s_C = \sqrt{\frac{1}{n-1}\sum_{i=1}^{n}(C_i - \hat{C}_n)^2} \quad (1.5)$$

denote the sample standard deviation of $C_1, \ldots, C_n$ and let z_δ denote the $1 - \delta$ quantile of the standard normal distribution (i.e., $\Phi(z_\delta) = 1 - \delta$). Then

$$\hat{C}_n \pm z_{\delta/2}\frac{s_C}{\sqrt{n}} \quad (1.6)$$

is an asymptotically (as $n \to \infty$) valid $1 - \delta$ confidence interval for C. (For a 95% confidence interval, $\delta = .05$ and $z_{\delta/2} \approx 1.96$.) Alternatively, because the standard deviation is estimated rather than known, we may prefer to replace $z_{\delta/2}$ with the corresponding quantile from the t distribution with $n-1$ degrees of freedom, which results in a slightly wider interval. In either case, the probability that the interval covers C approaches $1 - \delta$ as $n \to \infty$. (These ideas are reviewed in Appendix A.)

The problem of estimating $\mathsf{E}[e^{-rT}(S(T) - K)^+]$ by Monte Carlo is simple enough to be illustrated in a spreadsheet. Commercial spreadsheet software typically includes a method for sampling from the normal distribution and the mathematical functions needed to transform normal samples to terminal stock prices and then to discounted option payoffs. Figure 1.1 gives a schematic illustration. The Z_i are samples from the normal distribution; the comments in the spreadsheet illustrate the formulas used to transform these to arrive at the estimate $\hat{C}_n$. The spreadsheet layout in Figure 1.1 makes the method transparent but has the drawback that it requires storing all n replication in n rows of cells. It is usually possible to use additional spreadsheet commands to recalculate cell values n times without storing intermediate values.

Replication	Normals	Stock Price	Option Payoff		
1	Z_1	S_1	C_1		
2	Z_2	S_2	C_2		
3	Z_3	S_3	C_3		
4	Z_4	S_4			
5	Z_5	S_5	S_1=S(0)*exp((r-0.5*σ^2)*T+σ*sqrt(T)*Z_1)		
6	Z_6	S_6	C_6		
7	Z_7	S_7	C_7		
8	Z_8	S_8	C_8	C_8=exp(-rT)*max(0,S_8-K)	
9	Z_9	S_9	C_9		
10	Z_10	S_10	C_10		
11	Z_11	S_11	C_11		
$\vdots$	$\vdots$	$\vdots$	$\vdots$		
n	Z_n	S_n	C_n		
			$\hat{C}_n$ = AVERAGE(C_1,...,C_n)		
			s_C = STDEV(C_1,...,C_n)		

Fig. 1.1. A spreadsheet for estimating the expected present value of the payoff of a call option.

This simple example illustrates a general feature of Monte Carlo methods for valuing derivatives, which is that the simulation is built up in layers: each of the transformations

$$Z_i \longrightarrow S_i(T) \longrightarrow C_i$$

exemplifies a typical layer. The first transformation constructs a path of underlying assets from random variables with simpler distributions and the second calculates a discounted payoff from each path. In fact, we often have additional layers above and below these. At the lowest level, we typically start from independent random variables U_i uniformly distributed between 0 and 1, so we need a transformation taking the U_i to Z_i. The transformation taking the C_i to the sample mean $\hat{C}_n$ and sample standard deviation s_C may be viewed as another layer. We include another still higher level in, for example, valuing a portfolio of instruments, each of which is valued by Monte Carlo. Randomness (or apparent randomness) typically enters only at the lowest layer; the subsequent transformations producing asset paths, payoffs, and estimators are usually deterministic.

Path-Dependent Example

The payoff of a standard European call option is determined by the terminal stock price $S(T)$ and does not otherwise depend on the evolution of $S(t)$ between times 0 and T. In estimating $\mathsf{E}[e^{-rT}(S(T) - K)^+]$, we were able to jump directly from time 0 to time T using (1.3) to sample values of $S(T)$. Each simulated "path" of the underlying asset thus consists of just the two points $S(0)$ and $S(T)$.

In valuing more complicated derivative securities using more complicated models of the dynamics of the underlying assets, it is often necessary to simulate paths over multiple intermediate dates and not just at the initial and terminal dates. Two considerations may make this necessary:

o   the payoff of a derivative security may depend explicitly on the values of underlying assets at multiple dates;
o   we may not know how to sample transitions of the underlying assets exactly and thus need to divide a time interval $[0, T]$ into smaller subintervals to obtain a more accurate approximation to sampling from the distribution at time T.

In many cases, both considerations apply.

Before turning to a detailed example of the first case, we briefly illustrate the second. Consider a generalization of the basic model (1.1) in which the dynamics of the underlying asset $S(t)$ are given by

$$dS(t) = rS(t)\,dt + \sigma(S(t))S(t)\,dW(t). \tag{1.7}$$

In other words, we now let the volatility σ depend on the current level of S. Except in very special cases, this equation does not admit an explicit solution of the type in (1.2) and we do not have an exact mechanism for sampling from the distribution of $S(T)$. In this setting, we might instead partition $[0, T]$ into m subintervals of length $\Delta t = T/m$ and over each subinterval $[t, t + \Delta t]$ simulate a transition using a discrete (Euler) approximation to (1.7) of the form

$$S(t + \Delta t) = S(t) + rS(t)\Delta t + \sigma(S(t))S(t)\sqrt{\Delta t}Z,$$

with Z a standard normal random variable. This relies on the fact that $W(t + \Delta t) - W(t)$ has mean 0 and standard deviation $\sqrt{\Delta t}$. For each step, we would use an independent draw from the normal distribution. Repeating this for m steps produces a value of $S(T)$ whose distribution approximates the exact (unknown) distribution of $S(T)$ implied by (1.7). We expect that as m becomes larger (so that Δt becomes smaller) the approximating distribution of $S(T)$ draws closer to the exact distribution. In this example, intermediate times are introduced into the simulation to reduce *discretization error*, the topic of Chapter 6.

Even if we assume the dynamics in (1.1) of the Black-Scholes model, it may be necessary to simulate paths of the underlying asset if the payoff of a derivative security depends on the value of the underlying asset at intermediate dates and not just the terminal value. *Asian* options are arguably the simplest path-dependent options for which Monte Carlo is a competitive computational tool. These are options with payoffs that depend on the average level of the underlying asset. This includes, for example, the payoff $(\bar{S} - K)^+$ with

$$\bar{S} = \frac{1}{m} \sum_{j=1}^{m} S(t_j) \tag{1.8}$$

for some fixed set of dates $0 = t_0 < t_1 < \cdots < t_m = T$, with T the date at which the payoff is received.

To calculate the expected discounted payoff $\mathsf{E}[e^{-rT}(\bar{S} - K)^+]$, we need to be able to generate samples of the average $\bar{S}$. The simplest way to do this is to simulate the path $S(t_1), \ldots, S(t_m)$ and then compute the average along the path. We saw in (1.3) how to simulate $S(T)$ given $S(0)$; simulating $S(t_{j+1})$ from $S(t_j)$ works the same way:

$$S(t_{j+1}) = S(t_j) \exp \left([r - \tfrac{1}{2}\sigma^2](t_{j+1} - t_j) + \sigma \sqrt{t_{j+1} - t_j} Z_{j+1} \right) \tag{1.9}$$

where $Z_1, \ldots, Z_m$ are independent standard normal random variables. Given a path of values, it is a simple matter to calculate $\bar{S}$ and then the discounted payoff $e^{-rT}(\bar{S} - K)^+$.

The following algorithm illustrates the steps in simulating n paths of m transitions each. To be explicit, we use Z_{ij} to denote the jth draw from the normal distribution along the ith path. The $\{Z_{ij}\}$ are mutually independent.

> for $i = 1, \ldots, n$
>> for $j = 1, \ldots, m$
>>> generate Z_{ij}
>>> set $S_i(t_j) = S_i(t_{j-1}) \exp \left([r - \tfrac{1}{2}\sigma^2](t_j - t_{j-1}) + \sigma \sqrt{(t_j - t_{j-1})} Z_{ij} \right)$
>> set $\bar{S} = (S_i(t_1) + \cdots + S_i(t_m))/m$
>> set $C_i = e^{-rT}(\bar{S} - K)^+$
> set $\hat{C}_n = (C_1 + \cdots + C_n)/n$

Figure 1.2 gives a schematic illustration of a spreadsheet implementation of this method. The spreadsheet has n rows of standard normal random variables Z_{ij} with m variables in each row. These are mapped to n paths of the underlying asset, each path consisting of m steps. From each path, the spreadsheet calculates a value of the time average $\bar{S}_i$ and a value of the discounted payoff C_i. The C_i are averaged to produce the final estimate $\hat{C}_n$.

Path \ Step	1	2	3		m				
1	Z_11	Z_12	Z_13	···	Z_1m				
2	Z_21	Z_22	Z_23	···	Z_2m				
3	Z_31	Z_32	Z_33	···	Z_3m				
⋮	⋮	⋮	⋮		S_13=S12*exp((r-0.5*σ^2)*(t_3-t_2)+σ*sqrt(t_3-t_2)*Z_13)				
n	Z_n1	Z_n2	Z_n3		Z_nm				
1	S_11	S_12	S_13	···	S_1m	$\bar{S}_1$	C_1		
2	S_21	S_22	S_23	···	S_2m	$\bar{S}_2$	C_2		
3	S_31	S_32	S_33	···	S_3m	$\bar{S}_3$	C_3	C_2=exp(-rT)*max(0,$\bar{S}_2$-K)	
⋮	⋮	⋮	$\bar{S}_3$ =AVERAGE(S_31,S_32,...,S_3m)		⋮				
n	S_n1	S_n2	S_n3	···	S_nm	$\bar{S}_n$	C_n		
							$\hat{C}_n$ = AVERAGE(C_1,...,C_n)		

Fig. 1.2. A spreadsheet for estimating the expected present value of the payoff of an Asian call option.

1.1.3 Efficiency of Simulation Estimators

Much of this book is devoted to ways of improving Monte Carlo estimators. To discuss improvements, we first need to explain our criteria for comparing alternative estimators. Three considerations are particularly important: computing time, bias, and variance.

We begin by considering unbiased estimates. The two cases considered in Section 1.1.2 (the standard call and the Asian call) produced unbiased estimates in the sense that in both cases $E[\hat{C}_n] = C$, with $\hat{C}_n$ the corresponding estimator and C the quantity being estimated. Also, in both cases the estimator $\hat{C}_n$ was the mean of n independent and identically distributed samples. We proceed by continuing to consider estimators of this form because this setting is both simple and practically relevant.

Suppose, then, that

$$\hat{C}_n = \frac{1}{n} \sum_{i=1}^{n} C_i,$$

with C_i i.i.d., $E[C_i] = C$ and $Var[C_i] = \sigma_C^2 < \infty$. The central limit theorem asserts that as the number of replications n increases, the standardized estimator $(\hat{C}_n - C)/(\sigma_C/\sqrt{n})$ converges in distribution to the standard normal, a statement often abbreviated as

$$\frac{\hat{C}_n - C}{\sigma_C/\sqrt{n}} \Rightarrow N(0,1)$$

or, equivalently, as

$$\sqrt{n}[\hat{C}_n - C] \Rightarrow N(0, \sigma_C^2). \tag{1.10}$$

Here, $\Rightarrow$ denotes convergence in distribution and $N(a, b^2)$ denotes the normal distribution with mean a and variance b^2. The stated convergence in distribution means that

$$\lim_{n \to \infty} P\left(\frac{\hat{C}_n - C}{\sigma_C/\sqrt{n}} \leq x\right) = \Phi(x)$$

for all x, with Φ the cumulative normal distribution. The same limit holds if σ_C is replaced with the sample standard devation s_C (as in (1.5)); this is important because σ_C is rarely known in practice but s_C is easily calculated from the simulation output. The fact that we can replace σ_C with s_C without changing the limit in distribution follows from the fact that $s_C/\sigma_C \to 1$ as $n \to \infty$ and general results on convergence in distribution (cf. Appendix A).

The central limit theorem justifies the confidence interval (1.6): as $n \to \infty$, the probability that this interval straddles the true value C approaches $1 - \delta$. Put differently, the central limit theorem tells us something about the distribution of the error in our simulation estimate:

$$\hat{C}_n - C \approx N(0, \sigma_C^2/n),$$

meaning that the error on the left has approximately the distribution on the right. This makes precise the intuitively obvious notion that, other things being equal, in comparing two estimators of the same quantity we should prefer the one with lower variance.

But what if other things are not equal? In particular, suppose we have a choice between two unbiased estimators and that the one with smaller variance takes longer to compute. How should we balance variance reduction and computational effort? An informal answer was suggested by Hammersley and Handscomb [169]; Fox and Glynn [128] and Glynn and Whitt [160] develop a general framework for analyzing this issue and we now review some of its main conclusions.

Suppose that generating a replication C_i takes a fixed amount of computing time τ. Our objective is to compare estimators based on relative computational effort, so the units in which we measure computing time are unimportant. Let s denote our computational budget, measured in the same units as τ. Then the number of replications we can complete given the available budget is $\lfloor s/\tau \rfloor$, the integer part of s/τ, and the resulting estimator is $\hat{C}_{\lfloor s/\tau \rfloor}$. Directly from (1.10), we get

$$\sqrt{\lfloor s/\tau \rfloor}[\hat{C}_{\lfloor s/\tau \rfloor} - C] \Rightarrow N(0, \sigma_C^2)$$

as the computational budget s increases to infinity. Noting that $\lfloor s/\tau \rfloor / s \to 1/\tau$, it follows that $\sqrt{s}[\hat{C}_{\lfloor s/\tau \rfloor} - C]$ is also asymptotically normal but with an asymptotic variance of $\sigma_C^2 \tau$; i.e.,

$$\sqrt{s}[\hat{C}_{\lfloor s/\tau \rfloor} - C] \Rightarrow N(0, \sigma_C^2 \tau) \tag{1.11}$$

as $s \to \infty$. This limit normalizes the error in the estimator by the computing time s rather than by the number of replications. It tells us that, given a budget s, the error in our estimator will be approximately normally distributed with variance $\sigma_C^2 \tau / s$.

This property provides a criterion for comparing alternative unbiased estimators. Suppose, for example, that we have two unbiased estimators both of which are averages of independent replications, as above. Suppose the variance per replication σ_1^2 of the first estimator is larger than the variance per replication σ_2^2 of the second estimator, but the computing times per replication τ_i, $i = 1, 2$, of the two estimators satisfy $\tau_1 < \tau_2$. How should we choose between the faster, more variable estimator and the slower, less variable estimator? The formulation of the central limit theorem in (1.11) suggests that asymptotically (as the computational budget grows), we should prefer the estimator with the smaller value of $\sigma_i^2 \tau_i$, because this is the one that will produce the more precise estimate (and narrower confidence interval) from the budget s.

A feature of the product $\sigma^2 \tau$ (variance per replication times computer time per replication) as a measure of efficiency is that it is insensitive to bundling multiple replications into a single replication. Suppose, for example, that we simply redefine a replication to be the average of two independent copies of the original replications. This cuts the variance per replication in half but doubles the computing time per replication and thus leaves the product of the two unaltered. A purely semantic change in what we call a replication does not affect our measure of efficiency.

The argument leading to the work-normalized central limit theorem (1.11) requires that the computing time per replication be constant. This would be almost exactly the case in, for example, the simulation of the Asian option considered in Section 1.1.2: all replications require simulating the same number of transitions, and the time per transition is nearly constant. This feature is characteristic of many derivative pricing problems in which the time per replication is determined primarily by the number of time steps simulated. But there are also cases in which computing time can vary substantially across replications. In pricing a *barrier* option, for example (cf. Section 3.2.2), one might terminate a path the first time a barrier is crossed; the number of transitions until this happens is typically random. Sampling through acceptance-rejection (as discussed in Section 2.2.2) also introduces randomness in the time per replication.

To generalize (1.11) to these cases, we replace the assumption of a fixed computing time with the condition that $(C_1, \tau_1), (C_2, \tau_2), \dots$ are independent and identically distributed, with C_i as before and τ_i now denoting the computer time required for the ith replication. The number of replications that can be completed with a computing budget s is

$$N(s) = \sup\left\{n \geq 0 : \sum_{i=1}^{n} \tau_i \leq s\right\}$$

and is also random. Our estimator based on a budget s is $\hat{C}_{N(s)}$, the average of the first $N(s)$ replications. Our assumption of i.i.d. replications ensures that $N(s)/s \to 1/\mathsf{E}[\tau]$ with probability one (this is the elementary renewal theorem) and then that (1.11) generalizes to (cf. Appendix A.1)

$$\sqrt{s}[\hat{C}_{N(s)} - C] \Rightarrow N(0, \sigma_C^2 \mathsf{E}[\tau]). \tag{1.12}$$

This limit provides a measure of asymptotic relative efficiency when the computing time per replication is variable. It indicates that in comparing alternative estimators, each of which is the average of unbiased independent replications, we should prefer the one for which the product

(variance per replication) $\times$ (expected computing time per replication)

is smallest. This principle (an early version of which may be found in Hammersley and Handscomb [169], p.51) is a special case of a more general formulation developed by Glynn and Whitt [160] for comparing the efficiency of simulation estimators. Their results include a limit of the form in (1.12) that holds in far greater generality than the case of i.i.d. replications we consider here.

Bias

The efficiency comparisons above, based on the central limit theorems in (1.10) and (1.12), rely on the fact that the estimators to be compared are averages of unbiased replications. In the absence of bias, estimator variability and computational effort are the most important considerations. However, reducing variability or computing time would be pointless if it merely accelerated convergence to an incorrect value. While accepting bias in small samples is sometimes necessary, we are interested only in estimators for which any bias can be eliminated through increasing computational effort.

Some simulation estimators are biased for all finite sample sizes but become asymptotically unbiased as the number of replications increases. This is true of $\hat{C}_{N(s)}$, for example. When the τ_i are random, $\mathsf{E}[\hat{C}_{N(s)}] \neq C$, but the central limit theorem (1.12) shows that the bias in this case becomes negligible as s increases. Glynn and Heidelberger [155] show that it can be entirely eliminated by forcing completion of at least the first replication, because $\mathsf{E}[\hat{C}_{\max\{1,N(s)\}}] = C$.

Another example is provided by the problem of estimating a ratio of expectations $\mathsf{E}[X]/\mathsf{E}[Y]$ from i.i.d. replications (X_i, Y_i), $i = 1, \ldots, n$, of the pair (X, Y). The ratio of sample means $\bar{X}/\bar{Y}$ is biased for all n because

$$E\left[\frac{\bar{X}}{\bar{Y}}\right] \neq \frac{E[\bar{X}]}{E[\bar{Y}]};$$

but $\bar{X}/\bar{Y}$ clearly converges to $E[X]/E[Y]$ with probability 1 as $n \to \infty$. Moreover, the normalized error

$$\sqrt{n}\left(\frac{\bar{X}}{\bar{Y}} - \frac{E[X]}{E[Y]}\right)$$

is asymptotically normal, a point we return to in Section 4.3.3. Thus, the bias becomes negligible as the number of replications increases, and the convergence rate of the estimator is unaffected.

But not all types of bias vanish automatically in large samples — some require special effort. Three examples should help illustrate typical sources of non-negligible bias in financial engineering simulations. In each of these examples the bias persists as the number of replications increases, but the bias is nevertheless manageable in the sense that it can be made as small as necessary through additional computational effort.

Example 1.1.1 *Model discretization error.* In Section 1.1.2 we illustrated the use of Monte Carlo in estimating the expected present value of the payoff of a standard call option and an Asian call option under Black-Scholes assumptions on the dynamics of the underlying stock. We obtained unbiased estimates by simulating the underlying stock using (1.3) and (1.9). Suppose that instead of using (1.9) we divide the time horizon into small increments of length h and approximate changes in the underlying stock using the recursion

$$S((j+1)h) = S(jh) + rS(jh)h + \sigma S(jh)\sqrt{h}Z_{j+1},$$

with $Z_1, Z_2, \ldots$ independent standard normal random variables. The joint distribution of the values of the stock price along a path simulated using this rule will not be exactly the same as that implied by the Black-Scholes dynamics in (1.1). As a consequence, the expected present value of an option payoff estimated using this simulation rule will differ from the exact value — the simulation estimator is biased. This is an example of *discretization* bias because it results from time-discretization of the continuous-time dynamics of the underlying model.

Of course, in this example, the bias can be eliminated by using the exact method (1.9) to simulate values of the underlying stock at the relevant dates. But for many models, exact sampling of the continuous-time dynamics is infeasible and discretization error is inevitable. This is typically the case if, for example, the volatility parameter σ is a function of the stock price S, as in (1.7). The resulting bias can be managed because it typically vanishes as the time step h decreases. However, taking h smaller entails generating more transitions per path (assuming a fixed time horizon) and thus a higher computational burden. □

Example 1.1.2 *Payoff discretization error.* Suppose that in the definition of the Asian option in Section 1.1.2, we replace the discrete average in (1.8) with a continuous average

$$\bar{S} = \frac{1}{T} \int_0^T S(u)\,du.$$

In this case, even if we use (1.9) to generate values of $S(t_i)$ at a discrete set of dates t_i, we cannot calculate $\bar{S}$ exactly — we need to use a discrete approximation to the continuous average. A similar issue arises in estimating, e.g.,

$$\mathsf{E}[e^{-rT}(\max_{0 \le t \le T} S(t) - S(T))],$$

the expected present value of the payoff of a *lookback* option. Even if we simulate a path $S(0), S(t_1), \ldots, S(t_m)$ exactly (i.e., using (1.9)), the estimator

$$e^{-rT}(\max_{0 \le j \le m} S(t_j) - S(T))$$

is biased; in particular, the maximum over the $S(t_j)$ can never exceed and will almost surely underestimate the maximum of $S(t)$ over all t between 0 and T. In both cases, the bias can be made arbitrarily small by using a sufficiently small simulation time step, at the expense of increasing the computational cost per path. Notice that this example differs from Example 1.1.1 in that the source of discretization error is the form of the option payoff rather than the underlying model; the $S(t_i)$ themselves are sampled without discretization error. This type of bias is less common in practice than the model discretization error in Example 1.1.1 because option contracts are often sensitive to the value of the underlying asset at only a finite set of dates. □

Example 1.1.3 *Nonlinear functions of means.* Consider an option expiring at T_1 to buy a call option expiring at $T_2 > T_1$; this is an option on an option, sometimes called a *compound* option. Let $C^{(2)}(x)$ denote the expected discounted payoff of the option expiring at T_2 conditional on the underlying stock price equaling x at time T_1. More explicitly,

$$C^{(2)}(x) = \mathsf{E}[e^{-r(T_2 - T_1)}(S(T_2) - K_2)^+ | S(T_1) = x]$$

with K_2 the strike price. If the compound option has a strike of K_1, then the expected present value of its payoff is

$$C^{(1)} = \mathsf{E}[e^{-rT_1}(C^{(2)}(S(T_1)) - K_1)^+].$$

If the dynamics of the underlying stock are described by the Black-Scholes model (1.1), $C^{(2)}$ and $C^{(1)}$ can be evaluated explicitly. But consider the problem of estimating $C^{(1)}$ by simulation. To do this, we simulate n values $S_1(T_1), \ldots, S_n(T_1)$ of the stock at T_1 and then k values $S_{i1}(T_2), \ldots, S_{ik}(T_2)$ of the stock at T_2 from each $S_i(T_1)$, as illustrated in Figure 1.3. We estimate the inner option value at $S_i(T_1)$ using

$$\hat{C}_k^{(2)}(S_i(T_1)) = \frac{1}{k}\sum_{j=1}^{k} e^{-r(T_2-T_1)}(S_{ij}(T_2) - K_2)^+$$

and then estimate $C^{(1)}$ using

$$\hat{C}_n^{(1)} = \frac{1}{n}\sum_{i=1}^{n} e^{-rT_1}(\hat{C}_k^{(2)}(S_i(T_1)) - K_1)^+.$$

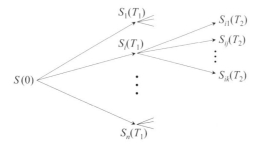

Fig. 1.3. Nested simulation used to estimate a function of a conditional expectation.

If we replaced the inner estimate $\hat{C}_k^{(2)}$ with its expectation, the result would be an unbiased estimator of $C^{(1)}$. But because we estimate the inner expectation, the overall estimator is biased high:

$$\begin{aligned}
\mathsf{E}[\hat{C}_n^{(1)}] &= \mathsf{E}[e^{-rT_1}(\hat{C}_k^{(2)}(S_i(T_1)) - K_1)^+]\\
&= \mathsf{E}[\mathsf{E}[e^{-rT_1}(\hat{C}_k^{(2)}(S_i(T_1)) - K_1)^+|S_i(T_1)]]\\
&\geq \mathsf{E}[e^{-rT_1}(\mathsf{E}[\hat{C}_k^{(2)}(S_i(T_1))|S_i(T_1)] - K_1)^+]\\
&= \mathsf{E}[e^{-rT_1}(C^{(2)}(S_i(T_1)) - K_1)^+]\\
&= C^{(1)}.
\end{aligned}$$

This follows from Jensen's inequality and the convexity of the function $y \mapsto (y - K_1)^+$. As the number k of samples of $S(T_2)$ generated per sample of $S(T_1)$ increases, the bias vanishes because $\hat{C}_k^{(2)}(S_i(T_1)) \to C^{(2)}(S_i(T_1))$ with probability one. The bias can therefore be managed, but once again only at the expense of increasing the computational cost per replication.

The source of bias in this example is the application of a nonlinear function (in this case, the option payoff) to an estimate of an expectation. Closely related biases arise in at least two important applications of Monte Carlo in financial engineering. In measuring portfolio risk over a fixed horizon, the value of the portfolio at the end of the horizon is a conditional expectation. In valuing American options by simulation, the option payoff at each exercise date must be compared with the conditionally expected discounted payoff from waiting to exercise. These topics are discussed in Chapters 8 and 9. □

Examples 1.1.1–1.1.3 share some important features. In each case, the relevant estimator is an average of independent replications; each replication is biased but the bias can be made arbitrarily small at the expense of increasing the computational cost per replication. Given a fixed computing budget, we therefore face a tradeoff in allocating the budget. Expending more effort per replication lowers bias, but it also decreases the number of replications that can be completed and thus tends to increase estimator variance.

We need a measure of estimator performance that balances bias and variance. A standard measure is mean square error, which equals the sum of bias squared and variance. More explicitly, if $\hat{\alpha}$ is an estimator of a quantity α, then

$$
\begin{aligned}
\text{MSE}(\hat{\alpha}) &= \mathsf{E}[(\hat{\alpha} - \alpha)^2] \\
&= (\mathsf{E}[\hat{\alpha}] - \alpha)^2 + \mathsf{E}[(\hat{\alpha} - \mathsf{E}[\hat{\alpha}])^2] \\
&= \text{Bias}^2(\hat{\alpha}) + \text{Variance}(\hat{\alpha}).
\end{aligned}
$$

While exact calculation of mean square error is generally impractical, it is often possible to compare estimators through their asymptotic MSE.

For simplicity, we restrict attention to estimators that are sample means of i.i.d. replications. Extending the notation used in the unbiased case, we write $\hat{C}(n, \delta)$ for the average of n independent replications with parameter δ. This parameter determines the bias: we assume $\mathsf{E}[\hat{C}(n, \delta)] = \alpha_\delta$ and $\alpha_\delta \to \alpha$ as $\delta \to 0$, with α the quantity to be estimated. In Examples 1.1.1 and 1.1.2, δ could be the simulation time increment along each path; in Example 1.1.3 we could take $\delta = 1/k$. We investigate the mean square error of $\hat{C}(n, \delta)$ as the computational budget grows.

Under reasonable additional conditions (in particular, uniform integrability), the central limit theorem in (1.12) for the asymptotically unbiased estimator $\hat{C}_{N(s)}$ implies

$$
s\text{Var}[\hat{C}_{N(s)}] \to \sigma_C^2 \mathsf{E}[\tau];
$$

equivalently,

$$
s^{1/2}\sqrt{\text{Var}[\hat{C}_{N(s)}]} \to \sigma_C \sqrt{\mathsf{E}[\tau]}. \tag{1.13}
$$

The power of s on the left tells us the rate at which the standard error of $\hat{C}_{N(s)}$ (the square root of its variance) decreases, and the limit on the right tells us the constant associated with this asymptotic rate. We proceed to derive similar information in the biased case, where the asymptotic rate of decrease of the mean square error depends, in part, on how computational effort is allocated to reducing bias and variance.

For this analysis, we need to make some assumptions about the estimator. Let τ_δ be the computer time per replication at parameter δ, which we assume to be nonrandom. For the estimator bias and computing time, we assume there are constants $\eta, \beta > 0$, b, and $c > 0$ such that, as $\delta \to 0$,

$$\alpha_\delta - \alpha = b\delta^\beta + o(\delta^\beta) \tag{1.14}$$

$$\tau_\delta = c\delta^{-\eta} + o(\delta^{-\eta}). \tag{1.15}$$

For Examples 1.1.1–1.1.3, it is reasonable to expect that (1.15) holds with $\eta = 1$ because in all three examples the work per path is roughly linear in $1/\delta$. The value of β can vary more from one problem to another, but typical values are $1/2$, 1, and 2. We will see in Chapter 6 that the value of β often depends on how one chooses to approximate a continuous-time process.

Given a computational budget s, we can specify an allocation of this budget to reducing bias and variance by specifying a rule $s \mapsto \delta(s)$ for selecting the parameter δ. The resulting number of replications is $N(s) = \lfloor s/\tau_{\delta(s)} \rfloor$ and the resulting estimator is $\hat{C}(s) \equiv \hat{C}(N(s), \delta(s))$; notice that the estimator is now indexed by the single parameter s whereas it was originally indexed by both the number of replications n and the bias parameter δ. We consider allocation rules $\delta(s)$ for which

$$\delta(s) = as^{-\gamma} + o(s^{-\gamma}) \tag{1.16}$$

for some constants $a, \gamma > 0$. A larger γ corresponds to a smaller $\delta(s)$ and thus greater effort allocated to reducing bias; through (1.15), smaller δ also implies greater computing time per replication, hence fewer replications and less effort allocated to reducing variance. Our goal is to relate the choice of γ to the rate at which the MSE of $\hat{C}(s)$ decreases as s increases.

For large s, we have $N(s) \approx s/\tau_{\delta(s)}$; (1.15) and (1.16) together imply that $\tau_{\delta(s)}$ is $O(s^{\gamma\eta})$ and hence that $N(s)$ is $O(s^{1-\gamma\eta})$. A minimal requirement on the allocation rule $\delta(s)$ is that the number of replications $N(s)$ increase with s. We therefore restrict γ to be less than $1/\eta$ so that $1 - \gamma\eta > 0$.

As a step in our analysis of the MSE, we write the squared bias as

$$
\begin{aligned}
(\alpha_{\delta(s)} - \alpha)^2 &= b^2 \delta(s)^{2\beta} + o(\delta(s)^{2\beta}) \\
&= b^2 a^{2\beta} s^{-2\beta\gamma} + o(s^{-2\beta\gamma}) \tag{1.17} \\
&= O(s^{-2\beta\gamma}) \tag{1.18}
\end{aligned}
$$

using (1.14) and (1.16).

Next we consider variance. Let σ_δ^2 denote the variance per replication at parameter δ. Then

$$\mathsf{Var}[\hat{C}(s)] = \frac{\sigma_{\delta(s)}^2}{\lfloor s/\tau_{\delta(s)} \rfloor}.$$

We assume that σ_δ^2 approaches a finite limit $\sigma^2 > 0$ as $\delta \to 0$. This is a natural assumption in the examples of this section: in Examples 1.1.1 and 1.1.2, σ^2 is the variance in the continuous-time limit; in Example 1.1.3, it is the variance that remains from the first simulation step after the variance in the second step is eliminated by letting $k \to \infty$. Under this assumption we have

$$\mathsf{Var}[\hat{C}(s)] = \frac{\sigma^2 \tau_{\delta(s)}}{s} + o(\tau_\delta(s)/s).$$

Combining this expression for the variance with (1.15) and (1.16), we get

$$\mathsf{Var}[\hat{C}(s)] = \frac{\sigma^2 c\delta(s)^{-\eta}}{s} + o(\delta(s)^{-\eta}/s)$$

$$= \sigma^2 ca^{-\eta}s^{\gamma\eta-1} + o(s^{\gamma\eta-1}) \tag{1.19}$$

$$= O(s^{\gamma\eta-1}). \tag{1.20}$$

The order of magnitude of the MSE is the sum of (1.18) and (1.20).

Consider the effect of different choices of γ. If $2\beta\gamma > 1 - \gamma\eta$ then the allocation rule drives the squared bias (1.18) to zero faster than the variance (1.20), so the MSE is eventually dominated by the variance. Conversely, if $2\beta\gamma < 1 - \gamma\eta$ then for large s the MSE is dominated by the squared bias. An optimal allocation rule selects γ to balance the two terms. Setting $2\beta\gamma = 1 - \gamma\eta$ means taking $\gamma = 1/(2\beta + \eta)$. Substituting this back into (1.17) and (1.19) results in

$$\mathrm{MSE}(\hat{C}(s)) = (b^2 a^{2\beta} + \sigma^2 ca^{-\eta})s^{-2\beta/(2\beta+\eta)} + o(s^{-2\beta/(2\beta+\eta)}) \tag{1.21}$$

and thus for the *root* mean square error we have

$$\mathrm{RMSE}(\hat{C}(s)) \equiv \sqrt{\mathrm{MSE}(\hat{C}(s))} = O(s^{-\beta/(2\beta+\eta)}). \tag{1.22}$$

The exponent of s in this approximation gives the convergence rate of the RMSE and should be contrasted with the convergence rate of $s^{-1/2}$ in (1.13). By minimizing the coefficient in (1.21) we can also find the optimal parameter a in the allocation rule (1.16),

$$a_* = \left(\frac{\eta\sigma^2 c}{2\beta b^2}\right)^{\frac{1}{2\beta+\eta}};$$

but this is of less immediate practical value than the convergence rate in (1.22).

A large β corresponds to a rapidly vanishing bias; as $\beta \to \infty$ we have $\beta/(2\beta + \eta) \to 1/2$, recovering the convergence rate of the standard error in the unbiased case. Similarly, when η is small it follows from (1.16) that the computational cost of reducing bias is small; in the limit as $\eta \to 0$ we again get $\beta/(2\beta + \eta) \to 1/2$. But for any finite β and positive η, (1.22) shows that we must expect a slower convergence rate using an estimator that is unbiased only asymptotically compared with one that is unbiased.

Under an allocation rule satisfying (1.16), taking $\gamma = 1/(2\beta + \eta)$ implies that the bias parameter δ should decrease rather slowly as the computational budget increases. Consider, for instance, bias resulting from model discretization error as in Example 1.1.1. In this setting, interpreting δ as the simulation time increment, the values $\beta = \eta = 1$ would often apply, resulting in $\gamma = 1/3$. Through (1.16), this implies that the time increment should be cut in half with an eight-fold increase in the computational budget.

In applications of Monte Carlo to financial engineering, estimator variance is typically larger than (squared) bias. With a few notable exceptions (including the pricing of American options), it is generally easier to implement a simulation with a comfortably small bias than with a comfortably small standard error. (For example, it is often difficult to measure the reduction in discretization bias achieved using the methods of Chapter 6 because the bias is overwhelmed by simulation variability.) This is consistent with the rather slow decrease in $\delta(s)$ recommended by the analysis above, but it may also in part reflect the relative magnitudes of the constants b, c, and σ. These constants may be difficult to determine; the order of magnitude in (1.21) can nevertheless provide useful insight, especially when very precise simulation results are required, for which the limit $s \to \infty$ is particularly relevant.

The argument above leading to (1.21) considers only the convergence of the mean square error. Glynn and Whitt [160] analyze asymptotic efficiency through the convergence rate of the limit *in distribution* of simulation estimators. Under uniform integrability conditions, a convergence rate in distribution implies a convergence rate for the MSE, but the limiting distribution also provides additional information, just as the central limit theorem (1.12) provides information beyond (1.13).

1.2 Principles of Derivatives Pricing

The mathematical theory of derivatives pricing is both elegant and remarkably practical. A proper development of the theory and of the tools needed even to state precisely its main results requires a book-length treatment; we therefore assume familiarity with at least the basic ideas of mathematical finance and refer the reader to Björk [48], Duffie [98], Hunt and Kennedy [191], Lamberton and Lapeyre [218], and Musiela and Rutkowski [275] for further background. We will, however, highlight some principles of the theory, especially those that bear on the applicability of Monte Carlo to the calculation of prices. Three ideas are particularly important:

1. If a derivative security can be perfectly *replicated* (equivalently, *hedged*) through trading in other assets, then the price of the derivative security is the cost of the replicating trading strategy.
2. Discounted (or *deflated*) asset prices are martingales under a probability measure associated with the choice of discount factor (or *numeraire*). Prices are expectations of discounted payoffs under such a martingale measure.
3. In a *complete* market, any payoff (satisfying modest regularity conditions) can be synthesized through a trading strategy, and the martingale measure associated with a numeraire is unique. In an *incomplete* market there are derivative securities that cannot be perfectly hedged; the price of such a derivative is not completely determined by the prices of other assets.

The rest of this chapter is devoted to explaining these principles and to developing enough of the underlying theory to indicate why, leaving technical issues aside, they ought to be true. A reader familiar with or uninterested in this background may want to skip to the recipe in Figure 1.4, with a warning that the overly simplified summary given there is at best a starting point for applying Monte Carlo to pricing.

The first of the principles above is the foundation of an industry. Financial intermediaries can sell options to their clients and then eliminate the risk from the resulting short position in the option through trading in other assets. They need to charge what it costs to implement the trading strategy, and competition ensures that they cannot charge (much) more. Their clients could in principle run the replicating trading strategy themselves instead of buying options, but financial institutions are better equipped to do this and can do it at lower cost. This role should be contrasted with that of the insurance industry. Insurers bear risk; derivative dealers transfer it.

The second principle is the main link between pricing and Monte Carlo. The first principle gives us a way of thinking about what the price of a derivative security ought to be, but it says little about how this price might be evaluated — it leaves us with the task of finding a hedging strategy and then determining the cost of implementing this strategy. But the second principle gives us a powerful shortcut because it tells us how to represent prices as expectations. Expectations (and, more generally, integrals) lend themselves to evaluation through Monte Carlo and other numerical methods. The subtlety in this approach lies in the fact that we must describe the dynamics of asset prices not as we observe them but as they would be under a *risk-adjusted* probability measure.

The third principle may be viewed as describing conditions under which the price of a derivative security is determined by the prices of other assets so that the first and second principles apply. A complete market is one in which all risks can be perfectly hedged. If all uncertainty in a market is generated by independent Brownian motions, then completeness roughly corresponds to the requirement that the number of traded assets be at least as large as the number of driving Brownian motions. Jumps in asset prices will often render a model incomplete because it may be impossible to hedge the effect of discontinuous movements. In an incomplete market, prices can still be represented as expectations in substantial generality, but the risk adjustment necessary for this representation may not be uniquely determined. In this setting, we need more economic information — an understanding of investor attitudes towards risk — to determine prices, so the machinery of derivatives pricing becomes less useful.

A derivative security introduced into a complete market is a redundant asset. It does not expand investment opportunities; rather, it packages the trading strategy (from the first principle above) investors could have used anyway to synthesize the security. In this setting, pricing a derivative (using the second principle) may be viewed as a complex form of interpolation: we use a model to determine the price of the derivative *relative* to the prices of other assets. On this point, mathematical theory and industry practice are remarkably well aligned. For a financial institution to create a new derivative security, it must determine how it will hedge (or synthesize) the security by trading in other, more liquid assets, and it must determine the cost of this trading strategy from the prices of these other assets.

1.2.1 Pricing and Replication

To further develop these ideas, we consider an economy with d assets whose prices $S_i(t)$, $i = 1, \ldots, d$, are described by a system of SDEs

$$\frac{dS_i(t)}{S_i(t)} = \mu_i(S(t), t) \, dt + \sigma_i(S(t), t)^\top \, dW^o(t), \qquad (1.23)$$

with W^o a k-dimensional Brownian motion, each σ_i taking values in $\Re^k$, and each μ_i scalar-valued. We assume that the μ_i and σ_i are deterministic functions of the current state $S(t) = (S_1(t), \ldots, S_d(t))^\top$ and time t, though the general theory allows these coefficients to depend on past prices as well. (See Appendix B for a brief review of stochastic differential equations and references for further background.) Let

$$\Sigma_{ij} = \sigma_i^\top \sigma_j, \quad i, j = 1, \ldots, d; \qquad (1.24)$$

this may be interpreted as the covariance between the instantaneous returns on assets i and j.

A portfolio is characterized by a vector $\theta \in \Re^d$ with θ_i representing the number of units held of the ith asset. Since each unit of the ith asset is worth $S_i(t)$ at time t, the value of the portfolio at time t is

$$\theta_1 S_1(t) + \cdots + \theta_d S_d(t),$$

which we may write as $\theta^\top S(t)$. A trading strategy is characterized by a stochastic process $\theta(t)$ of portfolio vectors. To be consistent with the intuitive notion of a trading strategy, we need to restrict $\theta(t)$ to depend only on information available at t; this is made precise through a measurability condition (for example, that θ be *predictable*).

If we fix the portfolio holdings at $\theta(t)$ over the interval $[t, t+h]$, then the change in value over this interval of the holdings in the ith asset is given by $\theta_i(t)[S_i(t+h) - S_i(t)]$; the change in the value of the portfolio is given by $\theta(t)^\top[S(t+h) - S(t)]$. This suggests that in the continuous-time limit we may describe the gains from trading over $[0, t]$ through the stochastic integral

$$\int_0^t \theta(u)^\top \, dS(u),$$

subject to regularity conditions on S and θ. Notice that we allow trading of arbitrarily large or small, positive or negative quantities of the underlying assets continuously in time; this is a convenient idealization that ignores constraints on real trading.

A trading strategy is *self-financing* if it satisfies

$$\theta(t)^\top S(t) - \theta(0)^\top S(0) = \int_0^t \theta(u)^\top dS(u) \tag{1.25}$$

for all t. The left side of this equation is the change in portfolio value from time 0 to time t and the right side gives the gains from trading over this interval. Thus, the self-financing condition states that changes in portfolio value equal gains from trading: no gains are withdrawn from the portfolio and no funds are added. By rewriting (1.25) as

$$\theta(t)^\top S(t) = \theta(0)^\top S(0) + \int_0^t \theta(u)^\top dS(u),$$

we can interpret it as stating that from an initial investment of $V(0) = \theta(0)^\top S(0)$ we can achieve a portfolio value of $V(t) = \theta(t)^\top S(t)$ by following the strategy θ over $[0, t]$.

Consider, now, a derivative security with a payoff of $f(S(T))$ at time T; this could be a standard European call or put on one of the d assets, for example, but the payoff could also depend on several of the underlying assets. Suppose that the value of this derivative at time t, $0 \leq t \leq T$, is given by some function $V(S(t), t)$. The fact that the dynamics in (1.23) depend only on $(S(t), t)$ makes it at least plausible that the same might be true of the derivative price. If we further conjecture that V is a sufficiently smooth function of its arguments, Itô's formula (see Appendix B) gives

$$V(S(t), t) = V(S(0), 0) + \sum_{i=1}^d \int_0^t \frac{\partial V(S(u), u)}{\partial S_i} dS_i(u) + \int_0^t \left[\frac{\partial V(S(u), u)}{\partial u} \right.$$

$$\left. + \frac{1}{2} \sum_{i,j=1}^d S_i(u) S_j(u) \Sigma_{ij}(S(u), u) \frac{\partial^2 V(S(u), u)}{\partial S_i \partial S_j} \right] du, \tag{1.26}$$

with Σ as in (1.24). If the value $V(S(t), t)$ can be achieved from an initial wealth of $V(S(0), 0)$ through a self-financing trading strategy θ, then we also have

$$V(S(t), t) = V(S(0), 0) + \sum_{i=1}^d \int_0^t \theta_i(u) \, dS_i(u). \tag{1.27}$$

Comparing terms in (1.26) and (1.27), we find that both equations hold if

$$\theta_i(u) = \frac{\partial V(S(u), u)}{\partial S_i}, \quad i = 1, \ldots, d, \tag{1.28}$$

and

$$\frac{\partial V(S,u)}{\partial u} + \frac{1}{2} \sum_{i,j=1}^{d} \Sigma_{ij}(S,u) S_i S_j \frac{\partial^2 V(S,u)}{\partial S_i \partial S_j} = 0. \tag{1.29}$$

Since we also have $V(S(t),t) = \theta^\top(t) S(t)$, (1.28) implies

$$V(S,t) = \sum_{i=1}^{d} \frac{\partial V(S,t)}{\partial S_i} S_i. \tag{1.30}$$

Finally, at $t = T$ we must have

$$V(S,T) = f(S) \tag{1.31}$$

if V is indeed to represent the value of the derivative security.

Equations (1.29) and (1.30), derived here following the approach in Hunt and Kennedy [191], describe V through a partial differential equation (PDE) with boundary condition (1.31). Suppose we could find a solution $V(S,t)$. In what sense would we be justified in calling this the price of the derivative security?

By construction, V satisfies (1.29) and (1.30), and then (1.26) implies that the (assumed) self-financing representation (1.27) indeed holds with the trading strategy defined by (1.28). Thus, we may sell the derivative security for $V(S(0),0)$ at time 0, use the proceeds to implement this self-financing trading strategy, and deliver the promised payoff of $f(S(T),T) = V(S(T),T)$ at time T *with no risk*. If anyone were willing to pay more than $V(S(0),0)$, we could sell the derivative and be guaranteed a riskless profit from a net investment of zero; if anyone were willing to sell the derivative for less than $V(S(0),0)$, we could buy it, implement the strategy $-\theta(t)$, and again be ensured a riskless profit without investment. Thus, $V(S(0),0)$ is the only price that rules out riskless profits from zero net investment.

From (1.30) we see that the trading strategy that replicates V holds $\partial V(S,t)/\partial S_i$ shares of the ith underlying asset at time t. This partial derivative is the *delta* of V with respect to S_i and the trading strategy is called *delta hedging*.

Inspection of (1.29) and (1.30) reveals that the drift parameters μ_i in the asset price dynamics (1.23) do not appear anywhere in the partial differential equation characterizing the derivative price V. This feature is sometimes paraphrased through the statement that the price of a derivative does not depend on the drifts of the underlying assets; it would be more accurate to say that the effect of the drifts on the price of a derivative is already reflected in the underlying asset prices S_i themselves, because V depends on the S_i and the S_i are clearly affected by the μ_i.

The drifts of the underlying asset prices reflect investor attitudes toward risk. In a world of risk-averse investors, we may expect riskier assets to grow at a higher rate of return, so larger values of σ_{ij} should be associated with larger values of μ_i. In a world of risk-neutral investors, all assets should grow at the same rate — investors will not demand higher returns for riskier assets. The fact that the μ_i do not appear in the equations for the derivative price V may therefore be interpreted as indicating that we can price the derivative without needing to know anything about investor attitudes toward risk. This relies critically on the existence of a self-financing trading strategy that replicates V: because we have assumed that V can be replicated by trading in the underlying assets, risk preferences are irrelevant; the price of the derivative is simply the minimal initial investment required to implement the replicating strategy.

Black-Scholes Model

As an illustration of the general formulation in (1.29) and (1.30), we consider the pricing of European options in the Black-Scholes model. The model contains two assets. The first (often interpreted as a stock price) is risky and its dynamics are represented through the scalar SDE

$$\frac{dS(t)}{S(t)} = \mu\,dt + \sigma\,dW^o(t)$$

with W^o a one-dimensional Brownian motion. The second asset (often called a savings account or a money market account) is riskless and grows deterministically at a constant, continuously compounded rate r; its dynamics are given by

$$\frac{d\beta(t)}{\beta(t)} = r\,dt.$$

Clearly, $\beta(t) = \beta(0)e^{rt}$ and we may assume the normalization $\beta(0) = 1$. We are interested in pricing a derivative security with a payoff of $f(S(T))$ at time T. For example, a standard call option pays $(S(T) - K)^+$, with K a constant.

If we were to formulate this model in the notation of (1.23), Σ would be a 2×2 matrix with only one nonzero entry, σ^2. Making the appropriate substitutions, (1.29) thus becomes

$$\frac{\partial V}{\partial t} + \tfrac{1}{2}\sigma^2 S^2 \frac{\partial^2 V}{\partial S^2} = 0. \tag{1.32}$$

Equation (1.30) becomes

$$V(S, \beta, t) = \frac{\partial V}{\partial S}S + \frac{\partial V}{\partial \beta}\beta. \tag{1.33}$$

These equations and the boundary condition $V(S, \beta, T) = f(S)$ determine the price V.

This formulation describes the price V as a function of the three variables S, β, and t. Because β depends deterministically on t, we are interested in values of V only at points (S, β, t) with $\beta = e^{rt}$. This allows us to eliminate one variable and write the price as $\tilde{V}(S,t) = V(S, e^{rt}, t)$, as in Hunt and Kennedy [191]. Making this substitution in (1.32) and (1.33), noting that

$$\frac{\partial \tilde{V}}{\partial t} = \frac{\partial V}{\partial \beta} r\beta + \frac{\partial V}{\partial t}$$

and simplifying yields

$$\frac{\partial \tilde{V}}{\partial t} + rS\frac{\partial \tilde{V}}{\partial S} + \frac{1}{2}\sigma^2 S^2 \frac{\partial^2 \tilde{V}}{\partial S^2} - r\tilde{V} = 0.$$

This is the Black-Scholes PDE characterizing the price of a European derivative security. For the special case of the boundary condition $\tilde{V}(S,T) = (S - K)^+$, the solution is given by $\tilde{V}(S,t) = \mathrm{BS}(S, \sigma, T - t, r, K)$, the Black-Scholes formula in (1.4).

1.2.2 Arbitrage and Risk-Neutral Pricing

The previous section outlined an argument showing how the existence of a self-financing trading strategy that replicates a derivative security determines the price of the derivative security. Under assumptions on the dynamics of the underlying assets, this argument leads to a partial differential equation characterizing the price of the derivative.

Several features may, however, limit the feasibility of calculating derivative prices by solving PDEs. If the asset price dynamics are sufficiently complex, a PDE characterizing the derivative price may be difficult to solve or may even fail to exist. If the payoff of a derivative security depends on the *paths* of the underlying assets and not simply their terminal values, the assumption that the price can be represented as a function $V(S,t)$ generally fails to hold. If the number of underlying assets required by the replicating strategy is large (greater than two or three), numerical solution of the PDE may be impractical. These are precisely the settings in which Monte Carlo simulation is likely to be most useful. However, to apply Monte Carlo we must first find a more convenient representation of derivative prices. In particular, we would like to represent derivative prices as expectations of random objects that we can simulate. This section develops such representations.

Arbitrage and Stochastic Discount Factors

We return to the general setting described by the asset price dynamics in (1.23), for emphasis writing P_o for the probability measure under which these dynamics are specified. (In particular, the process W^o in (1.23) is a standard Brownian motion under P_o.) The measure P_o is intended to describe objective ("real-world") probabilities and the system of SDEs in (1.23) thus describes the empirical dynamics of asset prices.

Recall the definition of a self-financing trading strategy $\theta(t)$ as given in (1.25). A self-financing trading strategy $\theta(t)$ is called an *arbitrage* if either of the following conditions holds for some fixed time t:

(i) $\theta(0)^\top S(0) < 0$ and $P_o(\theta(t)^\top S(t) \geq 0) = 1$;
(ii) $\theta(0)^\top S(0) = 0$, $P_o(\theta(t)^\top S(t) \geq 0) = 1$, and $P_o(\theta(t)^\top S(t) > 0) > 0$.

In (i), θ turns a negative initial investment into nonnegative final wealth with probability 1. In (ii), θ turns an initial net investment of 0 into nonnegative final wealth that is positive with positive probability. Each of these corresponds to an opportunity to create something from nothing and is incompatible with economic equilibrium. Precluding arbitrage is a basic consistency requirement on the dynamics of the underlying assets in (1.23) and on the prices of any derivative securities that can be synthesized from these assets through self-financing trading strategies.

Call a process $V(t)$ an *attainable* price process if $V(t) = \theta(t)^\top S(t)$ for some self-financing trading strategy θ. Thus, a European derivative security can be replicated by trading in the underlying assets precisely if its payoff at expiration T coincides with the value $V(T)$ of some attainable price process at time T. Each of the underlying asset prices $S_i(t)$ in (1.23) is attainable through the trivial strategy that sets $\theta_i \equiv 1$ and $\theta_j \equiv 0$ for all $j \neq i$.

We now introduce an object whose role may at first seem mysterious but which is central to asset pricing theory. Call a strictly positive process $Z(t)$ a *stochastic discount factor* (or a *deflator*) if the ratio $V(t)/Z(t)$ is a martingale for every attainable price process $V(t)$; i.e., if

$$\frac{V(t)}{Z(t)} = \mathsf{E}_o\left[\frac{V(T)}{Z(T)}|\mathcal{F}_t\right], \tag{1.34}$$

whenever $t < T$. Here, E_o denotes expectation under P_o and $\mathcal{F}_t$ represents the history of the Brownian motion W^o up to time t. We require that $Z(t)$ be adapted to $\mathcal{F}_t$, meaning that the value of $Z(t)$ is determined by the history of the Brownian motion up to time t. Rewriting (1.34) as

$$V(t) = \mathsf{E}_o\left[V(T)\frac{Z(t)}{Z(T)}|\mathcal{F}_t\right] \tag{1.35}$$

explains the term "stochastic discount factor": the price $V(t)$ is the expected discounted value of the price $V(T)$ if we discount using $Z(t)/Z(T)$. (It is more customary to refer to $1/Z(t)$ rather than $Z(t)$ as the stochastic discount factor, deflator, or *pricing kernel*; our use of the terminology is nonstandard but leads to greater symmetry when we discuss numeraire assets.) Notice that any constant multiple of a stochastic discount factor is itself a stochastic discount factor so we may adopt the normalization $Z(0) \equiv 1$. Equation (1.35) then specializes to

$$V(0) = \mathsf{E}_o\left[\frac{V(T)}{Z(T)}\right]. \tag{1.36}$$

Suppose, for example, that $V(t)$ represents the price at time t of a call option on the ith underlying asset with strike price K and expiration T. Then $V(T) = (S_i(T) - K)^+$; in particular, V is a known function of S_i at time T. Equation (1.36) states that the terminal value $V(T)$ determines the initial value $V(0)$ through stochastic discounting.

We may think of (1.36) as reflecting two ways in which the price $V(0)$ differs from the expected payoff $\mathsf{E}_o[V(T)]$. The first results from "the time value of money": the payoff $V(T)$ will not be received until T, and other things being equal we assume investors prefer payoffs received sooner rather than later. The second results from attitudes toward risk. In a world of risk-averse investors, risky payoffs should be more heavily discounted in valuing a security; this could not be accomplished through a deterministic discount factor.

Most importantly for our purposes, the existence of a stochastic discount factor rules out arbitrage. If θ is a self-financing trading strategy, then the process $\theta(t)^\top S(t)$ is an attainable price process and the ratio $\theta(t)^\top S(t)/Z(t)$ must be a martingale. In particular, then,

$$\theta(0)^\top S(0) = \mathsf{E}_o \left[\frac{\theta(T)^\top S(T)}{Z(T)} \right],$$

as in (1.36). Compare this with conditions (i) and (ii) above for an arbitrage, recalling that Z is nonnegative. If $\theta(T)^\top S(T)$ is almost surely positive, it is impossible for $\theta(0)^\top S(0)$ to be negative; if $\theta(T)^\top S(T)$ is positive with positive probability and almost surely nonnegative, then $\theta(0)^\top S(0) = 0$ is impossible. Thus, there can be no arbitrage if the attainable price processes admit a stochastic discount factor.

It is less obvious that the converse also holds: under a variety of technical conditions on asset price dynamics and trading strategies, it has been shown that the absence of arbitrage implies the existence of a stochastic discount factor (or the closely related concept of an equivalent martingale measure). We return to this point in Section 1.2.4. The equivalence of no-arbitrage to the existence of a stochastic discount factor is often termed the Fundamental Theorem of Asset Pricing, though it is not a single theorem but rather a body of results that apply under various sets of conditions. An essential early reference is Harrison and Kreps [170]; for further background and results, see Duffie [98] and Musiela and Rutkowski [275].

Risk-Neutral Pricing

Let us suppose that among the d assets described in (1.23) there is one that is risk-free in the sense that its coefficients σ_{ij} are identically zero. Let us further assume that its drift, which may be interpreted as a riskless interest rate, is a constant r. As in our discussion of the Black-Scholes model in Section 1.2.1, we denote this asset by $\beta(t)$ and refer to it as the money market account. Its dynamics are given by the equation $d\beta(t)/\beta(t) = r\,dt$, with solution $\beta(t) = \beta(0)\exp(rt)$; we fix $\beta(0)$ at 1.

Clearly, $\beta(t)$ is an attainable price process because it corresponds to the trading strategy that makes an initial investment of 1 in the money market account and continuously reinvests all gains in this single asset. Accordingly, if the market admits a stochastic discount factor $Z(t)$, the process $\beta(t)/Z(t)$ is a martingale. This martingale is positive because both $\beta(t)$ and $Z(t)$ are positive, and it has an initial value of $\beta(0)/Z(0) = 1$.

Any positive martingale with an initial value 1 defines a change of probability measure. For each fixed interval $[0, T]$, the process $\beta(t)/Z(t)$ defines a new measure P_β through the Radon-Nikodym derivative (or *likelihood ratio process*)

$$\left(\frac{dP_\beta}{dP_o}\right)_t = \frac{\beta(t)}{Z(t)}, \quad 0 \le t \le T. \tag{1.37}$$

More explicitly, this means (cf. Appendix B.4) that for any event $A \in \mathcal{F}_t$,

$$P_\beta(A) = \mathsf{E}_o\left[\mathbf{1}_A \cdot \left(\frac{dP_\beta}{dP_o}\right)_t\right] = \mathsf{E}_o\left[\mathbf{1}_A \cdot \frac{\beta(t)}{Z(t)}\right]$$

where $\mathbf{1}_A$ denotes the indicator of the event A. Similarly, expectation under the new measure is defined by

$$\mathsf{E}_\beta[X] = \mathsf{E}_o\left[X\frac{\beta(t)}{Z(t)}\right] \tag{1.38}$$

for any nonnegative X measurable with respect to $\mathcal{F}_t$. The measure P_β is called the *risk-neutral* measure; it is equivalent to P_o in the sense of measures, meaning that $P_\beta(A) = 0$ if and only if $P_o(A) = 0$. (Equivalent probability measures agree about which events are impossible.) The risk-neutral measure is a particular choice of *equivalent martingale measure*.

Consider, again, the pricing equation (1.36). In light of (1.38), we may rewrite it as

$$V(0) = \mathsf{E}_\beta\left[\frac{V(T)}{\beta(T)}\right] = e^{-rT}\mathsf{E}_\beta[V(T)]. \tag{1.39}$$

This simple transformation is the cornerstone of derivative pricing by Monte Carlo simulation. Equation (1.39) expresses the current price $V(0)$ as the expected present value of the terminal value $V(T)$ *discounted at the risk-free rate r* rather than through the stochastic discount factor Z. The expectation in (1.39) is taken with respect to P_β rather than P_o, so estimating the expectation by Monte Carlo entails simulating under P_β rather than P_o. These points are crucial to the applicability of Monte Carlo because

o   the dynamics of $Z(t)$ are generally unknown and difficult to model (since they embody time and risk preferences of investors);

o   the dynamics of the underlying asset prices are more easily described under the risk-neutral measure than under the objective probability measure.

The second point requires further explanation. Equation (1.39) generalizes to

$$V(t) = \mathsf{E}_\beta\left[V(T)\frac{\beta(t)}{\beta(T)}|\mathcal{F}_t\right], \quad t < T, \tag{1.40}$$

with $V(t)$ an attainable price process. In particular, then, since each $S_i(t)$ is an attainable price process, each ratio $S_i(t)/\beta(t)$ is a martingale under P_β. Specifying asset price dynamics under the risk-neutral measure thus entails specifying dynamics that make the ratios $S_i(t)/\beta(t)$ martingales. If the dynamics of the asset prices in (1.23) could be expressed as

$$\frac{dS_i(t)}{S_i(t)} = r\,dt + \sigma_i(S(t), t)^\top\,dW(t), \tag{1.41}$$

with W a standard k-dimensional Brownian motion under P_β, then

$$d\left(\frac{S_i(t)}{\beta(t)}\right) = \left(\frac{S_i(t)}{\beta(t)}\right)\sigma_i(S(t), t)^\top\,dW(t),$$

so $S_i(t)/\beta(t)$ would indeed be a martingale under P_β. Specifying a model of the form (1.41) is simpler than specifying the original equation (1.23) because all drifts in (1.41) are set equal to the risk-free rate r: the potentially complicated drifts in (1.23) are irrelevant to the asset price dynamics under the risk-neutral measure. Indeed, this explains the name "risk-neutral." In a world of risk-neutral investors, the rate of return on risky assets would be the same as the risk-free rate.

Comparison of (1.41) and (1.23) indicates that the two are consistent if

$$dW(t) = dW^\circ(t) + \nu(t)\,dt$$
$$\text{for some } \nu \text{ satisfying } \mu_i = r + \sigma_i^\top\nu,\ i = 1, \ldots, d, \tag{1.42}$$

because making this substitution in (1.41) yields

$$\frac{dS_i(t)}{S_i(t)} = r\,dt + \sigma_i(S(t), t)^\top\,[dW^\circ(t) + \nu(t)\,dt]$$
$$= (r + \sigma_i(S(t), t)^\top\nu(t))\,dt + \sigma_i(S(t), t)^\top\,dW^\circ(t)$$
$$= \mu_i(S(t), t)\,dt + \sigma_i(S(t), t)^\top\,dW^\circ(t),$$

as in (1.23). The condition in (1.42) states that the objective and risk-neutral measures are related through a change of drift in the driving Brownian motion. It follows from the Girsanov Theorem (see Appendix B) that any measure equivalent to P_o must be related to P_o in this way. In particular, the diffusion terms σ_{ij} in (1.41) and (1.23) must be the same. This is important because it ensures that the coefficients required to describe the dynamics of asset prices under the risk-neutral measure P_β can be estimated from data observed under the real-world measure P_o.

We now briefly summarize the pricing of derivative securities through the risk-neutral measure with Monte Carlo simulation. Consider a derivative security with a payoff at time T specified through a function f of the prices of the underlying assets, as in the case of a standard call or put. To price the derivative, we model the dynamics of the underlying assets under the risk-neutral measure, ensuring that discounted asset prices are martingales, typically through choice of the drift. The price of the derivative is then given by $E_\beta[e^{-rT}f(S(T))]$. To evaluate this expectation, we simulate paths of the underlying assets over the time interval $[0,T]$, simulating according to their risk-neutral dynamics. On each path we calculate the discounted payoff $e^{-rT}f(S(T))$; the average across paths is our estimate of the derivative's price. Figure 1.4 gives a succinct statement of these steps, but it should be clear that especially the first step in the figure is an oversimplification.

Monte Carlo Recipe for Cookbook Pricing

∘ replace drifts μ_i in (1.23) with risk-free interest rate and simulate paths;
∘ calculate payoff of derivative security on each path;
∘ discount payoffs at the risk-free rate;
∘ calculate average over paths.

Fig. 1.4. An overly simplified summary of risk-neutral pricing by Monte Carlo.

Black-Scholes Model

To illustrate these ideas, consider the pricing of a call option on a stock. Suppose the real-world dynamics of the stock are given by

$$\frac{dS(t)}{S(t)} = \mu(S(t), t)\, dt + \sigma\, dW^o(t),$$

with W^o a standard one-dimensional Brownian motion under P_o and σ a constant. Each unit invested in the money market account at time 0 grows to a value of $\beta(t) = e^{rt}$ at time t. Under the risk-neutral measure P_β, the stock price dynamics are given by

$$\frac{dS(t)}{S(t)} = r\, dt + \sigma\, dW(t)$$

with W a standard Brownian motion under P_β. This implies that

$$S(T) = S(0)e^{(r-\frac{1}{2}\sigma^2)T + \sigma W(T)}.$$

If the call option has strike K and expiration T, its price at time 0 is given by $E_\beta[e^{-rT}(S(T) - K)^+]$. Because $W(T)$ is normally distributed, this expectation can be evaluated explicitly and results in the Black-Scholes formula (1.4). In particular, pricing through the risk-neutral measure produces the same result as pricing through the PDE formulation in Section 1.2.1, as it must since in both cases the price is determined by the absence of arbitrage. This also explains why we are justified in equating the expected discounted payoff calculated in Section 1.1.2 with the price of the option.

Dividends

Thus far, we have implicitly assumed that the underlying assets S_i do not pay dividends. This is implicit, for example, in our discussion of the self-financing trading strategies. In the definition (1.25) of a self-financing strategy θ, we interpret $\theta_i(u)\,dS_i(u)$ as the trading gains from the ith asset over the time increment du. This, however, reflects only the capital gains resulting from the change in price in the ith asset. If each share pays dividends at rate $dD_i(u)$ over du, then the portfolio gains would also include terms of the form $\theta_i(u)\,dD_i(u)$.

In the presence of dividends, a simple strategy of holding a single share of a single asset is no longer self-financing, because it entails withdrawal of the dividends from the portfolio. In contrast, a strategy that continuously reinvests all dividends from an asset back into that asset is self-financing in the sense that it involves neither the withdrawal nor addition of funds from the portfolio. When dividends are reinvested, the number of shares held changes over time.

These observations suggest that we may accommodate dividends by re-defining the original assets to include the reinvested dividends. Let $\tilde{S}_i(t)$ be the ith asset price process with dividends reinvested, defined through the requirement

$$\frac{d\tilde{S}_i(t)}{\tilde{S}_i(t)} = \frac{dS_i(t) + dD_i(t)}{S_i(t)}. \tag{1.43}$$

The expression on the right is the instantaneous return on the ith original asset, including both capital gains and dividends; the expression on the left is the instantaneous return on the ith new asset in which all dividends are reinvested. For $\tilde{S}_i$ to support this interpretation, the two sides must be equal.

The new assets $\tilde{S}_i$ pay no dividends so we may apply the ideas developed above in the absence of dividends to these assets. In particular, we may rein-terpret the asset price dynamics in (1.23) as applying to the $\tilde{S}_i$ rather than to the original S_i. One consequence of this is that the $\tilde{S}_i$ will have continuous paths, so any discontinuities in the cumulative dividend process D_i must be offset by the original asset price S_i. For example, a discrete dividend corresponds to a positive jump in D_i and this must be accompanied by an offsetting negative jump in S_i.

For purposes of derivative pricing, the most important point is that the martingale property under the risk-neutral measure applies to $\tilde{S}_i(t)/\beta(t)$

rather than $S_i(t)/\beta(t)$. This affects how we model the dynamics of the S_i under P_β. Consider, for example, an asset paying a continuous *dividend yield* at rate δ, meaning that $dD_i(t) = \delta S_i(t)\, dt$. For $e^{-rt}\tilde{S}_i(t)$ to be a martingale, we require that the dt coefficient in $d\tilde{S}_i(t)/\tilde{S}_i(t)$ be r. Equating dt terms on the two sides of (1.43), we conclude that the coefficient on dt in the equation for $dS_i(t)/S_i(t)$ must be $r - \delta$. Thus, in modeling asset prices under the risk-neutral measure, the effect of a continuous dividend yield is to change the drift. The first step in Figure 1.4 is modified accordingly.

As a specific illustration, consider a version of the Black-Scholes model in which the underlying asset has dividend yield δ. The risk-neutral dynamics of the asset are given by

$$\frac{dS(t)}{S(t)} = (r - \delta)\, dt + \sigma\, dW(t)$$

with solution

$$S(t) = S(0)e^{(r-\delta-\frac{1}{2}\sigma^2)t+\sigma W(t)}.$$

The price of a call option with strike K and expiration T is given by the expectation $\mathsf{E}_\beta[e^{-rT}(S(T) - K)^+]$, which evaluates to

$$e^{-\delta T}S(0)\Phi(d) - e^{-rT}K\Phi(d - \sigma\sqrt{T}), \quad d = \frac{\log(S(0)/K) + (r - \delta + \frac{1}{2}\sigma^2)T}{\sigma\sqrt{T}},$$

$$(1.44)$$

with Φ the cumulative normal distribution.

1.2.3 Change of Numeraire

The risk-neutral pricing formulas (1.39) and (1.40) continue to apply if the constant risk-free rate r is replaced with a time-varying rate $r(t)$, in which case the money market account becomes

$$\beta(t) = \exp\left(\int_0^t r(u)\, du\right)$$

and the pricing formula becomes

$$V(t) = \mathsf{E}_\beta\left[\exp\left(-\int_t^T r(u)\, du\right)V(T)|\mathcal{F}_t\right].$$

The risk-neutral dynamics of the asset prices now take the form

$$\frac{dS_i(t)}{S_i(t)} = r(t)\, dt + \sigma_i(S(t), t)^\top dW(t),$$

with W a standard k-dimensional Brownian motion under P_β. Subject only to technical conditions, these formulas remain valid if the short rate $r(t)$ is a stochastic process.

Indeed, our choice of $\beta(t)$ as the asset through which to define a new probability measure in (1.38) was somewhat arbitrary. This choice resulted in pricing formulas with the appealing feature that they discount payoffs at the risk-free rate; it also resulted in a simple interpretation of the measure P_β as risk-neutral in the sense that all assets grow at the risk-free rate under this measure. Nevertheless, we could just as well have chosen a different asset as *numeraire*, meaning the asset relative to which all others are valued. As we explain next, all choices of numeraire result in analogous pricing formulas and the flexibility to change the numeraire is a useful modeling and computational tool.

Although we could start from the objective measure P_o as we did in Section 1.2.2, it may be simpler to start from the risk-neutral measure P_β, especially if we assume a constant risk-free rate r. Choosing asset S_d as numeraire means defining a new probability measure P_{S_d} through the likelihood ratio process (Radon-Nikodym derivative)

$$\left(\frac{dP_{S_d}}{dP_\beta} \right)_t = \frac{S_d(t)}{\beta(t)} \bigg/ \frac{S_d(0)}{\beta(0)}.$$

Recall that $S_d(t)/\beta(t)$ is a positive martingale under P_β; dividing it by its initial value produces a unit-mean positive martingale and thus defines a change of measure. Expectation under P_{S_d} is given by

$$\mathsf{E}_{S_d}[X] = \mathsf{E}_\beta \left[X \left(\frac{dP_{S_d}}{dP_\beta} \right)_t \right] = \mathsf{E}_\beta \left[X \frac{S_d(t)\beta(0)}{\beta(t)S_d(0)} \right]$$

for nonnegative $X \in \mathcal{F}_t$. The pricing formula (1.39) thus implies (recalling that $\beta(0) = 1$)

$$V(0) = \mathsf{E}_\beta \left[\frac{V(T)}{\beta(T)} \right] = S_d(0)\mathsf{E}_{S_d} \left[\frac{V(T)}{S_d(T)} \right]. \tag{1.45}$$

Equation (1.40) similarly implies

$$V(t) = S_d(t)\mathsf{E}_{S_d} \left[\frac{V(T)}{S_d(T)} |\mathcal{F}_t \right]. \tag{1.46}$$

Thus, to price under P_{S_d}, we discount the terminal value $V(T)$ by dividing by the terminal value of the numeraire and multiplying by the current value of the numeraire.

Some examples should help illustrate the potential utility of this transformation. Consider, first, an option to exchange one asset for another, with payoff $(S_1(T) - S_2(T))^+$ at time T. The price of the option is given by

$$e^{-rT}\mathsf{E}_\beta[(S_1(T) - S_2(T))^+]$$

but also by

$$S_2(0)\mathsf{E}_{S_2}\left[\frac{(S_1(T)-S_2(T))^+}{S_2(T)}\right] = S_2(0)\mathsf{E}_{S_2}\left[([S_1(T)/S_2(T)]-1)^+\right].$$

The expression on the right looks like the price of a standard call option on the ratio of the two assets with a strike of 1; it reveals that the price of the exchange option is sensitive to the dynamics of the ratio but not otherwise to the dynamics of the individual assets. In particular, if the ratio has a constant volatility (a feature invariant under equivalent changes of measure), then the option can be valued through a variant of the Black-Scholes formula due to Margrabe [247].

Consider, next, a call option on a foreign stock whose payoff will be converted to the domestic currency at the exchange rate prevailing at the expiration date T. Letting S_1 denote the stock price in the foreign currency and letting S_2 denote the exchange rate (expressed as number of domestic units per foreign unit), the payoff (in domestic currency) becomes $S_2(T)(S_1(T)-K)^+$ with price

$$e^{-rT}\mathsf{E}_\beta[S_2(T)(S_1(T)-K)^+].$$

Each unit of foreign currency earns interest at a risk-free rate r_f and this acts like a continuous dividend yield. Choosing $\tilde{S}_2(t) \equiv e^{r_f t}S_2(t)$ as numeraire, we may express the price as

$$e^{-r_f T}S_2(0)\mathsf{E}_{\tilde{S}_2}[(S_1(T)-K)^+],$$

noting that $S_2(0) = \tilde{S}_2(0)$. This expression involves the current exchange rate $S_2(0)$ but not the unknown future rate $S_2(T)$.

The flexibility to change numeraire can be particularly valuable in a model with stochastic interest rates, so our last example applies to this setting. Consider an interest rate derivative with a payoff of $V(T)$ at time T. Using the risk-neutral measure, we can express its price as

$$V(0) = \mathsf{E}_\beta\left[\exp\left(-\int_0^T r(u)\,du\right)V(T)\right].$$

The *forward measure* for maturity T_F is the measure associated with taking as numeraire a zero-coupon bond maturing at T_F with a face value of 1. We denote the time-t value of the bond by $B(t,T_F)$ (so $B(T_F,T_F) \equiv 1$) and the associated measure by P_{T_F}. Using this measure, we can write the price as

$$V(0) = B(0,T_F)\mathsf{E}_{T_F}\left[\frac{V(T)}{B(T,T_F)}\right].$$

With the specific choice $T_F = T$, we get

$$V(0) = B(0,T)\mathsf{E}_T[V(T)].$$

Observe that in this expression the discount factor (the initial bond price) is deterministic even though the interest rate $r(t)$ may be stochastic. This feature often leads to useful simplifications in pricing interest rate derivatives.

To use any of the price representations above derived through a change of numeraire, we need to know the dynamics of the underlying asset prices under the corresponding probability measure. For example, if in (1.45) the terminal value $V(T)$ is a function of the values $S_i(T)$ of the underlying assets, then to estimate the rightmost expectation through Monte Carlo we need to be able to simulate paths of the underlying assets according to their dynamics under P_{S_d}. We encountered the same issue in Section 1.2.2 in pricing under the risk-neutral measure P_β. There we noted that changing from the objective measure P_o to the risk-neutral measure had the effect of changing the drifts of all prices to the risk-free rate; an analogous change of drift applies more generally in changing numeraire.

Based on the dynamics in (1.41), we may write the asset price $S_d(t)$ as

$$S_d(t) = S_d(0) \exp \left(\int_0^t \left[r(u) - \tfrac{1}{2} \| \sigma_d(u) \|^2 \right] du + \int_0^t \sigma_d(u)^\top \, dW(u) \right), \quad (1.47)$$

with W a standard Brownian motion under P_β. Here, we have implicitly generalized the setting in (1.41) to allow the short rate to be time-varying and even stochastic; we have also abbreviated $\sigma_d(S(u), u)$ as $\sigma_d(u)$ to lighten notation. From this and the definition of P_{S_d}, we therefore have

$$\left(\frac{dP_{S_d}}{dP_\beta} \right)_t = \exp \left(\int_0^t -\tfrac{1}{2} \| \sigma_d(u) \|^2 \, du + \int_0^t \sigma_d(u)^\top \, dW(u) \right).$$

Through the Girsanov Theorem (see Appendix B), we find that changing measure from P_β to P_{S_d} has the effect of adding a drift to W. More precisely, the process W^d defined by

$$dW^d(t) = -\sigma_d(t) \, dt + dW(t) \qquad (1.48)$$

is a standard Brownian motion under P_{S_d}. Making this substitution in (1.41), we find that

$$\frac{dS_i(t)}{S_i(t)} = r(t) \, dt + \sigma_i(t)^\top \, dW(t)$$

$$= r(t) \, dt + \sigma_i(t)^\top \left[dW^d(t) + \sigma_d(t) \, dt \right]$$

$$= \left[r(t) + \sigma_i(t)^\top \sigma_d(t) \right] dt + \sigma_i(t)^\top \, dW^d(t)$$

$$= \left[r(t) + \Sigma_{id}(t) \right] dt + \sigma_i(t)^\top \, dW^d(t) \qquad (1.49)$$

with $\Sigma_{id}(t) = \sigma_i(t)^\top \sigma_d(t)$. Thus, when we change measures from P_β to P_{S_d}, an additional term appears in the drift of S_i reflecting the instantaneous covariance between S_i and the numeraire asset S_d.

The distinguishing feature of this change of measure is that it makes the ratios $S_i(t)/S_d(t)$ martingales. This is already implicit in (1.46) because each $S_i(t)$ is an attainable price process and thus a candidate for $V(t)$. To make the martingale property more explicit, we may use (1.47) for S_i and S_d and then simplify using (1.48) to write the ratio as

$$\frac{S_i(t)}{S_d(t)} =$$

$$\frac{S_i(0)}{S_d(0)} \exp\left(-\frac{1}{2} \int_0^t \|\sigma_i(u) - \sigma_d(u)\|^2 \, du + \int_0^t [\sigma_i(u) - \sigma_d(u)]^\top \, dW^d(u) \right).$$

This reveals that $S_i(t)/S_d(t)$ is an exponential martingale (see (B.21) in Appendix B) under P_{S_d} because W^d is a standard Brownian motion under that measure. This also provides a convenient way of thinking about asset price dynamics under the measure P_{S_d}: under this measure, the drifts of the asset prices make the ratios $S_i(t)/S_d(t)$ martingales.

1.2.4 The Market Price of Risk

In this section we conclude our overview of the principles underlying derivatives pricing by returning to the idea of a stochastic discount factor introduced in Section 1.2.1 and further developing its connections with the absence of arbitrage, market completeness, and dynamic hedging. Though not stricly necessary for the application of Monte Carlo (which is based on the pricing relations (1.39) and (1.45)), these ideas are important parts of the underlying theory.

We proceed by considering the dynamics of a stochastic discount factor $Z(t)$ as defined in Section 1.2.1. Just as the likelihood ratio process $(dP_\beta/dP_o)_t$ defined in (1.37) is a positive martingale under P_o, its reciprocal $(dP_o/dP_\beta)_t$ is a positive martingale under P_β; this is a general change of measure identity and is not specific to this context. From (1.37) we find that $(dP_o/dP_\beta)_t = Z(t)/\beta(t)$ and thus that $e^{-rt} Z(t)$ is a positive martingale under P_β. (For simplicity, we assume the short rate r is constant.) This suggests that $Z(t)$ should evolve according to an SDE of the form

$$\frac{dZ(t)}{Z(t)} = r \, dt + \nu(t)^\top \, dW(t), \tag{1.50}$$

for some process ν, with W continuing to be a standard Brownian motion under P_β. Indeed, under appropriate conditions, the martingale representation theorem (Appendix B) ensures that the dynamics of Z must have this form.

Equation (1.50) imposes a restriction on the dynamics of the underlying assets S_i under the objective probability measure P_o. The dynamics of the S_i under the risk-neutral measure are given in (1.41). Switching from P_β back to P_o is formally equivalent to applying a change of numeraire from $\beta(t)$ to $Z(t)$. The process $Z(t)$ may not correspond to an asset price, but this has no effect on the mechanics of the change of measure.

We saw in the previous section that switching from P_β to P_{S_d} had the effect of adding a drift to W; more precisely, the process W^d defined in (1.48) becomes a standard Brownian motion under P_{S_d}. We saw in (1.49) that this has the effect of adding a term to the drifts of the asset prices as viewed under P_{S_d}. By following exactly the same steps, we recognize that the likelihood ratio

$$\left(\frac{dP_o}{dP_\beta}\right)_t = e^{-rt} Z(t) = \exp\left(\int_0^t -\tfrac{1}{2}\|\nu(u)\|^2\, du + \int_0^t \nu(u)^\top dW(u)\right)$$

implies (through the Girsanov Theorem) that

$$dW^o = -\nu(t)\, dt + dW(t)$$

defines a standard Brownian motion under P_o and that the asset price dynamics can be expressed as

$$\begin{aligned}
\frac{dS_i(t)}{S_i(t)} &= r\, dt + \sigma_i(t)^\top dW(t) \\
&= r\, dt + \sigma_i(t)^\top [dW^o(t) + \nu(t)\, dt] \\
&= [r + \nu(t)^\top \sigma_i(t)]\, dt + \sigma_i(t)^\top dW^o(t).
\end{aligned} \tag{1.51}$$

Comparing this with our original specification in (1.23), we find that the existence of a stochastic discount factor implies that the drifts must have the form

$$\mu_i(t) = r + \nu(t)^\top \sigma_i(t). \tag{1.52}$$

This representation suggests an interpretation of ν as a *risk premium*. The components of ν determine the amount by which the drift of a risky asset will exceed the risk-free rate r. In the case of a scalar W^o and ν, from the equation $\mu_i = r + \nu\sigma_i$ we see that the excess return $\mu_i - r$ generated by a risky asset is proportional to its volatility σ_i, with ν the constant of proportionality. In this sense, ν is the *market price of risk*; it measures the excess return demanded by investors per unit of risk. In the vector case, each component ν_j may similarly be interpreted as the market price of risk associated with the jth risk factor — the jth component of W^o. It should also be clear that had we assumed the drifts in (1.23) to have the form in (1.52) (for some ν) from the outset, we could have defined a stochastic discount factor Z from ν and (1.50). Thus, the existence of a stochastic discount factor and a market price of risk vector are essentially equivalent.

An alternative line of argument (which we mention but do not develop) derives the market price of risk in a more fundamental way as the aggregate effect of the individual investment and consumption decisions of agents in an economy. Throughout this section, we have taken the dynamics of the asset prices to be specified exogenously. In a more general formulation, asset prices result from balancing supply and demand among agents who trade to optimize their lifetime investment and consumption; the market price of risk is then determined through the risk aversion of the agents as reflected in their utility for wealth and consumption. Thus, in a general equilibrium model of this type, the market price of risk emerges as a consequence of investor preferences and not just as a constraint to preclude arbitrage. For more on this approach, see Chapter 10 of Duffie [98].

Incomplete Markets

The economic foundation of the market price of risk and the closely related concept of a stochastic discount factor is particularly important in an *incomplete* market. A *complete* market is one in which all risks that affect asset prices can be perfectly hedged. Any new asset (such as an option on one of the existing assets) introduced into a complete market is redundant in the sense that it can be replicated by trading in the other assets. Derivative prices are thus determined by the absence of arbitrage. In an incomplete market, some risks cannot be perfectly hedged and it is therefore possible to introduce genuinely new assets that cannot be replicated by trading in existing assets. In this case, the absence of arbitrage constrains the price of a derivative security but may not determine it uniquely.

For example, market incompleteness may arise because there are fewer traded assets than driving Brownian motions. In this case, there may be infinitely many solutions to (1.52), and thus infinitely many choices of stochastic discount factor $Z(t)$ for which $S_i(t)/Z(t)$ will be martingales, $i = 1, \ldots, d$. Similarly, there are infinitely many possible risk-neutral measures, meaning measures equivalent to the original one under which $e^{-rt}S_i(t)$ are martingales. As a consequence of these indeterminacies, the price of a new security introduced into the market may not be uniquely determined by the prices of existing assets. The machinery of derivatives pricing is largely inapplicable in an incomplete market.

Market incompleteness can arise in various ways; a few examples should serve to illustrate this. Some assets are not traded, making them inaccessible for hedging. How would one eliminate the risk from an option on a privately held business, a parcel of land, or a work of art? Some sources of risk may not correspond to asset prices at all — think of hedging a weather derivative with a payoff tied to rainfall or temperature. Jumps in asset prices and stochastic volatility can often render a market model incomplete by introducing risks that cannot be eliminated through trading in other assets. In such cases, pricing derivatives usually entails making some assumptions, sometimes only implicitly, about the market price for bearing unhedgeable risks.

2

Generating Random Numbers and Random Variables

This chapter deals with algorithms at the core of Monte Carlo simulation: methods for generating uniformly distributed random variables and methods for transforming those variables to other distributions. These algorithms may be executed millions of times in the course of a simulation, making efficient implementation especially important.

Uniform and nonuniform random variate generation have each spawned a vast research literature; we do not attempt a comprehensive account of either topic. The books by Bratley, Fox, and Schrage [59], Devroye [95], Fishman [121], Gentle [136], Niederreiter [281], and others provide more extensive coverage of these areas. We treat the case of the normal distribution in more detail than is customary in books on simulation because of its importance in financial engineering.

2.1 Random Number Generation

2.1.1 General Considerations

At the core of nearly all Monte Carlo simulations is a sequence of apparently random numbers used to drive the simulation. In analyzing Monte Carlo methods, we will treat this driving sequence as though it were genuinely random. This is a convenient fiction that allows us to apply tools from probability and statistics to analyze Monte Carlo computations — convenient because modern *pseudorandom* number generators are sufficiently good at mimicking genuine randomness to make this analysis informative. Nevertheless, we should be aware that the apparently random numbers at the heart of a simulation are in fact produced by completely deterministic algorithms.

The objectives of this section are to discuss some of the primary considerations in the design of random number generators, to present a few simple generators that are good enough for practical use, and to discuss their implementation. We also provide references to a few more sophisticated (though not necessarily better) methods. Elegant theory has been applied to the problem of random number generation, but it is mostly unrelated to the tools we use elsewhere in the book (with the exception of Chapter 5), so we do not treat the topic in depth. The books of Bratley, Fox, and Schrage [59], Fishman [121], Gentle [136], Knuth [212], and Niederreiter [281], and the survey article of L'Ecuyer [223] provide detailed treatment and extensive references to the literature.

Before discussing sequences that appear to be random but are not, we should specify what we mean by a generator of genuinely random numbers: we mean a mechanism for producing a sequence of random variables $U_1, U_2, \ldots$ with the property that

(i) each U_i is uniformly distributed between 0 and 1;
(ii) the U_i are mutually independent.

Property (i) is a convenient but arbitrary normalization; values uniformly distributed between 0 and 1/2 would be just as useful, as would values from nearly any other simple distribution. Uniform random variables on the unit interval can be transformed into samples from essentially any other distribution using, for example, methods described in Section 2.2 and 2.3. Property (ii) is the more important one. It implies, in particular, that all pairs of values should be uncorrelated and, more generally, that the value of U_i should not be predictable from $U_1, \ldots, U_{i-1}$.

A random number generator (often called a *pseudorandom* number generator to emphasize that it only mimics randomness) produces a finite sequence of numbers $u_1, u_2, \ldots, u_K$ in the unit interval. Typically, the values generated depend in part on input parameters specified by the user. *Any* such sequence constitutes a set of possible outcomes of independent uniforms $U_1, \ldots, U_K$. A good random number generator is one that satisfies the admittedly vague requirement that small (relative to K) segments of the sequence $u_1, \ldots, u_K$ should be difficult to distinguish from a realization of independent uniforms.

An effective generator therefore produces values that appear consistent with properties (i) and (ii) above. If the number of values K is large, the fraction of values falling in any subinterval of the unit interval should be approximately the length of the subinterval — this is uniformity. Independence suggests that there should be no discernible pattern among the values. To put this only slightly more precisely, statistical tests for independence should not easily reject segments of the sequence $u_1, \ldots, u_K$.

We can make these and other considerations more concrete through examples. A *linear congruential generator* is a recurrence of the following form:

$$x_{i+1} = ax_i \bmod m \tag{2.1}$$

$$u_{i+1} = x_{i+1}/m \tag{2.2}$$

Here, the *multiplier* a and the *modulus* m are integer constants that determine the values generated, given an initial value (*seed*) x_0. The seed is an integer between 1 and $m - 1$ and is ordinarily specified by the user. The operation $y \bmod m$ returns the remainder of y (an integer) after division by m. In other words,

$$y \bmod m = y - \lfloor y/m \rfloor m, \qquad (2.3)$$

where $\lfloor x \rfloor$ denotes the greatest integer less than or equal to x. For example, 7 mod 5 is 2; 10 mod 5 is 0; 43 mod 5 is 3; and 3 mod 5 is 3. Because the result of the mod m operation is always an integer between 0 and $m - 1$, the output values u_i produced by (2.1)–(2.2) are always between 0 and $(m - 1)/m$; in particular, they lie in the unit interval.

Because of their simplicity and potential for effectiveness, linear congruential generators are among the most widely used in practice. We discuss them in detail in Section 2.1.2. At this point, we use them to illustrate some general considerations in the design of random number generators. Notice that the linear congruential generator has the form

$$x_{i+1} = f(x_i), \quad u_{i+1} = g(x_{i+1}), \qquad (2.4)$$

for some deterministic functions f and g. If we allow the x_i to be vectors, then virtually all random number generators fit this general form.

Consider the sequence of x_i produced in (2.1) by a linear congruential generator with $a = 6$ and $m = 11$. (In practice, m should be large; these values are solely for illustration.) Starting from $x_0 = 1$, the next value is 6 mod 11 = 6, followed by $(6 \cdot 6)$ mod 11 = 3. The seed $x_0 = 1$ thus produces the sequence

$$1, \ 6, \ 3, \ 7, \ 9, \ 10, \ 5, \ 8, \ 4, \ 2, \ 1, \ 6, \ \ldots.$$

Once a value is repeated, the entire sequence repeats. Indeed, since a computer can represent only a finite number of values, any recurrence of the form in (2.4) will eventually return to a previous x_i and then repeat all values that followed that x_i. Observe that in this example all ten distinct integers between 1 and $m - 1$ appeared in the sequence before a value was repeated. (If we were to start the sequence at 0, all subsequent values would be zero, so we do not allow $x_0 = 0$.) If we keep $m = 11$ but take $a = 3$, the seed $x_0 = 1$ yields

$$1, \ 3, \ 9, \ 5, \ 4, \ 1, \ \ldots,$$

whereas $x_0 = 2$ yields

$$2, \ 6, \ 7, \ 10, \ 8, \ 2, \ \ldots.$$

Thus, in this case, the possible values $\{1, 2, \ldots, 10\}$ split into two cycles. This means that regardless of what x_0 is chosen, a multiplier of $a = 3$ produces just five distinct numbers before it repeats, whereas a multiplier of $a = 6$ produces all ten distinct values before repeating. A linear congruential generator that produces all $m - 1$ distinct values before repeating is said to have *full period*. In practice we would like to be able to generate (at least) tens of millions of distinct values before repeating any. Simply choosing m to be very large does not ensure this property because of the possibility that a poor choice of parameters a and m may result in short cycles among the values $\{1, 2, \ldots, m-1\}$.

With these examples in mind, we discuss the following general considerations in the construction of a random number generator:

○ *Period length.* As already noted, any random number generator of the form (2.4) will eventually repeat itself. Other things being equal, we prefer generators with longer periods — i.e., generators that produce more distinct values before repeating. The longest possible period for a linear congruential generator with modulus m is $m - 1$. For a linear congruential generator with full period, the gaps between the values u_i produced are of width $1/m$; hence, the larger m is the more closely the values can approximate a uniform distribution.

○ *Reproducibility.* One might be tempted to look to physical devices — a computer's clock or a specially designed electronic mechanism — to generate true randomness. One drawback of a genuinely random sequence is that it cannot be reproduced easily. It is often important to be able to rerun a simulation using exactly the same inputs used previously, or to use the same inputs in two or more different simulations. This is easily accomplished with a linear congruential generator or any other procedure of the general form (2.4) simply by using the same seed x_0.

○ *Speed.* Because a random number generator may be called thousands or even millions of times in a single simulation, it must be fast. It is hard to imagine an algorithm simpler or faster than the linear congruential generator; most of the more involved methods to be touched on in Section 2.1.5 remain fast in absolute terms, though they involve more operations per value generated. The early literature on random number generation includes strategies for saving computing time through convenient parameter choices. For example, by choosing m to be a power of 2, the mod m operation can be implemented by shifting bits, without explicit division. Given current computing speeds, this incremental speed-up does not seem to justify choosing a generator with poor distributional properties.

○ *Portability.* An algorithm for generating random numbers should produce the same sequence of values on all computing platforms. The quest for speed and long periods occasionally leads to implementations that depend on machine-specific representations of numbers. Some implementations of linear congruential generators rely on the way overflow is handled on particular computers. We return to this issue in the next section.

○ *Randomness.* The most important consideration is the hardest to define or ensure. There are two broad aspects to constructing generators with apparent randomness: theoretical properties and statistical tests. Much is known about the structure of points produced by the most widely used generators and this helps narrow the search for good parameter values. Generators with good theoretical properties can then be subjected to statistical scrutiny to test for evident departures from randomness. Fortunately, the field is sufficiently well developed that for most applications one can comfortably use one of many generators in the literature that have survived rigorous tests and the test of time.

2.1.2 Linear Congruential Generators

The general linear congruential generator, first proposed by Lehmer [229], takes the form

$$x_{i+1} = (ax_i + c) \bmod m$$
$$u_{i+1} = x_{i+1}/m$$

This is sometimes called a *mixed* linear congruential generator and the multiplicative case in the previous section a *pure* linear congruential generator. Like a and m, the parameter c must be an integer.

Quite a bit is known about the structure of the sets of values $\{u_1, \ldots, u_K\}$ produced by this type of algorithm. In particular, simple conditions are available ensuring that the generator has full period — i.e., that the number of distinct values generated from any seed x_0 is $m - 1$. If $c \neq 0$, the conditions are (Knuth [212, p.17])

(a) c and m are relatively prime (their only common divisor is 1);
(b) every prime number that divides m divides $a - 1$;
(c) $a - 1$ is divisible by 4 if m is.

As a simple consequence, we observe that if m is a power of 2, the generator has full period if c is odd and $a = 4n + 1$ for some integer n.

If $c = 0$ and m is prime, full period is achieved from any $x_0 \neq 0$ if

○ $a^{m-1} - 1$ is a multiple of m;
○ $a^j - 1$ is not a multiple of m for $j = 1, \ldots, m - 2$.

A number a satisfying these two properties is called a *primitive root* of m. Observe that when $c = 0$ the sequence $\{x_i\}$ becomes

$$x_0, \ ax_0, \ a^2 x_0, \ a^3 x_0, \ \ldots \ (\bmod m).$$

The sequence first returns to x_0 at the smallest k for which $a^k x_0 \bmod m = x_0$. This is the smallest k for which $a^k \bmod m = 1$; i.e., the smallest k for which $a^k - 1$ is a multiple of m. So, the definition of a primitive root corresponds precisely to the requirement that the sequence not return to x_0 until $a^{m-1} x_0$. It can also be verified that when a is a primitive root of m, all x_i are nonzero if x_0 is nonzero. This is important because if some x_i were 0, then all subsequent values generated would be too.

Marsaglia [249] demonstrates that little additional generality is achieved by taking $c \neq 0$. Since a generator with a nonzero c is slower than one without, it is now customary to take $c = 0$. In this case, it is convenient to take m to be prime, since it is then possible to construct full-period generators simply by finding primitive roots of m.

Table 2.1 displays moduli and multipliers for seven linear congruential generators that have been recommended in the literature. In each case, the modulus m is a large prime not exceeding $2^{31} - 1$. This is the largest integer that can be represented in a 32-bit word (assuming one bit is used to determine the sign) and it also happens to be a prime — a Mersenne prime. Each multiplier a in the table is a primitive root of the corresponding modulus, so all generators in the table have full period. The first generator listed was dubbed the "minimal standard" by Park and Miller [294]; though widely used, it appears to be inferior to the others listed. Among the remaining generators, those identified by Fishman and Moore [123] appear to have slightly better uniformity while those from L'Ecuyer [222] offer a computational advantage resulting from having comparatively smaller values of a (in particular, $a < \sqrt{m}$). We discuss this computational advantage and the basis on which these generators have been compared next.

Generators with far longer periods are discussed in Section 2.1.5. L'Ecuyer, Simard, and Wegenkittl [228] reject all "small" generators like those in Table 2.1 as obsolete. Section 2.1.5 explains how they remain useful as components of combined generators.

Modulus m	Multiplier a	Reference
$2^{31} - 1$	16807	Lewis, Goodman, and Miller [234],
($= 2147483647$)		Park and Miller [294]
	39373	L'Ecuyer [222]
	742938285	Fishman and Moore [123]
	950706376	Fishman and Moore [123]
	1226874159	Fishman and Moore [123]
2147483399	40692	L'Ecuyer [222]
2147483563	40014	L'Ecuyer [222]

Table 2.1. Parameters for linear congruential generators. The generator in the first row appears to be inferior to the rest.

2.1.3 Implementation of Linear Congruential Generators

Besides speed, avoiding overflow is the main consideration in implementing a linear congruential generator. If the product ax_i can be represented exactly for every x_i in the sequence, then no overflow occurs. If, for example, every integer from 0 to $a(m - 1)$ can be represented exactly in double precision, then implementation in double precision is straightforward.

If the multiplier a is large, as in three of the generators of Table 2.1, even double precision may not suffice for an exact representation of every product

ax_i. In this case, the generator may be implemented by first representing the multiplier as $a = 2^\alpha a_1 + a_2$, with $a_1, a_2 < 2^\alpha$, and then using

$$ax_i \bmod m = (a_1(2^\alpha x_i \bmod m) + a_2 x_i \bmod m) \bmod m.$$

For example, with $\alpha = 16$ and $m = 2^{31} - 1$ this implementation never requires an intermediate value as large as 2^{47}, even though ax_i could be close to 2^{62}.

Integer arithmetic is sometimes faster than floating point arithmetic, in which case an implementation in integer variables is more appealing than one using double precision. Moreover, if variables y and m are represented as integers in a computer, the integer operation y/m produces $\lfloor y/m \rfloor$, so $y \bmod m$ can be implemented as $y - (y/m) * m$ (see (2.3)). However, working in integer variables restricts the magnitude of numbers that can be represented far more than does working in double precision. To avoid overflow, a straightforward implementation of a linear congruential generator in integer variables must be restricted to an unacceptably small modulus — e.g., $2^{15} - 1$. If a is not too large (say $a \le \sqrt{m}$, as in the first two and last two entries of Table 2.1), Bratley, Fox, and Schrage [59] show that a faster implementation is possible using only integer arithmetic, while still avoiding overflow.

Their method is based on the following observations. Let

$$q = \lfloor m/a \rfloor, \quad r = m \bmod a$$

so that the modulus can be represented as $m = aq + r$. The calculation to be carried out by the generator is

$$ax_i \bmod m = ax_i - \left\lfloor \frac{ax_i}{m} \right\rfloor m$$
$$= \left(ax_i - \left\lfloor \frac{x_i}{q} \right\rfloor m \right) + \left(\left\lfloor \frac{x_i}{q} \right\rfloor - \left\lfloor \frac{ax_i}{m} \right\rfloor \right) m. \qquad (2.5)$$

The first term on the right in (2.5) satisfies

$$ax_i - \left\lfloor \frac{x_i}{q} \right\rfloor m = ax_i - \left\lfloor \frac{x_i}{q} \right\rfloor (aq + r)$$
$$= a \left(x_i - \left\lfloor \frac{x_i}{q} \right\rfloor q \right) - \left\lfloor \frac{x_i}{q} \right\rfloor r$$
$$= a(x_i \bmod q) - \left\lfloor \frac{x_i}{q} \right\rfloor r.$$

Making this substitution in (2.5) yields

$$ax_i \bmod m = a(x_i \bmod q) - \left\lfloor \frac{x_i}{q} \right\rfloor r + \left(\left\lfloor \frac{x_i}{q} \right\rfloor - \left\lfloor \frac{ax_i}{m} \right\rfloor \right) m. \qquad (2.6)$$

To prevent overflow, we need to avoid calculation of the potentially large term ax_i on the right side of (2.6). In fact, we can entirely avoid calculation of

$$\left(\left\lfloor \frac{x_i}{q} \right\rfloor - \left\lfloor \frac{ax_i}{m} \right\rfloor \right) \tag{2.7}$$

if we can show that this expression takes only the values 0 and 1. For in this case, the last term in (2.6) is either 0 or m, and since the final calculation must result in a value in $\{0, 1, \ldots, m-1\}$, the last term in (2.6) is m precisely when

$$a(x_i \bmod q) - \left\lfloor \frac{x_i}{q} \right\rfloor r < 0.$$

Thus, the last term in (2.6) adds m to the first two terms precisely when not doing so would result in a value outside of $\{0, 1, \ldots, m-1\}$.

It remains to verify that (2.7) takes only the values 0 and 1. This holds if

$$\frac{x_i}{q} - \frac{ax_i}{m} \le 1. \tag{2.8}$$

But x_i never exceeds $m-1$, and

$$\frac{m-1}{q} - \frac{a(m-1)}{m} = \frac{r(m-1)}{qm}.$$

Thus, (2.8) holds if $r \le q$; a simple sufficient condition ensuring this is $a \le \sqrt{m}$.

The result of this argument is that (2.6) can be implemented so that every intermediate calculation results in an integer between $-(m-1)$ and $m-1$, allowing calculation of $ax_i \bmod m$ without overflow. In particular, explicit calculation of (2.7) is avoided by checking indirectly whether the result of this calculation would be 0 or 1. L'Ecuyer [222] gives a simple implementation of this idea, which we illustrate in Figure 2.1.

$$\boxed{\begin{array}{l}
(m, a \text{ integer constants} \\
q, r \text{ precomputed integer constants,} \\
\text{with } q = \lfloor m/a \rfloor, r = m \bmod a \\
x \text{ integer variable holding the current } x_i) \\
\quad k \leftarrow x/q \\
\quad x \leftarrow a * (x - k * q) - k * r \\
\quad \text{if } (x < 0) \; x \leftarrow x + m
\end{array}}$$

Fig. 2.1. Implementation of $ax \bmod m$ in integer arithmetic without overflow, assuming $r \le q$ (e.g., $a \le \sqrt{m}$).

The final step in using a congruential generator — converting the $x_i \in \{0, 1, \ldots, m-1\}$ to a value in the unit interval — is not displayed in Figure 2.1. This can be implemented by setting $u \leftarrow x * h$ where h is a precomputed constant equal to $1/m$.

Of the generators in Table 2.1, the first two and the last two satisfy $a \leq \sqrt{m}$ and thus may be implemented using Figure 2.1. L'Ecuyer [222] finds that the second, sixth, and seventh generators listed in the table have the best distributional properties among all choices of multiplier a that are primitive roots of m and satisfy $a \leq \sqrt{m}$, $m \leq 2^{31} - 1$. Fishman [121] recommends working in double precision in order to get the somewhat superior uniformity of the large multipliers in Table 2.1. We will see in Section 2.1.5 that by combining generators it is possible to maintain the computational advantage of having $a \leq \sqrt{m}$ without sacrificing uniformity.

Skipping Ahead

It is occasionally useful to be able to split a random number stream into apparently unrelated subsequences. This can be implemented by initializing the same random number to two or more distinct seeds. Choosing the seeds arbitrarily leaves open the possibility that the ostensibly unrelated subsequences will have substantial overlap. This can be avoided by choosing the seeds far apart along the sequence produced by a random number generator.

With a linear congruential generator, it is easy to skip ahead along the sequence without generating intermediate values. If $x_{i+1} = a x_i \bmod m$, then

$$x_{i+k} = a^k x_i \bmod m.$$

This in turn is equivalent to

$$x_{i+k} = ((a^k \bmod m) x_i) \bmod m.$$

Thus, one could compute the constant $a^k \bmod m$ just once and then easily produce a sequence of values spaced k apart along the generator's output. See L'Ecuyer, Simard, Chen, and Kelton [227] for an implementation.

Splitting a random number stream carefully is essential if the subsequences are to be assigned to parallel processors running simulations intended to be independent of each other. Splitting a stream can also be useful when simulation is used to compare results from a model at different parameter values. In comparing results, it is generally preferable to use the same random numbers for both sets of simulations, and to use them for the same purpose in both to the extent possible. For example, if the model involves simulating d asset prices, one would ordinarily want to arrange matters so that the random numbers used to simulate the ith asset at one parameter value are used to simulate the same asset at other parameter values. Dedicating a separate subsequence of the generator to each asset ensures this arrangement.

2.1.4 Lattice Structure

In discussing the generators of Table 2.1, we alluded to comparisons of their distributional properties. We now provide a bit more detail on how these comparisons are made. See Knuth [212] and Neiderreiter [281] for far more thorough treatments of the topic.

If the random variables $U_1, U_2, \ldots$ are independent and uniformly distributed over the unit interval, then (U_1, U_2) is uniformly distributed over the unit square, (U_1, U_2, U_3) is uniformly distributed over the unit cube, and so on. Hence, one way to evaluate a random number generator is to form points in $[0, 1]^d$ from consecutive output values and measure how uniformly these points fill the space.

The left panel of Figure 2.2 plots consecutive overlapping pairs (u_1, u_2), (u_2, u_3), ..., (u_{10}, u_{11}) produced by a linear congruential generator. The parameters of the generator are $a = 6$ and $m = 11$, a case considered in Section 2.1.1. The graph immediately reveals a regular pattern: the ten distinct points obtained from the full period of the generator lie on just two parallel lines through the unit square.

This phenomenon is characteristic of all linear congruential generators (and some other generators as well), though it is of course particularly pronounced in this simple example. Marsaglia [248] showed that overlapping d-tuples formed from consecutive outputs of a linear congruential generator with modulus m lie on at most $(d!m)^{1/d}$ hyperplanes in the d-dimensional unit cube. For $m = 2^{31} - 1$, this is approximately 108 with $d = 3$ and drops below 39 at $d = 10$. Thus, particularly in high dimensions, the lattice structure of even the best possible linear congruential generators distinguishes them from genuinely random numbers.

The right panel of Figure 2.2, based on a similar figure in L'Ecuyer [222], shows the positions of points produced by the first generator in Table 2.1. The figure magnifies the strip $\{(u_1, u_2) : u_1 < .001\}$ and plots the first 10,005 points that fall in this strip starting from a seed of $x_0 = 8835$. (These are all the points that fall in the strip out of the first ten million points generated by the sequence starting from that seed.) At this magnification, the lattice structure becomes evident, even in this widely used method.

The lattice structure of linear congruential generators is often used to compare their outputs and select parameters. There are many ways one might try to quantify the degree of equidistribution of points on a lattice. The most widely used in the analysis of random number generators is the *spectral test*, originally proposed by Coveyou and Macpherson [88]. For each dimension d and each set of parallel hyperplanes containing all points in the lattice, consider the distance between adjacent hyperplanes. The spectral test takes the maximum of these distances over all such sets of parallel hyperplanes.

To see why taking the maximum is appropriate, consider again the left panel of Figure 2.2. The ten points in the graph lie on two positively sloped lines. They also lie on five negatively sloped lines and ten vertical lines. Depending on which set of lines we choose, we get a different measure of distance between adjacent lines. The maximum distance is achieved by the two positively sloped lines passing through the points, and this measure is clearly the one that best captures the wide diagonal swath left empty by the generator.

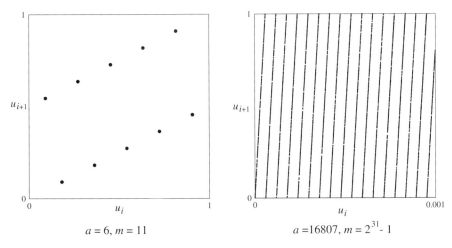

Fig. 2.2. Lattice structure of linear congruential generators.

Although the spectral test is an informative measure of uniformity, it does not provide a strict ranking of generators because it produces a separate value for each dimension d. It is possible for each of two generators to outperform the other at some values of d. Fishman and Moore [123] and L'Ecuyer [222] base their recommendations of the values in Table 2.1 on spectral tests up to dimension $d = 6$; computing the spectral test becomes increasingly difficult in higher dimensions. L'Ecuyer [222] combines results for $d = 2$–6 into a worst-case figure of merit in order to rank generators.

Niederreiter [281] analyzes the uniformity of point sets in the unit hypercube (including those produced by various random number generators) through *discrepancy* measures, which have some appealing theoretical features not shared by the spectral test. Discrepancy measures are particularly important in the analysis of *quasi-Monte Carlo* methods.

It is also customary to subject random number generators to various statistical tests of uniformity and independence. See, e.g., Bratley, Fox, and Schrage [59] or Knuth [212] for a discussion of some of the tests often used.

Given the inevitable shortcomings of any practical random number generator, it is advisable to use only a small fraction of the period of a generator. This again points to the advantage of generators with long periods — much longer than 2^{31}.

2.1.5 Combined Generators and Other Methods

We now turn to a discussion of a few other methods for random number generation. Methods that combine linear congruential generators appear to be particularly promising because they preserve attractive computational features of these generators while extending their period and, in some cases, attenuating their lattice structure. A combined generator proposed by L'Ecuyer [224] and discussed below appears to meet the requirements for speed, uniformity, and a long period of most current applications. We also note a few other directions of work in the area.

Combining Generators

One way to move beyond the basic linear congruential generator combines two or more of these generators through summation. Wichmann and Hill [355] propose summing values in the unit interval (i.e., after dividing by the modulus); L'Ecuyer [222] sums first and then divides.

To make this more explicit, consider J generators, the jth having parameters a_j, m_j:

$$x_{j,i+1} = a_j x_{j,i} \bmod m_j, \quad u_{j,i+1} = x_{j,i+1}/m_j, \quad j = 1, \ldots, J.$$

The Wichmann-Hill combination sets u_{i+1} equal to the fractional part of $u_{1,i+1} + u_{2,i+1} + \cdots + u_{J,i+1}$. L'Ecuyer's combination takes the form

$$x_{i+1} = \sum_{j=1}^{J} (-1)^{(j-1)} x_{j,i+1} \mod (m_1 - 1) \tag{2.9}$$

and

$$u_{i+1} = \begin{cases} x_{i+1}/m_1, & x_{i+1} > 0; \\ (m_1 - 1)/m_1, & x_{i+1} = 0. \end{cases} \tag{2.10}$$

This assumes that m_1 is the largest of the m_j.

A combination of generators can have a much longer period than any of its components. A long period can also be achieved in a single generator by using a larger modulus, but a larger modulus complicates the problem of avoiding overflow. In combining generators, it is possible to choose each multiplier a_j smaller than $\sqrt{m_j}$ in order to use the integer implementation of Figure 2.1 for each. The sum in (2.9) can then also be implemented in integer arithmetic, whereas the Wichmann-Hill summation of $u_{j,i}$ is a floating point operation. L'Ecuyer [222] gives a portable implementation of (2.9)–(2.10). He also examines a combination of the first and sixth generators of Table 2.1 and finds that the combination has no apparent lattice structure at a magnification at which each component generator has a very evident lattice structure. This suggests that combined generators can have superior uniformity properties as well as long periods and computational convenience.

Another way of extending the basic linear congruential generator uses a higher-order recursion of the form

$$x_i = (a_1 x_{i-1} + a_2 x_{i-2} + \cdots a_k x_{i-k}) \bmod m, \tag{2.11}$$

followed by $u_i = x_i/m$; this is called a *multiple recursive* generator, or MRG. A seed for this generator consists of initial values $x_{k-1}, x_{k-2}, \ldots, x_0$.

Each of the lagged values x_{i-j} in (2.11) can take up to m distinct values, so the vector $(x_{i-1}, \ldots, x_{i-k})$ can take up to m^k distinct values. The sequence x_i repeats once this vector returns to a previously visited value, and if the vector ever reaches $(0, \ldots, 0)$ all subsequent x_i are identically 0. Thus, the longest possible period for (2.11) is $m^k - 1$. Knuth [212] gives conditions on m and $a_1, \ldots, a_k$ under which this bound is achieved.

L'Ecuyer [224] combines MRGs using essentially the mechanism in (2.9)–(2.10). He shows that the combined generator is, in a precise sense, a close approximation to a single MRG with a modulus equal to the product of the moduli of the component MRGs. Thus, the combined generator has the advantages associated with a larger modulus while permitting an implementation using smaller values. L'Ecuyer's investigation further suggests that a combined MRG has a less evident lattice structure than the large-modulus MRG it approximates, indicating a distributional advantage to the method in addition to its computational advantages.

L'Ecuyer [224] analyzes and recommends a specific combination of two MRGs: the first has modulus $m = 2^{31} - 1 = 2147483647$ and coefficients $a_1 = 0$, $a_2 = 63308$, $a_3 = -183326$; the second has $m = 2145483479$ and $a_1 = 86098$, $a_2 = 0$, $a_3 = -539608$. The combined generator has a period close to 2^{185}. Results of the spectral tests in L'Ecuyer [224] in dimensions 4–20 indicate far superior uniformity for the combined generator than for either of its components. Because none of the coefficients a_i used in this method is very large, an implementation in integer arithmetic is possible. L'Ecuyer [224] gives an implementation in the C programming language which we reproduce in Figure 2.3. We have modified the introduction of the constants for the generator, using `#define` statements rather than variable declarations for greater speed, as recommended by L'Ecuyer [225]. The variables `x10,...,x22` must be initialized to an arbitrary seed before the first call to the routine.

Figure 2.4 reproduces an implementation from L'Ecuyer [225]. L'Ecuyer [225] reports that this combined generator has a period of approximately 2^{319} and good uniformity properties at least up to dimension 32. The variables `s10,...,s24` must be initialized to an arbitrary seed before the first call to the routine. The multipliers in this generator are too large to permit a 32-bit integer implementation using the method in Figure 2.3, so Figure 2.4 uses floating point arithmetic. L'Ecuyer [225] finds that the relative speeds of the two methods vary with the computing platform.

Other Methods

An alternative strategy for random number generation produces a stream of bits that are concatenated to produce integers and then normalized to produce points in the unit interval. Bits can be produced by linear recursions mod 2; e.g.,

$$b_i = (a_1 b_{i-1} + a_2 b_{i-2} + \cdots a_k b_{i-k}) \bmod 2,$$

```
#define m1 2147483647
#define m2 2145483479
#define a12 63308
#define a13 −183326
#define a21 86098
#define a23 −539608
#define q12 33921
#define q13 11714
#define q21 24919
#define q23 3976
#define r12 12979
#define r13 2883
#define r21 7417
#define r23 2071
#define Invmp1 4.656612873077393e−10;
int x10, x11, x12, x20, x21, x22;

int    Random()
       {
       int h, p12, p13, p21, p23;
       /* Component 1 */
       h = x10/q13; p13 = −a13*(x10−h*q13)−h*r13;
       h = x11/q12; p12 = a12*(x11−h*q12)−h*r12;
       if(p13<0) p13 = p13+m1; if(p12<0) p12 = p12+m1;
       x10 = x11; x11 = x12; x12 = p12−p13; if(x12<0) x12 = x12+m1;
       /* Component 2 */
       h = x20/q23; p23 = −a23*(x20−h*q23)−h*r23;
       h = x22/q21; p21 = a21*(x22−h*q21)−h*r21;
       if(p23<0) p23 = p23+m2; if(p21<0) p21 = p21+m2;
       /* Combination */
       if (x12<x22) return (x12−x22+m1); else return (x12−x22);
       }

double Uniform01()
       {
       int Z;
       Z=Random(); if(Z==0) Z=m1; return (Z*Invmp1);
       }
```

Fig. 2.3. Implementation in C of a combined multiple recursive generator using integer arithmetic. The generator and the implementation are from L'Ecuyer [224].

with all a_i equal to 0 or 1. This method was proposed by Tausworthe [346]. It can be implemented through a mechanism known as a *feedback shift register*. The implementation and theoretical properties of these generators (and also of *generalized* feedback shift register methods) have been studied extensively. Matsumoto and Nishimura [258] develop a generator of this type with a period of $2^{19937} − 1$ and apparently excellent uniformity properties. They provide C code for its implementation.

Inversive congruential generators use recursions of the form

$$x_{i+1} = (ax_i^- + c) \bmod m,$$

where the (mod m)-inverse x^- of x is an integer in $\{1, \ldots, m − 1\}$ (unique if it exists) satisfying $xx^- = 1 \bmod m$. This is an example of a nonlinear congruential generator. Inversive generators are free of the lattice structure

```
double s10, s11, s12, s13, s14, s20, s21, s22, s23, s24;

#define norm 2.3283163396834613e-10
#define m1 4294949027.0
#define m2 4294934327.0
#define a12 1154721.0
#define a14 1739991.0
#define a15n 1108499.0
#define a21 1776413.0
#define a23 865203.0
#define a25n 1641052.0

double MRG32k5a ()
    {
    long k;
    double p1, p2;
    /* Component 1 */
    p1 = a12 * s13 − a15n * s10;
    if (p1 > 0.0) p1 −= a14 * m1;
    p1 += a14 * s11; k = p1 / m1; p1 −= k * m1;
    if (p1 < 0.0) p1 += m1;
    s10 = s11; s11 = s12; s12 = s13; s13 = s14; s14 = p1;
    /* Component 2 */
    p2 = a21 * s24 − a25n * s20;
    if (p2 > 0.0) p2 −= a23 * m2;
    p2 += a23 * s22; k = p2 / m2; p2 −= k * m2;
    if (p2 < 0.0) p2 += m2;
    s20 = s21; s21 = s22; s22 = s23; s23 = s24; s24 = p2;
    /* Combination */
    if (p1 <= p2) return ((p1 − p2 + m1) * norm);
    else return ((p1 − p2) * norm);
    }
```

Fig. 2.4. Implementation in C of a combined multiple recursive generator using floating point arithmetic. The generator and implementation are from L'Ecuyer [225].

characteristic of linear congruential generators but they are much more computationally demanding. They may be useful for comparing results in cases where the deficiencies of a random number generator are cause for concern. See Eichenauer-Herrmann, Herrmann, and Wegenkittl [110] for a survey of this approach and additional references.

2.2 General Sampling Methods

With an introduction to random number generation behind us, we henceforth assume the availability of an ideal sequence of random numbers. More precisely, we assume the availability of a sequence $U_1, U_2, \ldots$ of independent random variables, each satisfying

$$P(U_i \leq u) = \begin{cases} 0, u < 0 \\ u, 0 \leq u \leq 1 \\ 1, u > 1 \end{cases} \tag{2.12}$$

i.e., each uniformly distributed between 0 and 1. A simulation algorithm transforms these independent uniforms into sample paths of stochastic processes.

Most simulations entail sampling random variables or random vectors from distributions other than the uniform. A typical simulation uses methods for transforming samples from the uniform distribution to samples from other distributions. There is a large literature on both general purpose methods and specialized algorithms for specific cases. In this section, we present two of the most widely used general techniques: the inverse transform method and the acceptance-rejection method.

2.2.1 Inverse Transform Method

Suppose we want to sample from a cumulative distribution function F; i.e., we want to generate a random variable X with the property that $P(X \leq x) = F(x)$ for all x. The inverse transform method sets

$$X = F^{-1}(U), \quad U \sim \text{Unif}[0, 1], \tag{2.13}$$

where F^{-1} is the inverse of F and Unif[0,1] denotes the uniform distribution on $[0, 1]$.

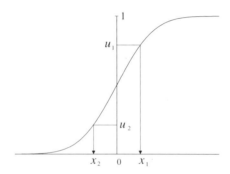

Fig. 2.5. Inverse transform method.

This transformation is illustrated in Figure 2.5 for a hypothetical cumulative distribution F. In the figure, values of u between 0 and $F(0)$ are mapped to negative values of x whereas values between $F(0)$ and 1 are mapped to positive values. The left panel of Figure 2.6 depicts a cumulative distribution function with a jump at x_0; i.e.,

$$\lim_{x \uparrow x_0} F(x) \equiv F(x-) < F(x+) \equiv \lim_{x \downarrow x_0} F(x).$$

Under the distribution F, the outcome x_0 has probability $F(x+) - F(x-)$. As indicated in the figure, all values of u between $u_1 = F(x-)$ and $u_2 = F(x+)$ are mapped to x_0.

The inverse of F is well-defined if F is strictly increasing; otherwise, we need a rule to break ties. For example, we may set

$$F^{-1}(u) = \inf\{x : F(x) \geq u\}; \tag{2.14}$$

if there are many values of x for which $F(x) = u$, this rule chooses the smallest.

We need a rule like (2.14) in cases where the cumulative distribution F has flat sections, because the inverse of F is not well-defined at such points; see, e.g., the right panel of Figure 2.6. Observe, however, that if F is constant over an interval $[a, b]$ and if X has distribution F, then

$$P(a < X \leq b) = F(b) - F(a) = 0,$$

so flat sections of F correspond to intervals of zero probability for the random variable. If F has a continuous density, then F is strictly increasing (and its inverse is well-defined) anywhere the density is nonzero.

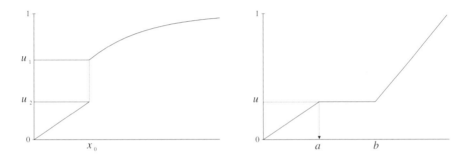

Fig. 2.6. Inverse transform for distributions with jumps (left) or flat sections (right).

To verify that the inverse transform (2.13) generates samples from F, we check the distribution of the X it produces:

$$\begin{aligned} P(X \leq x) &= P(F^{-1}(U) \leq x) \\ &= P(U \leq F(x)) \\ &= F(x). \end{aligned}$$

The second equality follows from the fact that, with F^{-1} as we have defined it, the events $\{F^{-1}(u) \leq x\}$ and $\{u \leq F(x)\}$ coincide for all u and x. The last equality follows from (2.12).

One may interpret the input U to the inverse transform method as a *random percentile*. If F is continuous and $X \sim F$, then X is just as likely to fall between, say, the 20th and 30th percentiles of F as it is to fall between the 85th and 95th. In other words, the percentile at which X falls (namely $F(X)$) is uniformly distributed. The inverse transform method chooses a percentile level uniformly and then maps it to a corresponding value of the random variable.

We illustrate the method with examples. These examples also show that a direct implementation of the inverse transform method can sometimes be made more efficient through minor modifications.

Example 2.2.1 *Exponential distribution.* The exponential distribution with mean θ has distribution

$$F(x) = 1 - e^{-x/\theta}, \quad x \geq 0.$$

This is, for example, the distribution of the times between jumps of a Poisson process with rate $1/\theta$. Inverting the exponential distribution yields the algorithm $X = -\theta \log(1 - U)$. This can also be implemented as

$$X = -\theta \log(U) \tag{2.15}$$

because U and $1 - U$ have the same distribution. □

Example 2.2.2 *Arcsine law.* The time at which a standard Brownian motion attains its maximum over the time interval $[0, 1]$ has distribution

$$F(x) = \frac{2}{\pi} \arcsin(\sqrt{x}), \quad 0 \leq x \leq 1.$$

The inverse transform method for sampling from this distribution is $X = \sin^2(U\pi/2)$, $U \sim \mathrm{Unif}[0,1]$. Using the identity $2\sin^2(t) = 1 - \cos(2t)$ for $0 \leq t \leq \pi/2$, we can simplify the transformation to

$$X = \tfrac{1}{2} - \tfrac{1}{2}\cos(U\pi), \quad U \sim \mathrm{Unif}[0, 1].$$

□

Example 2.2.3 *Rayleigh distribution.* If we condition a standard Brownian motion starting at the origin to be at b at time 1, then its maximum over $[0, 1]$ has the Rayleigh distribution

$$F(x) = 1 - e^{-2x(x-b)}, \quad x \geq b.$$

Solving the equation $F(x) = u$, $u \in (0, 1)$, results in a quadratic with roots

$$x = \frac{b}{2} \pm \frac{\sqrt{b^2 - 2\log(1 - u)}}{2}.$$

The inverse of F is given by the larger of the two roots — in particular, we must have $x \geq b$ since the maximum of the Brownian path must be at least as large as the terminal value. Thus, replacing $1 - U$ with U as we did in Example 2.2.1, we arrive at

$$X = \frac{b}{2} + \frac{\sqrt{b^2 - 2\log(U)}}{2}.$$

□

Even if the inverse of F is not known explicitly, the inverse transform method is still applicable through numerical evaluation of F^{-1}. Computing $F^{-1}(u)$ is equivalent to finding a root x of the equation $F(x) - u = 0$. For a distribution F with density f, Newton's method for finding roots produces a sequence of iterates

$$x_{n+1} = x_n - \frac{F(x_n) - u}{f(x_n)},$$

given a starting point x_0. In the next example, root finding takes a special form.

Example 2.2.4 *Discrete distributions.* In the case of a discrete distribution, evaluation of F^{-1} reduces to a table lookup. Consider, for example, a discrete random variable whose possible values are $c_1 < \cdots < c_n$. Let p_i be the probability attached to c_i, $i = 1, \ldots, n$, and set $q_0 = 0$,

$$q_i = \sum_{j=1}^{i} p_j, \quad i = 1, \ldots, n.$$

These are the cumulative probabilities associated with the c_i; that is, $q_i = F(c_i)$, $i = 1, \ldots, n$. To sample from this distribution,

(i)   generate a uniform U;
(ii) find $K \in \{1, \ldots, n\}$ such that $q_{K-1} < U \leq q_K$;
(iii) set $X = c_K$.

The second step can be implemented through binary search. Bratley, Fox, and Schrage [59], and Fishman [121] discuss potentially faster methods. $\square$

Our final example illustrates a general feature of the inverse transform method rather than a specific case.

Example 2.2.5 *Conditional distributions.* Suppose X has distribution F and consider the problem of sampling X conditional on $a < X \leq b$, with $F(a) < F(b)$. Using the inverse transform method, this is no more difficult than generating X unconditionally. If $U \sim \text{Unif}[0,1]$, then the random variable V defined by

$$V = F(a) + [F(b) - F(a)]U$$

is uniformly distributed between $F(a)$ and $F(b)$, and $F^{-1}(V)$ has the desired conditional distribution. To see this, observe that

$$\begin{aligned}
P(F^{-1}(V) \leq x) &= P(F(a) + [F(b) - F(a)]U \leq F(x)) \\
&= P(U \leq [F(x) - F(a)]/[F(b) - F(a)]) \\
&= [F(x) - F(a)]/[F(b) - F(a)],
\end{aligned}$$

and this is precisely the distribution of X given $a < X \leq b$. Either of the endpoints a, b could be infinite in this example. $\square$

The inverse transform method is seldom the fastest method for sampling from a distribution, but it has important features that make it attractive nevertheless. One is its use in sampling from conditional distributions just illustrated; we point out two others. First, the inverse transform method maps the input U monotonically and — if F is strictly increasing — continuously to the output X. This can be useful in the implementation of variance reduction techniques and in sensitivity estimation, as we will see in Chapters 4 and 7. Second, the inverse transform method requires just one uniform random variable for each sample generated. This is particularly important in using quasi-Monte Carlo methods where the *dimension* of a problem is often equal to the number of uniforms needed to generate one "path." Methods that require multiple uniforms per variable generated result in higher-dimensional representations for which quasi-Monte Carlo may be much less effective.

2.2.2 Acceptance-Rejection Method

The acceptance-rejection method, introduced by Von Neumann [353], is among the most widely applicable mechanisms for generating random samples. This method generates samples from a target distribution by first generating candidates from a more convenient distribution and then rejecting a random subset of the generated candidates. The rejection mechanism is designed so that the accepted samples are indeed distributed according to the target distribution. The technique is by no means restricted to univariate distributions.

Suppose, then, that we wish to generate samples from a density f defined on some set $\mathcal{X}$. This could be a subset of the real line, of $\Re^d$, or a more general set. Let g be a density on $\mathcal{X}$ from which we know how to generate samples and with the property that

$$f(x) \leq cg(x), \quad \text{for all } x \in \mathcal{X}$$

for some constant c. In the acceptance-rejection method, we generate a sample X from g and accept the sample with probability $f(X)/cg(X)$; this can be implemented by sampling U uniformly over $(0,1)$ and accepting X if $U \leq f(X)/cg(X)$. If X is rejected, a new candidate is sampled from g and the acceptance test applied again. The process repeats until the acceptance test is passed; the accepted value is returned as a sample from f. Figure 2.7 illustrates a generic implementation.

To verify the validity of the acceptance-rejection method, let Y be a sample returned by the algorithm and observe that Y has the distribution of X conditional on $U \leq f(X)/cg(X)$. Thus, for any $A \subseteq \mathcal{X}$,

$$P(Y \in A) = P(X \in A | U \leq f(X)/cg(X))$$
$$= \frac{P(X \in A, U \leq f(X)/cg(X))}{P(U \leq f(X)/cg(X))}. \tag{2.16}$$

Given X, the probability that $U \leq f(X)/cg(X)$ is simply $f(X)/cg(X)$ because U is uniform; hence, the denominator in (2.16) is given by

> 1. generate X from distribution g
> 2. generate U from Unif[0,1]
> 3. if $U \leq f(X)/cg(X)$
>   return X
>   otherwise
>   go to Step 1.

Fig. 2.7. The acceptance-rejection method for sampling from density f using candidates from density g.

$$P(U \leq f(X)/cg(X)) = \int_{\mathcal{X}} \frac{f(x)}{cg(x)} g(x)\, dx = 1/c \qquad (2.17)$$

(taking $0/0 = 1$ if $g(x) = 0$ somewhere on $\mathcal{X}$). Making this substitution in (2.16), we find that

$$P(Y \in A) = cP(X \in A, U \leq f(X)/cg(X)) = c \int_{A} \frac{f(x)}{cg(x)} g(x)\, dx = \int_{A} f(x)\, dx.$$

Since A is arbitrary, this verifies that Y has density f.

In fact, this argument shows more: Equation (2.17) shows that the probability of acceptance on each attempt is $1/c$. Because the attempts are mutually independent, the number of candidates generated until one is accepted is geometrically distributed with mean c. It is therefore preferable to have c close to 1 (it can never be less than 1 if f and g both integrate to 1). Tighter bounds on the target density f result in fewer wasted samples from g. Of course, a prerequisite for the method is the ability to sample from g; the speed of the method depends on both c and the effort involved in sampling from g.

We illustrate the method with examples.

Example 2.2.6 *Beta distribution.* The beta density on $[0, 1]$ with parameters $\alpha_1, \alpha_2 > 0$ is given by

$$f(x) = \frac{1}{B(\alpha_1, \alpha_2)} x^{\alpha_1 - 1} (1 - x)^{\alpha_2 - 1}, \quad 0 \leq x \leq 1,$$

with

$$B(\alpha_1, \alpha_2) = \int_0^1 x^{\alpha_1 - 1} (1 - x)^{\alpha_2 - 1}\, dx = \frac{\Gamma(\alpha_1)\Gamma(\alpha_2)}{\Gamma(\alpha_1 + \alpha_2)}$$

and Γ the gamma function. Varying the parameters α_1, α_2 results in a variety of shapes, making this a versatile family of distributions with bounded support. Among many other applications, beta distributions are used to model the random recovery rate (somewhere between 0 and 100%) upon default of a bond subject to credit risk. The case $\alpha_1 = \alpha_2 = 1/2$ is the arcsine distribution considered in Example 2.2.2.

If $\alpha_1, \alpha_2 \geq 1$ and at least one of the parameters exceeds 1, the beta density is unimodal and achieves its maximum at $(\alpha_1 - 1)/(\alpha_1 + \alpha_2 - 2)$. Let c be the value of the density f at this point. Then $f(x) \leq c$ for all x, so we may choose g to be the uniform density $(g(x) = 1, 0 \leq x \leq 1)$, which is in fact the beta density with parameters $\alpha_1 = \alpha_2 = 1$. In this case, the acceptance-rejection method becomes

> Generate U_1, U_2 from Unif[0,1] until $cU_2 \leq f(U_1)$
> Return U_1

This is illustrated in Figure 2.8 for parameters $\alpha_1 = 3$, $\alpha_2 = 2$.

As is clear from Figure 2.8, generating candidates from the uniform distribution results in many rejected samples and thus many evaluations of f. (The expected number of candidates generated for each accepted sample is $c \approx 1.778$ for the density in the figure.) Faster methods for sampling from beta distributions — combining more carefully designed acceptance-rejection schemes with the inverse transform and other methods — are detailed in Devroye [95], Fishman [121], Gentle [136], and Johnson, Kotz, and Balakrishnan [202]. □

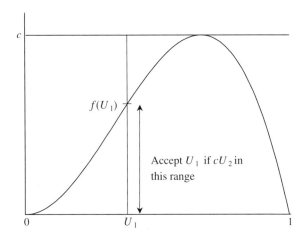

Fig. 2.8. Illustration of the acceptance-rejection method using uniformly distributed candidates.

Example 2.2.7 *Normal from double exponential.* Fishman [121, p.173] illustrates the use of the acceptance-rejection method by generating half-normal samples from the exponential distribution. (A half-normal random variable has the distribution of the absolute value of a normal random variable.) Fishman also notes that the method can be used to generate normal random variables and we present the example in this form. Because of its importance in financial applications, we devote all of Section 2.3 to the normal distribution; we include this example here primarily to further illustrate acceptance-rejection.

The double exponential density on $(-\infty, \infty)$ is $g(x) = \exp(-|x|)/2$ and the normal density is $f(x) = \exp(-x^2/2)/\sqrt{2\pi}$. The ratio is

$$\frac{f(x)}{g(x)} = \sqrt{\frac{2}{\pi}} e^{-\frac{1}{2}x^2 + |x|} \leq \sqrt{\frac{2e}{\pi}} \approx 1.3155 \equiv c.$$

Thus, the normal density is dominated by the scaled double exponential density $cg(x)$, as illustrated in Figure 2.9. A sample from the double exponential density can be generated using (2.15) to draw a standard exponential random variable and then randomizing the sign. The rejection test $u > f(x)/cg(x)$ can be implemented as

$$u > \exp(-\tfrac{1}{2}x^2 + |x| - \tfrac{1}{2}) = \exp(-\tfrac{1}{2}(|x| - 1)^2).$$

In light of the symmetry of both f and g, it suffices to generate positive samples and determine the sign only if the sample is accepted; in this case, the absolute value is unnecessary in the rejection test. The combined steps are as follows:

1. generate U_1, U_2, U_3 from Unif[0,1]
2. $X \leftarrow -\log(U_1)$
3. if $U_2 > \exp(-0.5(X-1)^2)$
      go to Step 1
4. if $U_3 \leq 0.5$
      $X \leftarrow -X$
5. return X

□

Example 2.2.8 *Conditional distributions.* Consider the problem of generating a random variable or vector X conditional on $X \in A$, for some set A. In the scalar case, this can be accomplished using the inverse transform method if A is an interval; see Example 2.2.5. In more general settings it may be difficult to sample directly from the conditional distribution. However, so long as it is possible to generate unconditional samples, one may always resort to the following crude procedure:

    Generate X until $X \in A$
    return X

This may be viewed as a degenerate form of acceptance-rejection. Let f denote the conditional density and let g denote the unconditional density; then

$$f(x)/g(x) = \begin{cases} 1/P(X \in A), & x \in A \\ 0, & x \notin A. \end{cases}$$

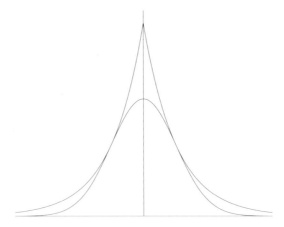

Fig. 2.9. Normal density and scaled double exponential.

Thus, $c = 1/P(X \in A)$ is an upper bound on the ratio. Moreover, since the ratio $f(x)/cg(x)$ is either 0 or 1 at every x, it is unnecessary to randomize the rejection decision: a candidate X is accepted precisely if $X \in A$. $\square$

Acceptance-rejection can often be accelerated through the *squeeze* method, in which simpler tests are applied before the exact acceptance threshold $f(x)/cg(x)$ is evaluated. The simpler tests are based on functions that bound $f(x)/cg(x)$ from above and below. The effectiveness of this method depends on the quality of the bounding functions and the speed with which they can be evaluated. See Fishman [121] for a detailed discussion.

Although we have restricted attention to sampling from densities, it should be clear that the acceptance-rejection method also applies when f and g are replaced with the mass functions of discrete distributions.

The best methods for sampling from a specific distribution invariably rely on special features of the distribution. Acceptance-rejection is frequently combined with other techniques to exploit special features — it is perhaps more a principle than a method.

At the end of Section 2.2.1 we noted that one attractive feature of the inverse transform method is that it uses exactly one uniform random variable per nonuniform random variable generated. When simulation problems are formulated as numerical integration problems, the dimension of the integrand is typically the maximum number of uniform variables needed to generate a simulation "path." The effectiveness of quasi-Monte Carlo and related integration methods generally deteriorates as the dimension increases, so in using those methods, we prefer representations that keep the dimension as small as possible. With an acceptance-rejection method, there is ordinarily no upper bound on the number of uniforms required to generate even a single nonuniform variable; simulations that use acceptance-rejection therefore correspond to infinite-dimensional integration problems. For this reason, acceptance-rejection methods are generally inapplicable with quasi-Monte Carlo methods. A further potential drawback of acceptance-rejection methods, compared with the inverse transform method, is that their outputs are generally neither continuous nor monotone functions of the input uniforms. This can diminish the effectiveness of the antithetic variates method, for example.

2.3 Normal Random Variables and Vectors

Normal random variables are the building blocks of many financial simulation models, so we discuss methods for sampling from normal distributions in detail. We begin with a brief review of basic properties of normal distributions.

2.3.1 Basic Properties

The standard univariate normal distribution has density

$$\phi(x) = \frac{1}{\sqrt{2\pi}} e^{-x^2/2}, \quad -\infty < x < \infty \tag{2.18}$$

and cumulative distribution function

$$\Phi(x) = \frac{1}{\sqrt{2\pi}} \int_{-\infty}^{x} e^{-u^2/2} \, du. \tag{2.19}$$

Standard indicates mean 0 and variance 1. More generally, the normal distribution with mean μ and variance σ^2, $\sigma > 0$, has density

$$\phi_{\mu,\sigma}(x) = \frac{1}{\sqrt{2\pi}\sigma} e^{-\frac{(x-\mu)^2}{2\sigma^2}}$$

and cumulative distribution

$$\Phi_{\mu,\sigma}(x) = \Phi\left(\frac{x-\mu}{\sigma}\right).$$

The notation $X \sim N(\mu, \sigma^2)$ abbreviates the statement that the random variable X is normally distributed with mean μ and σ^2.

If $Z \sim N(0, 1)$ (i.e., Z has the standard normal distribution), then

$$\mu + \sigma Z \sim N(\mu, \sigma^2).$$

Thus, given a method for generating samples $Z_1, Z_2, \ldots$ from the standard normal distribution, we can generate samples $X_1, X_2, \ldots$ from $N(\mu, \sigma^2)$ by setting $X_i = \mu + \sigma Z_i$. It therefore suffices to consider methods for sampling from $N(0, 1)$.

A d-dimensional normal distribution is characterized by a d-vector μ and a $d \times d$ covariance matrix Σ; we abbreviate it as $N(\mu, \Sigma)$. To qualify as a covariance matrix, Σ must be symmetric (i.e., Σ and its transpose $\Sigma^\top$ are equal) and positive semidefinite, meaning that

$$x^\top \Sigma x \geq 0 \qquad (2.20)$$

for all $x \in \Re^d$. This is equivalent to the requirement that all eigenvalues of Σ be nonnegative. (As a symmetric matrix, Σ automatically has real eigenvalues.) If Σ is positive definite (meaning that strict inequality holds in (2.20) for all nonzero $x \in \Re^d$ or, equivalently, that all eigenvalues of Σ are positive), then the normal distribution $N(\mu, \Sigma)$ has density

$$\phi_{\mu, \Sigma}(x) = \frac{1}{(2\pi)^{d/2}|\Sigma|^{1/2}} \exp\left(-\tfrac{1}{2}(x - \mu)^\top \Sigma^{-1}(x - \mu)\right), \quad x \in \Re^d, \quad (2.21)$$

with $|\Sigma|$ the determinant of Σ. The *standard* d-dimensional normal $N(0, I_d)$, with I_d the $d \times d$ identity matrix, is the special case

$$\frac{1}{(2\pi)^{d/2}} \exp\left(-\tfrac{1}{2}x^\top x\right).$$

If $X \sim N(\mu, \Sigma)$ (i.e., the random vector X has a multivariate normal distribution), then its ith component X_i has distribution $N(\mu_i, \sigma_i^2)$, with $\sigma_i^2 = \Sigma_{ii}$. The ith and jth components have covariance

$$\mathsf{Cov}[X_i, X_j] = \mathsf{E}[(X_i - \mu_i)(X_j - \mu_j)] = \Sigma_{ij},$$

which justifies calling Σ the covariance matrix. The correlation between X_i and X_j is given by

$$\rho_{ij} = \frac{\Sigma_{ij}}{\sigma_i \sigma_j}.$$

In specifying a multivariate distribution, it is sometimes convenient to use this definition in the opposite direction: specify the marginal standard deviations σ_i, $i = 1, \ldots, d$, and the correlations ρ_{ij} from which the covariance matrix

$$\Sigma_{ij} = \sigma_i \sigma_j \rho_{ij} \qquad (2.22)$$

is then determined.

If the $d \times d$ symmetric matrix Σ is positive semidefinite but not positive definite then the rank of Σ is less than d, Σ fails to be invertible, and there is no normal density with covariance matrix Σ. In this case, we can define the normal distribution $N(\mu, \Sigma)$ as the distribution of $X = \mu + AZ$ with $Z \sim N(0, I_d)$ for any $d \times d$ matrix A satisfying $AA^\top = \Sigma$. The resulting distribution is independent of which such A is chosen. The random vector X does not have a density in $\Re^d$, but if Σ has rank k then one can find k components of X with a multivariate normal density in $\Re^k$.

Three further properties of the multivariate normal distribution merit special mention:

Linear Transformation Property: Any linear transformation of a normal vector is again normal:

$$X \sim N(\mu, \Sigma) \Rightarrow AX \sim N(A\mu, A\Sigma A^{\top}), \tag{2.23}$$

for any d-vector μ, and $d \times d$ matrix Σ, and any $k \times d$ matrix A, for any k.

Conditioning Formula: Suppose the partitioned vector $(X_{[1]}, X_{[2]})$ (where each $X_{[i]}$ may itself be a vector) is multivariate normal with

$$\begin{pmatrix} X_{[1]} \\ X_{[2]} \end{pmatrix} \sim N\left(\begin{pmatrix} \mu_{[1]} \\ \mu_{[2]} \end{pmatrix}, \begin{pmatrix} \Sigma_{[11]} & \Sigma_{[12]} \\ \Sigma_{[21]} & \Sigma_{[22]} \end{pmatrix} \right), \tag{2.24}$$

and suppose $\Sigma_{[22]}$ has full rank. Then

$$(X_{[1]} | X_{[2]} = x) \sim N(\mu_{[1]} + \Sigma_{[12]}\Sigma_{[22]}^{-1}(x - \mu_{[2]}), \Sigma_{[11]} - \Sigma_{[12]}\Sigma_{[22]}^{-1}\Sigma_{[21]}). \tag{2.25}$$

In (2.24), the dimensions of the $\mu_{[i]}$ and $\Sigma_{[ij]}$ are consistent with those of the $X_{[i]}$. Equation (2.25) then gives the distribution of $X_{[1]}$ conditional on $X_{[2]} = x$.

Moment Generating Function: If $X \sim N(\mu, \Sigma)$ with X d-dimensional, then

$$\mathsf{E}[\exp(\theta^{\top} X)] = \exp\left(\mu^{\top}\theta + \tfrac{1}{2}\theta^{\top}\Sigma\theta \right) \tag{2.26}$$

for all $\theta \in \Re^d$.

2.3.2 Generating Univariate Normals

We now discuss algorithms for generating samples from univariate normal distributions. As noted in the previous section, it suffices to consider sampling from $N(0, 1)$. We assume the availability of a sequence $U_1, U_2, \ldots$ of independent random variables uniformly distributed on the unit interval $[0, 1]$ and consider methods for transforming these uniform random variables to normally distributed random variables.

Box-Muller Method

Perhaps the simplest method to implement (though not the fastest or necessarily the most convenient) is the Box-Muller [51] algorithm. This algorithm generates a sample from the bivariate standard normal, each component of which is thus a univariate standard normal. The algorithm is based on the following two properties of the bivariate normal: if $Z \sim N(0, I_2)$, then

(i) $R = Z_1^2 + Z_2^2$ is exponentially distributed with mean 2, i.e.,

$$P(R \leq x) = 1 - e^{-x/2};$$

(ii) given R, the point (Z_1, Z_2) is uniformly distributed on the circle of radius $\sqrt{R}$ centered at the origin.

Thus, to generate (Z_1, Z_2), we may first generate R and then choose a point uniformly from the circle of radius $\sqrt{R}$. To sample from the exponential distribution we may set $R = -2\log(U_1)$, with $U_1 \sim \text{Unif}[0,1]$, as in (2.15). To generate a random point on a circle, we may generate a random angle uniformly between 0 and 2π and then map the angle to a point on the circle. The random angle may be generated as $V = 2\pi U_2$, $U_2 \sim \text{Unif}[0,1]$; the corresponding point on the circle has coordinates $(\sqrt{R}\cos(V), \sqrt{R}\sin(V))$. The complete algorithm is given in Figure 2.10.

> generate U_1, U_2 independent $\text{Unif}[0,1]$
> $R \leftarrow -2\log(U_1)$
> $V \leftarrow 2\pi U_2$
> $Z_1 \leftarrow \sqrt{R}\cos(V),\ Z_2 \leftarrow \sqrt{R}\sin(V)$
> return Z_1, Z_2.

Fig. 2.10. Box-Muller algorithm for generating normal random variables.

Marsaglia and Bray [250] developed a modification of the Box-Muller method that reduces computing time by avoiding evaluation of the sine and cosine functions. The Marsaglia-Bray method instead uses acceptance-rejection to sample points uniformly in the unit disc and then transforms these points to normal variables.

The algorithm is illustrated in Figure 2.11. The transformation $U_i \leftarrow 2U_i - 1$, $i = 1, 2$, makes (U_1, U_2) uniformly distributed over the square $[-1,1] \times [-1,1]$. Accepting only those pairs for which $X = U_1^2 + U_2^2$ is less than or equal to 1 produces points uniformly distributed over the disc of radius 1 centered at the origin. Conditional on acceptance, X is uniformly distributed between 0 and 1, so the $\log X$ in Figure 2.11 has the same effect as the $\log U_1$ in Figure 2.10. Dividing each accepted (U_1, U_2) by $\sqrt{X}$ projects it from the unit disc to the unit circle, on which it is uniformly distributed. Moreover, $(U_1/\sqrt{X}, U_2/\sqrt{X})$ is independent of X conditional on $X \leq 1$. Hence, the justification for the last step in Figure 2.11 is the same as that for the Box-Muller method.

As is the case with most acceptance-rejection methods, there is no upper bound on the number of uniforms the Marsaglia-Bray algorithm may use to generate a single normal variable (or pair of variables). This renders the method inapplicable with quasi-Monte Carlo simulation.

```
while (X > 1)
    generate U₁, U₂ ~ Unif[0,1]
    U₁ ← 2 * U₁ - 1,   U₂ ← 2 * U₂ - 1
    X ← U₁² + U₂²
end
Y ← √(-2 log X/X)
Z₁ ← U₁Y,   Z₂ ← U₂Y
return Z₁, Z₂.
```

Fig. 2.11. Marsaglia-Bray algorithm for generating normal random variables.

Approximating the Inverse Normal

Applying the inverse transform method to the normal distribution entails evaluation of Φ^{-1}. At first sight, this may seem infeasible. However, there is really no reason to consider Φ^{-1} any less tractable than, e.g., a logarithm. Neither can be computed exactly in general, but both can be approximated with sufficient accuracy for applications. We discuss some specific methods for evaluating Φ^{-1}.

Because of the symmetry of the normal distribution,

$$\Phi^{-1}(1-u) = -\Phi^{-1}(u), \quad 0 < u < 1;$$

it therefore suffices to approximate Φ^{-1} on the interval $[0.5, 1)$ (or the interval $(0, 0.5]$) and then to use the symmetry property to extend the approximation to the rest of the unit interval. Beasley and Springer [43] provide a rational approximation

$$\Phi^{-1}(u) \approx \frac{\sum_{n=0}^{3} a_n (u - \frac{1}{2})^{2n+1}}{1 + \sum_{n=0}^{3} b_n (u - \frac{1}{2})^{2n}}, \tag{2.27}$$

for $0.5 \le u \le 0.92$, with constants a_n, b_n given in Figure 2.12; for $u > 0.92$ they use a rational function of $\sqrt{\log(1-u)}$. Moro [271] reports greater accuracy in the tails by replacing the second part of the Beasley-Springer approximation with a Chebyshev approximation

$$\Phi^{-1}(u) \approx g(u) = \sum_{n=0}^{8} c_n [\log(-\log(1-u))]^n, \quad 0.92 \le u < 1, \tag{2.28}$$

with constants c_n again given in Figure 2.12. Using the symmetry rule, this gives

$$\Phi^{-1}(u) \approx -g(1-u) \quad 0 < u \le .08.$$

With this modification, Moro [271] finds a maximum absolute error of 3×10^{-9} out to seven standard deviations (i.e., over the range $\Phi(-7) \le u \le \Phi(7)$). The combined algorithm from Moro [271] is given in Figure 2.13.

$$
\begin{array}{llll}
a_0 = & 2.50662823884 & b_0 = & \text{-8.47351093090} \\
a_1 = & \text{-18.61500062529} & b_1 = & 23.08336743743 \\
a_2 = & 41.39119773534 & b_2 = & \text{-21.06224101826} \\
a_3 = & \text{-25.44106049637} & b_3 = & 3.13082909833
\end{array}
$$

$$
\begin{array}{ll}
c_0 = 0.3374754822726147 & c_5 = 0.0003951896511919 \\
c_1 = 0.9761690190917186 & c_6 = 0.0000321767881768 \\
c_2 = 0.1607979714918209 & c_7 = 0.0000002888167364 \\
c_3 = 0.0276438810333863 & c_8 = 0.0000003960315187 \\
c_4 = 0.0038405729373609 &
\end{array}
$$

Fig. 2.12. Constants for approximations to inverse normal.

Input: u between 0 and 1
Output: x, approximation to $\Phi^{-1}(u)$.
$y \leftarrow u - 0.5$
if $|y| < 0.42$
  $r \leftarrow y * y$
  $x \leftarrow y * (((a_3 * r + a_2) * r + a_1) * r + a_0)/$
          $((((b_3 * r + b_2) * r + b_1) * r + b_0) * r + 1)$
else
  $r \leftarrow u;$
  if $(y > 0)$ $r \leftarrow 1 - u$
  $r \leftarrow \log(- \log(r))$
  $x \leftarrow c_0 + r * (c_1 + r * (c_2 + r * (c_3 + r * (c_4 +$
          $r * (c_5 + r * (c_6 + r * (c_7 + r * c_8))))))))$
  if $(y < 0)$ $x \leftarrow -x$
return x

Fig. 2.13. Beasley-Springer-Moro algorithm for approximating the inverse normal.

The problem of computing $\Phi^{-1}(u)$ can be posed as one of finding the root x of the equation $\Phi(x) = u$ and in principle addressed through any general root-finding algorithm. Newton's method, for example, produces the iterates

$$
x_{n+1} = x_n - \frac{\Phi(x_n) - u}{\phi(x_n)},
$$

or, more explicitly,

$$
x_{n+1} = x_n + (u - \Phi(x_n)) \exp(-0.5 x_n \cdot x_n + c), \quad c \equiv \log(\sqrt{2\pi}).
$$

Marsaglia, Zaman, and Marsaglia [251] recommend the starting point

$$
x_0 = \pm \sqrt{|-1.6 \log(1.0004 - (1 - 2u)^2)|},
$$

the sign depending on whether $u \geq 0$ or $u < 0$. This starting point gives a surprisingly good approximation to $\Phi^{-1}(u)$. A root-finding procedure is useful when extreme precision is more important than speed — for example, in tabulating "exact" values or evaluating approximations. Also, a small number of Newton steps can be appended to an approximation like the one in Figure 2.13 to further improve accuracy. Adding just a single step to Moro's [271] algorithm appears to reduce the maximum error to the order of 10^{-15}.

Approximating the Cumulative Normal

Of course, the application of Newton's method presupposes the ability to evaluate Φ itself quickly and accurately. Evaluation of the cumulative normal is necessary for many financial applications (including evaluation of the Black-Scholes formula), so we include methods for approximating this function. We present two methods; the first is faster and the second is more accurate, but both are probably fast enough and accurate enough for most applications.

The first method, based on work of Hastings [171], is one of several included in Abramowitz and Stegun [3]. For $x \geq 0$, it takes the form

$$\Phi(x) \approx 1 - \phi(x)(b_1 t + b_2 t^2 + b_3 t^3 + b_4 t^4 + b_5 t^5), \quad t = \frac{1}{1 + px},$$

for constants b_i and p. The approximation extends to negative arguments through the identity $\Phi(-x) = 1 - \Phi(x)$. The necessary constants and an explicit algorithm for this approximation are given in Figure 2.14. According to Hastings [171, p.169], this method has a maximum absolute error less than 7.5×10^{-8}.

$b_1 = 0.319381530 \qquad p = 0.2316419$
$b_2 = -0.356563782 \qquad c = \log(\sqrt{2\pi}) = 0.918938533204672$
$b_3 = 1.781477937$
$b_4 = -1.821255978$
$b_5 = 1.330274429$

Input: x
Output: y, approximation to $\Phi(x)$
$a \leftarrow |x|$
$t \leftarrow 1/(1 + a * p)$
$s \leftarrow ((((b_5 * t + b_4) * t + b_3) * t + b_2) * t + b_1) * t$
$y \leftarrow s * \exp(-0.5 * x * x - c)$
if $(x > 0)$ $y \leftarrow 1 - y$
return y;

Fig. 2.14. Hastings' [171] approximation to the cumulative normal distribution as modified in Abramowitz and Stegun [3].

The second method we include is from Marsaglia et al. [251]. Like the Hastings approximation above, this method is based on approximating the ratio $(1 - \Phi(x))/\phi(x)$. According to Marsaglia et al. [251], as an approximation to the tail probability $1 - \Phi(x)$ this method has a maximum *relative* error of 10^{-15} for $0 \leq x \leq 6.23025$ and 10^{-12} for larger x. (Relative error is much more stringent than absolute error in this setting; a small absolute error is easily achieved for large x using the approximation $1 - \Phi(x) \approx 0$.) This method takes about three times as long as the Hastings approximation, but both methods are very fast. The complete algorithm appears in Figure 2.15.

$$v_1 = 1.253314137315500 \qquad v_9 = 0.1231319632579329$$
$$v_2 = 0.6556795424187985 \qquad v_{10} = 0.1097872825783083$$
$$v_3 = 0.4213692292880545 \qquad v_{11} = 0.09902859647173193$$
$$v_4 = 0.3045902987101033 \qquad v_{12} = 0.09017567550106468$$
$$v_5 = 0.2366523829135607 \qquad v_{13} = 0.08276628650136917$$
$$v_6 = 0.1928081047153158 \qquad v_{14} = 0.0764757610162485$$
$$v_7 = 0.1623776608968675 \qquad v_{15} = 0.07106958053885211$$
$$v_8 = 0.1401041834530502$$
$$c = \log(\sqrt{2\pi}) = 0.918938533204672$$

Input: x between -15 and 15
Output: y, approximation to $\Phi(x)$.
$j \leftarrow \lfloor \min(|x| + 0.5, 14) \rfloor$
$z \leftarrow j, \quad h \leftarrow |x| - z, \quad a \leftarrow v_{j+1}$
$b \leftarrow z * a - 1, \quad q \leftarrow 1, \quad s \leftarrow a + h * b$
for $i = 2, 4, 6, \ldots, 24 - j$
$\quad a \leftarrow (a + z * b)/i$
$\quad b \leftarrow (b + z * a)/(i + 1)$
$\quad q \leftarrow q * h * h$
$\quad s \leftarrow s + q * (a + h * b)$
end
$y = s * \exp(-0.5 * x * x - c)$
if $(x > 0)$ $y \leftarrow 1 - y$
return y

Fig. 2.15. Algorithm of Marsaglia et al. [251] to approximate the cumulative normal distribution.

Marsaglia et al. [251] present a faster approximation achieving similar accuracy but requiring 121 tabulated constants. Marsaglia et al. also detail the use of accurate approximations to Φ in constructing approximations to Φ^{-1} by tabulating "exact" values at a large number of strategically chosen points. Their method entails the use of more than 2000 tabulated constants, but the constants can be computed rather than tabulated, given an accurate approximation to Φ.

Other methods for approximating Φ and Φ^{-1} found in the literature are often based on the *error function*

$$\text{Erf}(x) = \frac{2}{\sqrt{\pi}} \int_0^x e^{-t^2} \, dt$$

and its inverse. Observe that for $x \geq 0$,

$$\text{Erf}(x) = 2\Phi(x\sqrt{2}) - 1, \quad \Phi(x) = \tfrac{1}{2}[\text{Erf}(x/\sqrt{2}) + 1]$$

and

$$\text{Erf}^{-1}(u) = \frac{1}{\sqrt{2}}\Phi^{-1}(\frac{u+1}{2}), \quad \Phi^{-1}(u) = \sqrt{2}\text{Erf}^{-1}(2u - 1),$$

so approximations to Erf and its inverse are easily converted into approximations to Φ and its inverse. Hastings [171], in fact, approximates Erf, so the constants in Figure 2.14 (as modified in [3]) differ from his, with p smaller and the b_i larger by a factor of $\sqrt{2}$.

Devroye [95] discusses several other methods for sampling from the normal distribution, including some that may be substantially faster than evaluation of Φ^{-1}. Nevertheless, as discussed in Section 2.2.1, the inverse transform method has some advantages — particularly in the application of variance reduction techniques and low-discrepancy methods — that will often justify the additional computational effort. One advantage is that the inverse transform method requires just one uniform input per normal output: a relevant notion of the *dimension* of a Monte Carlo problem is often the maximum number of uniforms required to generate one sample path, so methods requiring more uniforms per normal sample implicitly result in higher dimensional representations. Another useful property of the inverse transform method is that the mapping $u \mapsto \Phi^{-1}(u)$ is both continuous and monotone. These properties can sometimes enhance the effectiveness of variance reduction techniques, as we will see in later sections.

2.3.3 Generating Multivariate Normals

A multivariate normal distribution $N(\mu, \Sigma)$ is specified by its mean vector μ and covariance matrix Σ. The covariance matrix may be specified implicitly through its diagonal entries σ_i^2 and correlations ρ_{ij} using (2.22); in matrix form,

$$\Sigma = \begin{pmatrix} \sigma_1 & & & \\ & \sigma_2 & & \\ & & \ddots & \\ & & & \sigma_d \end{pmatrix} \begin{pmatrix} \rho_{11} & \rho_{12} & \cdots & \rho_{1d} \\ \rho_{12} & \rho_{22} & & \rho_{2d} \\ \vdots & & \ddots & \vdots \\ \rho_{1d} & \rho_{2d} & \cdots & \rho_{dd} \end{pmatrix} \begin{pmatrix} \sigma_1 & & & \\ & \sigma_2 & & \\ & & \ddots & \\ & & & \sigma_d \end{pmatrix}.$$

From the Linear Transformation Property (2.23), we know that if $Z \sim N(0, I)$ and $X = \mu + AZ$, then $X \sim N(\mu, AA^\top)$. Using any of the methods discussed in Section 2.3.2, we can generate independent standard normal random variables $Z_1, \ldots, Z_d$ and assemble them into a vector $Z \sim N(0, I)$. Thus, the problem of sampling X from the multivariate normal $N(\mu, \Sigma)$ reduces to finding a matrix A for which $AA^\top = \Sigma$.

Cholesky Factorization

Among all such A, a lower triangular one is particularly convenient because it reduces the calculation of $\mu + AZ$ to the following:

$$X_1 = \mu_1 + A_{11}Z_1$$
$$X_2 = \mu_2 + A_{21}Z_1 + A_{22}Z_2$$
$$\vdots$$
$$X_d = \mu_d + A_{d1}Z_1 + A_{d2}Z_2 + \cdots + A_{dd}Z_d.$$

A full multiplication of the vector Z by the matrix A would require approximately twice as many multiplications and additions. A representation of Σ as $AA^\top$ with A lower triangular is a *Cholesky factorization* of Σ. If Σ is positive definite (as opposed to merely positive semidefinite), it has a Cholesky factorization and the matrix A is unique up to changes in sign.

Consider a 2×2 covariance matrix Σ, represented as

$$\Sigma = \begin{pmatrix} \sigma_1^2 & \sigma_1\sigma_2\rho \\ \sigma_1\sigma_2\rho & \sigma_2^2 \end{pmatrix}.$$

Assuming $\sigma_1 > 0$ and $\sigma_2 > 0$, the Cholesky factor is

$$A = \begin{pmatrix} \sigma_1 & 0 \\ \rho\sigma_2 & \sqrt{1-\rho^2}\,\sigma_2 \end{pmatrix},$$

as is easily verified by evaluating $AA^\top$. Thus, we can sample from a bivariate normal distribution $N(\mu, \Sigma)$ by setting

$$X_1 = \mu_1 + \sigma_1 Z_1$$
$$X_2 = \mu_2 + \sigma_2\rho Z_1 + \sigma_2\sqrt{1-\rho^2}\,Z_2,$$

with Z_1, Z_2 independent standard normals.

For the case of a $d \times d$ covariance matrix Σ, we need to solve

$$\begin{pmatrix} A_{11} & & & \\ A_{21} & A_{22} & & \\ \vdots & \vdots & \ddots & \\ A_{d1} & A_{d2} & \cdots & A_{dd} \end{pmatrix} \begin{pmatrix} A_{11} & A_{21} & \cdots & A_{d1} \\ & A_{22} & \cdots & A_{d2} \\ & & \ddots & \vdots \\ & & & A_{dd} \end{pmatrix} = \Sigma.$$

Traversing the Σ_{ij} by looping over $j = 1, \ldots, d$ and then $i = j, \ldots, d$ produces the equations

$$A_{11}^2 = \Sigma_{11}$$
$$A_{21}A_{11} = \Sigma_{21}$$
$$\vdots$$

$$A_{d1}A_{11} = \Sigma_{d1} \tag{2.29}$$
$$A_{21}^2 + A_{22}^2 = \Sigma_{22}$$
$$\vdots$$
$$A_{d1}^2 + \cdots + A_{dd}^2 = \Sigma_{dd}.$$

Exactly one new entry of the A matrix appears in each equation, making it possible to solve for the individual entries sequentially.

More compactly, from the basic identity

$$\Sigma_{ij} = \sum_{k=1}^{j} A_{ik}A_{jk}, \quad j \le i,$$

we get

$$A_{ij} = \left(\Sigma_{ij} - \sum_{k=1}^{j-1} A_{ik}A_{jk} \right) / A_{jj}, \quad j < i, \tag{2.30}$$

and

$$A_{ii} = \sqrt{\Sigma_{ii} - \sum_{k=1}^{i-1} A_{ik}^2}. \tag{2.31}$$

These expressions make possible a simple recursion to find the Cholesky factor. Figure 2.16 displays an algorithm based on one in Golub and Van Loan [162]. Golub and Van Loan [162] give several other versions of the algorithm and also discuss numerical stability.

Input: Symmetric positive definite matrix $d \times d$ matrix Σ
Output: Lower triangular A with $AA^\top = \Sigma$

$A \leftarrow 0 \ (d \times d$ zero matrix)
for $j = 1, \ldots, d$
    for $i = j, \ldots, d$
        $v_i \leftarrow \Sigma_{ij}$
        for $k = 1, \ldots, j - 1$
            $v_i \leftarrow v_i - A_{jk}A_{ik}$
        $A_{ij} \leftarrow v_i / \sqrt{v_j}$
return A

Fig. 2.16. Cholesky factorization.

The Semidefinite Case

If Σ is positive definite, an induction argument verifies that the quantity inside the square root in (2.31) is strictly positive so the A_{ii} are nonzero. This ensures that (2.30) does not entail division by zero and that the algorithm in Figure 2.16 runs to completion.

If, however, Σ is merely positive semidefinite, then it is rank deficient. It follows that any matrix A satisfying $AA^\top = \Sigma$ must also be rank deficient; for if A had full rank, then Σ would too. If A is lower triangular and rank deficient, at least one element of the diagonal of A must be zero. (The determinant of a triangular matrix is the product of its diagonal elements, and the determinant of A is zero if A is singular.) Thus, for semidefinite Σ, any attempt at Cholesky factorization must produce some $A_{jj} = 0$ and thus an error in (2.31) and the algorithm in Figure 2.16.

From a purely mathematical perspective, the problem is easily solved by making the jth column of A identically zero if $A_{jj} = 0$. This can be deduced from the system of equations (2.29): the first element of the jth column of A encountered in this sequence of equations is the diagonal entry; if $A_{jj} = 0$, all subsequent equations for the jth column of Σ may be solved with $A_{ij} = 0$. In the factorization algorithm of Figure 2.16, this is accomplished by inserting "if $v_j > 0$" before the statement "$A_{ij} \leftarrow v_i/\sqrt{v_j}$." Thus, if $v_j = 0$, the entry A_{ij} is left at its initial value of zero.

In practice, this solution may be problematic because it involves checking whether an intermediate calculation (v_j) is exactly zero, making the modified algorithm extremely sensitive to round-off error.

Rather than blindly subjecting a singular covariance matrix to Cholesky factorization, it is therefore preferable to use the structure of the covariance matrix to reduce the problem to one of full rank. If $X \sim N(0, \Sigma)$ and the $d \times d$ matrix Σ has rank $k < d$, it is possible to express all d components of X as linear combinations of just k of the components, these k components having a covariance matrix of rank k. In other words, it is possible to find a subvector $\tilde{X} = (X_{i_1}, \ldots, X_{i_k})$ and a $d \times k$ matrix D such that $D\tilde{X} \sim N(0, \Sigma)$ and for which the covariance matrix $\tilde{\Sigma}$ of $\tilde{X}$ has full rank k. Cholesky factorization can then be applied to $\tilde{\Sigma}$ to find $\tilde{A}$ satisfying $\tilde{A}\tilde{A}^\top = \tilde{\Sigma}$. The full vector X can be sampled by setting $X = D\tilde{A}Z$, $Z \sim N(0, I)$.

Singular covariance matrices often arise from factor models in which a vector of length d is determined by $k < d$ sources of uncertainty (factors). In this case, the prescription above reduces to using knowledge of the factor structure to generate X.

Eigenvector Factorization and Principal Components

The equation $AA^\top = \Sigma$ can also be solved by diagonalizing Σ. As a symmetric $d \times d$ matrix, Σ has d real eigenvalues $\lambda_1, \ldots, \lambda_d$, and because Σ must be positive definite or semidefinite the λ_i are nonnegative. Furthermore, Σ has an associated orthonormal set of eigenvectors $\{v_1, \ldots, v_d\}$; i.e., vectors satisfying

$$v_i^\top v_i = 1, \quad v_i^\top v_j = 0, \quad j \neq i, \quad i, j = 1, \ldots, d,$$

and

$$\Sigma v_i = \lambda_i v_i.$$

It follows that $\Sigma = V \Lambda V^\top$, where V is the orthogonal matrix ($V V^\top = I$) with columns $v_1, \ldots, v_d$ and Λ is the diagonal matrix with diagonal entries $\lambda_1, \ldots, \lambda_d$. Hence, if we choose

$$A = V \Lambda^{1/2} = V \begin{pmatrix} \sqrt{\lambda_1} & & & \\ & \sqrt{\lambda_2} & & \\ & & \ddots & \\ & & & \sqrt{\lambda_d} \end{pmatrix}, \tag{2.32}$$

then

$$A A^\top = V \Lambda V^\top = \Sigma.$$

Methods for calculating V and Λ are included in many mathematical software libraries and discussed in detail in Golub and Van Loan [162].

Unlike the Cholesky factor, the matrix A in (2.32) has no particular structure providing a computational advantage in evaluating AZ, nor is this matrix faster to compute than the Cholesky factorization. The eigenvectors and eigenvalues of a covariance matrix do however have a statistical interpretation that is occasionally useful. We discuss this interpretation next.

If $X \sim N(0, \Sigma)$ and $Z \sim N(0, I)$, then generating X as AZ for any choice of A means setting

$$X = a_1 Z_1 + a_2 Z_2 + \cdots + a_d Z_d$$

where a_j is the jth column of A. We may interpret the Z_j as independent factors driving the components of X, with A_{ij} the "factor loading" of Z_j on X_i. If Σ has rank 1, then X may be represented as $a_1 Z_1$ for some vector a_1, and in this case a single factor suffices to represent X. If Σ has rank k, then k factors $Z_1, \ldots, Z_k$ suffice.

If Σ has full rank and $AA^\top = \Sigma$, then A must have full rank and $X = AZ$ implies $Z = BX$ with $B = A^{-1}$. Thus, the factors Z_j are themselves linear combinations of the X_i. In the special case of A given in (2.32), we have

$$A^{-1} = \Lambda^{-1/2} V^\top \tag{2.33}$$

because $V^\top V = I$ (V is orthogonal). It follows that Z_j is proportional to $v_j^\top X$, where v_j is the jth column of V and thus an eigenvector of Σ.

The factors Z_j constructed proportional to the $v_j^\top X$ are optimal in a precise sense. Suppose we want to find the best single-factor approximation to X; i.e., the linear combination $w^\top X$ that best captures the variability of the components of X. A standard notion of optimality chooses w to maximize the variance of $w^\top X$, which is given by $w^\top \Sigma w$. Since this variance can be made arbitrarily large by multiplying any w by a constant, it makes sense to impose a normalization through a constraint of the form $w^\top w = 1$. We are thus led to the problem

$$\max_{w:w^\top w=1} w^\top \Sigma w.$$

If the eigenvalues of Σ are ordered so that

$$\lambda_1 \geq \lambda_2 \geq \cdots \geq \lambda_d,$$

then this optimization problem is solved by v_1, as is easily verified by appending the constraint with a Lagrange multiplier and differentiating. (This optimality property of eigenvectors is sometimes called Rayleigh's principle.) The problem of finding the next best factor orthogonal to the first reduces to solving

$$\max_{w:w^\top w=1, w^\top v_1=0} w^\top \Sigma w.$$

This optimization problem is solved by v_2. More generally, the best k-factor approximation chooses factors proportional to $v_1^\top X, v_2^\top X, \ldots, v_k^\top X$. Since

$$v_j^\top \Sigma v_j = \lambda_j,$$

normalizing the $v_j^\top X$ to construct unit-variance factors yields

$$Z_j = \frac{1}{\sqrt{\lambda_j}} v_j^\top X,$$

which coincides with (2.33). The transformation $X = AZ$ recovering X from the Z_j is precisely the A in (2.32).

The optimality of this representation can be recast in the following way. Suppose that we are given X and that we want to find vectors $a_1, \ldots, a_k$ in $\Re^d$ and unit-variance random variables $Z_1, \ldots, Z_k$ in order to approximate X by $a_1 Z_1 + \cdots + a_k Z_k$. For any $k = 1, \ldots, d$, the mean square approximation error

$$\mathsf{E}\left[\left\| X - \sum_{i=1}^k a_i Z_i \right\|^2 \right], \qquad (\|x\|^2 = x^\top x)$$

is minimized by taking the a_i to be the columns of A in (2.32) and setting $Z_i = v_i^\top X / \sqrt{\lambda_i}$.

In the statistical literature, the linear combinations $v_j^\top X$ are called the *principal components* of X (see, e.g., Seber [325]). We may thus say that the principal components provide an optimal lower-dimensional approximation to a random vector. The variance *explained* by the first k principal components is the ratio

$$\frac{\lambda_1 + \cdots + \lambda_k}{\lambda_1 + \cdots + \lambda_k + \cdots + \lambda_d}; \qquad (2.34)$$

in particular, the first principal component is chosen to explain as much variance as possible. In simulation applications, generating X from its principal components (i.e., using (2.32)) is sometimes useful in designing variance reduction techniques. In some cases, the principal components interpretation suggests that variance reduction should focus first on Z_1, then on Z_2, and so on. We will see examples of this in Chapter 4 and related ideas in Section 5.5.

3

Generating Sample Paths

This chapter develops methods for simulating paths of a variety of stochastic processes important in financial engineering. The emphasis in this chapter is on methods for *exact* simulation of continuous-time processes at a discrete set of dates. The methods are exact in the sense that the joint distribution of the simulated values coincides with the joint distribution of the continuous-time process on the simulation time grid. Exact methods rely on special features of a model and are generally available only for models that offer some tractability. More complex models must ordinarily be simulated through, e.g., discretization of stochastic differential equations, as discussed in Chapter 6.

The examples covered in this chapter are arranged roughly in increasing order of complexity. We begin with methods for simulating Brownian motion in one dimension or multiple dimensions and extend these to geometric Brownian motion. We then consider Gaussian interest rate models. Our first real break from Gaussian processes comes in Section 3.4, where we treat square-root diffusions. Section 3.5 considers processes with jumps as models of asset prices. Sections 3.6 and 3.7 treat substantially more complex models than the rest of the chapter; these are interest rate models that describe the term structure through a curve or vector of forward rates. Exact simulation of these models is generally infeasible; we have included them here because of their importance in financial engineering and because they illustrate some of the complexities of the use of simulation for derivatives pricing.

3.1 Brownian Motion

3.1.1 One Dimension

By a *standard* one-dimensional Brownian motion on $[0, T]$, we a mean a stochastic process $\{W(t), 0 \leq t \leq T\}$ with the following properties:

(i) $W(0) = 0$;

(ii) the mapping $t \mapsto W(t)$ is, with probability 1, a continuous function on $[0, T]$;

(iii) the increments $\{W(t_1) - W(t_0), W(t_2) - W(t_1), \ldots, W(t_k) - W(t_{k-1})\}$ are independent for any k and any $0 \leq t_0 < t_1 < \cdots < t_k \leq T$;

(iv) $W(t) - W(s) \sim N(0, t - s)$ for any $0 \leq s < t \leq T$.

In (iv) it would suffice to require that $W(t) - W(s)$ have mean 0 and variance $t - s$; that its distribution is in fact normal follows from the continuity of sample paths in (ii) and the independent increments property (iii). We include the condition of normality in (iv) because it is central to our discussion. A consequence of (i) and (iv) is that

$$W(t) \sim N(0, t), \tag{3.1}$$

for $0 < t \leq T$.

For constants μ and $\sigma > 0$, we call a process $X(t)$ a Brownian motion with drift μ and diffusion coefficient σ^2 (abbreviated $X \sim \mathrm{BM}(\mu, \sigma^2)$) if

$$\frac{X(t) - \mu t}{\sigma}$$

is a standard Brownian motion. Thus, we may construct X from a standard Brownian motion W by setting

$$X(t) = \mu t + \sigma W(t).$$

It follows that $X(t) \sim N(\mu t, \sigma^2 t)$. Moreover, X solves the stochastic differential equation (SDE)

$$dX(t) = \mu \, dt + \sigma \, dW(t).$$

The assumption that $X(0) = 0$ is a natural normalization, but we may construct a Brownian motion with parameters μ and σ^2 and initial value x by simply adding x to each $X(t)$.

For deterministic but time-varying $\mu(t)$ and $\sigma(t) > 0$, we may define a Brownian motion with drift μ and diffusion coefficient σ^2 through the SDE

$$dX(t) = \mu(t) \, dt + \sigma(t) \, dW(t);$$

i.e., through

$$X(t) = X(0) + \int_0^t \mu(s) \, ds + \int_0^t \sigma(s) \, dW(s),$$

with $X(0)$ an arbitrary constant. The process X has continuous sample paths and independent increments. Each increment $X(t) - X(s)$ is normally distributed with mean

$$\mathsf{E}[X(t) - X(s)] = \int_s^t \mu(u) \, du$$

and variance

$$\mathsf{Var}[X(t) - X(s)] = \mathsf{Var}\left[\int_s^t \sigma(u) \, dW(u)\right] = \int_s^t \sigma^2(u) \, du.$$

Random Walk Construction

In discussing the simulation of Brownian motion, we mostly focus on simulating values $(W(t_1), \ldots, W(t_n))$ or $(X(t_1), \ldots, X(t_n))$ at a fixed set of points $0 < t_1 < \cdots < t_n$. Because Brownian motion has independent normally distributed increments, simulating the $W(t_i)$ or $X(t_i)$ from their increments is straightforward. Let $Z_1, \ldots, Z_n$ be independent standard normal random variables, generated using any of the methods in Section 2.3.2, for example. For a standard Brownian motion set $t_0 = 0$ and $W(0) = 0$. Subsequent values can be generated as follows:

$$W(t_{i+1}) = W(t_i) + \sqrt{t_{i+1} - t_i} Z_{i+1}, \quad i = 0, \ldots, n - 1. \tag{3.2}$$

For $X \sim \mathrm{BM}(\mu, \sigma^2)$ with constant μ and σ and given $X(0)$, set

$$X(t_{i+1}) = X(t_i) + \mu(t_{i+1} - t_i) + \sigma\sqrt{t_{i+1} - t_i} Z_{i+1}, \quad i = 0, \ldots, n - 1. \tag{3.3}$$

With time-dependent coefficients, the recursion becomes

$$X(t_{i+1}) = X(t_i) + \int_{t_i}^{t_{i+1}} \mu(s)\, ds + \sqrt{\int_{t_i}^{t_{i+1}} \sigma^2(u)\, du} Z_{i+1}, \quad i = 0, \ldots, n - 1. \tag{3.4}$$

The methods in (3.2)–(3.4) are *exact* in the sense that the joint distribution of the simulated values $(W(t_1), \ldots, W(t_n))$ or $(X(t_1), \ldots, X(t_n))$ coincides with the joint distribution of the corresponding Brownian motion at $t_1, \ldots, t_n$. Of course, this says nothing about what happens between the t_i. One may extend the simulated values to other time points through, e.g., piecewise linear interpolation; but no deterministic interpolation method will give the extended vector the correct joint distribution. The methods in (3.2)–(3.4) are exact at the time points $t_1, \ldots, t_n$ but subject to *discretization error*, compared to a true Brownian motion, if deterministically interpolated to other time points. Replacing (3.4) with the Euler approximation

$$X(t_{i+1}) = X(t_i) + \mu(t_i)(t_{i+1} - t_i) + \sigma(t_i)\sqrt{t_{i+1} - t_i} Z_{i+1}, \quad i = 0, \ldots, n - 1,$$

will in general introduce discretization error even at $t_1, \ldots, t_n$, because the increments will no longer have exactly the right mean and variance. We return to the topic of discretization error in Chapter 6.

The vector $(W(t_1), \ldots, W(t_n))$ is a linear transformation of the the the vector of increments $(W(t_1), W(t_2) - W(t_1), \ldots, W(t_n) - W(t_{n-1}))$. Since these increments are independent and normally distributed, it follows from the Linear Transformation Property (2.23) that $(W(t_1), \ldots, W(t_n))$ has a multivariate normal distribution. Simulating $(W(t_1), \ldots, W(t_n))$ is thus a special case of the general problem, treated in Section 2.3.3, of generating multivariate normal vectors. While the random walk construction suffices for most applications, it is interesting and sometimes useful to consider alternative sampling methods.

To apply any of the methods considered in Section 2.3.3, we first need to find the mean vector and covariance matrix of $(W(t_1), \ldots, W(t_n))$. For a standard Brownian motion, we know from (3.1) that $\mathsf{E}[W(t_i)] = 0$, so the mean vector is identically 0. For the covariance matrix, consider first any $0 < s < t < T$; using the independence of the increments we find that

$$
\begin{aligned}
\mathsf{Cov}[W(s), W(t)] &= \mathsf{Cov}[W(s), W(s) + (W(t) - W(s))] \\
&= \mathsf{Cov}[W(s), W(s)] + \mathsf{Cov}[W(s), W(t) - W(s)] \\
&= s + 0 = s.
\end{aligned}
\tag{3.5}
$$

Letting C denote the covariance matrix of $(W(t_1), \ldots, W(t_n))$, we thus have

$$
C_{ij} = \min(t_i, t_j).
\tag{3.6}
$$

Cholesky Factorization

Having noted that the vector $(W(t_1), \ldots, W(t_n))$ has the distribution $N(0, C)$, with C as in (3.6), we may simulate this vector as AZ, where $Z = (Z_1, \ldots, Z_n)^\top \sim N(0, I)$ and A satisfies $AA^\top = C$. The Cholesky method discussed in Section 2.3.3 takes A to be lower triangular. For C in (3.6), the Cholesky factor is given by

$$
A = \begin{pmatrix}
\sqrt{t_1} & 0 & \cdots & 0 \\
\sqrt{t_1} & \sqrt{t_2 - t_1} & \cdots & 0 \\
\vdots & \vdots & \ddots & \vdots \\
\sqrt{t_1} & \sqrt{t_2 - t_1} & \cdots & \sqrt{t_n - t_{n-1}}
\end{pmatrix},
$$

as can be verified through calculation of $AA^\top$. In this case, generating $(W(t_1), \ldots, W(t_n))$ as AZ is simply a matrix-vector representation of the recursion in (3.2). Put differently, the random walk construction (3.2) may be viewed as an efficient implementation of the product AZ. Even exploiting the lower triangularity of A, evaluation of AZ is an $O(n^2)$ operation; the random walk construction reduces this to $O(n)$ by implicitly exploiting the fact that the nonzero entries of each column of A are identical.

For a $\mathrm{BM}(\mu, \sigma^2)$ process X, the mean vector of $(X(t_1), \ldots, X(t_n))$ has ith component μt_i and the covariance matrix is $\sigma^2 C$. The Cholesky factor is σA and we once again find that the Cholesky method coincides with the increment recursion (3.3).

Brownian Bridge Construction

The recursion (3.2) generates the vector $(W(t_1), \ldots, W(t_n))$ from left to right. We may however generate the $W(t_i)$ in any order we choose, provided that at each step we sample from the correct conditional distribution given the values already generated. For example, we may first generate the final value $W(t_n)$, then sample $W(t_{\lfloor n/2 \rfloor})$ conditional on the value of $W(t_n)$, and proceed by progressively filling in intermediate values. This flexibility can be useful in implementing variance reduction techniques and low-discrepancy methods. It follows from the Conditioning Formula (2.24) that the conditional distribution needed at each step is itself normal and this makes conditional sampling feasible.

Conditioning a Brownian motion on its endpoints produces a *Brownian bridge*. Once we determine $W(t_n)$, filling in intermediate values amounts to simulating a Brownian bridge from $0 = W(0)$ to $W(t_n)$. If we next sample $W(t_{\lfloor n/2 \rfloor})$, then filling in values between times $t_{\lfloor n/2 \rfloor}$ and t_n amounts to simulating a Brownian bridge from $W(t_{\lfloor n/2 \rfloor})$ to $W(t_n)$. This approach is thus referred to as a Brownian bridge construction.

As a first step in developing this construction, suppose $0 < u < s < t$ and consider the problem of generating $W(s)$ conditional on $W(u) = x$ and $W(t) = y$. We use the Conditioning Formula (2.24) to find the conditional distribution of $W(s)$. We know from (3.5) that the unconditional distribution is given by

$$\begin{pmatrix} W(u) \\ W(s) \\ W(t) \end{pmatrix} \sim N\left(0, \begin{pmatrix} u & u & u \\ u & s & s \\ u & s & t \end{pmatrix}\right).$$

The Conditioning Formula (2.24) gives the distribution of the second component of a partitioned vector conditional on a value of the first component. We want to apply this formula to find the distribution of $W(s)$ conditional on the value of $(W(u), W(t))$. We therefore first permute the entries of the vector to get

$$\begin{pmatrix} W(s) \\ W(u) \\ W(t) \end{pmatrix} \sim N\left(0, \begin{pmatrix} s & u & s \\ u & u & u \\ s & u & t \end{pmatrix}\right).$$

We now find from the Conditioning Formula that, given $(W(u) = x, W(t) = y)$, $W(s)$ is normally distributed with mean

$$\mathsf{E}[W(s)|W(u) = x, W(t) = y] =$$

$$0 - (u\ s)\begin{pmatrix} u & u \\ u & t \end{pmatrix}^{-1}\begin{pmatrix} x \\ y \end{pmatrix} = \frac{(t-s)x + (s-u)y}{(t-u)}, \qquad (3.7)$$

and variance

$$s - (u\ s)\begin{pmatrix} u & u \\ u & t \end{pmatrix}^{-1}\begin{pmatrix} u \\ s \end{pmatrix} = \frac{(s-u)(t-s)}{(t-u)}, \qquad (3.8)$$

since

$$\begin{pmatrix} u & u \\ u & t \end{pmatrix}^{-1} = \frac{1}{t-u}\begin{pmatrix} t/u & -1 \\ -1 & 1 \end{pmatrix}.$$

In particular, the conditional mean (3.7) is obtained by linearly interpolating between (u, x) and (t, y).

Suppose, more generally, that the values $W(s_1) = x_1$, $W(s_2) = x_2$, ...,
$W(s_k) = x_k$ of the Brownian path have been determined at the times $s_1 <$
$s_2 < \cdots < s_k$ and that we wish to sample $W(s)$ conditional on these values.
Suppose that $s_i < s < s_{i+1}$. Then

$$(W(s)|W(s_j) = x_j, j = 1, \ldots, k) = (W(s)|W(s_i) = x_i, W(s_{i+1}) = x_{i+1}),$$

in the sense that the two conditional distributions are the same. This can again
be derived from the Conditioning Formula (2.24) but is more immediate from
the Markov property of Brownian motion (a consequence of the independent
increments property): given $W(s_i)$, $W(s)$ is independent of all $W(t)$ with
$t < s_i$, and given $W(s_{i+1})$ it is independent of all $W(t)$ with $t > s_{i+1}$. Thus,
conditioning on all $W(s_j)$ is equivalent to conditioning on the values of the
Brownian path at the two times s_i and s_{i+1} closest to s. Combining these
observations with (3.7) and (3.8), we find that

$$(W(s)|W(s_1) = x_1, W(s_2) = x_2, \ldots, W(s_k) = x_k) =$$
$$N\left(\frac{(s_{i+1} - s)x_i + (s - s_i)x_{i+1}}{(s_{i+1} - s_i)}, \frac{(s_{i+1} - s)(s - s_i)}{(s_{i+1} - s_i)}\right).$$

This is illustrated in Figure 3.1. The conditional mean of $W(s)$ lies on the
line segment connecting (s_i, x_i) and (s_{i+1}, x_{i+1}); the actual value of $W(s)$ is
normally distributed about this mean with a variance that depends on $(s - s_i)$
and $(s_{i+1} - s)$. To sample from this conditional distribution, we may set

$$W(s) = \frac{(s_{i+1} - s)x_i + (s - s_i)x_{i+1}}{(s_{i+1} - s_i)} + \sqrt{\frac{(s_{i+1} - s)(s - s_i)}{(s_{i+1} - s_i)}}Z,$$

with $Z \sim N(0, 1)$ independent of all $W(s_1), \ldots, W(s_k)$.

By repeatedly using these observations, we may indeed sample the components of the vector $(W(t_1), \ldots, W(t_n))$ in any order. In particular, we may
start by sampling $W(t_n)$ from $N(0, t_n)$ and proceed by conditionally sampling
intermediate values, at each step conditioning on the two closest time points
already sampled (possibly including $W(0) = 0$).

If n is a power of 2, the construction can be arranged so that each $W(t_i)$,
$i < n$, is generated conditional on the values $W(t_\ell)$ and $W(t_r)$ with the
property that i is midway between ℓ and r. Figure 3.2 details this case. If, for
example, $n = 16$, the algorithm starts by sampling $W(t_{16})$; the first loop over
j samples $W(t_8)$; the second samples $W(t_4)$ and $W(t_{12})$; the third samples
$W(t_2)$, $W(t_6)$, $W(t_{10})$, and $W(t_{14})$; and the final loop fills in all $W(t_i)$ with
odd i. If n is not a power of 2, the algorithm could still be applied to a subset
of $2^m < n$ of the t_i, with the remaining points filled in at the end.

Our discussion of the Brownian bridge construction (and Figure 3.2 in
particular) has considered only the case of a standard Brownian motion. How
would the construction be modified for a Brownian motion with drift μ? Only

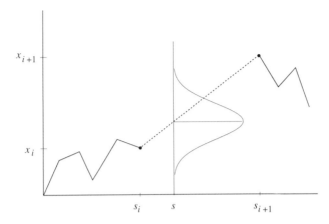

Fig. 3.1. Brownian bridge construction of Brownian path. Conditional on $W(s_i) = x_i$ and $W(s_{i+1}) = x_{i+1}$, the value at s is normally distributed. The conditional mean is obtained by linear interpolation between x_i and x_{i+1}; the conditional variance is obtained from (3.8).

Input: Time indices $(t_1, \ldots, t_{2^m})$
Output: Path $(w_1, \ldots, w_{2^m})$ with distribution of $(W(t_1), \ldots, W(t_{2^m}))$

    Generate $(Z_1, \ldots, Z_{2^m}) \sim N(0, I)$
    $h \leftarrow 2^m, \quad j_{\max} \leftarrow 1$
    $w_h \leftarrow \sqrt{t_h} Z_h$
    $t_0 \leftarrow 0, \quad w_0 \leftarrow 0$
    for $k = 1, \ldots, m$
        $i_{\min} \leftarrow h/2, \quad i \leftarrow i_{\min}$
        $\ell \leftarrow 0, \quad r \leftarrow h$
        for $j = 1, \ldots, j_{\max}$
            $a \leftarrow ((t_r - t_i)w_\ell + (t_i - t_\ell)w_r)/(t_r - t_\ell)$
            $b \leftarrow \sqrt{(t_i - t_\ell)(t_r - t_i)/(t_r - t_\ell)}$
            $w_i \leftarrow a + bZ_i$
            $i \leftarrow i + h; \quad \ell \leftarrow \ell + h, \quad r \leftarrow r + h$
        end
        $j_{\max} \leftarrow 2 * j_{\max};$
        $h \leftarrow i_{\min};$
    end
    return $(w_1, \ldots, w_{2^m})$

Fig. 3.2. Implementation of Brownian bridge construction when the number of time indices is a power of 2. The conditional standard deviations assigned to b could be precomputed and stored in an array $(b_1, \ldots, b_{2^m})$ if multiple paths are to be generated. The interpolation weights used in calculating the conditional mean a could also be precomputed.

the first step — sampling of the rightmost point — would change. Instead of sampling $W(t_m)$ from $N(0, t_m)$, we would sample it from $N(\mu t_m, t_m)$. The conditional distribution of $W(t_1), \ldots, W(t_{n-1})$ given $W(t_m)$ is the same *for all values of* μ. Put slightly differently, a Brownian bridge constructed from a Brownian motion with drift has the same law as one constructed from a standard Brownian motion. (For any finite set of points $t_1, \ldots, t_{n-1}$ this can be established from the Conditioning Formula (2.24).) Hence, to include a drift μ in the algorithm of Figure 3.2, it suffices to change just the third line, adding μt_h to w_h. For a Brownian motion with diffusion coefficient σ^2, the conditional mean (3.7) is unchanged but the conditional variance (3.8) is multiplied by σ^2. This could be implemented in Figure 3.2 by multiplying each b by σ (and setting $w_h \leftarrow \mu t_h + \sigma \sqrt{t_h} Z_h$ in the third line); alternatively, the final vector $(w_1, \ldots, w_{2m})$ could simply be multiplied by σ.

Why use a Brownian bridge construction? The algorithm in Figure 3.2 has no computational advantage over the simple random walk recursion (3.2). Nor does the output of the algorithm have any statistical feature not shared by the output of (3.2); indeed, the Brownian bridge construction is valid precisely because the distribution of the $(W(t_1), \ldots, W(t_m))$ it produces coincides with that resulting from (3.2). The potential advantage of the Brownian bridge construction arises when it is used with certain variance reduction techniques and low-discrepancy methods. We will return to this point in Section 4.3 and Chapter 5. Briefly, the Brownian bridge construction gives us greater control over the coarse structure of the simulated Brownian path. For example, it uses a single normal random variable to determine the endpoint of a path, which may be the most important feature of the path; in contrast, the endpoint obtained from (3.2) is the combined result of n independent normal random variables. The standard recursion (3.2) proceeds by evolving the path forward through time; in contrast, the Brownian bridge construction proceeds by adding increasingly fine detail to the path at each step, as illustrated in Figure 3.3. This can be useful in focusing variance reduction techniques on "important" features of Brownian paths.

Principal Components Construction

As just noted, under the Brownian bridge construction a single normal random variable (say Z_1) determines the endpoint of the path; conditional on the endpoint, a second normal random variable (say Z_2) determines the midpoint of the path, and so on. Thus, under this construction, much of the ultimate shape of the Brownian path is determined (or *explained*) by the values of just the first few Z_i. Is there a construction under which even more of the path is determined by the first few Z_i? Is there a construction that maximizes the variability of the path explained by $Z_1, \ldots, Z_k$ for all $k = 1, \ldots, n$?

This optimality objective is achieved for any normal random vector by the principal components construction discussed in Section 2.3.3. We now discuss its application to a discrete Brownian path $W(t_1), \ldots, W(t_n)$. It is useful to

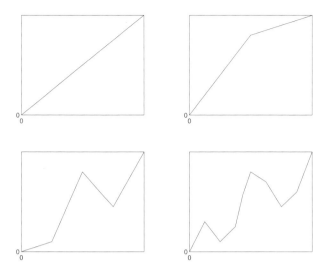

Fig. 3.3. Brownian bridge construction after 1, 2, 4, and 8 points have been sampled. Each step refines the previous path.

visualize the construction in vector form as

$$\begin{pmatrix} W(t_1) \\ W(t_2) \\ \vdots \\ W(t_n) \end{pmatrix} = \begin{pmatrix} a_{11} \\ a_{21} \\ \vdots \\ a_{n1} \end{pmatrix} Z_1 + \begin{pmatrix} a_{12} \\ a_{22} \\ \vdots \\ a_{n2} \end{pmatrix} Z_2 + \cdots + \begin{pmatrix} a_{1n} \\ a_{2n} \\ \vdots \\ a_{nn} \end{pmatrix} Z_n. \qquad (3.9)$$

Let $a_i = (a_{1i}, \ldots, a_{ni})^\top$ and let A be the $n \times n$ matrix with columns $a_1, \ldots, a_n$. We know from Section 2.3.3 that this is a valid construction of the discrete Brownian path if $AA^\top$ is the covariance matrix C of $W = (W(t_1), \ldots, W(t_n))^\top$, given in (3.6). We also know from the discussion of principal components in Section 2.3.3 that the approximation error

$$\mathsf{E}\left[\left\| W - \sum_{i=1}^{k} a_i Z_i \right\|^2 \right] \qquad (\|x\|^2 = x^\top x)$$

from using just the first k terms in (3.9) is minimized for all $k = 1, \ldots, n$ by using principal components. Specifically, $a_i = \sqrt{\lambda_i} v_i$, $i = 1, \ldots, n$, where $\lambda_1 > \lambda_2 > \cdots > \lambda_n > 0$ are the eigenvalues of C and the v_i are eigenvectors,

$$C v_i = \lambda_i v_i, \quad i = 1, \ldots, n,$$

normalized to have length $\|v_i\| = 1$.

Consider, for example, a 32-step discrete Brownian path with equal time increments $t_{i+1} - t_i = 1/32$. The corresponding covariance matrix has entries

$C_{ij} = \min(i,j)/32$, $i,j = 1,\ldots,32$. The magnitudes of the eigenvalues of this matrix drop off rapidly — the five largest are 13.380, 1.489, 0.538, 0.276, and 0.168. The variability explained by $Z_1,\ldots,Z_k$ (in the sense of (2.34)) is 81%, 90%, 93%, 95%, and 96%, for $k = 1,\ldots,5$; it exceeds 99% at $k = 16$. This indicates that although full construction of the 32-step path requires 32 normal random variables, most of the variability of the path can be determined using far fewer Z_i.

Figure 3.4 plots the normalized eigenvectors v_1, v_2, v_3, and v_4 associated with the four largest eigenvalues. (Each of these is a vector with 32 entries; they are plotted against the $j\Delta t$, $j = 1,\ldots,32$, with $\Delta t = 1/32$.) The v_i appear to be nearly sinusoidal, with frequencies that increase with i. Indeed, Åkesson and Lehoczky [8] show that for an n-step path with equal spacing $t_{i+1} - t_i = \Delta t$,

$$v_i(j) = \frac{2}{\sqrt{2n+1}} \sin\left(\frac{2i-1}{2n+1} j\pi \right), \quad j = 1,\ldots,n,$$

and

$$\lambda_i = \frac{\Delta t}{4} \sin^{-2}\left(\frac{2i-1}{2n+1} \frac{\pi}{2} \right),$$

for $i = 1,\ldots,n$. To contrast this with the Brownian bridge construction in Figure 3.3, note that in the principal components construction the v_i are multiplied by $\sqrt{\lambda_i} Z_i$ and then summed; thus, the discrete Brownian path may be viewed as a random linear combination of the vectors v_i, with random coefficients $\sqrt{\lambda_i} Z_i$. The coefficient on v_i has variance λ_i and we have seen that the λ_i drop off quickly. Thus, the first few v_i (and $\sqrt{\lambda_i} Z_i$) determine most of the shape of the Brownian path and the later v_i add high-frequency detail to the path. As in the Brownian bridge construction, these features can be useful in implementing variance reduction techniques by making it possible to focus on the most important Z_i. We return to this point in Sections 4.3.2 and 5.5.2.

Although the principal components construction is optimal with respect to explained variability, it has two drawbacks compared to the random walk and Brownian bridge constructions. The first is that it requires $O(n^2)$ operations to construct $W(t_1),\ldots,W(t_n)$ from $Z_1,\ldots,Z_n$, whereas the previous constructions require $O(n)$ operations. The second (potential) drawback is that with principal components none of the $W(t_i)$ is fully determined until all $Z_1,\ldots,Z_n$ have been processed — i.e., until all terms in (3.9) have been summed. In contrast, using either the random walk or Brownian bridge construction, exactly k of the $W(t_1),\ldots,W(t_n)$ are fixed by the first k normal random variables, for all $k = 1,\ldots,n$.

We conclude this discussion of the principal components construction with a brief digression into simulation of a continuous path $\{W(t), 0 \le t \le 1\}$. In the discrete case, the eigenvalue-eigenvector condition $Cv = \lambda v$ is (recall (3.6))

$$\sum_j \min(t_i, t_j) v(j) = \lambda v(i).$$

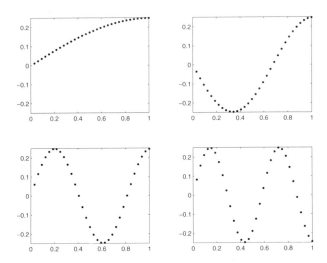

Fig. 3.4. First four eigenvectors of the covariance matrix of a 32-step Brownian path, ordered according to the magnitude of the corresponding eigenvalue.

In the continuous limit, the analogous property for an eigen*function* ψ on $[0,1]$ is

$$\int_0^1 \min(s,t)\psi(s)\,ds = \lambda\psi(t).$$

The solutions to this equation and the corresponding eigenvalues are

$$\psi_i(t) = \sqrt{2}\sin\left(\frac{(2i+1)\pi t}{2}\right), \quad \lambda_i = \left(\frac{2}{(2i+1)\pi}\right)^2, \quad i = 0,1,2,\ldots.$$

Note in particular that the ψ_i are periodic with increasing frequencies and that the λ_i decrease with i. The *Karhounen-Loève expansion* of Brownian motion is

$$W(t) = \sum_{i=0}^{\infty} \sqrt{\lambda_i}\psi_i(t)Z_i, \quad 0 \le t \le 1, \tag{3.10}$$

with $Z_0, Z_1,\ldots$ independent $N(0,1)$ random variables; see, e.g., Adler [5]. This infinite series is an exact representation of the continuous Brownian path. It may be viewed as a continuous counterpart to (3.9). By taking just the first k terms in this series, we arrive at an approximation to the continuous path $\{W(t), 0 \le t \le 1\}$ that is optimal (among all approximations that use just k standard normals) in the sense of explained variability. This approximation does not however yield the exact joint distribution for any subset $\{W(t_1),\ldots,W(t_n)\}$ except the trivial case $\{W(0)\}$.

The Brownian bridge construction also admits a continuous counterpart through a series expansion using *Schauder functions* in place of the $\sqrt{\lambda_i}\psi_i$

in (3.10). Lévy [233, pp.17–20] used the limit of the Brownian bridge construction to construct Brownian motion; the formulation as a series expansion is discussed in Section 1.2 of McKean [260]. Truncating the series after 2^m terms produces the piecewise linear interpolation of a discrete Brownian bridge construction of $W(0), W(2^{-m}), \ldots, W(1)$. See Acworth et al. [4] for further discussion with applications to Monte Carlo.

3.1.2 Multiple Dimensions

We call a process $W(t) = (W_1(t), \ldots, W_d(t))^\top$, $0 \leq t \leq T$, a standard Brownian motion on $\Re^d$ if it has $W(0) = 0$, continuous sample paths, independent increments, and

$$W(t) - W(s) \sim N(0, (t-s)I),$$

for all $0 \leq s < t \leq T$, with I the $d \times d$ identity matrix. It follows that each of the coordinate processes $W_i(t)$, $i = 1, \ldots, d$, is a standard one-dimensional Brownian motion and that W_i and W_j are independent for $i \neq j$.

Suppose μ is a vector in $\Re^d$ and Σ is a $d \times d$ matrix, positive definite or semidefinite. We call a process X a Brownian motion with drift μ and covariance Σ (abbreviated $X \sim \mathrm{BM}(\mu, \Sigma)$) if X has continuous sample paths and independent increments with

$$X(t) - X(s) \sim N((t-s)\mu, (t-s)\Sigma).$$

The initial value $X(0)$ is an arbitrary constant assumed to be 0 unless otherwise specified. If B is a $d \times k$ matrix satisfying $BB^\top = \Sigma$ and if W is a standard Brownian motion on $\Re^k$, then the process defined by

$$X(t) = \mu t + BW(t) \tag{3.11}$$

is a $\mathrm{BM}(\mu, \Sigma)$. In particular, the law of X depends on B only through $BB^\top$.

The process in (3.11) solves the SDE

$$dX(t) = \mu \, dt + B \, dW(t).$$

We may extend the definition of a multidimensional Brownian motion to deterministic, time-varying $\mu(t)$ and $\Sigma(t)$ through the solution to

$$dX(t) = \mu(t) \, dt + B(t) \, dW(t),$$

where $B(t)B^\top(t) = \Sigma(t)$. This process has continuous sample paths, independent increments, and

$$X(t) - X(s) \sim N\left(\int_s^t \mu(u) \, du, \int_s^t \Sigma(u) \, du \right).$$

A calculation similar to the one leading to (3.5) shows that if $X \sim \mathrm{BM}(\mu, \Sigma)$, then

$$\mathsf{Cov}[X_i(s), X_j(t)] = \min(s, t)\Sigma_{ij}. \tag{3.12}$$

In particular, given a set of times $0 < t_1 < t_2 < \cdots < t_n$, we can easily find the covariance matrix of

$$(X_1(t_1), \ldots, X_d(t_1), X_1(t_2), \ldots, X_d(t_2), \ldots, X_1(t_n), \ldots, X_d(t_n)) \tag{3.13}$$

along with its mean and reduce the problem of generating a discrete path of X to one of sampling this nd-dimensional normal vector. While there could be cases in which this is advantageous, it will usually be more convenient to use the fact that this nd-vector is the concatenation of d-vectors representing the state of the process at n distinct times.

Random Walk Construction

Let $Z_1, Z_2, \ldots$ be independent $N(0, I)$ random vectors in $\Re^d$. We can construct a standard d-dimensional Brownian motion at times $0 = t_0 < t_1 < \cdots < t_n$ by setting $W(0) = 0$ and

$$W(t_{i+1}) = W(t_i) + \sqrt{t_{i+1} - t_i}\, Z_{i+1}, \quad i = 0, \ldots, n - 1. \tag{3.14}$$

This is equivalent to applying the one-dimensional random walk construction (3.2) separately to each coordinate of W.

To simulate $X \sim \mathrm{BM}(\mu, \Sigma)$, we first find a matrix B for which $BB^\top = \Sigma$ (see Section 2.3.3). If B is $d \times k$, let $Z_1, Z_2, \ldots$ be independent standard normal random vectors in $\Re^k$. Set $X(0) = 0$ and

$$X(t_{i+1}) = X(t_i) + \mu(t_{i+1} - t_i) + \sqrt{t_{i+1} - t_i}\, BZ_i, \quad i = 0, \ldots, n - 1. \tag{3.15}$$

Thus, simulation of $\mathrm{BM}(\mu, \Sigma)$ is straightforward once Σ has been factored. For the case of time-dependent coefficients, we may set

$$X(t_{i+1}) = X(t_i) + \int_{t_i}^{t_{i+1}} \mu(s)\, ds + B(t_i, t_{i+1})Z_i, \quad i = 0, \ldots, n - 1,$$

with

$$B(t_i, t_{i+1})B(t_i, t_{i+1})^\top = \int_{t_i}^{t_{i+1}} \Sigma(u)\, du,$$

thus requiring n factorizations.

Brownian Bridge Construction

Application of a Brownian bridge construction to a standard d-dimensional Brownian motion is straightforward: we may simply apply independent one-dimensional constructions to each of the coordinates. To include a drift vector (i.e., for $\mathrm{BM}(\mu, I)$ process), it suffices to add $\mu_i t_n$ to $W_i(t_n)$ at the first step of the construction of the ith coordinate, as explained in Section 3.1.1. The rest of the construction is unaffected.

To construct $X \sim \mathrm{BM}(\mu, \Sigma)$, we may use the fact that X can be represented as $X(t) = \mu t + BW(t)$ with B a $d \times k$ matrix, $k \le d$, satisfying $BB^{\top} = \Sigma$, and W a standard k-dimensional Brownian motion. We may then apply a Brownian bridge construction to $W(t_1), \ldots, W(t_n)$ and recover $X(t_1), \ldots, X(t_n)$ through a linear transformation.

Principal Components Construction

As with the Brownian bridge construction, one could apply a one-dimensional principal components construction to each coordinate of a multidimensional Brownian motion. Through a linear transformation this then extends to the construction of $\mathrm{BM}(\mu, \Sigma)$. However, the optimality of principal components is lost in this reduction; to recover it, we must work directly with the covariance matrix of (3.13).

It follows from (3.12) that the covariance matrix of (3.13) can be represented as $(C \otimes \Sigma)$, where $\otimes$ denotes the Kronecker product producing

$$
(C \otimes \Sigma) = \begin{pmatrix}
C_{11}\Sigma & C_{12}\Sigma & \cdots & C_{1n}\Sigma \\
C_{21}\Sigma & C_{22}\Sigma & \cdots & C_{2n}\Sigma \\
\vdots & \vdots & \ddots & \vdots \\
C_{n1}\Sigma & C_{n2}\Sigma & \cdots & C_{nn}\Sigma
\end{pmatrix}.
$$

If C has eigenvectors $v_1, \ldots, v_n$ and eigenvalues $\lambda_1 \ge \cdots \ge \lambda_n$, and if Σ has eigenvectors $w_1, \ldots, w_d$ and eigenvalues $\eta_1 \ge \cdots \ge \eta_d$, then $(C \otimes \Sigma)$ has eigenvectors $(v_i \otimes w_j)$ and eigenvalues $\lambda_i \eta_j$, $i = 1, \ldots, n$, $j = 1, \times, d$. This special structure of the covariance matrix of (3.12) makes it possible to reduce the computational effort required to find all eigenvalues and eigenvectors from the $O((nd)^3)$ typically required for an $(nd \times nd)$ matrix to $O(n^3 + d^3)$.

If we rank the products of eigenvalues as

$$
(\lambda_i \eta_j)_{(1)} \ge (\lambda_i \eta_j)_{(2)} \ge \cdots (\lambda_i \eta_j)_{(nd)},
$$

then for any $k = 1, \ldots, n$,

$$
\frac{\sum_{r=1}^{k} (\lambda_i \eta_j)_{(r)}}{\sum_{r=1}^{nd} (\lambda_i \eta_j)_{(r)}} \le \frac{\sum_{i=1}^{k} \lambda_i}{\sum_{i=1}^{n} \lambda_i}.
$$

In other words, the variability explained by the first k factors is always smaller for a d-dimensional Brownian motion than it would be for a scalar Brownian motion over the same time points. This is to be expected since the d-dimensional process has greater total variability.

3.2 Geometric Brownian Motion

A stochastic process $S(t)$ is a *geometric Brownian motion* if $\log S(t)$ is a Brownian motion with initial value $\log S(0)$; in other words, a geometric Brownian motion is simply an exponentiated Brownian motion. Accordingly, all methods for simulating Brownian motion become methods for simulating geometric Brownian motion through exponentiation. This section therefore focuses more on modeling than on algorithmic issues.

Geometric Brownian motion is the most fundamental model of the value of a financial asset. In his pioneering thesis of 1900, Louis Bachelier developed a model of stock prices that in retrospect we describe as ordinary Brownian motion, though the mathematics of Brownian motion had not yet been developed. The use of geometric Brownian motion as a model in finance is due primarily to work of Paul Samuelson in the 1960s. Whereas ordinary Brownian motion can take negative values — an undesirable feature in a model of the price of a stock or any other limited liability asset — geometric Brownian motion is always positive because the exponential function takes only positive values. More fundamentally, for geometric Brownian motion the *percentage* changes

$$\frac{S(t_2) - S(t_1)}{S(t_1)}, \frac{S(t_3) - S(t_2)}{S(t_2)}, \ldots, \frac{S(t_n) - S(t_{n-1})}{S(t_{n-1})} \tag{3.16}$$

are independent for $t_1 < t_2 < \cdots < t_n$, rather than the absolute changes $S(t_{i+1}) - S(t_i)$. These properties explain the centrality of geometric rather than ordinary Brownian motion in modeling asset prices.

3.2.1 Basic Properties

Suppose W is a standard Brownian motion and X satisfies

$$dX(t) = \mu \, dt + \sigma \, dW(t),$$

so that $X \sim \mathrm{BM}(\mu, \sigma^2)$. If we set $S(t) = S(0) \exp(X(t)) \equiv f(X(t))$, then an application of Itô's formula shows that

$$
\begin{aligned}
dS(t) &= f'(X(t)) \, dX(t) + \tfrac{1}{2}\sigma^2 f''(X(t)) \, dt \\
&= S(0) \exp(X(t))[\mu \, dt + \sigma \, dW(t)] + \tfrac{1}{2}\sigma^2 S(0) \exp(X(t)) \, dt \\
&= S(t)(\mu + \tfrac{1}{2}\sigma^2) \, dt + S(t)\sigma \, dW(t).
\end{aligned} \tag{3.17}
$$

In contrast, a geometric Brownian motion process is often specified through an SDE of the form

$$\frac{dS(t)}{S(t)} = \mu \, dt + \sigma \, dW(t), \tag{3.18}$$

an expression suggesting a Brownian model of the "instantaneous returns" $dS(t)/S(t)$. Comparison of (3.17) and (3.18) indicates that the models are inconsistent and reveals an ambiguity in the role of "μ." In (3.17), μ is the drift of the Brownian motion we exponentiated to define $S(t)$ — the drift of $\log S(t)$. In (3.18), $S(t)$ has drift $\mu S(t)$ and (3.18) implies

$$d \log S(t) = (\mu - \tfrac{1}{2}\sigma^2)\, dt + \sigma\, dW(t), \qquad (3.19)$$

as can be verified through Itô's formula or comparison with (3.17).

We will use the notation $S \sim \mathrm{GBM}(\mu, \sigma^2)$ to indicate that S is a process of the type in (3.18). We will refer to μ in (3.18) as the *drift parameter* though it is not the drift of either $S(t)$ or $\log S(t)$. We refer to σ in (3.18) as the *volatility parameter* of $S(t)$; the diffusion coefficient of $S(t)$ is $\sigma^2 S^2(t)$.

From (3.19) we see that if $S \sim \mathrm{GBM}(\mu, \sigma^2)$ and if S has initial value $S(0)$, then

$$S(t) = S(0)\exp\left([\mu - \tfrac{1}{2}\sigma^2]t + \sigma W(t)\right). \qquad (3.20)$$

A bit more generally, if $u < t$ then

$$S(t) = S(u)\exp\left([\mu - \tfrac{1}{2}\sigma^2](t - u) + \sigma(W(t) - W(u))\right), \qquad (3.21)$$

from which the claimed independence of the returns in (3.16) becomes evident. Moreover, since the increments of W are independent and normally distributed, this provides a simple recursive procedure for simulating values of S at $0 = t_0 < t_1 < \cdots < t_n$:

$$S(t_{i+1}) = S(t_i)\exp\left([\mu - \tfrac{1}{2}\sigma^2](t_{i+1} - t_i) + \sigma\sqrt{t_{i+1} - t_i}\,Z_{i+1}\right), \qquad (3.22)$$
$$i = 0, 1, \ldots, n - 1,$$

with $Z_1, Z_2, \ldots, Z_n$ independent standard normals. In fact, (3.22) is equivalent to exponentiating both sides of (3.3) with μ replaced by $\mu - \tfrac{1}{2}\sigma^2$. This method is *exact* in the sense that the $(S(t_1), \ldots, S(t_n))$ it produces has the joint distribution of the process $S \sim \mathrm{GBM}(\mu, \sigma^2)$ at $t_1, \ldots, t_n$ — the method involves no discretization error. Time-dependent parameters can be incorporated by exponentiating both sides of (3.4).

Lognormal Distribution

From (3.20) we see that if $S \sim \mathrm{GBM}(\mu, \sigma^2)$, then the marginal distribution of $S(t)$ is that of the exponential of a normal random variable, which is called a lognormal distribution. We write $Y \sim LN(\mu, \sigma^2)$ if the random variable Y has the distribution of $\exp(\mu + \sigma Z)$, $Z \sim N(0, 1)$. This distribution is thus given by

$$P(Y \le y) = P(Z \le [\log(y) - \mu]/\sigma)$$
$$= \Phi\left(\frac{\log(y) - \mu}{\sigma}\right)$$

and its density by

$$\frac{1}{y\sigma}\phi\left(\frac{\log(y) - \mu}{\sigma}\right). \tag{3.23}$$

Moments of a lognormal random variable can be calculated using the basic identity

$$\mathsf{E}[e^{aZ}] = e^{\frac{1}{2}a^2}$$

for the moment generating function of a standard normal. From this it follows that $Y \sim LN(\mu, \sigma^2)$ has

$$\mathsf{E}[Y] = e^{\mu + \frac{1}{2}\sigma^2}, \quad \mathsf{Var}[Y] = e^{2\mu + \sigma^2}\left(e^{\sigma^2} - 1\right);$$

in particular, the notation $Y \sim LN(\mu, \sigma^2)$ does not imply that μ and σ^2 are the mean and variance of Y. From

$$P(Y \le e^\mu) = P(Z \le 0) = \tfrac{1}{2}$$

we see that e^μ is the median of Y. The mean of Y is thus larger than the median, reflecting the positive skew of the lognormal distribution.

Applying these observations to (3.20), we find that if $S \sim \mathrm{GBM}(\mu, \sigma^2)$ then $(S(t)/S(0)) \sim LN([\mu - \frac{1}{2}\sigma^2]t, \sigma^2 t)$ and

$$\mathsf{E}[S(t)] = e^{\mu t}S(0), \quad \mathsf{Var}[S(t)] = e^{2\mu t}S^2(0)\left(e^{\sigma^2 t} - 1\right).$$

In fact, we have

$$\mathsf{E}[S(t)|S(\tau), 0 \le \tau \le u] = \mathsf{E}[S(t)|S(u)] = e^{\mu(t-u)}S(u), \quad u < t, \tag{3.24}$$

and an analogous expression for the conditional variance. The first equality in (3.24) is the Markov property (which follows from the fact that S is a one-to-one transformation of a Brownian motion, itself a Markov process) and the second follows from (3.21).

Equation (3.24) indicates that μ acts as an average growth rate for S, a sort of average continuously compounded rate of return. Along a single sample path of S the picture is different. For a standard Brownian motion W, we have $t^{-1}W(t) \to 0$ with probability 1. For $S \sim \mathrm{GBM}(\mu, \sigma^2)$, we therefore find that

$$\frac{1}{t}\log S(t) \to \mu - \tfrac{1}{2}\sigma^2,$$

with probability 1, so $\mu - \frac{1}{2}\sigma^2$ serves as the growth rate along each path. If this expression is positive, $S(t) \to \infty$ as $t \to \infty$; if it is negative, then $S(t) \to 0$. In a model with $\mu > 0 > \mu - \frac{1}{2}\sigma^2$, we find from (3.24) that $\mathsf{E}[S(t)]$ grows exponentially although $S(t)$ converges to 0. This seemingly pathological behavior is explained by the increasing skew in the distribution of $S(t)$: although $S(t) \to 0$, rare but very large values of $S(t)$ are sufficiently likely to produce an increasing mean.

3.2.2 Path-Dependent Options

Our interest in simulating paths of geometric Brownian motion lies primarily in pricing options, particularly those whose payoffs depend on the path of an underlying asset S and not simply its value $S(T)$ at a fixed exercise date T. Through the principles of option pricing developed in Chapter 1, the price of an option may be represented as an expected discounted payoff. This price is estimated through simulation by generating paths of the underlying asset, evaluating the discounted payoff on each path, and averaging over paths.

Risk-Neutral Dynamics

The one subtlety in this framework is the probability measure with respect to which the expectation is taken and the nearly equivalent question of how the payoff should be discounted. This bears on how the paths of the underlying asset ought to be generated and more specifically in the case of geometric Brownian motion, how the drift parameter μ should be chosen.

We start by assuming the existence of a constant continuously compounded interest rate r for riskless borrowing and lending. A dollar invested at this rate at time 0 grows to a value of

$$\beta(t) = e^{rt}$$

at time t. Similarly, a contract paying one dollar at a future time t (a zero-coupon bond) has a value at time 0 of e^{-rt}. In pricing under the risk-neutral measure, we discount a payoff to be received at time t back to time 0 by dividing by $\beta(t)$; i.e., β is the numeraire asset.

Suppose the asset S pays no dividends; then, under the risk-neutral measure, the discounted price process $S(t)/\beta(t)$ is a martingale:

$$\frac{S(u)}{\beta(u)} = \mathsf{E}\left[\frac{S(t)}{\beta(t)}|\{S(\tau), 0 \leq \tau \leq u\}\right]. \tag{3.25}$$

Comparison with (3.24) shows that if S is a geometric Brownian motion under the risk-neutral measure, then it must have $\mu = r$; i.e.,

$$\frac{dS(t)}{S(t)} = r\,dt + \sigma\,dW(t). \tag{3.26}$$

As discussed in Section 1.2.2, this equation helps explain the name "risk-neutral." In a world of risk-neutral investors, all assets would have the same average rate of return — investors would not demand a higher rate of return for holding risky assets. In a risk-neutral world, the drift parameter for $S(t)$ would therefore equal the risk-free rate r.

In the case of an asset that pays dividends, we know from Section 1.2.2 that the martingale property (3.25) continues to hold but with S replaced by the sum of S, any dividends paid by S, and any interest earned from investing the dividends at the risk-free rate r. Thus, let $D(t)$ be the value of any dividends paid over $[0, t]$ and interest earned on those dividends. Suppose the asset pays a *continuous dividend yield* of δ, meaning that it pays dividends at rate $\delta S(t)$ at time t. Then D grows at rate

$$\frac{dD(t)}{dt} = \delta S(t) + rD(t),$$

the first term on the right reflecting the influx of new dividends and the second term reflecting interest earned on dividends already accumulated. If $S \sim \text{GBM}(\mu, \sigma^2)$, then the drift in $(S(t) + D(t))$ is

$$\mu S(t) + \delta S(t) + rD(t).$$

The martingale property (3.25), now applied to the combined process $(S(t) + D(t))$, requires that this drift equal $r(S(t) + D(t))$. We must therefore have $\mu + \delta = r$; i.e., $\mu = r - \delta$. The net effect of a dividend yield is to reduce the growth rate by δ.

We discuss some specific settings in which this formulation is commonly used:

○ *Equity Indices.* In pricing index options, the level of the index is often modeled as geometric Brownian motion. An index is not an asset and it does not pay dividends, but the individual stocks that make up an index may pay dividends and this affects the level of the index. Because an index may contain many stocks paying a wide range of dividends on different dates, the combined effect is often approximated by a continuous dividend yield δ.

○ *Exchange Rates.* In pricing currency options, the relevant underlying variable is an exchange rate. We may think of an exchange rate S (quoted as the number of units of domestic currency per unit of foreign currency) as the price of the foreign currency. A unit of foreign currency earns interest at some risk-free rate r_f, and this interest may be viewed as a dividend stream. Thus, in modeling an exchange rate using geometric Brownian motion, we set $\mu = r - r_f$.

○ *Commodities.* A physical commodity like gold or oil may in some cases behave like an asset that pays *negative* dividends because of the cost of storing the commodity. This is easily accommodated in the setting above by taking $\delta < 0$. There may, however, be some value in holding a physical commodity; for example, a party storing oil implicitly holds an option to sell or consume the oil in case of a shortage. This type of benefit is sometimes approximated through a hypothetical *convenience yield* that accrues from physical storage. The net dividend yield in this case is the difference between the convenience yield and the cost rate for storage.

○ *Futures Contracts.* A futures contract commits the holder to buying an underlying asset or commodity at a fixed price at a fixed date in the future. The *futures price* is the price specifed in a futures contract at which both the buyer and the seller would willingly enter into the contract without either party paying the other. A futures price is thus not the price of an asset but rather a price agreed upon for a transaction in the future.

Let $S(t)$ denote the price of the underlying asset (the spot price) and let $F(t,T)$ denote the futures prices at time t for a contract to be settled at a fixed time T in the future. Entering into a futures contract at time t to buy the underlying asset at time $T > t$ is equivalent to committing to exchange a known amount $F(t,T)$ for an uncertain amount $S(T)$. For this contract to have zero value at the inception time t entails

$$0 = e^{-r(T-t)}\mathsf{E}[(S(T) - F(t,T))|\mathcal{F}_t], \qquad (3.27)$$

where $\mathcal{F}_t$ is the history of market prices up to time t. At $t = T$ the spot and futures prices must agree, so $S(T) = F(T,T)$ and we may rewrite this condition as

$$F(t,T) = \mathsf{E}[F(T,T)|\mathcal{F}_t].$$

Thus, the futures price is a martingale (in its first argument) under the risk-neutral measure. It follows that if we choose to model a futures price (for fixed maturity T) using geometric Brownian motion, we should set its drift parameter to zero:

$$\frac{dF(t,T)}{F(t,T)} = \sigma\, dW(t).$$

Comparison of (3.27) and (3.25) reveals that

$$F(t,T) = e^{(r-\delta)(T-t)}S(t),$$

with δ the net dividend yield for S. If either process is a geometric Brownian motion under the risk-neutral measure then the other is as well and they have the same volatility σ.

This discussion blurs the distinction between futures and forward contracts. The argument leading to (3.27) applies more specifically to a forward price because a forward contract involves no intermediate cashflows. The holder of a futures contract typically makes or receives payments each day through a margin account; the discussion above ignores these cashflows. In a world with deterministic interest rates, futures and forward prices must be equal to preclude arbitrage so the conclusion in (3.27) is valid for both. With stochastic interest rates, it turns out that futures prices continue to be martingales under the risk-neutral measure but forward prices do not. The theoretical relation between futures and forward prices is investigated in Cox, Ingersoll, and Ross [90]; it is also discussed in many texts on derivative securities (e.g., Hull [189]).

Path-Dependent Payoffs

We turn now to some examples of path-dependent payoffs frequently encountered in option pricing. We focus primarily on cases in which the payoff depends on the values $S(t_1), \ldots, S(t_n)$ at a fixed set of dates $t_1, \ldots, t_n$; for these it is usually possible to produce an unbiased simulation estimate of the option price. An option payoff could in principle depend on the complete path $\{S(t), 0 \leq t \leq T\}$ over an interval $[0, T]$; pricing such an option by simulation will often entail some discretization bias. In the examples that follow, we distinguish between discrete and continuous monitoring of the underlying asset.

○ *Asian option: discrete monitoring.* An Asian option is an option on a time average of the underlying asset. Asian calls and puts have payoffs $(\bar{S} - K)^+$ and $(K - \bar{S})^+$ respectively, where the strike price K is a constant and

$$\bar{S} = \frac{1}{n} \sum_{i=1}^{n} S(t_i) \qquad (3.28)$$

is the average price of the underlying asset over the discrete set of monitoring dates $t_1, \ldots, t_n$. Other examples have payoffs $(S(T) - \bar{S})^+$ and $(\bar{S} - S(T))^+$. There are no exact formulas for the prices of these options, largely because the distribution of $\bar{S}$ is intractable.

○ *Asian option: continuous monitoring.* The continuous counterparts of the discrete Asian options replace the discrete average above with the continuous average

$$\bar{S} = \frac{1}{t - u} \int_u^t S(\tau) \, d\tau$$

over an interval $[u, t]$. Though more difficult to simulate, some instances of continuous-average Asian options allow pricing through the transform analysis of Geman and Yor [135] and the eigenfunction expansion of Linetsky [237].

○ *Geometric average option.* Replacing the arithmetic average $\bar{S}$ in (3.28) with

$$\left(\prod_{i=1}^{n} S(t_i) \right)^{1/n}$$

produces an option on the geometric average of the underlying asset price. Such options are seldom if ever found in practice, but they are useful as test cases for computational procedures and as a basis for approximating ordinary Asian options. They are mathematically convenient to work with because the geometric average of (jointly) lognormal random variables is itself lognormal. From (3.20) we find (with μ replaced by r) that

$$\left(\prod_{i=1}^{n} S(t_i) \right)^{1/n} = S(0) \exp \left([r - \tfrac{1}{2}\sigma^2] \frac{1}{n} \sum_{i=1}^{n} t_i + \frac{\sigma}{n} \sum_{i=1}^{n} W(t_i) \right).$$

From the Linear Transformation Property (2.23) and the covariance matrix (3.6), we find that

$$\sum_{i=1}^{n} W(t_i) \sim N\left(0, \sum_{i=1}^{n}(2i-1)t_{n+1-i}\right).$$

It follows that the geometric average of $S(t_1), \ldots, S(t_n)$ has the same distribution as the value at time T of a process $\mathrm{GBM}(r - \delta, \bar{\sigma}^2)$ with

$$T = \frac{1}{n}\sum_{i=1}^{n} t_i, \quad \bar{\sigma}^2 = \frac{\sigma^2}{n^2 T}\sum_{i=1}^{n}(2i-1)t_{n+1-i}, \quad \delta = \tfrac{1}{2}\sigma^2 - \tfrac{1}{2}\bar{\sigma}^2.$$

An option on the geometric average may thus be valued using the Black-Scholes formula (1.44) for an asset paying a continuous dividend yield. The expression

$$\exp\left(\int_u^t \log S(\tau)\, d\tau\right)$$

is a continuously monitored version of the geometric average and is also lognormally distributed. Options on a continuous geometric average can similarly be priced in closed form.

○ *Barrier options.* A typical example of a barrier option is one that gets "knocked out" if the underlying asset crosses a prespecified level. For instance, a *down-and-out call* with barrier b, strike K, and expiration T has payoff

$$\mathbf{1}\{\tau(b) > T\}(S(T) - K)^+,$$

where

$$\tau(b) = \inf\{t_i : S(t_i) < b\}$$

is the first time in $\{t_1, \ldots, t_n\}$ the price of the underlying asset drops below b (understood to be ∞ if $S(t_i) > b$ for all i) and $\mathbf{1}\{\ \}$ denotes the indicator of the event in braces. A down-and-*in* call has payoff $\mathbf{1}\{\tau(b) \leq T\}(S(T) - K)^+$: it gets "knocked in" only when the underlying asset crosses the barrier. Up-and-out and up-and-in calls and puts are defined analogously. Some knockout options pay a rebate if the underlying asset crosses the barrier, with the rebate possibly paid either at the time of the barrier crossing or at the expiration of the option.

These examples of discretely monitored barrier options are easily priced by simulation through sampling of $S(t_1), \ldots, S(t_n), S(T)$. A continuously monitored barrier option is knocked in or out the instant the underlying asset crosses the barrier; in other words, it replaces $\tau(b)$ as defined above with

$$\tilde{\tau}(b) = \inf\{t \geq 0 : S(t) \leq b\}.$$

Both discretely monitored and continuously monitored barrier options are found in practice. Many continuously monitored barrier options can be priced in closed form; Merton [261] provides what is probably the first such formula and many other cases can be found in, e.g., Briys et al. [62]. Discretely monitored barrier options generally do not admit pricing formulas and hence require computational procedures.

○ *Lookback options.* Like barrier options, lookback options depend on extremal values of the underlying asset price. Lookback puts and calls expiring at t_n have payoffs

$$(\max_{i=1,\ldots,n} S(t_i) - S(t_n)) \quad \text{and} \quad (S(t_n) - \min_{i=1,\ldots,n} S(t_i))$$

respectively. A lookback call, for example, may be viewed as the profit from buying at the lowest price over $t_1, \ldots, t_n$ and selling at the final price $S(t_n)$. Continuously monitored versions of these options are defined by taking the maximum or minimum over an interval rather than a finite set of points.

Incorporating a Term Structure

Thus far, we have assumed that the risk-free interest rate r is constant. This implies that the time-t price of a zero-coupon bond maturing (and paying 1) at time $T > t$ is

$$B(t, T) = e^{-r(T-t)}. \tag{3.29}$$

Suppose however that at time 0 we observe a collection of bond prices $B(0, T)$, indexed by maturity T, incompatible with (3.29). To price an option on an underlying asset price S consistent with the observed term structure of bond prices, we can introduce a deterministic but time-varying risk-free rate $r(u)$ by setting

$$r(u) = -\frac{\partial}{\partial T} \log B(0, T) \Big|_{T=u}.$$

Clearly, then,

$$B(0, T) = \exp\left(-\int_0^T r(u)\, du\right).$$

With a deterministic, time-varying risk-free rate $r(u)$, the dynamics of an asset price $S(t)$ under the risk-neutral measure (assuming no dividends) are described by the SDE

$$\frac{dS(t)}{S(t)} = r(t)\, dt + \sigma\, dW(t)$$

with solution

$$S(t) = S(0) \exp\left(\int_0^t r(u)\, du - \tfrac{1}{2}\sigma^2 t + \sigma W(t)\right).$$

This process can be simulated over $0 = t_0 < t_1 < \cdots < t_n$ by setting

$$S(t_{i+1}) = S(t_i) \exp\left(\int_{t_i}^{t_{i+1}} r(u)\, du - \tfrac{1}{2}\sigma^2(t_{i+1} - t_i) + \sigma\sqrt{t_{i+1} - t_i}\, Z_{i+1}\right),$$

with $Z_1, \ldots, Z_n$ independent $N(0,1)$ random variables.

If in fact we are interested only in values of $S(t)$ at $t_1, \ldots, t_n$, the simulation can be simplified, making it unnecessary to introduce a short rate $r(u)$ at all. If we observe bond prices $B(0, t_1), \ldots, B(0, t_n)$ (either directly or through interpolation from other observed prices), then since

$$\frac{B(0, t_i)}{B(0, t_{i+1})} = \exp\left(\int_{t_i}^{t_{i+1}} r(u)\, du\right),$$

we may simulate $S(t)$ using

$$S(t_{i+1}) = S(t_i) \frac{B(0, t_i)}{B(0, t_{i+1})} \exp\left(-\tfrac{1}{2}\sigma^2(t_{i+1} - t_i) + \sigma\sqrt{t_{i+1} - t_i}\, Z_{i+1}\right). \quad (3.30)$$

Simulating Off a Forward Curve

For some types of underlying assets, particularly commodities, we may observe not just a spot price $S(0)$ but also a collection of forward prices $F(0, T)$. Here, $F(0, T)$ denotes the price specified in a contract at time 0 to be paid at time T for the underlying asset. Under the risk-neutral measure, $F(0, T) = E[S(T)]$; in particular, the forward prices reflect the risk-free interest rate and any dividend yield (positive or negative) on the underlying asset. In pricing options, we clearly want to simulate price paths of the underlying asset consistent with the forward prices observed in the market.

The equality $F(0, T) = E[S(T)]$ implies

$$S(T) = F(0, T) \exp\left(-\tfrac{1}{2}\sigma^2 T + \sigma W(T)\right).$$

Given forward prices $F(0, t_1), \ldots, F(0, t_n)$, we can simulate using

$$S(t_{i+1}) = S(t_i) \frac{F(0, t_{i+1})}{F(0, t_i)} \exp\left(-\tfrac{1}{2}\sigma^2(t_{i+1} - t_i) + \sigma\sqrt{t_{i+1} - t_i}\, Z_{i+1}\right).$$

This generalizes (3.30) because in the absence of dividends we have $F(0, T) = S(0)/B(0, T)$. Alternatively, we may define $M(0) = 1$,

$$M(t_{i+1}) = M(t_i) \exp\left(-\tfrac{1}{2}\sigma^2(t_{i+1} - t_i) + \sigma\sqrt{t_{i+1} - t_i}\, Z_{i+1}\right), \quad i = 0, \ldots, n-1,$$

and set $S(t_i) = F(0, t_i)M(t_i)$, $i = 1, \ldots, n$.

Deterministic Volatility Functions

Although geometric Brownian motion remains an important benchmark, it has been widely observed across many markets that option prices are incompatible with a GBM model for the underlying asset. This has fueled research into alternative specifications of the dynamics of asset prices.

Consider a market in which several options with various strikes and maturities are traded simultaneously on the same underlying asset. Suppose the market is sufficiently liquid that we may effectively observe prices of the options without error. If the assumptions underlying the Black-Scholes formula held exactly, all of these option prices would result from using the same volatility parameter σ in the formula. In practice, one usually finds that this implied volatility actually varies with strike and maturity. It is therefore natural to seek a minimal modification of the Black-Scholes model capable of reproducing market prices.

Consider the extreme case in which we observe the prices $C(K,T)$ of call options on a single underlying asset for a continuum of strikes K and maturities T. Dupire [107] shows that, subject only to smoothness conditions on C as a function of K and T, it is possible to find a function $\sigma(S,t)$ such that the model

$$\frac{dS(t)}{S(t)} = r\,dt + \sigma(S(t),t)\,dW(t)$$

reproduces the given option prices, in the sense that

$$e^{-rT}\mathsf{E}[(S(T)-K)^+] = C(K,T)$$

for all K and T. This is sometimes called a *deterministic volatility function* to emphasize that it extends geometric Brownian motion by allowing σ to be a deterministic function of the current level of the underlying asset. This feature is important because it ensures that options can still be hedged through a position in the underlying asset, which would not be the case in a stochastic volatility model.

In practice, we observe only a finite set of option prices and this leaves a great deal of flexibility in specifying $\sigma(S,t)$ while reproducing market prices. We may, for example, impose smoothness constraints on the choice of volatility function. This function will typically be the result of a numerical optimization procedure and may never be given explicitly.

Once $\sigma(S,t)$ has been chosen to match a set of actively traded options, simulation may still be necessary to compute the prices of less liquid path-dependent options. In general, there is no exact simulation procedure for these models and it is necessary to use an Euler scheme of the form

$$S(t_{i+1}) = S(t_i)\left(1 + r(t_{i+1}-t_i) + \sigma(S(t_i),t_i)\sqrt{t_{i+1}-t_i}\,Z_{i+1}\right),$$

with $Z_1, Z_2, \ldots$ independent standard normals, or

$$S(t_{i+1}) = $$
$$S(t_i)\exp\left([r - \tfrac{1}{2}\sigma^2(S(t_i),t_i)](t_{i+1}-t_i) + \sigma(S(t_i),t_i)\sqrt{t_{i+1}-t_i}\,Z_{i+1}\right),$$

which is equivalent to an Euler scheme for $\log S(t)$.

3.2.3 Multiple Dimensions

A multidimensional geometric Brownian motion can be specified through a system of SDEs of the form

$$\frac{dS_i(t)}{S_i(t)} = \mu_i \, dt + \sigma_i \, dX_i(t), \quad i = 1, \ldots, d, \tag{3.31}$$

where each X_i is a standard one-dimensional Brownian motion and $X_i(t)$ and $X_j(t)$ have correlation ρ_{ij}. If we define a $d \times d$ matrix Σ by setting $\Sigma_{ij} = \sigma_i \sigma_j \rho_{ij}$, then $(\sigma_1 X_1, \ldots, \sigma_d X_d) \sim \mathrm{BM}(0, \Sigma)$. In this case we abbreviate the process $S = (S_1, \ldots, S_d)$ as $\mathrm{GBM}(\mu, \Sigma)$ with $\mu = (\mu_1, \ldots, \mu_d)$. In a convenient abuse of terminology, we refer to μ as the drift vector of S, to Σ as its covariance matrix and to the matrix with entries ρ_{ij} as its correlation matrix; the actual drift vector is $(\mu_1 S_1(t), \ldots, \mu_d S_d(t))$ and the covariances are given by

$$\mathsf{Cov}[S_i(t), S_j(t)] = S_i(0) S_j(0) e^{(\mu_i + \mu_j)t} \left(e^{\rho_{ij} \sigma_i \sigma_j} - 1 \right).$$

This follows from the representation

$$S_i(t) = S_i(0) e^{(\mu_i - \frac{1}{2}\sigma_i^2)t + \sigma_i X_i(t)}, \quad i = 1, \ldots, d.$$

Recall that a Brownian motion $\mathrm{BM}(0, \Sigma)$ can be represented as $AW(t)$ with W a standard Brownian motion $\mathrm{BM}(0, I)$ and A any matrix for which $AA^\top = \Sigma$. We may apply this to $(\sigma_1 X_1, \ldots, \sigma_d X_d)$ and rewrite (3.31) as

$$\frac{dS_i(t)}{S_i(t)} = \mu_i \, dt + a_i \, dW(t), \quad i = 1, \ldots, d, \tag{3.32}$$

with a_i the ith row of A. A bit more explicitly, this is

$$\frac{dS_i(t)}{S_i(t)} = \mu_i \, dt + \sum_{j=1}^{d} A_{ij} \, dW_j(t), \quad i = 1, \ldots, d.$$

This representation leads to a simple algorithm for simulating $\mathrm{GBM}(\mu, \Sigma)$ at times $0 = t_0 < t_1 < \cdots < t_n$:

$$S_i(t_{k+1}) = S_i(t_k) e^{(\mu_i - \frac{1}{2}\sigma_i^2)(t_{k+1} - t_k) + \sqrt{t_{k+1} - t_k} \sum_{j=1}^{d} A_{ij} Z_{k+1,j}}, \quad i = 1, \ldots, d, \tag{3.33}$$

$k = 0, \ldots, n-1$, where $Z_k = (Z_{k1}, \ldots, Z_{kd}) \sim N(0, I)$ and $Z_1, Z_2, \ldots, Z_n$ are independent. As usual, choosing A to be the Cholesky factor of Σ can reduce the number of multiplications and additions required at each step. Notice that (3.33) is essentially equivalent to exponentiating both sides of the recursion (3.15); indeed, all methods for simulating $\mathrm{BM}(\mu, \Sigma)$ provide methods for simulating $\mathrm{GBM}(\mu, \Sigma)$ (after replacement of μ_i by $\mu_i - \frac{1}{2}\sigma_i^2$).

The discussion of the choice of the drift parameter μ in Section 3.2.2 applies equally well to each μ_i in pricing options on multiple underlying assets. Often,

$\mu_i = r - \delta_i$ where r is the risk-free interest rate and δ_i is the dividend yield on the ith asset S_i.

We list a few examples of option payoffs depending on multiple assets:

o *Spread option.* A call option on the spread between two assets S_1, S_2 has payoff

$$([S_1(T) - S_2(T)] - K)^+$$

with K a strike price. For example, *crack spread* options traded on the New York Mercantile Exchange are options on the spread between heating oil and crude oil futures.

o *Basket option.* A basket option is an option on a portfolio of underlying assets and has a payoff of, e.g.,

$$([c_1 S_1(T) + c_2 S_2(T) + \cdots + c_d S_d(T)] - K)^+.$$

Typical examples would be options on a portfolio of related assets — bank stocks or Asian currencies, for instance.

o *Outperformance option.* These are options on the maximum or minimum of multiple assets and have payoffs of, e.g., the form

$$(\max\{c_1 S_1(T), c_2 S_2(T), \cdots, c_d S_d(T)\} - K)^+.$$

o *Barrier options.* A two-asset barrier option may have a payoff of the form

$$1\{\min_{i=1,\ldots,n} S_2(t_i) < b\}(K - S_1(T))^+;$$

This is a down-and-in put on S_1 that knocks in when S_2 drops below a barrier at b. Many variations on this basic structure are possible. In this example, one may think of S_1 as an individual stock and S_2 as the level of an equity index: the put on the stock is knocked in only if the market drops.

o *Quantos.* Quantos are options sensitive both to a stock price and an exchange rate. For example, consider an option to buy a stock denominated in a foreign currency with the strike price fixed in the foreign currency but the payoff of the option to be made in the domestic currency. Let S_1 denote the stock price and S_2 the exchange rate, expressed as the quantity of domestic currency required per unit of foreign currency. Then the payoff of the option in the domestic currency is given by

$$S_2(T)(S_1(T) - K)^+. \tag{3.34}$$

The payoff

$$\left(S_1(T) - \frac{K}{S_2(T)}\right)^+$$

corresponds to a quanto in which the level of the strike is fixed in the domestic currency and the payoff of the option is made in the foreign currency.

Change of Numeraire

The pricing of an option on two or more underlying assets can sometimes be transformed to a problem with one less underlying asset (and thus to a lower-dimensional problem) by choosing one of the assets to be the numeraire. Consider, for example, an option to exchange a basket of assets for another asset with payoff

$$\left(\sum_{i=1}^{d-1} c_i S_i(T) - c_d S_d(T) \right)^+,$$

for some constants c_i. The price of the option is given by

$$e^{-rT} \mathsf{E}\left[\left(\sum_{i=1}^{d-1} c_i S_i(T) - c_d S_d(T) \right)^+ \right], \tag{3.35}$$

the expectation taken under the risk-neutral measure. Recall that this is the measure associated with the numeraire asset $\beta(t) = e^{rt}$ and is characterized by the property that the processes $S_i(t)/\beta(t)$, $i = 1, \ldots, d$, are martingales under this measure.

As explained in Section 1.2.3, choosing a different asset as numeraire — say S_d — means switching to a probability measure under which the processes $S_i(t)/S_d(t)$, $i = 1, \ldots, d - 1$, and $\beta(t)/S_d(t)$ are martingales. More precisely, if we let P_β denote the risk-neutral measure, the new measure P_{S_d} is defined by the likelihood ratio process (cf. Appendix B.4)

$$\left(\frac{dP_{S_d}}{dP_\beta} \right)_t = \frac{S_d(t)}{\beta(t)} \frac{\beta(0)}{S_d(0)}. \tag{3.36}$$

Through this change of measure , the option price (3.35) can be expressed as

$$e^{-rT} \mathsf{E}_{S_d}\left[\left(\sum_{i=1}^{d-1} c_i S_i(T) - c_d S_d(T) \right)^+ \left(\frac{dP_\beta}{dP_{S_d}} \right)_T \right]$$

$$= e^{-rT} \mathsf{E}_{S_d}\left[\left(\sum_{i=1}^{d-1} c_i S_i(T) - c_d S_d(T) \right)^+ \left(\frac{\beta(T) S_d(0)}{S_d(T) \beta(0)} \right) \right]$$

$$= S_d(0) \mathsf{E}_{S_d}\left[\left(\sum_{i=1}^{d-1} c_i \frac{S_i(T)}{S_d(T)} - c_d \right)^+ \right],$$

with E_{S_d} denoting expectation under P_{S_d}. From this representation it becomes clear that only the $d - 1$ ratios $S_i(T)/S_d(T)$ (and the constant $S_d(0)$) are needed to price this option under the new measure. We thus need to determine the dynamics of these ratios under the new measure.

Using (3.32) and (3.36), we find that

$$\left(\frac{dP_{S_d}}{dP_\beta}\right)_t = \exp\left(-\tfrac{1}{2}\sigma_d^2 + a_d W(t)\right).$$

Girsanov's Theorem (see Appendix B.4) now implies that the process

$$W^d(t) = W(t) - a_d^\top t$$

is a standard Brownian motion under P_{S_d}. Thus, the effect of changing numeraire is to add a drift $a^\top$ to W. The ratio $S_i(t)/S_d(t)$ is given by

$$\begin{aligned}
\frac{S_i(t)}{S_d(t)} &= \frac{S_i(0)}{S_d(0)}\exp\left(-\tfrac{1}{2}\sigma_i^2 t + \tfrac{1}{2}\sigma_d^2 t + (a_i - a_d)W(t)\right) \\
&= \frac{S_i(0)}{S_d(0)}\exp\left(-\tfrac{1}{2}\sigma_i^2 t + \tfrac{1}{2}\sigma_d^2 t + (a_i - a_d)(W^d(t) + a_d^\top)\right) \\
&= \frac{S_i(0)}{S_d(0)}\exp\left(-\tfrac{1}{2}(a_i - a_d)(a_i - a_d)^\top + (a_i - a_d)W^d(t)\right),
\end{aligned}$$

using the identities $a_j a_j^\top = \sigma_j^2$, $j = 1,\ldots,d$, from the definition of the a_j in (3.32). Under P_{S_d}, the scalar process $(a_i - a_d)W^d(t)$ is a Brownian motion with drift 0 and diffusion coefficient $(a_i - a_d)(a_i - a_d)^\top$. This verifies that the ratios S_i/S_d are martingales under P_{S_d} and also that $(S_1/S_d,\ldots,S_{d-1}/S_d)$ remains a multivariate geometric Brownian motion under the new measure. It is thus possible to price the option by simulating just this $(d-1)$-dimensional process of ratios rather than the original d-dimensional process of asset prices.

This device would not have been effective in the example above if the payoff in (3.35) had instead been

$$\left(\sum_{i=1}^d c_i S_i(T) - K\right)^+$$

with K a constant. In this case, dividing through by $S_d(T)$ would have produced a term $K/S_d(T)$ and would thus have required simulating this ratio as well as S_i/S_d, $i = 1,\ldots,d-1$. What, then, is the scope of this method? If the payoff of an option is given by $g(S_1(T),\ldots,S_d(T))$, then the property we need is that g be homogeneous of degree 1, meaning that

$$g(\alpha x_1,\ldots,\alpha x_d) = \alpha g(x_1,\ldots,x_d)$$

for all scalars α and all $x_1,\ldots,x_d$. For in this case we have

$$\frac{g(S_1(T),\ldots,S_d(T))}{S_d(T)} = g(S_1(T)/S_d(T),\ldots,S_{d-1}(T)/S_d(T),1)$$

and taking one of the underlying assets as numeraire does indeed reduce by one the relevant number of underlying stochastic variables. See Jamshidian [197] for a more general development of this observation.

3.3 Gaussian Short Rate Models

This section and the next develop methods for simulating some simple but important stochastic interest rate models. These models posit the dynamics of an instantaneous continuously compounded short rate $r(t)$. An investment in a money market account earning interest at rate $r(u)$ at time u grows from a value of 1 at time 0 to a value of

$$\beta(t) = \exp\left(\int_0^t r(u)\, du\right)$$

at time t. Though this is now a stochastic quantity, it remains the numeraire for risk-neutral pricing. The price at time 0 of a derivative security that pays X at time T is the expectation of $X/\beta(T)$, i.e.,

$$\mathsf{E}\left[\exp\left(-\int_0^T r(u)\, du\right) X\right], \tag{3.37}$$

the expectation taken with respect to the risk-neutral measure. In particular, the time-0 price of a bond paying 1 at T is given by

$$B(0,T) = \mathsf{E}\left[\exp\left(-\int_0^T r(u)\, du\right)\right]. \tag{3.38}$$

We focus primarily on the dynamics of the short rate under the risk-neutral measure.

The Gaussian models treated in this section offer a high degree of tractability. Many simple instruments can be priced in closed form in these models or using deterministic numerical methods. Some extensions of the basic models and some pricing applications do, however, require simulation for the calculation of expressions of the form (3.37). The tractability of the models offers opportunities for increasing the accuracy of simulation.

3.3.1 Basic Models and Simulation

The classical model of Vasicek [352] describes the short rate through an Ornstein-Uhlenbeck process (cf. Karatzas and Shreve [207], p.358)

$$dr(t) = \alpha(b - r(t))\, dt + \sigma\, dW(t). \tag{3.39}$$

Here, W is a standard Brownian motion and α, b, and σ are positive constants. Notice that the drift in (3.39) is positive if $r(t) < b$ and negative if $r(t) > b$; thus, $r(t)$ is pulled toward level b, a property generally referred to as *mean reversion*. We may interpret b as a long-run interest rate level and α as the speed at which $r(t)$ is pulled toward b. The mean-reverting form of the drift is an essential feature of the Ornstein-Uhlenbeck process and thus of the Vasicek model.

The continuous-time Ho-Lee model [185] has

$$dr(t) = g(t)\,dt + \sigma\,dW(t) \tag{3.40}$$

with g a deterministic function of time. Both (3.39) and (3.40) define Gaussian processes, meaning that the joint distribution of $r(t_1), \ldots, r(t_n)$ is multivariate normal for any $t_1, \ldots, t_n$. Both define Markov processes and are special cases of the general Gaussian Markov process specified by

$$dr(t) = [g(t) + h(t)r(t)]\,dt + \sigma(t)\,dW(t), \tag{3.41}$$

with g, h, and σ all deterministic functions of time. Natural extensions of (3.39) and (3.40) thus allow σ, b, and α to vary with time. Modeling with the Vasicek model when b in particular is time-varying is discussed in Hull and White [190].

The SDE (3.41) has solution

$$r(t) = e^{H(t)}r(0) + \int_0^t e^{H(t)-H(s)}g(s)\,ds + \int_0^t e^{H(t)-H(s)}\sigma(s)\,dW(s),$$

with

$$H(t) = \int_0^t h(s)\,ds,$$

as can be verified through an application of Itô's formula. Because this produces a Gaussian process, simulation of $r(t_1), \ldots, r(t_n)$ is a special case of the general problem of sampling from a multivariate normal distribution, treated in Section 2.3. But it is a sufficiently interesting special case to merit consideration. To balance tractability with generality, we will focus on the Vasicek model (3.39) with time-varying b and on the Ho-Lee model (3.40). Similar ideas apply to the general case (3.41).

Simulation

For the Vasicek model with time-varying b, the general solution above specializes to

$$r(t) = e^{-\alpha t}r(0) + \alpha\int_0^t e^{-\alpha(t-s)}b(s)\,ds + \sigma\int_0^t e^{-\alpha(t-s)}\,dW(s). \tag{3.42}$$

Similarly, for any $0 < u < t$,

$$r(t) = e^{-\alpha(t-u)}r(u) + \alpha\int_u^t e^{-\alpha(t-s)}b(s)\,ds + \sigma\int_u^t e^{-\alpha(t-s)}\,dW(s).$$

From this it follows that, given $r(u)$, the value $r(t)$ is normally distributed with mean

$$e^{-\alpha(t-u)}r(u) + \mu(u,t), \quad \mu(u,t) \equiv \alpha \int_u^t e^{-\alpha(t-s)}b(s)\,ds \qquad (3.43)$$

and variance

$$\sigma_r^2(u,t) \equiv \sigma^2 \int_u^t e^{-2\alpha(t-s)}\,ds = \frac{\sigma^2}{2\alpha}\left(1 - e^{-2\alpha(t-u)}\right). \qquad (3.44)$$

To simulate r at times $0 = t_0 < t_1 < \cdots < t_n$, we may therefore set

$$r(t_{i+1}) = e^{-\alpha(t_{i+1}-t_i)}r(t_i) + \mu(t_i,t_{i+1}) + \sigma_r(t_i,t_{i+1})Z_{i+1}, \qquad (3.45)$$

with $Z_1,\ldots,Z_n$ independent draws from $N(0,1)$.

This algorithm is an exact simulation in the sense that the distribution of the $r(t_1),\ldots,r(t_n)$ it produces is precisely that of the Vasicek process at times $t_1,\ldots,t_n$ for the same value of $r(0)$. In contrast, the slightly simpler Euler scheme

$$r(t_{i+1}) = r(t_i) + \alpha(b(t_i) - r(t_i))(t_{i+1} - t_i) + \sigma\sqrt{t_{i+1}-t_i}Z_{i+1}$$

entails some discretization error. Exact simulation of the Ho-Lee process (3.40) is a special case of the method in (3.4) for simulating a Brownian motion with time-varying drift.

In the special case that $b(t) \equiv b$, the algorithm in (3.45) simplifies to

$$r(t_{i+1}) = e^{-\alpha(t_{i+1}-t_i)}r(t_i) + b(1 - e^{-\alpha(t_{i+1}-t_i)}) + \sigma\sqrt{\frac{1}{2\alpha}\left(1 - e^{-2\alpha(t_{i+1}-t_i)}\right)}Z_{i+1}.$$
$$(3.46)$$

The Euler scheme is then equivalent to making the approximation $e^x \approx 1 + x$ for the exponentials in this recursion.

Evaluation of the integral defining $\mu(t_i,t_{i+1})$ and required in (3.45) may seem burdensome. The effort involved in evaluating this integral clearly depends on the form of the function $b(t)$ so it is worth discussing how this function is likely to be specified in practice. Typically, the flexibility to make b vary with time is used to make the dynamics of the short rate consistent with an observed term structure of bond prices. The same is true of the function g in the Ho-Lee model (3.40). We return to this point in Section 3.3.2, where we discuss bond prices in Gaussian models.

Stationary Version

Suppose $b(t) \equiv b$ and $\alpha > 0$. Then from (3.43) we see that

$$\mathsf{E}[r(t)] = e^{-\alpha t}r(0) + (1 - e^{-\alpha t})b \to b \quad \text{as } t \to \infty,$$

so the process $r(t)$ has a limiting mean. It also has a limiting variance given (via (3.44)) by

$$\lim_{t \to \infty} \mathrm{Var}[r(t)] = \lim_{t \to \infty} \frac{\sigma^2}{2\alpha} \left(1 - e^{-2\alpha t}\right) = \frac{\sigma^2}{2\alpha}.$$

In fact, $r(t)$ converges in distribution to a normal distribution with this mean and variance, in the sense that for any $x \in \Re$

$$P(r(t) \le x) \to \Phi \left(\frac{x - b}{\sigma/\sqrt{2\alpha}} \right),$$

with Φ the standard normal distribution. The fact that $r(t)$ has a limiting distribution is a reflection of the stabilizing effect of mean reversion in the drift and contrasts with the long-run behavior of, for example, geometric Brownian motion.

The limiting distribution of $r(t)$ is also a stationary distribution in the sense that if $r(0)$ is given this distribution then every $r(t)$, $t > 0$, has this distribution as well. Because (3.46) provides an exact discretization of the process, the $N(b, \sigma^2/2\alpha)$ distribution is also stationary for the discretized process. To simulate a stationary version of the process, it therefore suffices to draw $r(0)$ from this normal distribution and then proceed as in (3.46).

3.3.2 Bond Prices

As already noted, time-dependent drift parameters are typically used to make a short rate model consistent with an observed set of bond prices. Implementation of the simulation algorithm (3.45) is thus linked to the *calibration* of the model through the choice of the function $b(t)$. The same applies to the function $g(t)$ in the Ho-Lee model and as this case is slightly simpler we consider it first.

Our starting point is the bond-pricing formula (3.38). The integral of $r(u)$ from 0 to T appearing in that formula is normally distributed because $r(u)$ is a Gaussian process. It follows that the bond price is the expectation of the exponential of a normal random variable. For a normal random variable $X \sim N(m, v^2)$, we have $\mathsf{E}[\exp(X)] = \exp(m + (v^2/2))$, so

$$\mathsf{E} \left[\exp \left(-\int_0^T r(t) \, dt \right) \right] = \exp \left(-\mathsf{E} \left[\int_0^T r(t) \, dt \right] + \tfrac{1}{2} \mathsf{Var} \left[\int_0^T r(t) \, dt \right] \right).$$

$$(3.47)$$

To find the price of the bond we therefore need to find the mean and variance of the integral of the short rate.

In the Ho-Lee model, the short rate is given by

$$r(t) = r(0) + \int_0^t g(s) \, ds + \sigma W(t)$$

and its integral by

$$\int_0^T r(u)\,du = r(0)T + \int_0^T \int_0^u g(s)\,ds\,du + \sigma \int_0^T W(u)\,du.$$

This integral has mean

$$r(0)T + \int_0^T \int_0^u g(s)\,ds\,du$$

and variance

$$\text{Var}\left[\sigma \int_0^T W(u)\,du\right] = 2\sigma^2 \int_0^T \int_0^t \text{Cov}[W(u), W(t)]\,du\,dt$$

$$= 2\sigma^2 \int_0^T \int_0^t u\,du\,dt$$

$$= \frac{1}{3}\sigma^2 T^3. \tag{3.48}$$

Substituting these expressions in (3.47), we get

$$B(0,T) = \mathsf{E}\left[\exp\left(-\int_0^T r(u)\,du\right)\right]$$

$$= \exp\left(-r(0)T - \int_0^T \int_0^u g(s)\,ds\,du + \frac{\sigma^2 T^3}{6}\right).$$

If we are given a set of bond prices $B(0,T)$ at time 0, our objective is to choose the function g so that this equation holds.

To carry this out we can write

$$B(0,T) = \exp\left(-\int_0^T f(0,t)\,dt\right),$$

with $f(0,t)$ the instantaneous forward rate for time t as of time 0 (cf. Appendix C). The initial forward curve $f(0,T)$ captures the same information as the initial bond prices. Equating the two expressions for $B(0,T)$ and taking logarithms, we find that

$$r(0)T + \int_0^T \int_0^u g(s)\,ds\,du - \frac{\sigma^2 T^3}{6} = \int_0^T f(0,t)\,dt.$$

Differentiating twice with respect to the maturity argument T, we find that

$$g(t) = \left.\frac{\partial}{\partial T} f(0,T)\right|_{T=t} + \sigma^2 t. \tag{3.49}$$

Thus, bond prices produced by the Ho-Lee model will match a given set of bond prices $B(0,T)$ if the function g is tied to the initial forward curve $f(0,T)$ in this way; i.e., if we specify

$$dr(t) = \left(\left. \frac{\partial}{\partial T} f(0, T) \right|_{T=t} + \sigma^2 t \right) dt + \sigma \, dW(t). \tag{3.50}$$

A generic simulation of the Ho-Lee model with drift function g can be written as

$$r(t_{i+1}) = r(t_i) + \int_{t_i}^{t_{i+1}} g(s) \, ds + \sigma \sqrt{t_{i+1} - t_i} Z_{i+1},$$

with $Z_1, Z_2, \ldots$ independent $N(0, 1)$ random variables. With g chosen as in (3.49), this simplifies to

$$r(t_{i+1}) = r(t_i) + [f(0, t_{i+1}) - f(0, t_i)] + \frac{\sigma^2}{2} [t_{i+1}^2 - t_i^2] + \sigma \sqrt{t_{i+1} - t_i} Z_{i+1}.$$

Thus, no integration of the drift function g is necessary; to put it another way, whatever integration is necessary must already have been dealt with in choosing the forward curve $f(0, t)$ to match a set of bond prices.

The situation is even simpler if we require that our simulated short rate be consistent only with bonds maturing at the simulation times $t_1, \ldots, t_n$. To satisfy this requirement we can weaken (3.49) to the condition that

$$\int_{t_i}^{t_{i+1}} g(s) \, ds = f(0, t_{i+1}) - f(0, t_i) + \frac{\sigma^2}{2} [t_{i+1}^2 - t_i^2].$$

Except for this constraint, the choice of g is immaterial — we could take it to be continuous and piecewise linear, for example. In fact, we never even need to specify g because only its integral over the intervals (t_i, t_{i+1}) influence the values of r on the time grid $t_1, \ldots, t_n$.

Bonds in the Vasicek Model

A similar if less explicit solution applies to the Vasicek model. The integral of the short rate is again normally distributed; we need to find the mean and variance of this integral to find the price of a bond using (3.47). Using (3.42), for the mean we get

$$\mathsf{E} \left[\int_0^T r(t) \, dt \right] = \int_0^T \mathsf{E}[r(t)] \, dt$$

$$= \frac{1}{\alpha} (1 - e^{-\alpha T}) r(0) + \alpha \int_0^T \int_0^t e^{-\alpha(t-s)} b(s) \, ds \, dt. \tag{3.51}$$

For the variance we have

$$\mathsf{Var} \left[\int_0^T r(t) \, dt \right] = 2 \int_0^T \int_0^t \mathsf{Cov}[r(t), r(u)] \, du \, dt. \tag{3.52}$$

From (3.42) we get, for $u \leq t$,

$$
\begin{aligned}
\mathsf{Cov}[r(t), r(u)] &= \sigma^2 \int_0^u e^{-\alpha(t-s)} e^{-\alpha(u-s)} \, ds \\
&= \frac{\sigma^2}{2\alpha} \left(e^{\alpha(u-t)} - e^{-\alpha(u+t)} \right).
\end{aligned} \tag{3.53}
$$

Integrating two more times as required for (3.52) gives

$$
\mathsf{Var}\left[\int_0^T r(t) \, dt \right] = \frac{\sigma^2}{\alpha^2} \left[T + \frac{1}{2\alpha} \left(1 - e^{-2\alpha T} \right) + \frac{2}{\alpha} \left(e^{-\alpha T} - 1 \right) \right]. \tag{3.54}
$$

By combining (3.51) and (3.54) as in (3.47), we arrive at an expression for the bond price $B(0, T)$.

Observe that (3.54) does not depend on $r(0)$ and (3.51) is a linear transformation of $r(0)$. If we set

$$
A(t, T) = \frac{1}{\alpha} \left(1 - e^{-\alpha(T-t)} \right)
$$

and

$$
\begin{aligned}
C(t, T) = &-\alpha \int_t^T \int_t^u e^{-\alpha(u-s)} b(s) \, ds \, du \\
&+ \frac{\sigma^2}{2\alpha^2} \left[(T-t) + \frac{1}{2\alpha} \left(1 - e^{-2\alpha(T-t)} \right) + \frac{2}{\alpha} \left(e^{-\alpha(T-t)} - 1 \right) \right],
\end{aligned}
$$

then substituting (3.51) and (3.54) in (3.47) produces

$$
B(0, T) = \exp(-A(0, T) r(0) + C(0, T)).
$$

In fact, the same calculations show that

$$
B(t, T) = \exp(-A(t, T) r(t) + C(t, T)). \tag{3.55}
$$

In particular, $\log B(t, T)$ is a linear transformation of $r(t)$. This feature has been generalized by Brown and Schaefer [71] and Duffie and Kan [101] to what is generally referred to as the *affine class* of interest rate models.

As in our discussion of the Ho-Lee model, the function $b(s)$ can be chosen to match a set of prices $B(0, T)$ indexed by T. If we are concerned only with matching a finite set of bond prices $B(0, t_1), \ldots, B(0, t_n)$, then only the values of the integrals

$$
\int_{t_i}^{t_{i+1}} e^{-\alpha(t_{i+1}-s)} b(s) \, ds
$$

need be specified. These are precisely the terms $\mu(t_i, t_{i+1})$ needed in the simulation algorithm (3.45). Thus, these integrals are by-products of fitting the model to a term structure and not additional computations required solely for the simulation.

Joint Simulation with the Discount Factor

Most applications that call for simulation of a short rate process $r(t)$ also require values of the discount factor

$$\frac{1}{\beta(t)} = \exp\left(-\int_0^t r(u)\,du\right)$$

or, equivalently, of

$$Y(t) = \int_0^t r(u)\,du.$$

Given values $r(0), r(t_1), \ldots, r(t_n)$ of the short rate, one can of course generate approximate values of $Y(t_i)$ using

$$\sum_{j=1}^{i} r(t_{j-1})[t_j - t_{j-1}], \quad t_0 = 0,$$

or some other approximation to the time integral. But in a Gaussian model, the pair $(r(t), Y(t))$ are jointly Gaussian and it is often possible to simulate paths of the pair without discretization error. To carry this out we simply need to find the means, variances, and covariance of the increments of $r(t)$ and $Y(t)$.

We have already determined (see (3.45)) that, given $r(t_i)$,

$$r(t_{i+1}) \sim N\left(e^{-\alpha(t_{i+1}-t_i)}r(t_i) + \mu(t_i, t_{i+1}), \sigma_r^2(t_i, t_{i+1})\right).$$

From the same calculations used in (3.51) and (3.54), we find that, given $r(t_i)$ and $Y(t_i)$,

$$Y(t_{i+1}) \sim N(Y(t_i) + \mu_Y(t_i, t_{i+1}), \sigma_Y^2(t_i, t_{i+1})),$$

with

$$\mu_Y(t_i, t_{i+1}) = \frac{1}{\alpha}\left(1 - e^{-\alpha(t_{i+1}-t_i)}\right)r(t_i) + \alpha\int_{t_i}^{t_{i+1}}\int_{t_i}^{u} e^{-\alpha(u-s)}b(s)\,ds\,du$$

and

$$\sigma_Y^2(t_i, t_{i+1}) =$$
$$\frac{\sigma^2}{\alpha^2}\left((t_{i+1} - t_i) + \frac{1}{2\alpha}\left(1 - e^{-2\alpha(t_{i+1}-t_i)}\right) + \frac{2}{\alpha}\left(e^{-\alpha(t_{i+1}-t_i)} - 1\right)\right).$$

It only remains to determine the conditional covariance between $r(t_{i+1})$ and $Y(t_{i+1})$ given $(r(t_i), Y(t_i))$. For this we proceed as follows:

$$\text{Cov}\left[r(t), Y(t)\right] = \int_0^t \text{Cov}[r(t), r(u)]\, du$$

$$= \frac{\sigma^2}{2\alpha} \int_0^t e^{\alpha(u-t)} - e^{-\alpha(u+t)}\, du$$

$$= \frac{\sigma^2}{2\alpha^2} \left[1 + e^{-2\alpha t} - 2e^{-\alpha t}\right].$$

The required covariance is thus given by

$$\sigma_{rY}(t_i, t_{i+1}) = \frac{\sigma^2}{2\alpha} \left[1 + e^{-2\alpha(t_{i+1}-t_i)} - 2e^{-\alpha(t_{i+1}-t_i)}\right].$$

The corresponding correlation is

$$\rho_{rY}(t_i, t_{i+1}) = \frac{\sigma_{rY}(t_i, t_{i+1})}{\sigma_r(t_i, t_{i+1})\sigma_Y(t_i, t_{i+1})}.$$

With this notation, the pair (r, Y) can be simulated at times $t_1, \ldots, t_n$ without discretization error using the following algorithm:

$$r(t_{i+1}) = e^{-\alpha(t_{i+1}-t_i)} r(t_i) + \mu(t_i, t_{i+1}) + \sigma_r(t_i, t_{i+1}) Z_1(i+1)$$
$$Y(t_{i+1}) = Y(t_i) + \mu_Y(t_i, t_{i+1}) + \sigma_Y(t_i, t_{i+1})[\rho_{rY}(t_i, t_{i+1}) Z_1(i+1)$$
$$+ \sqrt{1 - \rho_{rY}^2(t_i, t_{i+1})} Z_2(i+1)],$$

where $(Z_1(i), Z_2(i))$, $i = 1, \ldots, n$, are independent standard bivariate normal random vectors.

Change of Numeraire

Thus far, we have considered the dynamics of the short rate $r(t)$ only under the risk-neutral measure. Recall that the numeraire asset associated with the risk-neutral measure is $\beta(t) = \exp(\int_0^t r(u)\, du)$ and the defining feature of this probability measure is that it makes the discounted bond prices $B(t, T)/\beta(t)$ martingales. In fact, the dynamics of the bond prices under the Gaussian models we have considered are of the form (for fixed T)

$$\frac{dB(t, T)}{B(t, T)} = r(t)\, dt - A(t, T)\sigma\, dW(t) \tag{3.56}$$

with $A(t, T)$ deterministic; this follows from (3.55). The solution of this equation is

$$B(t, T) = B(0, T) \exp\left(\int_0^t [r(u) - \tfrac{1}{2}\sigma^2 A^2(u, T)]\, du - \sigma \int_0^t A(u, T)\, dW(u)\right),$$

from which it is evident that

$$\frac{B(t,T)}{\beta(t)} = B(0,T)\exp\left(-\tfrac{1}{2}\sigma^2\int_0^t A^2(u,T)\,du - \sigma\int_0^t A(u,T)\,dW(u)\right)$$

(3.57)

is an exponential martingale.

As discussed in Section 1.2.3, the *forward measure* for any date T_F is the measure associated with taking the T_F-maturity bond $B(t,T_F)$ as numeraire asset. The defining feature of the forward measure is that it makes the ratios $B(t,T)/B(t,T_F)$ martingales for $T < T_F$. It is defined by the likelihood ratio process

$$\left(\frac{dP_{T_F}}{dP_\beta}\right)_t = \frac{B(t,T_F)\beta(0)}{\beta(t)B(0,T_F)},$$

and this is given in (3.57) up to a factor of $1/B(0,T_F)$. From Girsanov's Theorem, it follows that the process W^{T_F} defined by

$$dW^{T_F}(t) = dW(t) + \sigma A(t,T_F)\,dt$$

is a standard Brownian motion under P_{T_F}. Accordingly, the dynamics of the Vasicek model become

$$\begin{aligned}
dr(t) &= \alpha(b(t) - r(t))\,dt + \sigma\,dW(t)\\
&= \alpha(b(t) - r(t))\,dt + \sigma\left(dW^{T_F}(t) - \sigma A(t,T_F)\,dt\right)\\
&= \alpha(b(t) - \sigma^2 A(t,T_F) - r(t))\,dt + \sigma\,dW^{T_F}(t).
\end{aligned}$$

(3.58)

Thus, under the forward measure, the short rate process remains a Vasicek process but the reversion level $b(t)$ becomes $b(t) - \sigma^2 A(t,T_F)$.

The process in (3.58) can be simulated using (3.45) with $b(t)$ replaced by $b(t) - \sigma^2 A(t,T_F)$. In particular, we simulate W^{T_F} the way we would simulate any other standard Brownian motion. The simulation algorithm does not "know" that it is simulating a Brownian motion under the forward measure rather than under the risk-neutral measure.

Suppose we want to price a derivative security making a payoff of $g(r(T_F))$ at time T_F. Under the risk-neutral measure, we would price the security by computing

$$\mathsf{E}\left[e^{-\int_0^{T_F} r(u)\,du} g(r(T_F))\right].$$

In fact, g could be a function of the path of $r(t)$ rather than just its terminal value. Switching to the forward measure, this becomes

$$\begin{aligned}
\mathsf{E}_{T_F}&\left[e^{-\int_0^{T_F} r(u)\,du} g(r(T_F))\left(\frac{dP_\beta}{dP_{T_F}}\right)_{T_F}\right]\\
&= \mathsf{E}_{T_F}\left[e^{-\int_0^{T_F} r(u)\,du} g(r(T_F))\left(\frac{\beta(T_F)B(0,T_F)}{B(T_F,T_F)\beta(0)}\right)\right]\\
&= B(0,T_F)\mathsf{E}_{T_F}\left[g(r(T_F))\right],
\end{aligned}$$

where E_{T_F} denotes expectation under the forward measure. Thus, we may price the derivative security by simulating $r(t)$ under the forward measure P_{T_F}, estimating the expectation of $g(r(T_F))$ and multiplying by $B(0, T_F)$. Notice that discounting in this case is deterministic — we do not need to simulate a discount factor. This apparent simplification results from inclusion of the additional term $-\sigma^2 A(t, T_F)$ in the drift of $r(t)$.

A consequence of working under the forward measure is that the simulation prices the bond maturing at T_F exactly: pricing this bond corresponds to taking $g(r(T_F)) \equiv 1$. Again, this apparent simplification is really a consequence of the form of the drift of $r(t)$ under the forward measure.

3.3.3 Multifactor Models

A general class of Gaussian Markov processes in $\Re^d$ have the form

$$dX(t) = C(b - X(t))\, dt + D\, dW(t) \tag{3.59}$$

where C and D are $d \times d$ matrices, b and $X(t)$ are in $\Re^d$, W is a standard d-dimensional Brownian motion, and $X(0)$ is Gaussian or constant. Such a process remains Gaussian and Markovian if the coefficients C, b, and D are made time-varying but deterministic. The solution of (3.59) is

$$X(t) = e^{-Ct}X(0) + \int_0^t e^{-C(t-s)}b\, ds + \int_0^t e^{-C(t-s)}D\, dW(s),$$

from which it is possible to define an exact time-discretization similar to (3.45).

A model of the short rate process can be specified by setting $r(t) = a^\top X(t)$ with $a \in \Re^d$ (or with a deterministically time-varying). The elements of $X(t)$ are then interpreted as "factors" driving the evolution of the short rate. Because each $X(t)$ is normally distributed, $r(t)$ is normally distributed. However, $r(t)$ is not in general a Markov process: to make the future evolution of r independent of the past, we need to condition on the full state information $X(t)$ and not merely $r(t)$.

Recall from (3.55) that in the Vasicek model (with constant or time-varying coefficients), bond prices are exponentials of affine functions of the short rate. A similar representation applies if the short rate has the form $r(t) = a^\top X(t)$ and $X(t)$ is as in (3.59); in particular, we have

$$B(t, T) = \exp(-A(t, T)^\top X(t) + C(t, T))$$

for some $\Re^d$-valued function $A(t, T)$ and some scalar function $C(t, T)$. In the single-factor setting, differentiating (3.55) and then simplifying leads to

$$\frac{dB(t, T)}{B(t, T)} = r(t)\, dt - A(t, T)\sigma\, dW(t),$$

with σ the diffusion parameter of $r(t)$. The instantaneous correlation between the returns on bonds with maturities T_1 and T_2 is therefore

$$\frac{A(t,T_1)\sigma \cdot A(t,T_2)\sigma}{\sqrt{A^2(t,T_1)\sigma^2}\sqrt{A^2(t,T_2)\sigma^2}} = 1.$$

In other words, all bonds are instantaneously perfectly correlated. In the multifactor setting, the bond price dynamics are given by

$$\frac{dB(t,T)}{B(t,T)} = r(t)\,dt - A(t,T)^\top D\,dW(t).$$

The instantaneous correlation for maturities T_1 and T_2 is

$$\frac{A(t,T_1)^\top DD^\top A(t,T_2)}{\|A(t,T_1)^\top D\|\|A(t,T_2)^\top D\|},$$

which can certainly take values other than 1. The flexibility to capture less than perfect instantaneous correlation between bond returns is the primary motivation for considering multifactor models.

Returning to the general formulation in (3.59), suppose that C can be diagonalized in the sense that $VCV^{-1} = \Lambda$ for some matrix V and diagonal matrix Λ with diagonal entries $\lambda_1, \ldots, \lambda_d$. Suppose further that C is nonsingular and define $Y(t) = VX(t)$. Then

$$
\begin{aligned}
dY(t) &= V\,dX(t)\\
&= V[C(b - X(t)\,dt + D\,dW(t)]\\
&= (VCb - \Lambda Y(t))\,dt + VD\,dW(t)\\
&= \Lambda(\Lambda^{-1}VCb - Y(t))\,dt + VD\,dW(t)\\
&= \Lambda(Vb - Y(t))\,dt + VD\,dW(t)\\
&\equiv \Lambda(\tilde{b} - Y(t))\,dt + d\tilde{W}(t)
\end{aligned}
$$

with $\tilde{W}$ a BM$(0,\Sigma)$ process, $\Sigma = VDD^\top V^\top$. It follows that the components of $(Y_1, \ldots, Y_d)$ satisfy

$$dY_j(t) = \lambda_j(\tilde{b}_j - Y_j(t))\,dt + d\tilde{W}_j(t), \quad j = 1, \ldots, d. \tag{3.60}$$

In particular, each Y_j is itself a Markov process. The Y_j remain coupled, however, through the correlation across the components of $\tilde{W}$. They can be simulated as in (3.46) by setting

$$Y_j(t_{i+1}) =$$

$$e^{\lambda_j(t_{i+1}-t_i)}Y_j(t_i) + (e^{\lambda_j(t_{i+1}-t_i)} - 1)\tilde{b}_j + \sqrt{\frac{1}{2\lambda_j}\left(1 - e^{-2\lambda_j(t_{i+1}-t_i)}\right)}\xi_j(i+1),$$

where $\xi(1), \xi(2), \ldots$ are independent $N(0,\Sigma)$ random vectors, $\xi(i) = (\xi_1(i), \ldots, \xi_d(i))$. Thus, when C is nonsingular and diagonalizable, simulation of (3.59) can be reduced to a system of scalar simulations.

As noted by Andersen and Andreasen [14], a similar reduction is possible even if C is not diagonalizable, but at the expense of making all coefficients time-dependent. If $V(t)$ is a deterministic $d \times d$ matrix-valued function of time and we set $Y(t) = V(t)X(t)$, then

$$
\begin{aligned}
dY(t) &= \dot{V}(t)X(t)\,dt + V(t)dX(t) \\
&= [\dot{V}(t)X(t) + V(t)C(b - X(t))]\,dt + V(t)D\,dW(t),
\end{aligned}
$$

where $\dot{V}(t)$ denotes the time derivative of $V(t)$. If we choose $V(t) = \exp([C - I]t)$, then

$$
\dot{V}(t) = V(t)C - V(t)
$$

and thus

$$
\begin{aligned}
dY(t) &= [V(t)Cb - V(t)X(t)]\,dt + V(t)D\,dW(t) \\
&= (\tilde{b}(t) - Y(t))\,dt + \tilde{D}(t)\,dW(t), \tag{3.61}
\end{aligned}
$$

with $\tilde{b}(t) = V(t)Cb$ and $\tilde{D}(t) = V(t)D$. Notice that the drift of each component $Y_i(t)$ depends only on that $Y_i(t)$. This transformation therefore decouples the drifts of the components of the state vector, making each Y_i a Markov process, though the components remain linked through the diffusion term. We can recover the original state vector by setting $X(t) = V(t)^{-1}Y(t)$ because $V(t)$ is always invertible. The seemingly special form of the dynamics in (3.61) is thus no less general than the dynamics in (3.59) with time-varying coefficients.

3.4 Square-Root Diffusions

Feller [118] studied a class of processes that includes the square-root diffusion

$$
dr(t) = \alpha(b - r(t))\,dt + \sigma\sqrt{r(t)}\,dW(t), \tag{3.62}
$$

with W a standard one-dimensional Brownian motion. We consider the case in which α and b are positive. If $r(0) > 0$, then $r(t)$ will never be negative; if $2\alpha b \geq \sigma^2$, then $r(t)$ remains strictly positive for all t, almost surely.

This process was proposed by Cox, Ingersoll, and Ross [91] as a model of the short rate, generally referred to as the CIR model. They developed a general equilibrium framework in which if the change in production opportunities is assumed to follow a process of this form, then the short rate does as well. As with the Vasicek model, the form of the drift in (3.62) suggests that $r(t)$ is pulled towards b at a speed controlled by α. In contrast to the Vasicek model, in the CIR model the diffusion term $\sigma\sqrt{r(t)}$ decreases to zero as $r(t)$ approaches the origin and this prevents $r(t)$ from taking negative values. This feature of (3.62) is attractive in modeling interest rates.

All of the coefficients in (3.62) could in principle be made time-dependent. In practice, it can be particularly useful to replace the constant b with a function of time and thus consider

$$dr(t) = \alpha(b(t) - r(t))\, dt + \sigma\sqrt{r(t)}\, dW(t). \qquad (3.63)$$

As with the Vasicek model, this extension is frequently used to make the bond price function

$$T \mapsto \mathsf{E}\left[\exp\left(-\int_0^T r(u)\, du\right)\right]$$

match a set of observed bond prices $B(0, T)$.

Although we stress the application of (3.63) to interest rate modeling, it should be noted that this process has other financial applications. For example, Heston [179] proposed a stochastic volatility model in which the price of an asset $S(t)$ is governed by

$$\frac{dS(t)}{S(t)} = \mu\, dt + \sqrt{V(t)}\, dW_1(t) \qquad (3.64)$$

$$dV(t) = \alpha(b - V(t))\, dt + \sigma\sqrt{V(t)}\, dW_2(t), \qquad (3.65)$$

where (W_1, W_2) is a two-dimensional Brownian motion. Thus, in Heston's model, the squared volatility $V(t)$ follows a square-root diffusion. In addition, the process in (3.63) is sometimes used to model a stochastic intensity for a jump process in, for example, modeling default.

A simple Euler discretization of (3.62) suggests simulating $r(t)$ at times $t_1, \ldots, t_n$ by setting

$$r(t_{i+1}) = r(t_i) + \alpha(b - r(t_i))[t_{i+1} - t_i] + \sigma\sqrt{r(t_i)^+}\sqrt{t_{i+1} - t_i}\, Z_{i+1}, \quad (3.66)$$

with $Z_1, \ldots, Z_n$ independent $N(0, 1)$ random variables. Notice that we have taken the positive part of $r(t_i)$ inside the square root; some modification of this form is necessary because the values of $r(t_i)$ produced by Euler discretization may become negative. We will see, however, that this issue can be avoided (along with any other discretization error) by sampling from the exact transition law of the process.

3.4.1 Transition Density

The SDE (3.62) is not explicitly solvable the way those considered in Sections 3.2 and 3.3 are; nevertheless, the transition density for the process is known. Based on results of Feller [118], Cox et al. [91] noted that the distribution of $r(t)$ given $r(u)$ for some $u < t$ is, up to a scale factor, a noncentral chi-square distribution. This property can be used to simulate the process (3.62). We follow the approach suggested by Scott [324].

A noncentral chi-square random variable $\chi_\nu'^2(\lambda)$ with ν degrees of freedom and noncentrality parameter λ has distribution

$$P(\chi_\nu'^2(\lambda) \le y) = F_{\chi_\nu'^2(\lambda)}(y)$$

$$\equiv e^{-\lambda/2} \sum_{j=0}^{\infty} \frac{(\frac{1}{2}\lambda)^j/j!}{2^{(\nu/2)+j}\Gamma(\frac{\nu}{2}+j)} \int_0^y z^{(\nu/2)+j-1} e^{-z/2}\, dz, \quad (3.67)$$

for $y > 0$. The transition law of $r(t)$ in (3.62) can be expressed as

$$r(t) = \frac{\sigma^2(1 - e^{-\alpha(t-u)})}{4\alpha} \chi_d'^2 \left(\frac{4\alpha e^{-\alpha(t-u)}}{\sigma^2(1 - e^{-\alpha(t-u)})} r(u) \right), \quad t > u, \quad (3.68)$$

where

$$d = \frac{4b\alpha}{\sigma^2}. \quad (3.69)$$

This says that, given $r(u)$, $r(t)$ is distributed as $\sigma^2(1 - e^{-\alpha(t-u)})/(4\alpha)$ times a noncentral chi-square random variable with d degrees of freedom and noncentrality parameter

$$\lambda = \frac{4\alpha e^{-\alpha(t-u)}}{\sigma^2(1 - e^{-\alpha(t-u)})} r(u); \quad (3.70)$$

equivalently,

$$P(r(t) \le y | r(u)) = F_{\chi_d'^2(\lambda)} \left(\frac{4\alpha y}{\sigma^2(1 - e^{-\alpha(t-u)})} \right),$$

with d as in (3.69), λ as in (3.70), and $F_{\chi_d'^2(\lambda)}$ as in (3.67). Thus, we can simulate the process (3.62) exactly on a discrete time grid provided we can sample from the noncentral chi-square distribution.

Like the Vasicek model, the square-root diffusion (3.62) has a limiting stationary distribution. If we let $t \to \infty$ in (3.68), we find that $r(t)$ converges in distribution to $\sigma^2/4\alpha$ times a noncentral chi-square random variable with d degrees of freedom and noncentrality parameter 0 (making it an ordinary chi-square random variable). This is a stationary distribution in the sense that if $r(0)$ is drawn from this distribution, then $r(t)$ has the same distribution for all t.

Chi-Square and Noncentral Chi-Square

If ν is a positive integer and $Z_1, \dots, Z_\nu$ are independent $N(0,1)$ random variables, then the distribution of

$$Z_1^2 + Z_2^2 + \cdots + Z_\nu^2$$

is called the chi-square distribution with ν degrees of freedom. The symbol χ_ν^2 denotes a random variable with this distribution; the prime in $\chi_\nu'^2(\lambda)$ emphasizes that this symbol refers to the noncentral case. The chi-square distribution is given by

$$P(\chi_\nu^2 \leq y) = \frac{1}{2^{\nu/2}\Gamma(\nu/2)} \int_0^y e^{-z/2} z^{(\nu/2)-1}\, dz, \qquad (3.71)$$

where $\Gamma(\cdot)$ denotes the gamma function and $\Gamma(n) = (n-1)!$ if n is a positive integer. This expression defines a valid probability distribution for all $\nu > 0$ and thus extends the definition of χ_ν^2 to non-integer ν.

For integer ν and constants $a_1, \ldots, a_\nu$, the distribution of

$$\sum_{i=1}^\nu (Z_i + a_i)^2 \qquad (3.72)$$

is noncentral chi-square with ν degrees of freedom and noncentrality parameter $\lambda = \sum_{i=1}^\nu a_i^2$. This representation explains the term "noncentral." The distribution in (3.67) extends the definition to non-integer ν.

It follows from the representation in (3.72) that if $\nu > 1$ is an integer, then

$$\chi_\nu'^2(\lambda) = \chi_1'^2(\lambda) + \chi_{\nu-1}^2,$$

meaning that the two sides have the same distribution when the random variables on the right are independent of each other. As discussed in Johnson et al. [202, p.436], this representation is valid even for non-integer $\nu > 1$. Thus, to generate $\chi_\nu'^2(\lambda)$, $\nu > 1$, it suffices to generate $\chi_{\nu-1}^2$ and an independent $N(0,1)$ random variable Z and to set

$$\chi_\nu'^2(\lambda) = (Z + \sqrt{\lambda})^2 + \chi_{\nu-1}^2. \qquad (3.73)$$

This reduces sampling of a noncentral chi-square to sampling of an ordinary chi-square (and an independent normal) when $\nu > 1$.

For any $\nu > 0$, (3.67) indicates that a noncentral chi-square random variable can be represented as an ordinary chi-square random variable with a random degrees-of-freedom parameter. In more detail, if N is a Poisson random variable with mean $\lambda/2$, then

$$P(N = j) = e^{-\lambda/2}\frac{(\lambda/2)^j}{j!}, \qquad j = 0, 1, 2, \ldots.$$

Consider now a random variable $\chi_{\nu+2N}^2$ with N having this Poisson distribution. Conditional on $N = j$, the random variable has an ordinary chi-square distribution with $\nu + 2j$ degrees of freedom:

$$P(\chi_{\nu+2N}^2 \leq y | N = j) = \frac{1}{2^{(\nu/2)+j}\Gamma((\nu/2)+j)} \int_0^y e^{-z/2} z^{(\nu/2)+j-1}\, dz.$$

The unconditional distribution is thus given by

$$\sum_{j=0}^\infty P(N = j)P(\chi_{\nu+2N}^2 \leq y | N = j) = \sum_{j=0}^\infty e^{-\lambda/2}\frac{(\lambda/2)^j}{j!} P(\chi_{\nu+2j}^2 \leq y),$$

which is precisely the noncentral chi-square distribution in (3.67). We may therefore sample $\chi_\nu'^2(\lambda)$ by first generating a Poisson random variable N and then, conditional on N, sampling a chi-square random variable with $\nu + 2N$ degrees of freedom. This reduces sampling of a noncentral chi-square to sampling of an ordinary chi-square and a Poisson random variable. We discuss methods for sampling from these distributions below. Figure 3.5 summarizes their use in simulating the square-root diffusion (3.62).

Simulation of $dr(t) = \alpha(b - r(t))\,dt + \sigma\sqrt{r(t)}\,dW(t)$
on time grid $0 = t_0 < t_1 < \cdots < t_n$ with $d = 4b\alpha/\sigma^2$

Case 1: $d > 1$
for $i = 0, \ldots, n - 1$
$\quad c \leftarrow \sigma^2(1 - e^{-\alpha(t_{i+1} - t_i)})/(4\alpha)$
$\quad \lambda \leftarrow r(t_i)(e^{-\alpha(t_{i+1} - t_i)})/c$
$\quad$generate $Z \sim N(0, 1)$
$\quad$generate $X \sim \chi_{d-1}^2$
$\quad r(t_{i+1}) \leftarrow c[(Z + \sqrt{\lambda})^2 + X]$
end

Case 2: $d \leq 1$
for $i = 0, \ldots, n - 1$
$\quad c \leftarrow \sigma^2(1 - e^{-\alpha(t_{i+1} - t_i)})/(4\alpha)$
$\quad \lambda \leftarrow r(t_i)(e^{-\alpha(t_{i+1} - t_i)})/c$
$\quad$generate $N \sim \text{Poisson}(\lambda/2)$
$\quad$generate $X \sim \chi_{d+2N}^2$
$\quad r(t_{i+1}) \leftarrow cX$
end

Fig. 3.5. Simulation of square-root diffusion (3.62) by sampling from the transition density.

Figure 3.6 compares the exact distribution of $r(t)$ with the distribution produced by the Euler discretization (3.66) after a single time step. The comparison is based on $\alpha = 0.2$, $\sigma = 0.1$, $b = 5\%$, and $r(0) = 4\%$; the left panel takes $t = 0.25$ and the right panel takes $t = 1$. These values for the model parameters are sensible for an interest rate model if time is measured in years, so the values of t should be interpreted as a quarter of a year and a full year, respectively. The figures suggest that the Euler discretization produces too many values close to or below 0 and a mode to the right of the true mode. The effect if particularly pronounced over the rather large time step $t = 1$.

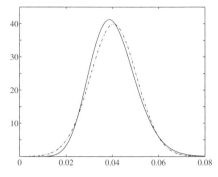 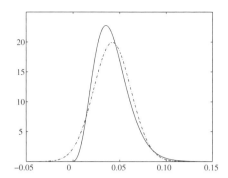

Fig. 3.6. Comparison of exact distribution (solid) and one-step Euler approximation (dashed) for a square-root diffusion with $\alpha = 0.2$, $\sigma = 0.1$, $b = 5\%$, and $r(0) = 4\%$. The left panel compares distributions at $t = 0.25$, the right panel at $t = 1$.

3.4.2 Sampling Gamma and Poisson

The discussion leading to Figure 3.5 reduces the problem of simulating the square-root diffusion (3.62) to one of sampling from a chi-square distribution and possibly also the normal and Poisson distributions. We discussed sampling from the normal distribution in Section 2.3; we now consider methods for sampling from the chi-square and Poisson distributions.

Gamma Distribution

The gamma distribution with shape parameter a and scale parameter β has density

$$f(y) = f_{a,\beta}(y) = \frac{1}{\Gamma(a)\beta^a} y^{a-1} e^{-y/\beta}, \quad y \geq 0. \tag{3.74}$$

It has mean $a\beta$ and variance $a\beta^2$. Comparison with (3.71) reveals that the chi-square distribution is the special case of scale parameter $\beta = 2$ and shape parameter $a = \nu/2$. We therefore consider the more general problem of generating samples from gamma distributions.

Methods for sampling from the gamma distribution typically distinguish the cases $a \leq 1$ and $a > 1$. For the application to the square-root diffusion (3.62), the shape parameter a is given by $d/2$ with d as in (3.69). At least in the case of an interest rate model, d would typically be larger than 2 so the case $a > 1$ is most relevant. We include the case $a \leq 1$ for completeness and other potential applications. There is no loss of generality in fixing the scale parameter β at 1: if X has the gamma distribution with parameters $(a, 1)$, then βX has the gamma distribution with parameters (a, β).

Cheng and Feast [83] develop a method based on a general approach to random variate generation known as the ratio-of-uniforms method. The ratio-of-uniforms method is closely related to the acceptance-rejection method discussed in Section 2.2.2. It exploits the following property. Suppose f is a nonnegative, integrable function on $[0, \infty)$; if (X, Y) is uniformly distributed over the set $A = \{(x, y) : x \leq \sqrt{f(y/x)}\}$, then the density of Y/X is proportional to f. (See p.180 of Fishman [121] or p.59 of Gentle [136].) Suppose A is contained in a bounded rectangle. Then to sample uniformly from A, we can repeatedly sample pairs (X, Y) uniformly over the rectangle and keep the first one that satisfies $X \leq \sqrt{f(Y/X)}$. The ratio-of-uniforms method delivers Y/X as a sample from the density proportional to f.

To sample from the gamma density with $a > 1$, define

$$A = \left\{ (x, y) : 0 \leq x \leq \sqrt{[(y/x)^{a-1} e^{-y/x}]} \right\}.$$

This set is contained in the rectangle $[0, \bar{x}] \times [0, \bar{y}]$ with $\bar{x} = [(a-1)/e]^{(a-1)/2}$ and $\bar{y} = [(a+1)/e]^{(a+1)/2}$. Sampling uniformly over this rectangle, the expected number of samples needed until one lands in A is given by the ratio of the area of A to that of the rectangle. As shown in Fishman [121], this ratio is $O(\sqrt{a})$, so the time required to generate a sample using this method grows with the shape parameter. Cheng and Feast [83] and Fishman [121] develop modifications of this basic approach that accelerate sampling. In Figure 3.7, which is Fishman's Algorithm GKM1, the first acceptance test is a fast check that reduces the number of logarithmic evaluations. When many samples are to be generated using the same shape parameter (as would be the case in the application to the square-root diffusion), the constants in the setup step in Figure 3.8 should be computed just once and then passed as arguments to the sampling routine. For large values of the shape parameter a, Algorithm GKM2 in Fishman [121] is faster than the method in Figure 3.7.

Setup: $\bar{a} \leftarrow a - 1$, $b \leftarrow (a - (1/(6a)))/\bar{a}$, $m \leftarrow 2/\bar{a}$, $d \leftarrow m + 2$
repeat
 generate $U_1, U_2 \sim \text{Unif}[0,1]$
 $V \leftarrow bU_2/U_1$
 if $mU_1 - d + V + (1/V) \leq 0$, accept
 elseif $m \log U_1 - \log V + V - 1 \leq 0$, accept
until accept
return $Z \leftarrow \bar{a}V$

Fig. 3.7. Algorithm GKM1 from Fishman [121], based on Cheng and Feast [83], for sampling from the gamma distribution with parameters $(a, 1)$, $a > 1$.

Ahrens and Dieter [6] provide a fast acceptance-rejection algorithm for the case $a \leq 1$. Their method generates candidates by sampling from distributions concentrated on $[0, 1]$ and $(1, \infty)$ with appropriate probabilities. In more detail, let $p = e/(a + e)$ ($e = \exp(1)$) and define

$$g(z) = \begin{cases} paz^{a-1}, & 0 \le z \le 1 \\ (1-p)e^{-z+1}, & z > 1. \end{cases}$$

This is a probability density; it is a mixture of the densities az^{a-1} on $[0,1]$ and e^{-z+1} on $(1,\infty)$, with weights p and $(1-p)$, respectively. We can sample from g by sampling from each of these densities with the corresponding probabilities. Each of these two densities is easily sampled using the inverse transform method : for the density az^{a-1} on $[0,1]$ we can use $U^{1/a}$, $U \sim \text{Unif}[0,1]$; for the density e^{-z+1} on $(1,\infty)$ we can use $1 - \log(U)$. Samples from g are suitable candidates for acceptance-rejection because the ratio $f_{a,1}(z)/g(z)$ with $f_{a,1}$ a gamma density as in (3.74) is bounded. Inspection of this ratio indicates that a candidate Z in $[0,1]$ is accepted with probability e^{-Z} and a candidate in $(1,\infty)$ is accepted with probability Z^{a-1}. A global bound on the ratio is given by

$$f_{a,1}(z)/g(z) \le \frac{a+e}{ae\Gamma(a)} \le 1.39;$$

recall from Section 2.2.2 that the upper bound on this ratio determines the expected number of candidates generated per accepted sample.

Figure 3.8 displays the method of Ahrens and Dieter [6]. The figure is based on Algorithm GS* in Fishman [121] but it makes the acceptance tests more explicit, if perhaps slightly slower. Notice that if the condition $Y \le 1$ fails to hold, then Y is uniformly distributed over $[1,b]$; this means that $(b-Y)/a$ has the distribution of U/e, $U \sim \text{Unif}[0,1]$ and thus $-\log((b-Y)/a)$ has the distribution of $1 - \log(U)$.

Setup: $b \leftarrow (a+e)/e$
repeat
    generate $U_1, U_2 \sim \text{Unif}[0,1]$; $Y \leftarrow bU_1$
    if $Y \le 1$
        $Z \leftarrow Y^{1/a}$
        if $U_2 < \exp(-Z)$, accept
    otherwise $Z \leftarrow -\log((b-Y)/a)$
        if $U_2 \le Z^{a-1}$, accept
until accept
return Z

Fig. 3.8. Ahrens-Dieter method for sampling from the gamma distribution with parameters $(a,1)$, $a \le 1$.

Poisson Distribution

The Poisson distribution with mean $\theta > 0$ is given by

$$P(N = k) = e^{-\theta}\frac{\theta^k}{k!}, \quad k = 0, 1, 2, \ldots. \tag{3.75}$$

We abbreviate this by writing $N \sim \text{Poisson}(\theta)$. This is the distribution of the number of events in $[0, 1]$ when the times between consecutive events are independent and exponentially distributed with mean $1/\theta$. Thus, a simple method for generating Poisson samples is to generate exponential random variables $X_i = -\log(U_i)/\theta$ from independent uniforms U_i and then take N to be the largest integer for which $X_1 + \cdots + X_N \leq 1$. This method is rather slow, especially if θ is large. In the intended application in Figure 3.5, the mean of the Poisson random variable — equal to half the noncentrality parameter in the transition density of the square-root diffusion — could be quite large for plausible parameter values.

An alternative is to use the inverse transform method. For discrete distributions, this amounts to a sequential search for the smallest n at which $F(n) \leq U$, where F denotes the cumulative distribution function and U is Unif$[0,1]$. In the case of a Poisson distribution, $F(n)$ is calculated as $P(N = 0) + \cdots + P(N = n)$; rather than calculate each term in this sum using (3.75), we can use the relation $P(N = k + 1) = P(N = k)\theta/(k + 1)$. Figure 3.9 illustrates the method.

$$
\begin{aligned}
&p \leftarrow \exp(-\theta), \; F \leftarrow p \\
&N \leftarrow 0 \\
&\text{generate } U \sim \text{Unif}[0,1] \\
&\text{while } U > F \\
&\qquad N \leftarrow N + 1 \\
&\qquad p \leftarrow p\theta/N \\
&\qquad F \leftarrow F + p \\
&\text{return } N
\end{aligned}
$$

Fig. 3.9. Inverse transform method for sampling from Poisson(θ), the Poisson distribution with mean θ.

3.4.3 Bond Prices

Cox, Ingersoll, and Ross [91] derived an expression for the price of a bond

$$B(t, T) = \mathsf{E}\left[\exp\left(-\int_t^T r(u)\, du\right) \Big| r(t)\right]$$

when the short rate evolves according to (3.62). The bond price has the exponential affine form

$$B(t,T) = e^{-A(t,T)r(t)+C(t,T)}$$

as in a Gaussian short rate model, but with

$$A(t,T) = \frac{2(e^{\gamma(T-t)} - 1)}{(\gamma + \alpha)(e^{\gamma(T-t)} - 1) + 2\gamma}$$

and

$$C(t,T) = \frac{2\alpha b}{\sigma^2} \log\left(\frac{2\gamma e^{(\alpha+\gamma)(T-t)/2}}{(\gamma + \alpha)(e^{\gamma(T-t)} - 1) + 2\gamma}\right),$$

and $\gamma = \sqrt{\alpha^2 + 2\sigma^2}$.

This expression for the bond price is a special case of a more general result, given as Proposition 6.2.5 in Lamberton and Lapeyre [218]. This result gives the bivariate Laplace transform of the short rate and its integral: for nonnegative λ, θ,

$$\mathsf{E}\left[\exp\left(-\lambda r(T) - \theta \int_t^T r(u)\,du\right)\Big|r(t)\right] = \exp(-\alpha b\psi_1(T-t) - r(t)\psi_2(T-t))$$

(3.76)

with

$$\psi_1(s) = -\frac{2}{\sigma^2} \log\left(\frac{2\gamma(\theta)e^{(\alpha+\gamma(\theta))s/2}}{\sigma^2\lambda(e^{\gamma(\theta)s} - 1) + \gamma(\theta) - \alpha + e^{\gamma(\theta)s}(\gamma(\theta) + \alpha)}\right),$$

and

$$\psi_2(s) = \frac{\lambda(\gamma(\theta) + \alpha + e^{\gamma(\theta)s}(\gamma(\theta) - \alpha)) + 2\theta(e^{\gamma(\theta)s} - 1)}{\sigma^2\lambda(e^{\gamma(\theta)s} - 1) + \gamma(\theta) - \alpha + e^{\gamma(\theta)s}(\gamma(\theta) + \alpha)}$$

and $\gamma(\theta) = \sqrt{\alpha^2 + 2\sigma^2\theta}$. The bond pricing formula is the special case $\lambda = 0$, $\theta = 1$.

The bivariate Laplace transform in (3.76) characterizes the joint distribution of the short rate and its integral. This makes it possible, at least in principle, to sample from the joint distribution of $(r(t_{i+1}), Y(t_{i+1}))$ given $(r(t_i), Y(t_i))$ with

$$Y(t) = \int_0^t r(u)\,du.$$

As explained in Section 3.3.2, this would allow exact simulation of the short rate and the discount factor on a discrete time grid. In the Gaussian setting, the joint distribution of $r(t)$ and $Y(t)$ is normal and therefore easy to sample; in contrast, the joint distribution determined by (3.76) is not explicitly available. Scott [324] derives the Laplace transform of the conditional distribution of $Y(t_{i+1}) - Y(t_i)$ given $r(t_i)$ and $r(t_{i+1})$, and explains how to use numerical transform inversion to sample from the conditional distribution. Through this method, he is able to simulate $(r(t_i), Y(t_i))$ without discretization error.

Time-Dependent Coefficients

As noted earlier, the parameter b is often replaced with a deterministic function of time $b(t)$ in order to calibrate the model to an initial term structure, resulting in the dynamics specified in (3.63). In this more general setting, a result of the form in (3.76) continues to hold but with functions ψ_1 and ψ_2 depending on both t and T rather than merely on $T-t$. Moreover, these functions will not in general be available in closed form, but are instead characterized by a system of ordinary differential equations. By solving these differential equations numerically, it then becomes possible to compute bond prices. Indeed, bond prices continue to have the exponential affine form, though the functions $A(t,T)$ and $C(t,T)$ in the exponent are no longer available explicitly but are also determined through ordinary differential equations (see Duffie, Pan, Singleton [105] and Jamshidian [195]). This makes it possible to use a numerical procedure to choose the function $b(t)$ to match an initial set of bond prices $B(0,T)$.

Once the constant b is replaced with a function of time, the transition density of the short rate process ceases to admit the relatively tractable form discussed in Section 3.4.1. One can of course simulate using an Euler scheme of the form

$$r(t_{i+1}) = r(t_i) + \alpha(b(t_i) - r(t_i))[t_{i+1} - t_i] + \sigma\sqrt{r(t_i)^+}\sqrt{t_{i+1} - t_i}Z_{i+1},$$

with independent $Z_i \sim N(0,1)$. However, it seems preferable (at least from a distributional perspective) to replace this normal approximation to the transition law with a noncentral chi-square approximation. For example, if we let

$$\bar{b}(t_i) = \frac{1}{t_{i+1} - t_i}\int_{t_i}^{t_{i+1}} b(s)\,ds$$

denote the average level of $b(t)$ over $[t_i, t_{i+1}]$ (assumed positive), then (3.68) suggests simulating by setting

$$r(t_{i+1}) = \frac{\sigma^2(1 - e^{-\alpha(t_{i+1}-t_i)})}{4\alpha}\chi_d'^2\left(\frac{4\alpha e^{-\alpha(t_{i+1}-t_i)}}{\sigma^2(1 - e^{-\alpha(t_{i+1}-t_i)})}r(t_i)\right), \qquad (3.77)$$

with $d = 4\bar{b}\alpha/\sigma^2$. We can sample from the indicated noncentral chi-square distribution using the methods discussed in Section 3.4.1. However, it must be stressed that whereas (3.68) is an exact representation in the case of constant coefficients, (3.77) is only an approximate procedure. If it suffices to choose the function $b(t)$ to match only bonds maturing at the simulation grid dates $t_1, \ldots, t_n$, then it may be possible to choose b to be constant over each interval $[t_i, t_{i+1}]$, in which case (3.77) becomes exact.

Jamshidian [195] shows that if α, b, and σ are all deterministic functions of time, the transition density of $r(t)$ can be represented through a noncentral chi-square distribution provided $\alpha(t)b(t)/\sigma^2(t)$ is independent of t. From (3.69) we see that this is equivalent to requiring that the degrees-of-freedom parameter $d = 4b\alpha/\sigma^2$ be constant. However, in this setting, the other parameters of the transition density are not given explicitly but rather as solutions to ordinary differential equations.

Change of Numeraire

Recall from Sections 1.2.3 and 3.3.2 that the forward measure for any date T_F is the measure associated with taking as numeraire asset the bond $B(t, T_F)$ maturing at T_F. We saw in Section 3.3.2 that if the short rate follows an Ornstein-Uhlenbeck process under the risk-neutral measure, then it continues to follow an OU process under a forward measure. An analogous property holds if the short rate follows a square-root diffusion.

Most of the development leading to (3.58) results from the exponential affine formula for bond prices and thus extends to the square-root model. In this setting, the bond price dynamics become

$$\frac{dB(t, T)}{B(t, T)} = r(t)\, dt - A(t, T)\sigma\sqrt{r(t)}\, dW(t);$$

in particular, the coefficient $\sigma\sqrt{r(t)}$ replaces the σ of the Gaussian case. Proceeding as in (3.56)–(3.58) but with this substitution, we observe that Girsanov's Theorem implies that the process W^{T_F} defined by

$$dW^{T_F}(t) = dW(t) + \sigma\sqrt{r(t)}A(t, T_F)\, dt$$

is a standard Brownian motion under the forward measure P_{T_F}. The dynamics of the short rate thus become

$$
\begin{aligned}
dr(t) &= \alpha(b(t) - r(t))\, dt + \sigma\sqrt{r(t)}\, dW(t) \\
&= \alpha(b(t) - r(t))\, dt + \sigma\sqrt{r(t)}[dW^{T_F}(t) - \sigma\sqrt{r(t)}A(t, T_F)dt] \\
&= \alpha(b(t) - (1 + \sigma^2 A(t, T_F))r(t)]\, dt + \sigma\sqrt{r(t)}\, dW^{T_F}(t).
\end{aligned}
$$

This can be written as

$$dr(t) = \alpha(1 + \sigma^2 A(t, T_F))\left(\frac{b(t)}{1 + \sigma^2 A(t, T_F)} - r(t)\right) dt + \sigma\sqrt{r(t)}\, dW^{T_F}(t),$$

which shows that under the forward measure the short rate is again a square-root diffusion but one in which both the level to which the process reverts and the speed with which it reverts are functions of time. The pricing of derivative securities through simulation in the forward measure works the same way here as in the Vasicek model.

3.4.4 Extensions

In this section we consider further properties and extensions of the square-root diffusion. We discuss multifactor models, a connection with squared Gaussian processes, and a connection with CEV (constant elasticity of variance) models.

Multifactor Models

The simplest multifactor extension of the CIR interest rate model defines independent processes

$$dX_i(t) = \alpha_i(b_i - X_i(t))\, dt + \sigma_i \sqrt{X_i(t)}\, dW_i(t), \quad i = 1, \ldots, d,$$

and takes the short rate to be $r(t) = X_1(t) + \cdots + X_d(t)$. Much as in the discussion of Section 3.3.3, this extension allows imperfect instantaneous correlation among bonds of different maturities. Each X_i can be simulated using the method developed in the previous sections for a single-factor model.

It is possible to consider more general models in which the underlying processes $X_1, \ldots, X_d$ are correlated. However, once one goes beyond the case of independent square-root factors, it seems more natural to move directly to the full generality of the affine models characterized by Duffie and Kan [101]. This class of models has a fair amount of tractability and computationally attractive features, but we will not consider it further here.

Squared Gaussian Models

We next point out a connection between the (single-factor) square-root diffusion and a Gaussian model of the type considered in Section 3.3. This connection is of intrinsic interest, it sheds further light on the simulation procedure of Section 3.4.1, and it suggests a wider class of interest rate models. The link between the CIR model and squared Gaussian models is noted in Rogers [307]; related connections are developed in depth by Revuz and Yor [306] in their discussion of Bessel processes.

Let $X_1(t), \ldots, X_d(t)$ be independent Ornstein-Uhlenbeck processes of the form

$$dX_i(t) = -\frac{\alpha}{2} X_i(t)\, dt + \frac{\sigma}{2}\, dW_i(t), \quad i = 1, \ldots, d,$$

for some constants α, σ, and independent Brownian motions $W_1, \ldots, W_d$. Let $Y(t) = X_1^2(t) + \cdots + X_d^2(t)$; then Itô's formula gives

$$dY(t) = \sum_{i=1}^{d}\left(2X_i(t)\, dX_i(t) + \frac{\sigma^2}{4}\, dt\right)$$

$$= \sum_{i=1}^{d}\left(-\alpha X_i^2(t) + \frac{\sigma^2}{4}\right) dt + \sigma \sum_{i=1}^{d} X_i(t)\, dW_i(t)$$

$$= \alpha\left(\frac{\sigma^2 d}{4\alpha} - Y(t)\right) dt + \sigma \sum_{i=1}^{d} X_i(t)\, dW_i(t).$$

If we now define

$$d\tilde{W}(t) = \sum_{i=1}^{d} \frac{X_i(t)}{\sqrt{Y(t)}}\, dW_i(t),$$

then $\tilde{W}(t)$ is a standard Brownian motion because the vector $(X_1(t), \ldots,$ $X_d(t))/\sqrt{Y(t)}$ multiplying $(dW_1(t), \ldots, dW_d(t))^\top$ has norm 1 for all t. Hence,

$$dY(t) = \alpha \left(\frac{\sigma^2 d}{4\alpha} - Y(t) \right) dt + \sigma \sqrt{Y(t)} \, d\tilde{W}(t),$$

which has the form of (3.62) with $b = \sigma^2 d/4\alpha$.

Starting from (3.62) and reversing these steps, we find that we can construct a square-root diffusion as a sum of squared independent Ornstein-Uhlenbeck processes provided $d = 4b\alpha/\sigma^2$ is an integer. Observe that this is precisely the degrees-of-freedom parameter in (3.69). In short, a square-root diffusion with an integer degrees-of-freedom parameter is a sum of squared Gaussian processes.

We can use this construction from Gaussian processes to simulate $r(t)$ in (3.62) if d is an integer. Writing $r(t_{i+1})$ as $\sum_{j=1}^{d} X_j^2(t_{i+1})$ and using (3.45) for the one-step evolution of the X_j, we arrive at

$$r(t_{i+1}) = \sum_{j=1}^{d} \left(e^{-\frac{1}{2}\alpha(t_{i+1}-t_i)} \sqrt{r(t_i)/d} + \frac{\sigma}{2}\sqrt{\frac{1}{\alpha}(1 - e^{-\alpha(t_{i+1}-t_i)})} Z_{i+1}^{(j)} \right)^2,$$

where $(Z_i^{(1)}, \ldots, Z_i^{(d)})$ are standard normal d-vectors, independent for different values of i. Comparison with (3.72) reveals that the expression on the right is a scalar multiple of a noncentral chi-square random variable, so this construction is really just a special case of the method in Section 3.4.1. It sheds some light on the appearance of the noncentral chi-square distribution in the law of $r(t)$.

This construction also points to another strategy for constructing interest rate models: rather than restricting ourselves to a sum of independent, identical squared OU processes, we can consider other quadratic functions of multivariate Gaussian processes. This idea has been developed in Beaglehole and Tenney [42] and Jamshidian [196]. The resulting models are closely related to the affine family.

CEV Process

We conclude this section with a digression away from interest rate models to consider a class of asset price processes closely related to the square-root diffusion.

Among the important alternatives to the lognormal model for an asset price considered in Section 3.2 is the *constant elasticity of variance* (CEV) process (see Cox and Ross [89], Schroder [322] and references there)

$$dS(t) = \mu S(t) \, dt + \sigma S(t)^{\beta/2} \, dW(t). \tag{3.78}$$

This includes geometric Brownian motion as the special case $\beta = 2$; some empirical studies have found that $\beta < 2$ gives a better fit to stock price data. If we write the model as

$$\frac{dS(t)}{S(t)} = \mu \, dt + \sigma S(t)^{(\beta-2)/2} \, dW(t),$$

we see that the instantaneous volatility $\sigma S(t)^{(\beta-2)/2}$ depends on the current level of the asset, and $\beta < 2$ implies a negative relation between the price level and volatility.

If we set $X(t) = S(t)^{2-\beta}$ and apply Itô's formula, we find that

$$dX(t) = \left[\frac{\sigma^2}{2}(2-\beta)(1-\beta) + \mu(2-\beta)X(t)\right] dt + \sigma(2-\beta)\sqrt{X(t)} \, dW(t),$$

revealing that $X(t)$ is a square-root diffusion. For $\mu > 0$ and $1 < \beta < 2$, we can use the method of the Section 3.4.1 to simulate $X(t)$ on a discrete time grid and then invert the transformation from S to X to get $S(t) = X(t)^{1/(2-\beta)}$. The case $\beta < 1$ presents special complications because of the behavior of S near 0; simulation of this case is investigated in Andersen and Andreasen [13].

3.5 Processes with Jumps

Although the vast majority of models used in derivatives pricing assume that the underlying assets have continuous sample paths, many studies have found evidence of the importance of jumps in prices and have advocated the inclusion of jumps in pricing models. Compared with a normal distribution, the logarithm of a price process with jumps is often *leptokurtotic*, meaning that it has a high peak and heavy tails, features typical of market data. In this section we discuss a few relatively simple models with jumps, highlighting issues that affect the implementation of Monte Carlo.

3.5.1 A Jump-Diffusion Model

Merton [263] introduced and analyzed one of the first models with both jump and diffusion terms for the pricing of derivative securities. Merton applied this model to options on stocks and interpreted the jumps as idiosyncratic shocks affecting an individual company but not the market as a whole. Similar models have subsequently been applied to indices, exchange rates, commodity prices, and interest rates.

Merton's jump-diffusion model can be specified through the SDE

$$\frac{dS(t)}{S(t-)} = \mu \, dt + \sigma \, dW(t) + dJ(t) \tag{3.79}$$

where μ and σ are constants, W is a standard one-dimensional Brownian motion, and J is a process independent of W with piecewise constant sample paths. In particular, J is given by

$$J(t) = \sum_{j=1}^{N(t)} (Y_j - 1) \tag{3.80}$$

where $Y_1, Y_2, \ldots$ are random variables and $N(t)$ is a *counting process*. This means that there are random arrival times

$$0 < \tau_1 < \tau_2 < \cdots$$

and

$$N(t) = \sup\{n : \tau_n \le t\}$$

counts the number of arrivals in $[0, t]$. The symbol $dJ(t)$ in (3.79) stands for the jump in J at time t. The size of this jump is $Y_j - 1$ if $t = \tau_j$ and 0 if t does not coincide with any of the τ_j.

In the presence of jumps, a symbol like $S(t)$ is potentially ambiguous: if it is possible for S to jump at t, we need to specify whether $S(t)$ means the value of S just before or just after the jump. We follow the usual convention of assuming that our processes are continuous from the right, so

$$S(t) = \lim_{u \downarrow t} S(u)$$

includes the effect of any jump at t. To specify the value just before a potential jump we write $S(t-)$, which is the limit

$$S(t-) = \lim_{u \uparrow t} S(u)$$

from the left.

If we write (3.79) as

$$dS(t) = \mu S(t-)\, dt + \sigma S(t-)\, dW(t) + S(t-)\, dJ(t),$$

we see that the increment $dS(t)$ in S at t depends on the value of S just before a potential jump at t and not on the value just after the jump. This is as it should be. The jump in S at time t is $S(t) - S(t-)$. This is 0 unless J jumps at t, which is to say unless $t = \tau_j$ for some j. The jump in S at τ_j is

$$S(\tau_j) - S(\tau_j-) = S(\tau_j-)[J(\tau_j) - J(\tau_j-)] = S(\tau_j-)(Y_j - 1),$$

hence

$$S(\tau_j) = S(\tau_j-)Y_j.$$

This reveals that the Y_j are the ratios of the asset price before and after a jump — the jumps are multiplicative. This also explains why we wrote $Y_j - 1$ rather than simply Y_j in (3.80).

By restricting the Y_j to be positive random variables, we ensure that $S(t)$ can never become negative. In this case, we see that

$$\log S(\tau_j) = \log S(\tau_j-) + \log Y_j,$$

so the jumps are additive in the logarithm of the price. Additive jumps are a natural extension of Brownian motion and multiplicative jumps (as in (3.79)) provide a more natural extension of geometric Brownian motion; see the discussion at the beginning of Section 3.2. The solution of (3.79) is given by

$$S(t) = S(0)e^{(\mu-\frac{1}{2}\sigma^2)t+\sigma W(t)} \prod_{j=1}^{N(t)} Y_j, \qquad (3.81)$$

which evidently generalizes the corresponding solution for geometric Brownian motion.

Thus far, we have not imposed any distributional assumptions on the jump process $J(t)$. We now consider the simplest model — the one studied by Merton [263] — which takes $N(t)$ to be a Poisson process with rate λ. This makes the interarrival times $\tau_{j+1} - \tau_j$ independent with a common exponential distribution,

$$P(\tau_{j+1} - \tau_j \le t) = 1 - e^{-\lambda t}, \quad t \ge 0.$$

We further assume that the Y_j are i.i.d. and independent of N (as well as W). Under these assumptions, J is called a *compound Poisson process*.

As noted by Merton [263], the model is particularly tractable when the Y_j are lognormally distributed, because a product of lognormal random variables is itself lognormal. In more detail, if $Y_j \sim LN(a, b^2)$ (so that $\log Y_j \sim N(a, b^2)$) then for any fixed n,

$$\prod_{j=1}^{n} Y_j \sim LN(an, b^2 n).$$

It follows that, conditional on $N(t) = n$, $S(t)$ has the distribution of

$$S(0)e^{(\mu-\frac{1}{2}\sigma^2)t+\sigma W(t)} \prod_{j=1}^{n} Y_j \sim S(0) \cdot LN((\mu - \tfrac{1}{2}\sigma^2)t, \sigma^2 t) \cdot LN(an, b^2 n)$$

$$= LN(\log S(0) + (\mu - \tfrac{1}{2}\sigma^2)t + an, \sigma^2 t + b^2 n),$$

using the independence of the Y_j and W. If we let $F_{n,t}$ denote this lognormal distribution (cf. Section 3.2.1) and recall that $N(t)$ has a Poisson distribution with mean λt, then from the Poisson probabilities (3.75) we find that the unconditional distribution of $S(t)$ is

$$P(S(t) \le x) = \sum_{n=0}^{\infty} e^{-\lambda t} \frac{(\lambda t)^n}{n!} F_{n,t}(x),$$

a Poisson mixture of lognormal distributions. Merton [263] used this property to express the price of an option on S as an infinite series, each term of which is the product of a Poisson probability and a Black-Scholes formula.

Recall that in the absence of jumps the drift μ in (3.79) would be the risk-free rate, assuming the asset pays no dividends and assuming the model represents the dynamics under the risk-neutral measure. Suppose, for simplicity, that the risk-free rate is a constant r; then the drift is determined by the condition that $S(t)e^{-rt}$ be a martingale. Merton [263] extends this principle to his jump-diffusion model under the assumption that jumps are specific to a single stock and can be diversified away; that is, by assuming that the market does not compensate investors for bearing the risk of jumps. We briefly describe how this assumption determines the drift parameter μ in (3.79).

A standard property of the Poisson process is that $N(t) - \lambda t$ is a martingale. A generalization of this property is that

$$\sum_{i=1}^{N(t)} h(Y_j) - \lambda \mathsf{E}[h(Y)]t$$

is a martingale for i.i.d. Y, Y_1, Y_2 and any function h for which $\mathsf{E}[h(Y)]$ is finite. Accordingly, the process

$$J(t) - \lambda m t$$

is a martingale if $m = \mathsf{E}[Y_j] - 1$. The choice of drift parameter in (3.79) that makes $S(t)e^{-rt}$ a martingale is therefore $\mu = r - \lambda m$. In this case, if we rewrite (3.79) as

$$\frac{dS(t)}{S(t-)} = r\,dt + \sigma\,dW(t) + [dJ(t) - \lambda m\,dt],$$

the last two terms on the right are martingales and the net growth rate in $S(t)$ is indeed r.

With this notation and with $\log Y_j \sim N(a, b^2)$, Merton's [263] option pricing formula becomes

$$e^{-rT}\mathsf{E}[(S(T) - K)^+] = \sum_{n=0}^{\infty} e^{-\lambda t}\frac{(\lambda t)^n}{n!}e^{-rT}\mathsf{E}[(S(T) - K)^+ | N(T) = n]$$

$$= \sum_{n=0}^{\infty} e^{-\lambda' t}\frac{(\lambda' t)^n}{n!}\mathrm{BS}(S(0), \sigma_n, T, r_n, K),$$

where $\lambda' = \lambda(1 + m)$, $\sigma_n^2 = \sigma^2 + b^2 n/T$, $r_n = r - \lambda m + n\log(1 + m)/T$, and $\mathrm{BS}(\cdot)$ denotes the Black-Scholes call option formula (1.4).

Simulating at Fixed Dates

We consider two approaches to simulating the jump-diffusion model (3.79), each of which is an instance of a more general strategy for simulating a broader class of jump-diffusion models. In the first method, we simulate the process at a fixed set of dates $0 = t_0 < t_1 < \cdots < t_n$ without explicitly distinguishing the effects of the jump and diffusion terms. In the second method, we simulate the jump times $\tau_1, \tau_2, \ldots$ explicitly.

We continue to assume that N is a Poisson process, that $Y_1, Y_2, \ldots$ are i.i.d., and that N, W, and $\{Y_1, Y_2, \ldots\}$ are mutually independent. We do not assume the Y_j are lognormally distributed, though that will constitute an interesting special case.

To simulate $S(t)$ at time $t_1, \ldots, t_n$, we generalize (3.81) to

$$S(t_{i+1}) = S(t_i) e^{(\mu - \frac{1}{2}\sigma^2)(t_{i+1} - t_i) + \sigma[W(t_{i+1}) - W(t_i)]} \prod_{j=N(t_i)+1}^{N(t_{i+1})} Y_j,$$

with the usual convention that the product over j is equal to 1 if $N(t_{i+1}) = N(t_i)$. We can simulate directly from this representation or else set $X(t) = \log S(t)$ and

$$X(t_{i+1}) = X(t_i) + (\mu - \tfrac{1}{2}\sigma^2)(t_{i+1} - t_i) + \sigma[W(t_{i+1}) - W(t_i)] + \sum_{j=N(t_i)+1}^{N(t_{i+1})} \log Y_j;$$

$$(3.82)$$

this recursion replaces products with sums and is preferable, at least if sampling $\log Y_j$ is no slower than sampling Y_j. We can exponentiate simulated values of the $X(t_i)$ to produce samples of the $S(t_i)$.

A general method for simulating (3.82) from t_i to t_{i+1} consists of the following steps:

1. generate $Z \sim N(0, 1)$
2. generate $N \sim \text{Poisson}(\lambda(t_{i+1} - t_i))$ (see Figure 3.9); if $N = 0$, set $M = 0$ and go to Step 4
3. generate $\log Y_1, \ldots, \log Y_N$ from their common distribution and set $M = \log Y_1 + \ldots + \log Y_N$
4. set

$$X(t_{i+1}) = X(t_i) + (\mu - \tfrac{1}{2}\sigma^2)(t_{i+1} - t_i) + \sigma\sqrt{t_{i+1} - t_i}\, Z + M.$$

This method relies on two properties of the Poisson process: the increment $N(t_{i+1}) - N(t_i)$ has a Poisson distribution with mean $\lambda(t_{i+1} - t_i)$, and it is independent of increments of N over $[0, t_i]$.

Under further assumptions on the distribution of the Y_j, this method can sometimes be simplified. If the Y_j have the lognormal distribution $LN(a, b^2)$, then $\log Y_j \sim N(a, b^2)$ and

$$\sum_{j=1}^{n} \log Y_j \sim N(an, b^2 n) = an + b\sqrt{n}N(0, 1).$$

In this case, we may therefore replace Step 3 with the following:

3'. generate $Z_2 \sim N(0,1)$; set $M = aN + b\sqrt{N}Z_2$

If the $\log Y_j$ have a gamma distribution with shape parameter a and scale parameter β (see (3.74)), then

$$\log Y_1 + \log Y_2 + \cdots + \log Y_n$$

has the gamma distribution with shape parameter an and scale parameter β. Consequently, in Step 3 above we may sample M directly from a gamma distribution, conditional on the value of N.

Kou [215] proposes and analyzes a model in which $|\log Y_j|$ has a gamma distribution (in fact exponential) and the sign of $\log Y_j$ is positive with probability q, negative with probability $1-q$. In this case, conditional on the Poisson random variable N taking the value n, the number of $\log Y_j$ with positive sign has a binomial distribution with parameters n and q. Step 3 can therefore be replaced with the following:

3a''. generate $K \sim \text{Binomial}(N,q)$
3b''. generate $R_1 \sim \text{Gamma}(Ka,\beta)$ and $R_2 \sim \text{Gamma}((N-K)a,\beta)$ and set $M = R_1 - R_2$

In 3b'', interpret a gamma random variable with shape parameter zero as the constant 0 in case $K = 0$ or $K = N$. In 3a'', conditional on $N = n$, the binomial distribution of K is given by

$$P(K = k) = \frac{n!}{k!(n-k)!}q^k(1-q)^{n-k}, \quad k = 0, 1, \ldots, n.$$

Samples from this distribution can be generated using essentially the same method used for the Poisson distribution in Figure 3.9 by changing just the first and sixth lines of that algorithm. In the first line, replace the mass at the origin $\exp(-\theta)$ for the Poisson distribution with the corresponding value $(1-q)^n$ for the binomial distribution. Observe that the ratio $P(K = k)/P(K = k - 1)$ is given by $q(n + 1 - k)/k(1 - q)$, so the sixth line of the algorithm becomes $p \leftarrow pq(n + 1 - N)/N(1 - q)$ (where N now refers to the binomial random variable produced by the algorithm).

Simulating Jump Times

Simulation methods based on (3.82) produce values $S(t_i) = \exp(X(t_i))$, $i = 1, \ldots, n$, with the exact joint distribution of the target process (3.79) at dates $t_1, \ldots, t_n$. Notice, however, that this approach does not identify the times at which $S(t)$ jumps; rather, it generates the total number of jumps in each interval $(t_i, t_{i+1}]$, using the fact that the number of jumps has a Poisson distribution.

An alternative approach to simulating (3.79) simulates the jump times $\tau_1, \tau_2, \ldots$ explicitly. From one jump time to the next, $S(t)$ evolves like an ordinary geometric Brownian motion because we have assumed that W and J in (3.79) are independent of each other. It follows that, conditional on the times $\tau_1, \tau_2, \ldots$ of the jumps,

$$S(\tau_{j+1}-) = S(\tau_j)e^{(\mu-\frac{1}{2}\sigma^2)(\tau_{j+1}-\tau_j)+\sigma[W(\tau_{j+1})-W(\tau_j)]}$$

and

$$S(\tau_{j+1}) = S(\tau_{j+1}-)Y_{j+1}.$$

Taking logarithms and combining these steps, we get

$$X(\tau_{j+1}) = X(\tau_j) + (\mu - \tfrac{1}{2}\sigma^2)(\tau_{j+1} - \tau_j) + \sigma[W(\tau_{j+1}) - W(\tau_j)] + \log Y_{j+1}.$$

A general scheme for simulating one step of this recursion now takes the following form:

1. generate R_{j+1} from the exponential distribution with mean $1/\lambda$
2. generate $Z_{j+1} \sim N(0,1)$
3. generate $\log Y_{j+1}$
4. set $\tau_{j+1} = \tau_j + R_{j+1}$ and

$$X(\tau_{j+1}) = X(\tau_j) + (\mu - \tfrac{1}{2}\sigma^2)R_{j+1} + \sigma\sqrt{R_{j+1}}Z_{j+1} + \log Y_{j+1}.$$

Recall from Section 2.2.1 that the exponential random variable R_{j+1} can be generated by setting $R_{j+1} = -\log(U)/\lambda$ with $U \sim \text{Unif}[0,1]$.

The two approaches to simulating $S(t)$ can be combined. For example, suppose we fix a date t in advance that we would like to include among the simulated dates. Suppose it happens that $\tau_j < t < \tau_{j+1}$ (i.e., $N(t) = N(t-) = j$). Then

$$S(t) = S(\tau_j)e^{(\mu-\frac{1}{2}\sigma^2)(t-\tau_j)+\sigma[W(t)-W(\tau_j)]}$$

and

$$S(\tau_{j+1}) = S(t)e^{(\mu-\frac{1}{2}\sigma^2)(\tau_{j+1}-t)+\sigma[W(\tau_{j+1})-W(t)]}Y_{j+1}.$$

Both approaches to simulating the basic jump-diffusion process (3.79) — simulating the number of jumps in fixed subintervals and simulating the times at which jumps occur — can be useful at least as approximations in simulating more general jump-diffusion models. Exact simulation becomes difficult when the times of the jumps and the evolution of the process between jumps are no longer independent of each other.

Inhomogeneous Poisson Process

A simple extension of the jump-diffusion model (3.79) replaces the constant jump intensity λ of the Poisson process with a deterministic (nonnegative) function of time $\lambda(t)$. This means that

$$P(N(t+h) - N(t) = 1|N(t)) = \lambda(t)h + o(h)$$

and $N(t)$ is called an *inhomogeneous* Poisson process. Like an ordinary Poisson process it has independent increments and these increments are Poisson distributed, but increments over different intervals of equal length can have different means. In particular, the number of jumps in an interval $(t_i, t_{i+1}]$ has a Poisson distribution with mean $\Lambda(t_{i+1}) - \Lambda(t_i)$, where

$$\Lambda(t) = \int_0^t \lambda(u) \, du.$$

Provided this function can be evaluated, simulation based on (3.82) generalizes easily to the inhomogeneous case: where we previously sampled from the Poisson distribution with mean $\lambda(t_{i+1} - t_i)$, we now sample from the Poisson distribution with mean $\Lambda(t_{i+1}) - \Lambda(t_i)$.

It is also possible to simulate the interarrival times of the jumps. The key property is

$$P(\tau_{j+1} - \tau_j \leq t | \tau_1, \dots, \tau_j) = 1 - \exp(-[\Lambda(\tau_j + t) - \Lambda(\tau_j)]), \quad t \geq 0,$$

provided $\Lambda(\infty) = \infty$. We can (at least in principle) sample from this distribution using the inverse transform method discussed in Section 2.2.1. Given τ_j, let

$$X = \inf \left\{ t \geq 0 : 1 - \exp\left(-\int_{\tau_j}^t \lambda(u) \, du \right) = U \right\}, \quad U \sim \text{Unif}[0,1]$$

then X has the required interarrival time distribution and we may set $\tau_{j+1} = \tau_j + X$. This is equivalent to setting

$$X = \inf \left\{ t \geq 0 : \int_{\tau_j}^t \lambda(u) \, du = \xi \right\} \tag{3.83}$$

where ξ is exponentially distributed with mean 1. We may therefore interpret the time between jumps as the time required to consume an exponential random variable if it is consumed at rate $\lambda(u)$ at time u.

If the time-varying intensity $\lambda(t)$ is bounded by a constant $\bar{\lambda}$, the jumps of the inhomogeneous Poisson process can be generated by *thinning* an ordinary Poisson process $\bar{N}$ with rate $\bar{\lambda}$, as in Lewis and Shedler [235]. In this procedure, the jump times of $\bar{N}$ become *potential* jump times of N; a potential jump at time t is accepted as an actual jump with probability $\lambda(t)/\bar{\lambda}$. A bit more explicitly, we have the following steps:

1. generate jump times $\bar{\tau}_j$ of $\bar{N}$ (the interarrival times $\bar{\tau}_{j+1} - \bar{\tau}_j$ are independent and exponentially distributed with mean $1/\bar{\lambda}$)
2. for each j generate $U_j \sim \text{Unif}[0,1]$; if $U_j \bar{\lambda} < \lambda(\bar{\tau}_j)$ then accept $\bar{\tau}_j$ as a jump time of N.

Figure 3.10 illustrates this construction.

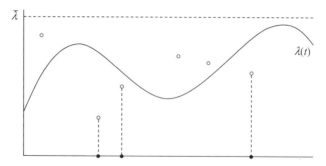

Fig. 3.10. Construction of an inhomogeneous Poisson process from an ordinary Poisson process by thinning. The horizontal coordinates of the open circles are the jump times of a Poisson process with rate $\bar{\lambda}$; each circle is raised to a height uniformly distributed between 0 and $\bar{\lambda}$. Circles below the curve $\lambda(t)$ are accepted as jumps of the inhomogeneous Poisson process. The times of the accepted jumps are indicated by the filled circles.

3.5.2 Pure-Jump Processes

If $S(t)$ is the jump-diffusion process in (3.79) with $J(t)$ a compound Poisson process, then $X(t) = \log S(t)$ is a process with independent increments. This is evident from (3.82) and the fact that both W and J have independent increments. Geometric Brownian motion also has the property that its logarithm has independent increments. It is therefore natural to ask what other potentially fruitful models of asset prices might arise from the representation

$$S(t) = S(0)\exp(X(t)) \tag{3.84}$$

with X having independent increments. Notice that we have adopted the normalization $X(0) = 0$.

The process X is a *Lévy process* if it has stationary, independent increments and satisfies the technical requirement that $X(t)$ converges in distribution to $X(s)$ as $t \to s$. Stationarity of the increments means that $X(t + s) - X(s)$ has the distribution of $X(t)$. Every Lévy process can be represented as the sum of a deterministic drift, a Brownian motion, and a pure-jump process independent of the Brownian motion (see, e.g., Chapter 4 of Sato [317]). If the number of jumps in every finite interval is almost surely finite, then the pure-jump component is a compound Poisson process. Hence, in constructing processes of the form (3.84) with X a Lévy process, the only way to move beyond the jump-diffusion process (3.79) is to consider processes with an infinite number of jumps in finite intervals. We will in fact focus on pure-jump processes of this type — that is, Lévy processes with no Brownian component. Several processes of this type have been proposed as models of asset prices, and we consider some of these examples. A more extensive discussion of the simulation of Lévy process can be found in Asmussen [21].

It should be evident that in considering processes with an infinite number of jumps in finite intervals, only the first of the two approaches developed in Section 3.5.1 is viable: we may be able to simulate the increments of such a process, but we cannot hope to simulate from one jump to the next. To simulate a pure-jump Lévy process we should therefore consider the distribution of its increments over a fixed time grid.

A random variable Y (more precisely, its distribution) is said to be *infinitely divisible* if for each $n = 2, 3, \ldots$, there are i.i.d. random variables $Y_1^{(n)}, \ldots, Y_n^{(n)}$ such that $Y_1^{(n)} + \cdots + Y_n^{(n)}$ has the distribution of Y. If X is a Lévy process $(X(0) = 0)$, then

$$X(t) = X(t/n) + [X(2t/n) - X(t/n)] + \cdots + [X(t) - X((n-1)t/n)]$$

decomposes $X(t)$ as the sum of n i.i.d. random variables and shows that $X(t)$ has an infinitely divisible distribution. Conversely, for each infinitely divisible distribution there is a Lévy process for which $X(1)$ has that distribution. Simulating a Lévy process on a fixed time grid is thus equivalent to sampling from infinitely divisible distributions.

A Lévy process with nondecreasing sample paths is called a *subordinator*. A large class of Lévy processes (sometimes called processes of type G) can be represented as $W(G(t))$ with W Brownian motion and G a subordinator independent of W. Several of the examples we consider belong to this class.

Gamma Processes

If $Y_1, \ldots, Y_n$ are independent with distribution Gamma$(a/n, \beta)$, then $Y_1 + \cdots + Y_n$ has distribution Gamma(a, β); thus, gamma distributions are infinitely divisible. For each choice of the parameters a and β there is a Lévy process (called a *gamma process*) such that $X(1)$ has distribution Gamma(a, β). We can simulate this process on a time grid $t_1, \ldots, t_n$ by sampling the increments

$$X(t_{i+1}) - X(t_i) \sim \text{Gamma}(a \cdot (t_{i+1} - t_i), \beta)$$

independently, using the methods of Sections 3.4.2.

A gamma random variable takes only positive values so a gamma process is nondecreasing. This makes it unsuitable as a model of (the logarithm of) a risky asset price. Madan and Seneta [243] propose a model based on (3.84) and $X(t) = U(t) - D(t)$, with U and D independent gamma processes representing the up and down moves of X. They call this the *variance gamma process*. Increments of X can be simulated through the increments of U and D.

If $U(1)$ and $D(1)$ have the same shape and scale parameters, then X admits an alternative representation as $W(G(t))$ where W is a standard Brownian motion and G is a gamma process. In other words, X can be viewed as the result of applying a random time-change to an ordinary Brownian motion: the deterministic time argument t has been replaced by the random time $G(t)$, which becomes the conditional variance of $W(G(t))$ given $G(t)$. This explains the name "variance gamma."

Madan et al. [242] consider the more general case $W(G(t))$ where W now has drift parameter μ and variance parameter σ^2. They restrict the shape parameter of $G(1)$ to be the reciprocal of its scale parameter β (so that $\mathsf{E}[G(t)] = t$) and show that this more general variance gamma process can still be represented as the difference $U(t) - D(t)$ of two independent gamma processes. The shape and scale parameters of $U(1)$ and $D(1)$ should be chosen to satisfy $a_U = a_D = 1/\beta$ and

$$\beta_U \beta_D = \frac{\sigma^2 \beta}{2}, \quad \beta_U - \beta_D = \mu \beta.$$

The general variance gamma process can therefore still be simulated as the difference between two independent gamma processes. Alternatively, we can use the representation $X(t) = W(G(t))$ for simulation. Conditional on the increment $G(t_{i+1}) - G(t_i)$, the increment $W(G(t_{i+1})) - W(G(t_i))$ has a normal distribution with mean $\mu[G(t_{i+1}) - G(t_i)]$ and variance $\sigma^2[G(t_{i+1}) - G(t_i)]$. Hence, we can simulate X as follows:

1. generate $Y \sim \text{Gamma}((t_{i+1} - t_i)/\beta, \beta)$ (this is the increment in G)
2. generate $Z \sim N(0,1)$
3. set $X(t_{i+1}) = X(t_i) + \mu Y + \sigma \sqrt{Y} Z$.

The relative merits of this method and simulation through the difference of U and D depend on the implementation details of the methods used for sampling from the gamma and normal distributions.

Figure 3.11 compares two variance gamma densities with a normal density; all three have mean 0 and standard deviation 0.4. The figure illustrates the much higher kurtosis that can be achieved within the variance gamma family. Although the examples in the figure are symmetric, positive and negative skewness can be introduced through the parameter μ.

Normal Inverse Gaussian Processes

This class of processes, described in Barndorff-Nielsen [36], has some similarities to the variance gamma model. It is a Lévy process whose increments have a *normal inverse Gaussian* distribution; it can also be represented through a random time-change of Brownian motion.

The inverse Gaussian distribution with parameters $\delta, \gamma > 0$ has density

$$f_{IG}(x) = \frac{\delta e^{\delta \gamma}}{\sqrt{2\pi}} x^{-3/2} \exp\left(-\tfrac{1}{2}(\delta^2 x^{-1} + \gamma^2 x)\right), \quad x > 0. \tag{3.85}$$

This is the density of the first passage time to level δ of a Brownian motion with drift γ. It has mean δ/γ and variance δ/γ^3. The inverse Gaussian distribution is infinitely divisible: if X_1 and X_2 are independent and have this density

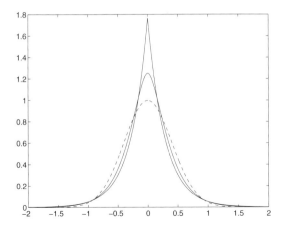

Fig. 3.11. Examples of variance gamma densities. The most peaked curve has $\mu = 0$, $\sigma = 0.4$, and $\beta = 1$ (and is in fact a double exponential density). The next most peaked curve has $\mu = 0$, $\sigma = 0.4$, and $\beta = 0.5$. The dashed line is the normal density with mean 0 and standard deviation 0.4.

with parameters (δ_1, γ) and (δ_2, γ), then it is clear from the first passage time interpretation that $X_1 + X_2$ has this density with parameters $(\delta_1 + \delta_2, \gamma)$. It follows that there is a Lévy process $Y(t)$ for which $Y(1)$ has density (3.85).

The *normal* inverse Gaussian distribution $NIG(\alpha, \beta, \mu, \delta)$ with parameters α, β, μ, δ can be described as the distribution of

$$\mu + \beta Y(1) + \sqrt{Y(1)}Z, \quad Z \sim N(0,1), \tag{3.86}$$

with $Y(1)$ having density (3.85), $\alpha = \sqrt{\beta^2 + \gamma^2}$, and Z independent of $Y(1)$. The mean and variance of this distribution are

$$\mu + \frac{\delta \beta}{\alpha \sqrt{1 - (\beta/\alpha)^2}} \quad \text{and} \quad \frac{\delta}{\alpha(1 - (\beta/\alpha)^2)^{3/2}},$$

respectively. The density is given in Barndorff-Nielsen [36] in terms of a modified Bessel function. Three examples are graphed in Figure 3.12; these illustrate the possibility of positive and negative skew and high kurtosis within this family of distributions.

Independent normal inverse Gaussian random variables add in the following way:

$$NIG(\alpha, \beta, \mu_1, \delta_1) + NIG(\alpha, \beta, \mu_2, \delta_2) = NIG(\alpha, \beta, \mu_1 + \mu_2, \delta_1 + \delta_2).$$

In particular, these distributions are infinitely divisible. Barndorff-Nielsen [36] studies Lévy processes with NIG increments. Such a process $X(t)$ can be represented as $W(Y(t))$ with $Y(t)$ the Lévy process defined from (3.85) and W a Brownian motion with drift β, unit variance, and initial value $W(0) = \mu$. At $t = 1$, this representation reduces to (3.86).

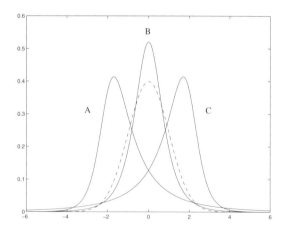

Fig. 3.12. Examples of normal inverse Gaussian densities. The parameters $(\alpha, \beta, \mu, \delta)$ are as follows: $(1, -0.75, 2, 1)$ for A, $(1, 0, 0, 1)$ for B, and $(1, 2, -0.75, 1)$ for C. The dashed line is the standard normal density and is included for comparison with case B, which also has mean 0 and standard deviation 1.

Eberlein [109] discusses the use of the NIG Lévy processes (in fact, a more general family called *generalized hyperbolic* Lévy processes) in modeling log returns. Barndorff-Nielsen [36] proposes several mechanisms for constructing models of price processes using NIG Lévy processes as a building block.

As with the variance gamma process of Madan and Seneta [243], there are in principle two strategies for simulating X on a discrete time grid. We can simulate the increments by sampling from the NIG distribution directly or we can use the representation as a time-changed Brownian motion (as in (3.86)). However, direct sampling from the NIG distribution does not appear to be particularly convenient, so we consider only the second of these two alternatives.

To simulate $X(t)$ as $W(Y(t))$ we need to be able to generate the increments of Y by sampling from the (ordinary) inverse Gaussian distribution. An interesting method for doing this was developed by Michael, Schucany, and Haas [264]. Their method uses the fact that if Y has the density in (3.85), then

$$\frac{(\gamma Y - \delta)^2}{Y} \sim \chi_1^2;$$

we may therefore sample Y by first generating $V \sim \chi_1^2$. Given a value of V, the resulting equation for Y has two roots,

$$y_1 = \frac{\delta}{\gamma} + \frac{V}{2\gamma^2} - \frac{1}{2\delta\gamma}\sqrt{4\delta^3 V/\gamma + \delta^2 V^2/\gamma}$$

and

$$y_2 = \delta^2/\gamma^2 y_1.$$

Michael et al. [264] show that the smaller root y_1 should be chosen with probability $\delta/(\delta + \gamma y_1)$ and the larger root y_2 with the complementary probability. Figure 3.13 illustrates the implementation of the method. The χ_1^2 random variable required for this algorithm can be generated as either a Gamma$(1/2, 2)$ or as the square of a standard normal.

Setup: $a \leftarrow 1/\gamma$, $b \leftarrow a * \delta$, $b \leftarrow b * b$

generate $V \sim \chi_1^2$
$\xi \leftarrow a * V$
$Y \leftarrow a * (\delta + (\xi/2) + \sqrt{\xi * (\delta + (\xi/4))})$
$p \leftarrow \delta/(\delta + \gamma * Y)$
generate $U \sim \mathrm{Unif}[0,1]$
if $U > p$ then $Y \leftarrow b/Y$
return Y

Fig. 3.13. Algorithm for sampling from the inverse Gaussian distribution (3.85), based on Michael et al. [264].

To simulate an increment of the NIG process $X(t) = W(Y(t))$ from t_i to t_{i+1}, we use the algorithm in Figure 3.13 to generate a sample Y from the inverse Gaussian distribution with parameters $\delta(t_{i+1} - t_i)$ and γ; we then set

$$X(t_{i+1}) = X(t_i) + \beta Y + \sqrt{Y} Z$$

with $Z \sim N(0, 1)$. (Recall that β is the drift of W in the NIG parameterization.)

Despite the evident similarity between this construction and the one used for the variance gamma process, a result of Asmussen and Rosiński [25] points to an important distinction between the two processes: the cumulative effect of small jumps can be well-approximated by Brownian motion in a NIG process but not in a variance gamma process. Loosely speaking, even the small jumps of the variance gamma process are too large or too infrequent to look like Brownian motion. Asmussen and Rosiński [25] discuss the use and applicability of a Brownian approximation to small jumps in simulating Lévy processes.

Stable Paretian Processes

A distribution is called *stable* if for each $n \geq 2$ there are constants $a_n > 0$ and b_n such that

$$X_1 + X_2 + \cdots + X_n =_{\mathrm{d}} a_n X + b_n,$$

where $X, X_1, \ldots, X_n$ are independent random variables with that distribution. (The symbol "$=_d$" indicates equality in distribution.) If $b_n = 0$ for all n, the distribution is *strictly* stable. The best known example is the standard normal distribution for which

$$X_1 + X_2 + \cdots + X_n =_d n^{1/2} X.$$

In fact, a_n must be of the form $n^{1/\alpha}$ for some $0 < \alpha \le 2$ called the *index* of the stable distribution. This is Theorem VI.1.1 of Feller [119]; for broader coverage of the topic see Samorodnitsky and Taqqu [316].

Stable random variables are infinitely divisible and thus define Lévy processes. Like the other examples in this section, these Lévy processes have no Brownian component (except in the case of Brownian motion itself) and are thus pure-jump processes. They can often be constructed by applying a random time change to an ordinary Brownian motion, the time change itself having stable increments.

Only the normal distribution has stable index $\alpha = 2$. Non-normal stable distributions (those with $\alpha < 2$) are often called *stable Paretian*. These are heavy-tailed distributions: if X has stable index $\alpha < 2$, then $E[|X|^p]$ is infinite for $p \ge \alpha$. In particular, all stable Paretian distributions have infinite variance and those with $\alpha \le 1$ have $E[|X|] = \infty$. Mandelbrot [246] proposed using stable Paretian distributions to model the high peaks and heavy tails (relative to the normal distribution) of market returns. Infinite variance suggests that the tails of these distributions may be *too* heavy for market data, but see Rachev and Mittnik [302] for a comprehensive account of applications in finance.

Stable random variables have probability densities but these are rarely available explicitly; stable distributions are usually described through their characteristic functions. The density is known for the normal case $\alpha = 2$; the Cauchy (or t_1) distribution, corresponding to $\alpha = 1$ and density

$$f(x) = \frac{1}{\pi} \frac{1}{1 + x^2}, \quad -\infty < x < \infty;$$

and the case $\alpha = 1/2$ with density

$$f(x) = \frac{1}{\sqrt{2\pi}} x^{-3/2} \exp(-1/(2x)), \quad x > 0.$$

This last example may be viewed as a limiting case of the inverse Gaussian distribution with $\gamma = 0$. Through a first passage time interpretation (see Feller [119], Example VI.2(f)), the Cauchy distribution may be viewed as a limiting case of the NIG distribution $\alpha = \beta = \mu = 0$. Both densities given above can be generalized by introducing scale and location parameters (as in [316], p.10). This follows from the simple observation that if X has a stable distribution then so does $\mu + \sigma X$, for any constants μ, σ.

As noted in Example 2.1.2, samples from the Cauchy distribution can be generated using the inverse transform method. If $Z \sim N(0, 1)$ then $1/Z^2$ has the stable density above with $\alpha = 1/2$, so this case is also straightforward. Perhaps surprisingly, it is also fairly easy to sample from other stable distributions even though their densities are unknown. An important tool in sampling from stable distributions is the following representation: if V is uniformly distributed over $[-\pi/2, \pi/2]$ and W is exponentially distributed with mean 1, then

$$\frac{\sin(\alpha V)}{(\cos(V))^{1/\alpha}} \left(\frac{\cos((1-\alpha)V)}{W} \right)^{(1-\alpha)/\alpha}$$

has a symmetric α-stable distribution; see p.42 of Samorodnitsky and Taqqu [316] for a proof. As noted there, this reduces to the Box-Muller method (see Section 2.3.2) when $\alpha = 2$. Chambers, Mallows, and Stuck [79] develop simulation procedures based on this representation and additional transformations. Samorodnitsky and Taqqu [316], pp.46-49, provide computer code for sampling from an arbitrary stable distribution, based on Chambers et al. [79].

Feller [119], p.336, notes that the Lévy process generated by a symmetric stable distribution can be constructed through a random time change of Brownian motion. This also follows from the observation in Samorodnitsky and Taqqu [316], p.21, that a symmetric stable random variable can be generated as the product of a normal random variable and a positive stable random variable, a construction similar to (3.86).

3.6 Forward Rate Models: Continuous Rates

The distinguishing feature of the models considered in this section and the next is that they explicitly describe the evolution of the full term structure of interest rates. This contrasts with the approach in Sections 3.3 and 3.4 based on modeling the dynamics of just the short rate $r(t)$. In a setting like the Vasicek model or the Cox-Ingersoll-Ross model, the current value of the short rate determines the current value of all other term structure quantities — forward rates, bond prices, etc. In these models, the state of the world is completely summarized by the value of the short rate. In multifactor extensions, like those described in Section 3.3.3, the state of the world is summarized by the current values of a finite number (usually small) of underlying factors; from the values of these factors all term structure quantities are determined, at least in principle.

In the framework developed by Heath, Jarrow, and Morton [174] (HJM), the state of the world is described by the full term structure and not necessarily by a finite number of rates or factors. The key contribution of HJM lies in identifying the restriction imposed by the absence of arbitrage on the evolution of the term structure.

At any point in time the term structure of interest rates can be described in various equivalent ways — through the prices or yields of zero-coupon bonds or par bonds, through forward rates, and through swap rates, to name just a few examples. The HJM framework models the evolution of the term structure through the dynamics of the forward rate curve. It could be argued that forward rates provide the most primitive description of the term structure (and thus the appropriate starting point for a model) because bond prices and yields reflect averages of forward rates across maturities, but it seems difficult to press this point too far.

From the perspective of simulation, this section represents a departure from the previous topics of this chapter. Thus far, we have focused on models that can be simulated exactly, at least at a finite set of dates. In the generality of the HJM setting, some discretization error is usually inevitable. HJM simulation might therefore be viewed more properly as a topic for Chapter 6; we include it here because of its importance and because of special simulation issues it raises.

3.6.1 The HJM Framework

The HJM framework describes the dynamics of the forward rate curve $\{f(t,T), 0 \leq t \leq T \leq T^*\}$ for some ultimate maturity T^* (e.g., 20 or 30 years from today). Think of this as a curve in the maturity argument T for each value of the time argument t; the length of the curve shrinks as time advances because $t \leq T \leq T^*$. Recall that the forward rate $f(t,T)$ represents the instantaneous continuously compounded rate contracted at time t for riskless borrowing or lending at time $T \geq t$. This is made precise by the relation

$$B(t,T) = \exp\left(-\int_t^T f(t,u)\, du\right)$$

between bond prices and forward rates, which implies

$$f(t,T) = -\frac{\partial}{\partial T} \log B(t,T). \tag{3.87}$$

The short rate is $r(t) = f(t,t)$. Figure 3.14 illustrates this notation and the evolution of the forward curve.

In the HJM setting, the evolution of the forward curve is modeled through an SDE of the form

$$df(t,T) = \mu(t,T)\, dt + \sigma(t,T)^\top\, dW(t). \tag{3.88}$$

In this equation and throughout, the differential df is with respect to time t and not maturity T. The process W is a standard d-dimensional Brownian motion; d is the number of *factors*, usually equal to 1, 2, or 3. Thus, while the forward rate curve is in principle an infinite-dimensional object, it is driven by a low-dimensional Brownian motion. The coefficients μ and σ in (3.88) (scalar and $\Re^d$-valued, respectively) could be stochastic or could depend on current

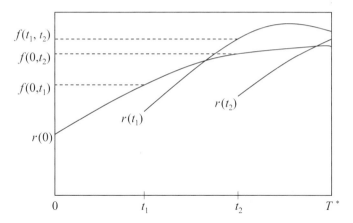

Fig. 3.14. Evolution of forward curve. At time 0, the forward curve $f(0, \cdot)$ is defined for maturities in $[0, T^*]$ and the short rate is $r(0) = f(0, 0)$. At $t > 0$, the forward curve $f(t, \cdot)$ is defined for maturities in $[t, T^*]$ and the short rate is $r(t) = f(t, t)$.

and past levels of forward rates. We restrict attention to the case in which μ and σ are deterministic functions of t, $T \geq t$, and the current forward curve $\{f(t, u), t \leq u \leq T^*\}$. Subject to technical conditions, this makes the evolution of the curve Markovian. We could make this more explicit by writing, e.g., $\sigma(f, t, T)$, but to lighten notation we omit the argument f. See Heath, Jarrow, and Morton [174] for the precise conditions needed for (3.88).

We interpret (3.88) as modeling the evolution of forward rates under the risk-neutral measure (meaning, more precisely, that W is a standard Brownian motion under that measure). We know that the absence of arbitrage imposes a condition on the risk-neutral dynamics of asset prices: the price of a (dividend-free) asset must be a martingale when divided by the numeraire

$$\beta(t) = \exp\left(\int_0^t r(u)\, du\right).$$

Forward rates are not, however, asset prices, so it is not immediately clear what restriction the absence of arbitrage imposes on the dynamics in (3.88). To find this restriction we must start from the dynamics of asset prices, in particular bonds. Our account is informal; see Heath, Jarrow, and Morton [174] for a rigorous development.

To make the discounted bond prices $B(t, T)/\beta(t)$ positive martingales, we posit dynamics of the form

$$\frac{dB(t, T)}{B(t, T)} = r(t)\, dt + \nu(t, T)^\top dW(t), \quad 0 \leq t \leq T \leq T^*. \tag{3.89}$$

The bond volatilities $\nu(t, T)$ may be functions of current bond prices (equivalently, of current forward rates since (3.87) makes a one-to-one correspondence between the two). Through (3.87), the dynamics in (3.89) constrain the evolution of forward rates. By Itô's formula,

$$d \log B(t,T) = [r(t) - \tfrac{1}{2}\nu(t,T)^\top \nu(t,T)] \, dt + \nu(t,T)^\top dW(t).$$

If we now differentiate with respect to T and then interchange the order of differentiation with respect to t and T, from (3.87) we get

$$\begin{aligned}
df(t,T) &= -\frac{\partial}{\partial T} d \log B(t,T) \\
&= -\frac{\partial}{\partial T}[r(t) - \tfrac{1}{2}\nu(t,T)^\top \nu(t,T)] \, dt - \frac{\partial}{\partial T}\nu(t,T)^\top dW(t).
\end{aligned}$$

Comparing this with (3.88), we find that we must have

$$\sigma(t,T) = -\frac{\partial}{\partial T}\nu(t,T)$$

and

$$\mu(t,T) = -\frac{\partial}{\partial T}[r(t) - \tfrac{1}{2}\nu(t,T)^\top \nu(t,T)] = \left(\frac{\partial}{\partial T}\nu(t,T)\right)^\top \nu(t,T).$$

To eliminate $\nu(t,T)$ entirely, notice that

$$\nu(t,T) = -\int_t^T \sigma(t,u) \, du + \text{constant}.$$

But because $B(t,T)$ becomes identically 1 as t approaches T (i.e., as the bond matures), we must have $\nu(T,T) = 0$ and thus the constant in this equation is 0. We can therefore rewrite the expression for μ as

$$\mu(t,T) = \sigma(t,T)^\top \int_t^T \sigma(t,u) \, du; \tag{3.90}$$

this is the risk-neutral drift imposed by the absence of arbitrage. Substituting in (3.88), we get

$$df(t,T) = \left(\sigma(t,T)^\top \int_t^T \sigma(t,u) \, du\right) dt + \sigma(t,T)^\top dW(t). \tag{3.91}$$

This equation characterizes the arbitrage-free dynamics of the forward curve under the risk-neutral measure; it is the centerpiece of the HJM framework.

Using a subscript $j = 1, \ldots, d$ to indicate vector components, we can write (3.91) as

$$df(t,T) = \sum_{j=1}^d \left(\sigma_j(t,T) \int_t^T \sigma_j(t,u) \, du\right) dt + \sum_{j=1}^d \sigma_j(t,T) \, dW_j(t). \tag{3.92}$$

This makes it evident that each factor contributes a term to the drift and that the combined drift is the sum of the contributions of the individual factors.

In (3.91), the drift is determined once σ is specified. This contrasts with the dynamics of the short rate models in Sections 3.3 and 3.4 where parameters of the drift could be specified independent of the diffusion coefficient without introducing arbitrage. Indeed, choosing parameters of the drift is essential in calibrating short rate models to an observed set of bond prices. In contrast, an HJM model is automatically calibrated to an initial set of bond prices $B(0,T)$ if the initial forward curve $f(0,T)$ is simply chosen consistent with these bond prices through (3.87). Put slightly differently, calibrating an HJM model to an observed set of bond prices is a matter of choosing an appropriate initial condition rather than choosing a parameter of the model dynamics. The effort in calibrating an HJM model lies in choosing σ to match market prices of interest rate *derivatives* in addition to matching bond prices.

We illustrate the HJM framework with some simple examples.

Example 3.6.1 *Constant σ.* Consider a single-factor $(d = 1)$ model in which $\sigma(t,T) \equiv \sigma$ for some constant σ. The interpretation of such a model is that each increment $dW(t)$ moves all points on the forward curve $\{f(t,u), t \leq u \leq T^*\}$ by an equal amount $\sigma \, dW(t)$; the diffusion term thus introduces only parallel shifts in the forward curve. But a model in which the forward curve makes only parallel shifts admits arbitrage opportunities: one can construct a costless portfolio of bonds that will have positive value under every parallel shift. From (3.90) we find that an HJM model with constant σ has drift

$$\mu(t,T) = \sigma \int_t^T \sigma \, du = \sigma^2(T - t).$$

In particular, the drift will vary (slightly, because σ^2 is small) across maturities, keeping the forward curve from making exactly parallel movements. This small adjustment to the dynamics of the forward curve is just enough to keep the model arbitrage-free. In this case, we can solve (3.91) to find

$$f(t,T) = f(0,T) + \int_0^t \sigma^2(T - u) \, du + \sigma W(t)$$
$$= f(0,T) + \tfrac{1}{2}\sigma^2[T^2 - (T - t)^2] + \sigma W(t).$$

In essentially any model, the identity $r(t) = f(t,t)$ implies

$$dr(t) = df(t,T)\bigg|_{T=t} + \frac{\partial}{\partial T} f(t,T)\bigg|_{T=t} dt.$$

In the case of constant σ, we can write this explicitly as

$$dr(t) = \sigma \, dW(t) + \left(\frac{\partial}{\partial T} f(0,T)\bigg|_{T=t} + \sigma^2 t \right) dt.$$

Comparing this with (3.50), we find that an HJM model with constant σ coincides with a Ho-Lee model with calibrated drift. □

Example 3.6.2 *Exponential* σ. Another convenient parameterization takes $\sigma(t,T) = \sigma \exp(-\alpha(T-t))$ for some constants $\sigma, \alpha > 0$. In this case, the diffusion term $\sigma(t,T)\,dW(t)$ moves forward rates for short maturities more than forward rates for long maturities. The drift is given by

$$\mu(t,T) = \sigma^2 e^{-\alpha(T-t)} \int_t^T e^{-\alpha(T-u)}\,du = \frac{\sigma^2}{\alpha}\left(e^{-2\alpha(T-t)} - e^{-\alpha(T-t)}\right).$$

An argument similar to the one used in Example 3.6.1 shows that the short rate in this case is described by the Vasicek model with time-varying drift parameters.

This example and the one that precedes it may be misleading. It would be incorrect to assume that the short rate process in an HJM setting will always have a convenient description. Indeed, such examples are exceptional. □

Example 3.6.3 *Proportional* σ. It is tempting to consider a specification of the form $\sigma(t,T) = \tilde{\sigma}(t,T)f(t,T)$ for some deterministic $\tilde{\sigma}$ depending only on t and T. This would make $\tilde{\sigma}(t,T)$ the volatility of the forward rate $f(t,T)$ and would suggest that the distribution of $f(t,T)$ is approximately lognormal. However, Heath et al. [174] note that this choice of σ is inadmissible: it produces forward rates that grow to infinity in finite time with positive probability. The difficulty, speaking loosely, is that if σ is proportional to the level of rates, then the drift is proportional to the rates *squared*. This violates the linear growth condition ordinarily required for the existence and uniqueness of solutions to SDEs (see Appendix B.2). Market conventions often presuppose the existence of a (proportional) volatility for forward rates, so the failure of this example could be viewed as a shortcoming of the HJM framework. We will see in Section 3.7 that the difficulty can be avoided by working with simple rather than continuously compounded forward rates. □

Forward Measure

Although the HJM framework is usually applied under the risk-neutral measure, only a minor modification is necessary to work in a forward measure. Fix a maturity T_F and recall that the forward measure associated with T_F corresponds to taking the bond $B(t,T_F)$ as numeraire asset. The forward measure P_{T_F} can be defined relative to the risk-neutral measure P_β through

$$\left(\frac{dP_{T_F}}{dP_\beta}\right)_t = \frac{B(t,T_F)\beta(0)}{\beta(t)B(0,T_F)}.$$

From the bond dynamics in (3.89), we find that this ratio is given by

$$\exp\left(-\frac{1}{2}\int_0^t \nu(u,T_F)^\top \nu(u,T_F)\,du + \int_0^t \nu(u,T_F)^\top dW(u)\right).$$

By the Girsanov Theorem, the process W^{T_F} defined by

$$dW^{T_F}(t) = -\nu(t, T_F)^\top \, dt + dW(t)$$

is therefore a standard Brownian motion under P_{T_F}. Recalling that $\nu(t, T)$ is the integral of $-\sigma(t, u)$ from $u = t$ to $u = T$, we find that the forward rate dynamics (3.91) become

$$
\begin{aligned}
df(t, T) &= -\sigma(t, T)^\top \nu(t, T) \, dt + \sigma(t, T)^\top [\nu(t, T_F)^\top \, dt + dW^{T_F}(t)] \\
&= -\sigma(t, T)^\top [\nu(t, T) - \nu(t, T_F)] \, dt + \sigma(t, T)^\top \, dW^{T_F}(t) \\
&= -\sigma(t, T)^\top \left(\int_T^{T_F} \sigma(t, u) \, du \right) dt + \sigma(t, T)^\top \, dW^{T_F}(t), \quad (3.93)
\end{aligned}
$$

for $t \leq T \leq T_F$. Thus, the HJM dynamics under the forward measure are similar to the dynamics under the risk-neutral measure, but where we previously integrated $\sigma(t, u)$ from t to T, we now integrate $-\sigma(t, u)$ from T to T_F. Notice that $f(t, T_F)$ is a martingale under P_{T_F}, though none of the forward rates is a martingale under the risk-neutral measure.

3.6.2 The Discrete Drift

Except under very special choices of σ, exact simulation of (3.91) is infeasible. Simulation of the general HJM forward rate dynamics requires introducing a discrete approximation. In fact, each of the two arguments of $f(t, T)$ requires discretization. For the first argument, fix a time grid $0 = t_0 < t_1 < \cdots < t_M$. Even at a fixed time t_i, it is generally not possible to represent the full forward curve $f(t_i, T)$, $t_i \leq T \leq T^*$, so instead we fix a grid of maturities and approximate the forward curve by its value for just these maturities. In principle, the time grid and the maturity grid could be different; however, assuming that the two sets of dates are the same greatly simplifies notation with little loss of generality.

We use hats to distinguish discretized variables from their exact continuous-time counterparts. Thus, $\hat{f}(t_i, t_j)$ denotes the discretized forward rate for maturity t_j as of time t_i, $j \geq i$, and $\hat{B}(t_i, t_j)$ denotes the corresponding bond price,

$$\hat{B}(t_i, t_j) = \exp\left(-\sum_{\ell=i}^{j-1} \hat{f}(t_i, t_\ell)[t_{\ell+1} - t_\ell] \right). \quad (3.94)$$

To avoid introducing any more discretization error than necessary, we would like the initial values of the discretized bonds $\hat{B}(0, t_j)$ to coincide with the exact values $B(0, t_j)$ for all maturities t_j on the discrete grid. Comparing (3.94) with the equation that precedes (3.87), we see that this holds if

$$\sum_{\ell=0}^{j-1} \hat{f}(0, t_\ell)[t_{\ell+1} - t_\ell] = \int_0^{t_j} f(0, u) \, du;$$

i.e., if

$$\hat{f}(0, t_\ell) = \frac{1}{t_{\ell+1} - t_\ell} \int_{t_\ell}^{t_{\ell+1}} f(0, u) \, du, \tag{3.95}$$

for all $\ell = 0, 1, \ldots, M - 1$. This indicates that we should initialize each $\hat{f}(0, t_\ell)$ to the *average* level of the forward curve $f(0, T)$ over the interval $[t_\ell, t_{\ell+1}]$ rather than, for example, initializing it to the value $f(0, t_\ell)$ at the left endpoint of this interval. The discretization (3.95) is illustrated in Figure 3.15.

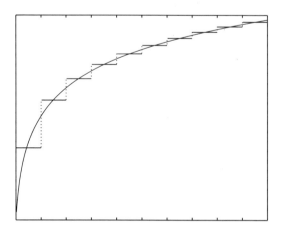

Fig. 3.15. Discretization of initial forward curve. Each discretized forward rate is the average of the underlying forward curve over the discretization interval.

Once the initial curve has been specified, a generic simulation of a single-factor model evolves like this: for $i = 1, \ldots, M$,

$$\hat{f}(t_i, t_j) = \hat{f}(t_{i-1}, t_j) +$$
$$\hat{\mu}(t_{i-1}, t_j)[t_i - t_{i-1}] + \hat{\sigma}(t_{i-1}, t_j)\sqrt{t_i - t_{i-1}} Z_i, \; j = i, \ldots, M, \tag{3.96}$$

where $Z_1, \ldots, Z_M$ are independent $N(0,1)$ random variables and $\hat{\mu}$ and $\hat{\sigma}$ denote discrete counterparts of the continuous-time coefficients in (3.91). We allow $\hat{\sigma}$ to depend on the current vector $\hat{f}$ as well as on time and maturity, though to lighten notation we do not include $\hat{f}$ as an explicit argument of $\hat{\sigma}$.

In practice, $\hat{\sigma}$ would typically be specified through a calibration procedure designed to make the simulated model consistent with market prices of actively traded derivative securities. (We discuss calibration of a closely related class of models in Section 3.7.4.) In fact, the continuous-time limit $\sigma(t, T)$ may never be specified explicitly because only the discrete version $\hat{\sigma}$ is used in the simulation. But the situation for the drift is different. Recall that in deriving (3.91) we chose the drift to make the model arbitrage-free; more precisely, we chose it to make the discounted bond prices martingales. There are many ways one might consider choosing the discrete drift $\hat{\mu}$ in (3.96) to approximate the continuous-time limit (3.90). From the many possible approximations, we choose the one that preserves the martingale property for the discounted bond prices.

Recalling that $f(s, s)$ is the short rate at time s, we can express the continuous-time condition as the requirement that

$$B(t, T) \exp\left(-\int_0^t f(s, s)\, ds\right)$$

be a martingale in t for each T. Similarly, in the discretized model we would like

$$\hat{B}(t_i, t_j) \exp\left(-\sum_{k=0}^{i-1} \hat{f}(t_k, t_k)[t_{k+1} - t_k]\right)$$

to be a martingale in i for each j. Our objective is to find a $\hat{\mu}$ for which this holds. For simplicity, we start by assuming a single-factor model.

The martingale condition can be expressed as

$$\mathsf{E}\left[\hat{B}(t_i, t_j) e^{-\sum_{k=0}^{i-1} \hat{f}(t_k, t_k)[t_{k+1} - t_k]} \Big| Z_1, \ldots, Z_{i-1}\right]$$
$$= \hat{B}(t_{i-1}, t_j) e^{-\sum_{k=0}^{i-2} \hat{f}(t_k, t_k)[t_{k+1} - t_k]}.$$

Using (3.94) and canceling terms that appear on both sides, this reduces to

$$\mathsf{E}\left[e^{-\sum_{\ell=i}^{j-1} \hat{f}(t_i, t_\ell)[t_{\ell+1} - t_\ell]} \Big| Z_1, \ldots, Z_{i-1}\right] = e^{-\sum_{\ell=i}^{j-1} \hat{f}(t_{i-1}, t_\ell)[t_{\ell+1} - t_\ell]}.$$

Now we introduce $\hat{\mu}$: on the left side of this equation we substitute for each $\hat{f}(t_i, t_\ell)$ according to (3.96). This yields the condition

$$\mathsf{E}\left[e^{-\sum_{\ell=i}^{j-1} \left(\hat{f}(t_{i-1}, t_\ell) + \hat{\mu}(t_{i-1}, t_\ell)[t_i - t_{i-1}] + \hat{\sigma}(t_{i-1}, t_\ell)\sqrt{t_i - t_{i-1}} Z_i\right)[t_{\ell+1} - t_\ell]} \Big| Z_1, \ldots, Z_{i-1}\right]$$
$$= e^{-\sum_{\ell=i}^{j-1} \hat{f}(t_{i-1}, t_\ell)[t_{\ell+1} - t_\ell]}.$$

Canceling terms that appear on both sides and rearranging the remaining terms brings this into the form

$$\mathsf{E}\left[e^{-\sum_{\ell=i}^{j-1} \hat{\sigma}(t_{i-1}, t_\ell)\sqrt{t_i - t_{i-1}}[t_{\ell+1} - t_\ell] Z_i} \Big| Z_1, \ldots, Z_{i-1}\right]$$
$$= e^{\sum_{\ell=i}^{j-1} \hat{\mu}(t_{i-1}, t_\ell)[t_i - t_{i-1}][t_{\ell+1} - t_\ell]}.$$

The conditional expectation on the left evaluates to

$$e^{\frac{1}{2}\left(\sum_{\ell=i}^{j-1} \hat{\sigma}(t_{i-1}, t_\ell)[t_{\ell+1} - t_\ell]\right)^2 [t_i - t_{i-1}]},$$

so equality holds if

$$\frac{1}{2}\left(\sum_{\ell=i}^{j-1}\hat{\sigma}(t_{i-1},t_\ell)[t_{\ell+1}-t_\ell]\right)^2 = \sum_{\ell=i}^{j-1}\hat{\mu}(t_{i-1},t_\ell)[t_{\ell+1}-t_\ell];$$

i.e., if

$$\hat{\mu}(t_{i-1},t_j)[t_{j+1}-t_j] =$$

$$\frac{1}{2}\left(\sum_{\ell=i}^{j}\hat{\sigma}(t_{i-1},t_\ell)[t_{\ell+1}-t_\ell]\right)^2 - \frac{1}{2}\left(\sum_{\ell=i}^{j-1}\hat{\sigma}(t_{i-1},t_\ell)[t_{\ell+1}-t_\ell]\right)^2 . \quad (3.97)$$

This is the discrete version of the HJM drift; it ensures that the discretized discounted bond prices are martingales.

To see the connection between this expression and the continuous-time drift (3.90), consider the case of an equally spaced grid, $t_i = ih$ for some increment $h > 0$. Fix a date t and maturity T and let $i, j \to \infty$ and $h \to 0$ in such a way that $jh = T$ and $ih = t$; each of the sums in (3.97) is then approximated by an integral. Dividing both sides of (3.97) by $t_{j+1} - t_j = h$, we find that for small h the discrete drift is approximately

$$\frac{1}{2h}\left[\left(\int_t^T \sigma(t,u)\,du\right)^2 - \left(\int_t^{T-h}\sigma(t,u)\,du\right)^2\right] \approx \frac{1}{2}\frac{\partial}{\partial T}\left(\int_t^T\sigma(t,u)\,du\right)^2,$$

which is

$$\sigma(t,T)\int_t^T\sigma(t,u)\,du.$$

This suggests that the discrete drift in (3.97) is indeed consistent with the continuous-time limit in (3.90).

In the derivation leading to (3.97) we assumed a single-factor model. A similar result holds with d factors. Let $\hat{\sigma}_k$ denote the kth entry of the d-vector $\hat{\sigma}$ and

$$\hat{\mu}_k(t_{i-1},t_j)[t_{j+1}-t_j] =$$

$$\frac{1}{2}\left(\sum_{\ell=i}^{j}\hat{\sigma}_k(t_{i-1},t_\ell)[t_{\ell+1}-t_\ell]\right)^2 - \frac{1}{2}\left(\sum_{\ell=i}^{j-1}\hat{\sigma}_k(t_{i-1},t_\ell)[t_{\ell+1}-t_\ell]\right)^2,$$

for $k = 1,\ldots,d$. The combined drift is given by the sum

$$\hat{\mu}(t_{i-1},t_j) = \sum_{k=1}^{d}\hat{\mu}_k(t_{i-1},t_j).$$

A generic multifactor simulation takes the form

$$\hat{f}(t_i,t_j) = \hat{f}(t_{i-1},t_j) + \hat{\mu}(t_{i-1},t_j)[t_i - t_{i-1}]$$

$$+ \sum_{k=1}^{d}\hat{\sigma}_k(t_{i-1},t_j)\sqrt{t_i - t_{i-1}}Z_{ik}, \quad j = i,\ldots,M, \quad (3.98)$$

where the $Z_i = (Z_{i1}, \ldots, Z_{id})$, $i = 1, \ldots, M$, are independent $N(0, I)$ random vectors.

We derived (3.97) by starting from the principle that the discretized discounted bond prices should be martingales. But what are the practical implications of using some other approximation to the continuous drift instead of this one? To appreciate the consequences, consider the following experiment. Imagine simulating paths of $\hat{f}$ as in (3.96) or (3.98). From a path of $\hat{f}$ we may extract a path

$$\hat{r}(t_0) = \hat{f}(t_0, t_0), \quad \hat{r}(t_1) = \hat{f}(t_1, t_1), \quad \ldots \quad \hat{r}(t_M) = \hat{f}(t_M, t_M),$$

of the discretized short rate $\hat{r}$. From this we can calculate a discount factor

$$\hat{D}(t_j) = \exp\left(-\sum_{i=0}^{j-1} \hat{r}(t_i)[t_{i+1} - t_i]\right) \tag{3.99}$$

for each maturity t_j. Imagine repeating this over n independent paths and let $\hat{D}^{(1)}(t_j), \ldots, \hat{D}^{(n)}(t_j)$ denote discount factors calculated over these n paths. A consequence of the strong law of large numbers, the martingale property, and the initialization in (3.95) is that, almost surely,

$$\frac{1}{n}\sum_{i=1}^{n} \hat{D}^{(i)}(t_j) \to \mathsf{E}[\hat{D}(t_j)] = \hat{B}(0, t_j) = B(0, t_j).$$

This means that if we simulate using (3.97) and then use the simulation to price a bond, the simulation price converges to the value to which the model was ostensibly calibrated. With some other choice of discrete drift, the simulation price would in general converge to something that differs from $B(0, t_j)$, even if only slightly. Thus, the martingale condition is not simply a theoretical feature — it is a prerequisite for internal consistency of the simulated model. Indeed, failure of this condition can create the illusion of arbitrage opportunities. If $\mathsf{E}[\hat{D}^{(1)}(t_j)] \neq B(0, t_j)$, the simulation would be telling us that the market has mispriced the bond.

The errors (or apparent arbitrage opportunities) that may arise from using a different approximation to the continuous-time drift may admittedly be quite small. But given that we have a simple way of avoiding such errors and given that the form of the drift is the central feature of the HJM framework, we may as well restrict ourselves to (3.97). This form of the discrete drift appears to be in widespread use in the industry; it is explicit in Andersen [11].

Forward Measure

Through an argument similar to the one leading to (3.97), we can find the appropriate form of the discrete drift under the forward measure. In continuous time, the forward measure for maturity T_F is characterized by the requirement that $B(t,T)/B(t,T_F)$ be a martingale, because the bond maturing at T_F is the numeraire asset associated with this measure. In the discrete approximation, if we take $t_M = T_F$, then we require that $\hat{B}(t_i,t_j)/\hat{B}(t_i,t_M)$ be a martingale in i for each j. This ratio is given by

$$\frac{\hat{B}(t_i,t_j)}{\hat{B}(t_i,t_M)} = \exp\left(\sum_{\ell=j}^{M-1} \hat{f}(t_i,t_\ell)[t_{\ell+1} - t_\ell]\right).$$

The martingale condition leads to a discrete drift $\hat{\mu}$ with

$$\hat{\mu}(t_{i-1},t_j)[t_{j+1} - t_j] =$$

$$\frac{1}{2}\left(\sum_{\ell=j+1}^{M-1} \hat{\sigma}(t_{i-1},t_\ell)[t_{\ell+1} - t_\ell]\right)^2 - \frac{1}{2}\left(\sum_{\ell=j}^{M-1} \hat{\sigma}(t_{i-1},t_\ell)[t_{\ell+1} - t_\ell]\right)^2. \quad (3.100)$$

The relation between this and the risk-neutral discrete drift (3.97) is, not surprisingly, similar to the relation between their continuous-time counterparts in (3.91) and (3.93).

3.6.3 Implementation

Once we have identified the discrete form of the drift, the main consideration in implementing an HJM simulation is keeping track of indices. The notation $\hat{f}(t_i,t_j)$ is convenient in describing the discretized model — the first argument shows the current time, the second argument shows the maturity to which this forward rate applies. But in implementing the simulation we are not interested in keeping track of an $M \times M$ matrix of rates as the notation $\hat{f}(t_i,t_j)$ might suggest. At each time step, we need only the vector of current rates. To implement an HJM simulation we need to adopt some conventions regarding the indexing of this vector.

Recall that our time and maturity grid consists of a set of dates $0 = t_0 < t_1 < \cdots < t_M$. If we identify t_M with the ultimate maturity T^* in the continuous-time model, then t_M is the maturity of the longest-maturity bond represented in the model. In light of (3.94), this means that the last forward rate relevant to the model applies to the interval $[t_{M-1},t_M]$; this is the forward rate with maturity argument t_{M-1}. Thus, our initial vector of forward rates consists of the M components $\hat{f}(0,0), \hat{f}(0,t_1), \ldots, \hat{f}(0,t_{M-1})$, which is consistent with the initialization (3.95). At the start of the simulation we will represent this vector as $(f_1, \ldots, f_M)$. Thus, our first convention is to use 1 rather than 0 as the lowest index value.

As the simulation evolves, the number of relevant rates decreases. At time t_i, only the rates $\hat{f}(t_i,t_i), \ldots, \hat{f}(t_i,t_{M-1})$ are meaningful. We need to specify how these $M - i$ rates should be indexed, given that initially we had a vector of M rates: we can either pad the initial portion of the vector with irrelevant data or we can shorten the length of the vector. We choose the latter and represent the $M - i$ rates remaining at t_i as the vector $(f_1, \ldots, f_{M-i})$. Thus, our second convention is to index forward rates by *relative* maturity rather than absolute maturity. At time t_i, f_j refers to the forward rate $\hat{f}(t_i, t_{i+j-1})$. Under this convention f_1 always refers to the current level of the short rate because $\hat{r}(t_i) = \hat{f}(t_i, t_i)$.

Similar considerations apply to $\hat{\mu}(t_i, t_j)$ and $\hat{\sigma}_k(t_i, t_j)$, $k = 1, \ldots, d$, and we adopt similar conventions for the variables representing these terms. For values of $\hat{\mu}$ we use variables m_j and for values of $\hat{\sigma}_k$ we use variables $s_j(k)$; in both cases the subscript indicates a relative maturity and in the case of $s_j(k)$ the argument $k = 1, \ldots, d$ refers the factor index in a d-factor model. We design the indexing so that the simulation step from t_{i-1} to t_i indicated in (3.98) becomes

$$f_j \leftarrow f_{j+1} + m_j[t_i - t_{i-1}] + \sum_{k=1}^{d} s_j(k)\sqrt{t_i - t_{i-1}} Z_{ik}, \quad j = 1, \ldots, M - i.$$

Thus, in advancing from t_{i-1} to t_i we want

$$m_j = \hat{\mu}(t_{i-1}, t_{i+j-1}), \quad s_j(k) = \hat{\sigma}_k(t_{i-1}, t_{i+j-1}). \tag{3.101}$$

In particular, recall that $\hat{\sigma}$ may depend on the current vector of forward rates; as implied by (3.101), the values of all $s_j(k)$ should be determined before the forward rates are updated.

To avoid repeated calculation of the intervals between dates t_i, we introduce the notation

$$h_i = t_i - t_{i-1}, \quad , i = 1, \ldots, M.$$

These values do not change in the course of a simulation so we use the vector $(h_1, \ldots, h_M)$ to represent these same values at all steps of the simulation.

We now proceed to detail the steps in an HJM simulation. We separate the algorithm into two parts, one calculating the discrete drift parameter at a fixed time step, the other looping over time steps and updating the forward curve at each step. Figure 3.16 illustrates the calculation of

$$\hat{\mu}_k(t_{i-1}, t_j) =$$

$$\frac{1}{2h_j} \left[\sum_{k=1}^{d} \left(\sum_{\ell=i}^{j} \hat{\sigma}_k(t_{i-1}, t_\ell) h_{\ell+1} \right)^2 - \sum_{k=1}^{d} \left(\sum_{\ell=i}^{j-1} \hat{\sigma}_k(t_{i-1}, t_\ell) h_{\ell+1} \right)^2 \right]$$

in a way that avoids duplicate computation. In the notation of the algorithm, this drift parameter is evaluated as

$$\frac{1}{2(t_{j+1} - t_j)} [B_{\text{next}} - B_{\text{prev}}],$$

and each $A_{\text{next}}(k)$ records a quantity of the form

$$\sum_{\ell=i}^{j} \hat{\sigma}_k(t_{i-1}, t_\ell) h_{\ell+1}.$$

Inputs: $s_j(k)$, $j = 1, \ldots, M - i$, $k = 1, \ldots, d$ as in (3.101)
 and $h_1, \ldots, h_M$ ($h_\ell = t_\ell - t_{\ell-1}$)

$A_{\text{prev}}(k) \leftarrow 0$, $k = 1, \ldots, d$
for $j = 1, \ldots, M - i$
    $B_{\text{next}} \leftarrow 0$
    for $k = 1, \ldots, d$
        $A_{\text{next}}(k) \leftarrow A_{\text{prev}}(k) + s_j(k) * h_{i+j}$
        $B_{\text{next}} \leftarrow B_{\text{next}} + A_{\text{next}}(k) * A_{\text{next}}(k)$
        $A_{\text{prev}}(k) \leftarrow A_{\text{next}}(k)$
    end
    $m_j \leftarrow (B_{\text{next}} - B_{\text{prev}})/(2h_{i+j})$
    $B_{\text{prev}} \leftarrow B_{\text{next}}$
end
return $m_1, \ldots, m_{M-i}$.

Fig. 3.16. Calculation of discrete drift parameters $m_j = \hat{\mu}(t_{i-1}, t_{i+j-1})$ needed to simulate transition from t_{i-1} to t_i.

Figure 3.17 shows an algorithm for a single replication in an HJM simulation; the steps in the figure would naturally be repeated over many independent replications. This algorithm calls the one in Figure 3.16 to calculate the discrete drift for all remaining maturities at each time step. The two algorithms could obviously be combined, but keeping them separate should help clarify the various steps. In addition, it helps stress the point that in propagating the forward curve from t_{i-1} to t_i, we first evaluate the $s_j(k)$ and m_j using the forward rates at step $i - 1$ and then update the rates to get their values at step i.

To make this point a bit more concrete, suppose we specified a single-factor model with $\hat{\sigma}(t_i, t_j) = \tilde{\sigma}(i, j) \hat{f}(t_i, t_j)$ for some fixed values $\tilde{\sigma}(i, j)$. This makes each $\hat{\sigma}(t_i, t_j)$ proportional to the corresponding forward rate. We noted in Example 3.6.3 that this type of diffusion term is inadmissible in the continuous-time limit, but it can be (and often is) used in practice so long as the increments h_i are kept bounded away from zero. In this model it should be clear that in updating $\hat{f}(t_{i-1}, t_j)$ to $\hat{f}(t_i, t_j)$ we need to evaluate $\tilde{\sigma}(i-1, j) \hat{f}(t_{i-1}, t_j)$ before we update the forward rate.

Since an HJM simulation is typically used to value interest rate derivatives, we have included in Figure 3.17 a few additional generic steps illustrating how

---

Inputs: initial curve $(f_1, \ldots, f_M)$ and intervals $(h_1, \ldots, h_M)$

$D \leftarrow 1, P \leftarrow 0, C \leftarrow 0.$
for $i = 1, \ldots, M - 1$
    $D \leftarrow D * \exp(-f_1 * h_i)$
    evaluate $s_j(k)$, $j = 1, \ldots, M - i$, $k = 1, \ldots, d$
        (recall that $s_j(k) = \hat{\sigma}_k(t_{i-1}, t_{i+j-1})$)
    evaluate $m_1, \ldots, m_{M-i}$ using Figure 3.16
    generate $Z_1, \ldots, Z_d \sim N(0, 1)$
    for $j = 1, \ldots, M - i$
        $S \leftarrow 0$
        for $k = 1, \ldots, d$ $S \leftarrow S + s_j(k) * Z_k$
        $f_j \leftarrow f_{j+1} + m_j * h_i + S * \sqrt{h_i}$
    end
    $P \leftarrow$ cashflow at t_i (depending on instrument)
    $C \leftarrow C + D * P$
end
return C.

Fig. 3.17. Algorithm to simulate evolution of forward curve over $t_0, t_1, \ldots, t_{M-1}$ and calculate cumulative discounted cashflows from an interest rate derivative.

a path of the forward curve is used both to compute and to discount the payoff of a derivative. The details of a particular instrument are subsumed in the placeholder "cashflow at t_i." This cashflow is discounted through multiplication by D, which is easily seen to contain the simulated value of the discount factor $\hat{D}(t_i)$ as defined in (3.99). (When D is updated in Figure 3.17, before the forward rates are updated, f_1 records the short rate for the interval $[t_{i-1}, t_i]$.) To make the pricing application more explicit, we consider a few examples.

Example 3.6.4 *Bonds.* There is no reason to use an HJM simulation to price bonds — if properly implemented, the simulation will simply return prices that could have been computed from the initial forward curve. Nevertheless, we consider this example to help fix ideas. We discussed the pricing of a zero-coupon bond following (3.99); in Figure 3.17 this corresponds to setting $P \leftarrow 1$ at the maturity of the bond and $P \leftarrow 0$ at all other dates. For a coupon paying bond with a face value of 100 and a coupon of c, we would set $P \leftarrow c$ at the coupon dates and $P \leftarrow 100 + c$ at maturity. This assumes, of course, that the coupon dates are among the $t_1, \ldots, t_M$. $\square$

Example 3.6.5 *Caps.* A *caplet* is an interest rate derivative providing protection against an increase in an interest rate for a single period; a *cap* is a portfolio of caplets covering multiple periods. A caplet functions almost like a call option on the short rate, which would have a payoff of the form

$(r(T) - K)^+$ for some strike K and maturity T. In practice, a caplet differs from this in some small but important ways. (For further background, see Appendix C.)

In contrast to the instantaneous short rate $r(t)$, the underlying rate in a caplet typically applies over an interval and is based on discrete compounding. For simplicity, suppose the interval is of the form $[t_i, t_{i+1}]$. At t_i, the continuously compounded rate for this interval is $\hat{f}(t_i, t_i)$; the corresponding discretely compounded rate $\hat{F}$ satisfies

$$\frac{1}{1 + \hat{F}(t_i)[t_{i+1} - t_i]} = e^{-\hat{f}(t_i, t_i)[t_{i+1} - t_i]};$$

i.e.,

$$\hat{F}(t_i) = \frac{1}{t_{i+1} - t_i} \left(e^{\hat{f}(t_i, t_i)[t_{i+1} - t_i]} - 1 \right).$$

The payoff of the caplet would then be $(\hat{F}(t_i) - K)^+$ (or a constant multiple of this). Moreover, this payment is ordinarily made at the end of the interval, t_{i+1}. To discount it properly we should therefore simulate to t_i and set

$$P \leftarrow \frac{1}{1 + \hat{F}(t_i)[t_{i+1} - t_i]} (\hat{F}(t_i) - K)^+; \qquad (3.102)$$

in the notation of Figure 3.17, this is

$$P \leftarrow e^{-f_1 h_{i+1}} \left(\frac{1}{h_{i+1}} \left(e^{f_1 h_{i+1}} - 1 \right) - K \right)^+.$$

Similar ideas apply if the caplet covers an interval longer than a single simulation interval. Suppose the caplet applies to an interval $[t_i, t_{i+n}]$. Then (3.102) still applies at t_i, but with t_{i+1} replaced by t_{i+n} and $\hat{F}(t_i)$ redefined to be

$$\hat{F}(t_i) = \frac{1}{t_{n+i} - t_i} \left(\exp \left(\sum_{\ell=0}^{n-1} \hat{f}(t_i, t_{i+\ell})[t_{i+\ell+1} - t_{i+\ell}] \right) - 1 \right).$$

In the case of a cap consisting of caplets for, say, the periods $[t_{i_1}, t_{i_2}]$, $[t_{i_2}, t_{i_3}]$, $\ldots$, $[t_{i_k}, t_{i_{k+1}}]$, for some $i_1 < i_2 < \cdots < i_{k+1}$, this calculation would be repeated and a cashflow recorded at each t_{i_j}, $j = 1, \ldots, k$. □

Example 3.6.6 *Swaptions.* Consider, next, an option to swap fixed-rate payments for floating-rate payments. (See Appendix C for background on swaps and swaptions.) Suppose the underlying swap begins at t_{j_0} with payments to be exchanged at dates $t_{j_1}, \ldots, t_{j_n}$. If we denote the fixed rate in the swap by R, then the fixed-rate payment at t_{j_k} is $100R[t_{j_k} - t_{j_{k-1}}]$, assuming a principal or *notional* amount of 100. As explained in Section C.2 of Appendix C, the value of the swap at t_{j_0} is

$$\hat{V}(t_{j_0}) = 100 \left(R \sum_{\ell=1}^{n} \hat{B}(t_{j_0}, t_{j_\ell})[t_{j_\ell} - t_{j_{\ell-1}}] + \hat{B}(t_{j_0}, t_{j_n}) - 1 \right).$$

The bond prices $\hat{B}(t_{j_0}, t_{j_\ell})$ can be computed from the forward rates at t_{j_0} using (3.94).

The holder of an option to enter this swap will exercise the option if $\hat{V}(t_{j_0}) > 0$ and let it expire otherwise. (For simplicity, we are assuming that the option expires at t_{j_0} though similar calculations apply for an option to enter into a *forward* swap, in which case the option expiration date would be prior to t_{j_0}.) Thus, we may view the swaption as having a payoff of $\max\{0, \hat{V}(t_{j_0})\}$ at t_{j_0}. In a simulation, we would therefore simulate the forward curve to the option expiration date t_{j_0}; at that date, calculate the prices of the bonds $\hat{B}(t_{j_0}, t_{j_\ell})$ maturing at the payment dates of the swaps; from the bond prices calculate the value of the swap $\hat{V}(t_{j_0})$ and thus the swaption payoff $\max\{0, \hat{V}(t_{j_0})\}$; record this as the cashflow P in the algorithm of Figure 3.17 and discount it as in the algorithm.

This example illustrates a general feature of the HJM framework that contrasts with models based on the short rate as in Sections 3.3 and 3.4. Consider valuing a 5-year option on a 20-year swap. This instrument involves maturities as long as 25 years, so valuing it in a model of the short rate could involve simulating paths over a 25-year horizon. In the HJM framework, if the initial forward curve extends for 25 years, then we need to simulate only for 5 years; at the expiration of the option, the remaining forward rates contain all the information necessary to value the underlying swap. Thus, although the HJM setting involves updating many more variables at each time step, it may also require far fewer time steps. □

3.7 Forward Rate Models: Simple Rates

The models considered in this section are closely related to the HJM framework of the previous section in that they describe the arbitrage-free dynamics of the term structure of interest rates through the evolution of forward rates. But the models we turn to now are based on *simple* rather than continuously compounded forward rates. This seemingly minor shift in focus has surprisingly far-reaching practical and theoretical implications. This modeling approach has developed primarily through the work of Miltersen, Sandmann, and Sondermann [268], Brace, Gatarek, and Musiela [56], Musiela and Rutkowski [274], and Jamshidian [197]; it has gained rapid acceptance in the financial industry and stimulated a growing stream of research into what are often called *LIBOR market models*.

3.7.1 LIBOR Market Model Dynamics

The basic object of study in the HJM framework is the forward rate curve $\{f(t,T), t \leq T \leq T^*\}$. But the instantaneous, continuously compounded forward rates $f(t,T)$ might well be considered mathematical idealizations — they are not directly observable in the marketplace. Most market interest rates are based on simple compounding over intervals of, e.g., three months or six months. Even the instantaneous short rate $r(t)$ treated in the models of Sections 3.3 and 3.4 is a bit of a mathematical fiction because short-term rates used for pricing are typically based on periods of one to three months. The term "market model" is often used to describe an approach to interest rate modeling based on observable market rates, and this entails a departure from instantaneous rates.

Among the most important benchmark interest rates are the London Inter-Bank Offered Rates or LIBOR. LIBOR is calculated daily through an average of rates offered by banks in London. Separate rates are quoted for different maturities (e.g., three months and six months) and different currencies. Thus, each day new values are calculated for three-month Yen LIBOR, six-month US dollar LIBOR, and so on.

LIBOR rates are based on *simple* interest. If L denotes the rate for an accrual period of length δ (think of δ as $1/4$ or $1/2$ for three months and six months respectively, with time measured in years), then the interest earned on one unit of currency over the accrual period is δL. For example, if three-month LIBOR is 6%, the interest earned at the end of three months on a principal of 100 is $0.25 \cdot 0.06 \cdot 100 = 1.50$.

A *forward* LIBOR rate works similarly. Fix δ and consider a maturity T. The forward rate $L(0,T)$ is the rate set at time 0 for the interval $[T, T+\delta]$. If we enter into a contract at time 0 to borrow 1 at time T and repay it with interest at time $T + \delta$, the interest due will be $\delta L(0,T)$. As shown in Appendix C (specifically equation (C.5)), a simple replication argument leads to the following identity between forward LIBOR rates and bond prices:

$$L(0,T) = \frac{B(0,T) - B(0, T+\delta)}{\delta B(0, T+\delta)}. \tag{3.103}$$

This further implies the relation

$$L(0,T) = \frac{1}{\delta}\left(\exp\left(\int_T^{T+\delta} f(0,u)\,du\right) - 1\right) \tag{3.104}$$

between continuous and simple forwards, though it is not necessary to introduce the continuous rates to build a model based on simple rates.

It should be noted that, as is customary in this literature, we treat the forward LIBOR rates as though they were risk-free rates. LIBOR rates are based on quotes by banks which could potentially default and this risk is presumably reflected in the rates. US Treasury bonds, in contrast, are generally considered to have a negligible chance of default. The argument leading to (3.103) may not hold exactly if the bonds on one side and the forward rate on the other reflect different levels of creditworthiness. We will not, however, attempt to take account of these considerations.

Although (3.103) and (3.104) apply in principle to a continuum of maturities T, we consider a class of models in which a finite set of maturities or *tenor dates*

$$0 = T_0 < T_1 < \cdots < T_M < T_{M+1}$$

are fixed in advance. As argued in Jamshidian [197], many derivative securities tied to LIBOR and swap rates are sensitive only to a finite set of maturities and it should not be necessary to introduce a continuum to price and hedge these securities. Let

$$\delta_i = T_{i+1} - T_i, \quad i = 0, \ldots, M,$$

denote the lengths of the intervals between tenor dates. Often, these would all be equal to a nominally fixed interval of a quarter or half year; but even in this case, day-count conventions would produce slightly different values for the fractions δ_i.

For each date T_n we let $B_n(t)$ denote the time-t price of a bond maturing at T_n, $0 \le t \le T_n$. In our usual notation this would be $B(t, T_n)$, but writing $B_n(t)$ and restricting n to $\{1, 2, \ldots, M+1\}$ emphasizes that we are working with a finite set of bonds. Similarly, we write $L_n(t)$ for the forward rate as of time t for the accrual period $[T_n, T_{n+1}]$; see Figure 3.18. This is given in terms of the bond prices by

$$L_n(t) = \frac{B_n(t) - B_{n+1}(t)}{\delta_n B_{n+1}(t)}, \quad 0 \le t \le T_n, \quad n = 0, 1, \ldots, M. \tag{3.105}$$

After T_n, the forward rate L_n becomes meaningless; it sometimes simplifies notation to extend the definition of $L_n(t)$ beyond T_n by setting $L_n(t) = L_n(T_n)$ for all $t \ge T_n$.

From (3.105) we know that bond prices determine the forward rates. At a tenor date T_i, the relation can be inverted to produce

$$B_n(T_i) = \prod_{j=i}^{n-1} \frac{1}{1 + \delta_j L_j(T_i)}, \quad n = i+1, \ldots, M+1. \tag{3.106}$$

However, at an arbitrary date t, the forward LIBOR rates do not determine the bond prices because they do not determine the discount factor for intervals shorter than the accrual periods. Suppose for example that $T_i < t < T_{i+1}$ and we want to find the price $B_n(t)$ for some $n > i+1$. The factor

$$\prod_{j=i+1}^{n-1} \frac{1}{1 + \delta_j L_j(t)}$$

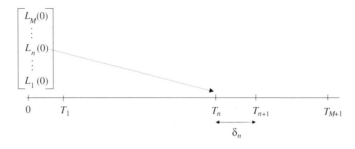

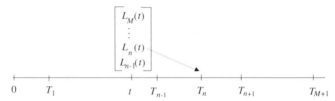

Fig. 3.18. Evolution of vector of forward rates. Each $L_n(t)$ is the forward rate for the interval $[T_n, T_{n+1}]$ as of time $t \leq T_n$.

discounts the bond's payment at T_n back to time T_{i+1}, but the LIBOR rates do not specify the discount factor from T_{i+1} to t.

Define a function $\eta : [0, T_{M+1}) \to \{1, \ldots, M+1\}$ by taking $\eta(t)$ to be the unique integer satisfying

$$T_{\eta(t)-1} \leq t < T_{\eta(t)};$$

thus, $\eta(t)$ gives the index of the next tenor date at time t. With this notation, we have

$$B_n(t) = B_{\eta(t)}(t) \prod_{j=\eta(t)}^{n-1} \frac{1}{1+\delta_j L_j(t)}, \quad 0 \leq t < T_n; \qquad (3.107)$$

the factor $B_{\eta(t)}(t)$ (the current price of the shortest maturity bond) is the missing piece required to express the bond prices in terms of the forward LIBOR rates.

Spot Measure

We seek a model in which the evolution of the forward LIBOR rates is described by a system of SDEs of the form

$$\frac{dL_n(t)}{L_n(t)} = \mu_n(t)\, dt + \sigma_n(t)^\top dW(t), \quad 0 \leq t \leq T_n, \quad n = 1, \ldots, M, \qquad (3.108)$$

with W a d-dimensional standard Brownian motion. The coefficients μ_n and σ_n may depend on the current vector of rates $(L_1(t), \ldots, L_M(t))$ as well as the current time t. Notice that in (3.108) σ_n is the (proportional) volatility because we have divided by L_n on the left, whereas in the HJM setting (3.91) we took $\sigma(t, T)$ to be the absolute level of volatility. At this point, the distinction is purely one of notation rather than scope because we allow $\sigma_n(t)$ to depend on the current level of rates.

Recall that in the HJM setting we derived the form of the drift of the forward rates from the absence of arbitrage. More specifically, we derived the drift from the condition that bond prices be martingales when divided by the numeraire asset. The numeraire we used is the usual one associated with the risk-neutral measure, $\beta(t) = \exp(\int_0^t r(u)\, du)$. But introducing a short-rate process $r(t)$ would undermine our objective of developing a model based on the simple (and thus more realistic) rates $L_n(t)$. We therefore avoid the usual risk-neutral measure and instead use a numeraire asset better suited to the tenor dates T_i.

A simply compounded counterpart of $\beta(t)$ works as follows. Start with 1 unit of account at time 0 and buy $1/B_1(0)$ bonds maturing at T_1. At time T_1, reinvest the funds in bonds maturing at time T_2 and proceed this way, at each T_i putting all funds in bonds maturing at time T_{i+1}. This trading strategy earns (simple) interest at rate $L_i(T_i)$ over each interval $[T_i, T_{i+1}]$, just as in the continuously compounded case a savings account earns interest at rate $r(t)$ at time t. The initial investment of 1 at time 0 grows to a value of

$$B^*(t) = B_{\eta(t)}(t) \prod_{j=0}^{\eta(t)-1} [1 + \delta_j L_j(T_j)]$$

at time t. Following Jamshidian [197], we take this as numeraire asset and call the associated measure the *spot measure*.

Suppose, then, that (3.108) holds under the spot measure, meaning that W is a standard Brownian motion under that measure. The absence of arbitrage restricts the dynamics of the forward LIBOR rates through the condition that bond prices be martingales when *deflated* by the numeraire asset. (We use the term "deflated" rather than "discounted" to emphasize that we are dividing by the numeraire asset and not discounting at a continuously compounded rate.) From (3.107) and the expression for B^*, we find that the deflated bond price $D_n(t) = B_n(t)/B^*(t)$ is given by

$$D_n(t) = \left(\prod_{j=0}^{\eta(t)-1} \frac{1}{1 + \delta_j L_j(T_j)} \right) \prod_{j=\eta(t)}^{n-1} \frac{1}{1 + \delta_j L_j(t)}, \quad 0 \le t \le T_n. \quad (3.109)$$

Notice that the spot measure numeraire B^* cancels the factor $B_{\eta(t)}(t)$ used in (3.107) to discount between tenor dates. We are thus left in (3.109) with an expression defined purely in terms of the LIBOR rates. This would not have been the case had we divided by the risk-neutral numeraire asset $\beta(t)$.

We require that the deflated bond prices D_n be positive martingales and proceed to derive the restrictions this imposes on the LIBOR dynamics (3.108). If the deflated bonds are indeed positive martingales, we may write

$$\frac{dD_{n+1}(t)}{D_{n+1}(t)} = \nu_{n+1}(t)^{\top} dW(t), \quad n = 1, \ldots, M,$$

for some $\Re^d$-valued processes ν_{n+1} which may depend on the current level of $(D_2, \ldots, D_{M+1})$ (equivalently, of $(L_1, \ldots, L_M)$). By Itô's formula,

$$d \log D_{n+1}(t) = -\tfrac{1}{2} \|\nu_{n+1}(t)\| dt + \nu_{n+1}^{\top}(t) dW(t).$$

We may therefore express ν_{n+1} by finding the coefficient of dW in

$$d \log D_{n+1}(t) = -\sum_{j=\eta(t)}^{n} d \log(1 + \delta_j L_j(t));$$

notice that the first factor in (3.109) is constant between maturities T_i. Applying Itô's formula and (3.108), we find that

$$\nu_{n+1}(t) = -\sum_{j=\eta(t)}^{n} \frac{\delta_j L_j(t)}{1 + \delta_j L_j(t)} \sigma_j(t). \tag{3.110}$$

We now proceed by induction to find the μ_n in (3.108). Setting $D_1(t) \equiv B_1(0)$, we make D_1 constant and hence a martingale without restrictions on any of the LIBOR rates. Suppose now that $\mu_1, \ldots, \mu_{n-1}$ have been chosen consistent with the martingale condition on D_n. From the identity $D_n(t) = D_{n+1}(1 + \delta_n L_n(t))$, we find that $\delta_n L_n(t) D_{n+1}(t) = D_n(t) - D_{n+1}(t)$, so D_{n+1} is a martingale if and only if $L_n D_{n+1}$ is a martingale. Applying Itô's formula, we get

$$
\begin{aligned}
d(L_n D_{n+1}) \\
= D_{n+1} \, dL_n + L_n \, dD_{n+1} + L_n D_{n+1} \nu_{n+1}^{\top} \sigma_n \, dt \\
= \left(D_{n+1} \mu_n L_n + L_n D_{n+1} \nu_{n+1}^{\top} \sigma_n \right) dt + L_n D_{n+1} \sigma_n^{\top} \, dW + L_n \, dD_{n+1}.
\end{aligned}
$$

(We have suppressed the time argument to lighten the notation.) To be consistent with the martingale restriction on D_{n+1} and $L_n D_{n+1}$, the dt coefficient must be zero, and thus

$$\mu_n = -\sigma_n^{\top} \nu_{n+1};$$

notice the similarity to the HJM drift (3.90). Combining this with (3.110), we arrive at

$$\mu_n(t) = \sum_{j=\eta(t)}^{n} \frac{\delta_j L_j(t) \sigma_n(t)^{\top} \sigma_j(t)}{1 + \delta_j L_j(t)} \tag{3.111}$$

as the required drift parameter in (3.108), so

$$\frac{dL_n(t)}{L_n(t)} = \sum_{j=\eta(t)}^{n} \frac{\delta_j L_j(t) \sigma_n(t)^{\top} \sigma_j(t)}{1 + \delta_j L_j(t)} dt + \sigma_n(t)^{\top} dW(t), \quad 0 \leq t \leq T_n, \tag{3.112}$$

$n = 1, \ldots, M$, describes the arbitrage-free dynamics of forward LIBOR rates under the spot measure. This formulation is from Jamshidian [197], which should be consulted for a rigorous and more general development.

Forward Measure

As in Musiela and Rutkowski [274], we may alternatively formulate a LIBOR market model under the forward measure P_{M+1} for maturity T_{M+1} and take the bond B_{M+1} as numeraire asset. In this case, we redefine the deflated bond prices to be the ratios $D_n(t) = B_n(t)/B_{M+1}(t)$, which simplify to

$$D_n(t) = \prod_{j=n+1}^{M} (1 + \delta_j L_j(t)). \tag{3.113}$$

Notice that the numeraire asset has once again canceled the factor $B_{\eta(t)}(t)$, leaving an expression that depends solely on the forward LIBOR rates.

We could derive the dynamics of the forward LIBOR rates under the forward measure through the Girsanov Theorem and (3.112), much as we did in the HJM setting to arrive at (3.93). Alternatively, we could start from the requirement that the D_n in (3.113) be martingales and proceed by induction (backwards from $n = M$) to derive restrictions on the evolution of the L_n. Either way, we find that the arbitrage-free dynamics of the L_n, $n = 1, \ldots, M$, under the forward measure P_{M+1} are given by

$$\frac{dL_n(t)}{L_n(t)} = -\sum_{j=n+1}^{M} \frac{\delta_j L_j(t)\sigma_n(t)^\top \sigma_j(t)}{1 + \delta_j L_j(t)}\, dt + \sigma_n(t)^\top\, dW^{M+1}(t), \quad 0 \le t \le T_n, \tag{3.114}$$

with W^{M+1} a standard d-dimensional Brownian motion under P_{M+1}. The relation between the drift in (3.114) and the drift in (3.112) is analogous to the relation between the risk-neutral and forward-measure drifts in the HJM setting; compare (3.90) and (3.93).

If we take $n = M$ in (3.114), we find that

$$\frac{dL_M(t)}{L_M(t)} = \sigma_M(t)^\top\, dW^{M+1}(t),$$

so that, subject only to regularity conditions on its volatility, L_M is a martingale under the forward measure for maturity T_{M+1}. Moreover, if σ_M is deterministic then $L_M(t)$ has lognormal distribution $LN(-\bar\sigma_M^2(t)/2, \bar\sigma_M^2(t))$ with

$$\bar\sigma_M(t) = \sqrt{\frac{1}{t}\int_0^t \|\sigma_M(u)\|^2\, du}. \tag{3.115}$$

In fact, the choice of M is arbitrary: each L_n is a martingale (lognormal if σ_n is deterministic) under the forward measure P_{n+1} associated with T_{n+1}.

These observations raise the question of whether we may in fact take the coefficients σ_n to be deterministic in (3.112) and (3.114). Recall from Example 3.6.3 that this choice (deterministic proportional volatility) is inadmissible in the HJM setting, essentially because it makes the HJM drift quadratic in the current level of rates. To see what happens with simple compounding, rewrite (3.112) as

$$dL_n(t) = \sum_{j=\eta(t)}^{n} \frac{\delta_j L_j(t) L_n(t) \sigma_n(t)^\top \sigma_j(t)}{1 + \delta_j L_j(t)} \, dt + L_n(t) \sigma_n(t)^\top \, dW(t) \quad (3.116)$$

and consider the case of deterministic σ_i. The numerators in the drift are quadratic in the forward LIBOR rates, but they are stabilized by the terms $1 + \delta_j L_j(t)$ in the denominators; indeed, because $L_j(t) \geq 0$ implies

$$\left| \frac{\delta_j L_j(t)}{1 + \delta_j L_j(t)} \right| \leq 1,$$

the drift is linearly bounded in $L_n(t)$, making deterministic σ_i admissible. This feature is lost in the limit as the compounding period δ_j decreases to zero. Thus, the distinction between continuous and simple forward rates turns out to have important mathematical as well as practical implications.

3.7.2 Pricing Derivatives

We have noted two important features of LIBOR market models: they are based on observable market rates, and (in contrast to the HJM framework) they admit deterministic volatilities σ_j. A third important and closely related feature arises in the pricing of interest rate caps.

Recall from Example 3.6.5 (or Appendix C.2) that a cap is a collection of caplets and that each caplet may be viewed as a call option on a simple forward rate. Consider, then, a caplet for the accrual period $[T_n, T_{n+1}]$. The underlying rate is L_n and the value $L_n(T_n)$ is fixed at T_n. With a strike of K, the caplet's payoff is $\delta_n(L_n(T_n) - K)^+$; think of the caplet as refunding the amount by which interest paid at rate $L_n(T_n)$ exceeds interest paid at rate K. This payoff is made at T_{n+1}.

Let $C_n(t)$ denote the price of this caplet at time t; we know the terminal value $C_n(T_{n+1}) = \delta_n(L_n(T_n) - K)^+$ and we want to find the initial value $C_n(0)$. Under the spot measure, the deflated price $C_n(t)/B^*(t)$ must be a martingale, so

$$C_n(0) = B^*(0)\mathsf{E}^* \left[\frac{\delta_n(L_n(T_n) - K)^+}{B^*(T_{n+1})} \right],$$

where we have written E^* for expectation under the spot measure. Through $B^*(T_{n+1})$, this expectation involves the joint distribution of $L_1(T_1)$, ..., $L_n(T_n)$, making its value difficult to discern. In contrast, under the forward measure P_{n+1} associated with maturity T_{n+1}, the martingale property applies to $C_n(t)/B_{n+1}(t)$. We may therefore also write

$$C_n(0) = B_{n+1}(0)\mathsf{E}_{n+1}\left[\frac{\delta_n(L_n(T_n) - K)^+}{B_{n+1}(T_{n+1})}\right],$$

with E_{n+1} denoting expectation under P_{n+1}. Conveniently, $B_{n+1}(T_{n+1}) \equiv 1$, so this expectation depends only on the marginal distribution of $L_n(T_n)$. If we take σ_n to be deterministic, then $L_n(T_n)$ has the lognormal distribution $LN(-\bar{\sigma}_n^2(T_n)/2, \bar{\sigma}_n^2(T_n))$, using the notation in (3.115). In this case, the caplet price is given by the *Black formula* (after Black [49]),

$$C_n(0) = \mathrm{BC}(L_n(0), \bar{\sigma}_n(T_n), T_n, K, \delta_n B_{n+1}(0)),$$

with

$$\mathrm{BC}(F, \sigma, T, K, b) =$$
$$b\left(F\Phi\left(\frac{\log(F/K) + \sigma^2 T/2}{\sigma\sqrt{T}}\right) - K\Phi\left(\frac{\log(F/K) - \sigma^2 T/2}{\sigma\sqrt{T}}\right)\right) \quad (3.117)$$

and Φ the cumulative normal distribution. Thus, under the assumption of deterministic volatilities, caplets are priced in closed form by the Black formula.

This formula is frequently used in the reverse direction. Given the market price of a caplet, one can solve for the "implied volatility" that makes the formula match the market price. This is useful in calibrating a model to market data, a point we return to in Section 3.7.4.

Consider, more generally, a derivative security making a payoff of $g(L(T_n))$ at T_k, with $L(T_n) = (L_1(T_1), \ldots, L_{n-1}(T_{n-1}), L_n(T_n), \ldots, L_M(T_n))$ and $k \geq n$. The price of the derivative at time 0 is given by

$$\mathsf{E}^*\left[\frac{g(L(T_n))}{B^*(T_k)}\right]$$

(using the fact that $B^*(0) = 1$), and also by

$$B_m(0)\mathsf{E}_m\left[\frac{g(L(T_n))}{B_m(T_k)}\right]$$

for every $m \geq k$. Which measure and numeraire are most convenient depends on the payoff function g. However, in most cases, the expectation cannot be evaluated explicitly and simulation is required.

As a further illustration, we consider the pricing of a swaption as described in Example 3.6.6 and Appendix C.2. Suppose the underlying swap begins at T_n with fixed- and floating-rate payments exchanged at $T_{n+1}, \ldots, T_{M+1}$. From equation (C.7) in Appendix C, we find that the forward swap rate at time t is given by

$$S_n(t) = \frac{B_n(t) - B_{M+1}(t)}{\sum_{j=n+1}^{M+1} \delta_j B_j(t)}. \quad (3.118)$$

Using (3.107) and noting that $B_{\eta(t)}(t)$ cancels from the numerator and denominator, this swap rate can be expressed purely in terms of forward LIBOR rates.

Consider, now, an option expiring at time $T_k \leq T_n$ to enter into the swap over $[T_n, T_{M+1}]$ with fixed rate T. The value of the option at expiration can be expressed as (cf. equation (C.11))

$$\sum_{j=n+1}^{M+1} \delta_j B_j(T_k)(R - S_n(T_k))^+.$$

This can be written as a function $g(L(T_k))$ of the LIBOR rates. The price at time zero can therefore be expressed as an expectation using the general expressions above.

By applying Itô's formula to the swap rate (3.118), it is not difficult to conclude that if the forward LIBOR rates have deterministic volatilities, then the forward swap rate cannot also have a deterministic volatility. In particular, then, the forward swap rate cannot be geometric Brownian motion under any equivalent measure. Brace et al. [56] nevertheless use a lognormal approximation to the swap rate to develop a method for pricing swaptions; their approximation appears to give excellent results. An alternative approach has been developed by Jamshidian [197]. He develops a model in which the term structure is described through a vector $(S_0(t), \ldots, S_M(t))$ of forward swap rates. He shows that one may choose the volatilities of the forward swap rates to be deterministic, and that in this case swaption prices are given by a variant of the Black formula. However, in this model, the LIBOR rates cannot also have deterministic volatilities, so caplets are no longer priced by the Black formula. One must therefore choose between the two pricing formulas.

3.7.3 Simulation

Pricing derivative securities in LIBOR market models typically requires simulation. As in the HJM setting, exact simulation is generally infeasible and some discretization error is inevitable. Because the models of this section deal with a finite set of maturities from the outset, we need only discretize the time argument, whereas in the HJM setting both time and maturity required discretization.

We fix a time grid $0 = t_0 < t_1 < \cdots < t_m < t_{m+1}$ over which to simulate. It is sensible to include the tenor dates $T_1, \ldots, T_{M+1}$ among the simulation dates. In practice, one would often even take $t_i = T_i$ so that the simulation evolves directly from one tenor date to the next. We do not impose any restrictions on the volatilities σ_n, though the deterministic case is the most widely used. The only other specific case that has received much attention takes $\sigma_n(t)$ to be the product of a deterministic function of time and a function of $L_n(t)$ as proposed in Andersen and Andreasen [13]. For example, one may take $\sigma_n(t)$ proportional to a power of $L_n(t)$, resulting in a CEV-type of volatility. In either this extension or in the case of deterministic volatilities, it often suffices to restrict the dependence on time to piecewise constant functions that change values only at the T_i. We return to this point in Section 3.7.4.

Simulation of forward LIBOR rates is a special case of the general problem of simulating a system of SDEs. One could apply an Euler scheme or a higher-order method of the type discussed in Chapter 6. However, even if we restrict ourselves to Euler schemes (as we do here), there are countless alternatives. We have many choices of variables to discretize and many choices of probability measure under which to simulate. Several strategies are compared both theoretically and numerically in Glasserman and Zhao [151], and the discussion here draws on that investigation.

The most immediate application of the Euler scheme under the spot measure discretizes the SDE (3.116), producing

$$\hat{L}_n(t_{i+1}) = \hat{L}_n(t_i) + \mu_n(\hat{L}(t_i), t_i)\hat{L}_n(t_i)[t_{i+1} - t_i]$$
$$+ \hat{L}_n(t_i)\sqrt{t_{i+1} - t_i}\sigma_n(t_i)^\top Z_{i+1} \qquad (3.119)$$

with

$$\mu_n(\hat{L}(t_i), t_i) = \sum_{j=\eta(t_i)}^{n} \frac{\delta_j \hat{L}_j(t_i)\sigma_n(t_i)^\top \sigma_j(t_i)}{1 + \delta_j \hat{L}_j(t_i)}$$

and $Z_1, Z_2, \ldots$ independent $N(0, I)$ random vectors in $\Re^d$. Here, as in Section 3.6.2, we use hats to identify discretized variables. We assume that we are given an initial set of bond prices $B_1(0), \ldots, B_{M+1}(0)$ and initialize the simulation by setting

$$\hat{L}_n(0) = \frac{B_n(0) - B_{n+1}(0)}{\delta_n B_{n+1}(0)}, \quad n = 1, \ldots, M,$$

in accordance with (3.105).

An alternative to (3.119) approximates the LIBOR rates under the spot measure using

$$\hat{L}_n(t_{i+1}) = \hat{L}_n(t_i) \times$$
$$\exp\left(\left[\mu_n(\hat{L}(t_i), t_i) - \tfrac{1}{2}\|\sigma_n(t_i)\|^2\right][t_{i+1} - t_i] + \sqrt{t_{i+1} - t_i}\sigma_n(t_i)^\top Z_{i+1}\right).$$
$$(3.120)$$

This is equivalent to applying an Euler scheme to $\log L_n$; it may also be viewed as approximating L_n by geometric Brownian motion over $[t_i, t_{i+1}]$, with drift and volatility parameters fixed at t_i. This method seems particularly attractive in the case of deterministic σ_n, since then L_n is close to lognormal. A further property of (3.120) is that it keeps all $\hat{L}_n$ positive, whereas (3.119) can produce negative rates.

For both of these algorithms it is important to note that our definition of η makes η right-continuous. For the original continuous-time processes we could just as well have taken η to be left-continuous, but the distinction is important in the discrete approximation. If $t_i = T_k$, then $\eta(t_i) = k+1$ and the sum in each $\mu_n(\hat{L}(t_i), t_i)$ starts at $k+1$. Had we taken η to be left-continuous, we would have $\eta(T_i) = k$ and thus an additional term in each μ_n. It seems intuitively more natural to omit this term as time advances beyond T_k since L_k ceases to be meaningful after T_k. Glasserman and Zhao [151] and Sidenius [330] both find that omitting it (i.e., taking η right-continuous) results in smaller discretization error.

Both (3.119) and (3.120) have obvious counterparts for simulation under the forward measure P_{M+1}. The only modification necessary is to replace $\mu_n(\hat{L}(t_i), t_i)$ with

$$\mu_n(\hat{L}(t_i), t_i) = -\sum_{j=n+1}^{M} \frac{\delta_j \hat{L}_j(t_i) \sigma_n(t_i)^\top \sigma_j(t_i)}{1 + \delta_j \hat{L}_j(t_i)}.$$

Notice that $\mu_M \equiv 0$. It follows that if the σ_M is deterministic and constant between the t_i (for example, constant between tenor dates), then the log Euler scheme (3.120) with $\mu_M = 0$ simulates L_M without discretization error under the forward measure P_{M+1}. None of the L_n is simulated without discretization error under the spot measure, but we will see that the spot measure is nevertheless generally preferable for simulation.

Martingale Discretization

In our discussion of simulation in the HJM setting, we devoted substantial attention to the issue of choosing the discrete drift to keep the model arbitrage-free even after discretization. It is therefore natural to examine whether an analogous choice of drift can be made in the LIBOR rate dynamics. In the HJM setting, we derived the discrete drift from the condition that the discretized discounted bond prices must be martingales. In the LIBOR market model, the corresponding requirement is that

$$\hat{D}_n(t_i) = \prod_{j=0}^{n-1} \frac{1}{1 + \delta_j \hat{L}_j(t_i \wedge T_j)} \tag{3.121}$$

be a martingale (in i) for each n under the spot measure; see (3.109). Under the forward measure, the martingale condition applies to

$$\hat{D}_n(t_i) = \prod_{j=n}^{M} \left(1 + \delta_j \hat{L}_j(t_i)\right); \tag{3.122}$$

see (3.113).

Consider the spot measure first. We would like, as a special case of (3.121), for $1/(1 + \delta_1 \hat{L}_1)$ to be a martingale. Using the Euler scheme (3.119), this requires

$$\mathsf{E}\left[\frac{1}{1+\delta_1(\hat{L}_1(0)[1+\mu_1 t_1 + \sqrt{t_1}\sigma_1^\top Z_1)]}\right] = \frac{1}{1+\delta_1\hat{L}_1(0)},$$

the expectation taken with respect to $Z_1 \sim N(0, I)$. However, because the denominator inside the expectation has a normal distribution, the expectation is infinite no matter how we choose μ_1. There is no discrete drift that preserves the martingale property. If, instead, we use the method in (3.120), the condition becomes

$$\mathsf{E}\left[\frac{1}{1+\delta_1(\hat{L}_1(0)\exp([\mu_1 - \|\sigma_1\|^2/2]t_1 + \sqrt{t_1}\sigma_1^\top Z_1))}\right] = \frac{1}{1+\delta_1\hat{L}_1(0)}.$$

In this case, there is a value of μ_1 for which this equation holds, but there is no explicit expression for it. The root of the difficulty lies in evaluating an expression of the form

$$\mathsf{E}\left[\frac{1}{1+\exp(a+bZ)}\right], \quad Z \sim N(0,1),$$

which is effectively intractable. In the HJM setting, calculation of the discrete drift relies on evaluating far more convenient expressions of the form $\mathsf{E}[\exp(a+bZ)]$; see the steps leading to (3.97).

Under the forward measure, it is feasible to choose μ_1 so that $\hat{D}_2$ in (3.122) is a martingale using an Euler scheme for either L_1 or $\log L_1$. However, this quickly becomes cumbersome for $\hat{D}_n$ with larger values of n. As a practical matter, it does not seem feasible under any of these methods to adjust the drift to make the deflated bond prices martingales. A consequence of this is that if we price bonds in the simulation by averaging replications of (3.121) or (3.122), the simulation price will not converge to the corresponding $B_n(0)$ as the number of replications increases.

An alternative strategy is to discretize and simulate the deflated bond prices themselves, rather than the forward LIBOR rates. For example, under the spot measure, the deflated bond prices satisfy

$$\frac{dD_{n+1}(t)}{D_{n+1}(t)} = -\sum_{j=\eta(t)}^{n}\left(\frac{\delta_j L_j(t)}{1+\delta_j L_j(t)}\right)\sigma_j^\top(t)\,dW(t)$$

$$= \sum_{j=\eta(t)}^{n}\left(\frac{D_{j+1}(t)}{D_j(t)} - 1\right)\sigma_j^\top(t)\,dW(t). \qquad (3.123)$$

An Euler scheme for $\log D_{n+1}$ therefore evolves according to

$$\hat{D}_{n+1}(t_{i+1}) =$$
$$\hat{D}_{n+1}(t_i)\exp\left(-\tfrac{1}{2}\|\hat{\nu}_{n+1}(t_i)\|^2[t_{i+1} - t_i] + \sqrt{t_{i+1} - t_i}\,\hat{\nu}_{n+1}(t_i)^\top Z_{i+1}\right) \qquad (3.124)$$

with

$$\hat{\nu}_{n+1}(t_i) = \sum_{j=\eta(t_i)}^{n} \left(\frac{\hat{D}_{j+1}(t_i)}{\hat{D}_j(t_i)} - 1 \right) \sigma_j(t_i). \tag{3.125}$$

In either case, the discretized deflated bond prices are automatically martingales; in (3.124) they are positive martingales and in this sense the discretization is arbitrage-free. From the simulated $\hat{D}_n(t_i)$ we can then *define* the discretized forward LIBOR rates by setting

$$\hat{L}_n(t_i) = \frac{1}{\delta_n} \left(\frac{\hat{D}_n(t_i) - \hat{D}_{n+1}(t_i)}{\hat{D}_{n+1}(t_i)} \right),$$

for $n = 1, \ldots, M$. Any other term structure variables (e.g., swap rates) required in the simulation can then be defined from the $\hat{L}_n$.

Glasserman and Zhao [151] recommend replacing

$$\left(\frac{\hat{D}_{j+1}(t_i)}{\hat{D}_j(t_i)} - 1 \right) \quad \text{with} \quad \min \left\{ \left(\frac{\hat{D}_{j+1}(t_i)}{\hat{D}_j(t_i)} \right)^+ - 1, 0 \right\}. \tag{3.126}$$

This modification has no effect in the continuous-time limit because $0 \leq D_{j+1}(t) \leq D_j(t)$ (if $L_j(t) \geq 0$). But in the discretized process the ratio $\hat{D}_{j+1}/\hat{D}_j$ could potentially exceed 1.

Under the forward measure P_{M+1}, the deflated bond prices (3.113) satisfy

$$\frac{dD_{n+1}(t)}{D_{n+1}(t)} = \sum_{j=n+1}^{M} \frac{\delta_j L_j(t)}{1 + \delta_j L_j(t)} \sigma_j(t)^{\top} \, dW^{M+1}(t)$$

$$= \sum_{j=n+1}^{M} \left(1 - \frac{D_{j+1}(t)}{D_j(t)} \right) \sigma_j^{\top}(t) \, dW^{M+1}(t). \tag{3.127}$$

We can again apply an Euler discretization to the logarithm of these variables to get (3.124), except that now

$$\hat{\nu}_{n+1}(t_i) = \sum_{j=n+1}^{M} \left(1 - \frac{\hat{D}_{j+1}(t_i)}{\hat{D}_j(t_i)} \right) \sigma_j(t_i),$$

possibly modified as in (3.126).

Glasserman and Zhao [151] consider several other choices of variables for discretization, including (under the spot measure) the normalized differences

$$V_n(t) = \frac{D_n(t) - D_{n+1}(t)}{B_1(0)}, \quad n = 1, \ldots, M;$$

these are martingales because the deflated bond prices are martingales. They satisfy

$$\frac{dV_n}{V_n} =$$

$$\left[\left(\frac{V_n + V_{n-1} + \cdots + V_1 - 1}{V_{n-1} + \cdots + V_1 - 1} \right) \sigma_n^\top + \sum_{j=\eta}^{n-1} \left(\frac{V_j}{V_{j-1} + \cdots + V_1 - 1} \right) \sigma_j^\top \right] dW,$$

with the convention $\sigma_{M+1} \equiv 0$. Forward rates are recovered using

$$\delta_n L_n(t) = \frac{V_n(t)}{V_{n+1}(t) + \cdots + V_{M+1}(t)}.$$

Similarly, the variables

$$\delta_n X_n(t) = \delta_n L_n(t) \prod_{j=n+1}^{M} (1 + \delta_j L_j(t))$$

are differences of deflated bond prices under the forward measure P_{M+1} and thus martingales under that measure. The X_n satisfy

$$\frac{dX_n}{X_n} = \left(\sigma_n^\top + \sum_{j=n+1}^{M} \frac{\delta_j X_j \sigma_j^\top}{1 + \delta_j X_j + \cdots + \delta_M X_M} \right) dW^{M+1}.$$

Forward rates are recovered using

$$L_n = \frac{X_n}{1 + \delta_{n+1} X_{n+1} + \cdots + \delta_M X_M}.$$

Euler discretizations of $\log V_n$ and $\log X_n$ preserve the martingale property and thus keep the discretized model arbitrage-free.

Pricing Derivatives

The pricing of a derivative security in a simulation proceeds as follows. Using any of the methods considered above, we simulate paths of the discretized variables $\hat{L}_1, \ldots, \hat{L}_M$. Suppose we want to price a derivative with a payoff of $g(L(T_n))$ at time T_n. Under the spot measure, we simulate to time T_n and then calculate the deflated payoff

$$g(\hat{L}(T_n)) \cdot \prod_{j=0}^{n-1} \frac{1}{1 + \delta_j \hat{L}_j(T_j)}.$$

Averaging over independent replications produces an estimate of the derivative's price at time 0. If we simulate under the forward measure, the estimate consists of independent replications of

$$g(\hat{L}(T_n)) \cdot B_{M+1}(0) \prod_{j=1}^{n-1} (1 + \delta_j \hat{L}_j(T_j)).$$

Glasserman and Zhao [151] compare various simulation methods based, in part, on their discretization error in pricing caplets. For the case of a caplet over $[T_{n-1}, T_n]$, take $g(x) = \delta_{n-1}(x - K)^+$ in the expressions above. If the σ_j are deterministic, the caplet price is given by the Black formula, as explained in Section 3.7.2. However, because of the discretization error, the simulation price will not in general converge exactly to the Black price as the number of replications increase. The bias in pricing caplets serves as a convenient indication of the magnitude of the discretization error.

Figure 3.19, reproduced from Glasserman and Zhao [151], graphs biases in caplet pricing as a function of caplet maturity for various simulation methods. The horizontal line through the center of each panel corresponds to zero bias. The error bars around each curve have halfwidths of one standard error, indicating that the apparent biases are statistically significant. Details of the parameters used for these experiments are reported in Glasserman and Zhao [151] along with several other examples.

These and other experiments suggest the following observations. The smallest biases are achieved by simulating the differences of deflated bond prices (the V_n in the spot measure and the X_n in the forward measure) using an Euler scheme for the logarithms of these variables. (See Glasserman and Zhao [151] for an explanation of the modified V_n method.) An Euler scheme for $\log D_n$ is nearly indistinguishable from an Euler scheme for L_n. Under the forward measure P_{M+1}, the final caplet is priced without discretization error by the Euler schemes for $\log X_n$ and $\log L_n$; these share the feature that they make the discretized rate $\hat{L}_M$ lognormal.

The graphs in Figure 3.19 compare discretization biases but say nothing about the relative variances of the methods. Glasserman and Zhao [151] find that simulating under the spot measure usually results in smaller variance than simulating under the forward measure, especially at high levels of volatility. An explanation for this is suggested by the expressions (3.109) and (3.113) for the deflated bond prices under the two measures: whereas (3.109) always lies between 0 and 1, (3.113) can take arbitrarily large values. This affects derivatives pricing through the discounting of payoffs.

3.7.4 Volatility Structure and Calibration

In our discussion of LIBOR market models we have taken the volatility factors $\sigma_n(t)$ as inputs without indicating how they might be specified. In practice, these coefficients are chosen to calibrate a model to market prices of actively traded derivatives, especially caps and swaptions. (The model is automatically calibrated to bond prices through the relations (3.105) and (3.106).) Once the model has been calibrated to the market, it can be used to price less liquid

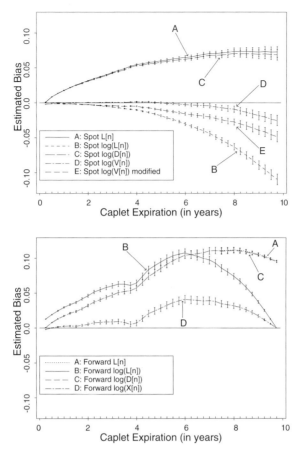

Fig. 3.19. Comparison of biases in caplet pricing for various simulation methods. Top panel uses spot measure; method A is an Euler scheme for L_n and methods B–E are Euler schemes for log variables. Bottom panel uses the forward measure P_{M+1}; method A is an Euler scheme for L_n and methods B–D are Euler schemes for log variables.

instruments for which market prices may not be readily available. Accurate and efficient calibration is a major topic in its own right and we can only touch on the key issues. For a more extensive treatment, see James and Webber [194] and Rebonato [303]. Similar considerations apply in both the HJM framework and in LIBOR market models; we discuss calibration in the LIBOR setting because it is somewhat simpler. Indeed, convenience in calibration is one of the main advantages of this class of models.

The variables $\sigma_n(t)$ are the primary determinants of both the level of volatility in forward rates and the correlations between forward rate. It is often useful to distinguish these two aspects and we will consider the overall level of volatility first. Suppose we are given the market price of a caplet for the interval $[T_n, T_{n+1}]$ and from this price we calculate an implied volatility v_n by inverting the Black formula (3.117). (We can assume that the other parameters of the formula are known.) If we choose σ_n to be any deterministic $\Re^d$-valued function satisfying

$$\frac{1}{T_n}\int_0^{T_n} \|\sigma_n(t)\|^2 \, dt = v_n^2,$$

then we know from the discussion in Section 3.7.2 that the model is calibrated to the market price of this caplet, because the model's caplet price is given by the Black formula with implied volatility equal to the square root of the expression on the left. By imposing this constraint on all of the σ_j, we ensure that the model is calibrated to all caplet prices. (As a practical matter, it may be necessary to infer the prices of individual caplets from the prices of caps, which are portfolios of caplets. For simplicity, we assume caplet prices are available.)

Because LIBOR market models do not specify interest rates over accrual periods shorter than the intervals $[T_i, T_{i+1}]$, it is natural and customary to restrict attention to functions $\sigma_n(t)$ that are constant between tenor dates. We take each σ_n to be right-continuous and thus denote by $\sigma_n(T_i)$ its value over the interval $[T_i, T_{i+1})$. Suppose, for a moment, that the model is driven by a scalar Brownian motion, so $d = 1$ and each σ_n is scalar valued. In this case, it is convenient to think of the volatility structure as specifed through a lower-triangular matrix of the following form:

$$\begin{pmatrix} \sigma_1(T_0) & & & \\ \sigma_2(T_0) & \sigma_2(T_1) & & \\ \vdots & \vdots & \ddots & \\ \sigma_M(T_0) & \sigma_M(T_1) & \cdots & \sigma_M(T_{M-1}) \end{pmatrix}.$$

The upper half of the matrix is empty (or irrelevant) because each $L_n(t)$ ceases to be meaningful for $t > T_n$. In this setting, we have

$$\int_0^{T_n} \sigma_n^2(t) \, dt = \sigma_n^2(T_0)\delta_0 + \sigma_n^2(T_1)\delta_1 + \cdots + \sigma_n^2(T_{n-1})\delta_{n-1},$$

so caplet prices impose a constraint on the sums of squares along each row of the matrix.

The volatility structure is *stationary* if $\sigma_n(t)$ depends on n and t only through the difference $T_n - t$. For a stationary, single-factor, piecewise constant volatility structure, the matrix above takes the form

$$\begin{pmatrix} \sigma(1) & & & \\ \sigma(2) & \sigma(1) & & \\ \vdots & \vdots & \ddots & \\ \sigma(M) & \sigma(M-1) & \cdots & \sigma(1) \end{pmatrix}.$$

for some values $\sigma(1), \ldots, \sigma(M)$. (Think of $\sigma(i)$ as the volatility of a forward rate i periods away from maturity.) In this case, the number of variables just equals the number of caplet maturities to which the model may be calibrated. Calibrating to additional instruments requires introducing nonstationarity or additional factors.

In a multifactor model (i.e., $d \geq 2$) we can think of replacing the entries $\sigma_n(T_i)$ in the volatility matrix with the norms $\|\sigma_n(T_i)\|$, since the $\sigma_n(T_i)$ are now vectors. With piecewise constant values, this gives

$$\int_0^{T_n} \|\sigma_n(t)\|^2 \, dt = \|\sigma_n(T_0)\|^2 \delta_0 + \|\sigma_n(T_1)\|^2 \delta_1 + \cdots + \|\sigma_n(T_{n-1})\|^2 \delta_{n-1},$$

so caplet implied volatilities continue to constrain the sums of squares along each row. This also indicates that taking $d \geq 2$ does not provide additional flexibility in matching these implied volatilities.

The potential value of a multifactor model lies in capturing correlations between forward rates of different maturities. For example, from the Euler approximation in (3.120), we see that over a short time interval the correlation between the increments of $\log L_j(t)$ and $\log L_k(t)$ is approximately

$$\frac{\sigma_k(t)^\top \sigma_j(t)}{\|\sigma_k(t)\| \, \|\sigma_j(t)\|}.$$

These correlations are often chosen to match market prices of swaptions (which, unlike caps, are sensitive to rate correlations) or to match historical correlations.

In the stationary case, we can visualize the volatility factors by graphing them as functions of time to maturity. This can be useful in interpreting the correlations they induce. Figure 3.20 illustrates three hypothetical factors in a model with $M = 15$. Because the volatility is assumed stationary, we may write $\sigma_n(T_i) = \sigma(n - i)$ for some vectors $\sigma(1), \ldots, \sigma(M)$. In a three-factor model, each $\sigma(i)$ has three components. The three curves in Figure 3.20 are graphs of the three components as functions of time to maturity. If we fix a time to maturity on the horizontal axis, the total volatility at that point is given by the sums of squares of the three components; the inner products of these three-dimensional vectors at different times determine the correlations between the forward rates.

Notice that the first factor in Figure 3.20 has the same sign for all maturities; regardless of the sign of the increment of the driving Brownian motion, this factor moves all forward rates in the same direction and functions approximately as a parallel shift. The second factor has values of opposite signs at short and long maturities and will thus have the effect of tilting the forward curve (up if the increment in the second component of the driving Brownian motion is positive and down if it is negative). The third factor bends the forward curve by moving intermediate maturities in the opposite direction of long and short maturities, the direction depending on the sign of the increment of the third component of the driving Brownian motion.

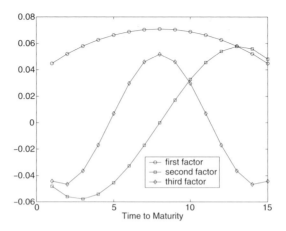

Fig. 3.20. Hypothetical volatility factors.

The hypothetical factors in Figure 3.20 are the first three principal components of the matrix

$$0.12^2 \exp((-0.8\sqrt{|i - j|})), \quad i, j = 1, \ldots, 15.$$

More precisely, they are the first three eigenvectors of this matrix as ranked by their eigenvalues, scaled to have length equal to their eigenvalues. It is common in practice to use the principal components of either the covariance matrix or the correlation matrix of changes in forward rates in choosing a factor structure. Principal components analysis typically produces the qualitative features of the hypothetical example in Figure 3.20; see, e.g., the examples in James and Webber [194] or Rebonato [304].

An important feature of LIBOR market models is that a good deal of calibration can be accomplished through closed form expressions or effective approximations for the prices of caps and swaptions. This makes calibration fast. In the absence of formulas or approximations, calibration is an iterative procedure requiring repeated simulation at various parameter values until the model price matches the market. Because each simulation can be quite time consuming, calibration through simulation can be onerous.

4

Variance Reduction Techniques

This chapter develops methods for increasing the efficiency of Monte Carlo simulation by reducing the variance of simulation estimates. These methods draw on two broad strategies for reducing variance: taking advantage of tractable features of a model to adjust or correct simulation outputs, and reducing the variability in simulation inputs. We discuss control variates, antithetic variates, stratified sampling, Latin hypercube sampling, moment matching methods, and importance sampling, and we illustrate these methods through examples. Two themes run through this chapter:

○ The greatest gains in efficiency from variance reduction techniques result from exploiting specific features of a problem, rather than from generic applications of generic methods.
○ Reducing simulation error is often at odds with convenient estimation of the simulation error itself; in order to supplement a reduced-variance estimator with a valid confidence interval, we sometimes need to sacrifice some of the potential variance reduction.

The second point applies, in particular, to methods that introduce dependence across replications in the course of reducing variance.

4.1 Control Variates

4.1.1 Method and Examples

The method of control variates is among the most effective and broadly applicable techniques for improving the efficiency of Monte Carlo simulation. It exploits information about the errors in estimates of known quantities to reduce the error in an estimate of an unknown quantity.

To describe the method, we let $Y_1, \ldots, Y_n$ be outputs from n replications of a simulation. For example, Y_i could be the discounted payoff of a derivative security on the ith simulated path. Suppose that the Y_i are independent and identically distributed and that our objective is to estimate $\mathsf{E}[Y_i]$. The usual estimator is the sample mean $\bar{Y} = (Y_1 + \cdots + Y_n)/n$. This estimator is unbiased and converges with probability 1 as $n \to \infty$.

Suppose, now, that on each replication we calculate another output X_i along with Y_i. Suppose that the pairs (X_i, Y_i), $i = 1, \ldots, n$, are i.i.d. and that the expectation $\mathsf{E}[X]$ of the X_i is known. (We use (X, Y) to denote a generic pair of random variables with the same distribution as each (X_i, Y_i).) Then for any fixed b we can calculate

$$Y_i(b) = Y_i - b(X_i - \mathsf{E}[X])$$

from the ith replication and then compute the sample mean

$$\bar{Y}(b) = \bar{Y} - b(\bar{X} - \mathsf{E}[X]) = \frac{1}{n} \sum_{i=1}^{n} (Y_i - b(X_i - \mathsf{E}[X])). \qquad (4.1)$$

This is a control variate estimator; the observed error $\bar{X} - \mathsf{E}[X]$ serves as a control in estimating $\mathsf{E}[Y]$.

As an estimator of $\mathsf{E}[Y]$, the control variate estimator (4.1) is unbiased because

$$\mathsf{E}[\bar{Y}(b)] = \mathsf{E}\left[\bar{Y} - b(\bar{X} - \mathsf{E}[X])\right] = \mathsf{E}[\bar{Y}] = \mathsf{E}[Y]$$

and it is consistent because, with probability 1,

$$\lim_{n \to \infty} \frac{1}{n} \sum_{i=1}^{n} Y_i(b) = \lim_{n \to \infty} \frac{1}{n} \sum_{i=1}^{n} (Y_i - b(X_i - \mathsf{E}[X]))$$
$$= \mathsf{E}\left[Y - b(X - \mathsf{E}[X])\right]$$
$$= \mathsf{E}[Y].$$

Each $Y_i(b)$ has variance

$$\mathsf{Var}[Y_i(b)] = \mathsf{Var}\left[Y_i - b(X_i - \mathsf{E}[X])\right]$$
$$= \sigma_Y^2 - 2b\sigma_X\sigma_Y\rho_{XY} + b^2\sigma_X^2 \equiv \sigma^2(b), \qquad (4.2)$$

where $\sigma_X^2 = \mathsf{Var}[X]$, $\sigma_Y^2 = \mathsf{Var}[Y]$, and ρ_{XY} is the correlation between X and Y. The control variate estimator $\bar{Y}(b)$ has variance $\sigma^2(b)/n$ and the ordinary sample mean $\bar{Y}$ (which corresponds to $b = 0$) has variance σ_Y^2/n. Hence, the control variate estimator has smaller variance than the standard estimator if $b^2\sigma_X < 2b\sigma_Y\rho_{XY}$.

The optimal coefficient b^* minimizes the variance (4.2) and is given by

$$b^* = \frac{\sigma_Y}{\sigma_X}\rho_{XY} = \frac{\mathsf{Cov}[X, Y]}{\mathsf{Var}[X]}. \qquad (4.3)$$

Substituting this value in (4.2) and simplifying, we find that the ratio of the variance of the optimally controlled estimator to that of the uncontrolled estimator is

$$\frac{\text{Var}[\bar{Y} - b^*(\bar{X} - \mathsf{E}[X])]}{\text{Var}[\bar{Y}]} = 1 - \rho_{XY}^2. \tag{4.4}$$

A few observations follow from this expression:

○ With the optimal coefficient b^*, the effectiveness of a control variate, as measured by the variance reduction ratio (4.4), is determined by the strength of the correlation between the quantity of interest Y and the control X. The sign of the correlation is irrelevant because it is absorbed in b^*.

○ If the computational effort per replication is roughly the same with and without a control variate, then (4.4) measures the computational speed-up resulting from the use of a control. More precisely, the number of replications of the Y_i required to achieve the same variance as n replications of the control variate estimator is $n/(1 - \rho_{XY}^2)$.

○ The variance reduction factor $1/(1 - \rho_{XY}^2)$ increases very sharply as $|\rho_{XY}|$ approaches 1 and, accordingly, it drops off quickly as $|\rho_{XY}|$ decreases away from 1. For example, whereas a correlation of 0.95 produces a ten-fold speed-up, a correlation of 0.90 yields only a five-fold speed-up; at $|\rho_{XY}| = 0.70$ the speed-up drops to about a factor of two. This suggests that a rather high degree of correlation is needed for a control variate to yield substantial benefits.

These remarks and equation (4.4) apply if the optimal coefficient b^* is known. In practice, if $\mathsf{E}[Y]$ is unknown it is unlikely that σ_Y or ρ_{XY} would be known. However, we may still get most of the benefit of a control variate using an estimate of b^*. For example, replacing the population parameters in (4.3) with their sample counterparts yields the estimate

$$\hat{b}_n = \frac{\sum_{i=1}^n (X_i - \bar{X})(Y_i - \bar{Y})}{\sum_{i=1}^n (X_i - \bar{X})^2}. \tag{4.5}$$

Dividing numerator and denominator by n and applying the strong law of large numbers shows that $\hat{b}_n \to b^*$ with probability 1. This suggests using the estimator $\bar{Y}(\hat{b}_n)$, the sample mean of $Y_i(\hat{b}_n) = Y_i - \hat{b}_n(X_i - \mathsf{E}[X])$, $i = 1, \ldots, n$. Replacing b^* with $\hat{b}_n$ introduces some bias; we return to this point in Section 4.1.3.

The expression in (4.5) is the slope of the least-squares regression line through the points (X_i, Y_i), $i = 1, \ldots, n$. The link between control variates and regression is useful in the statistical analysis of control variate estimators and also permits a graphical interpretation of the method. Figure 4.1 shows a hypothetical scatter plot of simulation outputs (X_i, Y_i) and the estimated regression line for these points, which passes through the point $(\bar{X}, \bar{Y})$. In the figure, $\bar{X} < \mathsf{E}[X]$, indicating that the n replications have underestimated $\mathsf{E}[X]$. If the X_i and Y_i are positively correlated, this suggests that the simulation estimate $\bar{Y}$ likely underestimates $\mathsf{E}[Y]$. This further suggests that we should adjust the estimator upward. The regression line determines the magnitude of the adjustment; in particular, $\bar{Y}(\hat{b}_n)$ is the value fitted by the regression line at the point $\mathsf{E}[X]$.

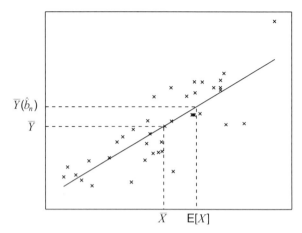

Fig. 4.1. Regression interpretation of control variate method. The regression line through the points (X_i, Y_i) has slope $\hat{b}_n$ and passes through $(\bar{X}, \bar{Y})$. The control variate estimator $\bar{Y}(\hat{b}_n)$ is the value fitted by the line at $\mathsf{E}[X]$. In the figure, the sample mean $\bar{X}$ underestimates $\mathsf{E}[X]$ and $\bar{Y}$ is adjusted upward accordingly.

Examples

To make the method of control variates more tangible, we now illustrate it with several examples.

Example 4.1.1 *Underlying assets.* In derivative pricing simulations, underlying assets provide a virtually universal source of control variates. We know from Section 1.2.1 that the absence of arbitrage is essentially equivalent to the requirement that appropriately discounted asset prices be martingales. Any martingale with a known initial value provides a potential control variate precisely because its expectation at any future time is its initial value. To be concrete, suppose we are working in the risk-neutral measure and suppose the interest rate is a constant r. If $S(t)$ is an asset price, then $\exp(-rt)S(t)$ is a martingale and $\mathsf{E}[\exp(-rT)S(T)] = S(0)$. Suppose we are pricing an option on S with discounted payoff Y, some function of $\{S(t), 0 \le t \le T\}$. From independent replications S_i, $i = 1, \ldots, n$, each a path of S over $[0, T]$, we can form the control variate estimator

$$\frac{1}{n}\sum_{i=1}^{n}(Y_i - b[S_i(T) - e^{rT}S(0)]),$$

or the corresponding estimator with b replaced by $\hat{b}_n$. If $Y = e^{-rT}(S(T)-K)^+$, so that we are pricing a standard call option, the correlation between Y and $S(T)$ and thus the effectiveness of the control variate depends on the strike K. At $K = 0$ we would have perfect correlation; for an option that is deep out-of-the-money (i.e., with large K), the correlation could be quite low. This is illustrated in Table 4.1 for the case of $S \sim \text{GBM}(r, \sigma^2)$ with parameters $r = 5\%$, $\sigma = 30\%$, $S(0) = 50$, and $T = 0.25$. This example shows that the effectiveness of a control variate can vary widely with the parameters of a problem. $\square$

K	40	45	50	55	60	65	70
$\hat{\rho}$	0.995	0.968	0.895	0.768	0.604	0.433	0.286
$\hat{\rho}^2$	0.99	0.94	0.80	0.59	0.36	0.19	0.08

Table 4.1. Estimated correlation $\hat{\rho}$ between $S(T)$ and $(S(T) - K)^+$ for various values of K, with $S(0) = 50$, $\sigma = 30\%$, $r = 5\%$, and $T = 0.25$. The third row measures the fraction of variance in the call option payoff eliminated by using the underlying asset as a control variate.

Example 4.1.2 *Tractable options.* Simulation is sometimes used to price complex options in a model in which simpler options can be priced in closed form. For example, even under Black-Scholes assumptions, some path-dependent options require simulation for pricing even though formulas are available for simpler options. A tractable option can sometimes provide a more effective control than the underlying asset.

A particularly effective example of this idea was suggested by Kemna and Vorst [209] for the pricing of Asian options. Accurate pricing of an option on the arithmetic average

$$\bar{S}_A = \frac{1}{n} \sum_{i=1}^{n} S(t_i)$$

requires simulation, even if S is geometric Brownian motion. In contrast, calls and puts on the geometric average

$$\bar{S}_G = \left(\prod_{i=1}^{n} S(t_i) \right)^{1/n}$$

can be priced in closed form, as explained in Section 3.2.2. Thus, options on $\bar{S}_G$ can be used as control variates in pricing options on $\bar{S}_A$.

Figure 4.2 shows scatter plots of simulated values of $(\bar{S}_A - K)^+$ against the terminal value of the underlying asset $S(T)$, a standard call payoff $(S(T) - K)^+$, and the geometric call payoff $(\bar{S}_G - K)^+$. The figures are based on $K = 50$ and thirteen equally spaced averaging dates; all other parameters are as in Example 4.1.1. The leftmost panel shows that the weak correlation between $\bar{S}_A$ and $S(T)$ is further weakened by applying the call option payoff to $\bar{S}_A$, which projects negative values of $\bar{S}_A - K$ to zero; the resulting correlation is approximately 0.79. The middle panel shows the effect of applying the call option payoff to $S(T)$ as well; in this case the correlation increases to approximately 0.85. The rightmost panel illustrates the extremely strong relation between the payoffs on the arithmetic and geometric average call options. The correlation in this case is greater then 0.99. A similar comparison is made in Broadie and Glasserman [67]. □

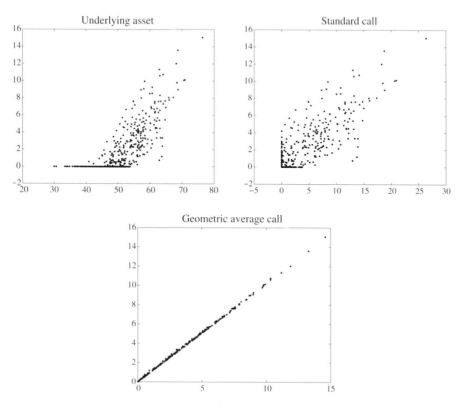

Fig. 4.2. Scatter plots of payoff of call option on arithmetic average against the underlying asset, the payoff of a standard call, and the payoff of a call on the geometric average.

Example 4.1.3 *Bond prices.* In a model with stochastic interest rates, bond prices often provide a convenient source of control variates. As emphasized in Sections 3.3–3.4 and Sections 3.6–3.7, an important consideration in implementing an interest rate simulation is ensuring that the simulation correctly prices bonds. While this is primarily important for consistent pricing, as a byproduct it makes bonds available as control variates. Bonds may be viewed as the underlying assets of an interest rate model, so in a sense this example is a special case of Example 4.1.1.

In a model of the short rate $r(t)$, a bond maturing at time T has initial price

$$B(0,T) = \mathsf{E}\left[\exp\left(-\int_0^T r(u)\,du\right)\right],$$

the expectation taken with respect to the risk-neutral measure. Since we may assume that $B(0,T)$ is known, the quantity inside the expectation provides a potential control variate. But even if r is simulated without discretization error at dates $t_1,\ldots,t_n = T$, the difference

$$\exp\left(-\frac{1}{n}\sum_{i=1}^n r(t_i)\right) - B(0,T)$$

will not ordinarily have mean 0 because of the error in approximating the integral. Using this difference in a control variate estimator could therefore introduce some bias, though the bias can be made as small as necessary by taking a sufficiently fine partition of $[0,T]$. In our discussion of the Vasicek model in Section 3.3, we detailed the exact joint simulation of $r(t_i)$ and its time integral

$$Y(t_i) = \int_0^{t_i} r(u)\,du.$$

This provides a bias-free control variate because

$$\mathsf{E}\left[\exp(-Y(T))\right] = B(0,T).$$

Similar considerations apply to the forward rate models of Sections 3.6 and 3.7. In our discussion of the Heath-Jarrow-Morton framework, we devoted considerable attention to deriving the appropriate discrete drift condition. Using this drift in a simulation produces unbiased bond price estimates and thus makes bonds available as control variates. In our discussion of LIBOR market models, we noted that discretizing the system of SDEs for the LIBOR rates L_n would not produce unbiased bond estimates; in contrast, the methods in Section 3.7 based on discretizing deflated bonds or their differences do produce unbiased estimates and thus allow the use of bonds as control variates.

Comparison of this discussion with the one in Section 3.7.3 should make clear that the question of whether or not asset prices can be used as control variates is closely related to the question of whether a simulated model is arbitrage-free. □

Example 4.1.4 *Tractable dynamics.* The examples discussed thus far are all based on using one set of prices in a model as control variates for some other price in the same model. Another strategy for developing effective control variates uses prices in a simpler model. We give two illustrations of this idea.

Consider, first, the pricing of an option on an asset whose dynamics are modeled by

$$\frac{dS(t)}{S(t)} = r\,dt + \sigma(t)\,dW(t),$$

where $\sigma(t)$ may be function of $S(t)$ or may be stochastic and described by a second SDE. We might simulate S at dates $t_1, \ldots, t_n$ using an approximation of the form

$$S(t_{i+1}) = S(t_i) \exp\left([r - \tfrac{1}{2}\sigma(t_i)^2](t_{i+1} - t_i) + \sigma(t_i)\sqrt{t_{i+1} - t_i}\, Z_{i+1} \right),$$

where the Z_i are independent $N(0,1)$ variables. In a stochastic volatility model, a second recursion would determine the evolution of $\sigma(t_i)$. Suppose the option we want to price would be tractable if the underlying asset were geometric Brownian motion. Then along with S we could simulate

$$\tilde{S}(t_{i+1}) = \tilde{S}(t_i) \exp\left([r - \tfrac{1}{2}\tilde{\sigma}^2](t_{i+1} - t_i) + \tilde{\sigma}\sqrt{t_{i+1} - t_i}\, Z_{i+1} \right)$$

for some constant $\tilde{\sigma}$, the *same* sequence Z_i, and with initial condition $\tilde{S}(0) = S(0)$. If, for example, the option is a standard call with strike K and expiration t_n, we could form a controlled estimator using independent replications of

$$(S(t_n) - K)^+ - b\left((\tilde{S}(t_n) - K)^+ - \mathsf{E}\left[(\tilde{S}(t_n) - K)^+ \right] \right).$$

Except for a discount factor, the expectation on the right is given by the Black-Scholes formula. For effective variance reduction, the constant $\tilde{\sigma}$ should be chosen close to a typical value of σ.

As a second illustration of this idea, recall that the dynamics of forward LIBOR under the spot measure in Section 3.7 are given by

$$\frac{dL_n(t)}{L_n(t)} = \sum_{j=\eta(t)}^{n} \frac{\sigma_j(t)^\top \sigma_n(t)\delta_j L_j(t)}{1 + \delta_j L_j(t)}\, dt + \sigma_n(t)^\top dW(t), \quad n = 1, \ldots, M. \quad (4.6)$$

Suppose the σ_n are deterministic functions of time. Along with the forward LIBOR rates, we could simulate auxiliary processes

$$\frac{dS_n(t)}{S_n(t)} = \sigma_n^\top(t)\, dW(t), \quad n = 1, \ldots, M. \quad (4.7)$$

These form a multivariate geometric Brownian motion and lend themselves to tractable pricing and thus to control variates. Alternatively, we could use

$$\frac{d\tilde{L}_n(t)}{\tilde{L}_n(t)} = \sum_{j=\eta(t)}^{n} \frac{\sigma_j(t)^\top \sigma_n(t)\delta_j L_j(0)}{1 + \delta_j L_j(0)}\, dt + \sigma_n(t)^\top dW(t). \quad (4.8)$$

Notice that the drift in this expression is a function of the constants $L_j(0)$ rather than the stochastic processes $L_j(t)$ appearing in the drift in (4.6). Hence, $\tilde{L}_n$ is also a geometric Brownian motion though with time-varying drift. Even if an option on the S_n or $\tilde{L}_n$ cannot be valued in closed form, if it can be valued quickly using a numerical procedure it may yield an effective control variate.

The evolution of L_n, S_n, and $\tilde{L}_n$ is illustrated in Figure 4.3. This example initializes all rates at 6%, takes $\delta_j \equiv 0.5$ (corresponding to semi-annual rates), and assumes a stationary specification of the volatility functions in which $\sigma_n(t) \equiv \sigma(n - \eta(t) + 1)$ with σ increasing linearly from 0.15 to 0.25. The figure plots the evolution of L_{40}, S_{40}, and $\tilde{L}_{40}$ using a log-Euler approximation of the type in (3.120). The figure indicates that $\tilde{L}_{40}$ tracks L_{40} quite closely and that even S_{40} is highly correlated with L_{40}.

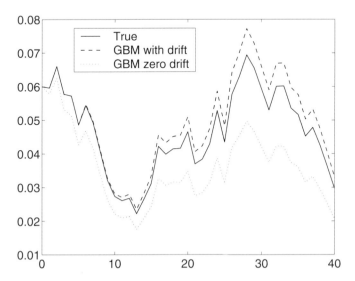

Fig. 4.3. A sample path of L_{40} using the true dynamics in (4.6), the geometric Brownian motion approximation $\tilde{L}_{40}$ with time-varying drift, and the driftless geometric Brownian motion S_{40}.

It should be noted that simulating an auxiliary process as suggested here may substantially increase the time required per replication — perhaps even doubling it. As with any variance reduction technique, the benefit must be weighed against the additional computing time required, using the principles in Section 1.1.3. □

Example 4.1.5 *Hedges as controls.* There is a close link between the selection of control variates and the selection of hedging instruments. If Y is a discounted payoff and we are estimating $\mathsf{E}[Y]$, then any instrument that serves as an effective hedge for Y also serves as an effective control variate if it can be easily priced. Indeed, the calculation of the optimal coefficient b^* is identical to the calculation of the optimal hedge ratio in minimum-variance hedging (see, e.g., Section 2.9 of Hull [189]).

Whereas a static hedge (a fixed position in another asset) may provide significant variance reduction, a dynamic hedge can, in principle, remove all variance — at least under the assumptions of a complete market and continuous trading as discussed in Section 1.2.1. Using the notation of Section 1.2.1, let $V(t)$ denote the value at time t of a price process that can be replicated through a self-financing trading strategy applied to a set of underlying assets S_j, $j = 1, \ldots, d$. As in Section 1.2.1, under appropriate conditions we have

$$V(T) = V(0) + \int_0^T \sum_{j=1}^d \frac{\partial V(t)}{\partial S_j} \, dS_j(t).$$

In other words, V is replicated through a delta-hedging strategy that holds $\partial V/\partial S_j$ shares of asset S_j at each instant. This suggests that $V(T)$ should be highly correlated with

$$\sum_{i=1}^m \sum_{j=1}^d \frac{\partial V(t_{i-1})}{\partial S_j} [S_j(t_i) - S_j(t_{i-1})] \tag{4.9}$$

where $0 = t_0 < t_1 < \cdots < t_m \equiv T$; this is a discrete-time approximation to the dynamic hedging strategy. Of course, in practice if $V(t)$ is unknown then its derivatives are likely to be unknown. One may still obtain an effective control variate by using a rough approximation to the $\partial V/\partial S_j$; for example, one might calculate these deltas as though the underlying asset prices followed geometric Brownian motions, even if their actual dynamics are more complex. See Clewlow and Carverhill [87] for examples of this.

Using an expression like (4.9) as a control variate is somewhat similar to using all the increments $S_j(t_i) - S_j(t_{i-1})$ as controls or, more conveniently, the increments of the discounted asset prices since these have mean zero. The main difference is that the coefficients $\partial V/\partial S_j$ in (4.9) will not in general be constants but will depend on the S_j themselves. We may therefore interpret (4.9) as using the $S_j(t_i) - S_j(t_{i-1})$ as *nonlinear* control variates. We discuss nonlinear controls more generally in Section 4.1.4. □

Example 4.1.6 *Primitive controls.* In the examples above, we have stressed the use of special features of derivative pricing models in identifying potential control variates. Indeed, significant variance reduction is usually achieved only by taking advantage of special properties of a model. It is nevertheless worth mentioning that many generic (and thus typically not very effective) control variates are almost always available in a simulation. For example, most of the models discussed in Chapter 3 are simulated from a sequence $Z_1, Z_2, \ldots$ of independent standard normal random variables. We know that $\mathsf{E}[Z_i] = 0$ and $\mathsf{Var}[Z_i] = 1$, so the sample mean and sample variance of the Z_i are available control variates. At a still more primitive level, most simulations start from a sequence $U_1, U_2, \ldots$ of independent Unif[0,1] random variables. Sample moments of the U_i can also be used as controls. □

Later in this chapter we discuss other techniques for reducing variance. In a sense, all of these can be viewed as strategies for selecting control variates.

For suppose we want to estimate $\mathsf{E}[Y]$ and in addition to the usual sample mean $\bar{Y}$ we have available an alternative unbiased estimator $\tilde{Y}$. The difference $(\bar{Y} - \tilde{Y})$ has (known) expectation zero and can thus be used to form a control variate estimate of the form

$$\bar{Y} - b(\bar{Y} - \tilde{Y}).$$

The special cases $b = 0$ and $b = 1$ correspond to using just one of the two estimators; by optimizing b, we obtain a combined estimator that has lower variance than either of the two.

Output Analysis

In analyzing variance reduction techniques, along with the effectiveness of a technique it is important to consider how the technique affects the statistical interpretation of simulation outputs. So long as we deal with unbiased, independent replications, computing confidence intervals for expectations is a simple matter, as noted in Section 1.1 and explained in Appendix A. But we will see that some variance reduction techniques complicate interval estimation by introducing dependence across replications. This issue arises with control variates if we use the estimated coefficient $\hat{b}_n$ in (4.5). It turns out that in the case of control variates the dependence can be ignored in large samples; a more careful consideration of small-sample issues will be given in Section 4.1.3.

For any fixed b, the control variate estimator $\bar{Y}(b)$ in (4.1) is the sample mean of independent replications $Y_i(b)$, $i = 1, \ldots, n$. Accordingly, an asymptotically valid $1 - \delta$ confidence interval for $\mathsf{E}[Y]$ is provided by

$$\bar{Y}(b) \pm z_{\delta/2} \frac{\sigma(b)}{\sqrt{n}}, \qquad (4.10)$$

where $z_{\delta/2}$ is the $1 - \delta/2$ quantile from the normal distribution ($\Phi(z_{\delta/2}) = 1 - \delta/2$) and $\sigma(b)$ is the standard deviation per replication, as in (4.2).

In practice, $\sigma(b)$ is typically unknown but can be estimated using the sample standard deviation

$$s(b) = \sqrt{\frac{1}{n-1} \sum_{i=1}^{n} (Y_i(b) - \bar{Y}(b))^2}.$$

The confidence interval (4.10) remains asymptotically valid if we replace $\sigma(b)$ with $s(b)$, as a consequence of the limit in distribution

$$\frac{\bar{Y}(b) - \mathsf{E}[Y]}{\sigma(b)/\sqrt{n}} \Rightarrow N(0, 1)$$

and the fact that $s(b)/\sigma(b) \to 1$; see Appendix A.2. If we use the estimated coefficient $\hat{b}_n$ in (4.5), then the estimator

$$\bar{Y}(\hat{b}_n) = \frac{1}{n} \sum_{i=1}^{n} (Y_i - \hat{b}_n(X_i - \mathsf{E}[X]))$$

is not quite of the form $\bar{Y}(b)$ because we have replaced the constant b with the random quantity $\hat{b}_n$. Nevertheless, because $\hat{b}_n \to b^*$, we have

$$\sqrt{n}(\bar{Y}(\hat{b}_n) - \bar{Y}(b^*)) = (\hat{b}_n - b^*) \cdot \sqrt{n}(\bar{X} - \mathsf{E}[X]) \Rightarrow 0 \cdot N(0, \sigma_X^2) = 0,$$

so $\bar{Y}(\hat{b}_n)$ satisfies the same central limit theorem as $\bar{Y}(b^*)$. This means that $\bar{Y}(\hat{b}_n)$ is asymptotically as precise as $\bar{Y}(b^*)$. Moreover, the central limit theorem applies in the form

$$\frac{\bar{Y}(\hat{b}_n) - \mathsf{E}[Y]}{s(\hat{b}_n)/\sqrt{n}} \Rightarrow N(0,1),$$

with $s(\hat{b}_n)$ the sample standard deviation of the $Y_i(\hat{b}_n)$, $i = 1, \ldots, n$, because $s(\hat{b}_n)/\sigma(b^*) \to 1$. In particular, the confidence interval (4.10) remains asymptotically valid if we replace $\bar{Y}(b)$ and $\sigma(b)$ with $\bar{Y}(\hat{b}_n)$ and $s(\hat{b}_n)$, and confidence intervals estimated using $\hat{b}_n$ are asymptotically no wider than confidence intervals estimated using the optimal coefficient b^*.

We may summarize this discussion as follows. It is a simple matter to estimate asymptotically valid confidence intervals for control variate estimators. Moreover, for large n, we get all the benefit of the optimal coefficient b^* by using the estimate $\hat{b}_n$. However, for finite n, there may still be costs to using an estimated rather than a fixed coefficient; we return to this point in Section 4.1.3.

4.1.2 Multiple Controls

We now generalize the method of control variates to the case of multiple controls. Examples 4.1.1–4.1.6 provide ample motivation for considering this extension.

○ If $e^{-rt}S(t)$ is a martingale, then $e^{-rt_1}S(t_1), \ldots, e^{-rt_d}S(t_d)$ all have expectation $S(0)$ and thus provide d controls on each path.
○ If the simulation involves d underlying assets, the terminal values of all assets may provide control variates.
○ Rather than use a single option as a control variate, we may want to use options with multiple strikes and maturities.
○ In an interest rate simulation, we may choose to use d bonds of different maturities as controls.

Suppose, then, that each replication i of a simulation produces outputs Y_i and $X_i = (X_i^{(1)}, \ldots, X_i^{(d)})^\top$ and suppose the vector of expectations $\mathsf{E}[X]$ is known. We assume that the pairs (X_i, Y_i), $i = 1, \ldots, n$, are i.i.d. with covariance matrix

$$\begin{pmatrix} \Sigma_X & \Sigma_{XY} \\ \Sigma_{XY}^\top & \sigma_Y^2 \end{pmatrix}. \tag{4.11}$$

Here, Σ_X is $d \times d$, Σ_{XY} is $d \times 1$, and, as before the scalar σ_Y^2 is the variance of the Y_i. We assume that Σ_X is nonsingular; otherwise, some $X^{(k)}$ is a linear combination of the other $X^{(j)}$s and may be removed from the set of controls.

Let $\bar{X}$ denote the vector of sample means of the controls. For fixed $b \in \Re^d$, the control variate estimator $\bar{Y}(b)$ is

$$\bar{Y}(b) = \bar{Y} - b^\top (\bar{X} - \mathsf{E}[X]).$$

Its variance per replication is

$$\mathsf{Var}[Y_i - b^\top (X_i - \mathsf{E}[X])] = \sigma_Y^2 - 2b^\top \Sigma_{XY} + b^\top \Sigma_{XX} b. \tag{4.12}$$

This is minimized at

$$b^* = \Sigma_X^{-1} \Sigma_{XY}. \tag{4.13}$$

As in the case of a single control variate, this is also the slope (more precisely, the vector of coefficients) in a regression of Y against X.

As is customary in regression analysis, define

$$R^2 = \Sigma_{XY}^\top \Sigma_X^{-1} \Sigma_{XY} / \sigma_Y^2; \tag{4.14}$$

this generalizes the squared correlation coefficient between scalar X and Y and measures the strength of the linear relation between the two. Substituting b^* into the expression for the variance per replication and simplifying, we find that the minimal variance (that is, the variance of $Y_i(b^*)$) is

$$\sigma_Y^2 - \Sigma_{XY}^\top \Sigma_X^{-1} \Sigma_{XY} = (1 - R^2) \sigma_Y^2. \tag{4.15}$$

Thus, R^2 measures the fraction of the variance of Y that is removed in optimally using X as a control.

In practice, the optimal vector of coefficients b^* is unknown but may be estimated. The standard estimator replaces Σ_X and Σ_{XY} in (4.13) with their sample counterparts to get

$$\hat{b}_n = S_X^{-1} S_{XY}, \tag{4.16}$$

where S_X is the $d \times d$ matrix with jk entry

$$\frac{1}{n-1} \left(\sum_{i=1}^n X_i^{(j)} X_i^{(k)} - n\bar{X}^{(j)} \bar{X}^{(k)} \right) \tag{4.17}$$

and S_{XY} is the d-vector with jth entry

$$\frac{1}{n-1} \left(\sum_{i=1}^n X_i^{(j)} Y_i - n\bar{X}^{(j)} \bar{Y} \right).$$

The number of controls d is ordinarily not very large so size is not an obstacle in inverting S_X, but if linear combinations of some of the controls are highly correlated this matrix may be nearly singular. This should be considered in choosing multiple controls.

A simple estimate of the variance of $\bar{Y}(\hat{b}_n)$ is provided by $s_n/\sqrt{n}$ where s_n is the sample standard deviation of the adjusted replications

$$Y_i(\hat{b}_n) = Y_i - \hat{b}_n^\top (X_i - \mathsf{E}[X]).$$

The estimator s_n ignores the fact that $\hat{b}_n$ is itself estimated from the replications, but it is nevertheless a consistent estimator of $\sigma_Y(b^*)$, the optimal standard deviation in (4.15). An asymptotically valid $1-\delta$ confidence interval is thus provided by

$$\bar{Y}(\hat{b}_n) \pm z_{\delta/2} \frac{s_n}{\sqrt{n}}. \tag{4.18}$$

The connection between control variates and regression analysis suggests an alternative way of forming confidence intervals; under additional assumptions about the joint distribution of the (X_i, Y_i) the alternative is preferable, especially if n is not very large. We return to this point in Section 4.1.3.

Variance Decomposition

In looking for effective control variates, it is useful to understand what part of the variance of an ordinary estimator is removed through the use of controls. We now address this point.

Let (X, Y) be any random vector with Y scalar and X d-dimensional. Let (X, Y) have the partitioned covariance matrix in (4.11). For any b, we can write

$$Y = \mathsf{E}[Y] + b^\top (X - \mathsf{E}[X]) + \epsilon,$$

simply by defining ϵ so that equality holds. If $b = b^* = \Sigma_X^{-1} \Sigma_{XY}$, then in fact ϵ is uncorrelated with X; i.e., $Y - b^{*\top}(X - \mathsf{E}[X])$ is uncorrelated with X so

$$\mathsf{Var}[Y] = \mathsf{Var}[b^{*\top} X] + \mathsf{Var}[\epsilon] = \mathsf{Var}[b^{*\top} X] + \mathsf{Var}[Y - b^{*\top} X].$$

In this decomposition, the part of $\mathsf{Var}[Y]$ eliminated by using X as a control is $\mathsf{Var}[b^{*\top} X]$ and the remaining variance is $\mathsf{Var}[\epsilon]$.

The optimal vector b^* makes $b^{*\top}(X - \mathsf{E}[X])$ the projection of $Y - \mathsf{E}[Y]$ onto $X - \mathsf{E}[X]$; the residual ϵ may be interpreted as the part of $Y - \mathsf{E}[Y]$ orthogonal to $X - \mathsf{E}[X]$, orthogonal here meaning uncorrelated. The smaller this orthogonal component (as measured by its variance), the greater the variance reduction achieved by using X as a control for Y. If, in particular, Y is a linear combination of the components of X, then using X as a control eliminates all variance. Of course, in this case $\mathsf{E}[Y]$ is a linear combination of the (known) components of $\mathsf{E}[X]$, so simulation would be unnecessary.

Consider, again, the examples with which we opened this section. If we use multiple path values $S(t_i)$ of an underlying asset as control variates, we eliminate all variance in estimating the expected value of any instrument whose payoff is a linear combination of the $S(t_i)$. (In particular, each $\mathsf{E}[S(t_i)]$ is trivially estimated without error if we use $S(t_i)$ as a control.) The variance that remains in estimating an expected payoff while using the $S(t_i)$ as controls is attributable to the part of the payoff that is uncorrelated with the $S(t_i)$. Similarly, if we use bond prices as control variates in pricing an interest rate derivative, the remaining variance is due to the part of the derivative's payoff that is uncorrelated with any linear combination of bond prices.

Control Variates and Weighted Monte Carlo

In introducing the idea of a control variate in Section 4.1.1, we explained that the observed error in estimating a known quantity could be used to adjust an estimator of an unknown quantity. But the technique has an alternative interpretation as a method for assigning weights to replications. This alternative perspective is sometimes useful, particularly in relating control variates to other methods.

For simplicity, we start with the case of a single control; thus, Y_i and X_i are scalars and the pairs (X_i, Y_i) are i.i.d. The control variate estimator with estimated optimal coefficient $\hat{b}_n$ is $\bar{Y}(\hat{b}_n) = \bar{Y} - \hat{b}_n(\bar{X} - \mathsf{E}[X])$. As in (4.5), the estimated coefficient is given by

$$\hat{b}_n = \frac{\sum_{i=1}^{n}(X_i - \bar{X})(Y_i - \bar{Y})}{\sum_{i=1}^{n}(X_i - \bar{X})^2}.$$

By substituting this expression into $\bar{Y}(\hat{b}_n)$ and simplifying, we arrive at

$$\bar{Y}(\hat{b}_n) = \sum_{i=1}^{n}\left(\frac{1}{n} + \frac{(\bar{X} - X_i)(\bar{X} - \mathsf{E}[X])}{\sum_{i=1}^{n}(X_i - \bar{X})^2}\right)Y_i \equiv \sum_{i=1}^{n}w_iY_i. \qquad (4.19)$$

In other words, the control variate estimator is a weighted average of the replications $Y_1, \ldots, Y_n$. The weights w_i are completely determined by the observations $X_1, \ldots, X_n$ of the control.

A similar representation applies with multiple controls. Using the estimated vector of coefficients in (4.16), the sample covariance matrix S_X in (4.17) and simplifying, we get

$$\bar{Y}(\hat{b}_n) = \sum_{i=1}^{n}\left(\frac{1}{n} + \frac{1}{n-1}(\bar{X} - X_i)^\top S_X^{-1}(\bar{X} - \mathsf{E}[X])\right)Y_i, \qquad (4.20)$$

which is again a weighted average of the Y_i. Here, as before, X_i denotes the vector of controls $(X_i^{(1)}, \ldots, X_i^{(d)})^\top$ from the ith replication, and $\bar{X}$ and $\mathsf{E}[X]$ denote the sample mean and expectation of the X_i, respectively.

The representation in (4.20) is a special case of a general feature of regression — namely, that a fitted value of Y is a weighted average of the observed values of the Y_i with weights determined by the X_i. One consequence of this representation is that if we want to estimate multiple quantities (e.g., prices of various derivative securities) from the same set of simulated paths using the same control variates, the weights can be calculated just once and then applied to all the outputs. Hesterberg and Nelson [178] also show that (4.20) is useful in applying control variates to quantile estimation. They indicate that although it is possible for some of the weights in (4.19) and (4.20) to be negative, the probability of negative weights is small in a sense they make precise.

4.1.3 Small-Sample Issues

In our discussion (following (4.10)) of output analysis with the method of control variates, we noted that because $\hat{b}_n$ converges to b^*, we obtain an asymptotically valid confidence interval if we ignore the randomness of $\hat{b}_n$ and the dependence it introduces among $Y_i(\hat{b}_n)$, $i = 1, \ldots, n$. Moreover, we noted that as $n \to \infty$, the variance reduction achieved using $\hat{b}_n$ approaches the variance reduction that would be achieved if the optimal b^* were known.

In this section, we supplement these large-sample properties with a discussion of statistical issues that arise in analyzing control variate estimators based on a finite number of samples. We note that stronger distributional assumptions on the simulation output lead to confidence intervals valid for all n. Moreover, it becomes possible to quantify the loss in efficiency due to estimating b^*. This offers some guidance in deciding how many control variates to use in a simulation. This discussion is based on results of Lavenberg, Moeller, and Welch [221] and Nelson [277].

For any fixed b, the control variate estimator $\bar{Y}(b)$ is unbiased. But using $\hat{b}_n$, we have

$$\text{Bias}(\bar{Y}(\hat{b}_n)) = \mathsf{E}[\bar{Y}(\hat{b}_n)] - \mathsf{E}[Y] = -\mathsf{E}[\hat{b}_n^\top(\bar{X} - \mathsf{E}[X])],$$

which need not be zero because $\hat{b}_n$ and $\bar{X}$ are not independent. A simple way to eliminate bias is to use n_1 replications to compute an estimate $\hat{b}_{n_1}$ and to then apply this coefficient with the remaining $n - n_1$ replications of (X_i, Y_i). This makes the coefficient estimate independent of $\bar{X}$ and thus makes $\mathsf{E}[\hat{b}_{n_1}^\top \bar{X}] = \mathsf{E}[\hat{b}_{n_1}^\top]\mathsf{E}[\bar{X}]$. In practice, the bias produced by estimating b^* is usually small so the cost of estimating coefficients through separate replications is unattractive. Indeed, the bias is typically $O(1/n)$, whereas the standard error is $O(1/\sqrt{n})$.

Lavenberg, Moeller, and Welch [221] and Nelson [277] note that even if $\hat{b}_n$ is estimated from the same replications used to compute $\bar{Y}$ and $\bar{X}$, the control variate estimator is unbiased if the regression of Y on X is linear. More precisely, if

$$E[Y|X] = c_0 + c_1 X^{(1)} + \cdots + c_d X^{(d)} \quad \text{for some constants } c_0, c_1, \ldots, c_d, \quad (4.21)$$

then $E[\bar{Y}(\hat{b}_n)] = E[Y]$.

Under the additional assumption that $\text{Var}[Y|X]$ does not depend on X, Nelson [277] notes that an *unbiased* estimator of the variance of $\bar{Y}(\hat{b}_n)$, $n > d - 1$, is provided by

$$\hat{s}_n^2 = \left(\frac{1}{n - d - 1} \sum_{i=1}^{n} [Y_i - \bar{Y}(\hat{b}_n) - \hat{b}_n^\top (X_i - E[X])]^2 \right)$$
$$\times \left(\frac{1}{n} + \frac{1}{n - 1} (\bar{X} - E[X])^\top S_X^{-1} (\bar{X} - E[X]) \right).$$

The first factor in this expression is the sample variance of the regression residuals, the denominator $n - d - 1$ reflecting the loss of $d + 1$ degrees of freedom in estimating the regression coefficients in (4.21). The second factor inflates the variance estimate when $\bar{X}$ is far from $E[X]$.

As in Lavenberg et al. [221] and Nelson [277], we now add a final assumption that (X, Y) has a multivariate normal distribution, from which two important consequences follow. The first is that this provides an exact confidence interval for $E[Y]$ for all n: the interval

$$\bar{Y}(\hat{b}_n) \pm t_{n-d-1,\delta/2} \hat{s}_n \tag{4.22}$$

covers $E[Y]$ with probability $1 - \delta$, where $t_{n-d-1,\delta/2}$ denotes the $1 - \delta/2$ quantile of the t distribution with $n - d - 1$ degrees of freedom. This confidence interval may have better coverage than the crude interval (4.18), even if the assumptions on which it is based do not hold exactly.

A second important conclusion that holds under the added assumption of normality is an exact expression for the variance of $\bar{Y}(\hat{b}_n)$. With d controls and $n > d + 2$ replications,

$$\text{Var}[\bar{Y}(\hat{b}_n)] = \frac{n - 2}{n - d - 2} (1 - R^2) \frac{\sigma_Y^2}{n}. \tag{4.23}$$

Here, as before, $\sigma_Y^2 = \text{Var}[Y]$ is the variance per replication without controls and R^2 is the squared multiple correlation coefficient defined in (4.14). As noted in (4.15), $(1 - R^2)\sigma_Y^2$ is the variance per replication of the control variate estimator with known optimal coefficient. We may thus write (4.23) as

$$\text{Var}[\bar{Y}(\hat{b}_n)] = \frac{n - 2}{n - d - 2} \text{Var}[\bar{Y}(b^*)]. \tag{4.24}$$

In light of this relation, Lavenberg et al. [221] call $(n - 2)/(n - d - 2)$ the *loss factor* measuring the loss in efficiency due to using the estimate $\hat{b}_n$ rather than the exact value b^*.

Both (4.22) and (4.24) penalize the use of too many controls — more precisely, they penalize the use of control variates that do not provide a sufficiently large reduction in variance. In (4.22), a larger d results in a loss of degrees of freedom, a larger multiplier $t_{n-d-1,\delta/2}$, and thus a wider confidence interval unless the increase in d is offset by a sufficient decrease in $\hat{s}_n$. In (4.24), a larger d results in a larger loss factor and thus a greater efficiency cost from using estimated coefficients. In both cases, the cost from using more controls is eventually overwhelmed by increasing the sample size n; what constitutes a reasonable number of control variates thus depends in part on the intended number of replications.

The validity of the confidence interval (4.22) and the loss factor in (4.24) depends on the distributional assumptions on (X, Y) introduced above leading up to (4.22); in particular, these results depend on the assumed normality of (X, Y). (Loh [239] provides extensions to more general distributions but these seem difficult to use in practice.) In pricing applications, Y would often be the discounted payoff of an option contract and thus highly skewed and distinctly non-normal. In this case, application of (4.22) and (4.24) lacks theoretical support.

Nelson [277] analyzes the use of various remedies for control variate estimators when the distributional assumptions facilitating their statistical analysis fail to hold. Among the methods he examines is *batching*. This method groups the replications (X_i, Y_i), $i = 1, \ldots, n$, into k disjoint batches of n/k replications each. It then calculates sample means of the (X_i, Y_i) within each batch and applies the usual control variate procedure to the k sample means of the k batches. The appeal of this method lies in the fact that the batch means should be more nearly normally distributed than the original (X_i, Y_i). The cost of batching lies in the loss of degrees of freedom: it reduces the effective sample size from n to k. Based on a combination of theoretical and experimental results, Nelson [277] recommends forming 30 to 60 batches if up to five controls are used. With a substantially larger number of controls, the cost of replacing the number of replications n with $k = 30$–60 in (4.22) and (4.24) would be more significant; this would argue in favor of using a larger number of smaller batches.

Another strategy for potentially improving the performance of control variate estimators replaces the estimated covariance matrix S_X with its true value Σ_X in estimating b^*. This is feasible if Σ_X is known, which would be the case in at least some of the examples introduced in Section 4.1.1. Nelson [277] and Bauer, Venkatraman, and Wilson [40] analyze this alternative; perhaps surprisingly, they find that it generally produces estimators inferior to the usual method based on $\hat{b}_n$.

4.1.4 Nonlinear Controls

Our discussion of control variates has thus far focused exclusively on *linear* controls, meaning estimators of the form

$$\bar{Y} - b^\top(\bar{X} - \mathsf{E}[X]), \tag{4.25}$$

with the vector b either known or estimated. There are, however, other ways one might use the discrepancy between $\bar{X}$ and $\mathsf{E}[X]$ to try to improve the estimator $\bar{Y}$ in estimating $\mathsf{E}[Y]$. For example, in the case of scalar X, the estimator

$$\bar{Y}\frac{\mathsf{E}[X]}{\bar{X}}$$

adjusts $\bar{Y}$ upward if $0 < \bar{X} < \mathsf{E}[X]$, downward if $0 < \mathsf{E}[X] < \bar{X}$, and thus may be attractive if X_i and Y_i are positively correlated. Similarly, the estimator $\bar{Y}\bar{X}/\mathsf{E}[X]$ may have merit if the X_i and Y_i are negatively correlated. Other estimators of this type include

$$\bar{Y}\exp\left(\bar{X} - \mathsf{E}[X]\right) \quad \text{and} \quad \bar{Y}^{(\bar{X}/\mathsf{E}[X])}.$$

In each case, the convergence of $\bar{X}$ to $\mathsf{E}[X]$ ensures that the adjustment to $\bar{Y}$ vanishes as the sample size increases, just as in (4.25). But for any finite number of replications, the variance of the adjusted estimator could be larger or smaller than that of $\bar{Y}$.

These are examples of *nonlinear* control variate estimators. They are all special cases of estimators of the form $h(\bar{X}, \bar{Y})$ for functions h satisfying

$$h(\mathsf{E}[X], y) = y \text{ for all } y.$$

The difference between the controlled estimator $h(\bar{X}, \bar{Y})$ and $\bar{Y}$ thus depends on the deviation of $\bar{X}$ from $\mathsf{E}[X]$.

Although the introduction of nonlinear controls would appear to substantially enlarge the class of candidate estimators, it turns out that in large samples, a nonlinear control variate estimator based on a smooth h is equivalent to an ordinary linear control variate estimator. This was demonstrated in Glynn and Whitt [159], who note a related observation in Cheng and Feast [84]. We present the analysis leading to this conclusion and then discuss its implications.

Delta Method

The main tool for the large-sample analysis of nonlinear control variate estimators is the *delta method*. This is a result providing a central limit theorem for functions of sample means. To state it generally, we let ξ_i, $i = 1, 2, \ldots$ be i.i.d. random vectors in $\Re^k$ with mean vector μ and covariance matrix Σ. The sample mean $\bar{\xi}$ of $\xi_1, \ldots, \xi_n$ satisfies the central limit theorem

$$\sqrt{n}[\bar{\xi} - \mu] \Rightarrow N(0, \Sigma).$$

Now let $h : \Re^k \to \Re$ be continuously differentiable in a neighborhood of μ and suppose the partial derivatives of h at μ are not all zero. For sufficiently large n, a Taylor approximation gives

$$h(\bar{\xi}) = h(\mu) + \nabla h(\zeta_n)[\bar{\xi} - \mu],$$

with ∇h the gradient of h (a row vector) and ζ_n a point on the line segment joining μ and $\bar{\xi}$. As $n \to \infty$, $\bar{\xi} \to \mu$ and thus $\zeta_n \to \mu$ as well; continuity of the gradient implies $\nabla h(\zeta_n) \to \nabla h(\mu)$. Thus, for large n, the error $h(\bar{\xi}) - h(\mu)$ is approximately the inner product of the constant vector $\nabla h(\mu)$ and the asymptotically normal vector $\bar{\xi} - \mu$, and is itself asymptotically normal. More precisely,

$$\sqrt{n}[h(\bar{\xi}) - h(\mu)] \Rightarrow N(0, \nabla h(\mu)\Sigma\nabla h(\mu)^{\top}). \tag{4.26}$$

See also Section 3.3 of Serfling [326], for example.

For the application to nonlinear controls, we replace ξ_i with (X_i, Y_i), μ with $(\mathsf{E}[X], \mathsf{E}[Y])$, and Σ with

$$\Sigma = \begin{pmatrix} \Sigma_X & \Sigma_{XY} \\ \Sigma_{XY}^{\top} & \sigma_Y^2 \end{pmatrix},$$

the covariance matrix in (4.11). From the delta method, we know that the nonlinear control variate estimator is asymptotically normal with

$$\sqrt{n}[h(\bar{X}, \bar{Y}) - \mathsf{E}[Y]] \Rightarrow N(0, \sigma_h^2),$$

(recall that $h(\mathsf{E}[X], \mathsf{E}[Y]) = \mathsf{E}[Y]$) and

$$\sigma_h^2 = \left(\frac{\partial h}{\partial y}\right)^2 \sigma_Y^2 + 2\left(\frac{\partial h}{\partial y}\right)\nabla_x h\Sigma_{XY} + \nabla_x h\Sigma_X\nabla_x h^{\top},$$

with $\nabla_x h$ denoting the gradient of h with respect to the elements of X and with all derivatives evaluated at $(\mathsf{E}[X], \mathsf{E}[Y])$. Because $h(\mathsf{E}[X], \cdot)$ is the identity, the partial derivative of h with respect to its last argument equals 1 at $(\mathsf{E}[X], \mathsf{E}[Y])$, so

$$\sigma_h^2 = \sigma_Y^2 + 2\nabla_x h\Sigma_{XY} + \nabla_x h\Sigma_X\nabla_x h^{\top}.$$

But this is precisely the variance of

$$Y_i - b^{\top}(X_i - \mathsf{E}[X])$$

with $b = -\nabla_x h(\mathsf{E}[X], \mathsf{E}[Y])$; see (4.12). Thus, the distribution of the nonlinear control variate estimator using $\bar{X}$ is asymptotically the same as the distribution of an ordinary linear control variate estimator using $\bar{X}$ and a specific vector of coefficients b. In particular, the limiting variance parameter σ_h^2 can be no smaller than the optimal variance that would be derived from using the optimal vector b^*.

A negative reading of this result leads to the conclusion that nonlinear controls add nothing beyond what can be achieved using linear controls. A somewhat more positive and more accurate interpretation would be that whatever advantages a nonlinear control variate estimator may have must be limited to small samples. "Small" may well include all relevant sample sizes in specific applications. The delta method tells us that asymptotically only the linear part of h matters, but if h is highly nonlinear a very large sample may be required for this asymptotic conclusion to be relevant. For fixed n, each of the examples with which we opened this section may perform rather differently from a linear control.

It should also be noted that in the linear control variate estimator to which any nonlinear control variate estimator is ultimately equivalent, the coefficient b is implicitly determined by the function h. In particular, using a nonlinear control does not entail estimating this coefficient. In some cases, a nonlinear control may be effective because $-\nabla_x h$ is close to optimal but need not be estimated.

4.2 Antithetic Variates

The method of *antithetic variates* attempts to reduce variance by introducing negative dependence between pairs of replications. The method can take various forms; the most broadly applicable is based on the observation that if U is uniformly distributed over $[0, 1]$, then $1 - U$ is too. Hence, if we generate a path using as inputs $U_1, \ldots, U_n$, we can generate a second path using $1 - U_1, \ldots, 1 - U_n$ without changing the law of the simulated process. The variables U_i and $1 - U_i$ form an antithetic pair in the sense that a large value of one is accompanied by a small value of the other. This suggests that an unusually large or small output computed from the first path may be balanced by the value computed from the antithetic path, resulting in a reduction in variance.

These observations extend to other distributions through the inverse transform method: $F^{-1}(U)$ and $F^{-1}(1 - U)$ both have distribution F but are antithetic to each other because F^{-1} is monotone. For a distribution symmetric about the origin, $F^{-1}(1 - u)$ and $F^{-1}(u)$ have the same magnitudes but opposite signs. In particular, in a simulation driven by independent standard normal random variables, antithetic variates can be implemented by pairing a sequence $Z_1, Z_2, \ldots$ of i.i.d. $N(0, 1)$ variables with the sequence $-Z_1, -Z_2, \ldots$ of i.i.d. $N(0, 1)$ variables, whether or not they are sampled through the inverse transform method. If the Z_i are used to simulate the increments of a Brownian path, then the $-Z_i$ simulate the increments of the reflection of the path about the origin. This again suggests that running a pair of simulations using the original path and then its reflection may result in lower variance.

To analyze this approach more precisely, suppose our objective is to estimate an expectation $\mathsf{E}[Y]$ and that using some implementation of antithetic sampling produces a sequence of pairs of observations $(Y_1, \tilde{Y}_1), (Y_2, \tilde{Y}_2), \ldots, (Y_n, \tilde{Y}_n)$. The key features of the antithetic variates method are the following:

○ the *pairs* $(Y_1, \tilde{Y}_1), (Y_2, \tilde{Y}_2), \ldots, (Y_n, \tilde{Y}_n)$ are i.i.d.;
○ for each i, Y_i and $\tilde{Y}_i$ have the same distribution, though ordinarily they are not independent.

We use Y generically to indicate a random variable with the common distribution of the Y_i and $\tilde{Y}_i$.

The antithetic variates estimator is simply the average of all $2n$ observations,

$$\hat{Y}_{AV} = \frac{1}{2n} \left(\sum_{i=1}^{n} Y_i + \sum_{i=1}^{n} \tilde{Y}_i \right) = \frac{1}{n} \sum_{i=1}^{n} \left(\frac{Y_i + \tilde{Y}_i}{2} \right). \tag{4.27}$$

The rightmost representation in (4.27) makes it evident that $\hat{Y}_{AV}$ is the sample mean of the n *independent* observations

$$\left(\frac{Y_1 + \tilde{Y}_1}{2} \right), \left(\frac{Y_2 + \tilde{Y}_2}{2} \right), \dots, \left(\frac{Y_n + \tilde{Y}_n}{2} \right). \tag{4.28}$$

The central limit theorem therefore applies and gives

$$\frac{\hat{Y}_{AV} - E[Y]}{\sigma_{AV}/\sqrt{n}} \Rightarrow N(0, 1)$$

with

$$\sigma_{AV}^2 = \text{Var} \left[\frac{Y_i + \tilde{Y}_i}{2} \right].$$

As usual, this limit in distribution continues to hold if we replace σ_{AV} with s_{AV}, the sample standard deviation of the n values in (4.28). This provides asymptotic justification for a $1 - \delta$ confidence interval of the form

$$\hat{Y}_{AV} \pm z_{\delta/2} \frac{s_{AV}}{\sqrt{n}},$$

where $1 - \Phi(z_{\delta/2}) = \delta/2$.

Under what conditions is an antithetic variates estimator to be preferred to an ordinary Monte Carlo estimator based on independent replications? To make this comparison, we assume that the computational effort required to generate a pair $(Y_i, \tilde{Y}_i)$ is approximately twice the effort required to generate Y_i. In other words, we ignore any potential computational savings from, for example, flipping the signs of previously generated $Z_1, Z_2, \dots$ rather than generating new normal variables. This is appropriate if the computational cost of generating these inputs is a small fraction of the total cost of simulating Y_i. Under this assumption, the effort required to compute $\hat{Y}_{AV}$ is approximately that required to compute the sample mean of $2n$ independent replications, and it is therefore meaningful to compare the variances of these two estimators. Using antithetics reduces variance if

$$\text{Var} \left[\hat{Y}_{AV} \right] < \text{Var} \left[\frac{1}{2n} \sum_{i=1}^{2n} Y_i \right];$$

i.e., if

$$\mathsf{Var}\left[Y_i + \tilde{Y}_i\right] < 2\mathsf{Var}[Y_i].$$

The variance on the left can be written as

$$\mathsf{Var}\left[Y_i + \tilde{Y}_i\right] = \mathsf{Var}[Y_i] + \mathsf{Var}[\tilde{Y}_i] + 2\mathsf{Cov}[Y_i, \tilde{Y}_i]$$
$$= 2\mathsf{Var}[Y_i] + 2\mathsf{Cov}[Y_i, \tilde{Y}_i],$$

using the fact that Y_i and $\tilde{Y}_i$ have the same variance if they have the same distribution. Thus, the condition for antithetic sampling to reduce variance becomes

$$\mathsf{Cov}\left[Y_i, \tilde{Y}_i\right] < 0. \qquad (4.29)$$

Put succinctly, this condition requires that negative dependence in the inputs (whether U and $1 - U$ or Z and $-Z$) produce negative correlation between the outputs of paired replications. A simple sufficient condition ensuring this is monotonicity of the mapping from inputs to outputs defined by a simulation algorithm. To state this precisely and to give a general formulation, suppose the inputs to a simulation are independent random variables $X_1, \ldots, X_m$. Suppose that Y is an increasing function of these inputs and $\tilde{Y}$ is a decreasing function of the inputs; then

$$\mathsf{E}[Y\tilde{Y}] \leq \mathsf{E}[Y]\mathsf{E}[\tilde{Y}].$$

This is a special case of more general properties of *associated* random variables, in the sense of Esary, Proschan, and Walkup [113]. Observe that if $Y = f(U_1, \ldots, U_d)$ or $Y = f(Z_1, \ldots, Z_d)$ for some increasing function f, then $\tilde{Y} = f(1 - U_1, \ldots, 1 - U_d)$ and $\tilde{Y} = f(-Z_1, \ldots, -Z_d)$ are decreasing functions of $(U_1, \ldots, U_d)$ and $(Z_1, \ldots, Z_d)$, respectively. The requirement that the simulation map inputs to outputs monotonically is rarely satisfied exactly, but provides some qualitative insight into the scope of the method.

The antithetic pairs $(U, 1 - U)$ with $U \sim \text{Unif}[0,1]$ and $(Z, -Z)$ with $Z \sim N(0, 1)$ share an additional relevant property: in each case, the average of the paired values is the population mean, because

$$\frac{U + (1 - U)}{2} = 1/2 \quad \text{and} \quad \frac{Z + (-Z)}{2} = 0.$$

It follows that if the output Y is a linear function of inputs $(U_1, \ldots, U_d)$ or $(Z_1, \ldots, Z_d)$, then antithetic sampling results in a zero-variance estimator. Of course, in the linear case simulation would be unnecessary, but this observation suggests that antithetic variates will be very effective if the mapping from inputs to outputs is close to linear.

Variance Decomposition

Antithetic variates eliminate the variance due to the *antisymmetric* part of an integrand, in a sense we now develop. For simplicity, we restrict attention to the case of standard normal inputs, but our observations apply equally well to any other distribution symmetric about the origin and apply with minor modifications to uniformly distributed inputs.

Suppose, then, that $Y = f(Z)$ with $Z = (Z_1, \ldots, Z_d) \sim N(0, I)$. Define the symmetric and antisymmetric parts of f, respectively, by

$$f_0(z) = \frac{f(z) + f(-z)}{2} \quad \text{and} \quad f_1(z) = \frac{f(z) - f(-z)}{2}.$$

Clearly, $f = f_0 + f_1$; moreover, this gives an orthogonal decomposition of f in the sense that $f_0(Z)$ and $f_1(Z)$ are uncorrelated:

$$\begin{aligned}
\mathsf{E}[f_0(Z)f_1(Z)] &= \frac{1}{4}\mathsf{E}[f^2(Z) - f^2(-Z)] \\
&= 0 \\
&= \mathsf{E}[f_0(Z)]\mathsf{E}[f_1(Z)].
\end{aligned}$$

It follows that
$$\mathsf{Var}[f(Z)] = \mathsf{Var}[f_0(Z)] + \mathsf{Var}[f_1(Z)]. \tag{4.30}$$

The first term on the right is the variance of an estimate of $\mathsf{E}[f(Z)]$ based on an antithetic pair $(Z, -Z)$. Thus, antithetic sampling eliminates all variance if f is antisymmetric ($f = f_1$) and it eliminates no variance if f is symmetric ($f = f_0$).

Fox [127] advocates the use of antithetic sampling as the first step of a more elaborate framework, in order to eliminate the variance due to the linear (or, more generally, the antisymmetric) part of f.

Systematic Sampling

Antithetic sampling pairs a standard normal vector $Z = (Z_1, \ldots, Z_d)$ with its reflection $-Z = (-Z_1, \ldots, -Z_d)$, but it is natural to consider other vectors formed by changing the signs of the components of Z. Generalizing still further leads us to consider transformations $T : \Re^d \to \Re^d$ (such as multiplication by an orthogonal matrix) with the property that $TZ \sim N(0, I)$ whenever $Z \sim N(0, I)$. This property implies that the iterated transformations T^2Z, $T^3Z, \ldots$ will also have standard normal distributions. Suppose that T^k is the identity for some k. The usual antithetic transformation has $k = 2$, but by considering other rotations and reflections of $\Re^d$, it is easy to construct examples with larger values of k.

Define

$$f_0(z) = \frac{1}{k}\sum_{i=1}^{k} f(T^i Z) \quad \text{and} \quad f_1(z) = f(z) - f_0(z).$$

We clearly have $\mathsf{E}[f_0(Z)] = \mathsf{E}[f(Z)]$. The estimator $f_0(Z)$ generalizes the antithetic variates estimator; in the survey sampling literature, methods of this type are called *systematic* sampling because after the initial random drawing of Z, the $k - 1$ subsequent points are obtained through deterministic transformations of Z.

The representation $f(Z) = f_0(Z) + f_1(Z)$ again gives an orthogonal decomposition. To see this, first observe that

$$
\begin{aligned}
\mathsf{E}[f_0(Z)^2] &= \frac{1}{k} \sum_{i=1}^{k} \mathsf{E}\left[f(T^i Z) \cdot \frac{1}{k} \sum_{j=1}^{k} f(T^j T^i Z) \right] \\
&= \frac{1}{k} \sum_{i=1}^{k} \mathsf{E}\left[f(T^i Z) \cdot f_0(T^i Z) \right] \\
&= \mathsf{E}[f(Z) f_0(Z)],
\end{aligned}
$$

so

$$
\mathsf{E}[f_0(Z) f_1(Z)] = \mathsf{E}[f_0(Z)(f(Z) - f_0(Z))] = 0.
$$

Thus, (4.30) continues to hold under the new definitions of f_0 and f_1. Assuming the $f(T^i Z)$, $i = 1, \ldots, k$, require approximately equal computing times, the estimator $f_0(Z)$ beats ordinary Monte Carlo if

$$
\mathsf{Var}[f_0(Z)] < \frac{1}{k} \mathsf{Var}[f(Z)].
$$

The steps leading to (4.29) generalize to the requirement

$$
\sum_{j=1}^{k-1} \mathsf{Cov}[f(Z), f(T^j Z)] < 0.
$$

This condition is usually at least as difficult to satisfy as the simple version (4.29) for ordinary antithetic sampling.

For a more general formulation of antithetic sampling and for historical remarks, see Hammersley and Handscomb [169]. Boyle [52] is an early application in finance. Other work on antithetic variates includes Fishman and Huang [122] and Rubinstein, Samorodnitsky, and Shaked [312].

4.3 Stratified Sampling

4.3.1 Method and Examples

Stratified sampling refers broadly to any sampling mechanism that constrains the fraction of observations drawn from specific subsets (or *strata*) of the sample space. Suppose, more specifically, that our goal is to estimate $\mathsf{E}[Y]$ with Y real-valued, and let $A_1, \ldots, A_K$ be disjoint subsets of the real line for which $P(Y \in \cup_i A_i) = 1$. Then

$$\mathsf{E}[Y] = \sum_{i=1}^{K} P(Y \in A_i)\mathsf{E}[Y|Y \in A_i] = \sum_{i=1}^{K} p_i\mathsf{E}[Y|Y \in A_i] \qquad (4.31)$$

with $p_i = P(Y \in A_i)$. In random sampling, we generate independent $Y_1, \ldots, Y_n$ having the same distribution as Y. The fraction of these samples falling in A_i will not in general equal p_i, though it would approach p_i as the sample size n increased. In stratified sampling, we decide in advance what fraction of the samples should be drawn from each stratum A_i; each observation drawn from A_i is constrained to have the distribution of Y conditional on $Y \in A_i$.

The simplest case is proportional sampling, in which we ensure that the fraction of observations drawn from stratum A_i matches the theoretical probability $p_i = P(Y \in A_i)$. If the total sample size is n, this entails generating $n_i = np_i$ samples from A_i. (To simplify the discussion, we ignore rounding and assume np_i is an integer instead of writing $\lfloor np_i \rfloor$.) For each $i = 1, \ldots, K$, let $Y_{ij}, j = 1, \ldots, n_i$ be independent draws from the conditional distribution of Y given $Y \in A_i$. An unbiased estimator of $\mathsf{E}[Y|Y \in A_i]$ is provided by the sample mean $(Y_{i1} + \cdots + Y_{in_i})/n_i$ of observations from the ith stratum. It follows from (4.31) that an unbiased estimator of $\mathsf{E}[Y]$ is provided by

$$\hat{Y} = \sum_{i=1}^{K} p_i \cdot \frac{1}{n_i} \sum_{j=1}^{n_i} Y_{ij} = \frac{1}{n} \sum_{i=1}^{K} \sum_{j=1}^{n_i} Y_{ij}. \qquad (4.32)$$

This estimator should be contrasted with the usual sample mean $\bar{Y} = (Y_1 + \cdots + Y_n)/n$ of a random sample of size n. Compared with $\bar{Y}$, the stratified estimator $\hat{Y}$ eliminates sampling variability across strata without affecting sampling variability within strata.

We generalize this formulation in two simple but important ways. First, we allow the strata to be defined in terms of a second variable X. This *stratification variable* could take values in an arbitrary set; to be concrete we assume it is $\Re^d$-valued and thus take the strata A_i to be disjoint subsets of $\Re^d$ with $P(X \in \cup_i A_i) = 1$. The representation (4.31) generalizes to

$$\mathsf{E}[Y] = \sum_{i=1}^{K} P(X \in A_i)\mathsf{E}[Y|X \in A_i] = \sum_{i=1}^{K} p_i\mathsf{E}[Y|X \in A_i], \qquad (4.33)$$

where now $p_i = P(X \in A_i)$. In some applications, Y is a function of X (for example, X may be a discrete path of asset prices and Y the discounted payoff of a derivative security), but more generally they may be dependent without either completely determining the other. To use (4.33) for stratified sampling, we need to generate pairs $(X_{ij}, Y_{ij}), j = 1, \ldots, n_i$, having the conditional distribution of (X, Y) given $X \in A_i$.

As a second extension of the method, we allow the stratum allocations $n_1, \ldots, n_K$ to be arbitrary (while summing to n) rather than proportional to

$p_1, \ldots, p_K$. In this case, the first representation in (4.32) remains valid but the second does not. If we let $q_i = n_i/n$ be the fraction of observations drawn from stratum i, $i = 1, \ldots, K$, we can write

$$\hat{Y} = \sum_{i=1}^{K} p_i \cdot \frac{1}{n_i} \sum_{j=1}^{n_i} Y_{ij} = \frac{1}{n} \sum_{i=1}^{K} \frac{p_i}{q_i} \sum_{j=1}^{n_i} Y_{ij}. \tag{4.34}$$

By minimizing the variance of this estimator over the q_i, we can find an allocation rule that is at least as effective as a proportional allocation. We return to this point later in this section.

From this introduction it should be clear that the use of stratified sampling involves consideration of two issues:

○ choosing the stratification variable X, the strata $A_1, \ldots, A_K$, and the allocation $n_1, \ldots, n_K$;
○ generating samples from the distribution of (X, Y) conditional on $X \in A_i$.

In addressing the first issue we will see that stratified sampling is most effective when the variability of Y within each stratum is small. Solutions to the second issue are best illustrated through examples.

Example 4.3.1 *Stratifying uniforms.* Perhaps the simplest application of stratified sampling stratifies the uniformly distributed random variables that drive a simulation. Partition the unit interval $(0, 1)$ into the n strata

$$A_1 = \left(0, \frac{1}{n}\right], \quad A_2 = \left(\frac{1}{n}, \frac{2}{n}\right], \quad \ldots, \quad A_n = \left(\frac{n-1}{n}, 1\right).$$

Each of these intervals has probability $1/n$ under the uniform distribution, so in a proportional allocation we should draw one sample from each stratum. (The sample size n and the number of strata K are equal in this example.) Let $U_1, \ldots, U_n$ be independent and uniformly distributed between 0 and 1 and let

$$V_i = \frac{i-1}{n} + \frac{U_i}{n}, \quad i = 1, \ldots, n. \tag{4.35}$$

Each V_i is uniformly distributed between $(i-1)/n$ and i/n, which is to say that V_i has the conditional distribution of U given $U \in A_i$ for $U \sim \text{Unif}[0,1]$. Thus, $V_1, \ldots, V_n$ constitute a stratified sample from the uniform distribution. (In working with the unit interval and the subintervals A_i, we are clearly free to define these to be open or closed on the left or right; in each setting, we adopt whatever convention is most convenient.)

Suppose $Y = f(U)$ so that $\mathsf{E}[Y]$ is simply the integral of f over the unit interval. Then the stratified estimator

$$\hat{Y} = \frac{1}{n} \sum_{i=1}^{n} f(V_i)$$

is similar to the deterministic midpoint integration rule

$$\frac{1}{n}\sum_{i=1}^{n} f\left(\frac{2i-1}{2n}\right)$$

based on the value of f at the midpoints of the A_i. A feature of the randomization in the stratified estimator is that it makes $\hat{Y}$ unbiased.

This example easily generalizes to partitions of $(0,1)$ into intervals of unequal lengths. If $A_i = (a_i, b_i]$, then the conditional distribution of U given $U \in A_i$ is uniform between a_i and b_i; we can sample from this conditional distribution by setting $V = a_i + U(b_i - a_i)$. $\square$

Example 4.3.2 *Stratifying nonuniform distributions.* Let F be a cumulative distribution function on the real line and let

$$F^{-1}(u) = \inf\{x : F(x) \le u\}$$

denote its inverse as defined in Section 2.2.1. Given probabilities $p_1, \ldots, p_K$ summing to 1, define $a_0 = -\infty$,

$$a_1 = F^{-1}(p_1), \; a_2 = F^{-1}(p_1 + p_2), \; \ldots, \; a_K = F^{-1}(p_1 + \cdots + p_K) = F^{-1}(1).$$

Define strata

$$A_1 = (a_0, a_1], \; A_2 = (a_1, a_2], \; \ldots, \; A_K = (a_{K-1}, a_K]$$

or with $A_K = (a_{K-1}, a_K)$ if $a_K = \infty$. By construction, each stratum A_i has probability p_i under F; for if Y has distribution F, then

$$P(Y \in A_i) = F(a_i) - F(a_{i-1}) = p_i.$$

Thus, defining strata for F with specified probabilities is straightforward, provided one can find the quantiles a_i. Figure 4.4 displays ten equiprobable $(p_i = 1/K)$ strata for the standard normal distribution.

To use the sets $A_1, \ldots, A_K$ for stratified sampling, we need to be able to generate samples of Y conditional on $Y \in A_i$. As demonstrated in Example 2.2.5, this is easy using the inverse transform method. If $U \sim \text{Unif}[0,1]$, then

$$V = a_{i-1} + U(a_i - a_{i-1})$$

is uniformly distributed between a_{i-1} and a_i and then $F^{-1}(V)$ has the distribution of Y conditional on $Y \in A_i$.

Figure 4.5 illustrates the difference between stratified and random sampling from the standard normal distribution. The left panel is a histogram of 500 observations, five from each of 100 equiprobable strata; the right panel is a histogram of 500 independent draws from the normal distribution. Stratification clearly produces a better approximation to the underlying distribution.

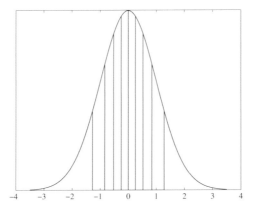

Fig. 4.4. A partition of the real line into ten intervals of equal probability under the standard normal distribution. The area under the normal density over each interval is $1/10$.

How might we use stratified samples from the normal distribution in simulating paths of a stochastic process? It would *not* be legitimate to use one value from each of 100 strata to generate 100 steps of a single Brownian path: the increments of Brownian motion are independent but the stratified values are not, and ignoring this dependence would produce nonsensical results. In contrast, we could validly use the stratified values to generate the first increment of 100 replications of a single Brownian path (or the terminal values of the paths, as explained in Section 4.3.2). In short, in using stratified sampling or any other variance reduction technique, we are free to introduce dependence *across* replications but not *within* replications. □

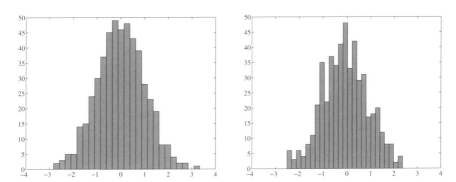

Fig. 4.5. Comparison of stratified sample (left) and random sample (right). The stratified sample uses 100 equiprobable strata with five samples from each stratum; the random sample consists of 500 independent draws from the normal distribution. Both histograms use 25 bins.

Example 4.3.3 *Stratification through acceptance-rejection.* A crude but al-most universally applicable method for generating samples conditional on a stratum generates unconditional samples and keeps those that fall in the target set. This is the method described in Example 2.2.8 for conditional sampling, and may be viewed as a form of acceptance-rejection in which the acceptance probability is always 0 or 1.

To describe this in more detail, we use the notation of (4.33). Our goal is to generate samples of the pair (X, Y) using strata $A_1, \ldots, A_K$ for X, with n_i samples to be generated conditional on $X \in A_i$, $i = 1, \ldots, K$. Given a mechanism for generating unconditional samples from the distribution of (X, Y), we can repeatedly generate such samples until we have produced n_i samples with $X \in A_i$ for each $i = 1, \ldots, K$; any extra samples generated from a stratum are simply rejected.

The efficiency of this method depends on the computational cost of gen-erating pairs (X, Y) and determining the stratum in which X falls. It also depends on the stratum probabilities: if $P(X \in A_i)$ is small, a large number of candidates may be required to produce n_i samples from A_i. These com-putational costs must be balanced against the reduction in variance achieved through stratification. Glasserman, Heidelberger, and Shahabuddin [143] an-alyze the overhead from rejected samples in this method based on a Poisson approximation to the arrival of samples from each stratum. □

Example 4.3.4 *Stratifying the unit hypercube.* The methods described in Ex-amples 4.3.1 and 4.3.2 extend, in principle, to multiple dimensions. Using the inverse transform method, a vector $(X_1, \ldots, X_d)$ of independent random vari-ables can be represented as $(F_1^{-1}(U_1), \ldots, F_d^{-1}(U_d))$ with F_i the distribution of X_i and $U_1, \ldots, U_d$ independent and uniform over $[0, 1)$. In this sense, it suf-fices to consider the uniform distribution over the d-dimensional hypercube $[0, 1)^d$. (In the case of dependent $X_1, \ldots, X_d$, replace F_i with the conditional distribution of X_i given $X_1, \ldots, X_{i-1}$.) In stratifying the unit hypercube with respect to the uniform distribution, it is convenient to take the strata to be products of intervals because the probability of such a set is easily calculated and because it is easy to sample uniformly from such a set by applying a transformation like (4.35) to each coordinate.

Suppose, for example, that we stratify the jth coordinate of the hypercube into K_j intervals of equal length. Each stratum of the hypercube has the form

$$\prod_{j=1}^{d} \left[\frac{i_j - 1}{K_j}, \frac{i_j}{K_j} \right), \quad i_j \in \{1, \ldots, K_j\}$$

and has probability $1/(K_1 \cdots K_d)$. To generate a vector V uniformly distrib-uted over this set, generate $U_1, \ldots, U_d$ independently from Unif[0,1) and define the jth coordinate of V to be

$$V_j = \frac{i_j - 1 + U_j}{K_j}, \quad j = 1, \ldots, d. \tag{4.36}$$

In this example, the total number of strata is $K_1 \cdots K_d$. Generating at least one point from each stratum therefore requires a sample size at least this large. Unless the K_j are quite small (in which case stratification may provide little benefit), this is likely to be prohibitive for d larger than 5, say. In Section 4.4 and Chapter 5, we will see methods related to stratified sampling that are better suited to higher dimensions. □

Output Analysis

We now turn to the problem of interval estimation for $\mu \overset{\triangle}{=} E[Y]$ using stratified sampling. As in (4.33), let $A_1, \ldots, A_K$ denote strata for a stratification variable X and let Y_{ij} have the distribution of Y conditional on $X \in A_i$. For $i = 1, \ldots, K$, let

$$\mu_i = E[Y_{ij}] = E[Y|X \in A_i] \tag{4.37}$$
$$\sigma_i^2 = \text{Var}[Y_{ij}] = \text{Var}[Y|X \in A_i]. \tag{4.38}$$

Let $p_i = P(X \in A_i)$, $i = 1, \ldots, K$, denote the stratum probabilities; we require these to be strictly positive and to sum to 1. Fix an allocation $n_1, \ldots, n_K$ with all $n_i \geq 1$ and $n_1 + \cdots + n_K = n$. Let $q_i = n_i/n$ denote the fraction of samples allocated to the ith stratum. For any such allocation the estimator $\hat{Y}$ in (4.33) is unbiased because

$$E[\hat{Y}] = \sum_{i=1}^{K} p_i \cdot \frac{1}{n_i} \sum_{j=1}^{n_i} E[Y_{ij}] = \sum_{i=1}^{K} p_i \mu_i = \mu.$$

The variance of $\hat{Y}$ is given by

$$\text{Var}[\hat{Y}] = \sum_{i=1}^{K} p_i^2 \text{Var}\left[\frac{1}{n_i} \sum_{j=1}^{n_i} Y_{ij} \right] = \sum_{i=1}^{K} p_i^2 \frac{\sigma_i^2}{n_i} = \frac{\sigma^2(q)}{n},$$

with

$$\sigma^2(q) = \sum_{i=1}^{K} \frac{p_i^2}{q_i} \sigma_i^2. \tag{4.39}$$

For each stratum A_i, the samples $Y_{i1}, Y_{i2}, \ldots$ are i.i.d. with mean μ_i and variance σ_i^2 and thus satisfy

$$\frac{1}{\sqrt{\lfloor nq_i \rfloor}} \sum_{j=1}^{\lfloor nq_i \rfloor} (Y_{ij} - \mu_i) \Rightarrow N(0, \sigma_j^2),$$

as $n \to \infty$ with $q_1, \ldots, q_K$ fixed. The centered and scaled estimator $\sqrt{n}(\hat{Y} - \mu)$ can be written as

$$\sqrt{n}(\hat{Y} - \mu) = \sqrt{n} \sum_{i=1}^{K} p_i \left(\frac{1}{\lfloor nq_i \rfloor} \sum_{j=1}^{\lfloor nq_i \rfloor} (Y_{ij} - \mu_i) \right)$$

$$\approx \sum_{i=1}^{K} \frac{p_i}{\sqrt{q_i}} \left(\frac{1}{\sqrt{\lfloor nq_i \rfloor}} \sum_{j=1}^{\lfloor nq_i \rfloor} (Y_{ij} - \mu_i) \right),$$

the approximation holding in the sense that the ratio of the two expressions approaches 1 as $n \to \infty$. This shows that $\sqrt{n}(\hat{Y} - \mu)$ is asymptotically a linear combination (with coefficients $p_i/\sqrt{q_i}$) of independent normal random variables (with mean 0 and variances σ_i^2). It follows that

$$\sqrt{n}(\hat{Y} - \mu) \Rightarrow N(0, \sigma^2(q))$$

with $\sigma^2(q)$ as defined in (4.39). This limit holds as the sample size n increases with the number of strata K held fixed.

A consequence of this central limit theorem for $\hat{Y}$ is the asymptotic validity of

$$\hat{Y} \pm z_{\delta/2} \frac{\sigma(q)}{\sqrt{n}} \tag{4.40}$$

as a $1 - \delta$ confidence interval for μ, with $z_{\delta/2} = \Phi^{-1}(1 - \delta/2)$. In practice, $\sigma^2(q)$ is typically unknown but can be consistently estimated using

$$s^2(q) = \sum_{i=1}^{K} \frac{p_i^2}{q_i} s_i^2,$$

where s_i^2 is the sample standard deviation of $Y_{i1}, \ldots, Y_{in_i}$.

Alternatively, one can estimate $\sigma^2(q)$ through independent replications of $\hat{Y}$. More precisely, suppose the sample size n can be expressed as mk with m and k integers and $m \geq 2$. Suppose $k_i = q_i k$ is an integer for all $i = 1, \ldots, K$ and note that $n_i = mk_i$. Then $\hat{Y}$ is the average of m independent stratified estimators $\hat{Y}_1, \ldots, \hat{Y}_m$, each of which allocates a fraction q_i of observations to stratum i and has a total sample size of k. Each $\hat{Y}_j$ thus has variance $\sigma^2(q)/k$; because $\hat{Y}$ is the average of the $\hat{Y}_1, \ldots, \hat{Y}_m$, an asymptotically (as $m \to \infty$) valid confidence interval for μ is provided by

$$\hat{Y} \pm z_{\delta/2} \frac{\sigma(q)/\sqrt{k}}{\sqrt{m}}. \tag{4.41}$$

This reduces to (4.40), but $\sigma(q)/\sqrt{k}$ can now be consistently estimated using the sample standard deviation of $\hat{Y}_1, \ldots, \hat{Y}_m$. This is usually more convenient than estimating all the stratum variances σ_i^2, $i = 1, \ldots, K$.

In this formulation, each $\hat{Y}_j$ may be thought of as a batch with sample size k and the original estimator $\hat{Y}$ as the sample mean of m independent batches. Given a total sample size n, is it preferable to have at least m observations from each stratum, as in this setting, or to increase the number of strata so that only one observation is drawn from each? A larger m should improve our estimate of $\sigma(q)$ and the accuracy of the normal approximation implicit in the confidence intervals above. However, we will see below (cf. (4.46)) that taking finer strata reduces variance. Thus, as is often the case, we face a tradeoff between reducing variance and accurately measuring variance.

Optimal Allocation

In the case of a proportional allocation of samples to strata, $q_i = p_i$ and the variance parameter $\sigma^2(q)$ simplifies to

$$\sum_{i=1}^{K} \frac{p_i^2}{q_i} \sigma_i^2 = \sum_{i=1}^{K} p_i \sigma_i^2. \tag{4.42}$$

To compare this to the variance without stratification, observe that

$$\mathsf{E}[Y^2] = \sum_{i=1}^{K} p_i \mathsf{E}[Y^2 | X \in A_i] = \sum_{i=1}^{K} p_i(\sigma_i^2 + \mu_i^2),$$

so using $\mu = \sum_{i=1}^{K} p_i \mu_i$ we get

$$\mathsf{Var}[Y] = \mathsf{E}[Y^2] - \mu^2 = \sum_{i=1}^{K} p_i \sigma_i^2 + \sum_{i=1}^{K} p_i \mu_i^2 - \left(\sum_{i=1}^{K} p_i \mu_i \right)^2. \tag{4.43}$$

By Jensen's inequality,

$$\sum_{i=1}^{K} p_i \mu_i^2 \geq \left(\sum_{i=1}^{K} p_i \mu_i \right)^2$$

with strict inequality unless all μ_i are equal. Thus, comparing (4.42) and (4.43), we conclude that *stratified sampling with a proportional allocation can only decrease variance.*

Optimizing the allocation can produce further variance reduction. Minimizing $\sigma^2(q)$ subject to the constraint that $(q_1, \ldots, q_K)$ be a probability vector yields the optimal allocation

$$q_i^* = \frac{p_i \sigma_i}{\sum_{\ell=1}^{K} p_\ell \sigma_\ell}, \quad i = 1, \ldots, K.$$

In other words, the optimal allocation for each stratum is proportional to the product of the stratum probability and the stratum standard deviation. The optimal variance is thus

$$\sigma^2(q^*) = \sum_{i=1}^{K} \frac{p_i^2}{q_i^*} \sigma_i^2 = \left(\sum_{i=1}^{K} p_i \sigma_i \right)^2 .$$

Comparison with (4.42) indicates that the additional reduction in variance from optimizing the allocation is greatest when the stratum standard deviations vary widely.

In practice, the σ_i are rarely known so the optimal fractions q_i^* are not directly applicable. Nevertheless, it is often practical to use pilot runs to get estimates of the σ_i and thus of the q_i^*. The estimated optimal fractions can then be used to allocate samples to strata in a second (typically larger) set of runs.

In taking the optimal allocation to be the one that minimizes variance, we are implicitly assuming that the computational effort required to generate samples is the same across strata. But this assumption is not always appropriate. For example, in sampling from strata through acceptance-rejection as described in Example 4.3.3, the expected time required to sample from A_i is proportional to $1/p_i$. A more complete analysis should therefore account for differences in computational costs across strata.

Suppose, then, that τ_i denotes the expected computing time required to sample (X, Y) conditional on $X \in A_i$ and let s denote the total computing budget. Let $\hat{Y}(s)$ denote the stratified estimator produced with a budget s, assuming the fraction of samples allocated to stratum i is q_i. (This is asymptotically equivalent to assuming the fraction of the computational budget allocated to stratum i is proportional to $q_i \tau_i$.) Arguing much as in Section 1.1.3, we find that

$$\sqrt{s}[\hat{Y}(s) - \mu] \Rightarrow N\left(0, \sigma^2(q, \tau)\right),$$

with

$$\sigma^2(q, \tau) = \left(\sum_{i=1}^{K} \frac{p_i^2 \sigma_i^2}{q_i} \right) \left(\sum_{i=1}^{K} q_i \tau_i \right).$$

By minimizing this work-normalized variance parameter we find that the optimal allocation is

$$q_i^* = \frac{p_i \sigma_i / \sqrt{\tau_i}}{\sum_{\ell=1}^{K} p_\ell \sigma_\ell / \sqrt{\tau_\ell}},$$

which now accounts for differences in computational costs across strata. Like the σ_i, the τ_i can be estimated through pilot runs.

Variance Decomposition

The preceding discussion considers the allocation of samples to given strata. In order to consider the question of how strata should be selected in the first place, we now examine what part of the variance of Y is removed through stratification of X.

As before, let $A_1, \ldots, A_K$ be strata for X. Let $\eta \equiv \eta(X) \in \{1, \ldots, K\}$ denote the index of the stratum containing X, so that $X \in A_\eta$. We can always write

$$Y = \mathsf{E}[Y|\eta] + \epsilon \qquad (4.44)$$

simply by defining the residual ϵ so that equality holds. It is immediate that $\mathsf{E}[\epsilon|\eta] = 0$ and that ϵ is uncorrelated with $\mathsf{E}[Y|\eta]$ because

$$\mathsf{E}[\epsilon\,(\mathsf{E}[Y|\eta] - \mathsf{E}[Y])] = 0,$$

as can be seen by first conditioning on η. Because (4.44) decomposes Y into the sum of uncorrelated terms, we have

$$\mathsf{Var}[Y] = \mathsf{Var}[\mathsf{E}[Y|\eta]] + \mathsf{Var}[\epsilon].$$

We will see that stratified sampling with proportional allocation eliminates the first term on the right, leaving only the variance of the residual term and thus guaranteeing a variance reduction.

The residual variance is $\mathsf{E}[\epsilon^2]$ because $\mathsf{E}[\epsilon] = 0$. Also,

$$\mathsf{E}[\epsilon^2|\eta] = \mathsf{E}\left[(Y - \mathsf{E}[Y|\eta])^2|\eta\right] = \mathsf{Var}[Y|\eta].$$

We thus arrive at the familiar decomposition

$$\mathsf{Var}[Y] = \mathsf{Var}[\mathsf{E}[Y|\eta]] + \mathsf{E}\left[\mathsf{Var}[Y|\eta]\right]. \qquad (4.45)$$

The conditional expectation of Y given $\eta = i$ is μ_i, and the probability that $\eta = i$ is p_i. The first term on the right side of (4.45) is thus

$$\mathsf{Var}[\mathsf{E}[Y|\eta]] = \sum_{i=1}^{K} p_i \mu_i^2 - \left(\sum_{i=1}^{K} p_i \mu_i\right)^2.$$

Comparing this with (4.43), we conclude from (4.44) and (4.45) that

$$\mathsf{Var}[\epsilon] = \mathsf{E}\left[\mathsf{Var}[Y|\eta]\right] = \sum_{i=1}^{K} p_i \sigma_i^2,$$

which is precisely the variance parameter in (4.42) for stratified sampling with proportional allocation. This confirms that the variance parameter of the stratified estimator is the variance of the residual of Y after conditioning on η.

Consider now the effect of alternative choices of strata. The total variance $\mathsf{Var}[Y]$ in (4.45) is constant, so making the residual variance small is equivalent to making $\mathsf{Var}[\mathsf{E}[Y|\eta]]$ large — i.e., to making $\mathsf{Var}[\mu_\eta]$ large. This indicates that we should try to choose strata to achieve a high degree of variability across the stratum means $\mu_1, \ldots, \mu_K$ and low variability within each stratum. Indeed, from (4.45) we find that stratification eliminates inter-stratum variability, leaving only intra-stratum variability.

Another consequence of (4.45) is that further stratification results in further variance reduction. More precisely, suppose the partition $\{\tilde{A}_1, \ldots, \tilde{A}_{\tilde{K}}\}$ refines the partition $\{A_1, \ldots, A_K\}$, in the sense that the stratum index $\tilde{\eta}$ of the new partition completely determines η. Then $\mathsf{E}[Y|\eta] = \mathsf{E}[\mathsf{E}[Y|\tilde{\eta}]|\eta]$ and Jensen's inequality yields

$$\mathsf{Var}[\mathsf{E}[Y|\eta]] \le \mathsf{Var}[\mathsf{E}[Y|\tilde{\eta}]], \tag{4.46}$$

from which it follows that the residual variance from the refined strata cannot exceed the residual variance from the original strata.

The decomposition (4.44) invites a comparison between stratified sampling and control variates. Consider the case of real-valued X. Using the method of Example 4.3.2, we can in principle stratify X using an arbitrarily large number of equiprobable intervals. As we refine the stratification, it is reasonable to expect that $\mathsf{E}[Y|\eta]$ will approach $\mathsf{E}[Y|X]$. (For a specific result of this type see Lemma 4.1 of Glasserman, Heidelberger, and Shahabuddin [139].) The decomposition (4.44) becomes

$$Y = \mathsf{E}[Y|X] + \epsilon = g(X) + \epsilon,$$

with $g(x) = \mathsf{E}[Y|X = x]$. If g is linear, then the variance removed through (infinitely fine) stratification of X is precisely the same as the variance that would be removed using X as a control variate. But in the general case, using X as a control variate would remove only the variance associated with the linear part of g near $\mathsf{E}[X]$; see the discussion in Section 4.1.4. In contrast, infinitely fine stratification of X removes all the variance of $g(X)$ leaving only the variance of the residual ϵ. In this sense, using X as a stratification variable is more effective than using it as a control variate. However, it should also be noted that using X as a control requires knowledge only of $\mathsf{E}[X]$ and not the full distribution of X; moreover, it is often easier to use X as a control than to generate samples from the conditional law of (X, Y) given the stratum containing X.

4.3.2 Applications

This section illustrates the application of stratified sampling in settings somewhat more complex than those in Examples 4.3.1–4.3.4. As noted in Example 4.3.4, fully stratifying a random vector becomes infeasible in high dimensions. We therefore focus primarily on methods that stratify a scalar projection in a multidimensional setting. This can be effective in valuing a derivative security if its discounted payoff is highly variable along the selected projection.

Terminal Stratification

In the pricing of options, the most important feature of the path of an underlying asset is often its value at the option expiration; much of the variability in the option's payoff can potentially be eliminated by stratifying the terminal value. As a step in this direction, we detail the stratification of Brownian motion along its terminal value. In the special case of an asset described by geometric Brownian motion with constant volatility, this is equivalent to stratifying the terminal value of the asset price itself.

Suppose, then, that we need to generate a discrete Brownian path $W(t_1)$, $\ldots$, $W(t_m)$ and that we want to stratify the terminal value $W(t_m)$. We can accomplish this through a variant of the Brownian bridge construction of Brownian motion presented in Section 3.1. Using the inverse transform method as in Example 4.3.2 we can stratify $W(t_m)$, and then conditional on each value of $W(t_m)$ we can generate the intermediate values $W(t_1), \ldots, W(t_{m-1})$.

Consider, in particular, the case of K equiprobable strata and a proportional allocation. Let $U_1, \ldots, U_K$ be independent Unif[0,1] random variables and set

$$V_i = \frac{i-1}{K} + \frac{U_i}{K}, \quad i = 1, \ldots, K.$$

Then $\Phi^{-1}(V_1), \ldots, \Phi^{-1}(V_K)$ form a stratified sample from the standard normal distribution and $\sqrt{t_m}\Phi^{-1}(V_1), \ldots, \sqrt{t_m}\Phi^{-1}(V_m)$ form a stratified sample from $N(0, t_m)$, the distribution of $W(t_m)$. To fill in the path leading to each $W(t_m)$, we recall from Section 3.1 that the conditional distribution of $W(t_j)$ given $W(t_{j-1})$ and $W(t_m)$ is

$$N\left(\frac{t_m - t_j}{t_m - t_{j-1}}W(t_{j-1}) + \frac{t_j - t_{j-1}}{t_m - t_{j-1}}W(t_m), \frac{(t_m - t_j)(t_j - t_{j-1})}{t_m - t_{j-1}}\right),$$

with $t_0 = 0$ and $W(0) = 0$.

The following algorithm implements this idea to generate K Brownian paths stratified along $W(t_m)$:

for $i = 1, \ldots, K$
    generate $U \sim$ Unif[0,1]
    $V \leftarrow (i - 1 + U)/K$
    $W(t_m) \leftarrow \sqrt{t_m}\Phi^{-1}(V)$
    for $j = 1, \ldots, m - 1$
        generate $Z \sim N(0, 1)$
        $W(t_j) \leftarrow \frac{t_m - t_j}{t_m - t_{j-1}}W(t_{j-1}) + \frac{t_j - t_{j-1}}{t_m - t_{j-1}}W(t_m) + \sqrt{\frac{(t_m - t_j)(t_j - t_{j-1})}{t_m - t_{j-1}}}Z$

Figure 4.6 illustrates the construction. Of the $K = 10$ paths generated by this algorithm, exactly one terminates in each of the $K = 10$ strata defined for $W(t_m)$, here with $t_m = 1$.

If the underlying asset price $S(t)$ is modeled by geometric Brownian motion, then driving the simulation of S with these Brownian paths stratifies the terminal asset price $S(t_m)$; this is a consequence of the fact that $S(t_m)$ is a monotone transformation of $W(t_m)$. In valuing an option on S, rather than constructing equiprobable strata over all possible terminal values, we may combine all values of $S(t_m)$ that result in zero payoff into a single stratum

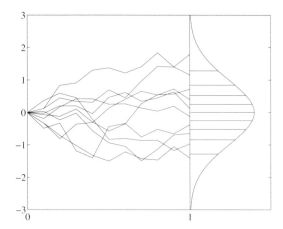

Fig. 4.6. Simulation of K Brownian paths using terminal stratification. One path reaches each of the K strata. The strata are equiprobable under the distribution of $W(1)$.

and create a finer stratification of terminal values that potentially produce a nonzero payoff. (The payoff will not be completely determined by $S(t_m)$ if it is path-dependent.)

As an example of how a similar construction can be used in a more complex example, consider the dynamics of a forward LIBOR rate L_n as in (4.6). Consider a single-factor model (so that W is a scalar Brownian motion) with deterministic but time-varying volatility $\sigma_n \equiv \sigma$. Without the drift term in (4.6), the terminal value $L_n(t_m)$ would be determined by

$$\int_0^{t_m} \sigma(u)\, dW(u)$$

rather than $W(t_m)$, so we may prefer to stratify this integral instead. If σ is constant over each interval $[t_i, t_{i+1})$, this integral simplifies to

$$W(t_m)\sigma(t_{m-1}) + \sum_{i=1}^{m-1} W(t_i)[\sigma(t_{i-1}) - \sigma(t_i)]. \tag{4.47}$$

Similarly, for some path-dependent options one may want to stratify the average

$$\frac{1}{m} \sum_{i=1}^{m} W(t_i). \tag{4.48}$$

In both cases, the stratification variable is a linear combination of the $W(t_i)$ and is thus a special case of the general problem treated next.

Stratifying a Linear Projection

Generating $(W(t_1), \ldots, W(t_m))$ stratified along $W(t_m)$ or (4.47) or (4.48) are all special cases of the problem of generating a multivariate normal random vector stratified along some projection. We now turn to this general formulation of the problem.

Suppose, then, that $\xi \sim N(\mu, \Sigma)$ in $\Re^d$ and that we want to generate ξ with $X \equiv v^\top \xi$ stratified for some fixed vector $v \in \Re^d$. Suppose the $d \times d$ matrix Σ has full rank. We may take μ to be the zero vector because stratifying $v^\top \xi$ is equivalent to stratifying $v^\top (\xi - \mu)$ since $v^\top \mu$ is a constant. Also, stratifying X is equivalent to stratifying any multiple of X; by scaling v if necessary, we may therefore assume that $v^\top \Sigma v = 1$. Thus,

$$ X = v^\top \xi \sim N(0, v^\top \Sigma v) = N(0, 1), $$

so we know how to stratify X using the method in Example 4.3.1.

The next step is to generate ξ conditional on the value of X. First observe that ξ and X are jointly normal with

$$ \begin{pmatrix} \xi \\ X \end{pmatrix} \sim N \left(0, \begin{pmatrix} \Sigma & \Sigma v \\ v^\top \Sigma & v^\top \Sigma v \end{pmatrix} \right). $$

Using the Conditioning Formula (2.25), we find that

$$ (\xi | X = x) \sim N \left(\frac{\Sigma v}{v^\top \Sigma v} x, \Sigma - \frac{\Sigma v v^\top \Sigma}{v^\top \Sigma v} \right) = N \left(\Sigma v x, \Sigma - \Sigma v v^\top \Sigma \right). $$

Observe that the conditional covariance matrix does not depend on x; this is important because it means that only a single factorization is required for the conditional sampling. Let A be any matrix for which $AA^\top = \Sigma$ (such as the one found by Cholesky factorization) and observe that

$$ \begin{aligned} (A - \Sigma v v^\top A)(A - \Sigma v v^\top A)^\top \\ = AA^\top - AA^\top v v^\top \Sigma - \Sigma v v^\top AA^\top + \Sigma v v^\top \Sigma v v^\top \Sigma \\ = \Sigma - \Sigma v v^\top \Sigma, \end{aligned} $$

again using the fact that $v^\top \Sigma v = 1$. Thus, we can use the matrix $A - \Sigma v v^\top A$ to sample from the conditional distribution of ξ given X.

The following algorithm generates K samples from $N(0, \Sigma)$ stratified along the direction determined by v:

> for $i = 1, \ldots, K$
> generate $U \sim \text{Unif}[0,1]$
> $V \leftarrow (i - 1 + U)/K$
> $X \leftarrow \Phi^{-1}(V)$
> generate $Z \sim N(0, I)$ in $\Re^d$
> $\xi \leftarrow \Sigma v X + (A - \Sigma v v^\top A)Z$

By construction, of the K values of X generated by this algorithm, exactly one will fall in each of K equiprobable strata for the standard normal distribution. But observe that under this construction,

$$v^\top \xi = v^\top \Sigma v X + v^\top (A - \Sigma v v^\top A) Z = X.$$

Thus, of the K values of ξ generated, exactly one has a projection $v^\top \xi$ falling into each of K equiprobable strata. In this sense, the algorithm generates samples from $N(0, \Sigma)$ stratified along the direction determined by v.

To apply this method to generate a Brownian path with the integral in (4.47) stratified, take Σ to be the covariance matrix of the Brownian path ($\Sigma_{ij} = \min(t_i, t_j)$, as in (3.6)) and

$$v \propto (\sigma(0) - \sigma(1), \sigma(1) - \sigma(2), \ldots, \sigma(m-2) - \sigma(m-1), \sigma(m-1))^\top,$$

normalized so that $v^\top \Sigma v = 1$. To generate the path with its average (4.48) stratified, take v to be the vector with all entries equal to the square root of the sum of the entries of Σ. Yet another strategy for choosing stratification directions is to use the principal components of Σ (cf. Section 2.3.3).

Further simplification is possible in stratifying a sample from the *standard* multivariate normal distribution $N(0, I)$. In this case, the construction above becomes

$$\xi = vX + (I - vv^\top)Z, \quad X \sim N(0,1), \quad Z \sim N(0, I),$$

with v now normalized so that $v^\top v = 1$. Since $X = v^\top \xi$, by stratifying X we stratify the projection of ξ onto v. The special feature of this setting is that the matrix-vector product $(I - vv^\top)Z$ can be evaluated as $Z - v(v^\top Z)$, which requires $O(d)$ operations rather than $O(d^2)$.

This construction extends easily to allow stratification along multiple directions simultaneously. Let B denote a $d \times m$ matrix, $m \le d$, whose columns represent the stratification directions. Suppose B has been normalized so that $B^\top \Sigma B = I$. If Σ itself is the identity matrix, this says that the m columns of B form a set of orthonormal vectors in $\Re^d$. Where we previously stratified the scalar projection $X = v^\top \xi$, we now stratify the m-vector $X = B^\top \xi$, noting that $X \sim N(0, I)$. For this, we first stratify the m-dimensional hypercube as in Example 4.3.4 and then set $X_j = \Phi^{-1}(V_j)$, $j = 1, \ldots, m$, with $(V_1, \ldots, V_m)$ sampled from a stratum of the hypercube as in (4.36). This samples X from the m-dimensional standard normal distribution with each of its components stratified. We then set

$$\xi = \Sigma BX + (A - \Sigma B B^\top A)Z, \quad Z \sim N(0, I),$$

for any $d \times d$ matrix A satisfying $AA^\top = \Sigma$. The projection $B^\top \xi$ of ξ onto the columns of B returns X and is stratified by construction.

To illustrate this method, we apply it to the LIBOR market model discussed in Section 3.7, using the notation and terminology of that section. We use accrual intervals of length $\delta = 1/2$ and set all forward rates initially equal to 6%. We consider a single-factor model (i.e., one driven by a scalar Brownian motion) with a piecewise constant stationary volatility, meaning that $\sigma_n(t)$ depends on n and t only through $n - \eta(t)$, the number of maturity dates remaining until T_n. We consider a model in which volatility decreases linearly from 0.20 to 0.10 over a 20-year horizon, and a model in which all forward rate volatilities are 0.15.

Our simulation uses a time increment equal to δ, throughout which volatilities are constant. We therefore write

$$\int_0^{T_n} \sigma_n(t) \, dW(t) = \sqrt{\delta} \sum_{i=1}^{n} \sigma_n(T_{i-1}) Z_i, \tag{4.49}$$

with $Z_1, Z_2, \ldots$ independent $N(0,1)$ variables. This suggests using the vector $(\sigma_n(0), \sigma_n(T_1), \ldots, \sigma_n(T_{n-1}))$ as the stratification direction in sampling $(Z_1, \ldots, Z_n)$ from the standard normal distribution in $\Re^n$.

Table 4.2 reports estimated variance reduction ratios for pricing various options in this model. Each entry in the table gives an estimate of the ratio of the variance using ordinary Monte Carlo to the variance using a stratified sample of equal size. The results are based on 40 strata (or simply 40 independent samples for ordinary Monte Carlo); the estimated ratios are based 1000 replications, each replication using a sample size of 40. The 1000 replications merely serve to make the ratio estimates reliable; the ratios themselves should be interpreted as the variance reduction achieved by using 40 strata.

The results shown are for a caplet with a maturity of 20 years, a caplet with a maturity of 5 years, bonds with maturities 20.5 and 5.5 years, and a swaption maturing in 5 years to enter into a 5-year, fixed-for-floating interest rate swap. The options are all at-the-money. The results are based on simulation in the spot measure using the log-Euler scheme in (3.120), except for the last row which applies to the forward measure for maturity 20.5. In each case, the stratification direction is based on the relevant portion of the volatility vector — forty components for a 20-year simulation, ten components for a 5-year simulation. (Discounting a payment to be received at T_{n+1} requires simulating to T_n.)

The results in the table indicate that the variance reduction achieved varies widely but can be quite substantial. With notable exceptions, we generally see greater variance reduction at shorter maturities, at least in part because of the discount factor (see, e.g., (3.109)). The stratification direction we use is tailored to a particular rate L_n (through (4.49)), but not necessarily to the discount factor. The discount factor becomes a constant under the forward measure and, accordingly, we see a greater variance reduction in this case, at the same maturity. Surprisingly, we find the greatest improvement in the case of the swaption, even though the stratification direction is not specifically tailored to the swap rate.

	Linear volatility	Constant volatility
Spot Measure		
Caplet, $T = 20$	2	8
Caplet, $T = 5$	26	50
Swaption, $T = 5$	38	79
Bond, $T = 20.5$	12	4
Bond, $T = 5.5$	5	4
Forward Measure		
Caplet, $T = 20$	11	11

Table 4.2. Variance reduction factors using one-dimensional stratified sampling in a single-factor LIBOR market model. The stratification direction is determined by the vector of volatilities. The results are based on 1000 replications of samples (stratified or independent) of size 40. For each instrument, the value of T indicates the maturity.

Optimal Directions

In estimating $\mathsf{E}[f(\xi)]$ with $\xi \sim N(\mu, \Sigma)$ and f a function from $\mathfrak{R}^d$ to $\mathfrak{R}$, it would be convenient to know the stratification direction v for which stratifying $v^\top \xi$ would produce the greatest reduction in variance. Finding this optimal v is rarely possible; we give a few examples for which the optimal direction is available explicitly.

With no essential loss of generality, we restrict attention to the case of $\mathsf{E}[f(Z)]$ with $Z \sim N(0, I)$. From the variance decomposition (4.45) and the surrounding discussion, we know that the residual variance after stratifying a linear combination $v^\top Z$ is $\mathsf{E}[\mathsf{Var}[f(Z)|\eta]]$, where η is the (random) index of the stratum containing $v^\top Z$. If we use equiprobable strata and let the number of strata grow (with each new set of strata refining the previous set), this residual variance converges to $\mathsf{E}[\mathsf{Var}[f(Z)|v^\top Z]]$ (cf. Lemma 4.1 of [139]). We will therefore compare alternative choices of v through this limiting value.

In the linear case $f(z) = b^\top z$, it is evident that the optimal direction is $v = b$. Next, let $f(z) = z^\top A z$ for some $d \times d$ matrix A. We may assume that A is symmetric and thus that it has real eigenvalues $\lambda_1 \geq \lambda_2 \geq \cdots \geq \lambda_d$ and associated orthonormal eigenvectors $v_1, \ldots, v_d$. Minimizing $\mathsf{E}[\mathsf{Var}[f(Z)|v^\top Z]]$ over vectors v for which $v^\top v = 1$ is equivalent to maximizing $\mathsf{Var}[\mathsf{E}[f(Z)|v^\top Z]]$ over the same set because the two terms sum to $\mathsf{Var}[f(Z)]$ for any v. In the quadratic case, some matrix algebra shows that $v^\top v = 1$ implies

$$\mathsf{Var}[\mathsf{E}[Z^\top A Z|v^\top Z]] = (v^\top A v)^2.$$

This is maximized by v_1 if $\lambda_1^2 \geq \lambda_d^2$ and by v_d if $\lambda_d^2 \geq \lambda_1^2$. In other words, the optimal stratification direction is an eigenvector of A associated with an eigenvalue of largest absolute value. The effect of optimal stratification is to reduce variance from $\sum_i \lambda_i^2$ to $\sum_i \lambda_i^2 - \max_i \lambda_i^2$.

As a final case, let $f(z) = \exp(\frac{1}{2}z^\top Az)$. For $f(Z)$ to have finite second moment, we now require that $\lambda_1 < 1/2$. Theorem 4.1 of [139] shows that the optimal stratification direction in this case is an eigenvector v_{j^*} where j^* satisfies

$$\left(\frac{\lambda_{j^*}}{1 - \lambda_{j^*}}\right)^2 = \max_{i=1,\dots,d}\left(\frac{\lambda_i}{1 - \lambda_i}\right)^2. \qquad (4.50)$$

As in the previous case, this criterion will always select either λ_1 or λ_d, but it will not necessarily select the one with largest absolute value.

Simulation is unnecessary for evaluation of $\mathsf{E}[f(Z)]$ in each of these examples. Nevertheless, a linear, quadratic, or exponential-quadratic function may be useful as an approximation to a more general f and thus as a guide in selecting stratification directions. Fox [127] uses quadratic approximations for related purposes in implementing quasi-Monte Carlo methods. Glasserman, Heidelberger, and Shahabuddin [139] use an exponential-quadratic approximation for stratified sampling in option pricing; in their application, A is the Hessian of the logarithm of an option's discounted payoff. We discuss this method in Section 4.6.2.

Radial Stratification

The symmetry of the standard multivariate normal distribution makes it possible to draw samples from this distribution with stratified norm. For $Z \sim N(0, I)$ in $\Re^d$, let

$$X = Z_1^2 + \cdots + Z_d^2,$$

so that $\sqrt{X} = \|Z\|$ is the radius of the sphere on which Z falls. The distribution of X is chi-square with d degrees of freedom (abbreviated χ_d^2) and is given explicitly in (3.71). Section 3.4.2 discusses efficient methods for sampling from χ_d^2, but for stratification it is more convenient to use the inverse transform method as explained in Example 4.3.2. There is no closed-form expression for the inverse of the χ_d^2 distribution, but the inverse can be evaluated numerically and methods for doing this are available in many statistical software libraries (see, e.g., the survey in Section 18.5 of [201]). Hence, by generating a stratified sample from Unif[0,1] and then applying the inverse of the χ_d^2 distribution, we can generate stratified values of X.

The next step is to sample Z conditional on the value of the stratification variable X. Because of the symmetry of the normal distribution, given X the vector Z is uniformly distributed on the sphere of radius $\sqrt{X}$. This is the basis of the Box-Muller method (cf. Section 2.3) in dimension 2 but it holds for all dimensions d. To sample uniformly from the sphere of radius $R = \sqrt{X}$ in $\Re^d$, we can extend the Box-Muller construction as follows: sample $U_1, \dots, U_{d-1}$ independently from Unif[0,1] and set

$$Z_1 = R\cos(2\pi U_1)$$
$$Z_2 = R\sin(2\pi U_1)\cos(2\pi U_2)$$

$$\vdots \quad \vdots \qquad \vdots$$

$$Z_{d-1} = R\sin(2\pi U_1)\sin(2\pi U_2)\cdots\sin(2\pi U_{d-2})\cos(2\pi U_{d-1})$$
$$Z_d = R\sin(2\pi U_1)\sin(2\pi U_2)\cdots\sin(2\pi U_{d-2})\sin(2\pi U_{d-1}).$$

Alternatively, given a method for generating standard normal random variables we can avoid the evaluation of sines and cosines. If $\xi_1,\dots,\xi_d$ are independent $N(0,1)$ random variables and $\xi = (\xi_1,\dots,\xi_d)^\top$, then $\xi/\|\xi\|$ is uniformly distributed over the unit sphere and

$$Z = R\frac{\xi}{\|\xi\|}$$

is uniformly distributed over the sphere of radius of R.

It should be noted that neither of these constructions extends easily to stratified sampling from $N(0,\Sigma)$ for general Σ. If $\zeta \sim N(0,\Sigma)$ and $X = \zeta\Sigma^{-1}\zeta$, then $X \sim \chi_d^2$ and we can stratify X just as before; moreover, given X, ζ is uniformly distributed over the ellipsoid

$$\mathcal{H}_X = \{x \in \Re^d : x^\top\Sigma^{-1}x = X\}.$$

The difficulty lies in sampling uniformly from the ellipsoid. Extending the Box-Muller construction entails replacing the sines and cosines with elliptic functions. The second construction does not appear to generalize at all: if $\xi \sim N(0,\Sigma)$, the vector $\sqrt{X}\xi/\sqrt{\xi^\top\Sigma^{-1}\xi}$ lies on the ellipsoid $\mathcal{H}_X$ but is not uniformly distributed over the ellipsoid.

The construction does, however, generalize beyond the standard normal to the class of *spherically contoured* distributions. The random vector Y is said to have a spherically contoured distribution if its conditional distribution given $\|Y\|$ is uniform over the sphere of radius $\|Y\|$; see Fang, Kotz, and Ng [114]. To stratify Y along its radius, we must therefore stratify $X = \|Y\|$, which will not be χ_d^2 except in the normal case. Given X, we can sample Y uniformly from the sphere of radius $\|X\|$ using either of the methods described above for the normal distribution.

Radial stratification is proposed and applied in [142] as a method for reducing variance in estimating the risk in a portfolio of options for which losses result from large moves of the underlying assets in any direction.

Stratifying a Poisson Process

In this example, we generate a Poisson process on $[0,T]$ with the total number of jumps in this interval stratified. Let λ denote the arrival rate for the Poisson process and N the number of jumps in $[0,T]$. Then N is a Poisson random variable with distribution

$$P(N = k) = e^{-\lambda T}\frac{(\lambda T)^k}{k!}, \quad k = 0,1,2,\dots.$$

We can sample from this distribution using the inverse transform method, as in Figure 3.9, and thus generate a stratified sample of values as in Example 4.3.2.

For each value of N in the stratified sample, we need to generate the arrival times of the jumps in $[0, T]$ conditional on the number of jumps N. For this we use a standard property of the Poisson process: given $N = k$, the arrival times of the jumps have the joint distribution of the order statistics of k independent random variables uniformly distributed over $[0, T]$. Thus, we may start from Unif[0,1] random variables $U_1, \ldots, U_k$, multiply them by T to make them uniform over $[0, T]$, and then sort them in ascending order to obtain the arrival times.

An alternative to sorting, detailed in Fox [127], samples directly from the joint distribution of the order statistics. Let $V_1, \ldots, V_k$ and $U_1, \ldots, U_k$ denote independent Unif[0,1] random variables. Then

$$V_1^{1/k} V_2^{1/(k-1)} \cdots V_k, \quad \ldots, \quad V_1^{1/k} V_2^{1/(k-1)}, \quad V_1^{1/k} \qquad (4.51)$$

have the joint distribution of the ascending order statistics of $U_1, \ldots, U_k$. For example,

$$P(\max(U_1, \ldots, U_k) \le x) = P(U_1 \le x) \cdots P(U_k \le x) = x^k, \quad x \in [0, 1],$$

and the last term in (4.51) simply samples from this distribution by applying its inverse to V_1. An induction argument verifies correctness of the remaining terms in (4.51). The products in (4.51) can be evaluated recursively from right to left. (To reduce round-off error, Fox [127] recommends recursively summing the logarithms of the $V_i^{1/(k-i+1)}$ and then exponentiating.) The ith arrival time, $i = 1, \ldots, k$, can then be generated as

$$\tau_i = T V_1^{1/k} V_2^{1/(k-1)} \cdots V_{k-i+1}^{1/i}; \qquad (4.52)$$

i.e., by rescaling from $[0, 1]$ to $[0, T]$. Because the terms in (4.51) are generated from right to left, Fox [127] instead sets

$$\tau_i = T \cdot (1 - V_1^{1/k} V_2^{1/(k-1)} \cdots V_i^{1/(k-i+1)});$$

this has the same distribution as (4.52) and allows generation of the arrival times in a single pass. (Subtracting the values in (4.51) from 1 maps the ith largest value to the ith smallest.)

This method of stratification extends, in principle, to inhomogeneous Poisson processes. The number of arrivals in $[0, T]$ continues to be a Poisson random variable in this case. Conditional on the number of arrivals, the times of the arrivals continue to be distributed as order statistics, but now of random variables with a density proportional to the arrival rate $\lambda(t)$, $t \in [0, T]$, rather than a uniform density.

Terminal Stratification in a Binomial Lattice

A binomial lattice provides a discrete-time, discrete-space approximation to the evolution of a diffusion process. Each node in the lattice (see Figure 4.7) is associated with a level of the underlying asset (or rate) S; over a single time step, the movement of the asset is restricted to two successor nodes, usually corresponding to a move up and a move down. By varying the spacing of the nodes and the transition probabilities, it is possible to vary the conditional mean and variance of the change in the underlying asset over a single time step, and thus to approximate virtually any diffusion processes.

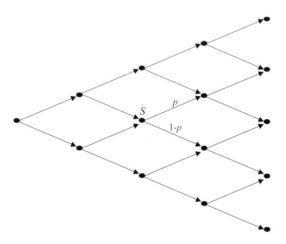

Fig. 4.7. A four-step binomial lattice. Each node has an associated value S of the underlying asset. Each node has two successor nodes, corresponding to a move up and a move down.

Binomial lattices are widely used for numerical option pricing. A typical algorithm proceeds by backward induction: an option contract determines the payoffs at the terminal nodes (which correspond to the option expiration); the option value at any other node is determined by discounting the values at its two successor nodes.

Consider, for example, the pricing of a put with strike K. Each terminal node corresponds to some level S of the underlying asset at expiration and thus to an option value $(K - S)^+$. A generic node in the lattice has an "up" successor node and a "down" successor node; suppose the option values V_u and V_d, respectively, at the two successor nodes have already been calculated. If the probability of a move up is p, and if the discount factor over a single step is $1/(1 + R)$, then the value at the current node is

$$V = \frac{1}{1 + R} \left(pV_u + (1 - p)V_d \right).$$

In pricing an *American* put, the backward induction rule is

$$V = \max\left(\frac{1}{1+R}(pV_u + (1-p)V_d), K - S\right).$$

Binomial option pricing is ordinarily a deterministic calculation, but it can be combined with Monte Carlo. Some path-dependent options, for example, are more easily valued through simulation than through backward induction. In some cases, there are advantages to sampling paths through the binomial lattice rather than sampling paths of a diffusion. For example, in an interest rate lattice, it is possible to compute bond prices at every node. The availability of these bond prices can be useful in pricing path-dependent options on, e.g., bonds or swaps through simulation.

An ordinary simulation through a binomial lattice starts at the root node and generates moves up or down using the appropriate probabilities for each node. As a further illustration of stratified sampling, we show how to simulate paths through a binomial lattice with the terminal value stratified.

Consider, first, the case of a binomial lattice for which the probability of an up move has the same value p at all nodes. In this case, the total number of up moves N through an m-step lattice has the binomial distribution

$$P(N = k) = \binom{m}{k}p^k(1-p)^{m-k}, \quad k = 0, 1, \ldots, m.$$

Samples from this distribution can be generated using the inverse transform method for discrete distributions, as in Example 2.2.4, much as in the case of the Poisson distribution in Figure 3.9. As explained in Example 4.3.2, it is a simple matter to generate stratified samples from a distribution using the inverse transform method. Thus, we have a mechanism for stratifying the total number of up moves through the lattice. Since the terminal node is determined by the difference $N - (m - N) = 2N - m$ between the number of up and down moves, stratifying N is equivalent to stratifying the terminal node.

The next step is to sample a path through the lattice *conditional* on the terminal node — equivalently, conditional on the number of up moves N. The key observation for this procedure is that, given N, all paths through the lattice with N up moves (hence $m - N$ down moves) are equally likely. Generating a path conditional on N is simply a matter of randomly distributing N "ups" among m moves. At each step, the probability of a move up is the ratio of the number of remaining up moves to the number of remaining steps. The following algorithm implements this idea:

$k \leftarrow N$ (total number of up moves to be made)
for $i = 0, \ldots, m - 1$
    if $k = 0$ move down
    if $k \geq m - i$ move up
    if $0 < k < m - i$

> generate $U \sim \text{Unif}[0,1]$
> if $(m-i)U < k$
> $k \leftarrow k - 1$
>   move up
> else move down

The variable k records the number of remaining up moves and $m-i$ is the number of remaining steps. The condition $(m-i)U < k$ is satisfied with probability $k/(m-i)$. This is the ratio of the number of remaining up moves to the number of remaining steps, and is thus the conditional probability of an up move on the next step. Repeating this algorithm for each of the stratified values of N produces a set of paths through the lattice with stratified terminal node.

This method extends to lattices in which the probability of an up move varies from node to node, though this extension requires substantial additional computing. The first step is to compute the distribution of the terminal node, which is no longer binomial. The probability of reaching a node can be calculated using the lattice itself: this probability is the "price" (without discounting) of a security that pays 1 in that node and 0 everywhere else. Through forward induction, the probabilities of all terminal nodes can be found in $O(m^2)$ operations. Once these are computed, it becomes possible to use the discrete inverse transform method to generate stratified samples from the terminal distribution.

The next step is to simulate paths through the lattice conditional on a terminal node. For this, let p denote the unconditional probability of an up move at the current node. Let h_u denote the unconditional probability of reaching the given terminal node from the up successor of the current node; let h_d denote the corresponding probability from the down successor. Then the *conditional* probability of an up move at the current node (given the terminal node) is $ph_u/(ph_u+(1-p)h_d)$ and the conditional probability of a down move is $(1-p)h_d/(ph_u+(1-p)h_d)$. Once the h_u and h_d have been calculated at every node, it is therefore a simple matter to simulate paths conditional on a given terminal node by applying these conditional probabilities at each step. Implementing this requires calculation of $O(m)$ conditional probabilities at every node, corresponding to the $O(m)$ terminal nodes. These can be calculated with a total effort of $O(m^3)$ using backward induction.

4.3.3 Poststratification

As should be evident from our discussion thus far, implementation of stratified sampling requires knowledge of stratum probabilities and a mechanism for conditional sampling from strata. Some of the examples discussed in Section 4.3.2 suggest that conditional sampling may be difficult even when computing stratum probabilities is not. *Poststratification* combines knowledge of stratum probabilities with ordinary independent sampling to reduce variance, at least asymptotically. It can therefore provide an attractive alternative to genuine stratification when conditional sampling is costly.

As before, suppose our objective is to estimate $E[Y]$. We have a mechanism for generating independent replications $(X_1, Y_1), \ldots, (X_n, Y_n)$ of the pair (X, Y); moreover, we know the probabilities $p_i = P(X \in A_i)$ for strata $A_1, \ldots, A_K$. As usual, we require that these be positive and sum to 1. For $i = 1, \ldots, K$, let

$$N_i = \sum_{j=1}^{n} \mathbf{1}\{X_j \in A_i\}$$

denote the number of samples that fall in stratum i and note that this is now a random variable. Let

$$S_i = \sum_{j=1}^{n} \mathbf{1}\{X_j \in A_i\}Y_j$$

denote the sum of those Y_j for which X_i falls in stratum i, for $i = 1, \ldots, K$.

The usual sample mean $\bar{Y} = (Y_1 + \cdots + Y_n)/n$ can be written as

$$\bar{Y} = \frac{S_1 + \cdots + S_K}{n} = \sum_{i=1}^{K} \frac{N_i}{n} \cdot \frac{S_i}{N_i},$$

at least if all N_i are nonzero. By the strong law of large numbers, $N_i/n \to p_i$ and $S_i/N_i \to \mu_i$, with probability 1, where $\mu_i = E[Y|X \in A_i]$ denotes the stratum mean, as in (4.37). Poststratification replaces the random fraction N_i/n with its expectation p_i to produce the estimator

$$\hat{Y} = \sum_{i=1}^{K} p_i \frac{S_i}{N_i}. \tag{4.53}$$

Whereas the sample mean $\bar{Y}$ assigns weight $1/n$ to every observation, the poststratified estimator weights values falling in stratum i by the ratio p_i/N_i. Thus, values from undersampled strata $(N_i < np_i)$ get more weight and values from oversampled strata $(N_i > np_i)$ get less weight. To cover the possibility that none of the n replications falls in the ith stratum, we replace S_i/N_i with zero in (4.53) if $N_i = 0$.

It is immediate from the almost sure convergence of S_i/N_i to μ_i that the poststratified estimator $\hat{Y}$ is a consistent estimator of $E[Y]$. Less clear are its merits relative to the ordinary sample mean or a genuinely stratified estimator. We will see that, asymptotically as the sample size grows, poststratification is as effective as stratified sampling in reducing variance. To establish this result, we first consider properties of ratio estimators more generally.

Ratio Estimators

We digress briefly to derive a central limit theorem for ratio estimators. For this discussion, let (R_i, Q_i), $i = 1, 2, \ldots$, be independent and identically distributed pairs of random variables with

$$\mathsf{E}[R_i] = \mu_R, \ \mathsf{E}[Q_i] = \mu_Q, \ \mathsf{Var}[R_i] = \sigma_R^2, \ \mathsf{Var}[Q_i] = \sigma_Q^2, \ \mathsf{Cov}[R_i, Q_i] = \sigma_{RQ},$$

and $\mu_Q \neq 0$. The sample means of the first n values are

$$\bar{R} = \frac{1}{n} \sum_{i=1}^{n} R_i, \quad \bar{Q} = \frac{1}{n} \sum_{i=1}^{n} Q_i.$$

By the strong law of large numbers, the ratio $\bar{R}/\bar{Q}$ converges with probability 1 to μ_R/μ_Q.

By applying the delta method introduced in Section 4.1.4 to the function $h(x, y) = x/y$, we obtain a central limit theorem of the form

$$\sqrt{n} \left(\frac{\bar{R}}{\bar{Q}} - \frac{\mu_R}{\mu_Q} \right) \Rightarrow N(0, \sigma^2)$$

for the ratio estimator. The variance parameter σ^2 is given by the general expression in (4.26) for the delta method and simplifies in this case to

$$\sigma^2 = \frac{\mu_R^2}{\mu_Q^4} \sigma_R^2 - \frac{2\mu_R}{\mu_Q^3} \sigma_{RQ} + \frac{\sigma_R^2}{\mu_Q^2} = \frac{\mathsf{Var}[R - \frac{\mu_R}{\mu_Q} Q]}{\mu_Q^2}. \tag{4.54}$$

This parameter is consistently estimated by

$$s^2 = \sum_{i=1}^{n} \left(R_i - \bar{R} Q_i / \bar{Q} \right)^2 \Big/ n(\bar{Q})^2,$$

from which we obtain an asymptotically valid $1 - \delta$ confidence interval

$$\frac{\bar{R}}{\bar{Q}} \pm z_{\delta/2} \frac{s}{\sqrt{n}},$$

with $z_{\delta/2} = -\Phi^{-1}(\delta/2)$.

For fixed n, $\bar{R}/\bar{Q}$ is a biased estimator of μ_R/μ_Q. The bias has the form

$$\mathsf{E}\left[\frac{\bar{R}}{\bar{Q}} - \frac{\mu_R}{\mu_Q} \right] = \frac{(\mu_R \sigma_Q^2 / \mu_Q^3) - (\sigma_{RQ}/\mu_Q^2)}{n} + O(1/n^2);$$

see, e.g., Fishman [121], p.109. Subtracting an estimate of the leading term can reduce the bias to $O(1/n^2)$.

Poststratification: Asymptotic Variance

We now apply this analysis of ratio estimators to derive a central limit theorem for the poststratified estimator $\hat{Y}$, which is a linear combination of ratio estimators. A straightforward extension of the result for a single ratio gives

$$\sqrt{n}\left(\frac{S_1}{N_1} - \mu_1, \ldots, \frac{S_K}{N_K} - \mu_K\right) \Rightarrow N(0, \Sigma),$$

with the limiting matrix Σ again determined by the delta method. For the diagonal entries of Σ, (4.54) gives

$$\Sigma_{ii} = \frac{\mathsf{Var}[Y\mathbf{1}\{X \in A_i\} - \mu_i\mathbf{1}\{X \in A_i\}]}{p_i^2} = \frac{\sigma_i^2}{p_i},$$

with σ_i^2 the stratum variance defined in (4.38). A similar calculation for $j \neq i$ gives

$$\Sigma_{ij} = \frac{\mathsf{Cov}[(Y - \mu_i)\mathbf{1}\{X \in A_i\}, (Y - \mu_j)\mathbf{1}\{X \in A_j\}]}{p_i p_j} = 0$$

because A_i and A_j are disjoint.

The poststratified estimator satisfies

$$\hat{Y} - \mu = \sum_{i=1}^{K} p_i\left(\frac{S_i}{N_i} - \mu_i\right)$$

and therefore

$$\sqrt{n}[\hat{Y} - \mu] \Rightarrow N(0, \sigma^2)$$

with

$$\sigma^2 = \sum_{i,j=1}^{K} p_i\Sigma_{ij}p_j = \sum_{i=1}^{K} p_i\sigma_i^2.$$

This is precisely the asymptotic variance for the stratified estimator based on proportional allocation of samples to strata; see (4.42). It can be estimated consistently by replacing each σ_i^2 with the sample variance of the observations falling in the ith stratum.

From this result we see that *in the large-sample limit*, we can extract all the variance reduction of stratified sampling without having to sample conditionally from the strata by instead weighting each observation according to its stratum. How large the sample needs to be for the two methods to give similar results depends in part on the number of strata and their probabilities. There is no simple way to determine at what sample size this limit becomes relevant without experimentation. But stratified sampling is generally preferable and poststratification is best viewed as an alternative for settings in which conditional sampling from the strata is difficult.

4.4 Latin Hypercube Sampling

Latin hypercube sampling is an extension of stratification for sampling in multiple dimensions. Recall from the discussion in Example 4.3.4 that stratified sampling in high dimensions is possible in principle but often infeasible in practice. The difficulty is apparent even in the simple case of sampling from the d-dimensional hypercube $[0, 1)^d$. Partitioning each coordinate into K strata produces K^d strata for the hypercube, thus requiring a sample size of at least K^d to ensure that each stratum is sampled. For even moderately large d, this may be prohibitive unless K is small, in which case stratification provides little benefit. For this reason, in Section 4.3.2 we focused on methods for stratifying a small number of important directions in multidimensional problems.

Latin hypercube sampling treats all coordinates equally and avoids the exponential growth in sample size resulting from full stratification by stratifying only the one-dimensional marginals of a multidimensional joint distribution. The method, introduced by McKay, Conover, and Beckman [259] and further analyzed in Stein [337], is most easily described in the case of sampling from the uniform distribution over the unit hypercube. Fix a dimension d and a sample size K. For each coordinate $i = 1, \ldots, d$, independently generate a stratified sample $V_i^{(1)}, \ldots, V_i^{(K)}$ from the unit interval using K equiprobable strata; each $V_i^{(j)}$ is uniformly distributed over $[(j-1)/K, j/K)$. If we arrange the d stratified samples in columns,

$$
\begin{array}{cccc}
V_1^{(1)} & V_2^{(1)} & \cdots & V_d^{(1)} \\
V_1^{(2)} & V_2^{(2)} & \cdots & V_d^{(2)} \\
\vdots & \vdots & & \vdots \\
V_1^{(K)} & V_2^{(K)} & \cdots & V_d^{(K)}
\end{array}
$$

then each row gives the coordinates of a point in $[0, 1)^d$. The first row identifies a point in $[0, 1/K)^d$, the second a point in $[1/K, 2/K)^d$, and so on, corresponding to K points falling in subcubes along the diagonal of the unit hypercube. Now randomly permute the entries in each column of the array. More precisely, let $\pi_1, \ldots, \pi_d$ be permutations of $\{1, \ldots, K\}$, drawn independently from the distribution that makes all $K!$ such permutations equally likely. Let $\pi_j(i)$ denote the value to which i is mapped by the jth permutation. The rows of the array

$$
\begin{array}{cccc}
V_1^{\pi_1(1)} & V_2^{\pi_2(1)} & \cdots & V_d^{\pi_d(1)} \\
V_1^{\pi_1(2)} & V_2^{\pi_2(2)} & \cdots & V_d^{\pi_d(2)} \\
\vdots & \vdots & & \vdots \\
V_1^{\pi_1(K)} & V_2^{\pi_2(K)} & \cdots & V_d^{\pi_d(K)}
\end{array}
\tag{4.55}
$$

continue to identify points in $[0, 1)^d$, but they are no longer restricted to the diagonal. Indeed, each row is a point *uniformly distributed* over the unit hypercube. The K points determined by the K rows are not independent: if we project the K points onto their ith coordinates, the resulting set of values $\{V_i^{\pi_i(1)}, \ldots, V_i^{\pi_i(K)}\}$ is the same as the set $\{V_i^{(1)}, \ldots, V_i^{(K)}\}$, and thus forms a stratified sample from the unit interval.

The "marginal" stratification property of Latin hypercube sampling is illustrated in Figure 4.8. The figure shows a sample of size $K = 8$ in dimension $d = 2$. Projecting the points onto either of their two coordinates shows that exactly one point falls in each of the eight bins into which each axis is partitioned. Stratified sampling would require drawing a point from each square and thus a sample size of 64.

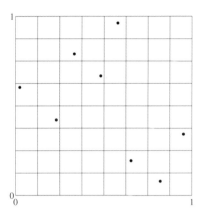

Fig. 4.8. A Latin hypercube sample of size $K = 8$ in dimension $d = 2$.

To generate a Latin hypercube sample of size K in dimension d, let $U_i^{(j)}$ be independent Unif[0,1) random variables for $i = 1, \ldots, d$ and $j = 1, \ldots, K$. Let $\pi_1, \ldots, \pi_d$ be independent random permutations of $\{1, \ldots, K\}$ and set

$$V_i^{(j)} = \frac{\pi_i(j) - 1 + U_i^{(j)}}{K}, \quad i = 1, \ldots, d, \quad j = 1, \ldots, K. \tag{4.56}$$

The sample consists of the K points $(V_1^{(j)}, \ldots, V_d^{(j)})$, $j = 1, \ldots, K$. To generate a random permutation, first sample uniformly from $\{1, \ldots, K\}$, then sample uniformly from the remaining values, and continue until only one value remains. In (4.56) we may choose one of the permutations (π_d, say) to be the identity, $\pi_d(i) \equiv i$ without affecting the joint distribution of the sample.

Using the inverse transform method, this construction easily extends to nonuniform distributions. For example, to generate a Latin hypercube sample of size K from $N(0, I)$ in $\Re^d$, set

$$Z_i^{(j)} = \Phi^{-1}(V_i^{(j)}), \quad i = 1, \ldots, d, \quad j = 1, \ldots, K$$

with Φ the cumulative normal and $V_i^{(j)}$ as in (4.56). The sample consists of the vectors

$$Z^{(j)} = (Z_1^{(j)}, \ldots, Z_d^{(j)}), \quad j = 1, \ldots, K, \qquad (4.57)$$

in $\Re^d$. Projecting these K points onto any axis produces a stratified sample of size K from the standard univariate normal distribution. Even if the inverse transform is inconvenient or if the marginals have different distributions, the construction in (4.55) continues to apply, provided we have some mechanism for stratifying the marginals: generate a stratified sample from each marginal using K equiprobable strata for each, then randomly permute these d stratified samples.

This construction does rely crucially on the assumption of independent marginals, and transforming variables to introduce dependence can affect the partial stratification properties of Latin hypercube samples in complicated ways. This is evident in the case of a multivariate normal distribution $N(0, \Sigma)$. To sample from $N(0, \Sigma)$, we set $X = AZ$ with $Z \sim N(0, I)$ and $AA^\top = \Sigma$. Replacing independently generated Zs with the Latin hypercube sample (4.57) produces points $X^{(j)} = AZ^{(j)}$, $j = 1, \ldots, K$; but the marginals of the $X^{(j)}$ so constructed will not in general be stratified. Rather, the marginals of the $A^{-1}X^{(j)}$ are stratified.

Example 4.4.1 *Brownian paths.* As a specific illustration, consider the simulation of Brownian paths at times $0 = t_0 < t_1 \cdots < t_d$. As in (4.57), let $Z^{(1)}, \ldots, Z^{(K)}$ denote a Latin hypercube sample from $N(0, I)$ in d dimensions. From these K points in $\Re^d$, we generate K discrete Brownian paths $W^{(1)}, \ldots, W^{(K)}$ by setting

$$W^{(j)}(t_n) = \sum_{i=1}^n \sqrt{t_i - t_{i-1}} Z_i^{(j)}, \quad n = 1, \ldots, d.$$

If we fix a time t_n, $n \geq 2$, and examine the K values $W^{(1)}(t_n), \ldots, W^{(K)}(t_n)$, these will *not* form a stratified sample from $N(0, t_n)$. It is rather the *increments* of the Brownian paths that would be stratified.

These K Brownian paths could be used to generate K paths of a process driven by a single Brownian motion. It would *not* be appropriate to use the K Brownian paths to generate a single path of a process driven by a K-dimensional Brownian motion. Latin hypercube sampling introduces dependence between elements of the sample, whereas the coordinates of a K-dimensional (standard) Brownian motion are independent. Using $(W^{(1)}, \ldots, W^{(K)})$ in place of a K-dimensional Brownian motion would thus change the law of the simulated process and could introduce severe bias. In contrast, the *marginal* law of each $W^{(j)}$ coincides with that of a scalar Brownian motion. Put succinctly, in implementing a variance reduction technique we are free to introduce dependence *across* paths but not *within* paths. □

Example 4.4.2 *Paths through a lattice.* To provide a rather different example, we apply Latin hypercube sampling to the problem of simulating paths through a binomial lattice. (See Figure 4.7 and the surrounding discussion for background.) Consider an m-step lattice with fixed probabilities p and $1 - p$. The "marginals" in this example correspond to the m time steps, so the dimension d equals m, and the sample size K is the number of paths. We encode a move up as a 1 and a move down as a 0. For each $i = 1, \ldots, m$ we generate a stratified sample of 1s and 0s: we "generate" $\lfloor pK \rfloor$ 1s and $\lfloor K - pK \rfloor$ 0s, and if pK is not an integer, the Kth value in the sample is 1 with probability p and 0 with probability $1 - p$. For example, with $d = 4$, $K = 8$, and $p = 0.52$, we might get

$$
\begin{array}{cccc}
1 & 1 & 1 & 1 \\
1 & 1 & 1 & 1 \\
1 & 1 & 1 & 1 \\
1 & 1 & 1 & 1 \\
0 & 0 & 0 & 0 \\
0 & 0 & 0 & 0 \\
0 & 0 & 0 & 0 \\
0 & 1 & 0 & 0 \\
\end{array}
$$

the columns corresponding to the $d = 4$ stratified samples. Applying a random permutation to each column produces, e.g.,

$$
\begin{array}{cccc}
0 & 1 & 0 & 1 \\
0 & 0 & 1 & 1 \\
1 & 1 & 0 & 1 \\
1 & 1 & 1 & 0 \\
0 & 1 & 0 & 0 \\
1 & 0 & 1 & 1 \\
0 & 0 & 1 & 0 \\
1 & 1 & 0 & 0 \\
\end{array}
$$

Each row now encodes a path through the lattice. For example, the last row corresponds to two consecutive moves up followed by two consecutive moves down. Notice that, for each time step i, the fraction of paths on which the ith step is a move up is very nearly p. This is the property enforced by Latin hypercube sampling.

One could take this construction a step further to enforce the (nearly) correct fraction of up moves at each *node* rather than just at each time step. For simplicity, suppose Kp^d is an integer. To the root node, assign Kp 1s and $K(1 - p)$ 0s. To the node reached from the root by taking ℓ steps up and k steps down, $\ell + k < d$, assign $Kp^{\ell+1}(1 - p)^k$ ones and $Kp^{\ell}(1 - p)^{k+1}$ zeros. Randomly and independently permute the ones and zeros at all nodes. The result encodes K paths through the lattice with the fraction of up moves out of every node exactly equal to p.

Mintz [269] develops a simple way to implement essentially the same idea. His implementation eliminates the need to precompute and permute the outcomes at all nodes. Instead, it assigns to each node a counter that keeps track of the number of paths that have left that node by moving up. For example, consider again a node reached from the root by taking ℓ steps up and k steps down, $\ell + k < d$. Let K be the total number of paths to be generated and again suppose for simplicity that Kp^d is an integer. The number of paths reaching the designated node is

$$K_{\ell k} = Kp^\ell (1 - p)^k,$$

so the counter at that node counts from zero to $pK_{\ell k}$, the number of paths that exit that node by moving up. If a path reaches the node and finds the counter at i, it moves down with probability $i/pK_{\ell k}$ and moves up with the complementary probability. If it moves up, the counter is incremented to $i+1$. □

Variance Reduction and Variance Decomposition

We now state some properties of Latin hypercube sampling that shed light on its effectiveness. These properties are most easily stated in the context of sampling from $[0, 1)^d$. Thus, suppose our goal is to estimate

$$\alpha_f = \int_{[0,1)^d} f(u) \, du$$

for some square-integrable $f : [0, 1)^d \to \Re$. The standard Monte Carlo estimator of this integral can be written as

$$\bar{\alpha}_f = \frac{1}{K} \sum_{j=0}^{K-1} f(U_{jd+1}, U_{jd+2}, \ldots, U_{jd+d})$$

with $U_1, U_2, \ldots$ independent uniforms. The variance of this estimator is σ^2/K with $\sigma^2 = \mathsf{Var}[f(U_1, \ldots, U_d)]$. For $V^{(1)}, \ldots, V^{(K)}$ as in (4.56), define the estimator

$$\hat{\alpha}_f = \frac{1}{K} \sum_{j=1}^{K} f(V^{(j)}).$$

McKay et al. [259] show that

$$\mathsf{Var}[\hat{\alpha}_f] = \frac{\sigma^2}{K} + \frac{K-1}{K} \mathsf{Cov}[f(V^{(1)}), f(V^{(2)})],$$

which could be larger or smaller than the variance of the standard estimator $\bar{\alpha}_f$, depending on the covariance between distinct points in the Latin hypercube sample. By construction, $V^{(j)}$ and $V^{(k)}$ avoid each other — for example, their ith coordinates cannot fall in the same bin if $j \neq k$ — which suggests that the covariance will often be negative. This holds, in particular, if f is monotone in each coordinate, as shown by McKay et al. [259]. Proposition 3 of Owen [288] shows that for any (square-integrable) f and any $K \geq 2$,

$$\mathsf{Var}[\hat{\alpha}_f] \leq \frac{\sigma^2}{K-1},$$

so the variance produced by a Latin hypercube sample of size K is no larger than the variance produced by an i.i.d. sample of size $K-1$.

Stein [337] shows that as $K \to \infty$, Latin hypercube sampling eliminates the variance due to the *additive* part of f, in a sense we now explain. For each $i = 1, \ldots, d$, let

$$f_i(u) = \mathsf{E}[f(U_1, \ldots, U_{i-1}, u, U_i, \ldots, U_d)],$$

for $u \in [0, 1)$. Observe that each $f_i(U)$, $U \sim \mathrm{Unif}[0,1]$ has expectation α_f. The function

$$f_{\mathrm{add}}(u_1, \ldots, u_d) = \sum_{i=1}^{d} f_i(u_i) - (d-1)\alpha_f$$

also has expectation α_f and is the best additive approximation to f in the sense that

$$\int_{[0,1)^d} (f(u_1, \ldots, u_d) - f_{\mathrm{add}}(u_1, \ldots, u_d))^2 \, du_1 \cdots du_d$$

$$\leq \int_{[0,1)^d} \left(f(u_1, \ldots, u_d) - \sum_{i=1}^{d} h_i(u_i) \right)^2 \, du_1 \cdots du_d$$

for any univariate functions $h_1, \ldots, h_d$. Moreover, the residual

$$\epsilon = f(U_1, \ldots, U_d) - f_{\mathrm{add}}(U_1, \ldots, U_d)$$

is uncorrelated with $f(U_1, \ldots, U_d)$ and this allows us to decompose the variance σ^2 of $f(U_1, \ldots, U_d)$ as $\sigma^2 = \sigma_{\mathrm{add}}^2 + \sigma_\epsilon^2$ with σ_{add}^2 the variance of $f_{\mathrm{add}}(U_1, \ldots, U_d)$ and σ_ϵ^2 the variance of the residual. Stein [337] showed that

$$\mathsf{Var}[\hat{\alpha}_f] = \frac{\sigma_\epsilon^2}{K} + o(1/K). \tag{4.58}$$

Up to terms of order $1/K$, Latin hypercube sampling eliminates σ_{add}^2 — the variance due to the additive part of f — from the simulation variance. This further indicates that Latin hypercube sampling is most effective with integrands that nearly separate into a sum of one-dimensional functions.

Output Analysis

Under various additional conditions on f, Loh [238], Owen [285], and Stein [337] establish a central limit theorem for $\hat{Y}$ of the form

$$\sqrt{K}[\hat{\alpha}_f - \alpha_f] \Rightarrow N(0, \sigma_\epsilon^2),$$

which in principle provides the basis for a large-sample confidence interval for α_f based on Latin hypercube sampling. In practice, σ_ϵ^2 is neither known nor easily estimated, making this approach difficult to apply.

A simpler approach to interval estimation generates i.i.d. estimators $\hat{\alpha}_f(1), \ldots, \hat{\alpha}_f(n)$, each based on a Latin hypercube sample of size K. An asymptotically (as $n \to \infty$) valid $1 - \delta$ confidence interval for α_f is provided by

$$\left(\frac{1}{n} \sum_{i=1}^{n} \hat{\alpha}_f(i) \right) \pm z_{\delta/2} \frac{\hat{s}}{\sqrt{n}},$$

with $\hat{s}$ the sample standard deviation of $\hat{\alpha}_f(1), \ldots, \hat{\alpha}_f(n)$.

The only cost to this approach lies in foregoing the possibly greater variance reduction from generating a single Latin hypercube sample of size nK rather than n independent samples of size K. Stein [337] states that this loss is small if K/d is large.

A $K \times K$ array is called a Latin square if each of the symbols $1, \ldots, K$ appears exactly once in each row and column. This helps explain the name "Latin hypercube sampling." Latin squares are used in the design of experiments, along with the more general concept of an *orthogonal array*. Owen [286] extends Stein's [337] approach to analyze the variance of Monte Carlo estimates based on randomized orthogonal arrays. This method generalizes Latin hypercube sampling by stratifying low-dimensional (but not just one-dimensional) marginal distributions.

Numerical Illustration

We conclude this section with a numerical example. We apply Latin hypercube sampling to the pricing of two types of path-dependent options — an Asian option and a barrier option. The Asian option is a call on the arithmetic average of the underlying asset over a finite set of dates; the barrier option is a down-and-out call with a discretely monitored barrier. The underlying asset is $\text{GBM}(r, \sigma^2)$ with $r = 5\%$, $\sigma = 0.30$, and an initial value of 50. The barrier is fixed at 40. The option maturity is one year in all cases. We report results for 8 and 32 equally spaced monitoring dates; the number of dates is the dimension of the problem. With d monitoring dates, we may view each discounted option payoff as a function of a standard normal random vector in $\Re^d$ and apply Latin hypercube sampling to generate these vectors.

Table 4.3 reports estimated variance reduction factors. Each entry in the table is an estimate of the ratio of the variance using independent sampling to the variance using a Latin hypercube sample of the same size. Thus, larger ratios indicate greater variance reduction. The sample sizes displayed are 50, 200, and 800. The ratios are estimated based on 1000 replications of samples of the indicated sizes.

The most salient feature of the results in Table 4.3 is the effect of varying the strike: in all cases, the variance ratio increases as the strike decreases. This

is to be expected because at lower strikes the options are more nearly linear.
The variance ratios are nearly the same in dimensions 8 and 32 and show little
dependence on the sample size. We know that the variance of independent
replications (the numerators in these ratios) are inversely proportional to the
sample sizes. Because the ratios are roughly constant across sample sizes, we
may conclude that the variance using Latin hypercube sampling is nearly
inversely proportional to the sample size. This suggests that (at least in these
examples) the asymptotic result in (4.58) is relevant for sample sizes as small
as $K = 50$.

	Strike	8 steps			32 steps		
		50	200	800	50	200	800
Asian	45	7.5	8.6	8.8	7.1	7.6	8.2
Option	50	3.9	4.4	4.6	3.7	3.6	4.0
	55	2.4	2.6	2.8	2.3	2.1	2.5
Barrier	45	4.1	4.1	4.3	3.8	3.7	3.9
Option	50	3.2	3.2	3.4	3.0	2.9	3.1
	55	2.5	2.6	2.7	2.4	2.2	2.4

Table 4.3. Variance reduction factors using Latin hypercube sampling for two path-dependent options. Results are displayed for dimensions (number of monitoring dates) 8 and 32 using samples of size 50, 200, and 800. Each entry in the table is estimated from 1000 replications, each replication consisting of 50, 200, or 800 paths.

The improvements reported in Table 4.3 are mostly modest. Similar variance ratios could be obtained by using the underlying asset as a control variate; for the Asian option, far greater variance reduction could be obtained
by using a geometric average control variate as described in Example 4.1.2.
One potential advantage of Latin hypercube sampling is that it lends itself
to the use of a single set of paths to price many different types of options.
The marginal stratification feature of Latin hypercube sampling is beneficial
in pricing many different options, whereas control variates are ideally tailored
to a specific application.

4.5 Matching Underlying Assets

This section discusses a set of loosely related techniques with the common
objective of ensuring that certain sample means produced in a simulation exactly coincide with their population values (i.e., with the values that would
be attained in the limit of infinitely many replications). Although these techniques could be used in almost any application of Monte Carlo, they take on
special significance in financial engineering where matching sample and population means will often translate to ensuring exact finite-sample pricing of

underlying assets. The goal of derivatives pricing is to determine the value of a derivative security *relative* to its underlying assets. One could therefore argue that correct pricing of these underlying assets is a prerequisite for accurate valuation of derivatives.

The methods we discuss are closely related to control variates, which should not be surprising since we noted (in Example 4.1.1) that underlying assets often provide convenient controls. There is also a link with stratified sampling: stratification with proportional allocation ensures that the sample means of the stratum indicator functions coincide with their population means. We develop two types of methods: *moment matching* based on transformations of simulated paths, and methods that weight (but do not transform) paths in order to match moments. When compared with control variates or with each other, these methods may produce rather different small-sample properties while becoming equivalent as the number of samples grows. This makes it difficult to compare estimators on theoretical grounds.

4.5.1 Moment Matching Through Path Adjustments

The idea of transforming paths to match moments is most easily introduced in the setting of a single underlying asset $S(t)$ simulated under a risk-neutral measure in a model with constant interest rate r. If the asset pays no dividends, we know that $E[S(t)] = e^{rt}S(0)$. Suppose we simulate n independent copies $S_1, \ldots, S_n$ of the process and define the sample mean process

$$\bar{S}(t) = \frac{1}{n} \sum_{i=1}^{n} S_i(t).$$

For finite n, the sample mean will not in general coincide with $E[S(t)]$; the simulation could be said to misprice the underlying asset in the sense that

$$e^{-rt}\bar{S}(t) \neq S(0), \tag{4.59}$$

the right side being the current price of the asset and the left side its simulation estimate.

A possible remedy is to transform the simulated paths by setting

$$\tilde{S}_i(t) = S_i(t)\frac{E[S(t)]}{\bar{S}(t)}, \quad i = 1, \ldots, n, \tag{4.60}$$

or

$$\tilde{S}_i(t) = S_i(t) + E[S(t)] - \bar{S}(t), \quad i = 1, \ldots, n, \tag{4.61}$$

and then using the $\tilde{S}_i$ rather than the S_i to price derivatives. Using either the multiplicative adjustment (4.60) or the additive adjustment (4.61) ensures that the sample mean of $\tilde{S}_1(t), \ldots, \tilde{S}_n(t)$ exactly equals $E[S(t)]$.

These and related transformations are proposed and tested in Barraquand [37], Boyle et al. [53], and Duan and Simonato [96]. Duan and Simonato call

(4.60) *empirical martingale* simulation; Boyle et al. use the name *moment matching*. In other application domains, Hall [164] analyzes a related *centering* technique for bootstrap simulation and Gentle [136] refers briefly to *constrained* sampling. In many settings, making numerical adjustments to samples seems unnatural — some discrepancy between the sample and population mean is to be expected, after all. In the financial context, the error in (4.59) could be viewed as exposing the user to arbitrage through mispricing and this might justify attempts to remove the error completely.

A further consequence of matching the sample and population mean of the underlying asset is a finite-sample form of *put-call parity*. The algebraic identity

$$(a - b)^+ - (b - a)^+ = a - b$$

implies the constraint

$$e^{-rT}\mathsf{E}[(S(T) - K)^+] - e^{-rT}\mathsf{E}[(K - S(T))^+] = S(0) - e^{-rT}K$$

on the values of a call, a put, and the underlying asset. Any adjustment that equates the sample mean of $\tilde{S}_1(T), \ldots, \tilde{S}_n(T)$ to $\mathsf{E}[S(T)]$ ensures that

$$e^{-rT}\frac{1}{n}\sum_{i=1}^n(\tilde{S}_i(T) - K)^+ - e^{-rT}\frac{1}{n}\sum_{i=1}^n(K - \tilde{S}_i(T))^+ = S(0) - e^{-rT}K.$$

This, too, may be viewed as a type of finite-sample no-arbitrage condition.

Of (4.60) and (4.61), the multiplicative adjustment (4.60) seems preferable on the grounds that it preserves positivity whereas the additive adjustment (4.61) can make some $\tilde{S}_i$ negative even if $S_1(t), \ldots, S_n(t), \mathsf{E}[S(t)]$ are all positive. However, we get $\mathsf{E}[\tilde{S}_i(t)] = \mathsf{E}[S(t)]$ using (4.61) but not with (4.60). Indeed, (4.61) even preserves the martingale property in the sense that

$$\mathsf{E}[e^{-r(T-t)}\tilde{S}_i(T)|\tilde{S}_i(u), 0 \le u \le t] = \tilde{S}_i(t).$$

Both (4.60) and (4.61) change the law of the simulated process ($\tilde{S}_i(t)$ and $S_i(t)$ will not in general have the same distribution) and thus typically introduce some bias in estimates computed from the adjusted paths. This bias vanishes as the sample size n increases and is typically $O(1/n)$.

Large-Sample Properties

There is some similarity between the transformations in (4.60) and (4.61) and the nonlinear control variates discussed in Section 4.1.4. The current setting does not quite fit the formulation in Section 4.1.4 because the adjustments here affect individual observations and not just their means.

To extend the analysis in Section 4.1.4, we formulate the problem as one of estimating $\mathsf{E}[h_1(X)]$ with X taking values in $\Re^d$ and h_1 mapping $\Re^d$ into $\Re$. For example, we might have $X = S(T)$ and $h_1(x) = e^{-rT}(x - K)^+$ in the

case of pricing a standard call option. The moment matching estimator has
the form

$$\frac{1}{n}\sum_{i=1}^{n}h(X_i,\bar{X})$$

with $X_1,\ldots,X_n$ i.i.d. and $\bar{X}$ their sample mean. The function h is required to
satisfy $h(x,\mu_X)=h_1(x)$ with $\mu_X=\mathsf{E}[X]$. It is easy to see that an estimator
of the form

$$e^{-rT}\frac{1}{n}\sum_{i=1}^{n}(\tilde{S}_i(T)-K)^+,$$

with $\tilde{S}_i$ as in (4.60) or (4.61), fits in this framework. Notice, also, that by
including in the vector X powers of other components of X, we make this
formulation sufficiently general to include matching higher-order moments as
well as the mean.

Suppose now that $h(X_i,\cdot)$ is almost surely continuously differentiable in a
neighborhood of μ_X. Then

$$\frac{1}{n}\sum_{i=1}^{n}h(X_i,\bar{X})\approx\frac{1}{n}\sum_{i=1}^{n}h_1(X_i)+\frac{1}{n}\sum_{i=1}^{n}\nabla_\mu h(X_i,\mu_X)[\bar{X}-\mu_X],\qquad(4.62)$$

with $\nabla_\mu h$ denoting the gradient of h with respect to its second argument.
Because $\bar{X}\to\mu_X$, this approximation becomes increasingly accurate as n
increases. This suggests that, asymptotically in n, the moment matching
estimator is equivalent to a control variate estimator with control $\bar{X}$ and
coefficient vector

$$b_n^\top=\frac{1}{n}\sum_{i=1}^{n}\nabla_\mu h(X_i,\mu_X)\to\mathsf{E}\left[\nabla_\mu h(X,\mu_X)\right].\qquad(4.63)$$

Some specific results in this direction are established in Duan, Gauthier, and
Simonato [97] and Hall [164]. However, even under conditions that make this
argument rigorous, the moment matching estimator may perform either better
or worse in small samples than the approximating control variate estimator.

The dependence among the observations $h(X_i,\bar{X})$, $i=1,\ldots,n$, introduced
through use of a common sample mean $\bar{X}$ complicates output analysis. One
approach proceeds as though the approximation in (4.62) held exactly and
estimates a confidence interval the way one would with a linear control variate
(cf. Sections 4.1.1 and 4.1.3). An alternative is to generate k independent
batches, each of size m, and to apply moment matching separately to each
batch of m paths. A confidence interval can then be formed from the sample
mean and sample standard deviation of the k means computed from the k
batches. As with stratified sampling or Latin hypercube sampling, the cost of
batching lies in foregoing potentially greater variance reduction by applying
the method to all km paths.

Examples

We turn now to some more specific examples of moment matching transformations.

Example 4.5.1 *Brownian motion and geometric Brownian motion.* In the case of a standard one-dimensional Brownian motion W, the additive transformation

$$\tilde{W}_i(t) = W_i(t) - \bar{W}(t)$$

seems the most natural way to match the sample and population means — there is no reason to try to avoid negative values of $\tilde{W}_i$, and the mean of a normal distribution is a location parameter. The transformation

$$\tilde{W}_i(t) = \frac{W_i(t) - \bar{W}(t)}{s(t)/\sqrt{t}}, \qquad (4.64)$$

with $s(t)$ the sample standard deviation of $W_1(t), \ldots, W_n(t)$, matches both first and second moments. But for this it seems preferable to scale the increments of the Brownian motion: with

$$W_i(t_k) = \sum_{j=1}^{k} \sqrt{t_j - t_{j-1}} Z_{ij}$$

and $\{Z_{ij}\}$ independent $N(0,1)$ random variables, set

$$\bar{Z}_j = \frac{1}{n} \sum_{i=1}^{n} Z_{ij}, \quad s_j^2 = \frac{1}{n-1} \sum_{i=1}^{n} (Z_{ij} - \bar{Z}_j)^2$$

and

$$\tilde{W}_i(t_k) = \sum_{j=1}^{k} \sqrt{t_j - t_{j-1}} \frac{Z_{ij} - \bar{Z}_j}{s_j}.$$

This transformation preserves the independence of increments whereas (4.64) does not.

For geometric Brownian motion $S \sim \mathrm{GBM}(r, \sigma^2)$, the multiplicative transformation (4.60) is more natural. It reduces to

$$\tilde{S}_i(t) = S(0)e^{rT} \frac{ne^{\sigma W_i(t)}}{\sum_{j=1}^{n} e^{\sigma W_j(t)}}.$$

This transformation does not lend itself as easily to matching higher moments.

As a simple illustration, we apply these transformations to the pricing of a call option under Black-Scholes assumptions and compare results in Figure 4.9. An ordinary simulation generates replications of the terminal asset price using

$$S_i(T) = S(0) \exp\left((r - \tfrac{1}{2}\sigma^2)T + \sigma\sqrt{T}Z_i\right), \quad i = 1, \ldots, n.$$

Method Z1 replaces each Z_i with $Z_i - \bar{Z}$ and method Z2 uses $(Z_i - \bar{Z})/s$, with $\bar{Z}$ the sample mean and s the sample standard deviation of $Z_1, \ldots, Z_n$. Methods SM and SA use the multiplicative and additive adjustments (4.60) and (4.61). Method CV uses $\bar{S}$ as a control variate with the optimal coefficient estimated as in (4.5).

Figure 4.9 compares estimates of the absolute bias and standard error for these methods in small samples of size $n = 16, 64, 256$, and 1024. The model parameters are $S(0) = K = 50$, $r = 5\%$, $\sigma = 0.30$, and the option expiration is $T = 1$. The results are based on 5000 replications of each sample size. The graphs in Figure 4.9 are on log-log scales; the slopes in the top panel are consistent with a $O(1/n)$ bias for each method and those in the bottom panel are consistent with $O(1/\sqrt{n})$ standard errors. Also, the biases are about an order of magnitude smaller than the standard errors.

In this example, the control variate estimator has the highest bias — recall that bias in this method results from estimation of the optimal coefficient — though the bias is quite small at $n = 1024$. Interestingly, the standard errors for the CV and SM methods are virtually indistinguishable. This suggests that the implicit coefficient (4.63) in the linear approximation to the multiplicative adjustment coincides with the optimal coefficient. The CV and SM methods achieve somewhat smaller standard errors than SA and Z1, which may be considered suboptimal control variate estimators. The lowest variance is attained by Z2, but because this method adjusts both a sample mean and a sample variance, it should be compared with a control variate estimator using two controls. (Compared with ordinary simulation, the Z2 reduces variance by a factor of about 40; the other methods reduce variance by factors ranging from about 3 to 7.)

These results suggest that moment matching estimators are indeed closely related to control variate estimators. They can sometimes serve as an indirect way of implementing a control, potentially providing much of the variance reduction while reducing small-sample bias. Though we observe this in Figure 4.9, there is of course no guarantee that the same would hold in other examples. □

Example 4.5.2 *Short rate models.* We consider, next, finite-sample adjustments to short rate processes with the objective of matching bond prices. We begin with an idealized setting of continuous-time simulation of a short rate process $r(t)$ under the risk-neutral measure. From independent replications $r_1, \ldots, r_n$ of the process, suppose we can compute estimated bond prices

$$\bar{B}(0, T) = \frac{1}{n} \sum_{i=1}^{n} \exp\left(-\int_0^T r_i(t)\, dt\right). \tag{4.65}$$

From these we define empirical forward rates

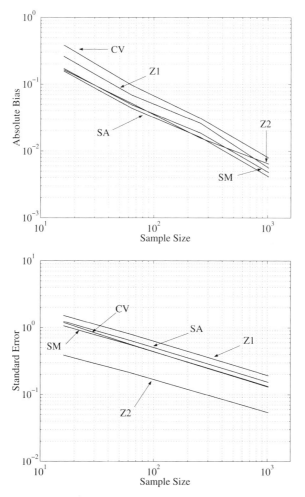

Fig. 4.9. Bias (top) and standard error (bottom) versus sample size in pricing a standard call option under Black-Scholes assumptions. The graphs compare moment matching based on the mean of Z_i (Z1), the mean and standard deviation of Z_i (Z2), multiplicative (SM) and additive (SA) adjustments based on $\bar{S}$, and an estimator using $\bar{S}$ as linear control variate (CV).

$$\hat{f}(0,T) = -\frac{\partial}{\partial T} \log \bar{B}(0,T). \qquad (4.66)$$

The model bond prices and forward rates are

$$B(0,T) = \mathsf{E}\left[\exp\left(-\int_0^T r(t)\,dt\right)\right]$$

and

$$f(0, T) = -\frac{\partial}{\partial T} \log B(0, T).$$

The adjustment

$$\tilde{r}_i(t) = r_i(t) + f(0, t) - \hat{f}(0, t)$$

results in

$$\frac{1}{n} \sum_{i=1}^{n} \exp\left(-\int_0^T \tilde{r}_i(t)\, dt\right) = B(0, T);$$

i.e., in exact pricing of bonds in finite samples.

Suppose, now, that we can simulate r exactly but only at discrete dates $0 = t_0, t_1, \ldots, t_m$, and our bond price estimates from n replications are given by

$$\bar{B}(0, t_m) = \frac{1}{n} \sum_{i=1}^{n} \exp\left(-\sum_{j=0}^{m-1} r_i(t_j)[t_{j+1} - t_j]\right).$$

The yield adjustment

$$\tilde{r}_i(t_j) = r_i(t_j) + \frac{\log \bar{B}(0, t_{j+1}) - \log \bar{B}(0, t_j)}{t_{j+1} - t_j} - \frac{\log B(0, t_{j+1}) - \log B(0, t_j)}{t_{j+1} - t_j}$$

results in

$$\frac{1}{n} \sum_{i=1}^{n} \exp\left(-\sum_{j=0}^{m-1} \tilde{r}_i(t_j)[t_{j+1} - t_j]\right) = B(0, t_m).$$

In this case, the adjustment corrects for discretization error as well as sampling variability. □

Example 4.5.3 *HJM framework.* Consider a continuous-time HJM model of the evolution of the forward curve $f(t, T)$, as in Section 3.6, and suppose we can simulate independent replications $f_1, \ldots, f_n$ of the paths of the curve. Because $f(t, t)$ is the short rate at time t, the previous example suggests the adjustment $\tilde{f}_i(t, t) = f_i(t, t) + f(0, t) - \hat{f}(0, t)$ with $\hat{f}$ as in (4.66) and $\bar{B}(0, T)$ as in (4.65) but with $r_i(t)$ replaced by $f_i(t, t)$. But the HJM setting provides additional flexibility to match additional moments by adjusting other rates $f(t, u)$.

We know that for any $0 < t < T$,

$$\mathsf{E}\left[\exp\left(-\int_0^t f(u, u)\, du - \int_t^T f(t, u)\, du\right)\right] = B(0, T);$$

this follows from the fact that discounted bond prices are martingales and is nearly the defining property of the HJM framework. We choose $\hat{f}$ to enforce the finite-sample analog of this property, namely

$$\frac{1}{n} \sum_{i=1}^{n} \exp\left(-\int_0^t \tilde{f}_i(u, u)\, du - \int_t^T \tilde{f}_i(t, u)\, du\right) = B(0, T).$$

This is accomplished by setting

$$\tilde{f}_i(t,T) = f_i(t,T) + f(0,T) - \frac{\sum_{i=1}^n f_i(t,T) H_i(t,T)}{\sum_{i=1}^n H_i(t,T)},$$

with

$$H_i(t,T) = \exp\left(-\int_0^t f_i(u,u)\,du - \int_t^T f_i(t,u)\,du\right).$$

In practice, one would simulate on a discrete grid of times and maturities as explained in Section 3.6.2 and this necessitates some modification. However, using the discretization in Section 3.6.2, the discrete discounted bond prices are martingales and thus lend themselves to a similar adjustment. □

Example 4.5.4 *Normal random vectors.* For i.i.d. normal random vectors, centering by the sample mean is equivalent to sampling *conditional* on the sample mean equaling the population mean. To see this, let $X_1, \ldots, X_n$ be independent $N(\mu, \Sigma)$ random vectors. The adjusted vectors

$$\tilde{X}_i = X_i - \bar{X} + \mu$$

have mean μ; moreover, they are jointly normal with

$$\begin{pmatrix} \tilde{X}_1 \\ \vdots \\ \tilde{X}_n \end{pmatrix} \sim N\left(\begin{pmatrix} \mu \\ \vdots \\ \mu \end{pmatrix}, \begin{pmatrix} (n-1)\Sigma/n & -\Sigma/n & \ldots & -\Sigma/n \\ -\Sigma/n & (n-1)\Sigma/n & & \vdots \\ \vdots & & \ddots & -\Sigma/n \\ -\Sigma/n & & -\Sigma/n & (n-1)\Sigma/n \end{pmatrix} \right),$$

as can be verified using the Linear Transformation Property (2.23). But this is also the joint distribution of $X_1, \ldots, X_n$ given $\bar{X} = \mu$, as can be verified using the Conditioning Formula (2.25). □

4.5.2 Weighted Monte Carlo

An alternative approach to matching underlying prices in finite samples assigns weights to the paths instead of shifting them. The weights are chosen to equate weighted averages over the paths to the corresponding population means.

Consider, again, the setting surrounding (4.59) with which we introduced the idea of moment matching. Suppose we want to price an option on the underlying asset $S(t)$; we note that the simulation misprices the underlying asset in finite samples in the sense that the sample mean $\bar{S}(t)$ deviates from $e^{rt}S(0)$. Rather than change the simulated values $S_1(t), \ldots, S_n(t)$, we may choose weights $w_1, \ldots, w_n$ satisfying

$$\sum_{i=1}^n w_i S_i(t) = e^{rt} S(0),$$

and then use these same weights in estimating the expected payoff of an option. For example, this yields the estimate

$$e^{-rt} \sum_{i=1}^{n} w_i (S_i(t) - K)^+$$

for the price of a call struck at K.

The method can be formulated more generically as follows. Suppose we want to estimate $\mathsf{E}[Y]$ and we know the mean vector $\mu_X = \mathsf{E}[X]$ for some random d-vector X. For example, X might record the prices of underlying assets at future dates, powers of those prices, or the discounted payoffs of tractable options. Suppose we know how to simulate i.i.d. replications (X_i, Y_i), $i = 1, \ldots, n$, of the pair (X, Y). In order to match the known mean μ_X in finite samples, we choose weights $w_1, \ldots, w_n$ satisfying

$$\sum_{i=1}^{n} w_i X_i = \mu_X \tag{4.67}$$

and then use

$$\sum_{i=1}^{n} w_i Y_i \tag{4.68}$$

to estimate $\mathsf{E}[Y]$. We may also want to require that the weights sum to 1:

$$\sum_{i=1}^{n} w_i = 1. \tag{4.69}$$

We can include this in (4.67) by taking one of the components of the X_i to be identically equal to 1.

The number of constraints d is typically much smaller than the number of replications n, so (4.67) does not determine the weights. We choose a particular set of weights by selecting an objective function $H : \Re^d \to \Re$ and solving the constrained optimization problem

$$\min H(w_1, \ldots, w_n) \quad \text{subject to (4.67).} \tag{4.70}$$

Because the replications are i.i.d., it is natural to restrict attention to functions H that are convex and symmetric in $w_1, \ldots, w_n$; the criterion in (4.70) then penalizes deviations from uniformity in the weights.

An approach of this type was proposed by Avellaneda et al. [26, 27], but more as a mechanism for model correction than variance reduction. In their setting, the constraint (4.67) uses a vector μ_o (to be interpreted as market prices) different from μ_X (the model prices). The weights thus serve to calibrate an imperfect model to the observed market prices of actively traded instruments in order to more accurately price a less liquid instrument. Broadie,

Glasserman, and Ha [68] use a related technique for pricing American options; we return to this in Chapter 8. It should be evident from the discussion leading to (4.68) and (4.70) that there is a close connection between this approach and using X as a control variate; we make the connection explicit in Example 4.5.6. First we treat the objective considered by Avellaneda et al. [26, 27].

Example 4.5.5 *Maximum entropy weights.* A particularly interesting and in some respects convenient objective H is the (negative) entropy function

$$H(w_1, \ldots, w_n) = \sum_{i=1}^{n} w_i \log w_i,$$

which we take to be $+\infty$ if any w_i is negative. (By convention, $0 \cdot \log 0 = 0$.) Using this objective will always produce positive weights, provided there is a positive feasible solution to (4.67). Such a solution exists whenever the convex hull of the points $X_1, \ldots, X_n$ contains μ_X, and this almost surely occurs for sufficiently large n.

We can solve for the optimal weights by first forming the Lagrangian

$$\sum_{i=1}^{n} w_i \log w_i - \nu \sum_{i=1}^{n} w_i - \lambda^\top \sum_{i=1}^{n} w_i X_i.$$

Here, ν is a scalar and λ is a d-vector. For this objective it turns out to be convenient to separate the constraint on the weight sum, as we have here. Setting the derivative with respect to w_i equal to zero and solving for w_i yields

$$w_i = e^{\nu - 1} \exp(\lambda^\top X_i).$$

Constraining the weights to sum to unity yields

$$w_i = \frac{\exp(\lambda^\top X_i)}{\sum_{j=1}^{n} \exp(\lambda^\top X_j)}. \tag{4.71}$$

The vector λ is then determined by the condition

$$\frac{\sum_{i=1}^{n} e^{\lambda^\top X_i} X_i}{\sum_{i=1}^{n} e^{\lambda^\top X_i}} = \mu_X,$$

which can be solved numerically.

Viewed as a probability distribution on $\{X_1, \ldots, X_n\}$, the $(w_1, \ldots, w_n)$ in (4.71) corresponds to an *exponential change of measure* applied to the uniform distribution $(1/n, \ldots, 1/n)$, in a sense to be further developed in Section 4.6. The solution in (4.71) may be viewed as the minimal adjustment to the uniform distribution needed to satisfy the constraints (4.67)–(4.69). □

Example 4.5.6 *Least-squares weights.* The simplest objective to consider in (4.70) is the quadratic

$$H(w_1, \ldots, w_n) = \tfrac{1}{2} w^\top w,$$

with w the vector of weights $(w_1, \ldots, w_n)^\top$. We will show that the estimator $w^\top Y = \sum w_i Y_i$ produced by these weights is identical to the control variate estimator $\bar{Y}(\hat{b}_n)$ defined by (4.16).

Define an $n \times (d + 1)$ matrix A whose ith row is $(1, X_i^\top - \mu_X^\top)$. (Here we assume that $X_1, \ldots, X_n$ do not contain an entry identically equal to 1.) Constraints (4.67)–(4.69) can be expressed as $w^\top A = (1, \mathbf{0})$, where $\mathbf{0}$ is a row vector of d zeros. The Lagrangian becomes

$$\tfrac{1}{2} w^\top w + w^\top A \lambda$$

with $\lambda \in \Re^d$. The first-order conditions are $w = -A\lambda$. From the constraint we get

$$(1, \mathbf{0}) = w^\top A = -\lambda^\top A^\top A \Rightarrow -\lambda^\top = (1, \mathbf{0})(A^\top A)^{-1}$$
$$\Rightarrow w^\top = (1, \mathbf{0})(A^\top A)^{-1} A^\top,$$

assuming the matrix A has full rank. The weighted Monte Carlo estimator of the expectation of the Y_i is thus

$$w^\top Y = (1, \mathbf{0})(A^\top A)^{-1} A^\top Y, \quad Y = (Y_1, \ldots, Y_n)^\top. \tag{4.72}$$

The control variate estimator is the first entry of the vector $\beta \in \Re^{d+1}$ that solves

$$\min_{\beta} \tfrac{1}{2} (Y - A\beta)^\top (Y - A\beta);$$

i.e., it is the value fitted at $(1, \mu_X)$ in a regression of the Y_i against the rows of A. From the first-order conditions $(Y - A\beta)^\top A = 0$, we find that the optimal β is $(A^\top A)^{-1} A^\top Y$. The control variate estimator is therefore

$$(1, \mathbf{0})\beta = (1, \mathbf{0})(A^\top A)^{-1} A^\top Y,$$

which coincides with (4.72). When written out explicitly, the weights in (4.72) take precisely the form displayed in (4.20), where we first noted the interpretation of a control variate estimator as a weighted Monte Carlo estimator. □

This link between the general strategy in (4.68) and (4.70) for constructing "moment-matched" estimators and the more familiar method of control variates suggests that (4.68) provides at best a small refinement of the control variate estimator. As the sample size n increases, the refinement typically vanishes and using knowledge of μ_X as a constraint in (4.67) becomes equivalent to using it in a control variate estimator. A precise result to this effect is proved in Glasserman and Yu [147]. This argues in favor of using control variate estimators rather than (4.68), because they are easier to implement and because more is known about their sampling properties.

4.6 Importance Sampling

4.6.1 Principles and First Examples

Importance sampling attempts to reduce variance by changing the probability measure from which paths are generated. Changing measures is a standard tool in financial mathematics; we encountered it in our discussion of pricing principles in Section 1.2.2 and several places in Chapter 3 in the guise of changing numeraire. Appendix B.4 reviews some of the underlying mathematical theory. When we switch from, say, the objective probability measure to the risk-neutral measure, our goal is usually to obtain a more convenient representation of an expected value. In importance sampling, we change measures to try to give more weight to "important" outcomes thereby increasing sampling efficiency.

To make this idea concrete, consider the problem of estimating

$$\alpha = \mathsf{E}[h(X)] = \int h(x) f(x)\, dx$$

where X is a random element of $\Re^d$ with probability density f, and h is a function from $\Re^d$ to $\Re$. The ordinary Monte Carlo estimator is

$$\hat{\alpha} = \hat{\alpha}(n) = \frac{1}{n} \sum_{i=1}^{n} h(X_i)$$

with $X_1, \ldots, X_n$ independent draws from f. Let g be any other probability density on $\Re^d$ satisfying

$$f(x) > 0 \Rightarrow g(x) > 0 \tag{4.73}$$

for all $x \in \Re^d$. Then we can alternatively represent α as

$$\alpha = \int h(x) \frac{f(x)}{g(x)} g(x)\, dx.$$

This integral can be interpreted as an expectation with respect to the density g; we may therefore write

$$\alpha = \tilde{\mathsf{E}}\left[h(X) \frac{f(X)}{g(X)} \right], \tag{4.74}$$

$\tilde{\mathsf{E}}$ here indicating that the expectation is taken with X distributed according to g. If $X_1, \ldots, X_n$ are now independent draws from g, the importance sampling estimator associated with g is

$$\hat{\alpha}_g = \hat{\alpha}_g(n) = \frac{1}{n} \sum_{i=1}^{n} h(X_i) \frac{f(X_i)}{g(X_i)}. \tag{4.75}$$

The weight $f(X_i)/g(X_i)$ is the *likelihood ratio* or *Radon-Nikodym derivative* evaluated at X_i.

It follows from (4.74) that $\tilde{\mathsf{E}}[\hat{\alpha}_g] = \alpha$ and thus that $\hat{\alpha}_g$ is an unbiased estimator of α. To compare variances with and without importance sampling it therefore suffices to compare second moments. With importance sampling, we have

$$\tilde{\mathsf{E}}\left[\left(h(X)\frac{f(X)}{g(X)}\right)^2\right] = \mathsf{E}\left[h(X)^2\frac{f(X)}{g(X)}\right].$$

This could be larger or smaller than the second moment $\mathsf{E}[h(X)^2]$ without importance sampling; indeed, depending on the choice of g it might even be *infinitely* larger or smaller. Successful importance sampling lies in the art of selecting an effective importance sampling density g.

Consider the special case in which h is nonnegative. The product $h(x)f(x)$ is then also nonnegative and may be normalized to a probability density. Suppose g is this density. Then

$$g(x) \propto h(x)f(x), \qquad (4.76)$$

and $h(X_i)f(X_i)/g(X_i)$ equals the constant of proportionality in (4.76) regardless of the value of X_i; thus, the importance sampling estimator $\hat{\alpha}_g$ in (4.75) provides a *zero-variance* estimator in this case. Of course, this is useless in practice: to normalize $h \cdot f$ we need to divide it by its integral, which is α; the zero-variance estimator is just α itself.

Nevertheless, this optimal choice of g does provide some useful guidance: in designing an effective importance sampling strategy, we should try to sample in proportion to the product of h and f. In option pricing applications, h is typically a discounted payoff and f is the risk-neutral density of a discrete path of underlying assets. In this case, the "importance" of a path is measured by the product of its discounted payoff and its probability density.

If h is the indicator function of a set, then the optimal importance sampling density is the original density conditioned on the set. In more detail, suppose $h(x) = \mathbf{1}\{x \in A\}$ for some $A \subset \Re^d$. Then $\alpha = P(X \in A)$ and the zero-variance importance sampling density $h(x)f(x)/\alpha$ is precisely the conditional density of X given $X \in A$ (assuming $\alpha > 0$). Thus, in applying importance sampling to estimate a probability, we should look for an importance sampling density that approximates the conditional density. This means choosing g to make the event $\{X \in A\}$ more likely, especially if A is a rare set under f.

Likelihood Ratios

In our discussion thus far we have assumed, for simplicity, that X is $\Re^d$-valued, but the ideas extend to X taking values in more general sets. Also, we have assumed that X has a density f, but the same observations apply if f is a probability mass function (or, more generally, a density with respect to some reference measure on $\Re^d$, possibly different from Lebesgue measure).

For option pricing applications, it is natural to think of X as a discrete path of underlying assets. The density of a path (if one exists) is ordinarily not specified directly, but rather built from more primitive elements. Consider, for example, a discrete path $S(t_i)$, $i = 0, 1, \ldots, m$, of underlying assets or state variables, and suppose that this process is Markov. Suppose the conditional distribution of $S(t_i)$ given $S(t_{i-1}) = x$ has density $f_i(x, \cdot)$. Consider a change of measure under which the transition densities f_i are replaced with transition densities g_i. The likelihood ratio for this change of measure is

$$\prod_{i=1}^{m} \frac{f_i(S(t_{i-1}), S(t_i))}{g_i(S(t_{i-1}), S(t_i))}.$$

More precisely, if E denotes expectation under the original measure and $\tilde{\mathsf{E}}$ denotes expectation under the new measure, then

$$\mathsf{E}[h(S(t_1), \ldots, S(t_m))] = \tilde{\mathsf{E}}\left[h(S(t_1), \ldots, S(t_m)) \prod_{i=1}^{m} \frac{f_i(S(t_{i-1}), S(t_i))}{g_i(S(t_{i-1}), S(t_i))}\right],$$
(4.77)

for all functions h for which the expectation on the left exists and is finite.

Here we have implicitly assumed that $S(t_0)$ is a constant. More generally, we could allow it to have density f_0 under the original measure and density g_0 under the new measure. This would result in an additional factor of $f_0(S(t_0))/g(S(t_0))$ in the likelihood ratio.

We often simulate a path $S(t_0), \ldots, S(t_m)$ through a recursion of the form

$$S(t_{i+1}) = G(S(t_i), X_{i+1}),$$
(4.78)

driven by i.i.d. random vectors $X_1, X_2, \ldots, X_m$. Many of the examples considered in Chapter 3 can be put in this form. The X_i will often be normally distributed, but for now let us simply assume they have common density f. If we apply a change of measure that preserves the independence of the X_i but changes their common density to g, then the corresponding likelihood ratio is

$$\prod_{i=1}^{m} \frac{f(X_i)}{g(X_i)}.$$

This means that

$$\mathsf{E}[h(S(t_1), \ldots, S(t_m))] = \tilde{\mathsf{E}}\left[h(S(t_1), \ldots, S(t_m)) \prod_{i=1}^{m} \frac{f(X_i)}{g(X_i)}\right],$$
(4.79)

where, again, E and $\tilde{\mathsf{E}}$ denote expectation under the original and new measures, respectively, and the expectation on the left is assumed finite. Equation (4.79) relies on the fact that $S(t_1), \ldots, S(t_m)$ are functions of $X_1, \ldots, X_m$.

Random Horizon

Identities (4.77) and (4.79) extend from a fixed number of steps m to a random number of steps, provided the random horizon is a stopping time. We demonstrate this in the case of i.i.d. inputs, as in (4.79). For each $n = 1, 2, \ldots$ let h_n be a function of n arguments and suppose we want to estimate

$$\mathsf{E}[h_N(S(t_1), \ldots, S(t_N))], \tag{4.80}$$

with N a random variable taking values in $\{1, 2, \ldots\}$. For example, in the case of a barrier option with barrier b, we might define N to be the index of the smallest t_i for which $S(t_i) > b$, taking $N = m$ if all $S(t_0), \ldots, S(t_m)$ lie below the barrier. We could then express the discounted payoff of an up-and-out put as $h_N(S(t_1), \ldots, S(t_N))$ with

$$h_n(S(t_1), \ldots, S(t_n)) = \begin{cases} e^{-rt_m}(K - S(t_m))^+, & n = m; \\ 0, & n = 0, 1, \ldots, m-1. \end{cases}$$

The option price then has the form (4.80).

Suppose that (4.78) holds and, as before, E denotes expectation when the X_i are i.i.d. with density f and $\tilde{\mathsf{E}}$ denotes expectation when they are i.i.d. with density g. For concreteness, suppose that $S(t_0)$ is fixed under both measures. Let N be a stopping time for the sequence $X_1, X_2, \ldots$; for example, N could be a stopping time for $S(t_1), S(t_2), \ldots$ as in the barrier option example. Then

$$\mathsf{E}[h_N(S(t_1), \ldots, S(t_N)) \mathbf{1}\{N < \infty\}]$$
$$= \tilde{\mathsf{E}}\left[h_N(S(t_1), \ldots, S(t_N)) \prod_{i=1}^{N} \frac{f(X_i)}{g(X_i)} \mathbf{1}\{N < \infty\} \right],$$

provided the expectation on the left is finite. This identity (sometimes called Wald's identity or the fundamental identity of sequential analysis — see, e.g., Asmussen [20]) is established as follows:

$$\mathsf{E}[h_N(S(t_1), \ldots, S(t_N)) \mathbf{1}\{N < \infty\}]$$
$$= \sum_{n=1}^{\infty} \mathsf{E}[h_n(S(t_1), \ldots, S(t_n)) \mathbf{1}\{N = n\}]$$
$$= \sum_{n=1}^{\infty} \tilde{\mathsf{E}}\left[h_n(S(t_1), \ldots, S(t_n)) \mathbf{1}\{N = n\} \prod_{i=1}^{n} \frac{f(X_i)}{g(X_i)} \right]$$
$$= \tilde{\mathsf{E}}\left[h_N(S(t_1), \ldots, S(t_N)) \prod_{i=1}^{N} \frac{f(X_i)}{g(X_i)} \mathbf{1}\{N < \infty\} \right]. \tag{4.81}$$

The second equality uses the stopping time property: because N is a stopping time the event $\{N = n\}$ is determined by $X_1, \ldots, X_n$ and this allows us to apply (4.79) to each term in the infinite sum. It is entirely possible for the event $\{N < \infty\}$ to have probability 1 under one of the measures but not the other; we will see an example of this in Example 4.6.3.

Long Horizon

We continue to consider two probability measures under which random vectors $X_1, X_2, \ldots$ are i.i.d., P giving the X_i density f, $\tilde{P}$ giving them density g.

It should be noted that even if f and g are mutually absolutely continuous, the probability measures P and $\tilde{P}$ will not be. Rather, absolute continuity holds for the restrictions of these measures to events defined by a finite initial segment of the infinite sequence. For $A \subseteq \Re^d$, the event

$$\left\{ \lim_{m \to \infty} \frac{1}{m} \sum_{i=1}^{m} \mathbf{1}\{X_i \in A\} = \int_A f(x)\, dx \right\}$$

has probability 1 under P; but some such event must have probability 0 under $\tilde{P}$ unless f and g are equal almost everywhere. In short, the strong law of large numbers forces P and $\tilde{P}$ to disagree about which events have probability 0.

This collapse of absolute continuity in the limit is reflected in the somewhat pathological behavior of the likelihood ratio as the number of terms grows, through an argument from Glynn and Iglehart [157]. Suppose that

$$\tilde{\mathsf{E}}\left[|\log(f(X_1)/g(X_1))|\right] < \infty;$$

then the strong law of large numbers implies that

$$\frac{1}{m} \sum_{i=1}^{m} \log(f(X_i)/g(X_i)) \to \tilde{\mathsf{E}}\left[\log(f(X_1)/g(X_1))\right] \equiv c \qquad (4.82)$$

with probability 1 under $\tilde{P}$. By Jensen's inequality,

$$c \leq \log \tilde{\mathsf{E}}[f(X_1)/g(X_1)] = \log \int \frac{f(x)}{g(x)} g(x)\, dx = 0,$$

with strict inequality unless $\tilde{P}(f(X_1) = g(X_1)) = 1$ because log is strictly concave. But if $c < 0$, (4.82) implies

$$\sum_{i=1}^{m} \log(f(X_i)/g(X_i)) \to -\infty;$$

exponentiating, we find that

$$\prod_{i=1}^{m} \frac{f(X_i)}{g(X_i)} \to 0$$

with $\tilde{P}$-probability 1. Thus, the likelihood ratio converges to 0 though its expectation equals 1 for all m. This indicates that the likelihood ratio becomes highly skewed, taking increasingly large values with small but non-negligible probability. This in turn can result in a large increase in variance if the change of measure is not chosen carefully.

Output Analysis

An importance sampling estimator does not introduce dependence between replications and is just an average of i.i.d. replications. We can therefore supplement an importance sampling estimator with a large-sample confidence interval in the usual way by calculating the sample standard deviation across replications and using it in (A.6). Because likelihood ratios are often highly skewed, the sample standard deviation will often underestimate the true standard deviation, and a very large sample size may be required for confidence intervals based on the central limit theorem to provide reasonable coverage. These features should be kept in mind in comparing importance sampling estimators based on estimates of their standard errors.

Examples

Example 4.6.1 *Normal distribution: change of mean.* Let f be the univariate standard normal density and g the univariate normal density with mean μ and variance 1. Then simple algebra shows that

$$\prod_{i=1}^{m} \frac{f(Z_i)}{g(Z_i)} = \exp\left(-\mu \sum_{i=1}^{m} Z_i + \frac{m}{2}\mu^2\right).$$

A bit more generally, if we let g_i have mean μ_i, then

$$\prod_{i=1}^{m} \frac{f(Z_i)}{g_i(Z_i)} = \exp\left(-\sum_{i=1}^{m} \mu_i Z_i + \tfrac{1}{2}\sum_{i=1}^{m} \mu_i^2\right). \tag{4.83}$$

If we simulate Brownian motion on a grid $0 = t_0 < t_1 < \cdots < t_m$ by setting

$$W(t_n) = \sum_{i=1}^{n} \sqrt{t_i - t_{i-1}} Z_i,$$

then (4.83) is the likelihood ratio for a change of measure that adds mean $\mu_i\sqrt{t_i - t_{i-1}}$ to the Brownian increment over $[t_{i-1}, t_i]$. □

Example 4.6.2 *Exponential change of measure.* The previous example is a special case of a more general class of convenient measure transformations. For a cumulative distribution function F on $\Re$, define

$$\psi(\theta) = \log \int_{-\infty}^{\infty} e^{\theta x} \, dF(x).$$

This is the *cumulant generating function* of F, the logarithm of the moment generating function of F. Let $\Theta = \{\theta : \psi(\theta) < \infty\}$ and suppose that Θ is nonempty. For each $\theta \in \Theta$, set

$$F_\theta(x) = \int_{-\infty}^{x} e^{\theta u - \psi(\theta)} \, dF(u);$$

each F_θ is a probability distribution, and $\{F_\theta, \theta \in \Theta\}$ form an *exponential family* of distributions. The transformation from F to F_θ is called exponential tilting, exponential twisting, or simply an exponential change of measure. If F has a density f, then F_θ has density

$$f_\theta(x) = e^{\theta x - \psi(\theta)} f(x).$$

Suppose that $X_1, \ldots, X_n$ are initially i.i.d. with distribution $F = F_0$ and that we apply a change of measure under which they become i.i.d. with distribution F_θ. The likelihood ratio for this transformation is

$$\prod_{i=1}^{n} \frac{dF_0(X_i)}{dF_\theta(X_i)} = \exp\left(-\theta \sum_{i=1}^{n} X_i + n\psi(\theta)\right). \tag{4.84}$$

The standard normal distribution has $\psi(\theta) = \theta^2/2$, from which we see that this indeed generalizes Example 4.6.1. A key feature of exponential twisting is that the likelihood ratio — which is in principle a function of all $X_1, \ldots, X_n$ — reduces to a function of the sum of the X_i. In statistical terminology, the sum of the X_i is a *sufficient statistic* for θ.

The cumulant generating function ψ records important information about the distributions F_θ. For example, $\psi'(\theta)$ is the mean of F_θ. To see this, let E_θ denote expectation with respect to F_θ and note that $\psi(\theta) = \log \mathsf{E}_0[\exp(\theta X)]$. Differentiation yields

$$\psi'(\theta) = \frac{\mathsf{E}_0[X e^{\theta X}]}{\mathsf{E}_0[e^{\theta X}]} = \mathsf{E}_0[X e^{\theta X - \psi(\theta)}] = \mathsf{E}_\theta[X].$$

A similar calculation shows that $\psi''(\theta)$ is the variance of F_θ. The function ψ passes through the origin; Hölder's inequality shows that it is convex, so that $\psi''(\theta)$ is indeed positive. For further theoretical background on exponential families see, e.g., Barndorff-Nielsen [35].

We conclude with some examples of exponential families. The normal distributions $N(\theta, \theta\sigma^2)$ form an exponential family in θ for all $\sigma > 0$. The gamma densities

$$\frac{1}{\Gamma(a)\theta^a} x^{a-1} e^{-x/\theta}, \quad x \geq 0,$$

form an exponential family in θ for each value of the shape parameter $a > 0$. With $a = 1$, this is the family of exponential distributions with mean θ. The Poisson distributions

$$e^{-\lambda} \frac{\lambda^k}{k!}, \quad k = 0, 1, \ldots,$$

form an exponential family in $\theta = \log \lambda$. The binomial distributions

$$\frac{n!}{k!(n-k)!}p^k(1-p)^{n-k}, \quad k=0,1,\ldots,n,$$

form an exponential family in $\theta = \log(p/(1-p))$. □

Example 4.6.3 *Ruin probabilities.* A classic application of importance sampling arises in estimating ruin probabilities in the theory of insurance risk. Consider an insurance firm earning premiums at a constant rate p per unit of time and paying claims that arrive at the jumps of a Poisson process with rate λ. Letting $N(t)$ denote the number of claims arriving in $[0,t]$ and Y_i the size of the ith claim, $i = 1,2,\ldots$, the net payout of the firm over $[0,t]$ is given by

$$\sum_{i=1}^{N(t)} Y_i - pt.$$

Suppose the firm has a reserve of x; then ruin occurs if the net payout ever exceeds x. We assume the claims are i.i.d. and independent of the Poisson process. We further assume that $\lambda\mathsf{E}[Y_i] < p$, meaning that premiums flow in at a faster rate than claims are paid out; this ensures that the probability of eventual ruin is less than 1.

If ruin ever occurs, it must occur at the arrival of a claim. It therefore suffices to consider the discrete-time process embedded at the jumps of the Poisson process. Let $\xi_1, \xi_2, \ldots$ be the interarrival times of the Poisson process; these are independent and exponentially distributed with mean $1/\lambda$. The net payout between the $(n-1)$th and nth claims (including the latter but not the former) is $X_n = Y_n - p\xi_n$. The net payout up to the nth claim is given by the random walk $S_n = X_1 + \cdots + X_n$. Ruin occurs at

$$\tau_x = \inf\{n \geq 0 : S_n > x\},$$

with the understanding that $\tau_x = \infty$ if S_n never exceeds x. The probability of eventual ruin is $P(\tau_x < \infty)$. Figure 4.10 illustrates the notation for this example.

The particular form of the increments X_n is not essential to the problem so we generalize the setting. We assume that $X_1, X_2, \ldots$ are i.i.d. with $0 < P(X_i > 0) < 1$ and $\mathsf{E}[X_i] < 0$, but we drop the specific form $Y_n - p\xi_n$. We add the assumption that the cumulant generating function ψ_X of the X_i (cf. Example 4.6.2) is finite in a neighborhood of the origin. This holds in the original model if the cumulant generating function ψ_Y of the claim sizes Y_i is finite in a neighborhood of the origin.

For any point θ in the domain of ψ_X, consider the exponential change of measure with parameter θ and let E_θ denote expectation under this measure. Because τ_x is a stopping time, we may apply (4.81) to write the ruin probability $P(\tau_x < \infty)$ as an E_θ-expectation. Because we have applied an exponential change of measure, the likelihood ratio simplifies as in (4.84); thus, the ruin probability becomes

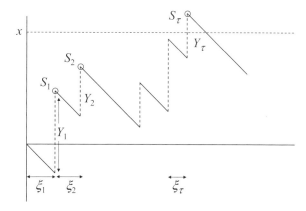

Fig. 4.10. Illustration of claim sizes Y_i, interarrival times ξ_i, and the random walk S_n. Ruin occurs at the arrival of the τth claim.

$$P(\tau_x < \infty) = \mathsf{E}_\theta \left[e^{-\theta S_{\tau_x} + \psi_X(\theta)\tau_x} \mathbf{1}\{\tau_x < \infty\} \right]. \qquad (4.85)$$

If $0 < \psi'_X(\theta) < \infty$ (which entails $\theta > 0$ because $\psi(0) = 0$ and $\psi'(0) = \mathsf{E}[X_n] < 0$), then the random walk has positive drift $\mathsf{E}_\theta[X_n] = \psi'_X(\theta)$ under the twisted measure, and this implies $P_\theta(\tau_x < \infty) = 1$. We may therefore omit the indicator inside the expectation on the right. It also follows that we may obtain an unbiased estimator of the ruin probability by simulating the random walk under P_θ until τ_x and returning the estimator $\exp(-\theta S_{\tau_x} + \psi_X(\theta)\tau_x)$. This would not be feasible under the original measure because of the positive probability that $\tau_x = \infty$.

Among all θ for which $\psi'_X(\theta) > 0$, one is particularly effective for simulation and indeed optimal in an asymptotic sense. Suppose there is a $\theta > 0$ at which $\psi_X(\theta) > 0$. There must then be a $\theta_* > 0$ at which $\psi_X(\theta_*) = 0$; convexity of ψ_X implies uniqueness of θ^* and positivity of $\psi'_X(\theta_*)$, as is evident from Figure 4.11. In the insurance risk model with $X_n = Y_n - p\xi_n$, θ_* is the unique positive solution to

$$\psi_Y(\theta) + \log\left(\frac{\lambda}{\lambda + p\theta}\right) = 0,$$

where ψ_Y is the cumulant generating function for the claim-size distribution.

With the parameter θ_*, (4.85) becomes

$$P(\tau_x < \infty) = \mathsf{E}_{\theta_*}\left[e^{-\theta_* S_{\tau_x}}\right] = e^{-\theta_* x}\mathsf{E}_{\theta_*}\left[e^{-\theta_*(S_{\tau_x} - x)}\right].$$

Because the overshoot $S_{\tau_x} - x$ is nonnegative, this implies the simple bound $P(\tau_x < \infty) \le e^{-\theta_* x}$ on the ruin probability. Under modest additional regularity conditions (for example, if the X_n have a density), $\mathsf{E}_{\theta_*}\left[e^{-\theta_*(S_{\tau_x} - x)}\right]$ converges to a constant c as $x \to \infty$, providing the classical approximation

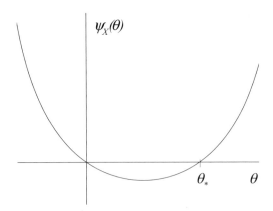

Fig. 4.11. Graph of a cumulant generating function ψ_X. The curve passes through the origin and has negative slope there because $\psi'_X(0) = \mathsf{E}[X] < 0$. At the positive root θ_*, the slope is positive.

$$P(\tau_x < \infty) \sim c e^{-\theta_* x},$$

meaning that the ratio of the two sides converges to 1 as $x \to \infty$. Further details and some of the history of this approximation are discussed in Asmussen [20] and in references given there.

From the perspective of simulation, the significance of θ_* lies in the variance reduction achieved by the associated importance sampling estimator. The unbiased estimator $\exp(-\theta_* S_{\tau_x})$, sampled under P_{θ_*}, has second moment

$$\mathsf{E}_{\theta_*}\left[e^{-2\theta_* S_{\tau_x}}\right] \le e^{-2\theta_* x}.$$

By Jensen's inequality, the second moment of any unbiased estimator must be at least as large as the square of the ruin probability, and we have seen that this probability is $O(e^{-\theta_* x})$. In this sense, the second moment of the importance sampling estimator based on θ_* is asymptotically optimal as $x \to \infty$.

This strategy for developing effective and even asymptotically optimal importance sampling estimators originated in Siegmund's [331] application in sequential analysis. It has been substantially generalized, particularly for queueing and reliability applications, as surveyed in Heidelberger [175]. □

Example 4.6.4 *A knock-in option.* As a further illustration of importance sampling through an exponential change of measure, we apply the method to a down-and-in barrier option. This example is from Boyle et al. [53]. The option is a digital knock-in option with payoff

$$\mathbf{1}\{S(T) > K\} \cdot \mathbf{1}\{\min_{1 \le k \le m} S(t_k) < H\},$$

with $0 < t_1 < \cdots < t_m = T$, S the underlying asset, K the strike, and H the barrier. If H is much smaller than $S(0)$, most paths of an ordinary simulation will result in a payoff of zero; importance sampling can potentially make knock-ins less rare.

Suppose the underlying asset is modeled through a process of the form

$$S(t_n) = S(0) \exp(L_n), \quad L_n = \sum_{i=1}^{n} X_i,$$

with $X_1, X_2, \ldots$ i.i.d. and $L_0 = 0$. This includes geometric Brownian motion but many other models as well; see Section 3.5. The option payoff is then

$$\mathbf{1}\{L_m > c, \tau < m\}$$

where $c = \log(K/S(0))$, τ is the first time the random walk L_n drops below $-b$, and $-b = \log(H/S(0))$. If b or c is large, the probability of a payoff is small. To increase the probability of a payoff, we need to drive L_n down toward $-b$ and then up toward c.

Suppose the X_i have cumulant generating function ψ and consider importance sampling estimators of the following form: exponentially twist the distribution of the X_i by some θ_- (with drift $\psi'(\theta_-) < 0$) until the barrier is crossed, then twist the remaining $X_{\tau+1}, \ldots, X_m$ by some θ_+ (with drift $\psi'(\theta_+) > 0$) to drive the process up toward the strike. On the event $\{\tau < m\}$, the likelihood ratio for this change of measure is (using (4.81) and (4.84))

$$\exp\left(-\theta_- L_\tau + \psi(\theta_-)\tau\right) \cdot \exp\left(-\theta_+[L_m - L_\tau] + \psi(\theta_+)[m - \tau]\right)$$
$$= \exp\left((\theta_+ - \theta_-)L_\tau - \theta_+ L_m + (\psi(\theta_-) - \psi(\theta_+))\tau + m\psi(\theta_+)\right).$$

The importance sampling estimator is the product of this likelihood ratio and the discounted payoff.

We now apply a heuristic argument to select the parameters θ_-, θ_+. We expect most of the variability in the estimator to result from the barrier crossing time τ, because for large b and c we expect $L_\tau \approx -b$ and $L_m \approx c$ on the event $\{\tau < m, L_m > c\}$. (In other words, the undershoot below $-b$ and the overshoot above c should be small.) If we choose θ_-, θ_+ to satisfy $\psi(\theta_-) = \psi(\theta_+)$, the likelihood ratio simplifies to

$$\exp\left((\theta_+ - \theta_-)L_\tau - \theta_+ L_m + m\psi(\theta_+)\right),$$

and we thus eliminate explicit dependence on τ.

To complete the selection of the parameters $\theta_\pm$, we impose the condition that traveling in a straight-line path from 0 to $-b$ at rate $|\psi'(\theta_-)|$ and then from $-b$ to c at rate $\psi'(\theta_+)$, the process should reach c at time m; i.e.,

$$\frac{-b}{\psi'(\theta_-)} + \frac{c+b}{\psi'(\theta_+)} = m.$$

These conditions uniquely determine $\theta_\pm$, at least if the domain of ψ is sufficiently large. This is illustrated in Figure 4.12.

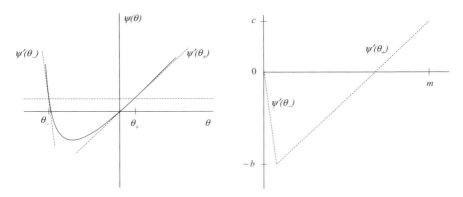

Fig. 4.12. Illustration of importance sampling strategy for a knock-in option. Twisting parameters $\theta_{\pm}$ are chosen so that (a) $\psi(\theta_-) = \psi(\theta_+)$ and (b) straight-line path with slopes $\psi'(\theta_-)$ and $\psi'(\theta_+)$ reaches $-b$ and then c in m steps.

In the case of geometric Brownian motion GBM(μ, σ^2) with equally spaced time points $t_n = nh$, we have

$$X_n \sim N((\mu - \tfrac{1}{2}\sigma^2)h, \sigma^2 h),$$

and the cumulant generating function is

$$\psi(\theta) = (\mu - \tfrac{1}{2}\sigma^2)h\theta + \tfrac{1}{2}\sigma^2 h\theta^2.$$

Because this function is quadratic in θ, it is symmetric about its minimum and the condition $\psi(\theta_-) = \psi(\theta_+)$ implies that $\psi'(\theta_-) = -\psi'(\theta_+)$. Thus, under our proposed change of measure, the random walk moves at a constant speed of $|\psi'(\theta_{\pm})|$. To traverse the path down to the barrier and up to the strike in m steps, we must have

$$|\psi'(\theta_{\pm})| = \frac{2b + c}{m}.$$

We can now solve for the twisting parameters to get

$$\theta_{\pm} = \left(\frac{1}{2} - \frac{\mu}{\sigma^2}\right) \pm \frac{2b + c}{m\sigma^2 h}.$$

The term in parentheses on the right is the point at which the quadratic ψ is minimized. The twisting parameters $\theta_{\pm}$ are symmetric about this point.

Table 4.4 reports variance ratios based on this method. The underlying asset $S(t)$ is GBM(r, σ^2) with $r = 5\%$, $\sigma = 0.15$, and initial value $S(0) = 95$. We consider an option paying 10,000 if not knocked out, hence having price $10,000 \cdot e^{-rT} P(\tau < m, S(T) > K)$. As above, m is the number of steps and $T \equiv t_m$ is the option maturity. The last column of the table gives the estimated ratio of the variance per replication using ordinary Monte Carlo to the variance using importance sampling. It is thus a measure of the speed-up produced by importance sampling. The estimates in the table are based on 100,000 replications for each case. The results suggest that the variance ratio depends primarily on the rarity of the payoff, and not otherwise on the maturity. The variance reduction can be dramatic for extremely rare payoffs.

An entirely different application of importance sampling to barrier options is developed in Glasserman and Staum [146]. In that method, at each step along a simulated path, the value of an underlying asset is sampled conditional on not crossing a knock-out barrier so that all paths survive to maturity. The one-step conditional distributions define the change of measure in this approach. □

	H	K	Price	Variance Ratio
$T = 0.25, m = 50$	94	96	3017.6	2
	90	96	426.6	10
	85	96	5.6	477
	90	106	13.2	177
$T = 1, m = 50$	90	106	664.8	6
	85	96	452.0	9
$T = 0.25, m = 100$	85	96	6.6	405
	90	106	15.8	180

Table 4.4. Variance reduction using importance sampling in pricing a knock-in barrier option with barrier H and strike K.

4.6.2 Path-Dependent Options

We turn now to a more ambitious application of importance sampling with the aim of reducing variance in pricing path-dependent options. We consider models of underlying assets driven by Brownian motion (or simply by normal random vectors after discretization) and change the drift of the Brownian motion to drive the underlying assets into "important" regions, with "importance" determined by the payoff of the option. We identify a specific change of drift through an optimization problem.

The method described in this section is from Glasserman, Heidelberger, and Shahabuddin (henceforth abbreviated GHS) [139], and that reference contains a more extensive theoretical development than we provide here. This method restricts itself to deterministic changes of drift over discrete time steps. It is theoretically possible to eliminate all variance through a stochastic change of drift in continuous time, essentially by taking the option being priced as the numeraire asset and applying the change of measure associated with this change of numeraire. This however requires knowing the price of the option in advance and is not literally feasible, though it potentially provides a basis for approximations. Related ideas are developed in Chapter 16 of Kloeden and Platen [211], Newton [278, 279], and Schoenmakers and Heemink [319].

We restrict ourselves to simulations on a discrete time grid $0 = t_0 < t_1 < \cdots < t_m = T$. We assume the only source of randomness in the simulated model is a d-dimensional Brownian motion. The increment of the Brownian motion from t_{i-1} to t_i is simulated as $\sqrt{t_i - t_{i-1}} Z_i$, where $Z_1, Z_2, \ldots, Z_m$ are independent d-dimensional standard normal random vectors. Denote by Z the concatenation of the Z_i into a single vector of length $n \equiv md$. Each outcome of Z determines a path of underlying assets or state variables, and each such path determines the discounted payoff of an option. If we let G denote the composition of these mappings, then $G(Z)$ is the discounted payoff derived from Z. Our task is to estimate $\mathsf{E}[G(Z)]$, the expectation taken with Z having the n-dimensional standard normal distribution.

An example will help fix ideas. Consider a single underlying asset modeled as geometric Brownian motion $\mathrm{GBM}(r, \sigma^2)$ and simulated using

$$S(t_i) = S(t_{i-1}) \exp\left([r - \tfrac{1}{2}\sigma^2](t_i - t_{i-1}) + \sigma\sqrt{t_i - t_{i-1}} Z_i \right), \quad i = 1, \ldots, m.$$
(4.86)

Consider an Asian call option on the arithmetic average $\bar{S}$ of the $S(t_i)$. We may view the payoff of the option as a function of the Z_i and thus write

$$G(Z) = G(Z_1, \ldots, Z_m) = e^{-rT}[\bar{S} - K]^+.$$

Pricing the option means evaluating $\mathsf{E}[G(Z)]$, the expectation taken with $Z \sim N(0, I)$.

Change of Drift: Linearization

Through importance sampling we can change the distribution of Z and still obtain an unbiased estimator of $\mathsf{E}[G(Z)]$, provided we weight each outcome by the appropriate likelihood ratio. We restrict ourselves to changes of distribution that change the mean of Z from 0 to some other vector μ. Let P_μ and E_μ denote probability and expectation when $Z \sim N(\mu, I)$. From the form of the likelihood ratio given in Example 4.6.1 for normal random vectors, we find that

$$\mathsf{E}[G(Z)] = \mathsf{E}_\mu\left[G(Z) e^{-\mu^\top Z + \frac{1}{2}\mu^\top \mu} \right]$$

for any $\mu \in \Re^n$. We may thus simulate as follows:

for replications $i = 1, \ldots, N$
    generate $Z^{(i)} \sim N(\mu, I)$
    $Y^{(i)} \leftarrow G(Z^{(i)}) \exp\left(-\mu^\top Z^{(i)} + \frac{1}{2}\mu^\top \mu \right)$
return $(Y^{(1)} + \cdots + Y^{(N)})/N$.

This estimator is unbiased for any choice of μ; we would like to choose a μ that produces a low-variance estimator.

If G takes only nonnegative values (as is typical of discounted option payoffs), we may write $G(z) = \exp(F(z))$, with the convention that $F(z) = -\infty$

if $G(z) = 0$. Also, note that taking an expectation over Z under P_μ is equivalent to replacing Z with $\mu + Z$ and taking the expectation under the original measure. (In the algorithm above, this simply means that we can sample from $N(\mu, I)$ by sampling from $N(0, I)$ and adding μ.) Thus,

$$\mathsf{E}[G(Z)] = \mathsf{E}\left[e^{F(Z)}\right] = \mathsf{E}_\mu\left[e^{F(Z)}e^{-\mu^\top Z + \frac{1}{2}\mu^\top \mu}\right]$$

$$= \mathsf{E}\left[e^{F(\mu+Z)}e^{-\mu^\top (\mu+Z)+\frac{1}{2}\mu^\top \mu}\right]$$

$$= \mathsf{E}\left[e^{F(\mu+Z)}e^{-\mu^\top Z - \frac{1}{2}\mu^\top \mu}\right]. \tag{4.87}$$

For any μ, the expression inside the expectation in (4.87) is an unbiased estimator with Z having distribution $N(0, I)$. To motivate a particular choice of μ, we now expand F to first order to approximate the estimator as

$$e^{F(\mu+Z)}e^{-\mu^\top Z - \frac{1}{2}\mu^\top \mu} \approx e^{F(\mu)+\nabla F(\mu)Z}e^{-\mu^\top Z - \frac{1}{2}\mu^\top \mu}, \tag{4.88}$$

with $\nabla F(\mu)$ the gradient of F at μ. If we can choose μ to satisfy the fixed-point condition

$$\nabla F(\mu) = \mu^\top, \tag{4.89}$$

then the expression on the right side of (4.88) collapses to a constant with no dependence on Z. Thus, applying importance sampling with μ satisfying (4.89) would produce a zero-variance estimator if (4.88) held exactly, and it should produce a low-variance estimator if (4.88) holds only approximately.

Change of Drift: Normal Approximation and Optimal Path

We now present an alternative argument leading to an essentially equivalent choice of μ. Recall from the discussion surrounding (4.76) that the optimal importance sampling density is the normalized product of the integrand and the original density. For the problem at hand, this means that the optimal density is proportional to

$$e^{F(z)-\frac{1}{2}z^\top z},$$

because $\exp(F(z))$ is the integrand and $\exp(-z^\top z/2)$ is proportional to the standard normal density. Normalizing this function by its integral produces a probability density but not, in general, a normal density. Because we have restricted ourselves to changes of mean, we may try to select μ so that $N(\mu, I)$ approximates the optimal distribution. One way to do this is to choose μ to be the mode of the optimal density; i.e., choose μ to solve

$$\max_z F(z) - \frac{1}{2}z^\top z. \tag{4.90}$$

The first-order condition for the optimum is $\nabla F(z) = z^\top$, which coincides with (4.89). If, for example, the objective in (4.90) is strictly concave, and if the first-order condition has a solution, this solution is the unique optimum.

We may interpret the solution z_* to (4.90) as an optimal *path*. Each $z \in \Re^n$ may be interpreted as a path because each determines a discrete Brownian path and thus a path of underlying assets. The solution to (4.90) is the most "important" path if we measure importance by the product of payoff $\exp(F(z))$ and probability density $\exp(-z^\top z/2)/(2\pi)^{n/2}$. In choosing $\mu = z_*$, we are therefore choosing the new drift to push the process along the optimal path.

GHS [139] give conditions under which this approach to importance sampling has an asymptotic optimality property. This property is based on introducing a parameter ϵ and analyzing the second moment of the estimator as ϵ approaches zero. From a practical perspective, a small ϵ should be interpreted as a nearly linear F.

Asian Option

We illustrate the selection and application of the optimal change of drift in the case of the Asian call defined above, following the discussion in GHS [139]. Solving (4.90) is equivalent to maximizing $G(z)\exp(-z^\top z/2)$ with G the discounted payoff of the Asian option. The discount factor e^{-rT} is a constant in this example, so for the purpose of optimization we may ignore it and redefine $G(z)$ to be $[\bar{S} - K]^+$. Also, in maximizing it clearly suffices to consider points z at which $\bar{S} > K$ and thus at which G is differentiable.

For the first-order conditions, we differentiate

$$[\bar{S} - K]e^{-z^\top z/2}$$

to get

$$\frac{\partial \bar{S}}{\partial z_j} - [\bar{S} - K]z_j = 0.$$

Using (4.86), we find that

$$\frac{\partial \bar{S}}{\partial z_j} = \frac{1}{m}\sum_{i=j}^{m}\frac{\partial S(t_i)}{\partial z_j} = \frac{1}{m}\sum_{i=j}^{m}\sigma\sqrt{t_i - t_{i-1}}S(t_i).$$

The first-order conditions thus become

$$z_j = \frac{\sum_{i=j}^{m}\sigma\sqrt{t_i - t_{i-1}}S(t_i)}{mG(z)}.$$

Now we specialize to the case of an equally spaced time grid with $t_i - t_{i-1} \equiv h$. This yields

$$z_1 = \frac{\sigma\sqrt{h}(G(z) + K)}{G(z)}, \quad z_{j+1} = z_j - \frac{\sigma\sqrt{h}S(t_j)}{mG(z)}, \quad j = 1, \ldots, m - 1. \quad (4.91)$$

Given the value of $G(z)$, (4.91) and (4.86) determine z. Indeed, if $y \equiv G(z)$, we could apply (4.91) to calculate z_1 from y, then (4.86) to calculate $S(t_1)$, then (4.91) to calculate z_2, and so on. Through this iteration, each value of y determines a $z(y)$ and path $S(t_i, y)$, $i = 1, \ldots, m$. Solving the first-order conditions reduces to finding the y for which the payoff at $S(t_1, y), \ldots, S(t_m, y)$ is indeed y; that is, it reduces to finding the root of the equation

$$\frac{1}{m}\sum_{j=1}^{m} S(t_j, y) - K - y = 0.$$

GHS [139] report that numerical examples suggest that this equation has a unique root. This root can be found very quickly through a one-dimensional search. Once the root y_* is found, the optimization problem is solved by $z_* = z(y_*)$. To simulate, we then set $\mu = z_*$ and apply importance sampling with mean μ.

Combined Importance Sampling and Stratification

In GHS [139], further (and in some cases enormous) variance reduction is achieved by combining importance sampling with stratification of a linear projection of Z. Recall from Section 4.3.2 that sampling Z so that $v^\top Z$ is stratified for some $v \in \Re^n$ is easy to implement. The change of mean does not affect this. Indeed, we may sample from $N(\mu, I)$ by sampling from $N(0, I)$ and then adding μ; we can apply stratified sampling to $N(0, I)$ before adding μ.

Two strategies for selecting the stratification direction v are considered in GHS [139]. One simply sets $v = \mu$ on the grounds that μ is an important path and thus a potentially important direction for stratification. The other strategy expands (4.88) to get

$$e^{F(\mu+Z)}e^{-\mu^\top Z - \frac{1}{2}\mu^\top \mu} \approx e^{F(\mu) + \nabla F(\mu)Z + \frac{1}{2}Z^\top H(\mu)Z}e^{-\mu^\top Z - \frac{1}{2}\mu^\top \mu},$$

with $H(\mu)$ the Hessian matrix of F at μ. Importance sampling with $\mu^\top = \nabla F(\mu)$ eliminates the linear term in the exponent, and this suggests that the stratification should be tailored to the quadratic term.

In Section 4.3.2, we noted that the optimal stratification direction for estimating an expression of the form $\mathsf{E}[\exp(\frac{1}{2}Z^\top AZ)]$ with A symmetric is an eigenvector of A. The optimal eigenvector is determined by the eigenvalues of A through the criterion in (4.50). This suggests that we should stratify along the optimal eigenvector of the Hessian of F at μ. This entails numerical calculation of the Hessian and its eigenvectors.

Table 4.5 shows results from GHS [139]. The table shows variance ratios (i.e., variance reduction factors) using importance sampling and two combinations of importance sampling with stratified sampling, using the two strategies just described for selecting a stratification direction. All results use $S(0) = 50$, $r = 0.05$, and $T = 1$ and are estimated from one million paths for each case. The results show that importance sampling by itself can produce noteworthy variance reduction (especially for out-of-the-money options) and that the combined impact with stratification can be astounding. The combination reduces variance by factors in the thousands.

n	σ	K	Price	Importance Sampling	IS & Strat. (μ)	IS & Strat. (v_{j*})
16	0.10	45	6.05	11	1,097	1,246
		50	1.92	7	4,559	5,710
		55	0.20	21	15,520	17,026
16	0.30	45	7.15	8	1,011	1,664
		50	4.17	9	1,304	1,899
		55	2.21	12	1,746	2,296
64	0.10	45	6.00	11	967	1,022
		50	1.85	7	4,637	5,665
		55	0.17	23	16,051	17,841
64	0.30	45	7.02	8	1,016	1,694
		50	4.02	9	1,319	1,971
		55	2.08	12	1,767	2,402

Table 4.5. Estimated variance reduction ratios for Asian options using importance sampling and combinations of importance sampling with stratified sampling, stratifying along the optimal μ or the optimal eigenvector v_{j*}. Stratified results use 100 strata.

The results in Table 4.5 may seem to suggest that stratification has a greater impact than importance sampling, and one may question the value of importance sampling in this example. But the effectiveness of stratification is indeed enhanced by the change in mean, which results in more paths producing positive payoffs. The positive-part operator $[\cdot]^+$ applied to $\bar{S} - K$ diminishes the effectiveness of stratified sampling, because it tends to produce many strata with a constant (zero) payoff — stratifying a region of constant payoff is useless. By shifting the mean of Z, we implicitly move more of the distribution of the stratification variable $v^\top Z$ (and thus more strata) into the region where the payoff varies. In this particular example, the region defined by $\bar{S} > K$ is reasonably easy to characterize and could be incorporated into the selection of strata; however, this is not the case in more complex examples.

A further notable feature of Table 4.5 is that the variance ratios are quite similar whether we stratify along μ or along the optimal eigenvector v_*. In fact, GHS [139] find that the vectors μ and v_* nearly coincide, once normalized to have the same length. They find similar patterns in other examples. This phenomenon can occur when G is well approximated by a nonlinear function of a linear combination of $z_1, \ldots, z_m$. For suppose $G(z) \approx g(v^\top z)$; then, the gradient of G is nearly proportional to $v^\top$ and so μ will be nearly proportional to v. Moreover, the Hessian of G will be nearly proportional to the rank-1 matrix $vv^\top$, whose only nonzero eigenvectors are multiples of v. Thus, in this setting, the optimal mean is proportional to the optimal eigenvector.

Application in the Heath-Jarrow-Morton Framework

GHS [140] apply the combination of importance sampling and stratified sampling in the Heath-Jarrow-Morton framework. The complexity of this setting necessitates some approximations in the calculation of the optimal path and eigenvector to make the method computationally feasible. We comment on these briefly.

We consider a three-factor model (so $d = 3$) discretized in time and maturity as detailed in Section 3.6.2. The discretization interval is a quarter of a year and we consider maturities up to 20 years, so $m = 80$ and the vector Z of random inputs has dimension $n = md = 240$. The factor loadings for each of the three factors are as displayed in Figure 4.13, where they are plotted against time to maturity.

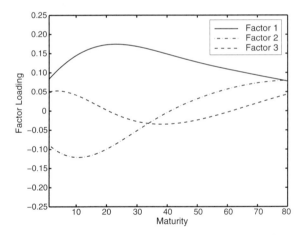

Fig. 4.13. Factor loadings in three-factor HJM model used to illustrate importance sampling.

As in Section 3.6.2, we use $\hat{f}(t_i, t_j)$ to denote the forward rate at time t_i for the interval $[t_j, t_{j+1}]$. We use an equally spaced time grid so $t_i = ih$ with h a quarter of a year. The initial forward curve is

$$\hat{f}(0, t_j) = \log(150 + 12j)/100, \quad j = 0, 1, \ldots, 80,$$

which increases gradually, approximately from 5% to 7%.

Among the examples considered in GHS [140] is the pricing of an interest rate caplet maturing in $T = t_m$ years with a strike of K. The caplet pays

$$\max \left(0, (e^{\hat{f}(t_m,t_m)h} - 1) - Kh) \right)$$

at t_{m+1}. For a maturity of $T = 5$ years, $m = 20$ and the optimal drift μ is a vector of dimension $md = 60$. We encode this vector in the following way: the first 20 components give the drift as a function of time for the first factor (i.e., the first component of the underlying three-dimensional Brownian motion); the next 20 components give the drift for the second factor; and the last 20 components give the drift for the third factor.

With this convention, the left panel of Figure 4.14 displays the optimal drift found through numerical solution of the optimization problem (4.90), with $\exp(F(z))$ the discounted caplet payoff determined by the input vector z. This optimal path gives a positive drift to the first and third factors and a negative drift to the second; all three drifts approach zero toward the end of the 20 steps (the caplet maturity).

The right panel of Figure 4.14 shows the net effect on the short rate of this change of drift. (Recall from Section 3.6.2 that the short rate at t_i is simply $\hat{f}(t_i, t_i)$.) If each of the three factors were to follow the mean paths in the left panel of Figure 4.14, the short rate would follow the dashed line in the right panel of the figure. This should be compared with the initial forward curve, displayed as the solid line: without a change of drift, the central path of the short rate would roughly follow this forward curve. Thus, the net effect of the change of drift in the underlying factors is to push the short rate higher. This is to be expected because the caplet payoff increases with the short rate. But it is not obvious that the "optimal" way to push the short rate has the factors follow the paths in the left panel of Figure 4.14.

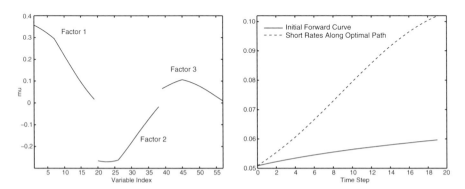

Fig. 4.14. Left panel shows optimal drift for factors in pricing a five-year caplet struck at 7%. Right panel compares path of short rate produced by optimal drift with the initial forward curve.

Table 4.6 compares the efficacy of four variance reduction techniques in pricing caplets: antithetic variates, importance sampling using the optimal drift, and two combinations of importance sampling with stratified sampling using either the optimal drift μ or the optimal eigenvector v_{j*} for the stratification direction. Calculation of the optimal eigenvector requires numerical calculation of the Hessian. The table (from [140]) displays estimated variance ratios based on 50,000 paths per method. The last two columns are based on 100 strata. The table indicates that importance sampling is most effective for deep out-of-the-money caplets, precisely where antithetics become least effective. Both strategies for stratification produce substantial additional variance reduction in these examples.

T	K	Antithetics	IS	IS & Strat. (μ)	IS & Strat. (v_{j*})
2.5	0.04	8	8	246	248
	0.07	1	16	510	444
	0.10	1	173	3067	2861
5.0	0.04	4	8	188	211
	0.07	1	11	241	292
	0.10	1	27	475	512
10.0	0.04	4	7	52	141
	0.07	1	8	70	185
	0.10	1	12	110	244
15.0	0.04	4	5	15	67
	0.07	2	6	22	112
	0.10	1	8	31	158

Table 4.6. Estimated variance ratios for caplets in three-factor HJM model.

Further numerical results for other interest rate derivatives are reported in GHS [140]. As in Table 4.6, the greatest variance reduction typically occurs for out-of-the-money options. The combination of importance sampling with stratification is less effective for options with discontinuous payoffs; a specific example of this in [140] is a "flex cap," in which only a subset of the caplets in a cap make payments even if all expire in-the-money.

In a complex, high-dimensional setting like an HJM model, the computational overhead involved in finding the optimal path or the optimal eigenvector can become substantial. These are fixed costs, in the sense that these calculations need only be done once for each pricing problem rather than once for each path simulated. In theory, as the number of paths increases (which is to say as the required precision becomes high), any fixed cost eventually becomes negligible, but this may not be relevant for practical sample sizes.

GHS [140] develop approximations to reduce the computational overhead in the optimization and eigenvector calculations. These approximations are based on assuming that the optimal drift (or optimal eigenvector) are piece-wise linear between a relatively small number of nodes. This reduces the dimension of the problem. Consider, for example, the computation of a 120-dimensional optimal drift consisting of three 40-dimensional segments. Making a piecewise linear approximation to each segment based on four nodes per segment reduces the problem to one of optimizing over the 12 nodes rather than all 120 components of the vector. Figure 4.15 illustrates the results of this approach in finding both the optimal drift and the optimal eigenvector for a ten-year caplet with a 7% strike. These calculations are explained in detail in GHS [140]. The figure suggests that the approach is quite effective. Simulation results reported in [140] based on using these approximations confirm the viability of the approach.

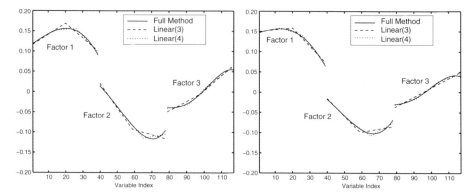

Fig. 4.15. Optimal drift (left) and eigenvector (right) calculated using a piecewise linear approximation with three or four nodes per segment.

4.7 Concluding Remarks

It is easier to survey the topic of variance reduction than to answer the question that brings a reader to such a survey: "Which technique should I use?" There is rarely a simple answer to this question. The most effective applications of variance reduction techniques take advantage of an understanding of the problem at hand. The choice of technique should depend on the available information and on the time available to tailor a general technique to a particular problem. An understanding of the strengths and weaknesses of alternative methods and familiarity with examples of effective applications are useful in choosing a technique.

Figure 4.16 provides a rough comparison of several techniques discussed in this chapter. The figure positions the methods according to their complexity and shows a range of typical effectiveness for each. We have not given the axes units or scales, nor have we defined what they mean or what counts as typical; the figure should clearly not be taken literally or too seriously. We include it to highlight some differences among the methods in the types of implementations and applications discussed in this chapter.

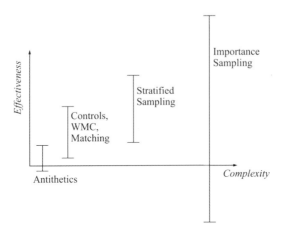

Fig. 4.16. Schematic comparison of some variance reduction techniques

The effectiveness of a method is the efficiency improvement it brings in the sense discussed in Section 1.1.3. Effectiveness below the horizontal axis in the figure is detrimental — worse than not using any variance reduction technique. In the complexity of a method we include the level of effort and detailed knowledge required for implementation. We now explain the positions of the methods in the figure:

o Antithetic sampling requires no specific information about a simulated model and is trivial to implement. It rarely provides much variance reduction. It can be detrimental (as explained in Section 4.2), but such cases are not common.

o Control variates and methods for matching underlying assets (including path adjustments and weighted Monte Carlo) give similar results. Of these methods, control variates are usually the easiest to implement and the best understood. Finding a good control requires knowing something about the simulated model — but not too much. A control variate implemented with an optimal coefficient is guaranteed not to increase variance. When the optimal coefficient is estimated rather than known, it is theoretically possible for a control variate to be detrimental at very small sample sizes, but this is not a practical limitation of the method.

o We have positioned stratified sampling at a higher level of complexity because we have in mind examples like those in Section 4.3.2 where the stratification variable is tailored to the model. In contrast, stratifying a uniform distribution (as in Example 4.3.1) is trivial but not very effective by itself. Using a variable for stratification requires knowing its distribution whereas using it as a control only requires knowing its mean. Stratified sampling is similar to using the distribution (more precisely, the stratum probabilities) as controls and is thus more powerful than using just the mean. With a proportional allocation, stratified sampling never increases variance; using any other allocation may be viewed as a form of importance sampling.

○ We have omitted Latin hypercube sampling from the figure. As a generalization of stratified sampling, it should lie upwards and to the right of that method. But a generic application of Latin hypercube sampling (in which the marginals simply correspond to the uniform random variables used to drive a simulation) is often both easier to implement and less effective than a carefully designed one-dimensional stratification. A problem for which a generic application of Latin hypercube sampling is effective, is also a good candidate for the quasi-Monte Carlo methods discussed in Chapter 5.

○ As emphasized in the figure, importance sampling is the most delicate of the methods discussed in this chapter. It has the capacity to exploit detailed knowledge about a model (often in the form of asymptotic approximations) to produce orders of magnitude variance reduction. But if the importance sampling distribution is not chosen carefully, this method can also increase variance. Indeed, it can even produce infinite variance.

There are counterexamples to nearly any general statement one could make comparing variance reduction techniques and one could argue against any of the comparisons implied by Figure 4.16. Nevertheless, we believe these comparisons to be indicative of what one finds in applying variance reduction techniques to the types of models and problems that arise in financial engineering.

We close this section with some further references to the literature on variance reduction. Glynn and Iglehart [156] survey the application of variance reduction techniques in queueing simulations and discuss some techniques not covered in this chapter. Schmeiser, Taaffe, and Wang [318] analyze biased control variates with coefficients chosen to minimize root mean square error; this is relevant to, e.g., Example 4.1.3 and to settings in which the mean of a control variate is approximated using a binomial lattice. The conditional sampling methods of Cheng [81, 82] are relevant to the methods discussed in Sections 4.3.2 and 4.5.1. Hesterberg [177] and Owen and Zhou [292] propose defensive forms of importance sampling that mix an aggressive change of distribution with a more conservative one to bound the worst-case variance. Dupuis and Wang [108] show that a dynamic exponential change of measure — in which the twisting parameter is recomputed at each step — can outperform a static twist. An adaptive importance sampling method for Markov chains is shown in Kollman et al. [214] to converge exponentially fast. An importance sampling method for stochastic volatility models is developed in Fournié, Lasry, and Touzi [125]. Schoenmakers and Heemink [319] apply importance sampling for derivatives pricing through an approximating PDE.

Shahabuddin [327] uses rare-transition asymptotics to develop an importance sampling procedure for reliability systems; see Heidelberger [175] and Shahabuddin [328] for more on this application area. Asmussen and Binswanger [22], Asmussen, Binswanger, and Højgaard [23], and Juneja and Shahabuddin [206] address difficulties in applying importance sampling to heavy-tailed distributions; see also Section 9.3.

Avramidis and Wilson [30] and Hesterberg and Nelson [178] analyze variance reduction techniques for quantile estimation. We return to this topic in Chapter 9.

Among the techniques not discussed in this chapter is *conditional Monte Carlo*, also called Rao-Blackwellization. This method replaces an estimator by its conditional expectation. Asmussen and Binswanger [22] give a particularly effective application of this idea to an insurance problem; Fox and Glynn [129] combine it with other techniques to estimate infinite horizon discounted costs; Boyle et al. [53] give some applications in option pricing.

5

Quasi-Monte Carlo

This chapter discusses alternatives to Monte Carlo simulation known as *quasi-Monte Carlo* or *low-discrepancy* methods. These methods differ from ordinary Monte Carlo in that they make no attempt to mimic randomness. Indeed, they seek to increase accuracy specifically by generating points that are too evenly distributed to be random. Applying these methods to the pricing of derivative securities requires formulating a pricing problem as the calculation of an integral and thus suppressing its stochastic interpretation as an expected value. This contrasts with the variance reduction techniques of Chapter 4, which take advantage of the stochastic formulation to improve precision.

Low-discrepancy methods have the potential to accelerate convergence from the $O(1/\sqrt{n})$ rate associated with Monte Carlo (n the number of paths or points generated) to nearly $O(1/n)$ convergence: under appropriate conditions, the error in a quasi-Monte Carlo approximation is $O(1/n^{1-\epsilon})$ for all $\epsilon > 0$. Variance reduction techniques, affecting only the implicit constant in $O(1/\sqrt{n})$, are not nearly so ambitious. We will see, however, that the ϵ in $O(1/n^{1-\epsilon})$ hides a dependence on problem *dimension*.

The tools used to develop and analyze low-discrepancy methods are very different from those used in ordinary Monte Carlo, as they draw on number theory and abstract algebra rather than probability and statistics. Our goal is therefore to present key ideas and methods rather than an account of the underlying theory. Niederreiter [281] provides a thorough treatment of the theory.

5.1 General Principles

This section presents definitions and results from the theory of quasi-Monte Carlo (QMC) methods. It is customary in this setting to focus on the problem of numerical integration over the unit hypercube. Recall from Section 1.1 and the many examples in Chapter 3 that each replication in a Monte Carlo simulation can be interpreted as the result of applying a series of transformations (implicit in the simulation algorithm) to an input sequence of independent uniformly distributed random variables $U_1, U_2, \ldots$. Suppose there is an upper bound d on the number of uniforms required to produce a simulation output and let $f(U_1, \ldots, U_d)$ denote this output. For example, f may be the result of transformations that convert the U_i to normal random variables, the normal random variables to paths of underlying assets, and the paths to the discounted payoff of a derivative security. We suppose the objective is to calculate

$$E[f(U_1, \ldots, U_d)] = \int_{[0,1)^d} f(x) \, dx. \tag{5.1}$$

Quasi-Monte Carlo approximates this integral using

$$\int_{[0,1)^d} f(x) \, dx \approx \frac{1}{n} \sum_{i=1}^n f(x_i), \tag{5.2}$$

for carefully (and deterministically) chosen points $x_1, \ldots, x_n$ in the unit hypercube $[0, 1)^d$.

A few issues require comment:

○ The function f need not be available in any explicit form; we merely require a method for evaluating f, and this is what a simulation algorithm does.
○ Whether or not we include the boundary of the unit hypercube in (5.1) and (5.2) has no bearing on the value of the integral and is clearly irrelevant in ordinary Monte Carlo. But some of the definitions and results in QMC require care in specifying the set to which points on a boundary belong. It is convenient and standard to take intervals to be closed on the left and open on the right, hence our use of $[0, 1)^d$ as the unit hypercube.
○ In ordinary Monte Carlo simulation, taking a scalar i.i.d. sequence of uniforms $U_1, U_2, \ldots$ and forming vectors $(U_1, \ldots, U_d), (U_{d+1}, \ldots, U_{2d}), \ldots$ produces an i.i.d. sequence of points from the d-dimensional hypercube. In QMC, the construction of the points x_i depends explicitly on the dimension of the problem — the vectors x_i in $[0, 1)^d$ cannot be constructed by taking sets of d consecutive elements from a scalar sequence.

The dependence of QMC methods on problem dimension is one of the features that most distinguishes them from Monte Carlo. If two different Monte Carlo algorithms corresponding to functions $f : [0, 1)^{d_1} \to \Re$ and $g : [0, 1)^{d_2} \to \Re$ resulted in $f(U_1, \ldots, U_{d_1})$ and $g(U_1, \ldots, U_{d_2})$ having the same distribution, then these two algorithms would have the same bias and variance properties. The preferred algorithm would be the one requiring less time to evaluate; the dimensions d_1, d_2 would be irrelevant except to the extent that they affect the computing times. In ordinary Monte Carlo one rarely even bothers to think about problem dimension, whereas in QMC the dimension must be identified explicitly before points can be generated. Lower-dimensional representations generally result in smaller errors. For some Monte Carlo algorithms, there is no upper bound d on the number of input uniforms required per output; this is true, for example, of essentially all simulations using acceptance-rejection methods, as noted in Section 2.2.2. Without an upper bound d, QMC methods are inapplicable.

For the rest of this chapter, we restrict attention to problems with a finite dimension d and consider approximations of the form in (5.2). The goal of low-discrepancy methods is to construct points x_i that make the error in (5.2) small for a large class of integrands f. It is intuitively clear (and, as we will see, correct in a precise sense) that this is equivalent to choosing the points x_i to fill the hypercube uniformly.

5.1.1 Discrepancy

A natural first attempt at filling the hypercube uniformly would choose the x_i to lie on a grid. But grids suffer from several related shortcomings. If the integrand f is nearly a separable function of its d arguments, the information contained in the values of f at n^d grid points is nearly the same as the information in just nd of these values. A grid leaves large rectangles within $[0, 1)^d$ devoid of any points. A grid requires specifying the total number of points n in advance. If one refines a grid by adding points, the number of points that must be added to reach the next favorable configuration grows very quickly. Consider, for example, a grid constructed as the Cartesian product of 2^k points along each of d dimensions for a total of 2^{kd} points. Now refine the grid by adding a point in each gap along each dimension; i.e., by doubling the number of points along each dimension. The total number of points added to the original grid to reach the new grid is $2^{(k+1)d} - 2^{kd}$, which grows very quickly with k. In contrast, there are low-discrepancy sequences with guarantees of uniformity over bounded-length extensions of an initial segment of the sequence.

To make these ideas precise, we need a precise notion of uniformity — or rather deviation from uniformity, which we measure through various notions of *discrepancy*. Given a collection $\mathcal{A}$ of (Lebesgue measurable) subsets of $[0, 1)^d$, the discrepancy of the point set $\{x_1, \ldots, x_n\}$ relative to $\mathcal{A}$ is

$$D(x_1, \ldots, x_n; \mathcal{A}) = \sup_{A \in \mathcal{A}} \left| \frac{\#\{x_i \in A\}}{n} - \text{vol}(A) \right|. \tag{5.3}$$

Here, $\#\{x_i \in A\}$ denotes the number of x_i contained in A and $\text{vol}(A)$ denotes the volume (measure) of A. Thus, the discrepancy is the supremum over errors in integrating the indicator function of A using the points $x_1, \ldots, x_n$. (In all interesting cases the x_i are distinct points, but to cover the possibility of duplication, count each point according to its multiplicity in the definition of discrepancy.)

Taking $\mathcal{A}$ to be the collection of all rectangles in $[0, 1)^d$ of the form

$$\prod_{j=1}^{d} [u_j, v_j), \quad 0 \le u_j < v_j \le 1,$$

yields the ordinary (or *extreme*) discrepancy $D(x_1, \ldots, x_n)$. Restricting $\mathcal{A}$ to rectangles of the form

$$\prod_{j=1}^{d} [0, u_j) \tag{5.4}$$

defines the *star* discrepancy $D^*(x_1, \ldots, x_n)$. The star discrepancy is obviously no larger than the ordinary discrepancy; Niederreiter [281], Proposition 2.4, shows that

$$D^*(x_1, \ldots, x_n) \leq D(x_1, \ldots, x_n) \leq 2^d D^*(x_1, \ldots, x_n),$$

so for fixed d the two quantities have the same order of magnitude.

Requiring each of these discrepancy measures to be small is consistent with an intuitive notion of uniformity. However, both measures focus on products of intervals and ignore, for example, a rotated subcube of the unit hypercube. If the integrand f represents a simulation algorithm, the coordinate axes may not be particularly meaningful. The asymmetry of the star discrepancy may seem especially odd: in a Monte Carlo simulation, we could replace any uniform input U_i with $1 - U_i$ and thus interchange 0 and 1 along one coordinate. If, as may seem more natural, we take $\mathcal{A}$ to be all convex subsets of $[0, 1]^d$, we get the *isotropic* discrepancy; but the magnitude of this measure can be as large as the dth root of the ordinary discrepancy (see p.17 of Niederreiter [281] and Chapter 3 of Matoušek [256]). We return to this point in Section 5.1.3.

We will see in Section 5.1.3 that these notions of discrepancy are indeed relevant to measuring the approximation error in (5.2). It is therefore sensible to look for points that achieve low values of these discrepancy measures, and that is what low-discrepancy methods do.

In dimension $d = 1$, Niederreiter [281], pp.23–24, shows that

$$D^*(x_1, \ldots, x_n) \geq \frac{1}{2n}, \quad D(x_1, \ldots, x_n) \geq \frac{1}{n}, \tag{5.5}$$

and that in both cases the minimum is attained by

$$x_i = \frac{2i - 1}{2n}, \quad i = 1, \ldots, n. \tag{5.6}$$

For this set of points, (5.2) reduces to the midpoint rule for integration over the unit interval. Notice that (5.6) does not define the first n points of an infinite sequence; in fact, the set of points defined by (5.6) has no values in common with the corresponding set for $n + 1$.

Suppose, in contrast, that we fix an infinite sequence $x_1, x_2, \ldots$ of points in $[0, 1)$ and measure the discrepancy of the first n points. From the perspective of numerical integration, this is a more relevant case if we hope to be able to increase the number of points in an approximation of the form (5.2). Niederreiter [281], p.24, cites references showing that in this case

$$D(x_1, \ldots, x_n) \geq D^*(x_1, \ldots, x_n) \geq \frac{c \log n}{n}$$

for infinitely many n, with c a constant. This situation is typical of low-discrepancy methods, even in higher dimensions: one can generally achieve a lower discrepancy by fixing the number of points n in advance; using the first n points of a sequence rather than a different set of points for each n typically increases discrepancy by a factor of $\log n$.

Much less is known about the best possible discrepancy in dimensions higher than 1. Niederreiter [281], p.32, states that "it is widely believed" that in dimensions $d \geq 2$, any point set $x_1, \ldots, x_n$ satisfies

$$D^*(x_1, \ldots, x_n) \geq c_d \frac{(\log n)^{d-1}}{n}$$

and the first n elements of any sequence $x_1, x_2, \ldots$ satisfy

$$D^*(x_1, \ldots, x_n) \geq c'_d \frac{(\log n)^d}{n},$$

for constants c_d, c'_d depending only on the dimension d. These order-of-magnitude discrepancies are achieved by explicit constructions (discussed in Section 5.2). It is therefore customary to reserve the informal term "low-discrepancy" for methods that achieve a star discrepancy of $O((\log n)^d/n)$. The logarithmic term can be absorbed into any power of n, allowing the looser bound $O(1/n^{1-\epsilon})$, for all $\epsilon > 0$.

Although any power of $\log n$ eventually becomes negligible relative to n, this asymptotic property may not be relevant at practical values of n if d is large. Accordingly, QMC methods have traditionally been characterized as appropriate only for problems of moderately high dimension, with some authors putting the upper limit at 40 dimensions, others putting it as low as 12 or 15. But in many recent applications of QMC to problems in finance, these methods have been found to be effective in much higher dimensions. We present some evidence of this in Section 5.5 and comment further in Section 5.6.

5.1.2 Van der Corput Sequences

Before proceeding with a development of further theoretical background, we introduce a specific class of one-dimensional low-discrepancy sequences called Van der Corput sequences. In addition to illustrating the general notion of discrepancy, this example provides the key element of many multidimensional constructions.

By a *base* we mean an integer $b \geq 2$. Every positive integer k has a unique representation (called its base-b or b-ary expansion) as a linear combination of nonnegative powers of b with coefficients in $\{0, 1, \ldots, b-1\}$. We can write this as

$$k = \sum_{j=0}^{\infty} a_j(k) b^j, \tag{5.7}$$

with all but finitely many of the coefficients $a_j(k)$ equal to zero. The *radical inverse function* ψ_b maps each k to a point in $[0, 1)$ by flipping the coefficients of k about the base-b "decimal" point to get the base-b fraction $.a_0a_1a_2\ldots$. More precisely,

$$\psi_b(k) = \sum_{j=0}^{\infty} \frac{a_j(k)}{b^{j+1}}. \tag{5.8}$$

The base-b Van der Corput sequence is the sequence $0 = \psi_b(0), \psi_b(1), \psi_b(2), \ldots$. Its calculation is illustrated in Table 5.1 for base 2.

k	k Binary	$\psi_2(k)$ Binary	$\psi_2(k)$
0	0	0	0
1	1	0.1	1/2
2	10	0.01	1/4
3	11	0.11	3/4
4	100	0.001	1/8
5	101	0.101	5/8
6	110	0.011	3/8
7	111	0.111	7/8

Table 5.1. Illustration of radical inverse function ψ_b in base $b = 2$.

Figure 5.1 illustrates how the base-2 Van der Corput sequence fills the unit interval. The kth row of the array in the figure shows the first k nonzero elements of the sequence; each row refines the previous one. The evolution of the point set is exemplified by the progression from the seventh row to the last row, in which the "sixteenths" are filled in. As these points are added, they appear on alternate sides of $1/2$: first $1/16$, then $9/16$, then $5/16$, and so on. Those that are added to the left of $1/2$ appear on alternate sides of $1/4$: first $1/16$, then $5/16$, then $3/16$, and finally $7/16$. Those on the right side of $1/2$ similarly alternate between the left and right sides of $3/4$. Thus, while a naive refinement might simply insert the new values in increasing order $1/16, 3/16, 5/16,\ldots, 15/16$, the Van der Corput inserts them in a maximally balanced way.

The effect of the size of the base b can be seen by comparing Figure 5.1 with the Van der Corput sequence in base 16. The first 15 nonzero elements of this sequence are precisely the values appearing the last row of Figure 5.1, but now they appear in increasing order. The first seven values of the base-16 sequence are all between 0 and $1/2$, whereas those of the base-2 sequence are spread uniformly over the unit interval. The larger the base, the greater the number of points required to achieve uniformity.

$\frac{1}{16}$	$\frac{1}{8}$	$\frac{3}{16}$	$\frac{1}{4}$	$\frac{5}{16}$	$\frac{3}{8}$	$\frac{7}{16}$	$\frac{1}{2}$	$\frac{9}{16}$	$\frac{5}{8}$	$\frac{11}{16}$	$\frac{3}{4}$	$\frac{13}{16}$	$\frac{7}{8}$	$\frac{15}{16}$
							$\frac{1}{2}$							
			$\frac{1}{4}$				$\frac{1}{2}$							
			$\frac{1}{4}$				$\frac{1}{2}$				$\frac{3}{4}$			
	$\frac{1}{8}$		$\frac{1}{4}$				$\frac{1}{2}$				$\frac{3}{4}$			
	$\frac{1}{8}$		$\frac{1}{4}$				$\frac{1}{2}$		$\frac{5}{8}$		$\frac{3}{4}$			
	$\frac{1}{8}$		$\frac{1}{4}$		$\frac{3}{8}$		$\frac{1}{2}$		$\frac{5}{8}$		$\frac{3}{4}$			
	$\frac{1}{8}$		$\frac{1}{4}$		$\frac{3}{8}$		$\frac{1}{2}$		$\frac{5}{8}$		$\frac{3}{4}$		$\frac{7}{8}$	
$\frac{1}{16}$	$\frac{1}{8}$		$\frac{1}{4}$		$\frac{3}{8}$		$\frac{1}{2}$		$\frac{5}{8}$		$\frac{3}{4}$		$\frac{7}{8}$	
$\frac{1}{16}$	$\frac{1}{8}$		$\frac{1}{4}$		$\frac{3}{8}$		$\frac{1}{2}$	$\frac{9}{16}$	$\frac{5}{8}$		$\frac{3}{4}$		$\frac{7}{8}$	
$\frac{1}{16}$	$\frac{1}{8}$		$\frac{1}{4}$	$\frac{5}{16}$	$\frac{3}{8}$		$\frac{1}{2}$	$\frac{9}{16}$	$\frac{5}{8}$		$\frac{3}{4}$		$\frac{7}{8}$	
$\frac{1}{16}$	$\frac{1}{8}$		$\frac{1}{4}$	$\frac{5}{16}$	$\frac{3}{8}$		$\frac{1}{2}$	$\frac{9}{16}$	$\frac{5}{8}$		$\frac{3}{4}$	$\frac{13}{16}$	$\frac{7}{8}$	
$\frac{1}{16}$	$\frac{1}{8}$	$\frac{3}{16}$	$\frac{1}{4}$	$\frac{5}{16}$	$\frac{3}{8}$		$\frac{1}{2}$	$\frac{9}{16}$	$\frac{5}{8}$		$\frac{3}{4}$	$\frac{13}{16}$	$\frac{7}{8}$	
$\frac{1}{16}$	$\frac{1}{8}$	$\frac{3}{16}$	$\frac{1}{4}$	$\frac{5}{16}$	$\frac{3}{8}$		$\frac{1}{2}$	$\frac{9}{16}$	$\frac{5}{8}$	$\frac{11}{16}$	$\frac{3}{4}$	$\frac{13}{16}$	$\frac{7}{8}$	
$\frac{1}{16}$	$\frac{1}{8}$	$\frac{3}{16}$	$\frac{1}{4}$	$\frac{5}{16}$	$\frac{3}{8}$	$\frac{7}{16}$	$\frac{1}{2}$	$\frac{9}{16}$	$\frac{5}{8}$	$\frac{11}{16}$	$\frac{3}{4}$	$\frac{13}{16}$	$\frac{7}{8}$	
$\frac{1}{16}$	$\frac{1}{8}$	$\frac{3}{16}$	$\frac{1}{4}$	$\frac{5}{16}$	$\frac{3}{8}$	$\frac{7}{16}$	$\frac{1}{2}$	$\frac{9}{16}$	$\frac{5}{8}$	$\frac{11}{16}$	$\frac{3}{4}$	$\frac{13}{16}$	$\frac{7}{8}$	$\frac{15}{16}$

Fig. 5.1. Illustration of the Van der Corput sequence in base 2. The kth row of this array shows the first k nonzero elements of the sequence.

Theorem 3.6 of Niederreiter [281] shows that all Van der Corput sequences are low-discrepancy sequences. More precisely, the star discrepancy of the first n elements of a Van der Corput sequence is $O(\log n/n)$, with an implicit constant depending on the base b.

5.1.3 The Koksma-Hlawka Bound

In addition to their intuitive appeal as indicators of uniformity, discrepancy measures play a central role in bounding the error in the approximation (5.2). The key result in this direction is generally known as the *Koksma-Hlawka inequality* after a one-dimensional result published by Jurjen Koksma in 1942 and its generalization by Edmund Hlawka in 1961. This result bounds the integration error in (5.2) by the product of two quantities, one depending only on the integrand f, the other — the star discrepancy of $x_1, \ldots, x_n$ — depending only on the point set used.

Finite Variation

The bound depends on the integrand f through its *Hardy-Krause variation*, which we now define, following Niederreiter [281]. For this we need f defined (and finite) on the closed unit hypercube $[0, 1]^d$. Consider a rectangle of the form

$$J = [u_1^-, u_1^+] \times [u_2^-, u_2^+] \times \cdots \times [u_d^-, u_d^+],$$

with $0 \leq u_i^- \leq u_i^+ \leq 1$, $i = 1, \ldots, d$. Each vertex of J has coordinates of the form $u_i^\pm$. Let $\mathcal{E}(J)$ be the set of vertices of J with an even number of $+$ superscripts and let $\mathcal{O}(J)$ contain those with an odd number of $+$ superscripts. Define

$$\Delta(f; J) = \sum_{u \in \mathcal{E}(J)} f(u) - \sum_{u \in \mathcal{O}(J)} f(u);$$

this is the sum of $f(u)$ over the vertices of J with function values at adjacent vertices given opposite signs.

The unit hypercube can be partitioned into a set $\mathcal{P}$ of rectangles of the form of J. Letting $\mathcal{P}$ range over all such partitions, define

$$V^{(d)}(f) = \sup_{\mathcal{P}} \sum_{J \in \mathcal{P}} |\Delta(f; J)|.$$

This is a measure of the variation of f. Niederreiter [281, p.19] notes that

$$V^{(d)}(f) = \int_0^1 \cdots \int_0^1 \left| \frac{\partial^d f}{\partial u_1 \cdots \partial u_d} \right| du_1 \cdots du_d,$$

if the partial derivative is continuous over $[0, 1]^d$. This expression makes the interpretation of $V^{(d)}$ more transparent, but it should be stressed that f need not be differentiable for $V^{(d)}(f)$ to be finite.

For any $1 \leq k \leq d$ and any $1 \leq i_1 < i_2 < \cdots < i_k \leq d$, consider the function on $[0, 1]^k$ defined by restricting f to points $(u_1, \ldots, u_d)$ with $u_j = 1$ if $j \notin \{i_1, \ldots, i_k\}$ and $(u_{i_1}, \ldots, u_{i_k})$ ranging over all of $[0, 1]^k$. Denote by $V^{(k)}(f; i_1, \ldots, i_k)$ the application of $V^{(k)}$ to this function. Finally, define

$$V(f) = \sum_{k=1}^{d} \sum_{1 \leq i_1 < \cdots < i_k \leq d} V^{(k)}(f; i_1, \ldots, i_k). \tag{5.9}$$

This is the variation of f in the sense of Hardy and Krause.

We can now state the Koksma-Hlawka bound: if the function f has finite Hardy-Krause variation $V(f)$, then for any $x_1, \ldots, x_n \in [0, 1)^d$,

$$\left| \frac{1}{n} \sum_{i=1}^{n} f(x_i) - \int_{[0,1]^d} f(u) \, du \right| \leq V(f) D^*(x_1, \ldots, x_n). \tag{5.10}$$

As promised, this result bounds the integration error through the product of two terms. The first term is a measure of the variation of the integrand; the second term is a measure of the deviation from uniformity of the points at which the integrand is evaluated.

Theorem 2.12 of Niederreiter [281] shows that (5.10) is a tight bound in the sense that for each $x_1, \ldots, x_n$ and $\epsilon > 0$ there is a function f for which the error on the left comes within ϵ of the bound on the right. The function can be chosen to be infinitely differentiable, so in this sense (5.10) is tight even for very smooth functions.

It is natural to contrast the Koksma-Hlwaka inequality with the error information available in ordinary Monte Carlo. To this end, let $U, U_1, U_2, \ldots$ be independent and uniformly distributed over the d-dimensional unit hypercube and let $\sigma_f^2 = \mathsf{Var}[f(U)]$. From the central limit theorem, we know that

$$\left| \frac{1}{n} \sum_{i=1}^{n} f(U_i) - \int_{[0,1]^d} f(u) \, du \right| \leq z_{\delta/2} \frac{\sigma_f}{\sqrt{n}}, \tag{5.11}$$

with probability approximately equal to $1 - \delta$, with $-z_{\delta/2}$ the $\delta/2$ quantile of the standard normal distribution. From Chebyshev's inequality we know that for any $\delta > 0$,

$$\left| \frac{1}{n} \sum_{i=1}^{n} f(U_i) - \int_{[0,1]^d} f(u) \, du \right| \leq \frac{\sigma_f}{\sqrt{\delta n}}, \tag{5.12}$$

with probability at least $1 - \delta$. The following observations are relevant in comparing the quasi-Monte Carlo and ordinary Monte Carlo error information:

○ The Koksma-Hlawka inequality (5.10) provides a strict bound on the integration error, whereas (5.11) and (5.12) are probabilistic bounds and (5.11) requires n to be large. In both (5.11) and (5.12) we may, however, choose $\delta > 0$ to bring the probability $1 - \delta$ arbitrarily close to 1.
○ Both of the terms $V(f)$ and $D^*(x_1, \ldots, x_n)$ appearing in the Koksma-Hlawka inequality are difficult to compute — potentially much more so than the integral of f. In contrast, the unknown parameter σ_f in (5.11) and (5.12) is easily estimated from $f(U_1), \ldots, f(U_n)$ with negligible additional computation.
○ In cases where $V(f)$ and $D^*(x_1, \ldots, x_n)$ are known, the Koksma-Hlawka bound is often found to grossly overestimate the true error of integration. In contrast, the central limit theorem typically provides a sound and informative measure of the error in a Monte Carlo estimate.
○ The condition that $V(f)$ be finite is restrictive. It requires, for example, that f be bounded, a condition often violated in option pricing applications.

In light of these observations, it seems fair to say that despite its theoretical importance the Koksma-Hlawka inequality has limited applicability as a practical error bound. This is a shortcoming of quasi-Monte Carlo methods in comparison to ordinary Monte Carlo methods, for which effective error information is readily available. The most important consequence of the Koksma-Hlawka inequality is that it helps guide the search for effective point sets and sequences by making precise the role of discrepancy.

The Koksma-Hlawka inequality is the best-known example of a set of related results. Hickernell [181] generalizes the inequality by extending both the star discrepancy and the Hardy-Krause variation using more general norms.

One such bound uses an L_2 discrepancy defined by replacing the maximum absolute deviation in (5.3) with a root-mean-square deviation. An analog of (5.10) then holds for this notion of discrepancy, with $V(f)$ replaced by an L_2 notion of variation. An advantage of the L_2 discrepancy is that it is comparatively easy to calculate — through simulation, for example.

Integrands arising in derivative pricing applications sometimes vanish off a subset of $[0, 1)^d$ (once formulated as functions on the hypercube) and may be discontinuous at the boundary of this domain. This is typical of barrier options, for example. Such integrands usually have infinite variation, as explained in Figure 5.2. The Koksma-Hlawka inequality is therefore uninformative for a large class of interesting integrands. An important variant of (5.10), one of several cited in Niederreiter [281], p.21, applies to integrals over arbitrary convex subsets of the hypercube. Thus, if our integrand would have finite variation but for the presence of the indicator function of a convex set, this result allows us to absorb the indicator into the integration domain and obtain a bound on the integration error. However, the bound in this case involves the isotropic discrepancy which, as we noted in Section 5.1.1, exhibits a much stronger dependence on dimension.

This points to another limitation of the Koksma-Hlawka bound, at least from the perspective of our intended application. The Koksma-Hlawka result is oriented to the axes of the hypercube, through the definitions of both $V(f)$ and the star discrepancy. The indicator of a rectangle, for example, has finite variation if the rectangle is parallel to the axes but infinite variation if the rectangle is rotated, as illustrated in Figure 5.2. This focus on a particular choice of coordinates seems unnatural if the function f is the result of transforming a simulation algorithm into a function on the unit hypercube; dropping this focus leads to much larger error bounds with a qualitatively different dependence on dimension (cf. Matoušek [257]). In Section 5.5.2, we discuss applications of QMC methods that take account of more specific features of integrands f arising in derivative pricing applications.

5.1.4 Nets and Sequences

Despite its possible shortcomings as a practical bound on integration error, the Koksma-Hlawka inequality (5.10) nevertheless suggests that constructing point sets and sequences with low discrepancy is a fruitful approach for numerical integration. A valuable tool for constructing and describing such point sets is the notion of a (t, m, d)-net and a (t, d)-sequence introduced by Niederreiter [280], extending ideas developed in base 2 by Sobol' [335]. These are more commonly referred to as (t, m, s)-nets and (t, s)-sequences; the parameter s in this terminology refers to the dimension, for which we have consistently used d. Briefly, a (t, m, d)-net is a finite set of points in $[0, 1)^d$ possessing a degree of uniformity quantified by t; a (t, d)-sequence is a sequence of points certain segments of which form (t, m, d)-nets.

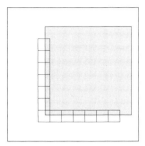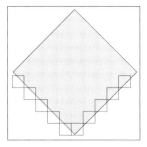

Fig. 5.2. Variation of the indicator function of the shaded square. In the left panel, each small box has a $\Delta(f; J)$ value of zero except the one containing the corner of the shaded square; the variation remains finite. In the right panel, each small box has a $|\Delta(f; J)|$ value of 1, except the one on the corner. Because the boxes can be made arbitrarily small, the indicator function on the right has infinite variation.

To formulate the defintions of these sets and sequences, we first need to define a *b-ary box*, also called an *elementary interval in base b*, with $b \geq 2$ an integer. This is a subset of $[0, 1)^d$ of the form

$$\prod_{i=1}^{d} \left[\frac{a_i}{b^{j_i}}, \frac{a_i + 1}{b^{j_i}} \right),$$

with $j_i \in \{0, 1, \ldots\}$ and $a_i \in \{0, 1, \ldots, b^{j_i} - 1\}$. The vertices of a b-ary box thus have coordinates that are multiples of powers of $1/b$, but with restrictions. In base 2, for example, $[3/4, 1)$ and $[3/4, 7/8)$ are admissible but $[5/8, 7/8)$ is not. The volume of a b-ary box is $1/b^{j_1 + \cdots + j_d}$.

For integers $0 \leq t \leq m$, a (t, m, d)-*net* in base b is a set of b^m points in $[0, 1)^d$ with the property that exactly b^t points fall in each b-ary box of volume b^{t-m}. Thus, the net correctly estimates the volume of each such b-ary box in the sense that the fraction of points b^t/b^m that lie in the box equals the volume of the box.

A sequence of points $x_1, x_2, \ldots$ in $[0, 1)^d$ is a (t, d)-*sequence* in base b if for all $m > t$ each segment $\{x_i : jb^m < i \leq (j + 1)b^m\}$, $j = 0, 1, \ldots$, is a (t, m, d)-net in base b.

In these definitions, it should be evident that smaller values of t are associated with greater uniformity: with smaller t, even small b-ary boxes contain the right number of points. It should also be clear that, other things being equal, a smaller base b is preferable because the uniformity properties of (t, m, d)-nets and (t, d)-sequences are exhibited in sets of b^m points. With larger b, more points are required for these properties to hold.

Figure 5.3 displays two nets. The 81 $(= 3^4)$ points in the left panel comprise a $(0, 4, 2)$-net in base 3. Dotted lines in the figure show 3-ary boxes with dimensions $1/9 \times 1/9$ and $1/27 \times 1/3$ containing one point each, as they must. (For points on the boundaries, recall our convention that intervals are closed on the left and open on the right.) The right panel shows a $(1, 7, 2)$-net in base 2 (with $2^7 = 128$ points) that is not a $(0, 7, 2)$-net. The dotted lines in the figure show that 2-ary boxes with area $1/64$ contain two points, but they also show boxes with dimensions $1/16 \times 1/8$ that do not contain any points.

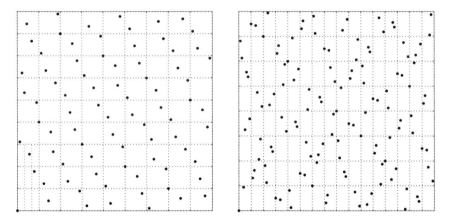

Fig. 5.3. Left panel shows 81 points comprising a $(0, 4, 2)$-net in base 3. Right panel shows 128 points comprising a $(1, 7, 2)$-net in base 2. Both include a point at the origin.

Niederreiter [281] contains an extensive analysis of discrepancy bounds for (t, m, d)-nets and (t, d)-sequences. Of his many results we quote just one, demonstrating that (t, d)-sequences are indeed low-discrepancy sequences. More precisely, Theorem 4.17 of Niederreiter [281] states that if $x_1, x_2, \ldots$ is a (t, d)-sequence in base b, then for $n \geq 2$,

$$D^*(x_1, \ldots, x_n) \leq C(d, b) b^t \frac{(\log n)^d}{n} + O\left(\frac{b^t (\log n)^{d-1}}{n} \right). \tag{5.13}$$

The factor $C(d, b)$ (for which Niederreiter provides an explicit expression) and the implicit constant in the $O(\cdot)$ term do not depend on n or t. Theorem 4.10 of Niederreiter [281] provides a similar bound for (t, m, d)-nets, but with each exponent of $\log n$ reduced by 1.

In the next section, we describe several specific constructions of low-discrepancy sequences. The simplest constructions, producing Halton sequences and Hammersley points, are the easiest to introduce, but they yield neither (t, d)-sequences nor (t, m, d)-nets. Faure sequences are $(0, d)$-sequences and thus optimize the uniformity parameter t; however, they require a base at least as large as the smallest prime greater than or equal to the dimension d. Sobol' sequences use base 2 regardless of the dimension (which has computational as well as uniformity advantages) but their t parameter grows with the dimension d.

5.2 Low-Discrepancy Sequences

We turn now to specific constructions of low-discrepancy sequences in arbitrary dimension d. We provide algorithms for the methods we consider and make some observations on the properties and relative merits of various sequences. All methods discussed in this section build on the Van der Corput sequences discussed in Section 5.1.2.

5.2.1 Halton and Hammersley

Halton [165], extending work of Hammersley [168], provides the simplest construction and first analysis of low-discrepancy sequences in arbitrary dimension d. The coordinates of a Halton sequence follow Van der Corput sequences in distinct bases. Thus, let $b_1, \ldots, b_d$ be relatively prime integers greater than 1, and set

$$x_k = (\psi_{b_1}(k), \psi_{b_2}(k), \ldots, \psi_{b_d}(k)), \quad k = 0, 1, 2, \ldots, \tag{5.14}$$

with ψ_b the radical inverse function defined in (5.8).

The requirement that the b_i be relatively prime is necessary for the sequence to fill the hypercube. For example, the two-dimensional sequence defined by $b_1 = 2$ and $b_2 = 6$ has no points in $[0, 1/2) \times [5/6, 1)$. Because we prefer smaller bases to larger bases, we therefore take $b_1, \ldots, b_d$ to be the first d prime numbers. The two-dimensional cases, using bases 2 and 3, is illustrated in Figure 5.4. With the convention that intervals are closed on the left and open on the right, each cell in the figure contains exactly one point.

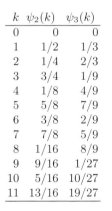

k	$\psi_2(k)$	$\psi_3(k)$
0	0	0
1	1/2	1/3
2	1/4	2/3
3	3/4	1/9
4	1/8	4/9
5	5/8	7/9
6	3/8	2/9
7	7/8	5/9
8	1/16	8/9
9	9/16	1/27
10	5/16	10/27
11	13/16	19/27

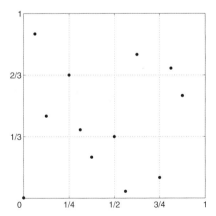

Fig. 5.4. First twelve points of two-dimensional Halton sequence.

A word about zero: Some properties of low-discrepancy sequences are most conveniently stated by including a 0th point, typically zero itself, as in (5.14).

When the points are fed into a simulation algorithm, there is often good reason to avoid zero — for example, $\Phi^{-1}(0) = -\infty$. In practice, we therefore omit it. Depending on whether or not x_0 is included, x_k is either the kth or $(k+1)$th point in the sequence, but we *always* take x_k to be the point constructed from the integer k, as in (5.14). Omission of x_0 has no bearing on asymptotic properties.

Halton points form an infinite sequence. We can achieve slightly better uniformity if we are willing to fix the number of points n in advance. The n points

$$\left\{ \left(k/n, \psi_{b_1}(k), \ldots, \psi_{b_{d-1}}(k)\right), k = 0, 1, \ldots, n-1 \right\}$$

with relatively prime $b_1, \ldots, b_{d-1}$ form a *Hammersley point set* in dimension d.

The star discrepancy of the first n Halton points in dimension d with relatively prime bases $b_1, \ldots, b_d$ satisfies

$$D^*(x_0, \ldots, x_{n-1}) \leq C_d(b_1, \ldots, b_d) \frac{(\log n)^d}{n} + O\left(\frac{(\log n)^{d-1}}{n}\right),$$

with $C_d(b_1, \ldots, b_d)$ independent of n; thus, Halton sequences are indeed low-discrepancy sequences. The corresponding n-element Hammersley point set satisfies

$$D^*(x_0, \ldots, x_{n-1}) \leq C_{d-1}(b_1, \ldots, b_{d-1}) \frac{(\log n)^{d-1}}{n} + O\left(\frac{(\log n)^{d-2}}{n}\right).$$

The leading orders of magnitude in these bounds were established in Halton [165] and subsequently refined through work reviewed in Niederreiter [281, p.44].

A formula for $C_d(b_1, \ldots, b_d)$ is given in Niederreiter [281]. This upper bound is minimized by taking the bases to be the first d primes. With C_d denoting this minimizing value, Niederreiter [281], p.47, observes that

$$\lim_{d \to \infty} \frac{\log C_d}{d \log d} = 1,$$

so the bounding constant C_d grows superexponentially. This indicates that while the Halton and Hammersley points exhibit good uniformity for fixed d as n increases, their quality degrades rapidly as d increases.

The deterioration of the Halton sequence and Hammersley points in high dimensions follows from the behavior of the Van der Corput sequence with a large base. The Van der Corput sequence in base b consists of consecutive monotone segments of length b. If the base is large, the sequence produces long monotone segments, and projections of a Halton sequence onto coordinates using large bases will have long diagonal segments in the projected hypercube.

This pattern is illustrated in Figure 5.5, which shows two projections of the first 1000 nonzero points of the Halton sequence in dimension 30. The left panel is the projection onto the first two coordinates, which use bases 2 and 3; the right panel is the projection onto the last two coordinates, which use bases 109 and 113, the 29th and 30th prime numbers. The impact of increasing the bases — and thus also of increasing the dimension — is evident from the figure.

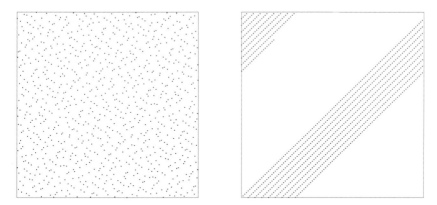

Fig. 5.5. First 1000 points of the Halton sequence in dimension 30. Left panel shows projection onto first two coordinates (bases 2 and 3); right panel shows projection onto last two coordinates (bases 109 and 113).

As a possible remedy for the problem illustrated in Figure 5.5, Kocis and Whiten [213] suggest using a *leaped* Halton sequence

$$x_k = (\psi_{b_1}(k\ell), \psi_{b_2}(k\ell), \ldots, \psi_{b_d}(k\ell)), \quad k = 0, 1, 2, \ldots,$$

for some integer $\ell \geq 2$. They recommend choosing ℓ to be relatively prime to the bases $b_1, \ldots, b_d$.

This idea is illustrated in Figure 5.6, where we have applied it to a two-dimensional Halton sequence with bases 109 and 113, the same bases used in the right panel of Figure 5.5. Each panel of Figure 5.6 shows 1000 points, using leaps $\ell = 3$, $\ell = 107$ (the prime that precedes 109), and $\ell = 127$ (the prime that succeeds 113). The figures suggest that leaping can indeed improve uniformity, but also that its effect is very sensitive to the choice of leap parameter ℓ.

The decline in uniformity of Halton sequences with increasing dimension is inherent to their construction. Several studies have concluded through numerical experiments that Halton sequences are not competitive with other methods in high dimensions; these studies include Fox [126], Kocis and Whiten [213], and, in financial applications, Boyle et al. [53] and Paskov [295]. An exception is Morokoff and Caflisch [272], where Halton sequences are found to be effective on a set of test problems; see also the comments of Matoušek [257, p.543] supporting randomized Halton sequences.

Fig. 5.6. First 1000 points of leaped Halton sequence with bases 109 and 113. From left to right, the leap parameters are $\ell = 3$, $\ell = 107$, and $\ell = 127$.

Implementation

Generating Halton points is essentially equivalent to generating a Van der Corput sequence, which in turn requires little more than finding base-b expansions. We detail these steps because they will be useful later as well.

Figure 5.7 displays an algorithm to compute the base-b expansion of an integer $k \geq 0$. The function returns an array **a** whose elements are the coefficients of the expansion ordered from most significant to least. Thus, B-ARY(6,2) is $(1, 1, 0)$ and B-ARY(135,10) is $(1, 3, 5)$.

B-ARY(k, b)
a $\leftarrow 0$
if $(k > 0)$
    $j_{\max} \leftarrow \lfloor \log(k) / \log(b) \rfloor$
    a $\leftarrow (0, 0, \ldots, 0)$ [length $j_{\max} + 1$]
    $q \leftarrow b^{j_{\max}}$
    for $j = 1, \ldots j_{\max} + 1$
        a$(j) \leftarrow \lfloor k/q \rfloor$
        $k \leftarrow k - q * \mathbf{a}(j)$
        $q \leftarrow q/b$
return **a**

Fig. 5.7. Function B-ARY(k, b) returns coefficients of base-b expansion of integer k in array **a**. Rightmost element of **a** is least significant digit in the expansion.

To generate elements $x_{n_1}, x_{n_1+1}, \ldots, x_{n_2}$ of the Van der Corput sequence in base b, we need the expansions of $k = n_1, n_1 + 1, \ldots, n_2$. But computing all of these through calls to B-ARY is wasteful. Instead, we can use B-ARY to expand n_1 and then update the expansion recursively. A function NEXTB-ARY that increments a base-b expansions by 1 is displayed in Figure 5.8.

$$
\begin{array}{l}
\text{NEXTB-ARY}(\mathbf{a}_{\text{in}}, b) \\
m \leftarrow \text{length}(\mathbf{a}_{\text{in}}),\ \text{carry} \leftarrow \text{TRUE} \\
\text{for } i = m, \ldots, 1 \\
\quad \text{if carry} \\
\quad\quad \text{if } (\mathbf{a}_{\text{in}}(i) = b - 1) \\
\quad\quad\quad \mathbf{a}_{\text{out}}(i) \leftarrow 0 \\
\quad\quad \text{else} \\
\quad\quad\quad \mathbf{a}_{\text{out}}(i) \leftarrow \mathbf{a}_{\text{in}}(i) + 1 \\
\quad\quad\quad \text{carry} \leftarrow \text{FALSE} \\
\quad \text{else} \\
\quad\quad \mathbf{a}_{\text{out}}(i) \leftarrow \mathbf{a}_{\text{in}}(i) \\
\text{if carry } \mathbf{a}_{\text{out}} \leftarrow (1, \mathbf{a}_{\text{out}}(1), \ldots, \mathbf{a}_{\text{out}}(m)) \\
\text{return } \mathbf{a}_{\text{out}}
\end{array}
$$

Fig. 5.8. Function NEXTB-ARY($\mathbf{a}_{\text{in}}, b$) returns coefficients of base-b expansion of integer $k + 1$ in array $\mathbf{a}_{\text{out}}$, given coefficients for integer k in array $\mathbf{a}_{\text{in}}$.

Elements $x_{n_1}, \ldots, x_{n_2}$ of the base-b Van der Corput sequence can now be calculated by making an initial call to B-ARY(n_1, b) to get the coefficients of the expansion of n_1 and then repeatedly applying NEXTB-ARY to get subsequent coefficients. If the array $\mathbf{a}$ for a point n has m elements, we compute x_n as follows:

$$
\begin{array}{l}
x_n \leftarrow 0,\ q \leftarrow 1/b \\
\text{for } j = 1, \ldots, m \\
\quad x_n \leftarrow x_n + q * \mathbf{a}(m - j + 1) \\
\quad q \leftarrow q/b
\end{array}
$$

This evaluates the radical-inverse function ψ_b. By applying the same procedure with prime bases $b_1, \ldots, b_d$, we construct Halton points in dimension d.

As noted by Halton [165], it is also easy to compute $\psi_b(k + 1)$ recursively from $\psi_b(k)$ without explicitly calculating the base-b expansion of either k or $k + 1$. Halton and Smith [166] provide a numerically stable implementation of this idea, also used in Fox [126]. We will, however, need the base-b expansions for other methods.

5.2.2 Faure

We noted in the previous section that the uniformity of Halton sequences degrades in higher dimensions because higher-dimensional coordinates are constructed from Van der Corput sequences with large bases. In particular, the dth coordinate uses a base at least as large as the dth prime, and this grows superexponentially with d. Faure [116] developed a different extension of Van der Corput sequences to multiple dimensions in which all coordinates use a common base. This base must be at least as large as the dimension itself, but can be much smaller than the largest base used for a Halton sequence of equal dimension.

In a d-dimensional Faure sequence, the coordinates are constructed by permuting segments of a single Van der Corput sequence. For the base b we choose the smallest prime number greater than or equal to d. As in (5.7), let $a_\ell(k)$ denote the coefficients in the base-b expansion of k, so that

$$k = \sum_{\ell=0}^{\infty} a_\ell(k)b^\ell. \tag{5.15}$$

The ith coordinate, $i = 1, \ldots, d$, of the kth point in the Faure sequence is given by

$$\sum_{j=1}^{\infty} \frac{y_j^{(i)}(k)}{b^j}, \tag{5.16}$$

where

$$y_j^{(i)}(k) = \sum_{\ell=0}^{\infty} \binom{\ell}{j-1}(i-1)^{\ell-j+1}a_\ell(k) \mod b, \tag{5.17}$$

with

$$\binom{m}{n} = \begin{cases} m!/(m-n)!n!, & m \geq n, \\ 0, & \text{otherwise,} \end{cases}$$

and $0! = 1$.

Each of the sums in (5.15)–(5.17) has only a finite number of nonzero terms. Suppose the base-b expansion of k in (5.15) has exactly r terms, meaning that $a_{r-1}(k) \neq 0$ and $a_\ell(k) = 0$ for all $\ell \geq r$. Then the summands in (5.17) vanish for $\ell \geq r$. If $j \geq r+1$, then the summands for $\ell = 0, \ldots, r-1$ also vanish, so $y_j^{(i)}(k) = 0$ if $j \geq r+1$, which implies that (5.16) has at most r nonzero terms. The construction may thus be viewed as the result of the matrix-vector calculation

$$\begin{pmatrix} y_1^{(i)}(k) \\ y_2^{(i)}(k) \\ \vdots \\ y_r^{(i)}(k) \end{pmatrix} = \mathbf{C}^{(i-1)} \begin{pmatrix} a_0(k) \\ a_1(k) \\ \vdots \\ a_{r-1}(k) \end{pmatrix} \mod b, \tag{5.18}$$

where $\mathbf{C}^{(i)}$ is the $r \times r$ matrix with entries

$$\mathbf{C}^{(i)}(m,n) = \binom{n-1}{m-1} i^{n-m} \tag{5.19}$$

for $n \geq m$ and zero otherwise. With the convention that $0^0 = 1$ and $0^j = 0$ for $j > 0$, this makes $\mathbf{C}^{(0)}$ the identity matrix. (Note that in (5.18) the coefficients $a_i(k)$ are ordered from least significant to most significant.)

These *generator* matrices have the following cyclic properties:

$$\mathbf{C}^{(i)} = \mathbf{C}^{(1)}\mathbf{C}^{(i-1)}, \quad i = 1, 2, \ldots,$$

and for $i \geq 2$, $\mathbf{C}^{(i)}$ mod i is the identity matrix.

To see the effect of the transformations (5.16)–(5.17), consider the integers from 0 to $b^r - 1$. These are the integers whose base-b expansions have r or fewer terms. As k varies over this range, the vector $\mathbf{a}(k) = (a_0(k), a_1(k), \ldots, a_{r-1}(k))^\top$ varies over all b^r vectors with elements in the set $\{0, 1, \ldots, b - 1\}$. The matrices $\mathbf{C}^{(i)}$ have the property that the product $\mathbf{C}^{(i)}\mathbf{a}(k)$ (taken mod b) ranges over exactly the same set. In other words,

$$\mathbf{C}^{(i)}\mathbf{a}(k) \bmod b, \quad 0 \leq k < b^r$$

is a permutation of $\mathbf{a}(k)$, $0 \leq k < b^r$. In fact, the same is true if we restrict k to any range of the form $jb^r \leq k < (j+1)b^r$, with $0 \leq j \leq b - 1$. It follows that the ith coordinate $x_k^{(i)}$ of the points x_k, for any such set of k, form a permutation of the segment $\psi_b(k)$, $jb^r \leq k < (j+1)b^{r+1}$, of the Van der Corput sequence.

As a simple illustration, consider the case $r = 2$ and $b = 3$. The generator matrices are

$$\mathbf{C}^{(1)} = \begin{pmatrix} 1 & 1 \\ 0 & 1 \end{pmatrix}, \quad \mathbf{C}^{(2)} = \mathbf{C}^{(1)}\mathbf{C}^{(1)} = \begin{pmatrix} 1 & 2 \\ 0 & 1 \end{pmatrix}.$$

For $k = 0, 1, \ldots, 8$, the vectors $\mathbf{a}(k)$ are

$$\begin{pmatrix} 0 \\ 0 \end{pmatrix}, \begin{pmatrix} 1 \\ 0 \end{pmatrix}, \begin{pmatrix} 2 \\ 0 \end{pmatrix}, \begin{pmatrix} 0 \\ 1 \end{pmatrix}, \begin{pmatrix} 1 \\ 1 \end{pmatrix}, \begin{pmatrix} 2 \\ 1 \end{pmatrix}, \begin{pmatrix} 0 \\ 2 \end{pmatrix}, \begin{pmatrix} 1 \\ 2 \end{pmatrix}, \begin{pmatrix} 2 \\ 2 \end{pmatrix}.$$

The vectors $\mathbf{C}^{(1)}\mathbf{a}(k)$ (mod b) are

$$\begin{pmatrix} 0 \\ 0 \end{pmatrix}, \begin{pmatrix} 1 \\ 0 \end{pmatrix}, \begin{pmatrix} 2 \\ 0 \end{pmatrix}, \begin{pmatrix} 1 \\ 1 \end{pmatrix}, \begin{pmatrix} 2 \\ 1 \end{pmatrix}, \begin{pmatrix} 0 \\ 1 \end{pmatrix}, \begin{pmatrix} 2 \\ 2 \end{pmatrix}, \begin{pmatrix} 0 \\ 2 \end{pmatrix}, \begin{pmatrix} 1 \\ 2 \end{pmatrix}.$$

And the vectors $\mathbf{C}^{(2)}\mathbf{a}(k)$ (mod b) are

$$\begin{pmatrix} 0 \\ 0 \end{pmatrix}, \begin{pmatrix} 1 \\ 0 \end{pmatrix}, \begin{pmatrix} 2 \\ 0 \end{pmatrix}, \begin{pmatrix} 2 \\ 1 \end{pmatrix}, \begin{pmatrix} 0 \\ 1 \end{pmatrix}, \begin{pmatrix} 1 \\ 1 \end{pmatrix}, \begin{pmatrix} 1 \\ 2 \end{pmatrix}, \begin{pmatrix} 2 \\ 2 \end{pmatrix}, \begin{pmatrix} 0 \\ 2 \end{pmatrix}.$$

Now we apply (5.16) to convert each of these sets of vectors into fractions, by premultiplying each vector by $(1/3, 1/9)$. Arranging these fractions into three rows, we get

$$\begin{array}{ccccccccc}
0 & 1/3 & 2/3 & 1/9 & 4/9 & 7/9 & 2/9 & 5/9 & 8/9 \\
0 & 1/3 & 2/3 & 4/9 & 7/9 & 1/9 & 8/9 & 2/9 & 5/9 \\
0 & 1/3 & 2/3 & 7/9 & 1/9 & 4/9 & 5/9 & 8/9 & 2/9
\end{array}$$

The first row gives the first nine elements of the base-3 Van der Corput sequence and the next two rows permute these elements. Finally, by taking each column in this array as a point in the three-dimensional unit hypercube, we get the first nine points of the three-dimensional Faure sequence.

Faure [116] showed that the discrepancy of the d-dimensional sequence constructed through (5.16)–(5.17) satisfies

$$D^*(x_1, \ldots, x_n) = F_d \frac{(\log n)^d}{n} + O\left(\frac{(\log n)^{d-1}}{n}\right),$$

with F_d depending on d but not n; thus, Faure sequences are indeed low-discrepancy sequences. In fact, $F_d \to 0$ quickly as $d \to \infty$, in marked contrast to the increase in the constant for Halton sequences.

In the terminology of Section 5.1.4, Faure sequences are $(0, d)$-sequences and thus achieve the best possible value of the uniformity parameter t. The example in the left panel of Figure 5.3 is the projection onto dimensions two and three of the first 81 points of the three-dimensional Faure sequence with base 3.

As a consequence of the definition of a (t, d)-sequence, any set of Faure points of the form $\{x_k : jb^m \le k < (j+1)b^m\}$ with $0 \le j \le b-1$ and $m \ge 1$ is a $(0, m, d)$-net, which we call a *Faure net*. (Recall that $x_0 = 0$.) The discussion following (5.19) may then be summarized as stating that, over a Faure net, all one-dimensional projections are the same, as each is a permutation of a segment of the Van der Corput sequence.

The cyclic properties of the generator matrices $\mathbf{C}^{(i)}$ have implications for higher dimensional projections as well. The projection of the points in a Faure net onto the ith and jth coordinates depends only on the distance $j-i$, modulo the base b. Thus, if the dimension equals the base, the $b(b-1)$ two-dimensional projections comprise at most $b-1$ distinct sets in $[0, 1)^2$. If we identify sets in $[0, 1)^2$ that result from interchanging coordinates, then there are at most $(b-1)/2$ distinct projections. Similar conclusions hold for projections onto more than two coordinates.

This phenomenon is illustrated in Figure 5.9, which is based on a Faure net in base 31, the 961 points constructed from $k = 5(31)^2, \ldots, 6(31)^2 - 1$. The projection onto coordinates 1 and 2 is identical to the projection onto coordinates 19 and 20. The projection onto coordinates 1 and 31 would look the same as the other two if plotted with the axes interchanged because modulo 31, 1 is the successor of 31.

Implementation

The construction of Faure points builds on the construction of Van der Corput sequences in Section 5.2.1. To generate the d coordinates of the Faure point x_k in base b, we record the base-b expansion of k in a vector, multiply the vector (mod b) by a generator matrix, and then convert the resulting vector to a point in the unit interval. We give a high-level description of an implementation, omitting many programming details. Fox [126] provides FORTRAN code to generate Faure points.

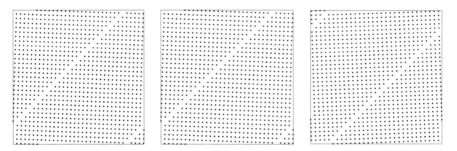

Fig. 5.9. Projections of 961 Faure points in 31 dimensions using base 31. From left to right, the figures show coordinates 1 and 2, 19 and 20, and 1 and 31. The first two are identical; the third would look the same as the others if the axes were interchanged.

A key step is the construction of the generator matrices (5.19). Because these matrices have the property that $\mathbf{C}^{(i)}$ is the ith power of $\mathbf{C}^{(1)}$, it is possible to construct just $\mathbf{C}^{(1)}$ and then recursively evaluate products of the form $\mathbf{C}^{(i)}\mathbf{a}$ as $\mathbf{C}^{(1)}\mathbf{C}^{(i-1)}\mathbf{a}$. However, to allow extensions to more general matrices, we do not take advantage of this in our implementation.

As noted by Fox [126] (for the case $i = 1$), in calculating the matrix entries in (5.19), evaluation of binomial coefficients can be avoided through the recursion

$$\binom{n+1}{k+1} = \binom{n}{k+1} + \binom{n}{k},$$

$n \geq k \geq 0$. Figure 5.10 displays a function FAUREMAT that uses this recursion to construct $\mathbf{C}^{(i)}$.

FAUREMAT(r, i)
$\mathbf{C}(1,1) \leftarrow 1$
for $m = 2, \ldots, r$
$\quad \mathbf{C}(m, m) \leftarrow 1$
$\quad \mathbf{C}(1, m) \leftarrow i * \mathbf{C}(1, m-1)$
for $n = 3, \ldots, r$
$\quad$ for $m = 2, \ldots, n-1$
$\quad\quad \mathbf{C}(m, n) = \mathbf{C}(m-1, n-1) + i * \mathbf{C}(m, n-1)$
return $\mathbf{C}$

Fig. 5.10. Function FAUREMAT(r, i) returns $r \times r$ generator matrix $\mathbf{C}^{(i)}$.

The function FAUREPTS in Figure 5.11 uses FAUREMAT to generate Faure points. The function takes as inputs the starting index n_0, the total number of points to generate n_{pts}, the dimension d, and the base b. The starting index n_0 must be greater than or equal to 1 and the base b must be a prime number at least as large as d. One could easily modify the function to include an array of prime numbers to save the user from having to specify the base. Calling FAUREPTS with $n_0 = 1$ starts the sequence at the first nonzero point. Fox [126] recommends starting at $n_0 = b^4 - 1$ to improve uniformity.

The advantage of generating n_{pts} points in a single call to the function lies in constructing the generator matrices just once. In FAUREPTS, r_{max} is the number of places in the base-b representation of $n_0 + n_{pts} - 1$, so the largest generator matrices needed are $r_{max} \times r_{max}$. We construct these and then use the submatrices needed to convert smaller integers. The variable r keeps track of the length of the expansion $\mathbf{a}$ of the current integer k, so we use the first r rows and columns of the full generator matrices to produce the required $r \times r$ generator matrices. The variable r increases by one each time k reaches q_{next}, the next power of b.

FAUREPTS(n_0, n_{pts}, d, b)
$n_{max} \leftarrow n_0 + n_{pts} - 1, \quad r_{max} \leftarrow 1 + \lfloor \log(n_{max})/\log(b) \rfloor$
$P \leftarrow 0 \ [n_{max} \times d], \quad \mathbf{y}(1, \ldots, r_{max}) \leftarrow 0$
$r \leftarrow 1 + \lfloor \log(\max(1, n_0 - 1))/\log(b) \rfloor, \quad q_{next} \leftarrow b^r$
$\mathbf{a} \leftarrow$ B-ARY($n_0 - 1, b, 1$)
for $i = 1, \ldots, d - 1$
 $\mathbf{C}^{(i)} \leftarrow$ FAUREMAT(r_{max}, i)
$b_{pwrs} \leftarrow (1/b, 1/b^2, \ldots, 1/b^{r_{max}})$
for $k = n_0, \ldots, n_{max}$
 $\mathbf{a} \leftarrow$ NEXTB-ARY($\mathbf{a}, b$)
 if ($k = q_{next}$)
  $r \leftarrow r + 1$
  $q_{next} \leftarrow b * q_{next}$
 for $j = 1, \ldots, r$
  $P(k - n_0 + 1, 1) \leftarrow P(k - n_0 + 1, 1) + b_{pwrs}(j) * \mathbf{a}(r - j + 1)$
 for $i = 2, \ldots, d$
  for $m = 1, \ldots, r$
   for $n = 1, \ldots, r$
    $\mathbf{y}(m) \leftarrow \mathbf{y}(m) + \mathbf{C}^{(i)}(m, n) * \mathbf{a}(r - n + 1)$
   $\mathbf{y}(m) \leftarrow \mathbf{y}(m) \bmod b$
   $P(k - n_0 + 1, i) \leftarrow P(k - n_0 + 1, i) + b_{pwrs}(m) * \mathbf{y}(m)$
   $\mathbf{y}(m) \leftarrow 0$
return P

Fig. 5.11. Function FAUREPTS(n_0, n_{pts}, d, b) returns $n_{pts} \times d$ array whose rows are coordinates of d-dimensional Faure points in base b, starting from the n_0th nonzero point.

The loop over m and n near the bottom of FAUREPTS executes the matrix-vector product in (5.18). The algorithm traverses the elements of the vector **a** from the highest index to the lowest because the variable **a** in the algorithm is flipped relative to the vector of coefficients in (5.18). This results from a conflict in two notational conventions: the functions B-ARY and NEXTB-ARY follow the usual convention of ordering digits from most significant to least, whereas in (5.18) and the surrounding discussion we prefer to start with the least significant digit.

In FAUREPTS, we apply the mod-b operation only after multiplying each vector of base-b coefficients by a generator matrix. We could also take the remainder mod b after each multiplication of an element of $\mathbf{C}^{(i)}$ by an element of **a**, or in setting up the matrices $\mathbf{C}^{(i)}$; indeed, we could easily modify FAUREMAT to return $\mathbf{C}^{(i)}$ mod b by including b as an argument of that function. Taking remainders at intermediate steps can eliminate problems from overflow, but requires additional calculation.

An alternative construction of Faure points makes it possible to replace the matrix-vector product in (5.18) (the loop over m and n in FAUREPTS) with a single vector addition. This alternative construction produces permutations of Faure points, rather than the Faure points themselves. It relies on the notion of a Gray code in base b; we return to it in the next section after a more general discussion of Gray codes.

5.2.3 Sobol'

Sobol' [335] gave the first construction of what is now known as a (t, d)-sequence (he used the name LP_τ-sequence). The methods of Halton and Hammersley have low discrepancy, but they are not (t, d)-sequences or (t, m, d)-nets. Sobol's construction can be succinctly contrasted with Faure's as follows: Whereas Faure points are $(0, d)$-sequences in a base at least as large as d, Sobol's points are (t, d)-sequences in base 2 for all d, with values of t that depend on d. Faure points therefore achieve the best value of the uniformity parameter t, but Sobol' points have the advantage of a much smaller base. Working in base 2 also lends itself to computational advantages through bit-level operations.

Like the methods of Halton, Hammersley, and Faure, Sobol' points start from the Van der Corput sequence, but now exclusively in base 2. The various coordinates of a d-dimensional Sobol' sequence result from permutations of segments of the Van der Corput sequence. As in Section 5.2.2, these permutations result from multiplying (binary) expansions of consecutive integers by a set of generator matrices, one for each dimension. The key difference lies in the construction of these generator matrices.

All coordinates of a Sobol' sequence follow the same construction, but each with its own generator. We may therefore begin by discussing the construction of a single coordinate based on a generator matrix **V**. The elements of **V** are equal to 0 or 1. Its columns are the binary expansions of a set of *direction* *numbers* $v_1, \ldots, v_r$. Here, r could be arbitrarily large; in constructing the kth point in the sequence, think of r as the number of terms in the binary expansion of k. The matrix $\mathbf{V}$ will be upper triangular, so regardless of the number of rows in the full matrix, it suffices to consider the square matrix consisting of the first r rows and columns.

Let $\mathbf{a}(k) = (a_0(k), \ldots, a_{r-1}(k))^\top$ denote the vector of coefficients of the binary representation of k, so that

$$k = a_0(k) + 2a_1(k) + \cdots + 2^{r-1} a_{r-1}(k).$$

Let

$$\begin{pmatrix} y_1(k) \\ y_2(k) \\ \vdots \\ y_r(k) \end{pmatrix} = \mathbf{V} \begin{pmatrix} a_0(k) \\ a_1(k) \\ \vdots \\ a_{r-1}(k) \end{pmatrix} \quad \text{mod } 2; \tag{5.20}$$

then $y_1(k), \ldots, y_r(k)$ are coefficients of the binary expansion of the kth point in the sequence; more explicitly,

$$x_k = \frac{y_1(k)}{2} + \frac{y_2(k)}{4} + \cdots + \frac{y_r(k)}{2^r}.$$

If $\mathbf{V}$ is the identity matrix, this produces the Van der Corput sequence in base 2.

The operation in (5.20) can be represented as

$$a_0(k)v_1 \oplus a_1(k)v_2 \oplus \cdots \oplus a_{r-1}(k)v_r, \tag{5.21}$$

where the v_j are the columns of (the $r \times r$ submatrix of) $\mathbf{V}$ and $\oplus$ denotes binary addition,

$$0 \oplus 0 = 0, \quad 0 \oplus 1 = 1 \oplus 0 = 1, \quad 1 \oplus 1 = 0.$$

This formulation is useful in computer implementation. If we reinterpret v_j as the computer representation of a number (i.e., as a computer word of bits rather than as an array) and implement $\oplus$ through a bitwise XOR operation, then (5.21) produces the computer representation of x_k.

We turn now to the heart of Sobol's method, which is the specification of the generator matrices — equivalently, of the direction numbers v_j. We use the same symbol v_j to denote the number itself (a binary fraction) as we use to denote the vector encoding its binary representation. For a d-dimensional Sobol' sequence we need d sets of direction numbers, one for each coordinate; for simplicity, we continue to focus on a single coordinate.

Sobol's method for choosing a set of direction numbers starts by selecting a *primitive polynomial* over binary arithmetic. This is a polynomial

$$x^q + c_1 x^{q-1} + \cdots + c_{q-1} x + 1, \tag{5.22}$$

with coefficients c_i in $\{0, 1\}$, satisfying the following two properties (with respect to binary arithmetic):

o it is irreducible (i.e., it cannot be factored);
o the smallest power p for which the polynomial divides $x^p + 1$ is $p = 2^q - 1$.

Irreducibility implies that the constant term 1 must indeed be present as implied by (5.22). The largest power q with a nonzero coefficient is the degree of the polynomial. The polynomials

$$x + 1, \quad x^2 + x + 1, \quad x^3 + x + 1, \quad x^3 + x^2 + 1$$

are all the primitive polynomials of degree one, two, or three.

Table 5.2 lists 53 primitive polynomials, including all those of degree 8 or less. (A list of 360 primitive polynomials is included in the implementation of Lemieux, Cieslak, and Luttmer [231].) Each polynomial in the table is encoded as the integer defined by interpreting the coefficients of the polynomial as bits. For example, the integer 37 in binary is 100101, which encodes the polynomial $x^5 + x^2 + 1$. Table 5.2 includes a polynomial of degree 0; this is a convenient convention that makes the construction of the first coordinate of a multidimensional Sobol' sequence consistent with the construction of the other coordinates.

Degree	Primitive Polynomials
0	1
1	3 $(x + 1)$
2	7 $(x^2 + x + 1)$
3	11 $(x^3 + x + 1)$, 13 $(x^3 + x^2 + 1)$
4	19, 25
5	37, 59, 47, 61, 55, 41
6	67, 97, 91, 109, 103, 115
7	131, 193, 137, 145, 143, 241, 157, 185, 167, 229, 171, 213, 191, 253, 203, 211, 239, 247
8	285, 369, 299, 425, 301, 361, 333, 357, 351, 501, 355, 397, 391, 451, 463, 487

Table 5.2. Primitive polynomials of degree 8 or less. Each number in the right column, when represented in binary, gives the coefficients of a primitive polynomial.

The polynomial (5.22) defines a recurrence relation

$$m_j = 2c_1 m_{j-1} \oplus 2^2 c_2 m_{j-2} \oplus \cdots \oplus 2^{q-1} c_{q-1} m_{j-q+1} \oplus 2^q m_{j-q} \oplus m_{j-q}. \quad (5.23)$$

The m_j are integers and $\oplus$ may again be interpreted either as binary addition of binary vectors (by identifying m_j with its vector of binary coefficients) or as bit-wise XOR applied directly to the computer representations of the operands. By convention, the recurrence relation defined by the degree-0 polynomial is $m_j \equiv 1$. From the m_j, the direction numbers are defined by setting

$$v_j = m_j / 2^j.$$

For this to fully define the direction numbers we need to specify initial values $m_1, \ldots, m_q$ for (5.23). A minimal requirement is that each initializing m_j be an odd integer less than 2^j; all subsequent m_j defined by (5.23) will then share this property and each v_j will lie strictly between 0 and 1. More can be said about the proper initialization of (5.23); we return to this point after considering an example.

Consider the primitive polynomial

$$x^3 + x^2 + 1$$

with degree $q = 3$. The recurrence (5.23) becomes

$$m_j = 2m_{j-1} \oplus 8m_{j-3} \oplus m_{j-3}$$

and suppose we initialize it with $m_1 = 1$, $m_2 = 3$, $m_3 = 3$. The next two elements in the sequence are as follows:

$$\begin{aligned}
m_4 &= (2 \cdot 3) \oplus (8 \cdot 1) \oplus 1 \\
&= 0110 \oplus 1000 \oplus 0001 \\
&= 1111 \\
&= 15 \\
m_5 &= (2 \cdot 15) \oplus (8 \cdot 3) \oplus 3 \\
&= 11110 \oplus 11000 \oplus 00011 \\
&= 00101 \\
&= 5
\end{aligned}$$

From these five values of m_j, we can calculate the corresponding values of v_j by dividing by 2^j. But dividing by 2^j is equivalent to shifting the "binary point" to the left j places in the representation of m_j. Hence, the first five direction numbers are

$$v_1 = 0.1, \ v_2 = 0.11, \ v_3 = 0.011, \ v_4 = 0.1111, \ v_5 = 0.00101,$$

and the corresponding generator matrix is

$$\mathbf{V} = \begin{pmatrix} 1 & 1 & 0 & 1 & 0 \\ 0 & 1 & 1 & 1 & 0 \\ 0 & 0 & 1 & 1 & 1 \\ 0 & 0 & 0 & 1 & 0 \\ 0 & 0 & 0 & 0 & 1 \end{pmatrix}. \tag{5.24}$$

Observe that taking $m_j = 1$, $j = 1, 2, \ldots$, (i.e., using the degree-0 polynomial) produces the identity matrix.

Finally, we illustrate the calculation of the sequence $x_1, x_2, \ldots$. For each k, we take the vector $\mathbf{a}(k)$ of binary coefficients of k and premultiply it (mod 2) by the matrix $\mathbf{V}$. The resulting vector gives the coefficients of a binary fraction. The first three vectors are

$$\mathbf{V}\begin{pmatrix}1\\0\\0\\0\\0\end{pmatrix}=\begin{pmatrix}1\\0\\0\\0\\0\end{pmatrix},\quad \mathbf{V}\begin{pmatrix}0\\1\\0\\0\\0\end{pmatrix}=\begin{pmatrix}1\\1\\0\\0\\0\end{pmatrix},\quad \mathbf{V}\begin{pmatrix}1\\1\\0\\0\\0\end{pmatrix}=\begin{pmatrix}0\\1\\0\\0\\0\end{pmatrix},$$

which produce the points $1/2$, $3/4$, and $1/4$. For $k = 29, 30, 31$ (the last three points that can be generated with a 5×5 matrix), we have

$$\mathbf{V}\begin{pmatrix}1\\0\\1\\1\\1\end{pmatrix}=\begin{pmatrix}0\\0\\1\\1\\1\end{pmatrix},\quad \mathbf{V}\begin{pmatrix}0\\1\\1\\1\\1\end{pmatrix}=\begin{pmatrix}0\\1\\1\\1\\1\end{pmatrix},\quad \mathbf{V}\begin{pmatrix}1\\1\\1\\1\\1\end{pmatrix}=\begin{pmatrix}1\\1\\1\\1\\1\end{pmatrix},$$

which produce $7/32$, $15/32$, and $31/32$.

Much as in the case of Faure points, this procedure produces a permutation of the segment $\psi_2(k)$, $2^{r-1} \le k < 2^r$, of the Van der Corput sequence when $\mathbf{V}$ is $r \times r$. This crucial property relies on the fact that $\mathbf{V}$ was constructed from a primitive polynomial.

Gray Code Construction

Antanov and Saleev [18] point out that Sobol's method simplifies if the usual binary representation $\mathbf{a}(k)$ is replaced with a Gray code representation. In a Gray code, k and $k+1$ have all but one coefficient in common, and this makes it possible to construct the values x_k recursively.

One way to define a Gray code is to take the bitwise binary sum of the usual binary representation of k, $a_{r-1}(k)\cdots a_1(k)a_0(k)$, with the shifted string $0a_{r-1}(k)\cdots a_1(k)$; in other words, take the $\oplus$-sum of the binary representations of k and $\lfloor k/2 \rfloor$. This encodes the numbers 1 to 7 as follows:

	1	2	3	4	5	6	7
Binary	001	010	011	100	101	110	111
Gray code	001	011	010	110	111	101	100

For example, the Gray code for 3 is calculated as $011 \oplus 001$.

Exactly one bit in the Gray code changes when k is incremented to $k+1$. The position of that bit is the position of the rightmost zero in the ordinary binary representation of k, taking this to mean an initial zero if the binary representation has only ones. For example, since the last bit of the binary representation of 4 is zero, the Gray codes for 4 and 5 differ in the last bit. Because the binary representation of 7 is displayed as 111, the Gray code for 8 differs from the Gray code for 7 through the insertion of an initial bit, which would produce 1100.

The binary strings formed by the Gray code representations of the integers $0, 1, \ldots, 2^r - 1$ are a permutation of the sequence of strings formed by the usual binary representations of the same integers, for any r. If in the definition of the radical inverse function ψ_2 in (5.8) we replaced the usual binary coefficients with Gray code coefficients, the first $2^r - 1$ values would be a permutation of the corresponding elements of the Van der Corput sequence. Hence, the two sequences have the same asymptotic discrepancy. Antanov and Saleev [18] show similarly that using a Gray code with Sobol's construction does not affect the asymptotic discrepancy of the resulting sequence.

Suppose, then, that in (5.21) we replace the binary coefficients $a_j(k)$ with Gray code coefficients $g_j(k)$ and redefine x_k to be

$$x_k = g_0(k)v_1 \oplus g_1(k)v_2 \oplus \cdots \oplus g_{r-1}(k)v_r.$$

Suppose that the Gray codes of k and $k+1$ differ in the ℓth bit; then

$$
\begin{aligned}
x_{k+1} &= g_0(k+1)v_1 \oplus g_1(k+1)v_2 \oplus \cdots \oplus g_{r-1}(k+1)v_r \\
&= g_0(k)v_1 \oplus g_1(k)v_2 \oplus \cdots \oplus (g_\ell(k) \oplus 1)v_\ell \oplus \cdots \oplus g_{r-1}(k)v_r \\
&= x_k \oplus v_\ell.
\end{aligned}
\tag{5.25}
$$

Rather than compute x_{k+1} from (5.21), we may thus compute it recursively from x_k through binary addition of a single direction number. The computer implementations of Bratley and Fox [57] and Press et al. [299] use this formulation.

It is worth noting that if we start the Sobol' sequence at $k = 0$, we never need to calculate a Gray code. To use (5.25), we need to know only ℓ, the index of the bit that would change if we calculated the Gray code. But, as explained above, ℓ is completely determined by the ordinary binary expansion of k. To start the Sobol' sequence at an arbitrary point x_n, we need to calculate the Gray code of n to initialize the recursion in (5.25).

The simplification in (5.25) extends to the construction of Faure points in any arbitrary (prime) base b through an observation of Tezuka [348]. We digress briefly to explain this extension. Let $a_0(k), a_1(k), \ldots, a_r(k)$ denote the coefficients in the base-b expansion of k. Setting

$$
\begin{pmatrix} g_0(k) \\ g_1(k) \\ \vdots \\ g_r(k) \end{pmatrix} = \begin{pmatrix} 1 & & & \\ b-1 & 1 & & \\ & & \ddots & \\ & & b-1 & 1 \\ & & & b-1 & 1 \end{pmatrix} \begin{pmatrix} a_0(k) \\ a_1(k) \\ \vdots \\ a_r(k) \end{pmatrix} \quad \bmod b
$$

defines a base-b Gray code, in the sense that the vectors thus calculated for k and $k+1$ differ in exactly one entry. The index of the entry that changes is the smallest ℓ for which $a_\ell(k) \neq b-1$ (padding the expansion of k with an initial zero, if necessary, to ensure it has the same length as the expansion of $k+1$). Moreover, $g_\ell(k+1) = g_\ell(k) + 1$ modulo b.

This simplifies the Faure construction. Instead of defining the coefficients $y_j^{(i)}(k)$ through the matrix-vector product (5.18), we may set

$$(y_1^{(i)}(k+1), \ldots, y_r^{(i)}(k+1)) =$$
$$(y_1^{(i)}(k), \ldots, y_r^{(i)}(k)) + (\mathbf{C}^{(i)}(1, \ell), \ldots, \mathbf{C}^{(i)}(r, \ell)) \bmod b,$$

once ℓ has been determined from the vector $\mathbf{a}(k)$. This makes it possible to replace the loop over both m and n near the bottom of FAUREPTS in Figure 5.11 with a loop over a single index. See Figure 5.13.

Choosing Initial Direction Numbers

In initializing the recurrence (5.23), we required only that each m_j be an odd integer less than 2^j. But suppose we initialize two different sequences (corresponding to two different coordinates of a d-dimensional sequence) with the same values $m_1, \ldots, m_r$. The first r columns of their generator matrices will then also be the same. The kth value generated in a Sobol' sequence depends only on as many columns as there are coefficients in the binary expansion of k; so, the first $2^r - 1$ values of the two sequences would be identical. Thus, whereas the choice of initial values may not be significant in constructing a one-dimensional sequence, it becomes important when d such sequences are yoked together to make a d-dimensional sequence.

Sobol' [336] provides some guidance for choosing initial values. He establishes two results on uniformity properties achieved by initial values satisfying additional conditions. A d-dimensional sequence $x_0, x_1, \ldots$ satisfies Sobol's Property A if for every $j = 0, 1, \ldots$ exactly one of the points x_k, $j2^d \le k < (j+1)2^d$ falls in each of the 2^d cubes of the form

$$\prod_{i=1}^{d} \left[\frac{a_i}{2}, \frac{a_i + 1}{2} \right), \quad a_i \in \{0, 1\}.$$

The sequence satisfies Sobol's Property A' if for every $j = 0, 1, \ldots$ exactly one of the points x_k, $j2^{2d} \le k < (j+1)2^{2d}$ falls in each of the 2^{2d} cubes of the form

$$\prod_{i=1}^{d} \left[\frac{a_i}{4}, \frac{a_i + 1}{4} \right), \quad a_i \in \{0, 1, 2, 3\}.$$

These properties bear some resemblance to the definition of a (t, d)-sequence but do not fit that definition because they restrict attention to specific equilateral boxes.

Let $v_j^{(i)}$ denote the jth direction number associated with the ith coordinate of the sequence. The generator matrix of the ith sequence is then

$$\mathbf{V}^{(i)} = [v_1^{(i)} | v_2^{(i)} | \cdots | v_r^{(i)}],$$

where we again use $v_j^{(i)}$ to denote the (column) vector of binary coefficients of a direction number as well as the direction number itself. Sobol' [336] shows that Property A holds if and only if the determinant constructed from the first d elements of the first row of each of the matrices satisfies

$$\begin{vmatrix} \mathbf{V}_{11}^{(1)} & \mathbf{V}_{12}^{(1)} & \cdots & \mathbf{V}_{1d}^{(1)} \\ \mathbf{V}_{11}^{(2)} & \mathbf{V}_{12}^{(2)} & \cdots & \mathbf{V}_{1d}^{(2)} \\ \vdots & \vdots & & \vdots \\ \mathbf{V}_{11}^{(d)} & \mathbf{V}_{12}^{(d)} & \cdots & \mathbf{V}_{1d}^{(d)} \end{vmatrix} \neq 0 \bmod 2.$$

Property A applies to sets of size 2^d and generating the first 2^d points involves exactly the first d columns of the matrices. Sobol' also shows that Property A' holds if and only if the determinant constructed from the first $2d$ elements of the first two rows of each of the matrices satisfies

$$\begin{vmatrix} \mathbf{V}_{11}^{(1)} & \mathbf{V}_{12}^{(1)} & \cdots & \mathbf{V}_{1,2d}^{(1)} \\ \vdots & \vdots & & \vdots \\ \mathbf{V}_{11}^{(d)} & \mathbf{V}_{12}^{(d)} & \cdots & \mathbf{V}_{1,2d}^{(d)} \\ \mathbf{V}_{21}^{(1)} & \mathbf{V}_{22}^{(1)} & \cdots & \mathbf{V}_{2,2d}^{(1)} \\ \vdots & \vdots & & \vdots \\ \mathbf{V}_{21}^{(d)} & \mathbf{V}_{22}^{(d)} & \cdots & \mathbf{V}_{2,2d}^{(d)} \end{vmatrix} \neq 0 \bmod 2.$$

To illustrate, suppose $d = 3$ and suppose the first three m_j values for the first three coordinates are as follows:

$$(m_1, m_2, m_3) = (1, 1, 1), \quad (m_1, m_2, m_3) = (1, 3, 5), \quad (m_1, m_2, m_3) = (1, 1, 7).$$

The first coordinate has $m_j = 1$ for all j. The second coordinate is generated by a polynomial of degree 1, so all subsequent values are determined by m_1; and because the third coordinate is generated by a polynomial of degree 2, that sequence is determined by the choice of m_1 and m_2.

From the first three m_j values in each coordinate, we determine the matrices

$$\mathbf{V}^{(1)} = \begin{pmatrix} 1 & 0 & 0 \\ 0 & 1 & 0 \\ 0 & 0 & 1 \end{pmatrix}, \quad \mathbf{V}^{(2)} = \begin{pmatrix} 1 & 1 & 1 \\ 0 & 1 & 0 \\ 0 & 0 & 1 \end{pmatrix}, \quad \mathbf{V}^{(3)} = \begin{pmatrix} 1 & 0 & 1 \\ 0 & 1 & 1 \\ 0 & 0 & 1 \end{pmatrix}.$$

From the first row of each of these we assemble the matrix

$$D = \begin{pmatrix} 1 & 0 & 0 \\ 1 & 1 & 1 \\ 1 & 0 & 1 \end{pmatrix}.$$

Because this matrix has a determinant of 1, Sobol's Property A holds.

The test matrix D can in fact be read directly from the m_j values: $D_{ij} = 1$ if the m_j value for the ith coordinate is greater than or equal to 2^{j-1}, and $D_{ij} = 0$ otherwise.

Table 5.3 displays initial values of the m_j for up to 20 dimensions. Recall that the number of initial values required equals the degree of the corresponding primitive polynomial, and this increases with the dimension. The values displayed in parentheses are determined by the initial values in each row. The values displayed are from Bratley and Fox [57], who credit unpublished work of Sobol' and Levitan (also cited in Sobol' [336]). These values satisfy Property A; Property A' holds for $d \leq 6$.

	m_1	m_2	m_3	m_4	m_5	m_6	m_7	m_8
1	1	(1)	(1)	(1)	(1)	(1)	(1)	(1)
2	1	(3)	(5)	(15)	(17)	(51)	(85)	(255)
3	1	1	(7)	(11)	(13)	(61)	(67)	(79)
4	1	3	7	(5)	(7)	(43)	(49)	(147)
5	1	1	5	(3)	(15)	(51)	(125)	(141)
6	1	3	1	1	(9)	(59)	(25)	(89)
7	1	1	3	7	(31)	(47)	(109)	(173)
8	1	3	3	9	9	(57)	(43)	(43)
9	1	3	7	13	3	(35)	(89)	(9)
10	1	1	5	11	27	(53)	(69)	(25)
11	1	3	5	1	15	(19)	(113)	(115)
12	1	1	7	3	29	(51)	(47)	(97)
13	1	3	7	7	21	(61)	(55)	(19)
14	1	1	1	9	23	37	(97)	(97)
15	1	3	3	5	19	33	(3)	(197)
16	1	1	3	13	11	7	(37)	(101)
17	1	1	7	13	25	5	(83)	(255)
18	1	3	5	11	7	11	(103)	(29)
19	1	1	1	3	13	39	(27)	(203)
20	1	3	1	15	17	63	13	(65)

Table 5.3. Initial values satisfying Sobol's Property A for up to 20 dimensions. In each row, values in parentheses are determined by the previous values in the sequence.

Bratley and Fox [57] include initializing values from the same source for up to 40 dimensions. A remark in Sobol' [336] indicates that the Sobol'-Levitan values should satisfy Property A for up to 51 dimensions; however, we find (as does Morland [270]) that this property does not consistently hold for $d > 20$. More precisely, for d ranging from 21 to 40, we find that Property A holds only at dimensions 23, 31, 33, 34, and 37. We have therefore limited Table 5.3 to the first 20 dimensions. The complete set of values used by Bratley and Fox [57] is in their FORTRAN program, available through the Collected Algorithms of the ACM.

Press et al. [299] give initializing values for up to six dimensions; their values fail the test for Property A in dimensions three, five, and six. A further distinction merits comment. We assume (as do Sobol' [336] and Bratley and Fox [57]) that the first coordinate uses $m_j \equiv 1$; this makes the first generator matrix the identity and thus makes the first coordinate the Van der Corput sequence in base 2. We use the polynomial $x + 1$ for the second coordinate, and so on. Press et al. [299] use $x + 1$ for the first coordinate. Whether or not Property A holds for a particular set of initializing values depends on whether the first row (with $m_j \equiv 1$) of Table 5.3 is included. Thus, one cannot interchange the initializing values used here with those used by Press et al. [299] for the same primitive polynomial, even in cases where both satisfy Property A.

The implementation of Lemieux, Cieslak, and Luttmer [231] includes initializing values for up to 360 dimensions. These values do not necessarily satisfy Sobol's property A, but they are the result of a search for good values based on a *resolution* criterion used in design of random number generators. See [231] and references cited there.

Discrepancy

In the terminology of Section 5.1.4, Sobol' sequences are (t, d)-sequences in base 2. The example in the right panel of Figure 5.3 is the projection onto dimensions four and five of the first 128 points of a five-dimensional Sobol' sequence.

Theorem 3.4 of Sobol' [335] provides a simple expression for the t parameter in a d-dimensional sequence as

$$t = q_1 + q_2 + \cdots + q_{d-1} - d + 1, \tag{5.26}$$

where $q_1 \leq q_2 \leq \cdots \leq q_{d-1}$ are the degrees of the primitive polynomials used to construct coordinates 2 through d. Recall that the first coordinate is constructed from the degenerate recurrence with $m_j \equiv 1$, which may be considered to have degree zero. If instead we used polynomials of degrees $q_1, \ldots, q_d$ for the d coordinates, the t value would be $q_1 + \cdots + q_d - d$. Sobol' [335] shows that while t grows faster than d, it does not grow faster than $d \log d$.

Although (5.26) gives a valid value for t, it does not always give the best possible value: a d-dimensional Sobol' sequence may be a (t', d)-sequence for some $t' < t$. Sobol [335] provides conditions under which (5.26) is indeed the minimum valid value of t.

Because they are (t, d)-sequences, Sobol' points are low-discrepancy sequences; see (5.13) and the surrounding discussion. Sobol' [335] provides more detailed discrepancy bounds.

Implementation

Bratley and Fox [57] and Press et al. [299] provide computer programs to generate Sobol' points in FORTRAN and C, respectively. Both take advantage of bit-level operations to increase efficiency. Timings reported in Bratley and Fox [57] indicate that using bit-level operations typically increases speed by a factor of more than ten, with greater improvements in higher dimensions.

We give a somewhat more schematic description of an implementation, suppressing programming details in the interest of transparency. Our description also highlights similarities between the construction of Sobol' points and Faure points.

As in the previous section, we separate the construction of generator matrices from the generation of the points themselves. Figure 5.12 displays a function SOBOLMAT to produce a generator matrix as described in the discussion leading to (5.24). The function takes as input a binary vector c_{vec} giving the coefficients of a primitive polynomial, a vector m_{init} of initializing values, and a parameter r determining the size of the matrix produced. For a polynomial of degree q, the vector c_{vec} has the form $(1, c_1, \ldots, c_{q-1}, 1)$; see (5.22). The vector m_{init} must then have q elements — think of using the row of Table 5.3 corresponding to the polynomial c_{vec}, including only those values in the row that do not have parentheses. The parameter r must be at least as large as q. Building an $r \times r$ matrix requires calculating $m_{q+1}, \ldots, m_r$ from the initial values $m_1, \ldots, m_q$ in m_{init}. These are ultimately stored in m_{vec} which (to be consistent with Table 5.3) orders $m_1, \ldots, m_r$ from left to right. In calling SOBOLMAT, the value of r is determined by the number of points to be generated: generating the point x_{2^k} requires $r = k + 1$.

The function SOBOLMAT could almost be substituted for FAUREMAT in the function FAUREPTS of Figure 5.11: the only modification required is passing the ith primitive polynomial and the ith set of initializing values, rather than just i itself. The result would be a legitimate algorithm to generate Sobol' points.

Rather than reproduce what we did in FAUREPTS, here we display an implementation using the Gray code construction of Antanov and Saleev [18]. The function SOBOLPTS in Figure 5.13 calls an undefined function GRAYCODE2 to find a binary Gray code representation. This can implemented as

$$\text{GRAYCODE2}(n) = \text{B-ARY}(n, 2) \oplus \text{B-ARY}(\lfloor n/2 \rfloor, 2),$$

after padding the second argument on the right with an initial zero to give the two arguments the same length.

In SOBOLPTS, a Gray code representation is explicitly calculated only for $n_0 - 1$. The Gray code vector $\mathbf{g}$ is subsequently incremented by toggling the ℓth bit, with ℓ determined by the usual binary representation $\mathbf{a}$, or by inserting a leading 1 at each power of 2 (in which case $\ell = 1$). As in (5.25), the value of ℓ is then the index of the column of $\mathbf{V}^{(i)}$ to be added (mod 2) to the previous point. The coefficients of the binary expansion of the ith

---

SOBOLMAT($c_{\text{vec}}, m_{\text{init}}, r$)
[c_{vec} has the form $(1, c_1, \ldots, c_{q-1}, 1)$, m_{init} has length $q \leq r$]
$q \leftarrow$ length(c_{vec}) $- 1$
if ($q = 0$) $\mathbf{V} \leftarrow I$ [$r \times r$ identity]
if ($q > 0$)
    $m_{\text{vec}} \leftarrow (m_{\text{init}}(1, \ldots, q), 0, \ldots, 0)$ [length r]
    $m_{\text{state}} \leftarrow m_{\text{init}}$
    for $i = q + 1, \ldots, r$
        $m_{\text{next}} \leftarrow 2c_1 m_{\text{state}}(q) \oplus 4c_2 m_{\text{state}}(q - 1) \oplus \cdots \oplus 2^q m_{\text{state}}(1) \oplus m_{\text{state}}(1)$
        $m_{\text{vec}}(i) \leftarrow m_{\text{next}}$
        $m_{\text{state}} \leftarrow (m_{\text{state}}(2, \ldots, q), m_{\text{next}})$
    for $j = 1, \ldots, r$
        $m_{\text{bin}} \leftarrow$ B-ARY($m_{\text{vec}}(j), 2$)
        $k \leftarrow$ length(m_{bin})
        for $i = 1, \ldots, k$
            $\mathbf{V}(j - i + 1, j) \leftarrow m_{\text{bin}}(k - i + 1)$
return $\mathbf{V}$

Fig. 5.12. Function SOBOLMAT(c_{vec}, m_{init}, r) returns $r \times r$ generator matrix $\mathbf{V}$ constructed from polynomial coefficients $(1, c_1, \ldots, c_{q-1}, 1)$ in c_{vec} and q initial values in array m_{init}.

coordinate of each point are held in the ith column of the array $\mathbf{y}$. Taking the inner product between this column and the powers of $1/2$ in b_{pwrs} maps the coefficients to $[0, 1)$. The argument p_{pvec} is an array encoding primitive polynomials using the numerical representation in Table 5.2, and m_{mat} is an array of initializing values (as in Table 5.3).

5.2.4 Further Constructions

The discussions in Sections 5.2.2 and 5.2.3 make evident similarities between the Faure and Sobol' constructions: both apply permutations to segments of the Van der Corput sequence, and these permutations can be represented through generator matrices. This strategy for producing low-discrepancy sequences has been given a very general formulation and analysis by Niederreiter [281]. Points constructed in this framework are called *digital* nets or sequences. Section 4.5 of Niederreiter [281] presents a special class of digital sequences, which encompass the constructions of Faure and Sobol'. Niederreiter shows how to achieve a t parameter through this construction (in base 2) strictly smaller than the best t parameter for Sobol' sequences in all dimensions greater than seven. Thus, these *Niederreiter sequences* have some theoretical superiority over Sobol' sequences. Larcher [219] surveys more recent theoretical developments in digital point sets.

Bratley, Fox, and Niederreiter [58] provide a FORTRAN generator for Niederreiter sequences. They note that for base 2 their program is "essentially

SOBOLPTS($n_0, n_{pts}, d, p_{vec}, m_{mat}$)
$n_{max} \leftarrow n_0 + n_{pts} - 1$
$r_{max} \leftarrow 1 + \lfloor (\log(n_{max})/\log(2)) \rfloor, \quad r \leftarrow 1$
$P \leftarrow 0 \; [n_{pts} \times d], \quad \mathbf{y} \leftarrow 0 \; [r_{max} \times d]$
if $(n_0 > 1) \; r \leftarrow 1 + \lfloor (\log(n_0 - 1)/\log(2)) \rfloor$
$q_{next} \leftarrow 2^r$
$\mathbf{a} \leftarrow$ B-ARY($n_0 - 1, 2$)
$\mathbf{g} \leftarrow$ GRAYCODE2($n_0 - 1$)
for $i = 1, \ldots, d$ [build matrices using polynomials in p_{vec}]
$\qquad q \leftarrow \lfloor (\log(p_{vec}(i))/\log(2)) \rfloor$
$\qquad c_{vec} \leftarrow$ B-ARY($p_{vec}(i), 2$)
$\qquad \mathbf{V}^{(i)} \leftarrow$ SOBOLMAT($c_{vec}, (m_{mat}(i,1), \ldots, m_{mat}(i,q)), r_{max}$)
$b_{pwrs} \leftarrow (1/2, 1/4, \ldots, 1/2^{r_{max}})$
for $i = 1, \ldots, d$ [Calculate point $n_0 - 1$ using Gray code]
$\qquad$ for $m = 1, \ldots, r$
$\qquad\qquad$ for $n = 1, \ldots, r$
$\qquad\qquad\qquad \mathbf{y}(m,i) \leftarrow \mathbf{y}(m,i) + \mathbf{V}^{(i)}(m,n) * \mathbf{g}(r - n + 1) \bmod 2$
for $k = n_0, \ldots, n_{max}$
$\qquad$ if $(k = q_{next})$
$\qquad\qquad r \leftarrow r + 1$
$\qquad\qquad \mathbf{g} \leftarrow (1, \mathbf{g})$ [insert 1 in Gray code at powers of 2]
$\qquad\qquad \ell \leftarrow 1$ [first bit changed]
$\qquad\qquad q_{next} \leftarrow 2 * q_{next}$
$\qquad$ else
$\qquad\qquad \ell \leftarrow$ index of rightmost zero in $\mathbf{a}$
$\qquad\qquad \mathbf{g}(\ell) \leftarrow 1 - \mathbf{g}(\ell)$ [increment Gray code]
$\qquad \mathbf{a} \leftarrow$ NEXTB-ARY($\mathbf{a}, 2$)
$\qquad$ for $i = 1, \ldots, d$ [Calculate point k recursively]
$\qquad\qquad$ for $m = 1, \ldots, r$
$\qquad\qquad\qquad \mathbf{y}(m,i) \leftarrow \mathbf{y}(m,i) + \mathbf{V}^{(i)}(m, r - \ell + 1) \bmod 2$
$\qquad\qquad\qquad$ for $j = 1, \ldots, r$
$\qquad\qquad\qquad\qquad P(k - n_0 + 1, i) \leftarrow P(k - n_0 + 1, i) + b_{pwrs}(j) * \mathbf{y}(j, i)$
return P

Fig. 5.13. Function SOBOLPTS($n_0, n_{pts}, d, p_{vec}, m_{mat}$) returns $n_{pts} \times d$ array whose rows are coordinates of d-dimensional Sobol' points, using polynomials encoded in p_{vec} and initializing values in the rows of m_{mat}.

identical" to one for generating Sobol' points, differing only in the choice of generator matrices. Their numerical experiments indicate roughly the same accuracy using Niederreiter and Sobol' points on a set of test integrals.

Tezuka [347] introduces a counterpart of the radical inverse function with respect to polynomial arithmetic. This naturally leads to a generalization of Halton sequences; Tezuka also extends Niederreiter's digital construction to this setting and calls the resulting points *generalized Niederreiter sequences*.

Tezuka and Tokuyama [349] construct $(0, d)$-sequences in this setting using generator matrices for which they give an explicit expression that generalizes the expression in (5.19) for Faure generator matrices. Tezuka [348] notes that these generator matrices have the form

$$\mathbf{A}^{(i)}(\mathbf{C}^{(1)})^{i-1}, \quad i = 1, \ldots, d, \tag{5.27}$$

with $\mathbf{C}^{(1)}$ as in (5.19), and $\mathbf{A}^{(i)}$ arbitrary nonsingular (mod b) lower triangular matrices. The method of Tezuka and Tokuyama [349] is equivalent to taking $\mathbf{A}^{(i)}$ to be the transpose of $(\mathbf{C}^{(1)})^{i-1}$. Tezuka [348] shows that all sequences constructed using generator matrices of the form (5.27) in a prime base $b \geq d$ are $(0, d)$-sequences. He calls these *generalized Faure sequences*; they are a special case of his generalized Niederreiter sequences and they include ordinary Faure sequences (take each $\mathbf{A}^{(i)}$ to be the identity matrix). Although the path leading to (5.27) is quite involved, the construction itself requires only minor modification of an algorithm to generate Faure points.

Faure [117] proposes an alternative method for choosing generator matrices to construct $(0, d)$-sequences and shows that these do not have the form in (5.27).

A series of theoretical breakthroughs in the construction of low-discrepancy sequences have been achieved by Niederreiter and Xing using ideas from algebraic geometry; these are reviewed in their survey article [282]. Their methods lead to (t, d)-sequences with theoretically optimal t parameters. Pirsic [298] provides a software implementation and some numerical tests. Further numerical experiments are reported in Hong and Hickernell [187].

5.3 Lattice Rules

The constructions in Sections 5.2.1–5.2.3 are all based on extending the Van der Corput sequence to multiple dimensions. The *lattice rules* discussed in this section provide a different mechanism for constructing low-discrepancy point sets. Some of the underlying theory of lattice methods suggests that they are particularly well suited to smooth integrands, but they are applicable to essentially any integrand.

Lattice methods primarily define fixed-size point sets, rather than infinite sequences. This is a shortcoming when the number of points required to achieve a satisfactory precision is not known in advance. We discuss a mechanism for extending lattice rules after considering the simpler setting of a fixed number of points.

A *rank-1* lattice rule of n points in dimension d is a set of the form

$$\left\{ \frac{k}{n}\mathbf{v} \bmod 1, \quad k = 0, 1, \ldots, n - 1 \right\}, \tag{5.28}$$

with $\mathbf{v}$ a d-vector of integers. Taking the remainder modulo 1 means taking the fractional part of a number ($x \bmod 1 = x - \lfloor x \rfloor$), and the operation is applied separately to each coordinate of the vector. To ensure that this set does indeed contain n distinct points (i.e., that no points are repeated), we require that n and the components of $\mathbf{v}$ have 1 as their greatest common divisor.

An n-point lattice rule of rank r takes the form

$$\left\{ \sum_{i=1}^{r} \frac{k_i}{n_i} \mathbf{v}_i \bmod 1, \quad k_i = 0, 1, \ldots, n_i - 1, \quad i = 1, \ldots, r \right\},$$

for linearly independent integer vectors $\mathbf{v}_1, \ldots, \mathbf{v}_r$ and integers $n_1, \ldots, n_r \geq 2$ with each n_i dividing n_{i+1}, $i = 1, \ldots, r - 1$, and $n_1 \cdots n_r = n$. As in the rank-1 case, we require that n_i and the elements of $\mathbf{v}_i$ have 1 as their greatest common divisor.

Among rank-1 lattices, a particularly simple and convenient class are the *Korobov rules*, which have a generating vector $\mathbf{v}$ of the form $(1, a, a^2, \ldots, a^{d-1})$, for some integer a. In this case, (5.28) can be described as follows: for each $k = 0, 1, \ldots, n - 1$, set $y_0 = k$, $u_0 = k/n$,

$$y_i = a y_{i-1} \bmod n, \quad i = 1, \ldots, d - 1, \quad u_i = y_i/n,$$

and set $x_k = (u_0, u_1, \ldots, u_{d-1})$. Comparison with Section 2.1.2 reveals that this is the set of vectors formed by taking d consecutive outputs from a multiplicative congruential generator, from all initial seeds y_0.

It is curious that the same mechanism used in Chapter 2 to mimic randomness is here used to try to produce low discrepancy. The apparent paradox is resolved by noting that here we intend to use the full period of the generator (we choose the modulus n equal to the number of points to be generated), whereas the algorithms of Section 2.1 are designed so that we use a small fraction of the period. To reconcile the two applications, we would like the discrepancy of the first N out of n points to be $O(1/\sqrt{N})$ for small N and $O((\log N)^d/N)$ for large N.

The connection between Korobov rules and multiplicative congruential generators has useful consequences. It simplifies implementation and it facilitates the selection of generating vectors by making relevant the extensively studied properties of random number generators; see Hellekalek [176], Hickernell, Hong, L'Ecuyer, and Lemieux [184], L'Ecuyer and Lemieux [226], Lemieux and L'Ecuyer [232], and Niederreiter [281]. Hickernell [182], Chapter 5 of Niederreiter [281], and Sloan and Joe [333] analyze the discrepancy of lattice rules and other measures of their quality.

Tables of good generating vectors $\mathbf{v}$ can be found in Fang and Wang [115] for up to 18 dimensions. Sloan and Joe [333] give tables for higher-rank lattice rules in up to 12 dimensions. L'Ecuyer and Lemieux [226] provide tables of multipliers a for Korobov rules passing tests of uniformity.

Integration Error

The Koksma-Hlawka bound (5.10) applies to lattice rules as it does to all point sets. But the special structure of lattice rules leads to a more explicit expression for the integration error using such a rule, and this in turn sheds light on both the design and scope of these methods.

Fix an integrand f on $[0,1]^d$, and for each d-vector of integers z define the Fourier coefficient

$$\hat{f}(z) = \int_{[0,1)^d} f(x)\, e^{-2\pi\sqrt{-1}x^\top z}\, dx.$$

The integral of f over the hypercube is $\hat{f}(0)$. Suppose that f is sufficiently regular to be represented by its Fourier series, in the sense that

$$f(x) = \sum_z \hat{f}(z) e^{2\pi\sqrt{-1}x^\top z}, \tag{5.29}$$

the sum ranging over all integer vectors z and converging absolutely.

A rank-1 lattice rule approximation to the integral of f is

$$\frac{1}{n}\sum_{k=0}^{n-1} f\left(\frac{k}{n}\mathbf{v} \bmod 1\right) = \frac{1}{n}\sum_{k=0}^{n-1}\sum_z \hat{f}(z)\exp\left(2\pi\sqrt{-1}k\mathbf{v}^\top z/n\right)$$

$$= \sum_z \hat{f}(z)\frac{1}{n}\sum_{k=0}^{n-1}\left[\exp\left(2\pi\sqrt{-1}\mathbf{v}^\top z/n\right)\right]^k. \tag{5.30}$$

The first equality follows from the Fourier representation of f and the periodicity of the function $u \mapsto \exp(2\pi\sqrt{-1}u)$, which allows us to omit the reduction modulo 1. For the second equality, the interchange in the order of summation is justified by the assumed absolute convergence of the Fourier series for f. Now the average over k in (5.30) simplifies to

$$\frac{1}{n}\sum_{k=0}^{n-1}\left[\exp\left(2\pi\sqrt{-1}\mathbf{v}^\top z/n\right)\right]^k = \begin{cases} 1, & \text{if } \mathbf{v}^\top z = 0 \bmod n, \\ 0, & \text{otherwise.} \end{cases}$$

To see why, observe that if $\mathbf{v}^\top z/n$ is an integer then each of the summands on the left is just 1; otherwise,

$$\sum_{k=0}^{n-1}\left[\exp\left(2\pi\sqrt{-1}\mathbf{v}^\top z/n\right)\right]^k = \frac{1 - \exp(2\pi\sqrt{-1}\mathbf{v}^\top z)}{1 - \exp(2\pi\sqrt{-1}\mathbf{v}^\top z/n)} = 0$$

because $\mathbf{v}^\top z$ is an integer. Using this in (5.30), we find that the lattice rule approximation simplifies to the sum of $\hat{f}(z)$ over all integer vectors z for which $\mathbf{v}^\top z = 0 \bmod n$. The correct value of the integral is $\hat{f}(0)$, so the error in the approximation is

$$\sum_{z \neq 0, \mathbf{v}^\top z = 0 \ \mathrm{mod} \ n} \hat{f}(z). \tag{5.31}$$

The values of $|\hat{f}(z)| = |\hat{f}(z_1, \ldots, z_d)|$ for large values of $|z| = |z_1| + \cdots + |z_d|$ reflect the smoothness of f, in the sense that large values of $|z|$ correspond to high-frequency oscillation terms in the Fourier representation of f. The expression in (5.31) for the integration error thus suggests the following:

(i) the generator $\mathbf{v}$ should be chosen so that $\mathbf{v}^\top z = 0 \ \mathrm{mod} \ n$ only if $|z|$ is large — vectors $\mathbf{v}$ with this property are known informally as *good lattice points*;

(ii) lattice rules are particularly well suited to integrands f that are smooth precisely in the sense that $\hat{f}(z)$ decreases quickly as $|z|$ increases.

The first of these observations helps guide the search for effective choices of $\mathbf{v}$. (Results showing the *existence* of good lattice points are detailed in Chapter 5 of Niederreiter [281].) Precise criteria for selecting $\mathbf{v}$ are related to the spectral test mentioned in the discussion of random number generators in Section 2.1. Recommended values are tabulated in Fang and Wang [115].

The direct applicability of observation (ii) seems limited, at least for the integrands implicit in derivative pricing. Bounding the Fourier coefficients of such functions is difficult, and there is little reason to expect these functions to be smooth. Moreover, the derivation leading to (5.31) obscures an important restriction: for the Fourier series to converge absolutely, f must be continuous on $[0, 1]^d$ and periodic at the boundaries (because absolute convergence makes f the uniform limit of functions with these properties). Sloan and Joe [333] advise caution in applying lattice rules to nonperiodic integrands; in their numerical results, they find Monte Carlo to be the best method for discontinuous integrands.

Extensible Lattice Rules

We conclude our discussion of lattice rules with a method of Hickernell et al. [184] for extending fixed-size lattice rules to infinite sequences.

Consider a rank-1 lattice rule with generating vector $\mathbf{v} = (v_1, \ldots, v_d)$. Suppose the number of points n equals b^r for some base b and integer r. Then the segment $\psi_b(0), \psi_b(1), \ldots, \psi_b(n-1)$ of the Van der Corput sequence in base b is a permutation of the coefficients k/n, $k = 0, 1, \ldots, n - 1$, appearing in (5.28). The point set is therefore unchanged if we represent it as

$$\{\psi_b(k)\mathbf{v} \ \mathrm{mod} \ 1, \ k = 0, 1, \ldots, n - 1\}.$$

We may now drop the upper limit on k to produce an infinite sequence.

The first b^r points in this sequence are the original lattice point set. Each of the next $(b-1)$ nonoverlapping segments of length b^r will be shifted versions of the original lattice. The first b^{r+1} points will again form a lattice rule of the type (5.28), but now with n replaced by b^{r+1}, and so on. In this way, the construction extends and refines the original lattice.

Hickernell et al. [184] give particular attention to extensible Korobov rules, which are determined by the single parameter a. They provide a table of values of this parameter that exhibit good uniformity properties when extended using $b = 2$. Their numerical results use $a = 17797$ and $a = 1267$.

5.4 Randomized QMC

We began this chapter with the suggestion that choosing points deterministically rather than randomly can reduce integration error. It may therefore seem odd to consider randomizing points chosen carefully for this purpose. There are, however, at least two good reasons for randomizing QMC.

The first reason is as immediately applicable as it is evident: by randomizing QMC points we open the possibility of measuring error through a confidence interval while preserving much of the accuracy of pure QMC. Randomized QMC thus seeks to combine the best features of ordinary Monte Carlo and quasi-Monte Carlo. The tradeoff it poses — sacrificing some precision to get a better measure of error — is essentially the same one we faced with several of the variance reduction techniques of Chapter 4.

The second reason to consider randomizing QMC is less evident and may also be less practically relevant: there are settings in which randomization actually improves accuracy. A particularly remarkable result of this type is a theorem of Owen [289] showing that the root mean square error of integration using a class of randomized nets is $O(1/n^{1.5-\epsilon})$, whereas the error without randomization is $O(1/n^{1-\epsilon})$. Owen's result applies to smooth integrands and may therefore be of limited applicability to pricing derivatives; nevertheless, it is notable that randomization takes advantage of the additional smoothness though QMC does not appear to. Hickernell [180], Matoušek [257], and L'Ecuyer and Lemieux [226] discuss other ways in which randomization can improve accuracy.

We describe four methods for randomizing QMC. For a more extensive treatment of the topic, see the survey article of L'Ecuyer and Lemieux [226] and Chapter 14 of Fox [127]. We limit the discussion to randomization of point sets of a fixed size n. We denote such a point set generically by

$$P_n = \{x_1, \ldots, x_n\},$$

each x_i an element of $[0, 1)^d$.

Random Shift

The simplest randomization of the point set P_n generates a random vector U uniformly distributed over the d-dimensional unit hypercube and shifts each point in P_n by U, modulo 1:

$$P_n(U) = \{x_i + U \bmod 1, i = 1, \ldots, n\}. \tag{5.32}$$

The reduction mod 1 applies separately to each coordinate. The randomized QMC estimate of the integral of f is

$$I_f(U) = \frac{1}{n} \sum_{i=1}^{n} f(x_i + U \bmod 1).$$

This mechanism was proposed by Cranley and Patterson [92] in the setting of a lattice rule, but can be applied with other low-discrepancy point sets. It should be noted, however, that the transformation changes the discrepancy of a point set and that a shifted (t, m, d)-net need not be a (t, m, d)-net.

For any $P_n \subseteq [0, 1)^d$, each element of $P_n(U)$ is uniformly distributed over the hypercube, though the points are clearly not independent. Repeating the randomization with independent replications of U produces independent batches of n points each. Each batch yields a QMC estimate of the form $I_f(U)$, and these estimates are independent and identically distributed. Moreover, each $I_f(U)$ is an unbiased estimate of the integral f, so computing an asymptotically valid confidence interval for the integral is straightforward.

L'Ecuyer and Lemieux [226] compare the variance of $I_f(U)$ and an ordinary Monte Carlo estimate with P_n a lattice rule. They show that either variance could be smaller, depending on the integrand f, but argue that $I_f(U)$ often has smaller variance in problems of practical interest.

The random-shift procedure may be viewed as an extreme form of systematic sampling (discussed in Section 4.2), in which a single point U is chosen randomly and n points are then chosen deterministically conditional on U. The variance calculation of L'Ecuyer and Lemieux [226] for a randomly shifted lattice rule has features in common with the calculation for systematic sampling in Section 4.2.

Random Permutation of Digits

Another mechanism for randomizing QMC applies a random permutation of $0, 1, \ldots, b - 1$ to the coefficients in the base-b expansion of the coordinates of each point. Consider, first, the one-dimensional case and write $x_k = 0.a_1(k)a_2(k) \ldots$ for a b-ary representation of x_k. Let π_j, $j = 1, 2, \ldots$ be independent random permutations of $\{0, 1, \ldots, b - 1\}$, uniformly distributed over all $b!$ permutations of the set. Randomize P_n by mapping each point x_k to the point $0.\pi_1(a_1(k))\pi_2(a_2(k)) \ldots$, applying the same permutations π_j to all points x_k. For a d-dimensional point set, randomize each coordinate in this way, using independent permutations for different coordinates.

Randomizing an arbitrary point $x \in [0, 1)^d$ in this way produces a random vector uniformly distributed over $[0, 1)^d$. Thus, for any P_n, the average of f over the randomization of P_n is an unbiased estimate of the integral of f. Independent randomizations produce independent estimates that can be combined to estimate a confidence interval.

Matoušek [257] analyzes the expected mean-square discrepancy for a general class of randomization procedures that includes this one. This randomization maps b-ary boxes to b-ary boxes of the same volume, so if P_n is a (t, m, d)-net, its randomization is too.

Scrambled Nets

Owen [287, 288, 289] introduces and analyzes a randomization mechanism that uses a hierarchy of permutations. This *scrambling* procedure permutes each digit of a b-ary expansion, but the permutation applied to the jth digit depends on the first $j - 1$ digits.

To make this more explicit, first consider the one-dimensional case. Suppose x has b-ary representation $0.a_1 a_2 a_3 \ldots$. The first coefficient a_1 is mapped to $\pi(a_1)$, with π a random permutation of $\{0, 1, \ldots, b - 1\}$. The second coefficient is mapped to $\pi_{a_1}(a_2)$, the third coefficient to $\pi_{a_1 a_2}(a_3)$, and so on; the random permutations $\pi, \pi_{a_1}, \pi_{a_1 a_2}, \ldots, a_j = 0, 1, \ldots, b - 1, j = 1, 2, \ldots$, are independent with each uniformly distributed over the set of all permutations of $\{0, 1, \ldots, b - 1\}$. To scramble a d-dimensional point set, apply this procedure to each coordinate, using independent sets of permutations for each coordinate.

Owen [290] describes scrambling as follows. In each coordinate, partition the unit interval into b subintervals of length $1/b$ and randomly permute those subintervals. Further partition each subinterval into b subintervals of length $1/b^2$ and permute those, randomly and independently, and so on. At the jth step, this procedure constructs b^{j-1} partitions, each consisting of b intervals, and permutes each partition independently. In contrast, Matoušek's random digit permutation applies the same permutation to all b^{j-1} partitions at each step j.

Owen [287] shows that a scrambled (t, m, d)-net is a (t, m, d)-net with probability one, and a scrambled (t, d)-sequence is a (t, d)-sequence with probability one. Owen [288, 289, 290] shows that the variance of a scrambled net estimator converges to zero faster than the variance of an ordinary Monte Carlo estimator does, while cautioning that the faster rate may not set in until the number of points becomes very large. For sufficiently smooth integrands, the variance is $O(1/n^{3-\epsilon})$ in the sample size n. The superior asymptotic performance with randomization results from cancellation of error terms. For fixed sample sizes, Owen [288, 290] bounds the amount by which the scrambled net variance can exceed the Monte Carlo variance. Hickernell and Hong [183] analyze the mean square discrepancy of scrambled nets.

Realizing the attractive features of scrambled nets in practice is not entirely straightforward because of the large number of permutations required for scrambling. Tan and Boyle [345] propose an approximate scrambling method based on permuting just the first few digits and find experimentally that it works well. Matoušek [257] outlines an implementation of full scrambling that reduces memory requirements at the expense of increasing computing time: rather than store a permutation, he stores the state of the random number generator and regenerates each permutation when it is needed. Hong and Hickernell [187] define a simplified form of scrambling and provide algorithms that generate scrambled points in about twice the time required for unscrambled points.

Linear Permutation of Digits

As an alternative to full scrambling, Matoušek [257] proposes a "linear" permutation method. This method maps a base-b expansion $0.a_1a_2\dots$ to $0.\tilde{a}_1\tilde{a}_2\dots$ using

$$\tilde{a}_j = \sum_{i=1}^{j} h_{ij}a_i + g_j \bmod b,$$

with the h_{ij} and g_j chosen randomly and independently from $\{0, 1, \dots, b-1\}$ and the h_{ii} required to be positive. This method is clearly easier to implement than full scrambling. Indeed, if the g_j were all 0, this would reduce to the generalized Faure method in (5.27) when applied to a Faure net P_n. The condition that the diagonal entries h_{ii} be positive ensures the nonsingularity required in (5.27).

All of the randomization methods described in this section produce points uniformly distributed over $[0, 1)^d$ and thus unbiased estimators of integrals over $[0, 1)^d$ when applied in the QMC approximation (5.2). Through independent replications of any of these it is a simple matter to construct asymptotically valid confidence intervals. The methods vary in evident ways in their computational requirements; the relative merits of the estimates they produce are less evident and warrant further investigation.

5.5 The Finance Setting

Our discussion of quasi-Monte Carlo has thus far been fairly abstract, dealing with the generic problem of numerical integration over $[0, 1)^d$. In this section, we deal more specifically with the application of QMC to the pricing of derivative securities. Section 5.5.1 discusses numerical results comparing QMC methods and ordinary Monte Carlo on some test problems. Section 5.5.2 discusses ways of taking advantage of the structure of financial models to enhance the effectiveness of QMC methods.

5.5.1 Numerical Examples

Several articles have reported numerical results obtained by applying QMC methods to financial problems. These include Acworth et al. [4], Berman [45],

Boyle et al. [53], Birge [47], Caflisch, Morokoff, and Owen [73], Joy, Boyle, and Tan [204], Ninomiya and Tezuka [283], Papageorgiou and Traub [293], Paskov [295], Paskov and Traub [296], Ross [309], and Tan and Boyle [345]. These investigations consider several different QMC methods applied to various pricing problems and find that they work well. We comment more generally on the numerical evidence after considering some examples.

A convenient set of problems for testing QMC methods are options on geometric averages of lognormally distributed asset prices. These options are tractable in arbitrarily high dimensions (and knowing the correct answer is useful in judging performance of numerical methods) while sharing features of more challenging multiple-asset and path-dependent pricing problems. We consider, then, options with payoffs $(\bar{S} - K)^+$ where either

$$\bar{S} = \prod_{i=1}^{d} S_i(T)^{1/d} \tag{5.33}$$

for multiple assets $S_1, \ldots, S_d$, or

$$\bar{S} = \prod_{i=1}^{d} S(iT/d)^{1/d}, \tag{5.34}$$

for a single asset S. The underlying assets $S_1, \ldots, S_d$ or S are modeled as geometric Brownian motion. Because $\bar{S}$ is lognormally distributed in both cases, the option price is given by a minor modification of the Black-Scholes formula, as noted in Section 3.2.2.

The two cases (5.33) and (5.34) reflect two potential sources of high dimensionality in financial problems: d is the number of underlying assets in (5.33) and it is the number of time steps in (5.34). Of course, in both cases $\bar{S}$ is the geometric average of (jointly) lognormal random variables so this distinction is purely a matter of interpretation. The real distinction is the correlation structure among the averaged random variables. In (5.34), the correlation is determined by the dynamics of geometric Brownian motion and is rather high; in (5.33), we are free to choose any correlation matrix for the (logarithms of the) d assets. Choosing a high degree of correlation would be similar to reducing the dimension of the problem; to contrast with (5.34), in (5.33) we choose the d assets to be independent of each other.

A comparison of methods requires a figure of merit. For Monte Carlo methods, variance is an appropriate figure of merit — at least for unbiased estimators with similar computing requirements, as argued in Section 1.1.3. The average of n independent replications has a variance exactly n times smaller than the variance of a single replication, so a comparison of variances is not tied to a particular sample size. In contrast, the integration error produced by a QMC method does depend on the number of points n, and often quite erratically. Moreover, the QMC error can be quite sensitive to problem parameters. This makes the comparison of QMC methods less straightforward.

As our figure of merit, we take the root mean square error or root mean square relative error over a fixed set of problem instances. This is somewhat arbitrary (especially in the choice of instances) but nevertheless informative. Given m problems with true values $C_1, \ldots, C_m$ and n-point QMC approximations $\hat{C}_1(n), \ldots, \hat{C}_m(n)$, the root mean square error is

$$\text{RMSE}(n) = \sqrt{\frac{1}{m} \sum_{i=1}^{m} (\hat{C}_i(n) - C_i)^2}$$

and the RMS relative error is

$$\sqrt{\frac{1}{m} \sum_{i=1}^{m} \left(\frac{\hat{C}_i(n) - C_i}{C_i} \right)^2}.$$

In order to compare QMC methods with Monte Carlo, we extend these definitions to random estimators $\hat{C}_i(n)$ by replacing $(\hat{C}_i(n) - C_i)^2$ with $E[(\hat{C}_i(n) - C_i)^2]$ in both cases.

Our first example is based on (5.33) with $d = 5$ assets; as this is a relatively low-dimensional problem, it should be particularly well suited to QMC methods. For simplicity, we take the five assets to be independent copies of the same process $\text{GBM}(r, \sigma^2)$ with an initial value of $S_i(0) = 100$. We fix r at 5%, and construct 500 problem instances through all combinations of the following parameters: the maturity T is 0.15, 0.25, 0.5, 1, or 2 years; the volatility σ varies from 0.21 to 0.66 in increments of 0.05; and the strike K varies from 94 to 103 in increments of 1. These 500 options range in price from 0.54 to 12.57; their average value is 5.62, and half lie between 4.06 and 7.02.

Figure 5.14 plots the RMSE against the number of points, using a log scale for both axes. For the QMC methods, the figure displays the exact number of points used. For the Sobol' points, we skipped the first 256 points and then chose the number of points to be powers of two. For the Faure points, we skipped the first 625 $(= 5^4)$ points and then chose the number of points to be powers of five (the base). These choices are favorable for each method. For the lattice rules the number of points is fixed. We used the following generating vectors from p.287 of Fang and Wang [115]:

n	$\mathbf{v}$				
1069	1,	63,	762,	970,	177
4001	1,	1534,	568,	3095,	2544
15019	1,	10641,	2640,	6710,	784
71053	1,	33755,	65170,	12740,	6878

These we implemented using the shifted points $k(\mathbf{v} - 0.5)/n \pmod 1$ as suggested in Fang and Wang [115], rather than (5.28). We also tested Korobov rules from L'Ecuyer and Lemieux [226] generated by $a = 331, 219, 1716, 7151, 665$, and 5693; these gave rather poor and erratic results and are therefore omitted from the figure. Using Monte Carlo, the RMSE scales exactly with $\sqrt{n}$, so we estimated it at $n = 64000$ and then extended this value to other values of n.

Figure 5.14 suggests several observations. The QMC methods produce root mean square errors three to ten times smaller than those of Monte Carlo over the range of sample sizes considered. Faure points appear to outperform the lattice rules and Sobol' points outperform the Faure points. In addition to producing smaller errors, the QMC methods appear to converge at a faster rate than Monte Carlo: their graphs are not only lower, they have a steeper slope. For Sobol' and Faure points, a convergence rate close to $O(1/n)$ (evidenced by a slope close to -1) sets in after a few thousand points. The slope for Monte Carlo is exactly $-1/2$ by construction.

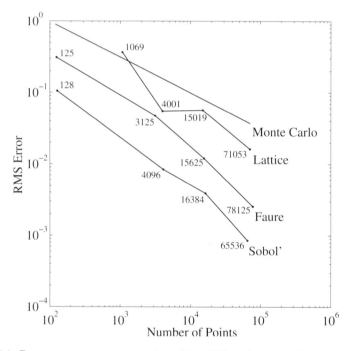

Fig. 5.14. Root mean square errors in pricing 500 options on the geometric mean of five underlying assets.

The relative smoothness of the convergence of the Faure and Sobol' approximations relies critically on our choice of favorable values of n for each method. For example, taking $n = 9000$ produces larger RMS errors than $n = 3125$ for the Faure sequence and larger than $n = 4096$ for the Sobol' sequence. The various points plotted for lattice rules are unrelated to each other because each value of n uses a different generating vector, whereas the Sobol' and Faure results use initial segments of infinite sequences.

Figures 5.15 and 5.16 examine the effect of increasing problem dimension while keeping the number of points nearly fixed. Figure 5.15 is based on (5.33) with $d = 10, 30, 40, 70, 100$, and 150. For this comparison we held the maturity T fixed at 0.25, let the strike K range from 94 to 102 in increments of 2, and let σ vary as before. Thus, we have fifty options for each value of d.

Because increasing d in the geometric mean (5.33) has the effect of reducing the volatility of $\bar{S}$, the average option price decreases with d, dropping from 3.46 at $d = 10$ to 1.85 at $d = 150$. Root mean square errors also decline, so to make the comparison more meaningful we look at relative errors. These increase with d for all three methods considered in Figure 5.15. The Monte Carlo results are estimated RMS relative errors for a sample size of 5000, but estimated from 64,000 replications. The Sobol' sequence results in all dimensions skip 4096 points and use $n = 5120$; this is $2^{12} + 2^{10}$ and should be favorable for a base-2 construction. For the Faure sequence, the base changes with dimension. For each d we chose a value of n near 5000 that should be favorable for the corresponding base: these values are $4 \cdot 11^3$, $5 \cdot 31^2 + 7 \cdot 31$, $3 \cdot 41^2$, 71^2, $50 \cdot 101$, and $34 \cdot 151$. In each case, we skipped the first b^4 points, with b the base. The figure suggests that the advantage of the QMC methods relative to Monte Carlo declines with increasing dimension but is still evident at $d = 150$.

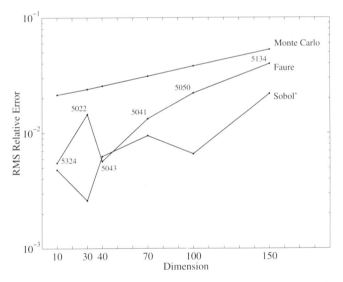

Fig. 5.15. Root mean square relative error in pricing options on the geometric average of d assets, with d the dimension.

The comparison in Figure 5.16 is similar but uses (5.34), so d now indexes the number of averaging dates along the path of a single asset. For this comparison we fixed T at 0.25, we let K vary from 96 to 104 in increments of 2, and let σ vary as before, to produce a total of fifty options. Each option price approaches a limit as d increases (the price associated with the continuous average), and the Monte Carlo RMSE is nearly constant across dimensions. The errors using Faure points show a sharp increase at $d = 100$ and $d = 150$. The errors using Sobol' points show a much less severe dependence on dimension. The number of points used for all three methods are the same in Figure 5.16 as Figure 5.15.

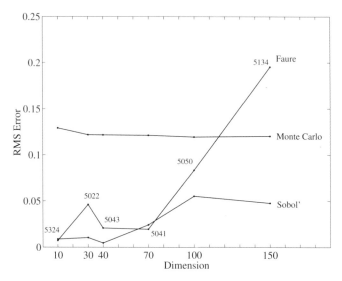

Fig. 5.16. Root mean square error in pricing options on the geometric time-average of d values of a single asset, with d the dimension.

Without experimentation, it is difficult to know how many QMC points to use to achieve a desired accuracy. In ordinary Monte Carlo, one can use a standard error estimated from a modest number of replications to determine the number of replications to undertake in a second stage of sampling to reach a target precision. Some authors have proposed stopping rules for QMC based on monitoring fluctuations in the approximation — rules that stop once the fluctuations are smaller than the required error tolerance. But such procedures are risky, as illustrated in Figure 5.17. The figure plots the running average of the estimated price of an option on the geometric average of 30 assets (with $T = 0.25$, $\sigma = 0.45$, and $K = 100$) using Faure points. An automatic stopping rule would likely detect convergence — erroneously — near 6000 points or 13000 points where the average plateaus. But in both cases, the QMC approximation remains far from the true value, which is not crossed until after more than 19000 points. These results use Faure points from the start of the sequence (in base 31); skipping an initial portion of the sequence would reduce the severity of this problem but would not eliminate it.

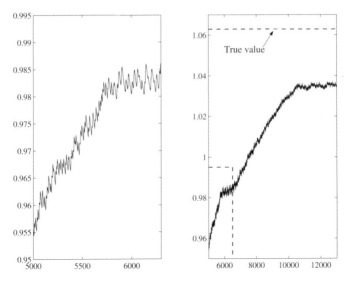

Fig. 5.17. Cumulative average approximation to a 30-dimensional option price using Faure points. The left panel magnifies the inset in the right panel. The approximation approaches the true value through plateaus that create the appearance of convergence.

Next we compare randomized QMC point sets using a random shift modulo 1 as in (5.32). For this comparison we consider a single option — a call on the geometric average of five assets, with $T = 0.25$, $K = 100$, and $\sigma = 0.45$. Because of the randomization, we can now compare methods based on their variances; these are displayed in Table 5.4. To compensate for differences in the cardinalities of the point sets, we report a product $n\sigma^2$, where n is the number of points in the set and σ^2 is the variance of the average value of the integrand over a randomly shifted copy of the point set. This measure makes the performance of ordinary Monte Carlo independent of the choice of n.

For the Faure and Sobol' results, we generated each point set of size n by starting at the nth point in the sequence; each n is a power of the corresponding base. The lattice rules are the same as those used for Figure 5.14. For the Korobov rules we display the number of points and the multiplier a; these values are from L'Ecuyer and Lemieux [226].

All the QMC methods show far smaller variance than ordinary Monte Carlo. Sobol' points generally appear to produce the smallest variance, but the smallest variance overall corresponds to a lattice rule. The Korobov rules have larger variances than the other methods.

The numerical examples considered here suggest some general patterns: the QMC methods produce substantially more precise values than ordinary Monte Carlo; this holds even at rather small values of n, before $O(1/n^{1-\epsilon})$

Lattice		Korobov			Faure		Sobol'		Monte Carlo
n	$n\sigma^2$	n	(a)	$n\sigma^2$	n	$n\sigma^2$	n	$n\sigma^2$	$n\sigma^2$
					125	11.9	128	5.9	34.3
1069	2.7	1021	(331)	2.7	3125	1.1	1024	2.0	34.3
4001	0.6	4093	(219)	1.5			4096	0.9	34.3
15019	0.3	16381	(665)	3.5	15625	0.4	16384	0.4	34.3

Table 5.4. Variance comparison for randomly shifted QMC methods and Monte Carlo.

convergence is evident; Sobol' points generally produce smaller errors than Faure points or lattice rules; the advantages of QMC persist even in rather high dimensions, especially for Sobol' points; randomized QMC point sets produce low-variance estimates.

The effectiveness of QMC methods in high-dimensional pricing problems runs counter to the traditional view that these methods are unsuitable in high dimensions. The traditional view is rooted in the convergence rate of $O((\log n)^d/n)$: if d is large then n must be *very* large for the denominator to overwhelm the numerator. The explanation for this apparent contradiction may lie in the structure of problems arising in finance — these high-dimensional integrals might be well-approximated by much lower-dimensional integrals, a possibility we exploit in Section 5.5.2.

Extrapolating from a limited set of examples (we have considered just one type of option and just one model of asset price dynamics) is risky, so we comment on results from other investigations. Acworth et al. [4] and Boyle et al. [53] find that Sobol' points outperform Faure points and that both outperform ordinary Monte Carlo in comparisons similar to those reported here. Morland [270] reports getting better results with Sobol' points than Niederreiter points (of the type generated in Bratley, Fox, and Niederreiter [58]). Joy et al. [204] test Faure sequences on several different types of options, including an HJM swaption pricing application, and find that they work well. Berman [45] compares methods on a broad range of options and models; he finds that Sobol' points give more precise results than ordinary Monte Carlo, but he also finds that with some simple variance reduction techniques the two methods perform very similarly. Paskov [295], Paskov and Traub [296], and Caflisch et al. [73] find that Sobol' points work well in pricing mortgage-backed securities formulated as 360-dimensional integrals. Papageorgiou and Traub [293] report improved results using a generalized Faure sequence and Ninomiya and Tezuka [283] report superior results for similar problems using a generalized Niederreiter sequence, but neither specifies the exact construction used.

For the most part, these comparisons (like those presented here) pit QMC methods against only the simplest form of Monte Carlo. Variance reduction techniques can of course improve the precision of Monte Carlo estimates; they provide a mechanism for taking advantage of special features of a model to a much greater extent than QMC. Indeed, the "black-box" nature of QMC methods is part of their appeal. As discussed in Section 5.1.3, the ready availability of error information through confidence intervals is an advantage of Monte Carlo methods.

For calculations that need to be repeated often with only minor changes in parameters — for example, options that need to be priced every day — this suggests the following approach: tailor a Monte Carlo method to the specific problem, using estimates of standard errors to compare algorithms, and determine the required sample size; once the problem and its solution are well understood, replace the random number generator with a quasi-Monte Carlo generator.

5.5.2 Strategic Implementation

QMC methods have the potential to improve accuracy for a wide range of integration problems without requiring an integrand-specific analysis. There are, however, two ways in which the application of QMC methods can be tailored to a specific problem to improve performance:

(i) changing the order in which coordinates of a sequence are assigned to arguments of an integrand;
(ii) applying a change of variables to produce a more tractable integrand.

The first of these transformations is actually a special case of the second but it merits separate consideration.

The strategy in (i) is relevant when some coordinates of a low-discrepancy sequence exhibit better uniformity properties than others. This holds for Halton sequences (in which coordinates with lower bases are preferable) and for Sobol' sequences (in which coordinates generated by lower-degree polynomials are preferable), but not for Faure sequences. As explained in Section 5.2.2, all coordinates of a Faure sequence are equally well distributed. But the more general strategy in (ii) is potentially applicable to all QMC methods.

As a simple illustration of (ii), consider the function on $[0, 1)^5$ defined by

$$f(u_1, u_2, u_3, u_4, u_5) = \mathbf{1}\{|\Phi^{-1}(u_4)| + |\Phi^{-1}(u_5)| \leq 2\sqrt{2}\},$$

with Φ^{-1} the inverse cumulative normal distribution. Although this reduces to a bivariate integrand, we have formulated it as a five-dimensional problem for purposes of illustration. The integral of this function is the probability that a pair of independent standard normal random variables fall in the square in $\Re^2$ with vertices $(0, \pm 2\sqrt{2})$ and $(\pm 2\sqrt{2}, 0)$, which is approximately 0.9111. Applying an orthogonal transformation to a pair of independent standard normal random variables produces another pair of independent standard normal random variables, so the same probability applies to the rotated square with vertices $(\pm 2, \pm 2)$. Thus, a change of variables transforms the integrand above to

$$\tilde{f}(u_1, u_2, u_3, u_4, u_5) = \mathbf{1}\{\max(|\Phi^{-1}(u_4)|, |\Phi^{-1}(u_5)|) \leq 2\}.$$

Figure 5.18 compares the convergence of QMC approximations to f (the dotted line) and $\tilde{f}$ (the solid line) using a five-dimensional Faure sequence starting at the 625th point. In this example, the rotation has an evident impact on the quality of the approximation: after 3125 points, the integration error for f is nearly four times as large as the error for $\tilde{f}$. That a rotation could affect convergence in this way is not surprising in view of the orientation displayed by Faure sequences, as in, e.g., Figure 5.3.

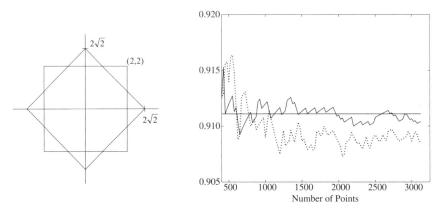

Fig. 5.18. Both squares on the left have probability 0.9111 under the bivariate standard normal distribution. The right panel shows the convergence of QMC approximations for the probabilities of the two squares. The solid horizontal line shows the exact value.

Assigning Coordinates

We proceed with an illustration of strategy (i) in which the form of the integrand is changed only through a permutation of its arguments. In the examples we considered in Section 5.5.1, the integrands are symmetric functions of their arguments because we took the underlying assets to be identical in (5.33) and we took the averaging dates to be equally spaced in (5.34). Changing the assignment of coordinates to variables would therefore have no effect on the value of a QMC approximation.

To break the symmetry of the multi-asset option in Section 5.5.1, we assign linearly increasing volatilities $\sigma_i = i\sigma_1$, $i = 1, \ldots, d$, to the d assets. We take the volatility of the ith asset as a rough measure of the importance of the ith coordinate (and continue to assume the assets are uncorrelated). With this interpretation, assigning the coordinates of a Sobol' sequence to the assets in reverse order should produce better results than assigning the ith coordinate to the ith asset.

To test this idea we take $d = 30$; the degrees of the primitive polynomials generating the coordinates then increase from 0 to 8. We compare straightforward application of Sobol' points with a reversed assignment of coordinates based on root mean square relative error. Given an average level of volatility $\bar{\sigma}$, we choose σ_1 so that $\bar{\sigma}^2 = (\sigma_1^2 + \cdots + \sigma_d^2)/d$, with $\sigma_i = i\sigma_1$. We let $\bar{\sigma}$ range from 0.21 to 0.66 in increments of 0.05, and we let K vary from 94 to 102 in increments of 2 to get the same fifty option values we used for Figure 5.15.

Table 5.5 displays the resulting RMS relative errors. The first row shows the number of points n. We specifically avoided values of n equal to powers of 2 in order to further differentiate the coordinates of the sequence; this makes the convergence of both methods erratic. In this example the reversed assignment usually produces smaller errors, but not always.

	750	1500	2500	3500	5000	7500	10000	12000
Sobol'	0.023	0.012	0.017	0.021	0.013	0.012	0.007	0.005
Reverse	0.020	0.021	0.010	0.015	0.009	0.007	0.005	0.003

Table 5.5. RMS relative errors for options on the geometric average of 30 assets with linearly increasing volatilities. Top row gives the number points. Second row is based on assigning ith coordinate to ith asset; last row uses reversed assignment.

Changing Variables

A general strategy for improving QMC approximations applies a change of variables to produce an integrand for which only a small number of arguments are "important" and then applies the lowest-indexed coordinates of a QMC sequence to those coordinates. Finding an effective transformation presents essentially the same challenge as finding good stratification variables, a topic treated in Section 4.3.2. As is the case in stratified sampling, the Gaussian setting offers particular flexibility.

In the application of QMC to derivatives pricing, the integrand f subsumes the dynamics of underlying assets as well as the form of the derivative contract. In the absence of specific information about the payoff of a derivative, one might consider transformations tied to the asset dynamics.

A simple yet effective example of this idea is the combination of Sobol' sequences with the Brownian bridge construction of Brownian motion developed in Section 3.1. In a straightforward application of Sobol' points to the generation of Brownian paths, the ith coordinate of each point would be transformed to a sample from the standard normal distribution (using Φ^{-1}), and these would be scaled and summed using the random walk construction (3.2). To the extent that the initial coordinates of a Sobol' sequence have uniformity superior to that of higher-indexed coordinates, this construction does a particularly good job of sampling the first few increments of the Brownian path.

However, many option contracts would be primarily sensitive to the *terminal* value of the Brownian path.

Through the Brownian bridge construction, the first coordinate of a Sobol' sequence determines the terminal value of the Brownian path, so this value should be particularly well distributed. Moreover, the first several coordinates of the Sobol' sequence determine the general shape of the Brownian path; the last few coordinates influence only the fine detail of the path, which is often less important. This combination of Sobol' points with the Brownian bridge construction was proposed by Moskowitz and Caflisch [273] and has been found by several authors (including Acworth et al. [4], Åkesson and Lehoczky [9], and Caflisch et al. [73]) to be highly effective in finance applications.

As discussed in Section 3.1, the principal components construction of a discrete Brownian path (or any other Gaussian vector) has an optimality property that maximizes the importance (in the statistical sense of explained variance) of any initial number of independent normals used to construct the vector. Though this property lacks a precise relation to discrepancy, it suggests a construction in which the ith coordinate of a Sobol' sequence is assigned to the ith principal component. Unlike the Brownian bridge construction, the principal components construction is applicable with any covariance matrix.

This construction was proposed and tested in Acworth et al. [4]. Tables 5.6 and 5.7 show some of their results. The tables report RMS relative errors comparing an ordinary application of Sobol' sequences with the Brownian bridge and principal components constructions. The errors are computed over 250 randomly generated problem instances as described in [4]. Table 5.6 reports results for barrier options and geometric average options on a single underlying asset. The results indicate that both the Brownian bridge (BB) and principal components (PC) constructions can produce substantial error reductions compared to straightforward application of Sobol' points in a random walk construction. This is particularly evident at smaller values of n.

Table 5.7 shows results for options on the geometric average of d assets. The Brownian bridge construction is inapplicable in this setting, so only an ordinary application of Sobol' points (using Cholesky factorization) and the principal components construction appear in the table. These methods are compared for uncorrelated assets and assets for which all correlations are 0.3. In the case of uncorrelated assets, the principal components construction simply permutes the coordinates of the Sobol' sequence, assigning the ith coordinate to the asset with the ith largest volatility. This suggests that the differences between the two methods should be greater in the correlated case, and this is borne out by the results in the table.

Neither the Brownian bridge nor the principal components construction is tailored to a particular type of option payoff. Given additional information about a payoff, we could try to find still better changes of variables. As an example, consider again an option on the geometric mean of d uncorrelated assets. A standard simulation would map a point $(u_1, \ldots, u_d) \in [0,1)^d$ to a value of the average $\bar{S}$ in (5.33) by first mapping each u_i to $\Phi^{-1}(u_i)$ and then

	Barrier Options			Average Options		
	Sobol'	BB	PC	Sobol'	BB	PC
$d = 10$, $n = 1{,}250$	1.32	0.78	0.97	2.14	0.71	0.32
5,000	0.75	0.41	0.49	0.18	0.24	0.11
20,000	0.48	0.53	0.50	0.08	0.08	0.02
80,000	0.47	0.47	0.47	0.03	0.03	0.01
$d = 50$, $n = 1{,}250$	7.10	1.14	1.18	4.24	0.53	0.33
5,000	1.10	0.87	0.59	0.61	0.16	0.11
20,000	0.30	0.25	0.31	0.24	0.05	0.02
80,000	0.22	0.12	0.08	0.06	0.03	0.01
$d = 100$, $n = 1{,}250$	9.83	1.32	1.41	10.12	0.63	0.33
5,000	1.70	0.91	0.46	1.27	0.18	0.11
20,000	0.62	0.23	0.28	0.24	0.04	0.02
80,000	0.19	0.09	0.11	0.05	0.03	0.01

Table 5.6. RMS relative errors (in percent) for single-asset options with d steps per path and n paths, using three different constructions of the underlying Brownian paths.

setting

$$\bar{S} = \left(\prod_{i=1}^{d} S_i(0) \right)^{1/d} \exp\left(rT - \frac{T}{2d} \sum_{i=1}^{d} \sigma_i^2 + \frac{\sqrt{T}}{d} \sum_{i=1}^{d} \sigma_i \Phi^{-1}(u_i) \right).$$

However, a simple change of variables allows us to replace

$$\sum_{i=1}^{d} \sigma_i \Phi^{-1}(u_i) \quad \text{with} \quad \sqrt{\sum_{i=1}^{d} \sigma_i^2} \, \Phi^{-1}(u_1).$$

This reduces the problem to a one-dimensional integral and uses the first coordinate u_1 for that integration. This example is certainly not typical, but it illustrates the flexibility available to change variables, particularly for models driven by normal random variables. All of the examples of stratified sampling in Section 4.3.2 can similarly be applied as changes of variables for QMC methods. Further strategies for improving the accuracy of QMC methods are developed in Fox [127].

5.6 Concluding Remarks

The preponderance of the experimental evidence amassed to date points to Sobol' sequences as the most effective quasi-Monte Carlo method for applications in financial engineering. They often produce more accurate results than other QMC and Monte Carlo methods, and they can be generated very quickly through the algorithms of Bratley and Fox [57] and Press et al. [299].

	Correlation 0		Correlation 0.3	
	Sobol'	PC	Sobol'	PC
$d = 10$, $n = 1{,}250$	1.20	1.01	1.03	0.23
5,000	0.37	0.50	0.17	0.06
20,000	0.19	0.20	0.06	0.02
80,000	0.06	0.03	0.04	0.01
$d = 50$, $n = 1{,}250$	3.55	2.45	1.58	0.16
5,000	0.50	0.34	0.21	0.05
20,000	0.18	0.08	0.05	0.02
80,000	0.08	0.04	0.04	0.01
$d = 100$, $n = 1{,}250$	3.18	3.59	2.15	0.16
5,000	0.53	0.56	0.34	0.04
20,000	0.13	0.10	0.06	0.02
80,000	0.07	0.02	0.03	0.00

Table 5.7. RMS relative errors (in percent) for options on the geometric average of d assets using n paths.

Although QMC methods are based on a deterministic perspective, the performance of Sobol' sequences in derivatives pricing can often be improved through examination of the underlying stochastic model. Because the initial coordinates of a Sobol' sequence are more uniform than later coordinates, a strategic assignment of coordinates to sources of randomness can improve accuracy. The combination of Sobol's points with the Brownian bridge construction is an important example of this idea, but by no means the only one. The applications of stratified sampling in Section 4.3.2 provide further examples, because good directions for stratification are also good candidates for effective use of the best Sobol' coordinates.

One might consider applying methods from Chapter 4 — a control variate, for example — in a QMC numerical integration. We prefer to take such combinations in the opposite order: first analyze a stochastic problem stochastically and use this investigation to find an effective variance reduction technique; then reformulate the variance-reduced simulation problem as an integration problem to apply QMC. Thus, one might develop an importance sampling technique and then implement it using QMC. It would be much more difficult to derive effective importance sampling methods of the type illustrated in Section 4.6 starting from a QMC integration problem.

Indeed, we view postponing the integration perspective as a good way to apply QMC techniques to stochastic problems more generally. The transformation to a Brownian bridge construction, for example, is easy to understand from a stochastic perspective but would be opaque if viewed as a change of variables for an integration problem. Also, the simple error estimates provided by Monte Carlo simulation are especially useful in developing and comparing algorithms. After finding a satisfactory algorithm one may apply QMC to try to further improve accuracy. This is particularly useful if similar problems need to be solved repeatedly, as is often the case in pricing applications. Randomized QMC methods make it possible to compute simple error estimates for QMC calculations and can sometimes reduce errors too.

The effectiveness of QMC methods in high-dimensional pricing problems cannot be explained by comparing the $O(1/\sqrt{n})$ convergence of Monte Carlo with the $O(1/n^{1-\epsilon})$ convergence of QMC because of the $(\log n)^d$ factor subsumed by the ϵ. An important part of the explanation must be that the main source of dimensionality in most finance problems is the number of time steps, and as the Brownian bridge and principal components constructions indicate, this may artificially inflate the nominal dimension. Recent work has identified abstract classes of integration problems for which QMC is provably effective in high dimensions because of the diminishing importance of higher dimensions; see Sloan and Wózniakowski [334] for a detailed analysis, Sloan [332] for an overview, and Larcher, Leobacher, and Scheicher [220] for an application of these ideas to the Brownian bridge construction. Owen [291] argues that the key requirement for the effectiveness of QMC in high dimensions is that the integrand be well-approximated by a sum of functions depending on a small number of variables each.

6

Discretization Methods

This chapter presents methods for reducing discretization error — the bias in Monte Carlo estimates that results from time-discretization of stochastic differential equations. Chapter 3 gives examples of continuous-time stochastic processes that can be simulated exactly at a finite set of dates, meaning that the joint distribution of the simulated values coincides with that of the continuous-time model at the simulated dates. But these examples are exceptional and most models arising in derivatives pricing can be simulated only approximately. The simplest approximation is the Euler scheme; this method is easy to implement and almost universally applicable, but it is not always sufficiently accurate. This chapter discusses methods for improving the Euler scheme and, as a prerequisite for this, discusses criteria for comparing discretization methods.

The issues addressed in this chapter are orthogonal to those in Chapters 4 and 5. Once a time-discretization method is fixed, applying a variance reduction technique or quasi-Monte Carlo method may improve precision in estimating an expectation at the fixed level of discretization, but it can do nothing to reduce discretization bias.

6.1 Introduction

We begin by discussing properties of the Euler scheme, the simplest method for approximate simulation of stochastic differential equations. We then undertake an expansion to refine the Euler scheme and present criteria for comparing methods.

6.1.1 The Euler Scheme and a First Refinement

We consider processes X satisfying a stochastic differential equation (SDE) of the form

$$dX(t) = a(X(t))\,dt + b(X(t))\,dW(t), \tag{6.1}$$

usually with $X(0)$ fixed. In the most general setting we consider, X takes values in $\Re^d$ and W is an m-dimensional standard Brownian motion, in which case a takes values in $\Re^d$ and b takes values in $\Re^{d \times m}$. Some of the methods in this chapter are most easily introduced in the simpler case of scalar X and W. The coefficient functions a and b are assumed to satisfy the conditions in Appendix B.2 for existence and uniqueness of a strong solution to the SDE (6.1); indeed, we will need to impose stronger conditions to reduce discretization error.

We use $\hat{X}$ to denote a time-discretized approximation to X. The Euler (or Euler-Maruyama, after [254]) approximation on a time grid $0 = t_0 < t_1 < \cdots < t_m$ is defined by $\hat{X}(0) = X(0)$ and, for $i = 0, \ldots, m - 1$,

$$\hat{X}(t_{i+1}) = \hat{X}(t_i) + a(\hat{X}(t_i))[t_{i+1} - t_i] + b(\hat{X}(t_i))\sqrt{t_{i+1} - t_i}\, Z_{i+1},$$

with $Z_1, Z_2, \ldots$ independent, m-dimensional standard normal random vectors. To lighten notation, we restrict attention to a grid with a fixed spacing h, meaning that $t_i = ih$. Everything we discuss carries over to the more general case provided the largest of the increments $t_{i+1} - t_i$ decreases to zero. Adaptive methods, in which the time steps depend on the evolution of $\hat{X}$ and are thus stochastic, require separate treatment; see, for example, Gaines and Lyons [133].

With a fixed time step $h > 0$, we may write $\hat{X}(ih)$ as $\hat{X}(i)$ and write the Euler scheme as

$$\hat{X}(i + 1) = \hat{X}(i) + a(\hat{X}(i))h + b(\hat{X}(i))\sqrt{h}\, Z_{i+1}. \tag{6.2}$$

Implementation of this method is straightforward, at least if a and b are easy to evaluate. Can we do better? And in what sense is one approximation better than another? These are the questions we address.

In the numerical solution of *ordinary* differential equations, methods of higher-order accuracy often rely on Taylor expansions. If b were identically zero (and thus (6.1) non-stochastic), (6.2) would reduce to a linear approximation, and a natural strategy for improving accuracy would include higher-order terms in a Taylor expansion of $a(X(t))$. A similar strategy applies to stochastic differential equations, but it must be carried out consistent with the rules of Itô calculus rather than ordinary calculus.

A First Refinement

Inspection of the Euler scheme (6.2) from the perspective of Taylor expansion suggests a possible inconsistency: this approximation expands the drift to $O(h)$ but the diffusion term only to $O(\sqrt{h})$. The approximation to the diffusion term omits $O(h)$ contributions, so including a term of order h in the drift looks like spurious accuracy. This discrepancy also suggests that to refine the Euler scheme we may want to focus on the diffusion term.

We now carry out this proposal. We will see, however, that whether or not it produces an improvement compared to the Euler scheme depends on how we measure error.

We start with the scalar case $d = m = 1$. Recall that the SDE (6.1) abbreviates the relation

$$X(t) = X(0) + \int_0^t a(X(u)) \, du + \int_0^t b(X(u)) \, dW(u). \tag{6.3}$$

The Euler scheme results from the approximations

$$\int_t^{t+h} a(X(u)) \, du \approx a(X(t))h \tag{6.4}$$

and

$$\int_t^{t+h} b(X(u)) \, dW(u) \approx b(X(t))[W(t+h) - W(t)]. \tag{6.5}$$

In both cases, an integrand over $[t, t+h]$ is approximated by its value at t. To improve the approximation of the diffusion term, we need a better approximation of $b(X(u))$ over an interval $[t, t+h]$. We therefore examine the evolution of $b(X(u))$.

From Itô's formula we get

$$
\begin{aligned}
db(X(t)) \\
&= b'(X(t)) \, dX(t) + \tfrac{1}{2} b''(X(t)) b^2(X(t)) \, dt \\
&= \left[b'(X(t)) a(X(t)) + \tfrac{1}{2} b''(X(t)) b^2(X(t)) \right] \, dt + b'(X(t)) b(X(t)) \, dW(t) \\
&\equiv \mu_b(X(t)) \, dt + \sigma_b(X(t)) \, dW(t),
\end{aligned}
$$

where b' and b'' are the first and second derivatives of b. Applying the Euler approximation to the process $b(X(t))$ results in the approximation of $b(X(u))$, $t \le u \le t+h$ by

$$
\begin{aligned}
b(X(u)) &\approx b(X(t)) + \mu_b(X(t))[u - t] + \sigma_b(X(t))[W(u) - W(t)] \\
&= b(X(t)) + \left(b'(X(t)) a(X(t)) + \tfrac{1}{2} b''(X(t)) b^2(X(t)) \right) [u - t] \\
&\quad + b'(X(t)) b(X(t))[W(u) - W(t)].
\end{aligned}
$$

Now $W(u) - W(t)$ is $O(\sqrt{u - t})$ (in probability) whereas the drift term in this approximation is $O(u-t)$ and thus of higher order. Dropping this higher-order term yields the simpler approximation

$$b(X(u)) \approx b(X(t)) + b'(X(t)) b(X(t))[W(u) - W(t)], \quad u \in [t, t+h]. \tag{6.6}$$

Armed with this approximation, we return to the problem of refining (6.5). Instead of freezing $b(X(u))$ at $b(X(t))$ over the interval $[t, t+h]$, as in (6.5), we use the approximation (6.6). Thus, we replace (6.5) with

$$\int_t^{t+h} b(X(u))\, dW(u)$$

$$\approx \int_t^{t+h} (b(X(t)) + b'(X(t))b(X(t))[W(u) - W(t)])\, dW(u)$$

$$= b(X(t))[W(t+h) - W(t)]$$

$$+ b'(X(t))b(X(t)) \left(\int_t^{t+h} [W(u) - W(t)]\, dW(u) \right). \tag{6.7}$$

The proposed refinement uses this expression in place of $b(\hat{X}(i))\sqrt{h}Z_{i+1}$ in the Euler scheme (6.2).

To make this practical, we need to simplify the remaining integral in (6.7). We can write this integral as

$$\int_t^{t+h} [W(u) - W(t)]\, dW(u)$$

$$= \int_t^{t+h} W(u)\, dW(u) - W(t) \int_t^{t+h} dW(u)$$

$$= Y(t+h) - Y(t) - W(t)[W(t+h) - W(t)] \tag{6.8}$$

with

$$Y(t) = \int_0^t W(t)\, dW(t);$$

i.e., $Y(0) = 0$ and

$$dY(t) = W(t)\, dW(t).$$

Itô's formula verifies that the solution to this SDE is

$$Y(t) = \tfrac{1}{2}W(t)^2 - \tfrac{1}{2}t.$$

Making this substitution in (6.8) and simplifying, we get

$$\int_t^{t+h} [W(u) - W(t)]\, dW(u) = \tfrac{1}{2}[W(t+h) - W(t)]^2 - \tfrac{1}{2}h. \tag{6.9}$$

Using this identity in (6.7), we get

$$\int_t^{t+h} b(X(u))\, dW(u) \approx b(X(t))[W(t+h) - W(t)]$$

$$+ \tfrac{1}{2}b'(X(t))b(X(t)) \left([W(t+h) - W(t)]^2 - h \right).$$

Finally, we use this approximation to approximate $X(t+h)$. We refine the one-step Euler approximation

$$X(t+h) \approx X(t) + a(X(t))h + b(X(t))[W(t+h) - W(t)]$$

to

$$X(t + h) \approx X(t) + a(X(t))h + b(X(t))[W(t + h) - W(t)]$$
$$+ \tfrac{1}{2}b'(X(t))b(X(t))\left([W(t + h) - W(t)]^2 - h\right).$$

In a simulation algorithm, we apply this recursively at $h, 2h, \ldots$, replacing the increments of W with $\sqrt{h}Z_{i+1}$; more explicitly, we have

$$\hat{X}(i + 1) = \hat{X}(i) + a(\hat{X}(i))h + b(\hat{X}(i))\sqrt{h}Z_{i+1}$$
$$+ \tfrac{1}{2}b'(\hat{X}(i))b(\hat{X}(i))h(Z_{i+1}^2 - 1). \tag{6.10}$$

This algorithm was derived by Milstein [266] through an analysis of partial differential equations associated with the diffusion X. It is sometimes called the Milstein scheme, but this terminology is ambiguous because there are several important methods due to Milstein.

The approximation method in (6.10) adds a term to the Euler scheme. It expands both the drift and diffusion terms to $O(h)$. Observe that, conditional on $\hat{X}(i)$, the new term

$$\tfrac{1}{2}b'(\hat{X}(i))b(\hat{X}(i))h(Z_{i+1}^2 - 1)$$

has mean zero and is uncorrelated with the Euler terms because $Z_{i+1}^2 - 1$ and Z_{i+1} are uncorrelated. The question remains, however, whether and in what sense (6.10) is an improvement over the Euler scheme. We address this in Section 6.1.2, after discussing the case of vector-valued X and W.

The Multidimensional Case

Suppose, now, that $X(t) \in \Re^d$ and $W(t) \in \Re^m$. Write X_i, W_i, and a_i for the ith components of X, W, and a, and write b_{ij} for the ij-entry of b. Then

$$X_i(t + h) = X_i(t) + \int_t^{t+h} a_i(X(u)) \, du + \sum_{j=1}^m \int_t^{t+h} b_{ij}(X(u)) \, dW_j(u),$$

and we need to approximate the integrals on the right. As in the Euler scheme, we approximate the drift term using

$$\int_t^{t+h} a_i(X(u)) \, du \approx a_i(X(t))h.$$

The argument leading to (6.7) yields

$$\int_t^{t+h} b_{ij}(X(u)) \, dW_j(u) \approx b_{ij}(X(t))[W_j(t + h) - W_j(t)]$$
$$+ \sum_{\ell=1}^d \sum_{k=1}^m \frac{\partial b_{ij}}{\partial x_\ell}(X(t))b_{\ell k}(X(t)) \int_t^{t+h} [W_k(u) - W_k(t)] \, dW_j(u). \tag{6.11}$$

For $k = j$, we can evaluate the integral in (6.11) as in the scalar case:

$$\int_t^{t+h} [W_j(u) - W_j(t)] \, dW_j(u) = \tfrac{1}{2}[W_j(t+h) - W_j(t)]^2 - \tfrac{1}{2}h.$$

However, there is no comparable expression for the off-diagonal terms

$$\int_t^{t+h} [W_k(u) - W_k(t)] \, dW_j(u), \quad k \neq j.$$

These mixed integrals (or more precisely their differences) are called *Lévy area* terms; see the explanation in Protter [300, p.82], for example. Generating samples from their distribution is a challenging simulation problem. Methods for doing so are developed in Gaines and Lyons [132] and Wiktorsson [356], but the difficulties involved limit the applicability of the expansion (6.11) in models driven by multidimensional Brownian motion. Fortunately, we will see that for the purpose of estimating an expectation it suffices to simulate rough approximations to these mixed Brownian integrals.

6.1.2 Convergence Order

Equation (6.10) displays a refinement of the Euler scheme based on expanding the diffusion term to $O(h)$ rather than just $O(\sqrt{h})$. To discuss the extent and the sense in which this algorithm is an improvment over the Euler scheme, we need to establish a figure of merit for comparing discretizations.

Two broad categories of error of approximation are commonly used in measuring the quality of discretization methods: criteria based on the pathwise proximity of a discretized process to a continuous process, and criteria based on the proximity of the corresponding distributions. These are generally termed *strong* and *weak* criteria, respectively.

Let $\{\hat{X}(0), \hat{X}(h), \hat{X}(2h), \ldots\}$ be any discrete-time approximation to a continuous-time process X. Fix a time T and let $n = \lfloor T/h \rfloor$. Typical strong error criteria are

$$\mathsf{E}\left[\|\hat{X}(nh) - X(T)\|\right], \quad \mathsf{E}\left[\|\hat{X}(nh) - X(T)\|^2\right],$$

and

$$\mathsf{E}\left[\sup_{0 \leq t \leq T} \|\hat{X}(\lfloor t/h \rfloor h) - X(t)\|\right],$$

for some vector norm $\|\cdot\|$. Each of these expressions measures the deviation between the individual values of X and the approximation $\hat{X}$.

In contrast, a typical weak error criterion has the form

$$\left|\mathsf{E}[f(\hat{X}(nh))] - \mathsf{E}[f(X(T))]\right|, \tag{6.12}$$

with f ranging over functions from $\Re^d$ to $\Re$ typically satisfying some smoothness conditions. Requring that an expression of the form (6.12) converge to zero as h decreases to zero imposes no constraint on the relation between the outcomes of $\hat{X}(nh)$ and $X(T)$; indeed, the two need not even be defined on the same probability space. Making the error criterion (6.12) small merely requires that the distributions of $\hat{X}(nh)$ and $X(T)$ be close.

For applications in derivatives pricing, weak error criteria are most relevant. We would like to ensure that prices (which are expectations) computed from $\hat{X}$ are close to prices computed from X; we are not otherwise concerned about the paths of the two processes. It is nevertheless useful to be aware of strong error criteria to appreciate the relative merits of alternative discretization methods.

Even after we fix an error criterion, it is rarely possible to ensure that the error using one discretization method will be smaller than the error using another in a specific problem. Instead, we compare methods based on their asymptotic performance for small h.

Under modest conditions, even the simple Euler scheme converges (with respect to both strong and weak criteria) as the time step h decreases to zero. We therefore compare discretization schemes based on the *rate* at which they converge. Following Kloeden and Platen [211], we say that a discretization $\hat{X}$ has *strong order of convergence* $\beta > 0$ if

$$\mathsf{E}\left[\|\hat{X}(nh) - X(T)\|\right] \leq ch^\beta \tag{6.13}$$

for some constant c and all sufficiently small h. The discretization scheme has *weak* order of convergence β if

$$\left|\mathsf{E}[f(\hat{X}(nh))] - \mathsf{E}[f(X(T))]\right| \leq ch^\beta \tag{6.14}$$

for some constant c and all sufficiently small h, for all f in a set $C_P^{2\beta+2}$. The set $C_P^{2\beta+2}$ consists of functions from $\Re^d$ to $\Re$ whose derivatives of order $0, 1, \ldots, 2\beta + 2$ are polynomially bounded. A function $g : \Re^d \to \Re$ is polynomially bounded if

$$|g(x)| \leq k(1 + \|x\|^q)$$

for some constants k and q and all $x \in \Re^d$. The constant c in (6.14) may depend on f.

In both (6.13) and (6.14), a larger value of β implies faster convergence to zero of the discretization error. The same scheme will often have a smaller strong order of convergence than its weak order of convergence. For example, the Euler scheme typically has a strong order of $1/2$, but it often achieves a weak order of 1.

Convergence Order of the Euler Scheme

In more detail, the Euler scheme has strong order $1/2$ under conditions only slightly stronger than those in Theorem B.2.1 of Appendix B.2 for existence and uniqueness of a (strong) solution to the SDE (6.1). We may generalize (6.1) by allowing the coefficient functions a and b to depend explicitly on time t as well as on $X(t)$. Because X is vector-valued, we could alternatively take t to be one of the components of $X(t)$; but that formulation leads to unnecessarily strong conditions for convergence because it requires that the coefficients be as smooth in t as they are in X. In addition to the conditions of Theorem B.2.1, suppose that

$$E\left[\|X(0) - \hat{X}(0)\|^2\right] \le K\sqrt{h} \tag{6.15}$$

and

$$\|a(x,s) - a(x,t)\| + \|b(x,s) - b(x,t)\| \le K(1 + \|x\|)\sqrt{|t-s|}, \tag{6.16}$$

for some constant K; then the Euler scheme has strong order $1/2$. (This is proved in Kloeden and Platen [211], pp.342–344. It is observed in Milstein [266] though without explicit hypotheses.) Condition (6.15) is trivially satisfied if $X(0)$ is known and we set $\hat{X}(0)$ equal to it.

Stronger conditions are required for the Euler scheme to have weak order 1. For example, Theorem 14.5.2 of Kloeden and Platen [211] requires that the functions a and b be four times continuously differentiable with polynomially bounded derivatives. More generally, the Euler scheme has weak order β if a and b are $2(\beta+1)$ times continuously differentiable with polynomially bounded derivatives; the condition (6.14) then applies only to functions f with the same degree of smoothness.

To see how smoothness can lead to a higher weak order than strong order, consider the following argument. Suppose, for simplicity, that $T = nh$ and that $X(0)$ is fixed so that $E[f(X(0))]$ is known. By writing

$$E[f(X(T))] = E[f(X(0))] + E\left[\sum_{i=0}^{n-1} E[f(X((i+1)h)) - f(X(ih))|X(ih)]\right],$$

we see that accurate estimation of $E[f(X(T))]$ follows from accurate estimation of the conditional expectations $E[f(X((i+1)h)) - f(X(ih))|X(ih)]$. Applying a Taylor approximation to f (and taking X scalar for simplicity), we get

$$E[f(X((i+1)h)) - f(X(ih))|X(ih)]$$
$$\approx \sum_{j=0}^{r} \frac{f^{(j)}(X(ih))}{j!} E[(X((i+1)h) - X(ih))^j|X(ih)]. \tag{6.17}$$

Thus, if f is sufficiently smooth, then to achieve a high order of weak convergence a discretization scheme need only approximate conditional *moments* of the increments of the process X. With sufficient smoothness in the coefficient functions a and b, higher conditional moments are of increasingly high order in h. Smoothness conditions on a, b, and f leading to a weak order of convergence β for the Euler scheme follow from careful accounting of the errors in expanding f and approximating the conditional moments; see Kloeden and Platen [211], Section 14.5, and Talay [340, 341].

The accuracy of a discretization scheme in estimating an expression of the form $\mathsf{E}[f(X(T))]$ does not necessarily extend to the simulation of other quantities associated with the same process. In Section 6.4 we discuss difficulties arising in simulating the maximum of a diffusion, for example. Talay and Zheng [344] analyze discretization error in estimating quantiles of the distribution of a component of $X(T)$. They provide very general conditions under which the bias in a quantile estimate computed from an Euler approximation is $O(h)$; but they also show that the implicit constant in this $O(h)$ error is large — especially in the tails of the distribution — and that this makes accurate quantile estimation difficult.

Convergence Order of the Refined Scheme

Theorem 10.3.5 of Kloeden and Platen [211] and Theorem 2-2 of Talay [340] provide conditions under which Milstein's refinement (6.10) and its multidimensional generalization based on (6.11) have strong order 1. The conditions required extend the linear growth, Lipschitz condition, and (6.16) to derivatives of the coefficient functions a and b. Thus, under these relatively modest additional conditions, expanding the diffusion term to $O(h)$ instead of just $O(\sqrt{h})$ through the derivation in Section 6.1.1 increases the order of strong convergence.

But the weak order of convergence of the refined scheme (6.10) is also 1, as it is for the Euler scheme. In this respect, including additional terms — as in (6.10) and (6.11) — does not result in greater accuracy. This should not be viewed as a deficiency of Milstein's method; rather, the Euler scheme is better than it "should" be, achieving order-1 weak convergence without expanding all terms to $O(h)$. This is in fact just the simplest example of a broader pattern of results on the number of terms required to achieve strong or weak convergence of a given order (to which we return in Section 6.3.1). In order to achieve a weak order greater than that of the Euler scheme, we need to expand dt-integrals to order h^2 and stochastic integrals to order h. We carry this out in the next section to arrive at a method with a higher weak order of convergence.

It is reassuring to know that a discretization scheme has a high order of convergence, but before venturing into our next derivation we should take note of the fact that good accuracy on smooth functions may not be directly relevant to our intended applications: option payoffs are typically nondifferentiable. Bally and Talay [34] show that the weak order of the Euler scheme holds for very general f and Yan [357] analyzes SDEs with irregular coefficients, but most of the literature requires significant smoothness assumptions.

When applying higher-order discretization methods, it is essential to test the methods numerically.

6.2 Second-Order Methods

We now proceed to further refine the Euler scheme to arrive at a method with weak order 2. The derivation follows the approach used in Section 6.1.1, expanding the integrals of $a(X(t))$ and $b(X(t))$ to refine the Euler approximations in (6.4) and (6.5), but now we keep more terms in the expansions. We begin by assuming that in the SDE (6.1) both X and W are scalar.

6.2.1 The Scalar Case

To keep the notation manageable, we adopt some convenient shorthand. With the scalar SDE (6.1) defining X, we associate the operators

$$\mathcal{L}^0 = a\frac{d}{dx} + \tfrac{1}{2}b^2\frac{d^2}{dx^2} \tag{6.18}$$

and

$$\mathcal{L}^1 = b\frac{d}{dx}, \tag{6.19}$$

meaning that for any twice differentiable f, we have

$$\mathcal{L}^0 f(x) = a(x)f'(x) + \tfrac{1}{2}b^2(x)f''(x)$$

and

$$\mathcal{L}^1 f(x) = b(x)f'(x).$$

This allows us to write Itô's formula as

$$df(X(t)) = \mathcal{L}^0 f(X(t))\,dt + \mathcal{L}^1 f(X(t))\,dW(t). \tag{6.20}$$

To accommodate functions $f(t, X(t))$ that depend explicitly on time, we would generalize (6.18) to

$$\mathcal{L}^0 = \frac{\partial}{\partial t} + a\frac{\partial}{\partial x} + \tfrac{1}{2}b^2\frac{\partial^2}{\partial x^2}.$$

As in Section 6.1.1, the key to deriving a discretization scheme lies in approximating the evolution of X over an interval $[t, t+h]$. We start from the representation

$$X(t+h) = X(t) + \int_t^{t+h} a(X(u))\,du + \int_t^{t+h} b(X(u))\,dW(u), \tag{6.21}$$

and approximate each of the two integrals on the right.

The Euler scheme approximates the first integral using the approximation $a(X(u)) \approx a(X(t))$ for $u \in [t, t+h]$. To derive a better approximation for $a(X(u))$, we start from the exact representation

$$a(X(u)) = a(X(t)) + \int_t^u \mathcal{L}^0 a(X(s)) \, ds + \int_t^u \mathcal{L}^1 a(X(s)) \, dW(s);$$

this is Itô's formula applied to $a(X(u))$. Next we apply the Euler approximation to each of the two integrals appearing in this representation; in other words, we set $\mathcal{L}^0 a(X(s)) \approx \mathcal{L}^0 a(X(t))$ and $\mathcal{L}^1 a(X(s)) \approx \mathcal{L}^1 a(X(t))$ for $s \in [t, u]$ to get

$$a(X(u)) \approx a(X(t)) + \mathcal{L}^0 a(X(t)) \int_t^u ds + \mathcal{L}^1 a(X(t)) \int_t^u dW(s).$$

Now we use this aproximation in the first integral in (6.21) to get

$$\int_t^{t+h} a(X(u)) \, du$$

$$\approx a(X(t))h + \mathcal{L}^0 a(X(t)) \int_t^{t+h} \int_t^u ds \, du + \mathcal{L}^1 a(X(t)) \int_t^{t+h} \int_t^u dW(s) \, du$$

$$\equiv a(X(t))h + \mathcal{L}^0 a(X(t)) I_{(0,0)} + \mathcal{L}^1 a(X(t)) I_{(1,0)}, \tag{6.22}$$

with $I_{(0,0)}$ and $I_{(1,0)}$ denoting the indicated double integrals. This gives us our approximation to the first term in integral in (6.21).

We use corresponding steps for the second integral in (6.21). We approximate the integrand $b(X(u))$, $u \in [t, t+h]$ using

$$b(X(u)) = b(X(t)) + \int_t^u \mathcal{L}^0 b(X(s)) \, ds + \int_t^u \mathcal{L}^1 b(X(s)) \, dW(s)$$

$$\approx b(X(t)) + \mathcal{L}^0 b(X(t)) \int_t^u ds + \mathcal{L}^1 b(X(t)) \int_t^u dW(s)$$

and thus approximate the integral as

$$\int_t^{t+h} b(X(u)) \, dW(u)$$

$$\approx b(X(t))[W(t+h) - W(t)] + \mathcal{L}^0 b(X(t)) \int_t^{t+h} \int_t^u ds \, dW(u)$$

$$+ \mathcal{L}^1 b(X(t)) \int_t^{t+h} \int_t^u dW(s) \, dW(u)$$

$$\equiv b(X(t))[W(t+h) - W(t)] + \mathcal{L}^0 b(X(t)) I_{(0,1)} + \mathcal{L}^1 b(X(t)) I_{(1,1)}. \tag{6.23}$$

Once again, the $I_{(i,j)}$ denote the indicated double integrals.

If we combine (6.22) and (6.23) and make explicit the application of the operators $\mathcal{L}^0$ and $\mathcal{L}^1$ to a and b, we arrive at the approximation

$$X(t + h) \approx X(t) + ah + b\Delta W + (aa' + \tfrac{1}{2}b^2 a'')I_{(0,0)}$$
$$+ (ab' + \tfrac{1}{2}b^2 b'')I_{(0,1)} + ba'I_{(1,0)} + bb'I_{(1,1)}, \qquad (6.24)$$

with $\Delta W = W(t + h) - W(t)$, and the functions a, b and their derivatives all evaluated at $X(t)$.

The Discretization Scheme

To turn the approximation in (6.24) into an implementable algorithm, we need to be able to simulate the double integrals $I_{(i,j)}$. Clearly,

$$I_{(0,0)} = \int_t^{t+h} \int_t^u ds\, du = \tfrac{1}{2}h^2.$$

From (6.9) we know that

$$I_{(1,1)} = \int_t^{t+h} [W(u) - W(t)]\, dW(u) = \tfrac{1}{2}[(\Delta W)^2 - h].$$

The term $I_{(0,1)}$ is

$$I_{(0,1)} = \int_t^{t+h} \int_t^u ds\, dW(u) = \int_t^{t+h} (u - t)\, dW(u).$$

Applying integration by parts (which can be justified by applying Itô's formula to $tW(t)$), we get

$$I_{(0,1)} = hW(t + h) - \int_t^{t+h} W(u)\, du$$
$$= h[W(t + h) - W(t)] - \int_t^{t+h} [W(u) - W(t)]\, du$$
$$= h\Delta W - I_{(1,0)}. \qquad (6.25)$$

So, it only remains to examine

$$I_{(1,0)} = \int_t^{t+h} [W(u) - W(t)]\, du.$$

Given $W(t)$, the area $I_{(1,0)}$ and the increment $\Delta W = W(t + h) - W(t)$ are jointly normal. Each has conditional mean 0; the conditional variance of ΔW is h and that of $I_{(1,0)}$ is $h^3/3$ (see (3.48)). For their covariance, notice first that

$$\mathsf{E}[I_{(1,0)}|W(t), \Delta W] = \tfrac{1}{2}h\Delta W \qquad (6.26)$$

(as illustrated in Figure 6.1), so $\mathsf{E}[I_{(1,0)}\Delta W] = \tfrac{1}{2}h^2$. We may therefore simulate $W(t + h) - W(t)$ and $I_{(1,0)}$ as

$$\begin{pmatrix} \Delta W \\ \Delta I \end{pmatrix} \sim N\left(0, \begin{pmatrix} h & \frac{1}{2}h^2 \\ \frac{1}{2}h^2 & \frac{1}{3}h^3 \end{pmatrix}\right). \tag{6.27}$$

This leads to the following second-order scheme:

$$\begin{aligned}
\hat{X}((i+1)h) = \hat{X}(ih) &+ ah + b\Delta W + (ab' + \tfrac{1}{2}b^2 b'')[\Delta W h - \Delta I] \\
&+ a'b\Delta I + \tfrac{1}{2}bb'[\Delta W^2 - h] \\
&+ (aa' + \tfrac{1}{2}b^2 a'')\tfrac{1}{2}h^2,
\end{aligned} \tag{6.28}$$

with the functions a, b and their derivatives all evaluated at $\hat{X}(ih)$.

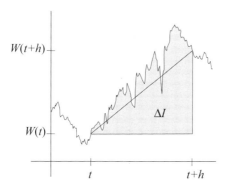

Fig. 6.1. The shaded area is ΔI. Given $W(t)$ and $W(t+h)$, the conditional expectation of W at any intermediate time lies on the straight line connecting these endpoints. The conditional expectation of ΔI is given by the area of the triangle with base h and height $\Delta W = W(t+h) - W(t)$.

This method was introduced by Milstein [267] in a slightly different form. Talay [341] shows that Milstein's scheme has weak order 2 under conditions on the coefficient functions a and b. These conditions include the requirement that the functions a and b be six times continuously differentiable with uniformly bounded derivatives. The result continues to hold if ΔI is replaced by its conditional expectation $\Delta W h/2$; this type of simplification becomes essential in the vector case, as we explain in the next section.

Implementation of (6.28) and similar methods requires calculation of the derivatives of the coefficient functions of a diffusion. Methods that use difference approximations to avoid derivative calculations without a loss in convergence order are developed in Milstein [267] and Talay [341]. These types of approximations are called Runge-Kutta methods in analogy with methods used in the numerical solution of ordinary differential equations.

6.2.2 The Vector Case

We now extend the scheme in (6.28) to d-dimensional X driven by m-dimensional W. Much as in the scalar case, we start from the representation

$$X_i(t+h) = X_i(t) + \int_t^{t+h} a_i(u)\,du + \sum_{k=1}^m \int_t^{t+h} b_{ik}(u)\,dW_k(u), \quad i = 1,\ldots,d,$$

and approximate each of the integrals on the right. In this setting, the relevant operators are

$$\mathcal{L}^0 = \frac{\partial}{\partial t} + \sum_{i=1}^d a_i \frac{\partial}{\partial x_i} + \frac{1}{2} \sum_{i,j=1}^d \sum_{k=1}^m b_{ik} b_{jk} \frac{\partial^2}{\partial x_i \partial x_j} \tag{6.29}$$

and

$$\mathcal{L}^k = \sum_{i=1}^d b_{ik} \frac{\partial}{\partial x_i}, \quad k = 1,\ldots,m. \tag{6.30}$$

The multidimensional Itô formula for twice continuously differentiable $f : \Re^d \to \Re$ becomes

$$df(X(t)) = \mathcal{L}^0 f(X(t))\,dt + \sum_{k=1}^m \mathcal{L}^k f(X(t))\,dW_k(t). \tag{6.31}$$

Applying (6.31) to a_i, we get

$$a_i(X(u)) = a_i(X(t)) + \int_t^u \mathcal{L}^0 a_i(X(s))\,ds + \sum_{k=1}^m \int_t^u \mathcal{L}^k a_i(X(s))\,dW_k(s).$$

The same steps leading to the approximation (6.22) in the scalar case now yield the approximation

$$\int_t^{t+h} a_i(X(u))\,du \approx a_i(X(t))h + \mathcal{L}^0 a_i(X(t))I_{(0,0)} + \sum_{k=1}^m \mathcal{L}^k a_i(X(t))I_{(k,0)},$$

with

$$I_{(k,0)} = \int_t^{t+h} \int_t^u dW_k(s)\,du, \quad k = 1,\ldots,m.$$

Similarly, the representation

$$b_{ik}(X(u)) = b_{ik}(X(t)) + \int_t^u \mathcal{L}^0 b_{ik}(X(s))\,ds + \sum_{j=1}^m \int_t^u \mathcal{L}^j b_{ik}(X(s))\,dW_j(s),$$

leads to the approximation

$$\int_t^{t+h} b_{ik}(X(u))\,dW_k(u)$$

$$\approx b_{ik}(X(t))h + \mathcal{L}^0 b_{ik}(X(t))I_{(0,k)} + \sum_{j=1}^m \mathcal{L}^j b_{ik}(X(t))I_{(j,k)},$$

with

$$I_{(0,k)} = \int_t^{t+h} \int_t^u ds\, dW_k(u), \quad k = 1, \ldots, m,$$

and

$$I_{(j,k)} = \int_t^{t+h} \int_t^u dW_j(u)\, dW_k(u), \quad j, k = 1, \ldots, m.$$

The notational convention for these integrals should be evident: in $I_{(j,k)}$ we integrate first over W_j and then over W_k. This interpretation extends to $j = 0$ if we set $W_0(t) \equiv t$.

By combining the expansions above for the integrals of a_i and b_{ik}, we arrive at the discretization

$$\hat{X}_i(t+h) = \hat{X}_i(t) + a_i(\hat{X}(t))h + \sum_{k=1}^m b_{ik}(\hat{X}(t))\Delta W_k$$

$$+ \tfrac{1}{2}\mathcal{L}^0 a_i(\hat{X}(t))h^2 + \sum_{k=1}^m \mathcal{L}^k a_i(\hat{X}(t))I_{(k,0)}$$

$$+ \sum_{k=1}^m \left(\mathcal{L}^0 b_{ik}(\hat{X}(t))I_{(0,k)} + \sum_{j=1}^m \mathcal{L}^j b_{ik}(\hat{X}(t))I_{(j,k)} \right), \quad (6.32)$$

for each $i = 1, \ldots, d$. Here we have substituted $h^2/2$ for $I_{(0,0)}$ and abbreviated $W_k(t+h) - W_k(t)$ as ΔW_k. The application of each of the operators $\mathcal{L}^j$ to any of the coefficient functions a_i, b_{ik} produces a polynomial in the coefficient functions and their derivatives; these expressions can be made explicit using (6.29) and (6.30). Using the identity

$$I_{(0,j)} + I_{(j,0)} = \Delta W_j h,$$

which follows from (6.25), we could rewrite all terms involving $I_{(0,j)}$ as multiples of $(\Delta W_j h - I_{(j,0)})$ instead. Thus, to implement (6.32) we need to sample, for each $j = 1, \ldots, m$, the Brownian increments ΔW_j together with the integrals $I_{(j,0)}$ and $I_{(j,k)}$, $k = 1, \ldots, m$. We address this issue next.

Commutativity Condition

As noted in Section 6.1.1, the mixed Brownian integrals $I_{(j,k)}$ with $j \neq k$ are difficult to simulate, so (6.32) does not provide a practical algorithm without further simplification. Simulation of the mixed integrals is obviated in models satisfying the *commutativity condition*

$$\mathcal{L}^k b_{ij} = \mathcal{L}^j b_{ik} \tag{6.33}$$

for all $i = 1, \ldots, d$. This is a rather artificial condition and is not often satisfied in practice, but it provides an interesting simplification of the second-order approximation.

When (6.33) holds, we may group terms in (6.32) involving mixed integrals $I_{(j,k)}$, $j, k \geq 1$, and write them as

$$\sum_{j=1}^{m}\sum_{k=1}^{m} \mathcal{L}^j b_{ik} I_{(j,k)} = \sum_{j=1}^{m} \mathcal{L}^j b_{ij} I_{(j,j)} + \sum_{j=1}^{m}\sum_{k=j+1}^{m} \mathcal{L}^j b_{ik}(I_{(j,k)} + I_{(k,j)}).$$

As in the scalar case (6.9), the diagonal term $I_{(j,j)}$ evaluates to $(\Delta W_j^2 - h)/2$ and is thus easy to simulate. The utility of the commutativity condition lies in the observation that even though each $I_{(j,k)}$, $j \neq k$, is difficult to simulate, the required sums simplify to

$$I_{(j,k)} + I_{(k,j)} = \Delta W_j \Delta W_k. \tag{6.34}$$

This follows from applying Itô's formula to $W_j(t)W_k(t)$ to get

$$W_j(t+h)W_k(t+h) - W_j(t)W_k(t) = \int_t^{t+h} W_k(u)\, dW_j(u) + \int_t^{t+h} W_j(u)\, dW_k(u)$$

and then subtracting $W_k(t)\Delta W_j + W_j(t)\Delta W_k$ from both sides.

When the commutativity condition is satisfied, the discretization scheme (6.32) thus simplifies to

$$\hat{X}_i(t + h) = \hat{X}_i(t) + a_i(\hat{X}(t))h + \sum_{k=1}^{m} b_{ik}(\hat{X}(t))\Delta W_k + \tfrac{1}{2}\mathcal{L}^0 a_i(\hat{X}(t))h^2$$

$$+ \sum_{k=1}^{m}\left(\left[\mathcal{L}^k a_i(\hat{X}(t)) - \mathcal{L}^0 b_{ik}(\hat{X}(t))\right]\Delta I_k + \mathcal{L}^0 b_{ik}(\hat{X}(t))\Delta W_k h\right)$$

$$+ \sum_{j=1}^{m}\left(\mathcal{L}^j b_{ij}(\hat{X}(t))\tfrac{1}{2}(\Delta W_j^2 - h) + \sum_{k=j+1}^{m} \mathcal{L}^j b_{ik}(\hat{X}(t))\Delta W_j \Delta W_k\right), \tag{6.35}$$

with $\Delta I_k = I_{(k,0)}$. Because the components of W are independent of each other, the pairs $(\Delta W_k, \Delta I_k)$, $k = 1, \ldots, m$, are independent of each other. Each such pair has the bivariate normal distribution identified in (6.27) and is thus easy to simulate.

Example 6.2.1 *LIBOR Market Model.* As an illustration of the commutativity condition (6.33), we consider the LIBOR market model of Section 3.7. Thus, take X_i to be the ith forward rate L_i in the spot measure dynamics in (3.112). This specifies that the evolution of L_i is governed by an SDE of the form

$$dL_i(t) = L_i(t)\mu_i(L(t), t)\, dt + L_i(t)\sigma_i(t)^\top\, dW(t),$$

with, for example, σ_i a deterministic function of time. In the notation of this section, $b_{ij} = L_i\sigma_{ij}$. The commutativity condition (6.33) requires

$$\sum_{r=1}^{d} b_{rk} \frac{\partial b_{ij}}{\partial x_r} = \sum_{r=1}^{d} b_{rj} \frac{\partial b_{ik}}{\partial x_r},$$

and this is satisfied because both sides evaluate to $\sigma_{ij}\sigma_{ik}L_i$. More generally, the commutativity condition is satisfied whenever $b_{ij}(X(t))$ factors as the product of a function of $X_i(t)$ and a deterministic function of time.

If we set $X_i(t) = \log L_i(t)$ then X solves an SDE of the form

$$dX_i(t) = \left(\mu_i(X(t),t) - \tfrac{1}{2}\|\sigma_i(t)\|^2\right) dt + \sigma_i(t)^{\top} dW(t).$$

In this case, $b_{ij} = \sigma_{ij}$ does not depend on X at all so the commutativity condition is automatically satisfied. □

A Simplified Scheme

Even when the commutativity condition fails, the discretization method (6.32) can be simplified for practical implementation. Talay [340] and Kloeden and Platen [211, p.465] show that the scheme continues to have weak order 2 if each ΔI_j is replaced with $\tfrac{1}{2}\Delta W_j h$. (Related simplifications are used in Milstein [267] and Talay [341].) Observe from (6.26) that this amounts to replacing ΔI_j with its conditional expectation given ΔW_j. As a consequence, $\tfrac{1}{2}\Delta W_j h$ has the same covariance with ΔW_j as ΔI_j does:

$$\mathsf{E}[\Delta W_j \cdot \tfrac{1}{2}\Delta W_j h] = \tfrac{1}{2}h\mathsf{E}[\Delta W_j^2] = \tfrac{1}{2}h^2.$$

It also has the same mean as ΔI_j but variance $h^3/4$ rather than $h^3/3$, an error of $O(h^3)$. This turns out to be close enough to preserve the order of convergence. In the scalar case (6.28), the simplified scheme is

$$
\begin{aligned}
\hat{X}(n+1) = {}& \hat{X}(n) + ah + b\Delta W \\
& + \tfrac{1}{2}(a'b + ab' + \tfrac{1}{2}b^2 b'')\Delta W h + \tfrac{1}{2}bb'[\Delta W^2 - h] \\
& + (aa' + \tfrac{1}{2}b^2 a'')\tfrac{1}{2}h^2,
\end{aligned}
\tag{6.36}
$$

with a, b, and their derivatives evaluated at $\hat{X}(n)$.

In the vector case, the simplified scheme replaces the double integrals in (6.32) with simpler random variables. As in the scalar case, $I_{(0,k)}$ and $I_{(k,0)}$ are approximated by $\Delta W_k h/2$. Each $I_{(j,j)}$, $j \neq 0$, evaluates to $(\Delta W_j^2 - h)/2$. For j, k different from zero and from each other, $I_{(j,k)}$ is approximated by (Talay [341], Kloeden and Platen [211], Section 14.2)

$$\tfrac{1}{2}(\Delta W_j \Delta W_k - V_{jk}), \tag{6.37}$$

with $V_{kj} = -V_{jk}$, and the V_{jk}, $j < k$, independent random variables taking values h and $-h$ each with probability $1/2$. Let $V_{jj} = h$. The resulting approximation is, for each coordinate $i = 1, \ldots, d$,

$$\hat{X}_i(n+1) =$$

$$\hat{X}_i(n) + a_i h + \sum_{k=1}^{m} b_{ik} \Delta W_k + \frac{1}{2} \mathcal{L}^0 a_i h^2 + \frac{1}{2} \sum_{k=1}^{m} \left(\mathcal{L}^k a_i + \mathcal{L}^0 b_{ik} \right) \Delta W_k h$$

$$+ \frac{1}{2} \sum_{k=1}^{m} \sum_{j=1}^{m} \mathcal{L}^j b_{ik} \left(\Delta W_j \Delta W_k - V_{jk} \right), \tag{6.38}$$

with all a_i, b_{ij}, and their derivatives evaluated at $\hat{X}(n)$.

In these simplified schemes, the ΔW can be replaced with other random variables $\widehat{\Delta W}$ with moments up to order 5 that are within $O(h^3)$ of those of ΔW. (See the discussion following (6.17) and, for precise results Kloeden and Platen [211, p.465] and Talay [341, 342].) This includes the three-point distributions

$$P(\widehat{\Delta W} = \pm\sqrt{3h}) = \frac{1}{6}, \quad P(\widehat{\Delta W} = 0) = \frac{2}{3}.$$

These are faster to generate, but using normally distributed ΔW will generally result in smaller bias. The justification for using (6.37) also lies in the fact that these simpler random variables have moments up to order five that are within $O(h^3)$ of those of the $I_{(j,k)}$; see Section 5.12 of Kloeden and Platen [211, p.465], Section 1.6 of Talay [341], or Section 5 of Talay [342]. Talay [341, 342] calls these "Monte Carlo equivalent" families of random variables.

Example 6.2.2 *Stochastic volatility model.* In Section 3.4, we noted that the square-root diffusion is sometimes used to model stochastic volatility. Heston's [179] model is

$$dS(t) = rS(t)\,dt + \sqrt{V(t)}S(t)\,dW_1(t)$$
$$dV(t) = \kappa(\theta - V(t))\,dt + \sqrt{V(t)}(\sigma_1\,dW_1(t) + \sigma_2\,dW_2(t)),$$

with S interpreted as, e.g., a stock price. The Brownian motions W_1 and W_2 are independent of each other. Heston [179] derives a formula for option prices in this setting using Fourier transform inversion. This provides a benchmark against which to compare simulation methods.

The simplified second-order scheme (6.38) for this model is as follows:

$$\hat{S}(i+1) = \hat{S}(i)(1 + rh + \sqrt{\hat{V}(i)}\Delta W_1) + \frac{1}{2}r^2\hat{S}(i)h^2$$

$$+ \left(\left[r + \frac{\sigma_1 - \kappa}{4} \right] \hat{S}(i)\sqrt{\hat{V}(i)} + \left[\frac{\kappa\theta}{4} - \frac{\sigma^2}{16} \right] \frac{\hat{S}(i)}{\sqrt{\hat{V}(i)}} \right) \Delta W_1 h$$

$$+ \frac{1}{2}\hat{S}(i)(\hat{V}(i) + \frac{\sigma_1}{2})(\Delta W_1^2 - h) + \frac{1}{4}\sigma_2\hat{S}(i)(\Delta W_2\Delta W_1 + \xi)$$

and

$$\hat{V}(i+1) =$$

$$\kappa\theta h + (1 - \kappa h)\hat{V}(i) + \sqrt{\hat{V}(i)}(\sigma_1 \Delta W_1 + \sigma_2 \Delta W_2) - \tfrac{1}{2}\kappa^2(\theta - \hat{V}(i))h^2$$

$$+ \left(\left[\frac{\kappa\theta}{4} - \frac{\sigma^2}{16}\right]\frac{1}{\sqrt{\hat{V}(i)}} - \frac{3\kappa}{2}\sqrt{\hat{V}(i)}\right)(\sigma_1 \Delta W_1 + \sigma_2 \Delta W_2)h$$

$$+ \tfrac{1}{4}\sigma_1^2(\Delta W_1^2 - h) + \tfrac{1}{4}\sigma_2^2(\Delta W_2^2 - h) + \tfrac{1}{2}\sigma_1\sigma_2\Delta W_1\Delta W_2,$$

with $\sigma^2 = \sigma_1^2 + \sigma_2^2$ and ξ taking the values h and $-h$ with probability $1/2$ independent of the Brownian increments. To avoid taking the square root of a negative number or dividing by zero, we replace $\hat{V}(i)$ by its absolute value before advancing these recursions.

Figure 6.2 displays numerical results using this scheme and a simple Euler approximation. We use parameters $S(0) = 100$, $V(0) = 0.04$, $r = 5\%$, $\kappa = 1.2$, $\theta = 0.04$, $\sigma = 0.30$, and $\sigma_1 = \rho\sigma$ with $\rho = -0.5$. Using Heston's [179] formula, the expectation $\mathsf{E}[e^{-rT}(S(T) - K)^+]$ with $T = 1$ and $K = 100$ evaluates to 10.3009. We compare our simulation results against this value to estimate bias. We use simulation time step $h = T/n$, with $n = 3$, 6, 12, 25, and 100 and run 2–4 million replications at each n for each method.

Figure 6.2 plots the estimated log absolute bias against $\log n$. The bias in the Euler scheme for this example falls below 0.01 at $n = 25$ steps per year, whereas the second-order method has a bias this small even at $n = 3$ steps per year. As n increases, the results for the Euler scheme look roughly consistent with first-order convergence; the second-order method produces smaller estimated biases but its convergence is much more erratic. In fact our use of (6.38) for this problem lacks theoretical support because the square-root functions in the model dynamics and the kink in the call option payoff violate the smoothness conditions required to ensure second-order convergence. The more regular convergence displayed by the Euler scheme in this example lends itself to the extrapolation method in Section 6.2.4.

6.2.3 Incorporating Path-Dependence

The error criterion in (6.14) applies to expectations of the form $\mathsf{E}[f(X(T))]$ with T fixed. Accurate estimation of $\mathsf{E}[f(X(T))]$ requires accurate approximation only of the distribution of $X(T)$. In many pricing problems, however, we are interested not only in the terminal state of an underlying process, but also in the path by which the terminal state is reached. The error criterion (6.14) does not appear to offer any guarantees on the approximation error in simulating functions of the path, raising the question of whether properties of the Euler and higher-order schemes extend to such functions.

One way to extend the framework of the previous sections to path-dependent quantities is to transform dependence on the past into dependence on supplementary state variables. This section illustrates this idea.

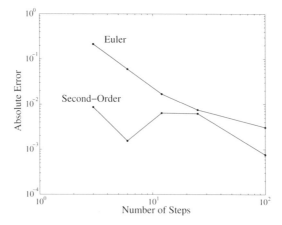

Fig. 6.2. Estimated bias versus number of steps in discretization of a stochastic volatility model.

Suppose we want to compute a bond price

$$\mathsf{E}\left[\exp\left(-\int_0^T r(t)\,dt\right)\right] \qquad (6.39)$$

with the (risk-neutral) dynamics of the short rate r described by the scalar SDE

$$dr(t) = \mu(r(t))\,dt + \sigma(r(t))\,dW(t).$$

If we simulate some discretization $\hat{r}(i) = \hat{r}(ih)$, $i = 0, 1, \ldots, n - 1$ with time step $h = T/n$, the simplest estimate of the bond price would be

$$\exp\left(-h\sum_{i=0}^{n-1}\hat{r}(i)\right). \qquad (6.40)$$

An alternative introduces the variable

$$D(t) = \exp\left(-\int_0^t r(u)\,du\right),$$

develops a discretization scheme for the bivariate diffusion

$$d\begin{pmatrix} r(t) \\ D(t) \end{pmatrix} = \begin{pmatrix} \mu(r(t)) \\ -r(t)D(t) \end{pmatrix} dt + \begin{pmatrix} \sigma(r(t)) \\ 0 \end{pmatrix} dW(t), \qquad (6.41)$$

and uses $\hat{D}(nh)$ as an estimate of the bond price (6.39). In (6.41), the driving Brownian motion is still one-dimensional, so we have not really made the problem any more difficult by enlarging the state vector. The difficulties addressed in Section 6.2.2 arise when W is vector-valued.

The Euler scheme for the bivariate diffusion is

$$\hat{r}(i+1) = \hat{r}(i) + \mu(\hat{r}(i))h + \sigma(\hat{r}(i))\Delta W$$
$$\hat{D}(i+1) = \hat{D}(i) - \hat{r}(i)\hat{D}(i)h.$$

Because of the smoothness of the coefficients of the SDE for $D(t)$, this discretization inherits whatever order of convergence the coefficients μ and σ ensure for $\hat{r}$. Beyond this guarantee, the bivariate formulation offers no clear advantage for the Euler scheme compared to simply using (6.40). Indeed, if we apply the Euler scheme to $\log D(t)$ rather than $D(t)$, we recover (6.40) exactly.

But we do find a difference when we apply a second-order discretization. The simplified second-order scheme for a generic bivariate diffusion X driven by a scalar Brownian motion has the form

$$\begin{pmatrix} \hat{X}_1(i+1) \\ \hat{X}_2(i+1) \end{pmatrix} = \text{Euler terms} + \frac{1}{2}\begin{pmatrix} \mathcal{L}^0 a_1(\hat{X}(i)) \\ \mathcal{L}^0 a_2(\hat{X}(i)) \end{pmatrix}h^2$$
$$+ \frac{1}{2}\begin{pmatrix} \mathcal{L}^1 b_1(\hat{X}(i)) \\ \mathcal{L}^1 b_2(\hat{X}(i)) \end{pmatrix}(\Delta W^2 - h) + \frac{1}{2}\begin{pmatrix} \mathcal{L}^1 a_1(\hat{X}(i)) + \mathcal{L}^0 b_1(\hat{X}(i)) \\ \mathcal{L}^1 a_2(\hat{X}(i)) + \mathcal{L}^0 b_2(\hat{X}(i)) \end{pmatrix}\Delta W h$$

with

$$\mathcal{L}^0 = a_1\frac{\partial}{\partial x_1} + a_2\frac{\partial}{\partial x_2} + \frac{1}{2}\left(b_1^2\frac{\partial^2}{\partial x_1^2} + 2b_1 b_2\frac{\partial^2}{\partial x_1 x_2} + b_2^2\frac{\partial^2}{\partial x_2^2}\right)$$

and

$$\mathcal{L}^1 = b_1\frac{\partial}{\partial x_1} + b_2\frac{\partial}{\partial x_2}.$$

When specialized to the bond-pricing setting, this discretizes $r(t)$ as

$$\hat{r}(i+1) = \hat{r}(i) + \mu h + \sigma\Delta W + \frac{1}{2}\sigma\sigma'[\Delta W^2 - h]$$
$$+ \frac{1}{2}(\sigma\mu' + \mu\sigma' + \frac{1}{2}\sigma^2\sigma'')\Delta W h + \frac{1}{2}(\mu'\mu + \frac{1}{2}\sigma^2\mu'')h^2$$

with μ, σ and their derivatives on the right evaluated at $\hat{r}(i)$. This is exactly the same as the scheme for $r(t)$ alone. But in discretizing $D(t)$, we get

$$\mathcal{L}^1(-r(t)D(t)) = -\sigma(r(t))D(t)$$

and

$$\mathcal{L}^0(-r(t)D(t)) = -\mu(r(t))D(t) + r(t)^2 D(t).$$

Hence, the scheme becomes

$$\hat{D}(i+1) = \hat{D}(i)\left(1 - \hat{r}(i)h + \frac{1}{2}[\hat{r}(i)^2 - \mu(\hat{r}(i))]h^2 - \frac{1}{2}\sigma(\hat{r}(i))\Delta W h\right),$$

which involves terms not reflected in (6.40).

Again because of the smoothness of the coefficients for $D(t)$, this method has weak order 2 if the scheme for $r(t)$ itself achieves this order. Thus, the weak error criterion extends to the bond price (6.39), though it would not necessarily extend if we applied the crude discretization in (6.40).

The same idea clearly applies in computing a function of, e.g.,

$$\left(X(T), \int_0^T X(t)\, dt \right),$$

as might be required in pricing an Asian option. In contrast, incorporating path-dependence through the maximum or minimum of a process (to price a barrier or lookback option, for example) is more delicate. We can define a supplementary variable of the form

$$M(t) = \max_{0 \leq s \leq t} X(t)$$

to remove dependence on the past of X, but the method applied above with the discount factor $D(t)$ does not extend to the bivariate process $(X(t), M(t))$. The difficulty lies in the fact that the running maximum $M(t)$ does not satisfy an SDE with smooth coefficients. For example, M remains constant except when $X(t) = M(t)$. Asmussen, Glynn, and Pitman [24] show that even when X is ordinary Brownian motion (so that the Euler scheme for X is exact), the Euler scheme for M has weak order $1/2$ rather than the weak order 1 associated with smooth coefficient functions. We return to the problem of discretizing the running maximum in Section 6.4.

6.2.4 Extrapolation

An alternative approach to achieving second-order accuracy applies *Richardson extrapolation* (also called *Romberg* extrapolation) to two estimates obtained from a first-order scheme at two different levels of discretization. This is easier to implement than a second-order scheme and usually achieves roughly the same accuracy — sometimes better, sometimes worse. The same idea can (under appropriate conditions) boost the order of convergence of a second-order or even higher-order scheme, but these extensions are not as effective in practice.

To emphasize the magnitude of the time increment, we write $\hat{X}^h$ for a discretized process with step size h. We write $\hat{X}^h(T)$ for the state of the discretized process at time T; more explicitly, this is $\hat{X}^h(\lfloor T/h \rfloor h)$.

As discussed in Section 6.1.2, the Euler scheme often has weak order 1, in which case

$$|\mathsf{E}[f(\hat{X}^h(T))] - \mathsf{E}[f(X(T))]| \leq Ch \tag{6.42}$$

for some constant C, for all sufficiently small h, for suitable f. Talay and Tubaro [343], Bally and Talay [34], and Protter and Talay [301] prove that the bound in (6.42) can often be strengthened to an equality of the form

$$E[f(\hat{X}^h(T))] = E[f(X(T))] + ch + o(h),$$

for some constant c depending on f. In this case, the discretization with time step $2h$ satisfies

$$E[f(\hat{X}^{2h}(T))] = E[f(X(T))] + 2ch + o(h),$$

with the same constant c.

By combining the approximations with time steps h and $2h$, we can eliminate the leading error term. More explicitly, from the previous two equations we get

$$2E[f(\hat{X}^h(T))] - E[f(\hat{X}^{2h}(T))] = E[f(X(T))] + o(h). \tag{6.43}$$

This suggests the following algorithm: simulate with time step h to estimate $E[f(\hat{X}^h(T))]$; simulate with time step $2h$ to estimate $E[f(\hat{X}^{2h}(T))]$; double the first estimate and subtract the second to estimate $E[f(X(T))]$. The bias in this combined estimate is of smaller order than the bias in either of its two components.

Talay and Tubaro [343] and Protter and Talay [301] give conditions under which the $o(h)$ term in (6.43) is actually $O(h^2)$ (and indeed under which the error can be expanded in arbitrarily high powers of h). This means that applying extrapolation to the Euler scheme produces an estimate with weak order 2. Because the Euler scheme is easy to implement, this offers an attractive alternative to the second-order schemes derived in Sections 6.2.1 and 6.2.2.

The variance of the extrapolated estimate is typically reduced if we use consistent Brownian increments in simulating paths of $\hat{X}^h$ and $\hat{X}^{2h}$. Each Brownian increment driving $\hat{X}^{2h}$ is the sum of two of the increments driving $\hat{X}^h$. If we use $\sqrt{h}Z_1, \sqrt{h}Z_2, \ldots$ as Brownian increments for $\hat{X}^h$, we should use $\sqrt{h}(Z_1 + Z_2), \sqrt{h}(Z_3 + Z_4), \ldots$ as Brownian increments for $\hat{X}^{2h}$. Whether or not we use this construction (as opposed to, e.g., simulating the two independently) has no bearing on the validity of (6.43) because (6.43) refers only to expectations and is unaffected by any dependence between $\hat{X}^h$ and $\hat{X}^{2h}$. Observe, however, that

$$\mathsf{Var}\left[2f(\hat{X}^h(T)) - f(\hat{X}^{2h}(T))\right] = 4\mathsf{Var}\left[f(\hat{X}^h(T))\right] + \mathsf{Var}\left[f(\hat{X}^{2h}(T))\right]$$
$$-4\mathsf{Cov}\left[f(\hat{X}^h(T)), f(\hat{X}^{2h}(T))\right].$$

Making $f(\hat{X}^h(T))$ and $f(\hat{X}^{2h}(T))$ positively correlated will therefore reduce variance, even though it has no effect on discretization bias. Using consistent Brownian increments will not always produce positive correlation, but it often will. Positive correlation can be guaranteed through monotonicity conditions, for example. This issue is closely related to the effectiveness of antithetic sampling; see Section 4.2, especially the discussion surrounding (4.29).

Extrapolation can theoretically be applied to a second-order scheme to further increase the order of convergence. Suppose that we start from a scheme having weak order 2, such as the simplified scheme (6.36) or (6.38). Suppose that in fact

$$E[f(\hat{X}^h(T))] = E[f(X(T))] + ch^2 + o(h^2).$$

Then

$$\frac{1}{3}(4E[f(\hat{X}^h(T))] - E[f(\hat{X}^{2h}(T))])$$

$$= \frac{1}{3}(\{4E[f(X(T))] + 4ch^2 + o(h^2)\} - \{E[f(X(T))] + 4ch^2 + o(h^2)\})$$

$$= E[f(X(T))] + o(h^2).$$

If the $o(h^2)$ error is in fact $O(h^3)$, then the combination

$$\frac{1}{21}(32E[f(\hat{X}^h(T))] - 12E[f(\hat{X}^{2h}(T))] + E[f(\hat{X}^{4h}(T))])$$

eliminates that term too. Notice that the correct weights to apply to $\hat{X}^h$, $\hat{X}^{2h}$, and any other discretization depend on the weak order of convergence of the scheme used.

6.3 Extensions

6.3.1 General Expansions

The derivations leading to the strong first-order scheme (6.10) and the weak second-order schemes (6.28) and (6.32) generalize to produce approximations that are, in theory, of arbitrarily high weak or strong order, under conditions on the coefficient functions. These higher-order methods can be cumbersome to implement and are of questionable practical significance; but they are of considerable theoretical interest and help underscore a distinction between weak and strong approximations.

We consider a d-dimensional process X driven by an m-dimensional standard Brownian motion W through an SDE of the form

$$dX(t) = b_0(X(t))\, dt + \sum_{j=1}^{d} b_j(X(t))^\top dW(t).$$

We have written the drift coefficient as b_0 rather than a to allow more compact notation in the expansions that follow. Let $\mathcal{L}^0$ be as in (6.29) but with a_i replaced by b_{0i} and let $\mathcal{L}^k$ be as in (6.30), $k = 1, \ldots, m$.

For any $n = 1, 2, \ldots$, and any $j_1, j_2, \ldots, j_n \in \{0, 1, \ldots, m\}$, define the multiple integrals

$$I_{(j_1, j_2, \ldots, j_n)} = \int_t^{t+h} \cdots \int_t^{u_3} \int_t^{u_2} dW_{j_1}(u_1)\, dW_{j_2}(u_2) \cdots dW_{j_n}(u_n),$$

with the convention that $dW_0(u) = du$. These integrals generalize those used in Section 6.2.2. Here, t and h are arbitrary positive numbers; as in Section 6.2.2, our objective is to approximate an arbitrary increment from $X(t)$ to $X(t + h)$.

The general weak expansion of order $\beta = 1, 2, \ldots$ takes the form

$$X(t + h) \approx X(t) + \sum_{n=1}^{\beta} \sum_{j_1, \ldots, j_n} \mathcal{L}^{j_1} \cdots \mathcal{L}^{j_{n-1}} b_{j_n} I_{(j_1, \ldots, j_n)}, \tag{6.44}$$

with each j_i ranging over $0, 1, \ldots, m$. (This approximation applies to each coordinate of the vectors X and b_{j_n}.) When $\beta = 2$, this reduces to the second-order scheme in (6.32). Kloeden and Platen [211] (Section 14.5) justify the general case and provide conditions under which this approximation produces a scheme with weak order β.

In contrast to (6.44), the general *strong* expansion of order $\beta = 1/2, 1, 3/2, \ldots$ takes the form (Kloeden and Platen [211], Section 14.5)

$$X(t + h) \approx X(t) + \sum_{(j_1, \ldots, j_n) \in \mathcal{A}_\beta} \mathcal{L}^{j_1} \cdots \mathcal{L}^{j_{n-1}} b_{j_n} I_{(j_1, \ldots, j_n)}. \tag{6.45}$$

The set $\mathcal{A}_\beta$ is defined as follows. A vector of indices $(j_1, \ldots, j_n)$ is in $\mathcal{A}_\beta$ if either (i) the number of indices n plus the number of indices that are 0 is less than or equal to 2β, or (ii) $n = \beta + \frac{1}{2}$ and all n indices are 0. Thus, when $\beta = 1$, the weak expansion sums over $j = 0$ and $j = 1$ (the Euler scheme), whereas the strong expansion sums over $j = 0$, $j = 1$, and $(j_1, j_2) = (1, 1)$ to get (6.10). Kloeden and Platen [211], Section 10.6, show that (6.45) indeed results in an approximation with strong order β.

Both expansions (6.44) and (6.45) follow from repeated application of the steps we used in (6.7) and (6.22)–(6.23). The distinction between the two expansions can be summarized as follows: the weak expansion treats terms ΔW_j, $j \neq 0$, as having the same order as h, whereas the strong expansion treats them as having order $h^{1/2}$. Thus, indices equal to zero (corresponding to "dt" terms rather than "dW_j" terms) count double in reckoning the number of terms to include in the strong expansion (6.45).

6.3.2 Jump-Diffusion Processes

Let $\{N(t), t \geq 0\}$ be a Poisson process, let $\{Y_1, Y_2, \ldots\}$ be i.i.d. random vectors, W a standard multidimensional Brownian motion with N, W, and $\{Y_1, Y_2, \ldots\}$ independent of each other. Consider jump-diffusion models of the form

$$dX(t) = a(X(t-)) \, dt + b(X(t-))^\top \, dW(t) + c(X(t-), Y_{N(t-)+1}) \, dN(t). \tag{6.46}$$

Between jumps of the Poisson process, X evolves like a diffusion with coefficient functions a and b; at the nth jump of the Poisson process, the jump in X is

$$X(t) - X(t-) = c(X(t-), Y_n),$$

a function of the state of X just before the jump and the random variable Y_n. We discussed a special case of this model in Section 3.5.1. Various processes of this type are used to model the dynamics of underlying assets in pricing derivative securities.

Mikulevicius and Platen [265] extend the general weak expansion (6.44) to processes of this type and analyze the discretization schemes that follow from this expansion. Their method uses a pure-diffusion discretization method between the jumps of the Poisson process and applies the function c to the discretized process to determine the jump in the discretized process at a jump of the Poisson process. The jump magnitudes are thus computed exactly, conditional on the value of the discretized process just before a jump. Mikulevicius and Platen [265] show, in fact, that the weak order of convergence of this method equals the order of the scheme used for the pure-diffusion part, under conditions on the coefficient functions a, b, and c.

In more detail, this method supplements the original time grid $0, h, 2h, \ldots$ with the jump times of the Poisson process. Because the Poisson process is independent of the Brownian motion, we can imagine generating all of these jump times at the start of a simulation. (See Section 3.5 for a discussion of the simulation of Poisson processes.) Let $0 = \tau_0, \tau_1, \tau_2, \ldots$ be the combined time grid, including both the multiples of h and the Poisson jump times. The discretization scheme proceeds by simulating $\hat{X}$ from τ_i to τ_{i+1}, $i = 0, 1, \ldots$. Given $\hat{X}(\tau_i)$, we apply an Euler scheme or higher-order scheme to generate $\hat{X}(\tau_{i+1}-)$, using the coefficient functions a and b. If τ_{i+1} is a Poisson jump time — the nth, say — we generate Y_n and set

$$\hat{X}(\tau_{i+1}) = \hat{X}(\tau_{i+1}-) + c(\hat{X}(\tau_{i+1}-), Y_n).$$

If τ_{i+1} is not a jump time, we set $\hat{X}(\tau_{i+1}) = \hat{X}(\tau_{i+1}-)$.

Glasserman and Merener [145] apply this method to a version of the LIBOR market model with jumps. The jump processes they consider are more general than Poisson processes, having arrival rates that depend on the current level of forward rates. They extend the method to this setting by using a bound on the state-dependent arrival rates to construct jumps by thinning a Poisson process. This requires relaxing the smoothness conditions imposed on c in Mikulevicius and Platen [265].

Maghsoodi [245] provides various alternative discretization schemes for jump-diffusion processes and considers both strong and weak error criteria. He distinguishes jump-adapted methods (like the one described above) that include the Poisson jump epochs in the time grid from those that use a fixed grid. A jump-adapted method may become computationally burdensome if the jump intensity is very high. Protter and Talay [301] analyze the Euler scheme for stochastic differential equations driven by Lévy processes, which include (6.46) as a special case. Among other results, they provide error expansions in powers of h justifying the use of Richardson extrapolation.

6.3.3 Convergence of Mean Square Error

The availability of discretization schemes of various orders poses a tradeoff: using a higher-order scheme requires more computing time per path and thus reduces the number of paths that can be completed in a fixed amount of time. The number of paths completed affects the standard error of any estimates we compute, but has no effect on discretization bias, which is determined by our choice of scheme. Thus, we face a tradeoff between reducing bias and reducing variance.

We discussed this tradeoff in a more general setting in Section 1.1.3 and the asymptotic conclusions reached there apply in the current setting. Here we present a slightly different argument to arrive at the same conclusion.

We suppose that our objective is to minimize mean square error (MSE), the sum of variance and squared bias. Using a discretization scheme of weak order β, we expect

$$\text{Bias} \approx c_1 h^\beta$$

for some constant c_1. For the variance based on n paths we expect

$$\text{Variance} \approx \frac{c_2}{n}$$

for some constant c_2. The time step h would generally have some effect on variance; think of c_2 as the limit as h decreases to zero of the variance per replication.

If we make the reasonable assumption that the computing time per path is proportional to the number of steps per path, then it is inversely proportional to h. The total computing time for n paths is then nc_3/h, for some constant c_3.

With these assumptions and approximations, we formulate the problem of minimizing MSE subject to a computational budget s as follows:

$$\min_{n,h} \left(c_1^2 h^{2\beta} + \frac{c_2}{n} \right) \quad \text{subject to} \quad \frac{nc_3}{h} = s.$$

Using the constraint to eliminate a variable, we put this in the form

$$\min_h \left(c_1^2 h^{2\beta} + \frac{c_2 c_3}{hs} \right),$$

which is minimized at

$$h = cs^{-\frac{1}{2\beta+1}} \tag{6.47}$$

with c a constant. Substituting this back into our expressions for the squared bias and the variance, we get

$$\text{MSE} \approx c_1' s^{-\frac{2\beta}{2\beta+1}} + c_2' s^{-\frac{2\beta}{2\beta+1}} = c' s^{-\frac{2\beta}{2\beta+1}},$$

for some constants c', c_1', c_2'. The optimal allocation thus balances variance and squared bias. Also, the optimal root mean square error becomes

$$\sqrt{\text{MSE}} \propto s^{-\frac{\beta}{2\beta+1}}. \tag{6.48}$$

This is what we found in Section 1.1.3 as well.

These calculations show how the order β of a scheme affects both the optimal allocation of effort and the convergence rate under the optimal allocation. As the convergence order β increases, the optimal convergence rate in (6.48) approaches $s^{-1/2}$, the rate associated with unbiased simulation. But for the important cases of $\beta = 1$ (first-order) and $\beta = 2$ (second-order) we get rates of $s^{-1/3}$ and $s^{-2/5}$. This makes precise the notion that simulating a process for which a discretization scheme is necessary is harder than simulating a solvable model. It also shows that when very accurate results are required (i.e., when s is large), a higher-order scheme will ultimately dominate a lower-order scheme.

Duffie and Glynn [100] prove a limit theorem that justifies the convergence rate implied by (6.48). They also report numerical results that are generally consistent with their theoretical predictions.

6.4 Extremes and Barrier Crossings: Brownian Interpolation

In Section 6.2.3 we showed that discretization methods can sometimes be extended to path-dependent payoffs through supplementary state variables. The additional state variables remove dependence on the past; standard discretization procedures can then be applied to the augmented state vector.

In option pricing applications, path-dependence often enters through the maximum or minimum of an underlying asset over the life of the option. This includes, for example, options whose payoffs depend on whether or not an underlying asset crosses a barrier. Here, too, path-dependence can be eliminated by including the running maximum or minimum in the state vector. However, this renders standard discretization procedures inapplicable because of the singular dynamics of these supplementary variables. The running maximum, for example, can increase only when it is equal to the underlying process.

This issue arises even when the underlying process X is a standard Brownian motion. Let

$$M(t) = \max_{0 \le u \le t} X(u)$$

and let

$$\hat{M}^h(n) = \max\{X(0), X(h), X(2h), \dots, X(nh)\}. \tag{6.49}$$

Then $\hat{M}^h(n)$ is the maximum of the Euler approximation to X over $[0, nh]$; the Euler approximation to X is exact for X itself because X is Brownian motion. Fix a time T and let $h = T/n$ so that $\hat{M}^h(n)$ is the discrete-time approximation of $M(T)$. Asmussen, Glynn, and Pitman [24] show that the normalized error

$$h^{-1/2}[\hat{M}^h(n) - M(T)]$$

has a limiting distribution as $h \to 0$. This result may be paraphrased as stating that the distribution of $\hat{M}^h(n)$ converges to that of $M(T)$ at rate $h^{1/2}$. It follows that the weak order of convergence (in the sense of (6.14)) cannot be greater than $1/2$. In contrast, we noted in Section 6.1.2 that for SDEs with smooth coefficient functions the Euler scheme has weak order of convergence 1. Thus, the singularity of the dynamics of the running maximum leads to a slower convergence rate.

In the case of Brownian motion, this difficulty can be circumvented by sampling $M(T)$ directly, rather than through (6.49). We can sample from the joint distribution of $X(T)$ and $M(T)$ as follows. First we generate $X(T)$ from $N(0,T)$. Conditional on $X(T)$ the process $\{X(t), 0 \le t \le T\}$ becomes a Brownian bridge, so we need to sample from the distribution of the maximum of a Brownian bridge. We discussed how to do this in Example 2.2.3. Given $X(T)$, set

$$M(T) = \frac{X(T) + \sqrt{X(T)^2 - 2T \log U}}{2}$$

with $U \sim \text{Unif}[0,1]$ independent of $X(T)$. The pair $(X(T), M(T))$ then has the joint distribution of the terminal and maximum value of the Brownian motion over $[0, T]$.

This procedure, exact for Brownian motion, suggests an approximation for more general processes. Suppose X is a diffusion satisfying the SDE (6.1) with scalar coefficient functions a and b. Let $\hat{X}(i) = \hat{X}(ih)$, $i = 0, 1, \ldots$, be a discrete-time approximation to X, such as one defined through an Euler or higher-order scheme. The simple estimate (6.49) applied to $\hat{X}$ is equivalent to taking the maximum over a piecewise linear interpolation of $\hat{X}$. We can expect to get a better approximation by interpolating over the interval $[ih, (i+1)h)$ using a Brownian motion with fixed parameters $a_i = a(\hat{X}(i))$ and $b_i = b(\hat{X}(i))$. Given the endpoints $\hat{X}(i)$ and $\hat{X}((i+1))$, the maximum of the interpolating Brownian bridge can be simulated using

$$\hat{M}_i = \frac{\hat{X}(i+1) + \hat{X}(i) + \sqrt{[\hat{X}(i+1) - \hat{X}(i)]^2 - 2b_i^2 h \log U_i}}{2}, \qquad (6.50)$$

with $U_0, U_1, \ldots$ independent $\text{Unif}[0,1]$ random variables. (The value of $a(\hat{X}(i))$ becomes immaterial once we condition on $\hat{X}(i+1)$.) The maximum of X over $[0, T]$ can then be approximated using

$$\max\{\hat{M}_0, \hat{M}_1, \ldots, \hat{M}_{n-1}\}.$$

Similar ideas are suggested in Andersen and Brotherton-Ratcliffe [16] and in Beaglehole, Dybvig, and Zhou [41] for pricing lookback options; their numerical results indicate that the approach can be very effective. Baldi [31] analyzes related techniques in a much more general setting.

In some applications, X may be better approximated by geometric Brownian motion than by ordinary Brownian motion. This can be accommodated by applying (6.50) to $\log \hat{X}$ rather than $\hat{X}$. This yields

$$\log \hat{M}_i = \frac{\log(\hat{X}(i+1)\hat{X}(i)) + \sqrt{[\log(\hat{X}(i+1)/\hat{X}(i))]^2 - 2(b_i/\hat{X}(i))^2 h \log U_i}}{2},$$

and exponentiating produces $\hat{M}_i$.

Barrier Crossings

Similar ideas apply in pricing barrier options with continuously monitored barriers. Suppose $B > X(0)$ and let

$$\tau = \inf\{t \geq 0 : X(t) > B\}.$$

A knock-out option might have a payoff of the form

$$(K - X(T))^+ \mathbf{1}\{\tau > T\}, \tag{6.51}$$

with K a constant. This requires simulation of $X(T)$ and the indicator $\mathbf{1}\{\tau > T\}$.

The simplest method sets

$$\hat{\tau} = \inf\{i : \hat{X}(i) > B\}$$

and approximates $(X(T), \mathbf{1}\{\tau > T\})$ by $(\hat{X}(n), \mathbf{1}\{\hat{\tau} > n\})$ with $h = T/n$, for some discretizaton $\hat{X}$. But even if we could simulate X exactly on the discrete grid $0, h, 2h, \ldots$, this would not sample $\mathbf{1}\{\tau > T\}$ exactly: it is possible for X to cross the barrier at some time t between grid points ih and $(i+1)h$ and never be above the barrier at any of the dates $0, h, 2h, \ldots$.

The method in (6.50) can be used to reduce discretization error in sampling the survival indicator $\mathbf{1}\{\tau > T\}$. Observe that the barrier is crossed in the interval $[ih, (i+1)h)$ precisely if the maximum over this interval exceeds B. Hence, we can approximate the survival indicator $\mathbf{1}\{\tau > T\}$ using

$$\prod_{i=0}^{n-1} \mathbf{1}\{\hat{M}_i \leq B\}, \tag{6.52}$$

with $nh = T$ and $\hat{M}_i$ as in (6.50).

This method can be simplified. Rather than generate $\hat{M}_i$, we can sample the indicators $\mathbf{1}\{\hat{M}_i \leq B\}$ directly. Given $\hat{X}(i)$ and $\hat{X}(i+1)$, this indicator takes the value 1 with probability

$$\hat{p}_i = P(\hat{M}_i \leq B | \hat{X}(i), \hat{X}(i+1)) = 1 - \exp\left(-\frac{2(B - \hat{X}(i))(B - \hat{X}(i+1))}{b(\hat{X}(i))^2 h}\right),$$

(assuming B is greater than both $\hat{X}(i)$ and $\hat{X}(i+1)$) and it takes the value 0 with probability $1 - \hat{p}_i$. Thus, we can approximate $\mathbf{1}\{\tau > T\}$ using

$$\prod_{i=0}^{n-1} \mathbf{1}\{U_i \le \hat{p}_i\}.$$

For fixed $U_0, U_1, \ldots, U_{n-1}$, this has the same value as (6.52) but is slightly simpler to evaluate. The probabilities $\hat{p}_i$ could alternatively be computed based on an approximating geometric (rather than ordinary) Brownian motion.

The discretized process $\hat{X}$ is often a Markov process and this leads to further simplification. Consider, for example, the payoff in (6.51). Using (6.52), we approximate the payoff as

$$(K - \hat{X}(n))^+ \prod_{i=0}^{n-1} \mathbf{1}\{\hat{M}_i \le B\}. \tag{6.53}$$

The conditional expectation of this expression given the values of $\hat{X}$ is

$$\mathsf{E}\left[(K - \hat{X}(n))^+ \prod_{i=0}^{n-1} \mathbf{1}\{\hat{M}_i \le B\} | \hat{X}(0), \hat{X}(1), \ldots, \hat{X}(n)\right] \tag{6.54}$$

$$= (K - \hat{X}(n))^+ \prod_{i=0}^{n-1} \mathsf{E}[\mathbf{1}\{\hat{M}_i \le B\} | \hat{X}(i), \hat{X}(i+1)]$$

$$= (K - \hat{X}(n))^+ \prod_{i=0}^{n-1} \hat{p}_i. \tag{6.55}$$

Thus, rather than generate the barrier-crossing indicators, we can just multiply by the probabilities $\hat{p}_i$.

Because (6.55) is the conditional expectation of (6.53), the two have the same expectation and thus the same discretization bias. By Jensen's inequality, the second moment of (6.53) is larger than the second moment of its conditional expectation (6.55), so using (6.55) rather than (6.53) reduces variance. (This is an instance of a more general strategy for reducing variance known as *conditional Monte Carlo*, based on replacing an estimator with its conditional expectation; see, Boyle et al. [53] for other applications in finance.) Using (6.53), we would stop simulating a path once some $\hat{M}_i$ exceeds B. Using (6.55), we never generate the $\hat{M}_i$ and must therefore simulate every path for n steps, unless some $\hat{X}(i)$ exceeds B (in which case $\hat{p}_i = 0$). So, although (6.55) has lower variance, it requires greater computational effort per path. A closely related tradeoff is investigated by Glasserman and Staum [146]; they consider estimators in which each transition of an underlying asset is sampled conditional on not crossing a barrier. In their setting, products of survival probabilities like those in (6.55) serve as likelihood ratios relating the conditional and unconditional evolution of the process.

Baldi, Caramellino, and Iovino [32] develop methods for reducing discretization error in a general class of barrier option simulation problems. They consider single- and double-barrier options with time-varying barriers and develop approximations to the one-step survival probabilities that refine the $\hat{p}_i$ above. The $\hat{p}_i$ are based on a single constant barrier and a Brownian approximation over a time interval of length h. Baldi et al. [32] derive asymptotics of the survival probabilities as $h \to 0$ for quite general diffusions based, in part, on a linear approximation to upper and lower barriers.

Averages Revisited

As already noted, the simulation estimators based on (6.50) or (6.52) can be viewed as the result of using Brownian motion to interpolate between the points $\hat{X}(i)$ and $\hat{X}(i+1)$ in a discretization scheme. The same idea can be applied in simulating other path-dependent quantities besides extremes and barrier-crossing indicators.

As an example, consider simulation of the pair

$$\left(X(T), \int_0^T X(t)\, dt \right)$$

for some scalar diffusion X. In Section 6.2.3, we suggested treating this pair as the state at time T of a bivariate diffusion and applying a discretization method to this augmented process. An alternative simulates a discretization $\hat{X}(i)$, $i = 0, 1, \ldots, n$, and uses Brownian interpolation to approximate the integral. More explicitly, the approximation is

$$\int_0^T X(t)\, dt \approx \sum_{i=0}^{n-1} b_i \hat{A}_i,$$

with $b_i = b(\hat{X}(i))$ and each $\hat{A}_i$ sampled from the distribution of

$$\int_t^{t+h} W(u)\, du, \quad W \sim \mathrm{BM}(0,1),$$

conditional on $W(t) = \hat{X}(i)$ and $W(t + h) = \hat{X}(i+1)$. The calculations used to derive (6.27) show that this conditional distribution is normal with mean $h(\hat{X}(i+1) + \hat{X}(i))/2$ and variance $h^3/3$. Thus, the $\hat{A}_i$ are easily generated.

This leads to a discretization scheme only slightly different from the one arrived at through the approach in Section 6.2.3. Because of the relative smoothness of the running integral, the effect of Brownian interpolation in this setting is minor compared to the benefit in simulating extremes or barrier crossings.

6.5 Changing Variables

We conclude our discussion of discretization methods by considering the flexibility to change variables through invertible transformations of a process. If X is a d-dimensional diffusion and $g : \Re^d \to \Re^d$ is a smooth, invertible transformation, we can define a process $Y(t) = g(X(t))$, simulate a discretization $\hat{Y}$, and define $\hat{X} = g^{-1}(\hat{Y})$ to get a discretization of the original process X. Thus, even if we restrict ourselves to a particular discretization method (an Euler or higher-order scheme), we have a great deal of flexibility in how we implement it. Changing variables has the potential to reduce bias and can also be useful in enforcing restrictions (such as nonnegativity) on simulated values. There is little theory available to guide such transformations; we illustrate the idea with some examples.

Taking Logarithms

Many of the stochastic processes that arise in mathematical finance take only positive values. This property often results from specifying that the diffusion term be proportional to the current level of the process, as in geometric Brownian motion and in the LIBOR market model of Section 3.7. If the coordinates of a d-dimensional process X are positive, we may define $Y_i(t) = \log X_i(t)$, $i = 1, \ldots, d$, apply Itô's formula to derive an SDE satisfied by $Y = (Y_1, \ldots, Y_d)$, simulate a discretization $\hat{Y}$ of Y, and then (if they are needed) approximate the original X_i with $\hat{X}_i = \exp(\hat{Y}_i)$. We encountered this idea in Section 3.7.3 in the setting of the LIBOR market model.

   Applying a logarithmic transformation can have several benefits. First, it ensures that the simulated $\hat{X}_i$ are positive because they result from exponentiation, whereas even a high-order scheme applied directly to the dynamics of X will produce some negative values. Keeping the variables positive can be important if the variables represent asset prices or interest rates.

   Second, a logarithmic transformation can enhance the numerical stability of a discretization method, meaning that it can reduce the propagation of round-off error. A process with "additive noise" can generally be simulated with less numerical error than a process with "multiplicative noise." Numerical stability is discussed in greater detail in Kloeden and Platen [211].

   Third, a logarithmic transformation can reduce discretization bias. For example, an Euler scheme applied to geometric Brownian motion becomes exact if we first take logarithms. More generally, if the coefficients of a diffusion are nearly proportional to the level of the process, then the coefficitions of the log process are nearly constant.

   This idea is illustrated in Figure 6.3, which is similar to examples in Glasserman and Merener [145] and is based on numerical results obtained by Nicolas Merener. The figure shows estimated biases in pricing a six-month caplet maturing in 20 years using the LIBOR market model with the decreasing volatility parameters used in Table 4.2. The largest practical time step in this setting is the length of the accrual period; the figure compares methods using one, two, and four steps per accrual period with between four and 20 million replications per method. In this example, taking logarithms cuts the absolute bias roughly in half for the Euler scheme. The figure also shows results using a second-order method for rates ($\times$) and log rates ($\circ$); even with 10 million replications, the mean errors in these methods are not statistically distinguishable from zero. In a LIBOR market model with jumps, Glasserman and Merener [145] find experimentally that a first-order scheme applied to log rates is as accurate as a second-order scheme applied to the rates themselves.

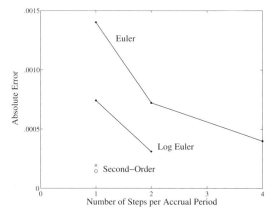

Fig. 6.3. Estimated bias versus number of steps in caplet pricing. The $\times$ and $\circ$ correspond to second-order schemes for rates and log rates, respectively.

As an aside, we note that the most time-consuming part of simulating an Euler approximation to a LIBOR market model is evaluating the drift coefficient. To save computing time, one might use the same value of the drift for multiple time steps, along the lines of Example 4.1.4, but with occasional updating of the drift. A similar idea is found to be effective in Hunter et al. [192] as part of a predictor-corrector method (of the type discussed in Chapter 15 of Kloeden and Platen [211]).

Imposing Upper and Lower Bounds

Taking logarithms enforces a lower bound at zero on a discretized process. Suppose the coordinates of X are known to evolve in the interval $(0, 1)$. How could this be imposed on a discretization of X? Enforcing such a condition could be important if, for example, the coordinates of X are zero-coupon bond prices, which should always be positive and should never exceed their face value of 1.

Glasserman and Wang [149] consider the transformations

$$Y_i = \Phi^{-1}(X_i) \quad \text{and} \quad Y_i = \log\left(\frac{X_i}{1 - X_i}\right), \tag{6.56}$$

both of which are increasing functions from $(0,1)$ onto the real line. Itô's formula produces an SDE for Y from an SDE for X in either case; the resulting SDE can be discretized and then the inverse transformations applied to produce

$$\hat{X}_i = \Phi(\hat{Y}_i) \quad \text{or} \quad \hat{X}_i = \frac{\exp(\hat{Y}_i)}{1 + \exp(\hat{Y}_i)}.$$

If the coordinates of X correspond to bonds of increasing maturities, we may also want to enforce an ordering of the form

$$1 \geq X_1(t) \geq X_2(t) \geq \cdots \geq X_d(t) \geq 1 \tag{6.57}$$

in the discretization. The method in [149] accomplishes this by defining $Y_1 = g(X_1)$, $Y_i = g(X_i/X_{i-1})$, $i = 2, \ldots, d$, with g an increasing function from $(0,1)$ to $\Re$ as in (6.56). Itô's formula gives the dynamics of $(Y_1, \ldots, Y_d)$ and then a discretization $(\hat{Y}_1, \ldots, \hat{Y}_d)$; applying the inverse of g produces a discretization $(\hat{X}_1, \ldots, \hat{X}_d)$ satsifying (6.57).

Constant Diffusion Transformation

Consider the SDE (6.1) in the case of scalar X and W. Suppose there exists an invertible, twice continuously differentiable transformation $g : \Re \rightarrow \Re$ for which $g'(x) = 1/b(x)$. With $Y(t) = g(X(t))$, Itô's formula gives

$$\begin{aligned}
dY(t) &= \left[a(X(t))g'(X(t)) + \tfrac{1}{2}b^2(X(t))g''(X(t))\right] dt + g'(X(t))b(X(t)) \, dW(t) \\
&= \tilde{a}(Y(t)) \, dt + dW(t),
\end{aligned}$$

with $\tilde{a}(y) = a(f(y))g'(f(y)) + \tfrac{1}{2}b^2(f(y))g''(f(y))$ and f the inverse of g. Changing variables from X to Y thus produces an SDE with a constant diffusion coefficient. This is potentially useful in reducing discretization error, though the impact on the drift cannot be disregarded. Moving all state-dependence from the diffusion to the drift is attractive in combination with a predictor-corrector method, which improves accuracy by averaging current and future levels of the drift coefficient.

To illustrate the constant diffusion transformation we apply it to the square-root diffusion

$$dX(t) = \alpha(\bar{x} - X(t)) \, dt + \sigma\sqrt{X(t)} \, dW(t).$$

Let $Y(t) = 2\sqrt{X}/\sigma$; then Y is a Bessel-like process,

$$dY(t) = \left[\left(\frac{4\alpha\bar{x} - \sigma^2}{2\sigma^2}\right)\frac{1}{Y(t)} - \frac{\alpha}{2}Y(t)\right] dt + dW(t).$$

In Section 3.4 we imposed the condition $2\alpha\bar{x} \geq \sigma^2$, and this is precisely the condition required to ensure that Y never reaches zero if $Y(0) > 0$.

Now suppose X and W take values in $\Re^d$ and suppose the $d \times d$ matrix $b(x)$ has inverse $c(x)$. Aït-Sahalia [7] shows that there is an invertible transformation $g : \Re^d \to \Re^d$ such that the diffusion matrix of $Y(t) = g(X(t))$ is the identity if and only if

$$\frac{\partial c_{ij}}{\partial x_k} = \frac{\partial c_{ik}}{\partial x_j},$$

for all $i, j, k = 1, \dots, d$. This condition implies the commutativity condition (6.33). It follows that processes X that can be transformed to have a constant diffusion matrix are also processes for which a second-order scheme is comparatively easy to implement.

Martingale Discretization

The requirement that discounted asset prices be martingales is central to the pricing of derivative securities. The martingale property is usually imposed in a continuous-time model; but as we have noted at several points, it is desirable to enforce the property on the simulated approximation as well. Doing so extends the no-arbitrage property to prices computed from a simulation. It also preserves the internal consistency of a simulated model by ensuring that prices of underlying assets computed in a simulation coincide with those used as inputs to the simulation; cf. the discussions in Sections 3.3.2, 3.6.2, 3.7.3, Example 4.1.3, and Section 4.5.

If the SDE (6.1) has drift coefficient a identically equal to zero, then the Euler and higher-order methods in Sections 6.1.1 and 6.2 all produce discretizations $\hat{X}$ that are discrete-time martingales. In this sense, the martingale property is almost trivially preserved by the discretization methods. However, the variables simulated do not always coincide with the variables to which the martingale property applies. For example, in pricing fixed-income derivatives we often simulate interest rates but the martingale property applies to discounted bonds and not to the interest rates themselves.

The simulation method developed for the Heath-Jarrow-Morton framework in Section 3.6.2 may be viewed as a nonstandard Euler scheme — nonstandard because of the modified drift coefficient. We modified the drift in forward rates precisely so that the discretized discounted bond prices would be martingales. In the LIBOR market models of Section 3.7, we argued that an analogous drift modification is infeasible. Instead, in Section 3.7.3 we discussed ways of preserving the martingale property through changes of variables. These methods include discretizing the discounted bond prices directly or discretizing their differences.

Preserving the martingale property is sometimes at odds with preserving bounds on variables. If X is a positive martingale then the Euler approximation to X is a martingale but not, in general, a positive process. Both properties can usually be enforced by discretizing $\log X$ instead. If X is a martingale taking values in $(0,1)$, then defining Y as in (6.56), discretizing Y, and then inverting the transformation preserves the bounds on X but not the martingale property. Using the transformation Φ^{-1}, a simple correction to the drift preserves the martingale property as well as the bounds. As shown in [149], when applied to the multidimensional constraint (6.57), this method preserves the constraints and limits departures from the martingale property to terms that are $o(h)$, with h the simulation time step.

6.6 Concluding Remarks

The benchmark method for reducing discretization error is the combination of the Euler scheme with two-point extrapolation, as in Section 6.2.4. This technique is easy to implement and is usually faster than a second-order expansion for comparable accuracy, especially in high dimensions. The smoothness conditions required on the coefficient functions to ensure the validity of extrapolation are not always satisfied in financial engineering applications, but similar conditions underlie the theoretical support for second-order approximations.

Second-order and higher-order schemes do have practical applications, but they should always be compared with the simpler alternative of an extrapolated Euler approximation. A theoretical comparison is usually difficult, but the magnitudes of the derivatives of the coefficient functions can provide some indication, with large derivatives unfavorable for higher-order methods. Talay and Tubaro [343] analyze specific models for which they derive explicit expressions for error terms. They show, for example, that the refinement in (6.10) can produce larger errors than an Euler approximation.

Of the many expansions discussed in this chapter, the most useful (in addition to the extrapolated Euler scheme) are (6.28) for scalar processes and (6.38) for vector processes.

For specific applications in financial engineering, a technique that uses information about the problem context is often more effective than a general-purpose expansion. For example, even a simple logarithmic change of variables can reduce discretization error, with the added benefit of ensuring that positive variables stay positive. The methods of Section 6.4 provide another illustration: they offer simple and effective ways of reducing discretization error in pricing options sensitive to barrier crossings and extreme values of the underlying assets. The simulation methods in Sections 3.6 and 3.7 also address discretization issues for specific models.

The literature on discretization methods for stochastic differential equations offers many modifications of the basic schemes discussed in Sections 6.1.1, 6.2, and 6.3.1. Kloeden and Platen [211] provide a comprehensive treatment of these methods. Another direction of research investigates the law of the error of a discretization method, as in Jacod and Protter [193].

This chapter has discussed the use of second-order and higher-order expansions in simulation, but the same techniques are sometimes useful in deriving approximations without recourse to simulation.

7

Estimating Sensitivities

Previous chapters have addressed various aspects of estimating expectations with a view toward computing the prices of derivative securities. This chapter develops methods for estimating *sensitivities* of expectations, in particular the derivatives of derivative prices commonly referred to as "Greeks." From the discussion in Section 1.2.1, we know that in an idealized setting of continuous trading in a complete market, the payoff of a contingent claim can be manufactured (or hedged) through trading in underlying assets. The risk in a short position in an option, for example, is offset by a delta-hedging strategy of holding delta units of each underlying asset, where delta is simply the partial derivative of the option price with respect to the current price of that underlying asset. Implementation of the strategy requires knowledge of these price sensitivities; sensitivities with respect to other parameters are also widely used to measure and manage risk. Whereas the prices themselves can often be observed in the market, their sensitivites cannot, so accurate calculation of sensitivities is arguably even more important than calculation of prices. We will see, however, that derivative estimation presents both theoretical and practical challenges to Monte Carlo simulation.

The methods for estimating sensitivities discussed in this chapter fall into two broad categories: methods that involve simulating at two or more values of the parameter of differentiation and methods that do not. The first category — finite-difference approximations — are at least superficially easier to understand and implement; but because they produce biased estimates their use requires balancing bias and variance. Methods in the second category, when applicable, produce unbiased estimates. They accomplish this by using information about the simulated stochastic process to replace numerical differentiation with exact calculations. The *pathwise method* differentiates each simulated outcome with respect to the parameter of interest; the *likelihood ratio method* differentiates a probability density rather than an outcome.

7.1 Finite-Difference Approximations

Consider a model that depends on a parameter θ ranging over some interval of the real line. Suppose that for each value of θ we have a mechanism for generating a random variable $Y(\theta)$, representing the output of the model at parameter θ. Let

$$\alpha(\theta) = \mathsf{E}[Y(\theta)].$$

The derivative estimation problem consists of finding a way to estimate $\alpha'(\theta)$, the derivative of α with respect to θ.

In the application to option pricing, $Y(\theta)$ is the discounted payoff of an option, $\alpha(\theta)$ is its price, and θ could be any of the many model or market parameters that influence the price. When θ is the initial price of an underlying asset, then $\alpha'(\theta)$ is the option's delta (with respect to that asset). The second derivative $\alpha''(\theta)$ is the option's gamma. When θ is a volatility parameter, $\alpha'(\theta)$ is often called "vega." For interest rate derivatives, the sensitivities of prices to the initial term structure (as represented by, e.g., a yield curve or forward curve) are important.

7.1.1 Bias and Variance

An obvious approach to derivative estimation proceeds as follows. Simulate independent replications $Y_1(\theta), \ldots, Y_n(\theta)$ of the model at parameter θ and n additional replications $Y_1(\theta + h), \ldots, Y_n(\theta + h)$ at $\theta + h$, for some $h > 0$. Average each set of replications to get $\bar{Y}_n(\theta)$ and $\bar{Y}_n(\theta + h)$ and form the forward-difference estimator

$$\hat{\Delta}_F \equiv \hat{\Delta}_F(n, h) = \frac{\bar{Y}_n(\theta + h) - \bar{Y}_n(\theta)}{h}. \qquad (7.1)$$

This estimator has expectation

$$\mathsf{E}[\hat{\Delta}_F] = h^{-1}[\alpha(\theta + h) - \alpha(\theta)]. \qquad (7.2)$$

We have not specified what relation, if any, holds between the outcomes $Y_i(\theta)$ and $Y_i(\theta + h)$; this will be important when we consider variance, but (7.2) is purely a property of the marginal distributions of the outcomes at θ and $\theta + h$.

If α is twice differentiable at θ, then

$$\alpha(\theta + h) = \alpha(\theta) + \alpha'(\theta)h + \tfrac{1}{2}\alpha''(\theta)h^2 + o(h^2).$$

In this case, it follows from (7.2) that the bias in the forward-difference estimator is

$$\mathrm{Bias}(\hat{\Delta}_F) = \mathsf{E}[\hat{\Delta}_F - \alpha'(\theta)] = \tfrac{1}{2}\alpha''(\theta)h + o(h). \qquad (7.3)$$

By simulating at $\theta - h$ and $\theta + h$, we can form a central-difference estimator

$$\hat{\Delta}_C \equiv \hat{\Delta}_C(n, h) = \frac{\bar{Y}_n(\theta + h) - \bar{Y}_n(\theta - h)}{2h}. \tag{7.4}$$

This estimator is often more costly than the forward-difference estimator in the following sense. If we are ultimately interested in estimating $\alpha(\theta)$ and $\alpha'(\theta)$, then we would ordinarily simulate at θ to estimate $\alpha(\theta)$, so the forward-difference estimator requires simulating at just one additional point $\theta + h$, whereas the central-difference estimator requires simulating at two additional points. But this additional computational effort yields an improvement in the convergence rate of the bias. If α is at least twice differentiable in a neighborhood of θ, then

$$\alpha(\theta + h) = \alpha(\theta) + \alpha'(\theta)h + \alpha''(\theta)h^2/2 + o(h^2)$$
$$\alpha(\theta - h) = \alpha(\theta) - \alpha'(\theta)h + \alpha''(\theta)h^2/2 + o(h^2),$$

so subtraction eliminates the second-order terms, leaving

$$\text{Bias}(\hat{\Delta}_C) = \frac{\alpha(\theta + h) - \alpha(\theta - h)}{2h} - \alpha'(\theta) = o(h), \tag{7.5}$$

which is of smaller order than (7.3). If α'' is itself differentiable at θ, we can refine (7.5) to

$$\text{Bias}(\hat{\Delta}_C) = \tfrac{1}{6}\alpha'''(\theta)h^2 + o(h^2). \tag{7.6}$$

The superior accuracy of a central-difference approximation compared with a forward-difference approximation is illustrated in Figure 7.1. The curve in the figure plots the Black-Scholes formula against the price of the underlying asset with volatility 0.30, an interest rate of 5%, a strike price of 100, and 0.04 years (about two weeks) to expiration. The figure compares the tangent line at 95 with a forward difference calculated from prices at 95 and 100 and a central difference using prices at 90 and 100. The slope of the central-difference line is clearly much closer to that of the tangent line.

In numerical differentiation of functions evaluated through deterministic algorithms, rounding errors resulting from small values of h limit the accuracy of finite-difference approximations. (See, for example, the discussion in Section 5.7 of Press et al. [299].) In applications of Monte Carlo, the variability in estimates of function values usually prevents us from taking very small values of h. So while it is advisable to be aware of possible round-off errors, this is seldom the main obstacle to accurate estimation of derivatives in simulation.

Variance

The form of the bias of the forward- and central-difference estimators would lead us to take ever smaller values of h to improve accuracy, at least if we ignore the limits of machine precision. But the effect of h on bias must be weighed against its effect on variance.

The variance of the forward-difference estimator (7.1) is

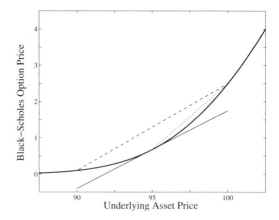

Fig. 7.1. Comparison of central-difference approximation (dashed) and forward-difference approximation (dotted) with exact tangent to the Black-Scholes formula.

$$\text{Var}[\hat{\Delta}_F(n, h)] = h^{-2}\text{Var}\left[\bar{Y}_n(\theta + h) - \bar{Y}_n(\theta)\right] \tag{7.7}$$

and a corresponding expression holds for the central-difference estimator (7.4). In both cases, the factor h^{-2} alerts us to possibly disastrous consequences of taking h to be very small. Equation (7.7) also makes clear that the dependence between values simulated at different values of θ affects the variance of a finite-difference estimator.

Suppose, for simplicity, that the pairs $(Y(\theta), Y(\theta + h))$ and $(Y_i(\theta), Y_i(\theta + h))$, $i = 1, 2, \ldots$, are i.i.d. so that

$$\text{Var}\left[\bar{Y}_n(\theta + h) - \bar{Y}_n(\theta)\right] = \frac{1}{n}\text{Var}\left[Y(\theta + h) - Y(\theta)\right].$$

How the variance in (7.7) changes with h is determined by the dependence of $\text{Var}[Y(\theta + h) - Y(\theta)]$ on h.

Three cases of primary importance arise in practice:

$$\text{Var}[Y(\theta + h) - Y(\theta)] = \begin{cases} O(1), & \text{Case (i)} \\ O(h), & \text{Case (ii)} \\ O(h^2), & \text{Case (iii)}. \end{cases} \tag{7.8}$$

Case (i) applies if we simulate $Y(\theta)$ and $Y(\theta + h)$ independently; for in this case we have

$$\text{Var}[Y(\theta + h) - Y(\theta)] = \text{Var}[Y(\theta + h)] + \text{Var}[Y(\theta)] \to 2\text{Var}[Y(\theta)],$$

under the minor assumption that $\text{Var}[Y(\theta)]$ is continuous in θ. Case (ii) is the typical consequence of simulating $Y(\theta + h)$ and $Y(\theta)$ using *common random numbers*; i.e., generating them from the same sequence $U_1, U_2, \ldots$ of Unif[0,1] random variables. (In practice, this is accomplished by starting the simulations at θ and $\theta + h$ with the same seed for the random number generator.) For Case (iii) to hold, we generally need not only that $Y(\theta)$ and $Y(\theta + h)$ use the same random numbers, but also that for (almost) all values of the random numbers, the output $Y(\cdot)$ is continuous in the input θ. We will have much more to say about Case (iii) in Section 7.2.2.

7.1.2 Optimal Mean Square Error

Because decreasing h can increase variance while decreasing bias, minimizing mean square error (MSE) requires balancing these two considerations. Increasing the number of replications n decreases variance with no effect on bias, whereas h affects both bias and variance; our objective is to find the optimal relation between the two. This tradeoff is analyzed by Glynn [153], Fox and Glynn [128], and Zazanis [358], and in early work by Frolov and Chentsov [130].

Consider the forward-difference estimator with independent simulation at θ and $\theta + h$. We can reasonably assume that Case (i) of (7.8) applies, and for emphasis we denote this estimator by $\hat{\Delta}_{F,i} = \hat{\Delta}_{F,i}(n, h)$. Squaring the bias in (7.3) and adding it to the variance in (7.7), we get

$$\mathrm{MSE}(\hat{\Delta}_{F,i}(n, h)) = O(h^2) + O(n^{-1}h^{-2}),$$

from which we see that minimal conditions for convergence are $h \to 0$ and $nh^2 \to \infty$.

To derive a more precise conclusion, we strengthen Cases (i) and (ii) of (7.8). We have four estimators to consider: forward-difference and central-difference using independent sampling or common random numbers at different values of θ. We can give a unified treatment of these cases by considering a generic estimator $\hat{\Delta} = \hat{\Delta}(n, h)$ for which

$$\mathsf{E}[\hat{\Delta} - \alpha'(\theta)] = bh^\beta + o(h^\beta), \quad \mathsf{Var}[\hat{\Delta}] = \frac{\sigma^2}{nh^\eta} + o(h^{-\eta}), \qquad (7.9)$$

for some positive β, η, σ, and some nonzero b. Forward- and central-difference estimators typically have $\beta = 1$ and $\beta = 2$; taking $\eta = 2$ sharpens Case (i) of (7.8) and taking $\eta = 1$ sharpens Case (ii).

Consider a sequence of estimators $\hat{\Delta}(n, h_n)$ with

$$h_n = h_* n^{-\gamma} \qquad (7.10)$$

for some positive h_* and γ. From our assumptions on bias and variance, we get

$$\mathrm{MSE}(\hat{\Delta}) = b^2 h_n^{2\beta} + \frac{\sigma^2}{nh_n^\eta}, \qquad (7.11)$$

up to terms that are of higher order in h_n. The value of γ that maximizes the rate of decrease of the MSE is $\gamma = 1/(2\beta + \eta)$, from which we draw the natural implication that with a smaller bias (larger β), we can use a larger increment h. If we substitute this value of γ into (7.11) and take the square root, we find that

$$\text{RMSE}(\hat{\Delta}) = O\left(n^{-\frac{\beta}{2\beta+\eta}}\right),$$

and this is a reasonable measure of the convergence rate of the estimator.

Taking this analysis one step further, we find that

$$n^{2\beta/(2\beta+\eta)}\text{MSE}(\hat{\Delta}) \to b^2 h_*^{2\beta} + \sigma^2 h_*^{-\eta};$$

minimizing over h_* yields an optimal value of

$$h_* = \left(\frac{\eta\sigma^2}{2\beta b^2}\right)^{\frac{1}{2\beta+\eta}}.$$

The results of this analysis are summarized in Table 7.1 for the forward- and central-difference estimators using independent sampling or common random numbers. The variance and bias columns display leading terms only and abbreviate the more complete expressions in (7.9). The table should be understood as follows: if the leading terms of the variance and bias are as indicated in the second and third columns, then the conclusions in the last three columns hold. The requirement that b not equal zero in (7.9) translates to $\alpha''(\theta) \neq 0$ and $\alpha'''(\theta) \neq 0$ for the forward and central estimators, respectively. The results in the table indicate that, at least asymptotically, $\hat{\Delta}_{C,ii}$ dominates the other three estimators because it exhibits the fastest convergence.

Estimator	Variance	Bias	Optimal h_n	Convergence	h_*
$\hat{\Delta}_{F,i}$	$\frac{\sigma^2_{F,i}}{nh^2}$	$\frac{1}{2}\alpha''(\theta)h$	$O(n^{-1/4})$	$O(n^{-1/4})$	$\left(\frac{4\sigma^2_{F,i}}{\alpha''(\theta)^2}\right)^{1/4}$
$\hat{\Delta}_{C,i}$	$\frac{\sigma^2_{C,i}}{nh^2}$	$\frac{1}{6}\alpha'''(\theta)h^2$	$O(n^{-1/6})$	$O(n^{-1/3})$	$\left(\frac{18\sigma^2_{C,i}}{\alpha'''(\theta)^2}\right)^{1/6}$
$\hat{\Delta}_{F,ii}$	$\frac{\sigma^2_{F,ii}}{nh}$	$\frac{1}{2}\alpha''(\theta)h$	$O(n^{-1/3})$	$O(n^{-1/3})$	$\left(\frac{2\sigma^2_{F,ii}}{\alpha''(\theta)^2}\right)^{1/3}$
$\hat{\Delta}_{C,ii}$	$\frac{\sigma^2_{C,ii}}{nh}$	$\frac{1}{6}\alpha'''(\theta)h^2$	$O(n^{-1/5})$	$O(n^{-2/5})$	$\left(\frac{9\sigma^2_{C,ii}}{\alpha'''(\theta)^2}\right)^{1/5}$

Table 7.1. Convergence rates of finite-difference estimators with optimal increment h_n. The estimators use either forward (F) or central (C) differences and either independent sampling (i) or common random numbers (ii).

Glynn [153] proves a central limit theorem for each of the cases in Table 7.1. These results take the form

$$n^{\frac{\beta}{2\beta+\eta}}[\hat{\Delta}(n,h_n) - \alpha'(\theta)] \Rightarrow N\left(bh_*^\beta, \frac{\sigma^2}{h_*^\eta}\right)$$

with h_n as in (7.10) and $\gamma = 1/(2\beta + \eta)$. The limit holds for any $h_* > 0$; the optimal values of h_* in Table 7.1 minimize the second moment of the limiting normal random variable.

For the forward-difference estimator with independent samples, the variance parameter $\sigma^2_{F,i}$ is given by

$$\sigma^2_{F,i} = \lim_{h \to 0} \left(\mathsf{Var}[Y(\theta + h)] + \mathsf{Var}[Y(\theta)] \right) = 2\mathsf{Var}[Y(\theta)]$$

under the minimal assumption of continuity of $\mathsf{Var}[Y(\theta)]$. Because the central-difference estimator has a denominator of $2h$, $\sigma^2_{C,i} = \mathsf{Var}[Y(\theta)]/2$. The parameters $\sigma^2_{F,ii}$ and $\sigma^2_{C,ii}$ do not admit a simple description. Their values depend on the joint distribution of $(Y(\theta - h), Y(\theta), Y(\theta + h))$, which depends in part on the particular algorithm used for simulation — different algorithms may respond differently to changes in an input parameter with the random numbers held fixed. In contrast, $\mathsf{Var}[Y(\theta)]$ is determined by the marginal distribution of $Y(\theta)$ so all algorithms that legitimately sample from this distribution produce the same variance. The variance of the finite-difference estimators appear in the optimal values of h_*; though these are unlikely to be known in advance, they can be estimated from preliminary runs and potentially combined with rough estimates of the derivatives of α to approximate h_*.

Case (iii) of (7.8) is not reflected in Table 7.1. When it applies, the mean square error takes the form

$$\text{MSE}(\hat{\Delta}) = b^2 h_n^{2\beta} + \frac{\sigma^2}{n},$$

and there is no tradeoff between bias and variance. We should take h_n as small as possible, and so long as $nh_n^{2\beta}$ is bounded, the RMSE is $O(n^{-1/2})$. This dominates all the convergence rates in Table 7.1.

The distinction between Cases (ii) and (iii) is illustrated in Figure 7.2. The figure compares RMS relative errors of forward-difference estimators of delta for a standard call option paying $(S(T) - K)^+$ and a digital option paying $\mathbf{1}\{S(T) > K\}$, with model parameters $K = S(0) = 100$, $\sigma = 0.30$, $r = 0.05$, and $T = 0.25$. The forward-difference estimators use common random numbers, which in this example simply means using the same draw from the normal distribution to generate $S(T)$ from both $S(0)$ and $S(0) + h$. This example is simple enough to allow exact calculation of the RMSE. To simplify comparison, in the figure we divide each RMSE by the true value of delta to get a relative error. The figure shows the effect of varying h with the number of replications n fixed at 5000.

The standard call option fits in Case (iii) of (7.8); this will be evident from the analysis in Section 7.2.2. As expected, its RMS relative error decreases with decreasing h. The digital option fits in Case (ii) so its relative error explodes as h approaches zero. From the figure we see that the relative error for the digital option is minimized at a surprisingly large value of about 4; the figure also shows that the cost of taking h too large is much smaller than the cost of taking it too small.

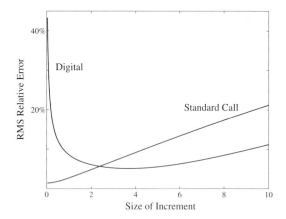

Fig. 7.2. RMS relative errors in forward-difference delta estimates for a standard call and a digital option as a function of the increment h, with $n = 5000$.

Extrapolation

For a smooth function $\alpha(\theta) = \mathsf{E}[Y(\theta)]$, the bias in finite-difference estimates can be reduced through extrapolation, much as in Section 6.2.4. This technique applies to all the finite-difference estimators considered above; we illustrate it in the case of $\hat{\Delta}_{C,ii}$, the central-difference estimator using common random numbers.

A Taylor expansion of $\alpha(\theta)$ shows that

$$\frac{\alpha(\theta + h) - \alpha(\theta - h)}{2h} = \alpha'(\theta) + \frac{1}{6}\alpha'''(\theta)h^2 + O(h^4);$$

odd powers of h are eliminated by the symmetry of the central-difference estimator. Similarly,

$$\frac{\alpha(\theta + 2h) - \alpha(\theta - 2h)}{4h} = \alpha'(\theta) + \frac{2}{3}\alpha'''(\theta)h^2 + O(h^4).$$

It follows that the bias in the combined estimator

$$\frac{4}{3}\hat{\Delta}_{C,ii}(n,h) - \frac{1}{3}\hat{\Delta}_{C,ii}(n,2h)$$

is $O(h^4)$. Accordingly, the RMSE of this estimator is $O(n^{-4/9})$ if h is taken to be $O(n^{-1/9})$. The ninth root of n varies little over practical sample sizes n, so this estimator achieves a convergence rate of nearly $n^{-1/2}$ with h_n nearly constant.

Second Derivatives

The analysis leading to Table 7.1 extends, with evident modifications, to finite-difference estimators of second derivatives. Consider a central-difference estimator of the form

$$\frac{\bar{Y}_n(\theta + h) - 2\bar{Y}_n(\theta) + \bar{Y}_n(\theta - h)}{h^2}. \tag{7.12}$$

Its expectation is

$$\frac{\alpha(\theta + h) - 2\alpha(\theta) + \alpha(\theta - h)}{h^2} = \alpha''(\theta) + O(h^2),$$

if α is four times differentiable. The bias is then $O(h^2)$.

If values at θ and $\theta \pm h$ are simulated independently of each other, the numerator in (7.12) has variance that is $O(1)$ in h, so the variance of the ratio is $O(h^{-4})$. By using common random numbers, we can often reduce the variance of the numerator to $O(h)$, but even in this case the estimator variance is $O(h^{-3})$. With h_n chosen optimally, this results in a covergence rate of $O(n^{-2/7})$ for the RMSE. This makes precise the idea that estimating second derivatives is fundamentally more difficult than estimating first derivatives.

For problems in which Case (iii) of (7.8) applies — the most favorable case for derivative estimation — the estimator in (7.12) often has variance $O(h^{-1})$. Because of the nondifferentiability of option payoffs, there are no interesting examples in which the variance of the numerator of (7.12) is smaller than $O(h^3)$, and there are plenty of examples in which it is $O(h)$.

Multiple Parameters

Suppose now that θ is a vector of parameters $(\theta_1, \ldots, \theta_m)$ and that we need to estimate the sensitivity of $\alpha(\theta) = \mathsf{E}[Y(\theta)]$ to each of these parameters. A straightforward approach selects an increment h_i for each θ_i and estimates $\partial\alpha/\partial\theta_i$ using

$$\frac{1}{2h_i} \left[\bar{Y}_n(\theta_1, \ldots, \theta_i + h_i, \ldots, \theta_m) - \bar{Y}_n(\theta_1, \ldots, \theta_i - h_i, \ldots, \theta_m) \right],$$

or the corresponding forward-difference estimator. In addition to the problem of selecting an appropriate h_i, this setting poses a further computational challenge. It requires estimation of $\alpha(\theta)$ at $2m + 1$ values of θ using central differences and $m+1$ values using forward differences. This difficulty becomes even more severe for second derivatives. Finite-difference estimation of all second derivatives $\partial^2\alpha/\partial\theta_i\partial\theta_j$ requires simulation at $O(m^2)$ parameter values. This may be onerous if m is large. In pricing interest rate derivatives, for example, we may be interested in sensitivities with respect to all initial forward rates or bond prices, in which case m could easily be 20 or more.

Techniques from the design of experiments and response surface methodology are potentially useful in reducing the number of parameter values at which one simulates. For this formulation, let ΔY denote the change in Y resulting from incrementing each θ_i by h_i, $i = 1, \ldots, m$. The problem of estimating sensitivities can be viewed as one of fitting a first-order model of the form

$$\Delta Y = \sum_{i=1}^{m} \beta_i h_i + \epsilon \tag{7.13}$$

or a second-order model of the form

$$\Delta Y = \sum_{i=1}^{m} \beta_i h_i + \sum_{i=1}^{m} \sum_{j=i}^{m} \beta_{ij} h_i h_j + \epsilon. \tag{7.14}$$

Each β_i approximates the partial derivative of α with respect to θ_i, $2\beta_{ii}$ approximates $\partial^2 \alpha / \partial \theta_i^2$, and each β_{ij}, $j \neq i$, approximates $\partial^2 \alpha / \partial \theta_i \partial \theta_j$. In both cases, ϵ represents a residual error.

   Given observations of Y at various values of θ, the coefficients β_i, β_{ij} can be estimated using, e.g., least squares or weighted least squares. Response surface methodology (as in, e.g., Khuri and Cornell [210]) provides guidance on choosing the values of θ at which to measure (i.e., simulate) Y.

   In the terminology of experimental design, simulating at all 2^m points defined by adding or subtracting h_i to each θ_i is a full factorial design. Simulating at fewer points may add bias to estimates of some coefficients. But by reducing the number of points at which we need to simulate, we can increase the number of replications at each point and reduce variance. A design with fewer points thus offers a tradeoff between bias and variance.

   Most of the literature on response surface methodology assumes independent residuals across observations. In simulation, we have the flexibility to introduce dependence between observations through the assignment of the output of the random number generator. Schruben and Margolin [323] analyze the design of simulation experiments in which the same random numbers are used at multiple parameter values. They recommend a combination of common random numbers and antithetic sampling across parameter values.

7.2 Pathwise Derivative Estimates

This section and the next develop alternatives to finite-difference methods that estimate derivatives directly, without simulating at multiple parameter values. They do so by taking advantage of additional information about the dynamics and parameter dependence of a simulated process.

7.2.1 Method and Examples

In our discussion of optimal mean square error in Section 7.1, we noted that in Case (iii) of (7.8) the MSE decreases with the parameter increment h. This suggests that we should let h decrease to zero and estimate the derivative of $\alpha(\theta) = \mathsf{E}[Y(\theta)]$ using

$$Y'(\theta) = \lim_{h \to 0} \frac{Y(\theta + h) - Y(\theta)}{h}. \tag{7.15}$$

This estimator has expectation $\mathsf{E}[Y'(\theta)]$. It is an unbiased estimator of $\alpha'(\theta)$ if

$$\mathsf{E}\left[\frac{d}{d\theta}Y(\theta)\right] = \frac{d}{d\theta}\mathsf{E}[Y(\theta)]; \tag{7.16}$$

i.e., if the interchange of differentiation and expectation is justified.

Even before discussing the validity of (7.16), we need to clarify what we mean by (7.15). Up until this point, we have not been explicit about what dependence, if any, the random variable Y has on the parameter θ. In our discussion of finite-difference estimators, the notation $Y(\theta)$ indicated an outcome of Y simulated at parameter value θ, but did not entail a functional relation between the two. Indeed, we noted that the relation between $Y(\theta)$ and $Y(\theta+h)$ could depend on whether the two outcomes were simulated using common random numbers.

To make (7.15) precise, we need a collection of random variables $\{Y(\theta), \theta \in \Theta\}$ defined on a single probability space $(\Omega, \mathcal{F}, P)$. In other words, $Y(\theta)$ is a stochastic process indexed by $\theta \in \Theta$. Take $\Theta \subseteq \Re$ to be an interval. We can fix $\omega \in \Omega$ and think of the mapping $\theta \mapsto Y(\theta, \omega)$ as a random function on Θ. We can then interpret $Y'(\theta) = Y'(\theta, \omega)$ as the derivative of the random function with respect to θ with ω held fixed. In (7.15), we implicitly assume that the derivative exists with probability one, and when this holds, we call $Y'(\theta)$ the *pathwise derivative* of Y at θ.

As a practical matter, we usually think of each ω as a realization of the output of an ideal random number generator. Each $Y(\theta, \omega)$ is then the output of a simulation algorithm at parameter θ with random number stream ω. Each $Y'(\theta, \omega)$ is the derivative of the simulation output with respect to θ with the random numbers held fixed. The value of this derivative depends, in part, on how we implement a simulation algorithm.

We will see through examples that the existence with probability 1 of the pathwise derivative $Y'(\theta)$ at each θ typically holds. This is not to say that, with probability 1, the mapping $\theta \mapsto Y(\theta)$ is a differentiable function on Θ. The distinction lies in the order of quantification: the exceptional set of probability 0 on which (7.15) fails to exist can and often does depend on θ. The union of these exceptional sets for θ ranging over Θ may well have positive probability.

There is a large literature on pathwise derivative estimation in the discrete-event simulation literature, where it is usually called infinitesimal perturbation analysis. This line of work stems primarily from Ho and Cao [186] and Suri and Zazanis [339]; a general framework is developed in Glasserman [138]. Broadie and Glasserman [64] apply the method to option pricing and we use some of their examples here. Chen and Fu [80] develop applications to mortgage-backed securities.

To illustrate the derivation and scope of pathwise derivative estimators, we now consider some examples.

Example 7.2.1 *Black-Scholes delta.* The derivative of the Black-Scholes formula with respect to the initial price $S(0)$ of the underlying asset can be calculated explicitly and is given by $\Phi(d)$ in the notation of (1.44). Calculation of the Black-Scholes delta does not require simulation, but nevertheless provides a useful example through which to introduce the pathwise method.

Let

$$Y = e^{-rT}[S(T) - K]^+$$

with

$$S(T) = S(0)e^{(r-\frac{1}{2}\sigma^2)T+\sigma\sqrt{T}Z}, \quad Z \sim N(0,1), \tag{7.17}$$

and take θ to be $S(0)$, with r, σ, T, and K positive constants. Applying the chain rule for differentiation, we get

$$\frac{dY}{dS(0)} = \frac{dY}{dS(T)}\frac{dS(T)}{dS(0)}. \tag{7.18}$$

For the first of these two factors, observe that

$$\frac{d}{dx}\max(0, x - K) = \begin{cases} 0, & x < K \\ 1, & x > K. \end{cases}$$

This derivative fails to exist at $x = K$. But because the event $\{S(T) = K\}$ has probability 0, Y is almost surely differentiable with respect to $S(T)$ and has derivative

$$\frac{dY}{dS(T)} = e^{-rT}\mathbf{1}\{S(T) > K\}. \tag{7.19}$$

For the second factor in (7.18), observe from (7.17) that $S(T)$ is linear in $S(0)$ with $dS(T)/dS(0) = S(T)/S(0)$. Combining the two factors in (7.18), we arrive at the pathwise estimator

$$\frac{dY}{dS(0)} = e^{-rT}\frac{S(T)}{S(0)}\mathbf{1}\{S(T) > K\}, \tag{7.20}$$

which is easily computed in a simulation of $S(T)$. The expected value of this estimator is indeed the Black-Scholes delta, so the estimator is unbiased.

A minor modification of this derivation produces the pathwise estimator of the Black-Scholes vega. Replace (7.18) with

$$\frac{dY}{d\sigma} = \frac{dY}{dS(T)}\frac{dS(T)}{d\sigma}.$$

The first factor is unchanged and the second is easily calculated from (7.17). Combining the two, we get the pathwise estimator

$$\frac{dY}{d\sigma} = e^{-rT}(-\sigma T + \sqrt{T}Z)S(T)\mathbf{1}\{S(T) > K\}.$$

The expected value of this expression is the Black-Scholes vega, so this estimator is unbiased.

Using (7.17) we can eliminate Z and write the vega estimator as

$$\frac{dY}{d\sigma} = e^{-rT} \left(\frac{\log(S(T)/S(0)) - (r + \frac{1}{2}\sigma^2)T}{\sigma} \right) S(T)\mathbf{1}\{S(T) > K\}.$$

This formulation has the feature that it does not rely on the particular expression in (7.17) for simulating $S(T)$. Although derived using (7.17), this estimator could be applied with any other mechanism for sampling $S(T)$ from its lognormal distribution. $\square$

This example illustrates a general point about differentiability made at the beginning of this section. Consider the discounted payoff Y as a function of $S(0)$. If we fix any $S(0) > 0$, then Y is differentiable at $S(0)$ with probability 1, because for each $S(0)$ the event $\{S(T) \neq K\}$ has probability 1. However, the probability that Y is a differentiable function of $S(0)$ throughout $(0, \infty)$ is zero: for each value of Z, there is some $S(0)$ at which Y fails to be differentiable — namely, the $S(0)$ that makes $S(T) = K$ for the given value of Z.

Example 7.2.2 *Path-dependent deltas.* As in the previous example, suppose the underlying asset is modeled by geometric Brownian motion, but now let the payoff be path-dependent. For example, consider an Asian option

$$Y = e^{-rT}[\bar{S} - K]^+, \quad \bar{S} = \frac{1}{m}\sum_{i=1}^{m} S(t_i),$$

for some fixed dates $0 < t_1 < \cdots < t_m \leq T$. Much as in Example 7.2.1,

$$\frac{dY}{dS(0)} = \frac{dY}{d\bar{S}}\frac{d\bar{S}}{dS(0)} = e^{-rT}\mathbf{1}\{\bar{S} > K\}\frac{d\bar{S}}{dS(0)}.$$

Also,

$$\frac{d\bar{S}}{dS(0)} = \frac{1}{m}\sum_{i=1}^{m} \frac{dS(t_i)}{dS(0)} = \frac{1}{m}\sum_{i=1}^{m} \frac{S(t_i)}{S(0)} = \frac{\bar{S}}{S(0)}.$$

The pathwise estimator of the option delta is

$$\frac{dY}{dS(0)} = e^{-rT}\mathbf{1}\{\bar{S} > K\}\frac{\bar{S}}{S(0)}.$$

This estimator is in fact unbiased; this follows from a more general result in Section 7.2.2. Because there is no formula for the price of an Asian option, this estimator has genuine practical value. Because $\bar{S}$ would be simulated anyway in estimating the price of the option, this estimator requires negligible additional effort. Compared with a finite-difference estimator, it reduces variance, eliminates bias, and cuts the computing time roughly in half.

Similar comments apply in the case of a lookback put with discounted payoff

$$Y = e^{-rT}\left(\max_{1\leq i\leq m} S(t_i) - S(t_m)\right).$$

The pathwise estimator of delta simplifies to $Y/S(0)$, which is also unbiased.

Using similar arguments, we could also derive pathwise estimators of delta and other derivatives for barrier options. However, for reasons discussed in Section 7.2.2, these are not in general unbiased: a small change in $S(0)$ can result in a large change in Y. $\square$

Example 7.2.3 *Path-dependent vega.* Now consider the sensitivity of the Asian option in the previous example to the volatility σ. Let

$$S(t_i) = S(t_{i-1})e^{(r-\frac{1}{2}\sigma^2)[t_i-t_{i-1}]+\sigma\sqrt{t_i-t_{i-1}}Z_i}, \quad i = 1,\ldots,m, \qquad (7.21)$$

with $Z_1,\ldots,Z_m$ independent $N(0,1)$ random variables. The parameter σ affects $S(t_i)$ explicitly through this functional relation but also implicitly through the dependence of $S(t_{i-1})$ on σ. By differentiating both sides we get a recursion for the derivatives along the path:

$$\frac{dS(t_i)}{d\sigma} = \frac{dS(t_{i-1})}{d\sigma}\frac{S(t_i)}{S(t_{i-1})} + S(t_i)[-\sigma(t_i-t_{i-1}) + \sqrt{t_i-t_{i-1}}Z_i].$$

With initial condition $dS(0)/d\sigma = 0$, this recursion is solved by

$$\frac{dS(t_i)}{d\sigma} = S(t_i)[-\sigma t_i + \sum_{j=1}^{i}\sqrt{t_j-t_{j-1}}Z_j],$$

which can also be written as

$$\frac{dS(t_i)}{d\sigma} = S(t_i)[\log(S(t_i)/S(0)) - (r + \tfrac{1}{2}\sigma^2)t_i]/\sigma.$$

The pathwise estimator of the Asian option vega is

$$e^{-rT}\frac{1}{m}\sum_{i=1}^{m}\frac{dS(t_i)}{d\sigma}\mathbf{1}\{\bar{S} > K\}.$$

This example illustrates a general technique for deriving pathwise estimators based on differentiating both sides of a recursive expression for the evolution of an underlying asset. This idea is further developed in Section 7.2.3. $\square$

Example 7.2.4 *Options on multiple assets.* Suppose the assets $S_1,\ldots,S_d$ are modeled by a multivariate geometric Brownian motion $GBM(r,\Sigma)$ as defined in Section 3.2.3. Their values can be generated by setting

$$S_i(T) = S_i(0)e^{(r-\frac{1}{2}\sigma_i^2)T+\sqrt{T}X_i}, \quad i = 1,\ldots,d,$$

with $(X_1, \ldots, X_d)$ drawn from $N(0, \Sigma)$. It follows that in this construction $dS_i(T)/dS_i(0) = S_i(T)/S_i(0)$, just as in the univariate case, and also $dS_i(T)/dS_j(0) = 0$, for $j \neq i$.

A spread option with discounted payoff

$$Y = e^{-rT}[(S_2(T) - S_1(T)) - K]^+$$

has a delta with respect to each underlying asset. Their pathwise estimators are

$$\frac{\partial Y}{\partial S_2(0)} = e^{-rT} \frac{S_2(T)}{S_2(0)} \mathbf{1}\{S_2(T) - S_1(T) > K\}$$

and

$$\frac{\partial Y}{\partial S_1(0)} = -e^{-rT} \frac{S_1(T)}{S_1(0)} \mathbf{1}\{S_2(T) - S_1(T) > K\}.$$

An option on the maximum of d assets with discounted payoff

$$Y = e^{-rT}[\max\{S_1(T), \ldots, S_d(T)\} - K]^+$$

has a delta with respect to each underlying asset. A small change in $S_i(T)$ affects Y only if the ith asset attained the maximum and exceeded the strike, so

$$\frac{\partial Y}{\partial S_i(T)} = e^{-rT} \mathbf{1}\{S_i(T) > \max_{j \neq i} S_j(T), S_i(T) > K\}.$$

Multiplying this expression by $S_i(T)/S_i(0)$ yields the pathwise estimator for the ith delta. $\square$

We introduced these examples in the setting of geometric Brownian motion for simplicity, but the expressions derived in the examples apply much more generally. Consider an underlying asset $S(t)$ described by an SDE

$$\frac{dS(t)}{S(t)} = \mu(t)\,dt + \sigma(t)\,dW(t),$$

in which $\mu(t)$ and $\sigma(t)$ could be stochastic but have no dependence on $S(0)$. The asset price at T is

$$S(T) = S(0)\exp\left(\int_0^T \left(\mu(t) - \tfrac{1}{2}\sigma^2(t)\right)dt + \int_0^T \sigma(t)\,dW(t)\right),$$

and we still have $dS(T)/dS(0) = S(T)/S(0)$. Indeed, this expression for the derivative is valid whenever $S(t)$ is given by $S(0)\exp(X(t))$ for some process X that does not depend on $S(0)$.

Example 7.2.5 *Square-root diffusion.* As an example of a process that is not linear in its initial state, consider a square-root diffusion

$$dX(t) = \alpha(b - X(t))\,dt + \sigma\sqrt{X(t)}\,dW(t).$$

From Section 3.4 (and (3.68) in particular) we know that $X(t)$ has the distribution of a multiple of a noncentral chi-square random variable with a noncentrality parameter proportional to $X(0)$:

$$X(t) \sim c_1 \chi_\nu'^2(c_2 X(0)).$$

See (3.68)–(3.70) for explicit expressions for c_1, c_2, and ν. As explained in Section 3.4, if $\nu > 1$ we can simulate $X(t)$ using

$$X(t) = c_1 \left(\left(Z + \sqrt{c_2 X(0)} \right)^2 + \chi_{\nu-1}^2 \right), \qquad (7.22)$$

with $Z \sim N(0, 1)$ and $\chi_{\nu-1}^2$ an ordinary chi-square random variable having $\nu - 1$ degrees of freedom, independent of Z. It follows that

$$\frac{dX(t)}{dX(0)} = c_1 c_2 \left(1 + \frac{Z}{\sqrt{c_2 X(0)}} \right).$$

This generalizes to a path simulated at dates $t_1 < t_2 < \cdots$ through the recursion

$$\frac{dX(t_{i+1})}{dX(0)} = c_1 c_2 \left(1 + \frac{Z_{i+1}}{\sqrt{c_2 X(t_i)}} \right) \frac{dX(t_i)}{dX(0)},$$

with Z_{i+1} used to generate $X(t_{i+1})$ from $X(t_i)$. The coefficients c_1, c_2 depend on the time increment $t_{i+1} - t_i$, as in (3.68)–(3.70). These expressions can be applied to the Cox-Ingersoll-Ross [91] interest rate model and to the Heston [179] stochastic volatility model in (3.65)–(3.65). □

Example 7.2.6 *Digital options and gamma.* Consider a digital option with discounted payoff

$$Y = e^{-rT} \mathbf{1}\{S(T) > K\},$$

and for concreteness let S be modeled by geometric Brownian motion. Viewed as a function of $S(T)$, the discounted payoff Y is differentiable except at $S(T) = K$, which implies that Y is differentiable with probability 1. But because Y is piecewise constant in $S(T)$, this derivative is 0 wherever it exists, and the same is true of Y viewed as a function of $S(0)$. Thus,

$$0 = \mathsf{E}\left[\frac{dY}{dS(0)} \right] \neq \frac{d}{dS(0)} \mathsf{E}[Y].$$

This is an example in which the pathwise derivative exists with probability 1 but is entirely uninformative. The change in $\mathsf{E}[Y]$ with a change in $S(0)$ is driven by the possibility that a change in $S(0)$ will cause $S(T)$ to cross the strike K, but this possibility is missed by the pathwise derivative. The pathwise derivative sees only the local insensitivity of Y to $S(0)$. This also explains the qualitatively different behavior as $h \to 0$ for the standard call and the digital option in Figure 7.2.

For much the same reason, the pathwise method is generally inapplicable to barrier options. On a fixed path, a sufficiently small change in the underlying asset will neither create nor eliminate a barrier crossing, so the effect of the barrier is missed by the pathwise derivative.

The example of a digital option also indicates that the pathwise method, at least in its simplest form, is generally inapplicable to estimating second derivatives. If, for example, Y is the discounted payoff of an ordinary call option, then its first derivative (in $S(T)$) has exactly the form of a digital payoff — see (7.19). The pathwise method faces the same difficulty in estimating the gamma of a standard call option as it does in estimating the delta of a digital option. $\square$

7.2.2 Conditions for Unbiasedness

Example 7.2.6 points to a limitation in the scope of the pathwise method: it is generally inapplicable to discontinuous payoffs. The possible failure of the interchange of derivative and expectation in (7.16) is a practical as well as theoretical issue. We therefore turn now to a discussion of conditions ensuring the validity of (7.16) and thus the unbiasedness of the pathwise method.

We continue to use θ to denote a generic parameter and assume the existence of a random function $\{Y(\theta), \theta \in \Theta\}$ with Θ an interval of the real line. As explained at the beginning of Section 7.2.1, this random function represents the output of a simulation algorithm as a function of θ with the simulation's random numbers held fixed. We consider settings in which the derivative $Y'(\theta)$ exists with probability 1 at each $\theta \in \Theta$. As the examples of Section 7.2.1 should make evident, this is by no means restrictive.

Given the existence of $Y'(\theta)$, the most important question is the validity of (7.16), which is a matter of interchanging a limit and an expectation to ensure

$$\mathsf{E}\left[\lim_{h \to 0} \frac{Y(\theta + h) - Y(\theta)}{h}\right] = \lim_{h \to 0} \mathsf{E}\left[\frac{Y(\theta + h) - Y(\theta)}{h}\right].$$

A necessary and sufficient condition for this is uniform integrability (see Appendix A) of the difference quotients $h^{-1}[Y(\theta + h) - Y(\theta)]$. Our objective is to provide sufficient conditions that are more readily verified in practice.

Inspection of the examples in Section 7.2.1 reveals a pattern in the derivation of the estimators: we apply the chain rule to express $Y'(\theta)$ as a product of terms, the first relating the discounted payoff to the path of the underlying asset, the second relating the path to the parameter. Because this is a natural and convenient way to derive estimators, we formulate conditions to fit this approach, following Broadie and Glasserman [64].

We restrict attention to discounted payoffs that depend on the value of an underlying asset or assets at a finite number of fixed dates. Rather than distinguish between multiple assets and values of a single asset at multiple dates, we simply suppose that the discounted payoff is a function of a random vector $X(\theta) = (X_1(\theta), \ldots, X_m(\theta))$, which is itself a function of the parameter θ. Thus,

$$Y(\theta) = f(X_1(\theta), \ldots, X_m(\theta)),$$

for some function $f : \Re^m \mapsto \Re$ depending on the specific derivative security. We require

(A1) At each $\theta \in \Theta$, $X_i'(\theta)$ exists with probability 1, for all $i = 1, \ldots, m$.

If f is differentiable, then Y inherits differentiability from X under condition (A1). As illustrated in the examples of Section 7.2.1, option payoffs often fail to be differentiable, but the points at which differentiability fails can often be ignored because they occur with probability 0. To make this precise, we let $D_f \subseteq \Re^m$ denote the set of points at which f is differentiable and require

(A2) $P(X(\theta) \in D_f) = 1$ for all $\theta \in \Theta$.

This then implies that $Y'(\theta)$ exists with probability 1 and is given by

$$Y'(\theta) = \sum_{i=1}^{m} \frac{\partial f}{\partial x_i}(X(\theta)) X_i'(\theta).$$

Comparison of Examples 7.2.1 and 7.2.6 indicates that even if points of nondifferentiability of f occur with probability 0, the behavior of f at these points is important. For both the standard call and the digital option, differentiability fails on the zero-probability event $\{S(T) = K\}$; but whereas the payoff of the call is continuous as $S(T)$ crosses the strike, the payoff of the digital option is not. This distinction leads to an unbiased estimator in the first case and not the second. In fact we need a bit more than continuity; a convenient condition is

(A3) There exists a constant k_f such that for all $x, y \in \Re^m$,

$$|f(x) - f(y)| \le k_f \|x - y\|;$$

i.e., f is Lipschitz.

The standard call, Asian option, lookback, spread option, and max option in Section 7.2.1 all satisfy this condition (as does any composition of linear transformations and the functions min and max); the digital payoff does not. We impose a related condition on $X(\theta)$:

(A4) There exist random variables κ_i, $i = 1, \ldots, m$, such that for all $\theta_1, \theta_2 \in \Theta$,

$$|X_i(\theta_2) - X_i(\theta_1)| \le \kappa_i |\theta_2 - \theta_1|,$$

and $\mathsf{E}[\kappa_i] < \infty$, $i = 1, \ldots, m$.

Conditions (A3) and (A4) together imply that Y is almost surely Lipschitz in θ because the Lipschitz property is preserved by composition. Thus,

$$|Y(\theta_2) - Y(\theta_1)| \leq \kappa_Y |\theta_2 - \theta_1|,$$

and for κ_Y we may take

$$\kappa_Y = k_f \sum_{i=1}^{m} \kappa_i.$$

It follows that $\mathsf{E}[\kappa_Y] < \infty$. Observing that

$$\left| \frac{Y(\theta + h) - Y(\theta)}{h} \right| \leq \kappa_Y,$$

we can now apply the dominated convergence theorem to interchange expectation and the limit as $h \to 0$ and conclude that $d\mathsf{E}[Y(\theta)]/d\theta$ exists and equals $\mathsf{E}[Y'(\theta)]$. In short, conditions (A1)–(A4) suffice to ensure that the pathwise derivative is an unbiased estimator.

Condition (A4) is satisfied in all the examples of Section 7.2.1, at least if Θ is chosen appropriately. No restrictions are needed when the dependence on the parameter is linear as in, e.g., the mapping from $S(0)$ to $S(T)$ for geometric Brownian motion. In the case of the mapping from σ to $S(T)$, (A4) holds if we take Θ to be a bounded interval; this is harmless because for the purposes of estimating a derivative at σ we may take an arbitrarily small neighborhood of σ. In Example 7.2.5, for the mapping from $X(0)$ to $X(t)$ to be Lipschitz, we need to restrict $X(0)$ to a set of values bounded away from zero.

If conditions (A1)–(A4) hold and we strengthen (A4) slightly to require $\mathsf{E}[\kappa_i^2] < \infty$, then $\mathsf{E}[\kappa_Y^2] < \infty$ and

$$\mathsf{E}[(Y(\theta + h) - Y(\theta))^2] \leq \mathsf{E}[\kappa_Y^2]h^2,$$

from which it follows that $\mathsf{Var}[Y(\theta + h) - Y(\theta)] = O(h^2)$, as in Case (iii) of (7.8). Conversely, if Case (iii) holds, if $Y'(\theta)$ exists with probability 1, and $\mathsf{E}[Y(\theta)]$ is differentiable, then

$$\mathsf{E}[(Y(\theta+h) - Y(\theta))^2] = \mathsf{Var}[Y(\theta+h) - Y(\theta)] + (\mathsf{E}[Y(\theta+h) - Y(\theta)])^2 = O(h^2).$$

Hence,

$$\mathsf{E}\left[\left(\frac{Y(\theta + h) - Y(\theta)}{h} \right)^2 \right]$$

remains bounded as $h \to 0$ and $[Y(\theta + h) - Y(\theta)]/h$ is uniformly integrable. Thus, the scope of the pathwise method is essentially the same as the scope of Case (iii) in (7.8).

Of conditions (A1)–(A4), the one that poses a practical limitation is (A3). As we noted previously, the payoffs of digital options and barrier options are not even continuous. Indeed, discontinuities in a payoff are the main obstacle to the applicability of the pathwise method. A simple rule of thumb states that the pathwise method applies when the payoff is continuous in the parameter of interest. Though this is not a precise guarantee (of the type provided by the conditions above), it provides sound guidance in most practical problems.

7.2.3 Approximations and Related Methods

This section develops further topics in sensitivity estimation using pathwise derivatives. Through various approximations and extensions, we can improve the efficiency or expand the scope of the method.

General Diffusion Processes

Consider a process X described by a stochastic differential equation

$$dX(t) = a(X(t))\,dt + b(X(t))\,dW(t) \tag{7.23}$$

with fixed initial condition $X(0)$. Suppose, for now, that this is a scalar process driven by a scalar Brownian motion. We might simulate the process using an Euler scheme with step size h; using the notation of Chapter 6, we write the discretization as

$$\hat{X}(i+1) = \hat{X}(i) + a(\hat{X}(i))h + b(\hat{X}(i))\sqrt{h}Z_{i+1}, \quad \hat{X}(0) = X(0),$$

with $\hat{X}(i)$ denoting the discretized approximation to $X(ih)$ and $Z_1, Z_2, \ldots$ denoting independent $N(0,1)$ random variables.

Let

$$\hat{\Delta}(i) = \frac{d\hat{X}(i)}{dX(0)}.$$

By differentiating both sides of the Euler scheme we get the recursion

$$\hat{\Delta}(i+1) = \hat{\Delta}(i) + a'(\hat{X}(i))\hat{\Delta}(i)h + b'(\hat{X}(i))\hat{\Delta}(i)\sqrt{h}Z_{i+1}, \quad \hat{\Delta}(0) = 1, \tag{7.24}$$

with a' and b' denoting derivatives of the coefficient functions. With

$$f(\hat{X}(1), \ldots, \hat{X}(m))$$

denoting the discounted payoff of some option, the pathwise derivative of the option's delta is

$$\sum_{i=1}^{m} \frac{\partial f}{\partial x_i}(\hat{X}(1), \ldots, \hat{X}(m))\hat{\Delta}(i).$$

A similar estimator could be derived by differentiating both sides of a higher-order discretization of X.

Under relatively minor conditions on the coefficient functions a and b, the (continuous-time) solution $X(t)$ to the SDE (7.23) is almost surely differentiable in $X(0)$. Moreover, its derivative $\Delta(t) = dX(t)/dX(0)$ satisfes

$$d\Delta(t) = a'(X(t))\Delta(t)\,dt + b'(X(t))\Delta(t)\,dW(t), \quad \Delta(0) = 1. \qquad (7.25)$$

See Section 4.7 of Kunita [217] or Section 5.7 of Protter [300] for precise results of this type. It follows that (7.24) — derived by differentiating the Euler scheme for X — is also the Euler scheme for Δ. Similar results apply to derivatives of $X(t)$ with respect to parameters of the coefficient functions a and b.

In light of these results, there is no real theoretical obstacle to developing pathwise estimators for general diffusions; through (7.25), the problem reduces to one of discretizing stochastic differential equations. There may, however, be practical obstacles to this approach. The additional effort required to simulate (7.25) may be roughly the same as the effort required to simulate a second path of X from a different initial state. (Contrast this with examples in Section 7.2.1 for which $dS(T)/dS(0)$ reduces to $S(T)/S(0)$ and is thus available at no additional cost.) Also, in the case of a d-dimensional diffusion X, we need to replace (7.25) with a $d \times d$ system of equations for

$$\Delta_{ij}(t) = \frac{\partial X_i(t)}{\partial X_j(0)}, \quad i, j = 1, \ldots, d.$$

Simulating a system of SDEs for this matrix of derivatives may be quite time-consuming.

These considerations motivate the use of approximations in simulating the derivatives of $X(t)$. One strategy is to use coarser time steps in simulating the derivatives than in simulating X. There are many ways of implementing this strategy; we discuss just one and for simplicity we consider only the scalar case.

Freezing $a'(X(t))$ and $b'(X(t))$ at their time-zero values transforms (7.24) into

$$\hat{\Delta}(i) = \hat{\Delta}(i-1)\left(1 + a'(X(0))h + b'(X(0))\sqrt{h}Z_i\right).$$

This is potentially much faster to simulate than (7.24), especially if evaluation of a' and b' is time-consuming. To improve the accuracy of the approximation, we might update the coefficients after every k steps. In this case, we would use the coefficiencts $a'(X(0))$ and $b'(X(0))$ for $i = 1, \ldots, k$, then use

$$\hat{\Delta}(i) = \hat{\Delta}(i-1)\left(1 + a'(\hat{X}(k))h + b'(\hat{X}(k))\sqrt{h}Z_i\right)$$

for $i = k+1, \ldots, 2k$, and so on.

If, for purposes of differentiation, we entirely ignored the dependence of a and b on the current state X, we would get simply $\hat{\Delta}(i) \equiv 1$. For $S(t) = S(0)\exp(X(t))$, this is equivalent to making the approximation $dS(t)/dS(0) \approx S(t)/S(0)$, which we know is exact in several cases.

LIBOR Market Model

Similar approximations are derived and tested in the setting of LIBOR market models by Glasserman and Zhao [150]. This application has considerable practical value and also serves to illustrate the approach. We use the notation of Section 3.7.

Consider, then, a system of SDEs of the form

$$\frac{dL_n(t)}{L_n(t)} = \mu_n(L(t), t)\, dt + \sigma_n(t)^\top dW(t), \quad n = 1, \ldots, M,$$

with $L(t)$ denoting the M-vector of all rates, W a d-dimensional Brownian motion, and each σ_n an $\Re^d$-valued deterministic function of time. We get the spot measure dynamics (3.112) by setting

$$\mu_n(L(t), t) = \sum_{j=\eta(t)}^{n} \frac{\delta_j L_j(t)\sigma_j(t)^\top \sigma_n(t)}{1 + \delta_j L_j(t)},$$

where, as in Section 3.7, $\eta(t)$ denotes the index of the next maturity date as of time t.

Let

$$\Delta_{nk}(t) = \frac{\partial L_n(t)}{\partial L_k(0)}$$

and write $\hat{\Delta}_{nk}$ for a discrete approximation to this process. From the spot measure dynamics it is evident that $\Delta_{nk}(t) \equiv 0$ unless $k \leq n$. Also, $\Delta_{nk}(0) = \mathbf{1}\{n = k\}$.

Suppose we simulate the forward rates L_n using a log-Euler scheme as in (3.120), with fixed time step h. By differentiating both sides, we get

$$\hat{\Delta}_{nk}(i + 1) = \hat{\Delta}_{nk}(i)\frac{\hat{L}_n(i+1)}{\hat{L}_n(i)} + \hat{L}_n(i+1) \sum_{j=1}^{M} \frac{\partial \hat{\mu}_n(i)}{\partial \hat{L}_j(i)} \hat{\Delta}_{jk}(i)h, \qquad (7.26)$$

with

$$\hat{\mu}_n(i) = \sum_{j=\eta(ih)}^{n} \frac{\delta_j \hat{L}_j(i)\sigma_j(ih)^\top \sigma_n(ih)}{1 + \delta_j \hat{L}_j(i)}.$$

Simulating all $O(M^2)$ derivatives using (7.26) is clearly quite time-consuming, and the main source of difficulty comes from the form of the drift.

If we replaced the $\mu_n(L(t), t)$ with

$$\mu_n^o(t) = \mu_n(L(0), t), \qquad (7.27)$$

we would make the drift time-varying but deterministic. (The deterministic dependence on t enters through η and the volatility functions σ_j.) We used this approximation in (4.8) to construct a control variate. Here we preserve the true drift μ_n in the dynamics of L_n but differentiate *as though* the drift were μ_n^o; i.e., as though the forward rates inside the drift function were frozen at their initial values. Differentiating the approximation

$$\hat{L}_n(i) \approx \hat{L}_n(0) \exp\left(\sum_{\ell=0}^{i-1} \left(\mu_n^o(\ell h)h - \tfrac{1}{2}\|\sigma_n(\ell h)\|^2 h + \sqrt{h}\sigma_n(\ell h)^\top Z_i \right) \right)$$

yields

$$\hat{\Delta}_{nk}(i) = \frac{\hat{L}_n(i)}{L_k(0)} \mathbf{1}\{n = k\} + \hat{L}_n(i) \sum_{\ell=0}^{i-1} \frac{\partial \mu_n^o(\ell h)}{\partial L_k(0)} h. \tag{7.28}$$

The derivatives of μ^o,

$$\frac{\partial \mu_n^o(\ell h)}{\partial L_k(0)} = \frac{\delta_k \sigma_j(\ell h)^\top \sigma_n(\ell h)}{(1 + \delta_k L_k(0))^2} \mathbf{1}\{\eta(\ell h) \le k \le n\},$$

are deterministic functions of time and can therefore be computed once, stored, and applied to each replication. Consequently, the computational cost of (7.28) is modest, especially when compared with the exact derivative recursion (7.26), ("exact" here meaning exact for the Euler scheme).

Glasserman and Zhao [150] report numerical results using these methods to estimate deltas of caplets. They find that the approximation (7.28) produces estimates quite close to the exact (for Euler) recursion (7.26), with some degradation in the approximation at longer maturities. Both methods are subject to discretization error but the discretization bias is small — less than 0.2% of the continuous-time delta obtained by differentiating the Black formula (3.117) for the continuous-time caplet price.

Smoothing

We noted in Section 7.2.2 that discontinuities in a discounted payoff generally make the pathwise method inapplicable. In some cases, this obstacle can be overcome by smoothing discontinuities through conditional expectations. We illustrate this idea with two examples.

Our first example is the discounted digital payoff $Y = e^{-rT}\mathbf{1}\{S(T) > K\}$, which also arises in the derivative of a standard call option; see Example 7.2.6. If $S \sim \text{GBM}(\mu, \sigma^2)$ and we fix $0 < \epsilon < T$, then

$$\mathsf{E}[Y|S(T-\epsilon)] = e^{-rT}\Phi\left(\frac{\log(S(T-\epsilon)/K) + (\mu - \tfrac{1}{2}\sigma^2)\epsilon}{\sigma\sqrt{\epsilon}} \right).$$

Differentiation yields

$$\frac{d\mathsf{E}[Y|S(T-\epsilon)]}{dS(T-\epsilon)} = \frac{e^{-rT}}{S(T-\epsilon)\sigma\sqrt{\epsilon}}\phi\left(\frac{\log(S(T-\epsilon)/K) + (\mu - \tfrac{1}{2}\sigma^2)\epsilon}{\sigma\sqrt{\epsilon}} \right),$$

$$\tag{7.29}$$

and then, through the chain rule, the unbiased delta estimator

$$\frac{d\mathsf{E}[Y|S(T-\epsilon)]}{dS(0)} = \frac{e^{-rT}}{S(0)\sigma\sqrt{\epsilon}}\phi\left(\frac{\log(S(T-\epsilon)/K)+(\mu-\frac{1}{2}\sigma^2)\epsilon}{\sigma\sqrt{\epsilon}}\right).$$

This calculation illustrates that by conditioning on the underlying asset ϵ time units before expiration, we can smooth the discontinuity in the digital payoff and then, by differentiating, arrive at an unbiased derivative estimator. Of course, the same argument allows exact calculation of the derivative of $\mathsf{E}[Y]$ in this example. The value of this derivation lies not so much in providing an exact expression for geometric Brownian motion as in providing an approximation for more general processes. Equation (7.29) could, for example, be applied to a stochastic volatility model by further conditioning on the volatility at $T-\epsilon$ and using this value in place of σ. This expression should then be multiplied by an exact or approximate expression for $dS(T-\epsilon)/dS(0)$. If the underlying asset is simulated using a log-Euler scheme with time step ϵ, then (7.29) does not introduce any error beyond that already inherent to the discretization method.

Our second example of smoothing applies to barrier options. We again introduce it through geometric Brownian motion though we intend it as an approximation for more general processes. Consider a discretely monitored knock-out option with discounted payoff

$$Y = e^{-rT}(S(T)-K)^+\mathbf{1}\{\min_{1\leq i\leq m} S(t_i) > b\},$$

for some fixed dates $0 < t_1 < \cdots < t_m = T$ and barrier $b < S(0)$. The knock-out feature makes Y discontinuous in the path of the underlying asset.

For each $i = 0, 1, \ldots, m-1$, define the one-step survival probability

$$p_i(x) = P(S(t_{i+1}) > b|S(t_i) = x) = \Phi\left(\frac{\log(x/b)+(\mu-\frac{1}{2}\sigma^2)(t_{i+1}-t_i)}{\sigma\sqrt{t_{i+1}-t_i}}\right).$$

From S construct a process $\tilde{S}$ having the same initial value as S but with state transitions generated conditional on survival: given $\tilde{S}(t_{i-1}) = x$, the next state $\tilde{S}(t_i)$ has the distribution of $S(t_i)$ conditional on $S(t_{i-1}) = x$ and $S(t_i) > b$. More explicitily,

$$\tilde{S}(t_i) = \tilde{S}(t_{i-1})e^{(\mu-\frac{1}{2}\sigma^2)[t_i-t_{i-1}]+\sigma\sqrt{t_i-t_{i-1}}Z_i}$$

where

$$Z_i = \Phi^{-1}\left(p_{i-1}(\tilde{S}(t_{i-1})) + (1-p_{i-1}(\tilde{S}(t_{i-1})))U_i\right),$$

with $U_1, \ldots, U_m$ independent Unif[0,1] random variables. This mechanism samples Z_i from the normal distribution conditional on $\tilde{S}(t_i) > b$; see Example 2.2.5.

Using the conditioned process, we can write

$$\mathsf{E}\left[e^{-rT}(S(T) - K)^+ \mathbf{1}\{\min_{1 \le i \le m} S(t_i) > b\}\right]$$

$$= \mathsf{E}\left[e^{-rT}(\tilde{S}(T) - K)^+ \prod_{i=0}^{m-1} p_i(\tilde{S}(t_i))\right]. \qquad (7.30)$$

Expressions of this form are derived in Glasserman and Staum [146] through the observation that switching from S to $\tilde{S}$ can be formulated as a change of measure and that the product of the survival probabilities is the appropriate likelihood ratio for this transformation. Because the conditioned process never crosses the barrier, we can omit the indicator from the expression on the right. The product of the survival probabilities smooths the indicator.

We could now differentiate with respect to $S(0)$ inside the expectation on the right side of (7.30) to get an estimator of the barrier option delta. Though straightforward in principle, this is rather involved in practice. The price we pay for switching from S to $\tilde{S}$ is the loss of linearity in $S(0)$. Each Z_i depends on $\tilde{S}(t_{i-1})$ and thus contributes a term when we differentiate $\tilde{S}(t_i)$. The practical value of this approach is therefore questionable, but it nevertheless serves to illustrate the flexibility we have to modify a problem before differentiating.

7.3 The Likelihood Ratio Method

As explained in several places in Section 7.2, the scope of the pathwise method is limited primarily by the requirement of continuity in the discounted payoff as a function of the parameter of differentiation. The *likelihood ratio method* provides an alternative approach to derivative estimation requiring no smoothness at all in the discounted payoff and thus complementing the pathwise method. It accomplishes this by differentiating probabilities rather than payoffs.

7.3.1 Method and Examples

As in Section 7.2.2, we consider a discounted payoff Y expressed as a function f of a random vector $X = (X_1, \ldots, X_m)$. The components of X could represent different underlying assets or values of a single asset at multiple dates. In our discussion of the pathwise method, we assumed the existence of a functional dependence of X (and then Y) on a parameter θ. In the likelihood ratio method, we instead suppose that X has a probability density g and that θ is a parameter of this density. We therefore write g_θ for the density, and to emphasize that an expectation is computed with respect to g_θ, we sometimes write it as E_θ.

In this formulation, the expected discounted payoff is given by

$$\mathsf{E}_\theta[Y] = \mathsf{E}_\theta[f(X_1,\ldots,X_m)] = \int_{\Re^m} f(x)g_\theta(x)\,dx.$$

To derive a derivative estimator, we suppose that the order of differentiation and integration can be interchanged to get

$$\frac{d}{d\theta}\mathsf{E}_\theta[Y] = \int_{\Re^m} f(x)\frac{d}{d\theta}g_\theta(x)\,dx. \tag{7.31}$$

If this indeed holds, then multiplying and dividing the integrand by g_θ yields

$$\frac{d}{d\theta}\mathsf{E}_\theta[Y] = \int_{\Re^m} f(x)\frac{\dot{g}_\theta(x)}{g_\theta(x)}g_\theta(x)\,dx = \mathsf{E}_\theta\left[f(X)\frac{\dot{g}_\theta(X)}{g_\theta(X)}\right],$$

where we have written $\dot{g}_\theta$ for $dg_\theta/d\theta$. It now follows from this equation that the expression

$$f(X)\frac{\dot{g}_\theta(X)}{g_\theta(X)} \tag{7.32}$$

is an unbiased estimator of the derivative of $\mathsf{E}_\theta[Y]$. This is a likelihood ratio method (LRM) estimator.

Three issues merit comment:

o As with the pathwise method, the validity of this approach relies on an interchange of differentation and integration. In practice, however, the interchange in (7.31) is relatively benign in comparison to (7.16), because probability densities are typically smooth functions of their parameters but option payoffs are not. Whereas the interchange (7.16) imposes practical limits on the use of the pathwise method, the validity of (7.31) is seldom an obstacle to the use of the likelihood ratio method.
o Whether we view θ as a parameter of the path X or of its density is largely a matter a choice. Suppose, for example, that X is a normal random variable with distribution $N(\theta,1)$ and $Y = f(X)$ for some function f. The density of X is $g_\theta(x) = \phi(x - \theta)$ with ϕ the standard normal density. By following the steps above, we arrive at the estimator

$$-f(X)\frac{\phi'(X-\theta)}{\phi(X-\theta)} = f(X)(X-\theta)$$

for the derivative of $\mathsf{E}_\theta[Y]$. But we could also write $X(\theta) = \theta + Z$ with $Z \sim N(0,1)$ and apply the pathwise method to get the estimator

$$Y'(\theta) = f'(X)\frac{dX}{d\theta} = f'(X).$$

This illustrates the flexibility we have in representing the dependence on a parameter through the path X or its density (or both); this flexibility should be exploited in developing derivative estimators.

○ In the statistical literature, an expression of the form $\dot{g}_\theta/g_\theta$, often written $d \log g_\theta/d\theta$, is called a *score function*. In the estimator (7.32), the score function is evaluated at the outcome X. We refer to the random variable $\dot{g}_\theta(X)/g_\theta(X)$ as the *score*. This term is short for "the differentiated log density evaluated at the simulated outcome" and also for "the expression that multiplies the discounted payoff in a likelihood ratio method estimator."

We illustrate the likelihood ratio method through examples. This approach has been developed primarily in the discrete-event simulation literature; important early references include Glynn [152], Reiman and Weiss [305], and Rubinstein [311]. It is sometimes called the score function method, as in Rubinstein and Shapiro [313]. Broadie and Glasserman [64] and Glasserman and Zhao [150] develop applications in finance, and we include some of their examples here.

Example 7.3.1 *Black-Scholes delta.* To estimate the Black-Scholes delta using the likelihood ratio method, we need to view $S(0)$ as a parameter of the density of $S(T)$. Using (3.23), we find that the lognormal density of $S(T)$ is given by

$$g(x) = \frac{1}{x\sigma\sqrt{T}}\phi(\zeta(x)), \quad \zeta(x) = \frac{\log(x/S(0)) - (r - \frac{1}{2}\sigma^2)T}{\sigma\sqrt{T}}, \quad (7.33)$$

with ϕ the standard normal density. Some algebra now shows that

$$\frac{dg(x)/dS(0)}{g(x)} = -\zeta(x)\frac{d\zeta(x)}{dS(0)} = \frac{\log(x/S(0)) - (r - \frac{1}{2}\sigma^2)T}{S(0)\sigma^2 T}.$$

We get the score by evaluating this expression at $S(T)$ and an unbiased estimator of delta by multiplying by the discounted payoff of the option:

$$e^{-rT}(S(T) - K)^+ \left(\frac{\log(S(T)/S(0)) - (r - \frac{1}{2}\sigma^2)T}{S(0)\sigma^2 T}\right).$$

If $S(T)$ is generated from $S(0)$ using a standard normal random variable Z as in (7.17), then $\zeta(S(T)) = Z$ and the estimator simplifies to

$$e^{-rT}(S(T) - K)^+ \frac{Z}{S(0)\sigma\sqrt{T}}. \quad (7.34)$$

The form of the option payoff in this example is actually irrelevant; any other function of $S(T)$ would result in an estimator of the same form. The delta of a digital option, for example, can be estimated using

$$e^{-rT}\mathbf{1}\{S(T) > K\}\frac{Z}{S(0)\sigma\sqrt{T}}.$$

This is a general feature of the likelihood ratio method that contrasts markedly with the pathwise method: the form of the estimator does not depend on the details of the discounted payoff. Once the score is calculated it can be multiplied by many different discounted payoffs to estimate their deltas.

For estimating vega, the score function is

$$\frac{dg(x)/d\sigma}{g(x)} = -\frac{1}{\sigma} - \zeta(x)\frac{d\zeta(x)}{d\sigma},$$

with

$$\frac{d\zeta(x)}{d\sigma} = \frac{\log(S(0)/x) + (r + \frac{1}{2}\sigma^2)T}{\sigma^2\sqrt{T}}.$$

The derivative estimator is the product of the discounted payoff and the score function evaluated at $x = S(T)$. After some algebraic simplification, the score can be expressed as

$$\frac{Z^2 - 1}{\sigma} - Z\sqrt{T},$$

with Z as in (7.17). □

Example 7.3.2 *Path-dependent deltas.* Consider an Asian option, as in Example 7.2.2. The payoff is a function of $S(t_1), \ldots, S(t_m)$, so we need the density of this path. Using the Markov property of geometric Brownian motion, we can factor this density as

$$g(x_1, \ldots, x_m) = g_1(x_1|S(0))g_2(x_2|x_1) \cdots g_m(x_m|x_{m-1}),$$

where each $g_j(x_j|x_{j-1})$, $j = 1, \ldots, m$, is the transition density from time t_{j-1} to time t_j,

$$g_j(x_j|x_{j-1}) = \frac{1}{x_j\sigma\sqrt{t_j - t_{j-1}}}\phi\left(\zeta_j(x_j|x_{j-1})\right),$$

with

$$\zeta_j(x_j|x_{j-1}) = \frac{\log(x_j/x_{j-1}) - (r - \frac{1}{2}\sigma^2)(t_j - t_{j-1})}{\sigma\sqrt{t_j - t_{j-1}}}.$$

It follows that $S(0)$ is a parameter of the first factor $g_1(x_1|S(0))$ but does not appear in any of the other factors. The score is thus given by

$$\frac{\partial \log g(S(t_1), \ldots, S(t_m))}{\partial S(0)} = \frac{\partial \log g_1(S(t_1)|S(0))}{\partial S(0)} = \frac{\zeta_1(S(t_1)|S(0))}{S(0)\sigma\sqrt{t_1}}.$$

This can also be written as

$$\frac{Z_1}{S(0)\sigma\sqrt{t_1}}, \tag{7.35}$$

with Z_1 the normal random variable used to generate $S(t_1)$ from $S(0)$, as in (7.21). The likelihood ratio method estimator of the Asian option delta is thus

$$e^{-rT}(\bar{S} - K)^+\frac{Z_1}{S(0)\sigma\sqrt{t_1}}. \tag{7.36}$$

Again, the specific form of the discounted payoff is irrelevant; the same derivation applies to any function of the path of S over the interval $[t_1, T]$, including
a discretely monitored barrier option.

Observe that the score has mean zero for all nonzero values of $S(0)$, σ,
and t_1, but its variance increases without bound as $t_1 \downarrow 0$. We discuss this
phenomenon in Section 7.3.2. $\square$

Example 7.3.3 *Path-dependent vega.* Whereas in the previous example the
parameter $S(0)$ appeared only in the distribution of $S(t_1)$, the parameter σ
influences every state transition. Accordingly, each transition along the path
contributes a term to the score. Following steps similar to those in the previous
two examples, we find that the score

$$\frac{\partial \log g(S(t_1), \ldots, S(t_m))}{\partial \sigma} = \sum_{j=1}^{m} \frac{\partial \log g(S(t_j)|S(t_{j-1}))}{\partial \sigma}$$

is given by

$$-\sum_{j=1}^{m} \left(\frac{1}{\sigma} + \zeta_j(S(t_j)|S(t_{j-1})) \frac{\partial \zeta_j}{\partial \sigma} \right).$$

This can be written as

$$\sum_{j=1}^{m} \left(\frac{Z_j^2 - 1}{\sigma} - Z_j \sqrt{t_j - t_{j-1}} \right), \tag{7.37}$$

using the normal random variables Z_j in (7.21). $\square$

Example 7.3.4 *Gaussian vectors and options on multiple assets.* The problem of LRM delta estimation for an option on multivariate geometric Brownian
motion can be reduced to one of differentiation with respect to a parameter of
the mean of a normal random vector. Suppose therefore that $X \sim N(\mu(\theta), \Sigma)$
with θ a scalar parameter of the d-dimensional mean vector μ, and Σ a $d \times d$
covariance matrix of full rank. Let $g = g_\theta$ denote the multivariate normal
density of X, as in (2.21). Differentiation reveals that the score is

$$\frac{d}{d\theta} \log g_\theta(X) = (X - \mu(\theta))^\top \Sigma^{-1} \dot{\mu}(\theta),$$

with $\dot{\mu}(\theta)$ the vector of derivatives of the components of μ with respect to the
parameter θ. If we simulate X as $\mu(\theta) + AZ$ with $Z \sim N(0, I)$ for some matrix
A satisfying $AA^\top = \Sigma$, then the score simplifies to

$$\frac{d}{d\theta} \log g_\theta(X) = Z^\top A^{-1} \dot{\mu}(\theta). \tag{7.38}$$

If θ is a parameter of Σ rather than μ, the score is

$d \log g_\theta(X)/d\theta$

$$= -\tfrac{1}{2}\mathrm{tr}\left(\Sigma^{-1}(\theta)\dot{\Sigma}(\theta)\right) + \tfrac{1}{2}(X-\mu)^\top\Sigma^{-1}(\theta)\dot{\Sigma}(\theta)\Sigma^{-1}(\theta)(X-\mu) \quad (7.39)$$

$$= -\tfrac{1}{2}\mathrm{tr}\left(A^{-1}(\theta)\dot{\Sigma}(\theta)A^{-1}(\theta)^\top\right) + \tfrac{1}{2}Z^\top A^{-1}(\theta)\dot{\Sigma}(\theta)A^{-1}(\theta)^\top Z, \quad (7.40)$$

where "tr" denotes trace, $A(\theta)A(\theta)^\top = \Sigma(\theta)$, and $\dot{\Sigma}(\theta)$ the matrix of derivatives of the elements of $\Sigma(\theta)$.

Now consider a d-dimensional geometric Brownian motion

$$\frac{dS_i(t)}{S_i(t)} = r\,dt + v_i^\top\,dW(t), \quad i = 1, \ldots, d,$$

with W a d-dimensional standard Brownian motion. Let A be the $d \times d$ matrix with rows $v_1^\top, \ldots, v_d^\top$ and suppose that A has full rank. Let $\Sigma = AA^\top$. Then $S_i(T)$ has the distribution of $\exp(X_i)$, $i = 1, \ldots, d$, with

$$X \sim N(\mu, T\Sigma), \quad \mu_i = \log S_i(0) + (r - \tfrac{1}{2}\|v_i\|^2)T.$$

We may view any discounted payoff $f(S_1(T), \ldots, S_d(T))$ as a function of the vector X. To calculate the delta with respect to the ith underlying asset, we take $S_i(0)$ as a parameter of the mean vector μ. The resulting likelihood ratio method estimator is

$$f(S_1(T), \ldots, S_d(T)) \cdot \frac{(Z^\top A^{-1})_i}{S_i(0)\sqrt{T}},$$

the numerator of the score given by the ith component of the vector $(Z^\top A^{-1})$. □

Example 7.3.5 *Square-root diffusion.* We use the notation of Example 7.2.5. We know from Section 3.4.1 that $X(t)/c_1$ has a noncentral chi-square density and that $X(0)$ appears in the noncentrality parameter. To estimate the sensitivity of some $\mathsf{E}[f(X(t))]$ with respect to $X(0)$, one could therefore calculate a score function by differentiating the log density (available from (3.67)) with respect to the noncentrality parameter. But the noncentral chi-square density is cumbersome and its score function still more cumbersome, so this is not a practical solution.

An alternative uses the decomposition (7.22). Any function of $X(t)$ can be represented as a function of a normal random variable with mean $\sqrt{c_2 X(0)}$ and an independent chi-square random variable. The expected value of the function is then an integral with respect to the bivariate density of these two random variables. Because the variables are independent, their joint density is just the product of their marginal normal and chi-square densities. Only the first of these densities depends on $X(0)$, and $X(0)$ enters through the mean. The score function thus reduces to

$$\frac{d}{dX(0)} \log \phi\left(x - \sqrt{c_2 X(0)}\right) = \frac{x\sqrt{c_2} - c_2\sqrt{X(0)}}{2\sqrt{X(0)}}.$$

Evaluating this at the normal random variable $Z + \sqrt{c_2 X(0)}$ (with Z as in (7.22)), we get the score

$$\frac{Z\sqrt{c_2}}{2\sqrt{X(0)}}.$$

Multiplying this by $f(X(t))$ produces an LRM estimator of the derivative $d\mathsf{E}[f(X(t))]/dX(0)$. For the derivative of a function of a discrete path $X(t_1)$, ..., $X(t_m)$, replace Z with Z_1 in the score, with Z_1 the normal random variable used in generating $X(t_1)$ from $X(0)$. $\square$

7.3.2 Bias and Variance Properties

The likelihood ratio method produces an unbiased estimator of a derivative when (7.31) holds; i.e., when the integral (over x) of the limit as $h \to 0$ of the functions

$$f(x)\frac{1}{h}\left(\frac{g_{\theta+h}(x)}{g_\theta(x)} - 1\right) g_\theta(x) \tag{7.41}$$

equals the limit of their integrals. Because probability densities tend to be smooth functions of their parameters, this condition is widely satisfied. Specific conditions for exponential families of distributions are given in, e.g., Barndorff-Nielsen [35], and this case is relevant to several of our examples. Glynn and L'Ecuyer [158] provide general conditions. In practice, the applicability of the likelihood ratio method is more often limited by either (i) the need for explicit knowledge of a density, or (ii) a large variance, than by the failure of (7.31).

Absolute Continuity

As suggested by (7.41), the likelihood ratio method is based on a limit of importance sampling estimators (see Section 4.6). For fixed h, the integral of (7.41) equals $1/h$ times the difference $\mathsf{E}_{\theta+h}[f(X)] - \mathsf{E}_\theta[f(X)]$, the first expectation relying on the importance sampling identity

$$\mathsf{E}_{\theta+h}[f(X)] = \mathsf{E}_\theta[f(X)g_{\theta+h}(X)/g_\theta(X)].$$

The validity of this identity relies on an implicit assumption of absolute continuity (cf. Appendix B.4): we need $g_{\theta+h}(x) > 0$ at all points x for which $g_\theta(x) > 0$.

For a simple example in which absolute continuity fails, suppose X is uniformly distributed over $(0, \theta)$. Its density is

$$g_\theta(x) = \frac{1}{\theta}\mathbf{1}\{0 < x < \theta\},$$

which is differentiable in θ at $x \in (0, \theta)$. The score $d\log g_\theta(X)/d\theta$ exists with probability 1 and equals $-1/\theta$. The resulting LRM estimator of the derivative of $\mathsf{E}_\theta[X]$ is

$$\mathsf{E}_\theta[X \cdot \frac{-1}{\theta}] = -\frac{\theta/2}{\theta} = -\tfrac{1}{2},$$

whereas the correct value is

$$\frac{d}{d\theta}\mathsf{E}_\theta[X] = \tfrac{1}{2}.$$

The estimator even fails to predict the direction of change. This failure results from the fact that $g_{\theta+h}$ is not absolutely continuous with respect to g_θ.

A related (and in practice more significant) limitation applies to the multivariate normal distribution. Suppose $X \sim N(\mu, \Sigma)$ and we are interested in the sensitivity of some expectation $\mathsf{E}[f(X)]$ to changes in θ, with θ a parameter of μ or Σ. If X is a vector of length d and the $d \times d$ matrix Σ has rank $k < d$, then X fails to have a density on $\Re^d$ and the likelihood ratio method is not directly applicable. As discussed in Section 2.3.3, in the rank-deficient case we could express $d - k$ components of X as linear transformations of $\tilde{X}$, a vector consisting of the other k components of X. Moreover, $\tilde{X}$ would then have a density in $\Re^k$. This suggests that we could write $f(X) = \tilde{f}(\tilde{X})$ for some function $\tilde{f}$ and then apply the likelihood ratio method.

This transformation may not, however, entirely remove the obstacle to using the method if it introduces explicit dependence on θ in the function $\tilde{f}$. A simple example illustrates this point. Suppose Z_1 and Z_2 are independent normal random variables and

$$\begin{pmatrix} X_1 \\ X_2 \\ X_3 \end{pmatrix} = \begin{pmatrix} \mu_1 \\ \mu_2 \\ \mu_3 \end{pmatrix} + \begin{pmatrix} 1 & 0 \\ 0 & 1 \\ a_1 & a_2 \end{pmatrix} \begin{pmatrix} Z_1 \\ Z_2 \end{pmatrix}. \tag{7.42}$$

We can reduce any function f of (X_1, X_2, X_3) to a function of just the first two components by defining

$$\tilde{f}(X_1, X_2) = f(X_1, X_2, a_1(X_1 - \mu_1) + a_2(X_2 - \mu_2) + \mu_3).$$

The pair (X_1, X_2) has a probability density in $\Re^2$. But if any of the μ_i or a_i depend on θ then $\tilde{f}$ does too. This dependence is not captured by differentiating the density of (X_1, X_2); capturing it requires differentiating $\tilde{f}$, in effect combining the likelihood ratio and pathwise methods. This may not be possible if f is discontinuous. The inability of the likelihood ratio method to handle this example again results from a failure of absolute continuity: (X_1, X_2, X_3) has a density on a two-dimensional subspace of $\Re^3$, but this subspace changes with θ, so the densities do not have common support.

Variance Properties

A near failure of absolute continuity is usually accompanied by an explosion in variance. This results from properties of the score closely related to the variance build-up in likelihood ratios discussed in Section 4.6.1 and Appendix B.4.

In (7.42), one could get around the lack of a density by adding $\sqrt{\epsilon}Z_3$ to X_3, with Z_3 independent of Z_1, Z_2. This would lead to a score of the form in (7.38) with

$$A^{-1} = \begin{pmatrix} 1 & 0 & 0 \\ 0 & 1 & 0 \\ -a_1/\sqrt{\epsilon} & -a_2/\sqrt{\epsilon} & 1/\sqrt{\epsilon} \end{pmatrix},$$

and thus infinite variance as $\epsilon \to 0$.

Something similar occurs with the score (7.35) for a path-dependent option sensitive to the values of $\mathrm{GBM}(r, \sigma^2)$ at dates $t_1, \ldots, t_m$. The score has expected value 0 for all $t_1 > 0$, but its variance grows without bound as t_1 approaches 0. This, too, is a consequence of a breakdown of absolute continuity. The score in (7.35) results from viewing a value of $S(t_1)$ generated from $S(0)$ as though it had been generated from $S(0) + h$. For $t_1 > 0$, all positive values of $S(t_1)$ are possible starting from both $S(0)$ and $S(0) + h$; the two distributions of $S(t_1)$ are mutually absolutely continuous and thus have a well-defined likelihood ratio. But viewed as probability measures on paths starting at time 0 rather t_1, the measures defined by $S(0)$ and $S(0) + h$ are mutually singular: no path that starts at one value can also start at the other. Absolute continuity thus fails as $t_1 \to 0$, and this is manifested by the increase in variance of the score.

Next we consider the effect of a long time horizon. The score in (7.37) for the vega of a function of the discrete path $(S(t_1), \ldots, S(t_m))$ is a sum of m independent random variables. Its variance grows linearly in m if the spacings $t_j - t_{j-1}$ are constant, faster if the spacings themselves are increasing. This suggests that the variance of an LRM estimator that uses this score will typically increase with the number of dates m. The same occurs if we fix a time horizon T, divide the interval $[0, T]$ into m equal time steps, and then let $m \to \infty$. Recall (cf. the discussion of Girsanov's Theorem in Appendix B.4) that the measures on paths for Brownian motions with different variance parameters are mutually singular, so this explosion in variance once again results from a breakdown of absolute continuity.

Figure 7.3 illustrates the growth in variance is estimating vega for an Asian option. The model parameters are $S(0) = K = 100$, $\sigma = 0.30$, $r = 5\%$; the option payoff is based on the average level of the underlying asset over m equally spaced dates with a spacing of $1/52$ (one week). The value of m varies along the horizontal access; the vertical axis shows variance (per replication) on a log scale, as estimated from one million replications. The figure shows that the variance grows with m for both the LRM and pathwise estimators. The variance of the LRM estimator starts higher (by more than a factor of 10 at $m = 2$) and grows faster as m increases.

The growth in variance exhibited by the vega score (7.37) is intrinsic to the method: subject only to modest technical conditions, the score is a martingale

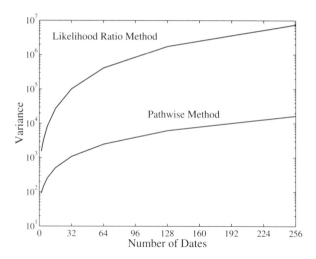

Fig. 7.3. Variance of vega estimators for an Asian option with weekly averaging as a function of the number of weeks in the average.

and thus a process of increasing variance. Consider, for example, a simulation (as in (4.78)) based on a recursion of the form

$$S(t_{i+1}) = G(S(t_i), X_{i+1}) \tag{7.43}$$

driven by i.i.d. random vectors $X_1, X_2, \ldots$. Suppose θ is a parameter of the density g_θ of the X_i. A likelihood ratio method estimator of the derivative with respect to θ of some expectation $\mathsf{E}[f(S(t_1), \ldots, S(t_m))]$ uses the score

$$\sum_{i=1}^{m} \frac{\dot{g}_\theta(X_i)}{g_\theta(X_i)}.$$

This is an i.i.d. sum and a martingale because

$$\mathsf{E}\left[\frac{\dot{g}_\theta(X_i)}{g_\theta(X_i)}\right] = \int \frac{\dot{g}_\theta(x)}{g_\theta(x)} g_\theta(x)\, dx$$

$$= \frac{d}{d\theta} \int g_\theta(x)\, dx = 0,$$

using the fact that g_θ integrates to 1 for all θ. Alternatively, if $g_\theta(y|x)$ is the transition density of the process $S(t_i)$, $i = 1, 2, \ldots$, we could use the score

$$\sum_{i=1}^{m} \frac{\dot{g}_\theta(S(t_i)|S(t_{i-1}))}{g_\theta(S(t_i)|S(t_{i-1}))}.$$

This is no longer an i.i.d. sum, but a similar argument shows that it is still a martingale and hence that its variance increases with m.

Because the score has mean zero, it provides a candidate control variate. Using it as a control removes some of the variance introduced by the score, but rarely changes the qualitative dependence on the number of steps m.

7.3.3 Gamma

We noted in Example 7.2.6 that the pathwise method does not readily extend to estimating second derivatives. With the likelihood ratio method, second derivatives are in principle no more difficult to estimate than first derivatives. Let $\ddot{g}_\theta$ denote the second derivative of g_θ in θ. The argument leading to (7.32) shows that

$$f(X)\frac{\ddot{g}_\theta(X)}{g_\theta(X)} \tag{7.44}$$

is an unbiased estimator of $d^2 E_\theta[f(X)]/d\theta^2$, subject to conditions permitting two interchanges of derivative and expectation. These conditions seldom limit the applicability of the method. However, just as the score $\dot{g}_\theta(X)/g_\theta(X)$ can lead to large variance in first-derivative estimates, its counterpart in (7.44) often produces even larger variance in second-derivative estimates.

We illustrate the method with some examples.

Example 7.3.6 *Black-Scholes gamma.* We apply (7.44) to estimate gamma — the second derivative with respect to the underlying price $S(0)$ — in the Black-Scholes setting. Using the notation of Example 7.3.1, we find that

$$\frac{d^2 g(S(T))/dS^2(0)}{g(S(T))} = \frac{\zeta(S(T))^2 - 1}{S(0)^2\sigma^2 T} - \frac{\zeta(S(T))}{S(0)^2\sigma\sqrt{T}}. \tag{7.45}$$

Multiplying this expression by the discounted payoff $e^{-rT}(S(T) - K)^+$ produces an unbiased estimator of the Black-Scholes gamma. If $S(T)$ is generated from $S(0)$ using (7.17), $\zeta(S(T))$ can be replaced with the standard normal random variable Z.

As with LRM first-derivative estimates, the form of the discounted payoff is unimportant; (7.45) could be applied with other functions of $S(T)$. Much as in Example 7.3.2, this expression also extends to path-dependent options. For a function of $S(t_1), \ldots, S(t_m)$, replace T with t_1 in (7.45); see the argument leading to (7.35). □

Example 7.3.7 *Second derivatives for Gaussian vectors.* If $X \sim N(\mu, \sigma^2)$ then differentiation with respect to μ yields

$$\frac{\dot{g}_\mu(X)}{g_\mu(X)} = \frac{X - \mu}{\sigma^2}$$

and

$$\frac{\ddot{g}_\mu(X)}{g_\mu(X)} = \frac{(X - \mu)^2 - \sigma^2}{\sigma^4}.$$

Higher orders of differentiation produce higher powers of X and often larger variance.

More generally, let X be a random vector with multivariate normal distribution $N(\mu(\theta), \Sigma)$, with θ a scalar parameter and Σ of full rank. Then

$$\frac{\ddot{g}_\theta(X)}{g_\theta(X)} = ([X - \mu(\theta)]^\top \Sigma^{-1} \dot{\mu}(\theta))^2 - \dot{\mu}(\theta)^\top \Sigma^{-1} \dot{\mu}(\theta) + [X - \mu(\theta)]^\top \Sigma^{-1} \ddot{\mu}(\theta),$$

where $\dot{\mu}$ and $\ddot{\mu}$ denote derivatives of μ with respect to θ. $\square$

An alternative approach to estimating second derivatives combines the pathwise and likelihood ratio methods, using each for one order of differentiation. We illustrate this idea by applying it to the Black-Scholes gamma. Suppose we apply the likelihood ratio method first and then the pathwise method. Multiplying the discounted payoff by the score yields (7.34), which we differentiate to get the "LR-PW" estimator

$$\frac{d}{dS(0)}\left(e^{-rT}(S(T) - K)^+ \frac{Z}{S(0)\sigma\sqrt{T}}\right) = e^{-rT}\frac{Z}{S(0)^2\sigma\sqrt{T}}\mathbf{1}\{S(T) > K\}K,$$
$$(7.46)$$

using $dS(T)/dS(0) = S(T)/S(0)$. If instead we apply pathwise differentiation first, we get the delta estimator (7.20). This expression has a functional dependence on $S(0)$ as well as a distributional dependence on $S(0)$ through the density of $S(T)$. Multiplying by the score captures the second dependence but for the first we need to take another pathwise derivative. The resulting "PW-LR" estimator is

$$e^{-rT}\left(\mathbf{1}\{S(T) > K\}\frac{S(T)}{S(0)} \cdot \frac{Z}{S(0)\sigma\sqrt{T}} + \mathbf{1}\{S(T) > K\}S(T)\frac{d}{dS(0)}\frac{1}{S(0)}\right)$$
$$= e^{-rT}\mathbf{1}\{S(T) > K\}\left(\frac{S(T)}{S(0)^2}\right)\left(\frac{Z}{\sigma\sqrt{T}} - 1\right).$$

The pure likelihood ratio method estimator of gamma is

$$e^{-rT}(S(T) - K)^+\left(\frac{Z^2 - 1}{S(0)^2\sigma^2 T} - \frac{Z}{S(0)^2\sigma\sqrt{T}}\right),$$

using (7.45).

Table 7.2 compares the variance per replication using these methods with parameters $S(0) = 100$, $\sigma = 0.30$, $r = 5\%$, three levels of the strike K, and two maturities T. The table also compares variances for the likelihood ratio method and pathwise estimators of delta; the exact values of delta and gamma are included in the table for reference. Each variance estimate is based on 1 million replications.

The results in the table indicate that the pathwise estimator of delta has lower variance than the likelihood ratio method estimator, especially at higher values of $S(0)/K$. The results also indicate that the mixed gamma estimators have substantially lower variance than the pure likelihood ratio method estimator. The two mixed estimators have similar variances, except at lower values of K where LR-PW shows a distinct advantage. This advantage is also evident from the expression in (7.46).

K	$T = 0.1$			$T = 0.5$		
	90	100	110	90	100	110
Delta	0.887	0.540	0.183	0.764	0.589	0.411
LR	3.4	1.5	0.5	2.7	2.0	1.4
PW	0.1	0.3	0.2	0.3	0.4	0.3
Gamma	0.020	0.042	0.028	0.015	0.018	0.018
LR	0.1232	0.0625	0.0265	0.0202	0.0154	0.0116
PW-LR	0.0077	0.0047	0.0048	0.0015	0.0013	0.0013
LR-PW	0.0052	0.0037	0.0045	0.0007	0.0007	0.0009

Table 7.2. Variance comparison for estimators of Black-Scholes deltas and gammas using the likelihood ratio (LR) and pathwise (PW) methods and combinations of the two.

Although generalizing from numerical results is risky (especially with results based on such a simple example), we expect the superior performance of the mixed gamma estimators campared with the pure likelihood ratio method estimator to hold more generally. It holds, for example, in the more complicated application to the LIBOR market model in Glasserman and Zhao [150].

Yet another strategy for estimating second derivatives applies a finite-difference approximation using two estimates of first derivatives. Glasserman and Zhao [150] report examples in which the best estimator of this type has smaller root mean square error than a mixed PW-LR estimator. A difficulty in this method as in any use of finite differences is finding an effective choice of parameter increment h.

7.3.4 Approximations and Related Methods

The main limitations on the use of the likelihood ratio method are (i) the method's reliance on explicit probability densities, and (ii) the large variance sometimes produced by the score. This section describes approximations and related methods that can in some cases address these limitations.

General Diffusion Processes

Several of the examples with which we introduced the likelihood ratio method in Section 7.3.1 rely on the simplicity of probability densities associated with geometric Brownian motion. In more complicated models, we rarely have explicit expressions for either marginal or transition densities, even in cases where these are known to exist. By the same token, in working with more complicated models we are often compelled to simulate approximations (using, e.g., the discretization methods of Chapter 6), and we may be able to develop derivative estimators for the approximating processes even if we cannot for the original process.

Consider, for example, a general diffusion of the type in (7.23). An Euler approximation to the process is a special case of the general recursion in (7.43) in which the driving random variables are normally distributed. In fact, whereas we seldom know the transition law of a general diffusion (especially in multiple dimensions), the Euler approximation has a Gaussian transition law and thus lends itself to use of the likelihood ratio method. If the method is unbiased for the Euler approximation, it does not introduce any discretization error beyond that already present in the Euler scheme. (Similar though more complicated techniques can potentially be developed using higher-order discretization methods; for example, in the second-order scheme (6.36), the state transition can be represented as a linear transformation of a noncentral chi-square random variable.)

We illustrate this idea through the Heston [179] stochastic volatility model, using the notation of Example 6.2.2. We consider an Euler approximation

$$\hat{S}(i+1) = (1+rh)\hat{S}(i) + \hat{S}(i)\sqrt{\hat{V}(i)h}Z_1(i+1)$$

$$\hat{V}(i+1) = \hat{V}(i) + \kappa(\theta - \hat{V}(i))h + \sigma\sqrt{\hat{V}(i)h}\left(\rho Z_1(i+1) + \sqrt{1-\rho^2}Z_2(i+1)\right)$$

with time step h and independent standard normal variables $(Z_1(i), Z_2(i))$, $i = 1, 2, \ldots$. The conditional distribution at step $i+1$ given the state at the ith step is normal,

$$N\left(\begin{pmatrix} (1+rh)\hat{S}(i) \\ \hat{V}(i) + \kappa(\theta - \hat{V}(i))h \end{pmatrix}, \begin{pmatrix} \hat{S}^2(i)\hat{V}(i)h & \rho\sigma\hat{S}(i)\hat{V}(i)h \\ \rho\sigma\hat{S}(i)\hat{V}(i)h & \sigma^2\hat{V}(i)h \end{pmatrix}\right). \quad (7.47)$$

Assume a fixed initial state $(\hat{S}(0), \hat{V}(0)) = (S(0), V(0))$ and $|\rho| \neq 1$.

Consider the estimation of delta, a derivative with respect to $S(0)$. Through the argument in Example 7.3.2, it suffices to consider the dependence of $(\hat{S}(1), \hat{V}(1))$ on $S(0)$. Because $S(0)$ is not a parameter of the distribution of $\hat{V}(1)$, it actually suffices to consider $\hat{S}(1)$. Observe that $S(0)$ appears in both the mean and variance of $\hat{S}(1)$, so we need to combine contributions from (7.38) and (7.40). After some algebraic simplification, the score becomes

$$\frac{Z_1(1)(1+rh)}{S(0)\sqrt{V(0)h}} + \frac{Z_1^2(1) - 1}{S(0)},$$

with $Z_1(1)$ the standard normal random variable used to generate $\hat{S}(1)$ in the Euler scheme.

Next consider estimation of sensitivity to the parameter σ. Because σ is a parameter of the transition law of the Euler scheme, every transition contributes a term to the score. For a generic transition out of a state $(\hat{S}, \hat{V})$, we use (7.40) with

$$A = \begin{pmatrix} \hat{S}\sqrt{\hat{V}h} & 0 \\ \rho\sigma\sqrt{\hat{V}h} & \sigma\sqrt{\hat{V}h(1-\rho^2)} \end{pmatrix}, \quad \dot{\Sigma} = \begin{pmatrix} 0 & \rho\hat{S}\hat{V}h \\ \rho\hat{S}\hat{V}h & 2\sigma\hat{V}h \end{pmatrix}.$$

For the ith transition, (7.40) then simplifies to

$$\frac{Z_2^2(i) - 1}{\sigma} + \frac{\rho Z_1(i)Z_2(i)}{\sigma\sqrt{1-\rho^2}}. \tag{7.48}$$

The score for a function of $(\hat{S}(i), \hat{V}(i))$, $i = 1, \ldots, m$, is then the sum of (7.48) for $i = 1, \ldots, m$.

LIBOR Market Model

The application of the likelihood ratio method to the Heston stochastic volatility model relies on the covariance matrix in (7.47) having full rank. But in an Euler scheme in which the dimension of the state vector exceeds the dimension of the driving Brownian motion, the covariance matrix of the transition law is singular. As explained in Section 7.3.2, this prevents a direct application of the likelihood ratio method. In LIBOR market models, the number of forward rates in the state vector typically does exceed the dimension of the driving Brownian motion (the number of factors), making this a practical as well as a theoretical complication.

One way to address this issue notes that even if the one-step transition law is singular, the k-step transition law may have a density for some $k > 1$. But a first attempt to use this observation raises another issue: in an Euler scheme for log forward rates $(\log \hat{L}_1, \ldots, \log \hat{L}_M)$, as in (3.120), the one-step transition law is normal but the k-step transition law is not, for $k > 1$, because of the dependence of the drift term on the current level of rates. To get around this problem, Glasserman and Zhao [150] make the approximation in (7.27) under which the drift becomes a function of the initial rates rather than the current rates. Using this approximation and a fixed time step h, the k-step evolution of each $\log \hat{L}_n$ takes the form

$$\log \hat{L}_n(k) = \log L_n(0) + h \sum_{i=0}^{k-1} [\mu_n^o(ih) - \tfrac{1}{2}\|\sigma_n(ih)\|^2]$$

$$+ \sqrt{h}\left[\sigma_n(0)^\top, \sigma_n(h)^\top, \ldots, \sigma_n((k-1)h)^\top\right] \begin{bmatrix} Z_1 \\ \vdots \\ Z_k \end{bmatrix}$$

where the row vectors $\sigma_n(ih)^\top$ have been concatenated into a single vector of length kd, the column vectors Z_i have been stacked into a column vector of the same length, and d is the number of factors. For sufficiently large k^*, the $M \times k^* d$ matrix

$$A(k^*) = \begin{pmatrix} \sigma_1(0)^\top & \sigma_1(h)^\top & \cdots & \sigma_1((k^*-1)h)^\top \\ \sigma_2(0)^\top & \sigma_2(h)^\top & \cdots & \sigma_2((k^*-1)h)^\top \\ \vdots & \vdots & & \vdots \\ \sigma_M(0)^\top & \sigma_M(h)^\top & \cdots & \sigma_M((k^*-1)h)^\top \end{pmatrix}$$

may have rank M even if the number of factors d is smaller than the number of rates M. Whether this occurs depends on how the volatilities $\sigma_n(ih)$ change with i. If it does occur, then the covariance matrix $A(k^*)A(k^*)^\top$ of the (approximate) k-step transition law is invertible and we may use formulas in Example 7.3.4 for an arbitrary Gaussian vector.

For example, to apply this idea to estimate a sensitivity with respect to some $L_j(0)$, we set

$$X = (\log \hat{L}_1(k^* h), \dots, \log \hat{L}_M(k^* h))^\top,$$

$$a_n = \log L_n(0) + h \sum_{\ell=0}^{k^*-1} [\hat{\mu}_n^o(\ell h) - \tfrac{1}{2}\|\sigma_n(\ell h)\|^2], \quad n = 1, \dots, M,$$

$$\dot{a}_n = \frac{\mathbf{1}\{n=j\}}{L_j(0)} + h \sum_{\ell=0}^{k^*-1} \frac{\partial \hat{\mu}_n^o(\ell h)}{\partial L_j(0)}, \quad n = 1, \dots, M,$$

and $\Sigma = A(k^*)A(k^*)^\top h$. The (approximate) LRM estimator for the sensitivity of an arbitrary discounted payoff $f(\hat{L}_1(i), \dots, \hat{L}_M(i))$, $i > k^*$, is then

$$f(\hat{L}_1(i), \dots, \hat{L}_M(i))(X - a)^\top \Sigma^{-1} \dot{a},$$

with $a = (a_1, \dots, a_M)^\top$ and $\dot{a} = (\dot{a}_1, \dots, \dot{a}_M)^\top$. Glasserman and Zhao [150] test this method on payoffs with discontinuities (digital caplets and knock-out caplets) and find that it substantially outperforms finite-difference estimation. They compare methods based on the root mean square error achieved in a fixed amount of computing time.

Mixed Estimators

Pathwise differentiation and LRM estimation can be combined to take advantage of the strengths of each approach. We illustrated one type of combination in discussing gamma estimation in Section 7.3.3. There we used each method for one order of differentiation. An alternative type of combination uses the likelihood ratio method near a discontinuity and pathwise differentiation everywhere else.

We illustrate this idea with an example from Fournié et al. [124]. The payoff of a digital option struck at K can be written as

$$\mathbf{1}\{x > K\} = f_\epsilon(x) + (\mathbf{1}\{x > K\} - f_\epsilon(x))$$
$$\equiv f_\epsilon(x) + h_\epsilon(x),$$

with

$$f_\epsilon(x) = \min\{1, \max\{0, x - K + \epsilon\}/2\epsilon\}.$$

The function f_ϵ makes a piecewise linear approximation to the step-function payoff of the digital option and h_ϵ corrects the approximation. We can apply pathwise differentiation to $f_\epsilon(S(T))$ and the likelihood ratio method to $h_\epsilon(S(T))$ to get the combined delta estimator

$$\frac{1}{2\epsilon}\mathbf{1}\{|S(T) - K| < \epsilon\}\frac{S(T)}{S(0)} + h_\epsilon(S(T))\frac{\zeta(S(T))}{S(0)\sigma\sqrt{T}},$$

assuming $S(t)$ is a geometric Brownian motion and using the notation of Example 7.3.1.

Figure 7.4 plots the variance of this estimator as a function of ϵ with parameters $S(0) = K = 100$, $\sigma = 0.30$, $r = 0.05$, and $T = 0.25$. The case $\epsilon = 0$ corresponds to using only the likelihood ratio method. A small $\epsilon > 0$ increases variance, because of the ϵ in the denominator of the estimator, but larger values of ϵ can substantially reduce variance. The optimum occurs at a surprisingly large value of $\epsilon \approx 35$.

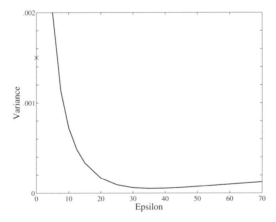

Fig. 7.4. Variance of mixed estimator as a function of linearization parameter ϵ.

Rewriting the combined estimator as

$$\mathbf{1}\{S(T) > K\}\frac{\zeta(S(T))}{S(0)\sigma\sqrt{T}} + \left(\mathbf{1}\{|S(T) - K| < \epsilon\}\frac{S(T)}{2\epsilon S(0)} - f_\epsilon(S(T))\frac{\zeta(S(T))}{S(0)\sigma\sqrt{T}}\right)$$

reveals an interpretation as a control variate estimator: the first term is the LRM estimator and the expression in parentheses has expected value zero. The implicit coefficient on the control is 1 and further variance reduction could be achieved by optimizing the coefficient.

Sensitivities to Calibration Instruments

Avellaneda and Gamba [28] consider the problem of estimating sensitivites of derivative securities with respect to the prices of securities used to calibrate a model, which may or may not be the underlying assets in the usual sense. These sensitivities are relevant if the securities used to calibrate a model to market prices are also the securities used for hedging, as is often the case.

The formulation in Avellaneda and Gamba [28] is based on weighted Monte Carlo (as in Section 4.5.2), but similar ideas apply in a simpler control variate setting. Denote by Y the discounted payoff of a derivative security and by X the discounted payoff of a hedging instrument. Suppose we estimate the price of the derivative security using an estimator of the form

$$\hat{Y} = Y - \beta(X - c), \tag{7.49}$$

with β interpreted as the slope of a regression line. The constant c is the observed market price of the hedging instrument, which may or may not equal $\mathsf{E}[X]$. The difference $\mathsf{E}[X] - c$ represents model error; if this is nonzero, then (7.49) serves primarily to correct for model error rather than to reduce variance. The coefficient β is the sensitivity of the corrected price to the market price of the hedging instrument and thus admits an interpretation as a hedge ratio. This coefficient can be estimated by applying, e.g., ordinary least squares regression to simulated values of (X, Y).

A link with regression can also be seen in the application of the likelihood ratio method with the normal distribution. Suppose $X \sim N(\theta, \sigma^2)$ and suppose Y depends on θ through X (for example, Y could be a function of X). As in (7.38), the score for estimating sensitivity to θ is $(X - \theta)/\sigma^2$ and the corresponding likelihood ratio method estimator has expectation

$$\mathsf{E}\left[Y\left(\frac{X - \theta}{\sigma^2}\right)\right] = \frac{\mathsf{Cov}[X, Y]}{\mathsf{Var}[X]}.$$

This is precisely the slope in a regression of Y against X. From this perspective, using regression to estimate a sensitivity can be interpreted as using an approximate LRM estimator based on the normal distribution.

7.4 Concluding Remarks

Extensive numerical evidence accumulated across many models and applications indicates that the pathwise method, when applicable, provides the best estimates of sensitivities. Compared with finite-difference methods, pathwise estimates require less computing time and they directly estimate derivatives rather than differences. Compared with the likelihood ratio method, pathwise estimates usually have smaller variance — often much smaller.

The application of the pathwise method requires interchanging the order of differentiation and integration. Sufficient conditions for this interchange, tailored to option pricing applications, are provided by conditions (A1)–(A4) in Section 7.2.2. A simple rule of thumb states that the pathwise method yields an unbiased estimate of the derivative of an option price if the option's discounted payoff is almost surely continuous in the parameter of differentiation. The discounted payoff is a stochastic quantity, so this rule of thumb requires continuity as the parameter varies with all random elements not depending on the parameter held fixed. (This should not be confused with continuity of the *expected* discounted payoff, which nearly always holds.) This rule excludes digital and barrier options, for example.

Finite-difference approximations are easy to implement and would appear to require less careful examination of the dependence of a model on its parameters. But when the pathwise method is inapplicable, finite-difference approximations have large mean square errors: the lack of continuity that may preclude the use of the pathwise method produces a large variance in a finite-difference estimate that uses a small parameter increment; a large increment leads to a large bias. The size of an effective parameter increment — one minimizing the mean square error — is very sensitive to the pathwise continuity of the discounted payoff. The question of continuity thus cannot be avoided by using finite-difference estimates.

In contrast, the likelihood ratio method does not require any smoothness in the discounted payoff because it is based on differentiation of probability densities instead. This makes LRM potentially attractive in exactly the settings in which the pathwise method fails. The application of LRM is, however, limited by two features: it requires explicit knowledge of the relevant probability densities, and its estimates often have large variance.

The variance of LRM estimates becomes problematic when the parameter of differentiation influences many of the random elements used to simulate a path. For example, consider the simulation of a discrete path $S(0), S(t_1), S(t_2), \ldots$ of geometric Brownian motion, and contrast the estimation of delta and vega. Only the distribution of $S(t_1)$ depends directly on $S(0)$; given $S(t_1)$, the subsequent $S(t_j)$ are independent of $S(0)$. But every transition depends on the volatility parameter σ. So, the score used to estimate delta has just a single term, whereas the score used to estimate vega has as many terms as there are transitions. Summing a large number of terms in the score produces an LRM estimator with a large variance.

Several authors (including Benhamou [44], Cvitanić et al. [94], Fournié et al. [124], and Gobet and Munos [161]) have proposed extensions of the likelihood ratio method using ideas from Malliavin calculus. These techniques reduce to LRM when the relevant densities are available; otherwise, they replace the score with a *Skorohod integral*, which is then computed numerically in the simulation itself. This integral may be viewed as a randomization of the score in the sense that the score is its conditional expectation. Further identities in this spirit are derived in Gobet and Munos [161], some leading to variance reduction. Except in special cases, evaluation of the Skorohod integral is computationally demanding. Estimators that require it should therefore be compared with the simpler alternative of applying LRM to an Euler approximation, for which the (Gaussian) transition laws are explicitly available.

Estimating second derivatives is fundamentally more difficult than estimating first derivatives, regardless of the method used. The pathwise method is generally inapplicable to second derivatives of option prices: a kink in an option payoff becomes a discontinuity in the derivative of the payoff. Combinations of the pathwise method and likelihood ratio method generally produce better gamma estimates than LRM alone. The two methods can also be combined to estimate first derivatives, with one method in effect serving as a control variate for the other.

8

Pricing American Options

Whereas a European option can be exercised only at a fixed date, an American option can be exercised any time up to its expiration. The value of an American option is the value achieved by exercising optimally. Finding this value entails finding the optimal exercise rule — by solving an optimal stopping problem — and computing the expected discounted payoff of the option under this rule. The embedded optimization problem makes this a difficult problem for simulation.

This chapter presents several methods that address this problem and discusses their strengths and weaknesses. The methods differ in the restrictions they impose on the number of exercise opportunities or the dimension of the underlying state vector, the information they require about the underlying processes, and the extent to which they seek to compute the exact price or just a reasonable approximation. Any general method for pricing American options by simulation requires substantial computational effort, and that is certainly the case for the methods we discuss here.

A theme of this chapter is the importance of understanding the sources of high and low bias that affect all methods for pricing American options by simulation. High bias results from using information about the future in making the decision to exercise, and this in turn results from applying backward induction to simulated paths. Low bias results from following a suboptimal exercise rule. Some methods mix the two sources of bias, but we will see that by separating them it is often possible to produce a pair of estimates straddling the optimal value.

8.1 Problem Formulation

A general class of continuous-time American option pricing problems can be formulated by specifying a process $U(t)$, $0 \leq t \leq T$, representing the discounted payoff from exercise at time t, and a class of admissible stopping times $\mathcal{T}$ with values in $[0, T]$. The problem, then, is to find the optimal expected discounted payoff

$$\sup_{\tau \in \mathcal{T}} \mathsf{E}[U(\tau)].$$

An arbitrage argument justifies calling this the option price under appropriate regularity conditions; see Duffie [98].

The process U is commonly derived from more primitive elements. With little loss of generality, we restrict attention to problems that can be formulated through an $\Re^d$-valued Markov process $\{X(t), 0 \leq t \leq T\}$ recording all necessary information about relevant financial variables, such as the current prices of the underlying assets. The Markov property can often be achieved by augmenting the state vector to include supplementary variables (such as stochastic volatility), if necessary. The payoff to the option holder from exercise at time t is $\tilde{h}(X(t))$ for some nonnegative payoff function $\tilde{h}$. If we further suppose the existence of an instantaneous short rate process $\{r(t), 0 \leq t \leq T\}$, the pricing problem becomes calculation of

$$\sup_{\tau \in \mathcal{T}} \mathsf{E} \left[e^{-\int_0^\tau r(u)\, du} \tilde{h}(X(\tau)) \right]. \tag{8.1}$$

It is implicit in the form of the discounting in this expression that the expectation is taken with respect to the risk-neutral measure. By taking the discount factor to be a component of X, we could absorb the discounting into the function $\tilde{h}$. In this Markovian setting, it is natural to take the admissible stopping rules $\mathcal{T}$ to be functions of the current state, augmenting the state vector if necessary. This means that the exercise decision at time t is determined by $X(t)$.

This formulation includes the classical American put as a special case. Consider a put struck at K on a single underlying asset $S(t)$. The risk-neutral dynamics of S are modeled as geometric Brownian motion $\mathrm{GBM}(r, \sigma^2)$, with r a constant risk-free interest rate. Suppose the option expires at T. Its value at time 0 is then

$$\sup_{\tau \in \mathcal{T}} \mathsf{E}[e^{-r\tau}(K - S(\tau))^+], \tag{8.2}$$

where the elements of $\mathcal{T}$ are stopping times (with respect to S) taking values in $[0, T]$. This supremum is achieved by an optimal stopping time τ^* that has the form

$$\tau^* = \inf\{t \geq 0 : S(t) \leq b^*(t)\}, \tag{8.3}$$

for some optimal *exercise boundary* b^*. This is illustrated in Figure 8.1.

We have written the payoff in (8.2) as $(K - S(\tau))^+$ rather than $(K - S(\tau))$ so that exercising an out-of-the-money option produces a zero payoff rather than a negative payoff. This allows us to include the possibility that the option expires worthless within the event $\{\tau = T\}$ rather than writing, e.g., $\tau = \infty$ for this case. For this reason, in (8.1) and throughout this chapter we take the payoff function $\tilde{h}$ to be nonnegative.

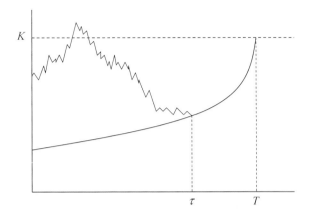

Fig. 8.1. Exercise boundary for American put with payoff $(K - S(t))^+$. The option is exercised at τ, the first time the underlying asset reaches the boundary.

In discussing simulation methods for pricing American options, we restrict ourselves to options that can be exercised only at a fixed set of exercise opportunities $t_1 < t_2 < \cdots < t_m$. In many cases, such a restriction is part of the option contract, and such options are often called "Bermudan" because they lie between European and continuously exercisable American options. In some cases, it is known in advance that exercise is suboptimal at all but a finite set of dates, such as dividend payment dates. One may alternatively view the restriction to a finite set of exercise dates as an approximation to a contract allowing continuous exercise, in which case one would want to consider the effect of letting m increase to infinity. Because even the finite-exercise problem poses a significant challenge to Monte Carlo, we focus exclusively on this case. We construe "American" to include both types of options.

To reduce notation, throughout this chapter we write $X(t_i)$ as X_i; this is the state of the underlying Markov process at the ith exercise opportunity. The discrete-time process $X_0 = X(0), X_1, \ldots, X_m$ is then a Markov chain on $\Re^d$. We discuss American option pricing based on simulation of this Markov chain. In practice, simulating X_{i+1} from X_i may entail simulating values of $X(t)$ with $t_i < t < t_{i+1}$. For example, it may be necessary to apply a time discretization using a time step smaller than the intervals $t_{i+1} - t_i$ between exercise dates. In this case, the simulated values of $X_1, \ldots, X_m$ are subject to discretization error. For the purposes of discussing American option pricing, we disregard this issue and focus on the challenge of solving the optimal stopping problem under the assumption that $X_1, \ldots, X_m$ can be simulated exactly.

Dynamic Programming Formulation

Working with a finite-exercise Markovian formulation lends itself to a characterization of the option value through dynamic programming. Let $\tilde{h}_i$ denote the payoff function for exercise at t_i, which we now allow to depend on i. Let $\tilde{V}_i(x)$ denote the value of the option at t_i given $X_i = x$, assuming the option has not previously been exercised. This can also be interpreted as the value of a newly issued option at t_i starting from state x. We are ultimately interested in $\tilde{V}_0(X_0)$. This value is determined recursively as follows:

$$\tilde{V}_m(x) = \tilde{h}_m(x) \tag{8.4}$$
$$\tilde{V}_{i-1}(x) = \max\{\tilde{h}_{i-1}(x), \mathsf{E}[D_{i-1,i}(X_i)\tilde{V}_i(X_i)|X_{i-1} = x]\}, \tag{8.5}$$
$$i = 1, \ldots, m.$$

Here we have introduced the notation $D_{i-1,i}(X_i)$ for the discount factor from t_{i-1} to t_i. Equation (8.4) states that the option value at expiration is given by the payoff function $\tilde{h}_m$; equation (8.5) states that at the $(i-1)$th exercise date the option value is the maximum of the immediate exercise value and the expected present value of continuing. (We usually exclude the current time 0 from the set of exercise opportunities; this can be accommodated in (8.5) by setting $\tilde{h}_0 \equiv 0$.)

Most methods for computing American option prices rely on the dynamic programming representation (8.4)–(8.5). This is certainly true of the binomial method in which the conditional expectation in (8.5) reduces to the average over the two successor nodes in the lattice. (See Figure 4.7 and the surrounding discussion.) Estimating these conditional expectations is the main difficulty in pricing American options by simulation. The difficulty is present even in the two-stage problem discussed in Example 1.1.3, and is compounded by the nesting of conditional expectations in (8.5) as i decreases from m to 1.

By augmenting the state vector if necessary, we assume that the discount factor $D_{i-1,i}$ is a deterministic function of X_i. The discount factor could have the form

$$D_{i-1,i}(X_i) = \exp\left(-\int_{t_{i-1}}^{t_i} r(u)\, du\right),$$

but this is not essential. The more general formulation frees us from reliance on the risk-neutral measure. We simply require that the expectation is taken with respect to the probability measure consistent with the choice of numeraire implicit in the discount factor; see the discussion in Section 1.2.3. If we simulate under the t_m-maturity forward measure, the discount factor $D_{i-1,i}$ becomes $B_m(t_{i-1})/B_m(t_i)$, with $B_m(t)$ denoting the time-t price of a bond maturing at t_m.

In discussing pricing algorithms in this chapter, we suppress explicit discounting and work with the following simplified dynamic programming recursion:

$$V_m(x) = h_m(x) \tag{8.6}$$
$$V_{i-1}(x) = \max\{h_{i-1}(x), \mathsf{E}[V_i(X_i)|X_{i-1} = x]\}. \tag{8.7}$$

This formulation is actually sufficiently general to include (8.4)–(8.5), as we now explain.

In (8.4)–(8.5), each $\hat{V}_i$ gives the value of the option in time-t_i dollars, but we could alternatively formulate the problem so that the value is always recorded in time-0 dollars. Let $D_{0,j}(X_j)$ denote the discount factor from time 0 to t_j, which we suppose is a deterministic function of X_j by augmenting the state vector if necessary. To be consistent with its interpretation as a discount factor, we require that $D_{0,j}(X_j)$ be nonnegative and that it satisfy $D_{0,0} \equiv 1$ and $D_{0,i-1}(X_{i-1})D_{i-1,i}(X_i) = D_{0,i}(X_i)$. Now let

$$h_i(x) = D_{0,i}(x)\tilde{h}_i(x), \quad i = 1, \ldots, m,$$

and

$$V_i(x) = D_{0,i}(x)\tilde{V}_i(x), \quad i = 0, 1, \ldots, m.$$

Then $V_0 = \tilde{V}_0$ and the V_i satisfy

$$V_m(x) = h_m(x)$$
$$V_{i-1}(x) = D_{0,i-1}(x)\tilde{V}_{i-1}(x)$$
$$= D_{0,i-1}(x)\max\{\tilde{h}_{i-1}(x), \mathsf{E}[D_{i-1,i}(X_i)\tilde{V}_i(X_i)|X_{i-1} = x]\}$$
$$= \max\{h_{i-1}(x), \mathsf{E}[D_{0,i-1}(x)D_{i-1,i}(X_i)\tilde{V}_i(X_i)|X_{i-1} = x]\}$$
$$= \max\{h_{i-1}(x), \mathsf{E}[V_i(X_i)|X_{i-1} = x]\}.$$

Thus, the discounted values V_i satisfy a dynamic programming recursion of exactly the form in (8.6)–(8.7). For the purpose of introducing simulation estimators, (8.7) is slightly simpler, though in practical implementation one usually uses (8.5).

Stopping Rules

The dynamic programming recursions (8.4)–(8.5) and (8.6)–(8.7) focus on option values, but it is also convenient to view the pricing problem through stopping rules and exercise regions. Any stopping time τ (for the Markov chain $X_0, X_1, \ldots, X_m$) determines a value (in general suboptimal) through

$$V_0^{(\tau)}(X_0) = \mathsf{E}[h_\tau(X_\tau)]. \tag{8.8}$$

Conversely, any rule assigning a "value" $\hat{V}_i(x)$ to each state $x \in \Re^d$ and exercise opportunity i, with $\hat{V}_m = h_m$, determines a stopping rule

$$\hat{\tau} = \min\{i \in \{1, \ldots, m\} : h_i(X_i) \geq \hat{V}_i(X_i)\}. \tag{8.9}$$

The exercise region determined by $\hat{V}_i$ at the ith exercise date is the set

$$\{x : h_i(x) \geq \hat{V}_i(x)\} \tag{8.10}$$

and the continuation region is the complement of this set. The stopping rule $\hat{\tau}$ can thus also be described as the first time the Markov chain X_i enters an exercise region. The value determined by using $\hat{\tau}$ in (8.8) does not in general coincide with $\hat{V}_0$, though the two would be equal if we started with the optimal value function.

Continuation Values

The continuation value of an American option with a finite number of exercise opportunities is the value of holding rather than exercising the option. The continuation value in state x at date t_i (measured in time-0 dollars) is

$$C_i(x) = \mathsf{E}[V_{i+1}(X_{i+1})|X_i = x], \tag{8.11}$$

$i = 0, \ldots, m-1$. These also satisfy a dynamic programming recursion: $C_m \equiv 0$ and

$$C_i(x) = \mathsf{E}\left[\max\{h_{i+1}(X_{i+1}), C_{i+1}(X_{i+1})\}|X_i = x\right], \tag{8.12}$$

for $i = 0, \ldots, m-1$. The option value is $C_0(X_0)$, the continuation value at time 0.

The value functions V_i determine the continuation values through (8.11). Conversely,

$$V_i(x) = \max\{h_i(x), C_i(x)\},$$

$i = 1, \ldots, m$, so the C_i determine the V_i. An approximation $\hat{C}_i$ to the continuation values determines a stopping rule through

$$\hat{\tau} = \min\{i \in \{1, \ldots, m\} : h_i(X_i) \geq \hat{C}_i(X_i)\}. \tag{8.13}$$

This is the same as the stopping rule in (8.9) if the approximations satisfy $\hat{V}_i(x) = \max\{h_i(x), \hat{C}_i(x)\}$.

8.2 Parametric Approximations

Genuine pricing of American options entails solving the dynamic programming problem (8.4)–(8.5) or (8.6)–(8.7). An alternative to trying to find the optimal value is to find the best value within a parametric class. This reduces the optimal stopping problem to a much more tractable finite-dimensional optimization problem. The reduction can be accomplished by considering a parametric class of exercise regions or a parametric class of stopping rules. The two approaches become equivalent through the correspondences in (8.8)–(8.10), so the distinction is primarily one of interpretation.

The motivation for considering parametric exercise regions is particularly evident in the case of an option on a single underlying asset. The curved exercise boundary in Figure 8.1, for example, is well-approximated by a piecewise linear function with three or four segments (or piecewise exponential as in Ju [205]), which could be specified with four or five parameters. The option value is usually not very sensitive to the exact position of the exercise boundary — the value is continuous across the boundary — suggesting that even a rough approximation to the boundary should produce a reasonable approximation to the optimal option value.

The one-dimensional setting is, however, somewhat misleading. In higher-dimensional problems (where Monte Carlo becomes relevant), the optimal exercise region need not have a simple structure and there may be no evident way of parameterizing a class of plausible approximations. Developing a parametric heuristic thus requires a deeper understanding of what drives early exercise. The financial interpretation of the optimal stopping problems that arise in pricing high-dimensional American options sometimes makes this possible.

To formalize this approach, consider a class of stopping rules τ_θ, $\theta \in \Theta$, with each $\tau_\theta \in \mathcal{T}$ and Θ a subset of some $\Re^M$. Let

$$V_0^\theta = \sup_{\theta \in \Theta} \mathsf{E}[h_{\tau(\theta)}(X_{\tau(\theta)})];$$

the objective of a parametric heuristic is to estimate V_0^θ, the optimal value within the parametric class. Because the supremum in this definition is taken over a subset of all admissible stopping rules $\mathcal{T}$, we have

$$V_0^\theta \le V_0 = \sup_{\tau \in \mathcal{T}} \mathsf{E}[h_\tau(X_\tau)], \qquad (8.14)$$

typically with strict inequality. Thus, a consistent estimator of V_0^θ underestimates the true option value.

A meta-algorithm for estimation of V_0^θ consists of the following steps:

Step 1: Simulate n_1 independent replications $X^{(j)}$, $j = 1, \ldots, n_1$, of the Markov chain $(X_0, X_1, \ldots, X_m)$;

Step 2: Find $\hat\theta$ maximizing

$$\hat{V}_0^{\hat\theta} = \frac{1}{n_1} \sum_{j=1}^{n_1} h_{\tau^{(j)}(\hat\theta)}(X_{\tau^{(j)}(\hat\theta)}^{(j)})$$

where for each $\theta \in \Theta$, $\tau^{(j)}(\theta)$ is the time of exercise on the jth replication at parameter θ.

Bias

Assuming Step 2 can be executed, the result is an estimator that is biased high relative to V_0^θ, in the sense that

$$\mathsf{E}[\hat{V}_0^{\hat\theta}] \ge V_0^\theta. \qquad (8.15)$$

This simply states that the expected value of the maximum over θ is at least as large as the maximum over θ of the expected values. This can be viewed as a consequence of Jensen's inequality. It also results from the fact that the in-sample optimum $\hat{\theta}$ implicitly uses information about the future evolution of the simulated replications in determining an exercise decision.

The combination of the in-sample bias in (8.15) and the suboptimality bias in (8.14) produces an unpredictable bias in $\hat{V}_0^{\hat{\theta}}$. One might hope that the high bias in (8.15) offsets the low bias in (8.14), but without further examination of specific cases this conjecture lacks support. Given that bias is inevitable in this setting, it is preferable to control the bias and determine its direction. This can be accomplished by adding the following to the meta-algorithm:

Step 3: Fix $\hat{\theta}$ at value found in Step 2. Simulate n_2 additional independent replications of the Markov chain using stopping rule $\tau_{\hat{\theta}}$ and compute the estimate

$$\hat{V}_0^{\hat{\theta}} = \frac{1}{n_2} \sum_{j=n_1+1}^{n_1+n_2} h_{\tau^{(j)}(\hat{\theta})}\big(X^{(j)}_{\tau^{(j)}(\hat{\theta})}\big).$$

Because the second set of replications is independent of the set used to determine $\hat{\theta}$, we now have

$$\mathsf{E}[\hat{V}_0^{\hat{\theta}}|\hat{\theta}] = V_0^{\hat{\theta}},$$

which is the true value at parameter $\hat{\theta}$ and cannot exceed V_0. Thus, taking the unconditional expectation we get

$$\mathsf{E}[\hat{V}_0^{\hat{\theta}}] \leq V_0,$$

from which we conclude that the estimator produced by Step 3 is biased low.

This is an instance of a more general strategy developed in Broadie and Glasserman [65, 66] for separating sources of high and low bias, to which we return in Sections 8.3 and 8.5.

Estimators like $\hat{\theta}$ defined as solutions to sample optimization problems have been studied extensively in other settings (including, for example, maximum likelihood estimation). There is a large literature establishing the convergence of such estimators to the true optimum and also the convergence of optimal-value estimators like $\hat{V}_0^{\hat{\theta}}$ to the true optimum V_0^{θ} over Θ. Some results of this type are presented in Chapter 6 of Rubinstein and Shapiro [313] and Chapter 7 of Serfling [326].

Optimization

The main difficulty in implementing Steps 1–2, beyond selection of the class of parametric rules, lies in the optimization problem of Step 2. Andersen [12] considers exercise rules defined by a single threshold at each exercise date and thus reduces the optimization problem to a sequence of one-dimensional searches. Fu and Hu [131] estimate derivatives with respect to parameters (through the pathwise method discussed in Section 7.2) and use these to search for optimal parameters. Garcia [134] uses a simplex method as in Press et al. [299] with the exercise region at each time step described by two parameters.

The optimization problem in Step 2 ordinarily decomposes into $m-1$ sub-problems, one for each exercise date except the last. This holds whenever the parameter vector θ decomposes into $m-1$ components with the ith component parameterizing the exercise region at t_i. This decomposition can be used to search for an optimal parameter vector by optimizing sequentially from the $(m-1)$th date to the first.

In more detail, suppose $\theta = (\theta_1, \ldots, \theta_{m-1})$ with θ_i parameterizing the exercise region at t_i. Each θ_i could itself be a vector. Now consider the following inductive procedure, applied to n independent replications of the Markov chain $X_0, X_1, \ldots, X_m$:

(2a) find the value $\hat{\theta}_{m-1}$ maximizing the average discounted payoff of the option over the n paths assuming exercise is possible only at the $(m-1)$th and mth dates;

(2b) with $\hat{\theta}_i, \ldots, \hat{\theta}_{m-1}$ fixed, find the value $\hat{\theta}_{i-1}$ maximizing the average discounted payoff of the option over the n paths, assuming exercise is possible only at $i-1, i, \ldots, m$, and following the exercise policy determined by $\hat{\theta}_i, \ldots, \hat{\theta}_{m-1}$ at $i, \ldots, m-1$.

If each θ_i is a scalar, then this procedure reduces to a sequence of one-dimensional optimization procedures. With a finite number of paths, each of these is typically a nonsmooth optimization problem and is best solved using an iterative search rather than a derivative-based method. Andersen [12] uses a golden section procedure, as in Section 10.1 of Press [299].

There is no guarantee that (2a)–(2b) will produce an optimum for the original problem in Step 2. In the decomposition, each $\hat{\theta}_i$ is optimized over all paths whereas in the original optimization problem only a subset of paths would survive until the ith date. One way to address this issue would be to repeat steps (2a)–(2b) as follows: working backward from dates $m-1$ to 1, update each $\hat{\theta}_i$ using parameter values from the previous iteration to determine the exercise decision at dates $1, \ldots, i-1$, and parameter values from the current iteration to determine the exercise decision at dates $i+1, \ldots, m-1$. Even this does not, however, guarantee an optimal solution to the problem in Step 2.

Andersen [12] uses an approach of the type in Steps 1–3 to value Bermudan swaptions in a LIBOR market model of the type discussed in Section 3.7. He considers threshold rules in which the option is exercised if some function of the state vector is above a threshold. He allows the threshold to vary with the exercise date and optimizes the vector of thresholds. In the notation of this section, each θ_i, $i = 1, \ldots, m-1$, is scalar and denotes the threshold for the ith exercise opportunity.

The simplest rule Andersen [12] considers exercises the swaption if the value of the underlying swap is above a time-varying threshold. All the threshold rules he tests try to capture the idea that exercise should occur when the underlying swap has sufficient value. His numerical results indicate that simple rules work well and that Bermudan swaption values are not very sensitive to the location of the exercise boundary in the parameterizations he uses.

This example illustrates a feature of many American option pricing problems. Although the problem is high-dimensional, it has a lot of structure, making approximate solution feasible using relatively simple exercise rules that tap into a financial understanding of what drives the exercise decision.

Parametric Value Functions

An alternative to specifying an approximate stopping rule or exercise regions uses a parametric approximation to the optimal value function. Although the two perspectives are ultimately equivalent, the interpretation and implementation are sufficiently different to merit separate consideration.

We work with the optimal continuation values $C_i(x)$ in (8.11) rather than the value function itself. Consider approximating each function $C_i(x)$ by a member $C_i(x, \theta_i)$ of a parametric family of functions. For example, we might take θ_i to be a vector with elements $\theta_{i1}, \ldots, \theta_{iM}$ and consider functions of the form

$$C_i(x, \theta_i) = \sum_{j=1}^{M} \theta_{ij} \psi_j(x),$$

for some set of basis functions $\psi_1, \ldots, \psi_M$. Our objective is to choose the parameters θ_i to approximate the recursion (8.12).

Proceeding backward by induction, this entails choosing $\hat{\theta}_{m-1}$ so that $C_{m-1}(x, \hat{\theta}_{m-1})$ approximates $\mathsf{E}[h_m(X_m)|X_{m-1} = x]$, with the conditional expectation estimated from simulated paths of the Markov chain. Given values of $\hat{\theta}_{i+1}, \ldots, \hat{\theta}_{m-1}$, we choose $\hat{\theta}_i$ so that $C_i(x, \hat{\theta}_i)$ approximates

$$\mathsf{E}\left[\max\{h_{i+1}(X_{i+1}), C_{i+1}(X_{i+1}, \hat{\theta}_{i+1})\}|\hat{\theta}_{i+1}, X_i = x\right],$$

again using simulated paths to estimate the conditional expectation.

Applying this type of approach in practice involves several issues. Choosing a parametric family of approximating functions is a problem-dependent modeling issue; finding the optimal parameters and, especially, estimating the conditional expectations present computational challenges. We return to this approach in Section 8.6.

8.3 Random Tree Methods

Whereas the approximations of the previous section search for the best solution within a parametric family, the random tree method of Broadie and

Glasserman [65] seeks to solve the full optimal stopping problem and estimate the genuine value of an American option. And whereas parametric approximations rely on insight into the form of a good stopping rule, the method of this section assumes little more than the ability to simulate paths of the underlying Markov chain. With only minimal conditions, the method produces two consistent estimators, one biased high and one biased low, and both converging to the true value. This combination makes it possible to measure and control error as the computational effort increases.

The main drawback of the random tree method is that its computational requirements grow exponentially in the number of exercise dates m, so the method is applicable only when m is small — not more than about 5, say. This substantially limits the scope of the method. Nevertheless, for problems with small m it is very effective, and it also serves to illustrate a theme of this chapter — managing sources of high and low bias.

Before discussing the details of the method, we explain how a combination of two biased estimators can be nearly as effective as a single unbiased estimator. Suppose, then, that $\hat{V}_n(b)$ and $\hat{v}_n(b)$ are each sample means of n independent replications, for each value of a simulation parameter b. Suppose that as estimators of some V_0 they are biased high and low, respectively, in the sense that

$$\mathsf{E}[\hat{V}_n(b)] \geq V_0 \geq \mathsf{E}[\hat{v}_n(b)]. \tag{8.16}$$

Suppose that, for some halfwidth $H_n(b)$,

$$\hat{V}_n(b) \pm H_n(b)$$

is a valid 95% confidence interval for $\mathsf{E}[\hat{V}_n(b)]$ in the sense that the interval contains this point with 95% probability; and suppose that

$$\hat{v}_n(b) \pm L_n(b)$$

is similarly a valid 95% confidence interval for $\mathsf{E}[\hat{v}_n(b)]$. Then by taking the lower confidence limit of the low estimator and the upper confidence interval of the high estimator, we get an interval

$$\left(\hat{v}_n(b) - L_n(b), \hat{V}_n(b) + H_n(b) \right) \tag{8.17}$$

containing the unknown value V_0 with probability at least 90% (at least 95% if $\hat{V}_n(b)$ and $\hat{v}_n(b)$ are symmetric about their means). Thus, we can produce a valid (though potentially very conservative) confidence interval by combining the two estimators; see Figure 8.2. In our application of this idea to the random tree method, the inequalities in (8.16) become equalities as $b \to \infty$ and the interval halfwidths $H_n(b)$ and $L_n(b)$ shrink to zero as $n \to \infty$. The interval in (8.17) can thus be made to shrink to the point V_0 in the limit as the computational effort grows.

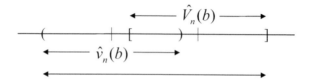

Fig. 8.2. Combining and high and low estimators to form a confidence interval.

8.3.1 High Estimator

As its name suggests, the random tree method is based on simulating a tree of paths of the underlying Markov chain $X_0, X_1, \ldots, X_m$. Fix a branching parameter $b \geq 2$. From the initial state X_0, simulate b independent successor states $X_1^1, \ldots, X_1^b$ all having the law of X_1. From each X_1^i, simulate b independent successors $X_2^{i1}, \ldots, X_2^{ib}$ from the conditional law of X_2 given $X_1 = X_1^i$. From each $X_2^{i_1 i_2}$, generate b successors $X_3^{i_1 i_2 1}, \ldots, X_3^{i_1 i_2 b}$, and so on. Figure 8.3 shows an example with $m = 2$ and $b = 3$. At the mth time step there are b^m nodes, and this is the source of the exponential growth in the computational cost of the method.

We denote a generic node in the tree at time step i by $X_i^{j_1 j_2 \cdots j_i}$. The superscript indicates that this node is reached by following the j_1th branch out of X_0, the j_2th branch out of the next node, and so on. Although it is not essential that the branching parameter remain fixed across time steps, this is a convenient simplification in discussing the method.

It should be noted that the random tree construction differs from what is sometimes called a "nonrecombining" tree in that successor nodes are sampled randomly. In a more standard nonrecombining tree, the placement of the nodes is deterministic, as it is in a binomial lattice; see, for example, Heath et al. [173].

From the random tree we define high and low estimators at each node by backward induction. We use the formulation in (8.6)–(8.7). Thus, h_i is the discounted payoff function at the ith exercise date, and the discounted option value satisfies $V_m \equiv h_m$,

$$V_i(x) = \max\{h_i(x), \mathsf{E}[V_{i+1}(X_{i+1})|X_i = x]\}, \tag{8.18}$$

$i = 0, \ldots, m - 1$.

Write $\hat{V}_i^{j_1 \cdots j_i}$ for the value of the high estimator at node $X_i^{j_1 \cdots j_i}$. At the terminal nodes we set

$$\hat{V}_m^{j_1 \cdots j_m} = h_m(X_m^{j_1 \cdots j_m}). \tag{8.19}$$

Working backward, we then set

$$\hat{V}_i^{j_1 \cdots j_i} = \max\left\{ h_i(X_i^{j_1 \cdots j_i}), \frac{1}{b} \sum_{j=1}^b \hat{V}_{i+1}^{j_1 \cdots j_i j} \right\}. \tag{8.20}$$

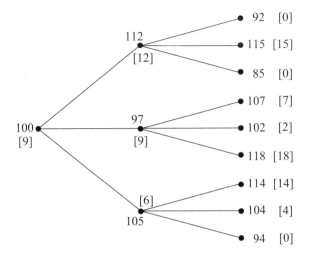

Fig. 8.3. Illustration of random tree and high estimator for a call option with a single underlying asset and a strike price of 100. Labels at each node show the level of the underlying asset and (in brackets) the value of the high estimator.

In other words, the high estimator is simply the result of applying ordinary dynamic programming to the random tree, assigning equal weight to each branch. Its calculation is illustrated in Figure 8.3 with $h_i(x) = (x - 100)^+$.

A simple induction argument demonstrates that the high estimator is indeed biased high at every node, in the sense that

$$\mathsf{E}[\hat{V}_i^{j_1 \cdots j_i} | X_i^{j_1 \cdots j_i}] \geq V_i(X_i^{j_1 \cdots j_i}). \tag{8.21}$$

First observe that this holds (with equality) at every terminal node because of (8.19) and the fact that $V_m = h_m$. Now we show that if (8.21) holds at $i+1$ it holds at i. From (8.20) we get

$$
\begin{aligned}
\mathsf{E}[\hat{V}_i^{j_1 \cdots j_i} | X_i^{j_1 \cdots j_i}] &= \mathsf{E}\left[\max\left\{h_i(X_i^{j_1 \cdots j_i}), \frac{1}{b}\sum_{j=1}^{b} \hat{V}_{i+1}^{j_1 \cdots j_i j}\right\} \Big| X_i^{j_1 \cdots j_i}\right] \\
&\geq \max\left\{h_i(X_i^{j_1 \cdots j_i}), \mathsf{E}\left[\frac{1}{b}\sum_{j=1}^{b} \hat{V}_{i+1}^{j_1 \cdots j_i j} | X_i^{j_1 \cdots j_i}\right]\right\} \\
&= \max\left\{h_i(X_i^{j_1 \cdots j_i}), \mathsf{E}\left[\hat{V}_{i+1}^{j_1 \cdots j_i 1} | X_i^{j_1 \cdots j_i}\right]\right\} \\
&\geq \max\left\{h_i(X_i^{j_1 \cdots j_i}), \mathsf{E}\left[V_{i+1}(X_{i+1}^{j_1 \cdots j_i 1}) | X_i^{j_1 \cdots j_i}\right]\right\} \\
&= V_i(X_i^{j_1 \cdots j_i}).
\end{aligned}
$$

The first equality uses (8.20), the next step applies Jensen's inequality, the third step uses the fact that the b successors of each node are conditionally i.i.d., the fourth step applies the induction hypothesis at $i + 1$, and the last step follows from (8.18).

A similar induction argument establishes that under modest moment conditions, each $\hat{V}_i^{j_1 \cdots j_i}$ converges in probability (and in norm) as $b \to \infty$ to the true value $V_i(X_i^{j_1 \cdots j_i})$, given $X_i^{j_1 \cdots j_i}$. This holds trivially at all terminal nodes because $\hat{V}_m$ is initialized to $h_m = V_m$. The continuation value at the $(m-1)$th step is the average of i.i.d. replications and converges by the law of large numbers. This convergence extends to the option value — the greater of the immediate exercise and continuation values — by the continuity of the max operation. A key step in the induction is the "contraction" property

$$|\max(a, c_1) - \max(a, c_2)| \leq |c_1 - c_2|, \tag{8.22}$$

which, together with (8.18) and (8.20), gives

$$|\hat{V}_i^{j_1 \cdots j_i} - V_i(X_i^{j_1 \cdots j_i})| \leq \frac{1}{b} \sum_{j=1}^{b} |\hat{V}_{i+1}^{j_1 \cdots j_i j} - \mathsf{E}[V_{i+1}(X_{i+1}^{j_1 \cdots j_i 1})|X_i^{j_1 \cdots j_i}]|.$$

This allows us to deduce convergence at step i from convergence at step $i+1$. For details, see Theorem 1 of Broadie and Glasserman [65].

We are primarily interested in $\hat{V}_0$, the estimate of the option price at the current time and state. Theorem 1 of [65] shows that if $\mathsf{E}[h_i^2(X_i)]$ is finite for all $i = 1, \ldots, m$, then $\hat{V}_0$ converges in probability to the true value $V_0(X_0)$; moreover, it is asymptotically unbiased in the sense that $\mathsf{E}[\hat{V}_0] \to V_0(X_0)$.

These properties hold as the branching parameter b increases to infinity. A simple way to compute a confidence interval fixes a value of b and replicates the random tree n times. (This is equivalent to generating nb branches out of X_0 and b branches out of all subsequent nodes.) Let $\bar{V}_0(n, b)$ denote the sample mean of the n replications of $\hat{V}_0$ generated this way, and let $s_V(n, b)$ denote their sample standard deviation. Then with $z_{\delta/2}$ denoting the $1 - \delta/2$ quantile of the normal distribution,

$$\bar{V}_0(n, b) \pm z_{\delta/2} \frac{s_V(n, b)}{\sqrt{n}}$$

provides an asymptotically valid (for large n) $1 - \delta$ confidence interval for $\mathsf{E}[\hat{V}_0]$. This is half of what we need for the interval in (8.17). The next section provides the other half.

8.3.2 Low Estimator

The high bias of the high estimator may be attributed to its use of the same information in deciding whether to exercise as in estimating the continuation value. This is implicit in the dynamic programming recursion (8.20): the first term inside the maximum is the immediate-exercise value, the second term is the estimated continuation value, and in choosing the maximum the estimator is deciding whether to exercise or continue. But the estimated continuation value is based on successor nodes, so the estimator is unfairly peeking into the future in making its decision.

To remove this source of bias, we need to separate the exercise decision from the value received upon continuation. This is the key to removing high bias in all Monte Carlo methods for pricing American options. (See, for example, the explanation of Step 3 in Section 8.2.) There are several ways of accomplishing this in the random tree method.

To simplify the discussion, consider the related problem of estimating

$$\max(a, \mathsf{E}[Y])$$

from i.i.d. replications $Y_1, \ldots, Y_b$, for some constant a and random variable Y. This is a simplified version of the problem we face at each node in the tree, with a corresponding to the immediate exercise value and $\mathsf{E}[Y]$ the continuation value. The estimator $\max(a, \bar{Y})$, with $\bar{Y}$ the sample mean of the Y_i, is biased high:

$$\mathsf{E}[\max(a, \bar{Y})] \geq \max(a, \mathsf{E}[\bar{Y}]) = \max(a, \mathsf{E}[Y]).$$

This corresponds to the high estimator of the previous section.

Suppose that we instead separate the Y_i into two disjoint subsets and calculate their sample means $\bar{Y}_1$ and $\bar{Y}_2$; these are independent of each other. Now set

$$\hat{v} = \begin{cases} a, & \text{if } \bar{Y}_1 \leq a; \\ \bar{Y}_2, & \text{otherwise.} \end{cases}$$

This estimator uses $\bar{Y}_1$ to decide whether to "exercise," and if it decides not to, it uses $\bar{Y}_2$ to estimate the "continuation" value. Its expectation is

$$\mathsf{E}[\hat{v}] = P(\bar{Y}_1 \leq a)a + (1 - P(\bar{Y}_1 \leq a))\mathsf{E}[Y] \leq \max(a, \mathsf{E}[Y]), \qquad (8.23)$$

so the estimator is indeed biased low. If $a \neq \mathsf{E}[Y]$, then $P(\bar{Y}_1 \leq a) \to \mathbf{1}\{\mathsf{E}[Y] < a\}$ and $\mathsf{E}[\hat{v}] \to \max(a, \mathsf{E}[Y])$ as the number of replications used to calculate $\bar{Y}_1$ increases. If the number of replications used to calculate $\bar{Y}_2$ also increases then $\hat{v} \to \max(a, \mathsf{E}[Y])$. Thus, in this simplified setting we can easily produce a consistent estimator that is biased low.

Broadie and Glasserman [65] use a slightly different estimator. They use all but one of the Y_i to calculate $\bar{Y}_1$ and use the remaining one for $\bar{Y}_2$; they then average the result over all b ways of leaving out one of the Y_i. In more detail, the estimator is defined as follows. At all terminal nodes, set the estimator equal to the payoff at that node:

$$\hat{v}_m^{j_1 j_2 \cdots j_m} = h_m(X_m^{j_1 j_2 \cdots j_m}).$$

At node $j_1 j_2 \cdots j_i$ at time step i, and for each $k = 1, \ldots, b$, set

$$\hat{v}_{ik}^{j_1 j_2 \cdots j_i} = \begin{cases} h_i(X_i^{j_1 j_2 \cdots j_i}) & \text{if } \frac{1}{b-1} \sum_{j=1, j \neq k}^{b} \hat{v}_{i+1}^{j_1 j_2 \cdots j_i j} \leq h_i(X_i^{j_1 j_2 \cdots j_i}); \\ \hat{v}_{i+1}^{j_1 j_2 \cdots j_i k} & \text{otherwise;} \end{cases} \qquad (8.24)$$

then set

$$\hat{v}_i^{j_1 j_2 \cdots j_i} = \frac{1}{b} \sum_{k=1}^{b} \hat{v}_{ik}^{j_1 j_2 \cdots j_i}. \qquad (8.25)$$

The estimator of the option price at the current time and state is $\hat{v}_0$.

The calculation of the low estimator is illustrated in Figure 8.4. Consider the third node at the first exercise date. When we leave out the first successor we estimate a continuation value of $(4 + 0)/2 = 2$ so we exercise and get 5. If we leave out the second successor node we continue (because $7 > 5$) and get 4. In the third case we continue and get 0. Averaging the three payoffs 5, 4, and 0 yields a low estimate of 3 at that node.

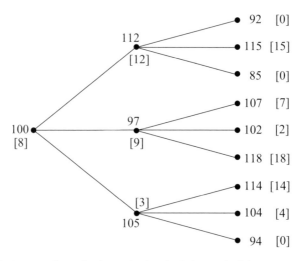

Fig. 8.4. Labels at each node show the level of the underlying asset and the value of the low estimator in brackets.

An induction argument similar to the one used for the high estimator in Section 8.3.1 and using the observation in (8.23) verifies that $\hat{v}_0$ is indeed biased low. Theorem 3 of Broadie and Glasserman [65] establishes the convergence in probability and in norm of $\hat{v}_0$ to the true value $V_0(X_0)$.

From n independent replications of the random tree we can calculate the sample mean $\bar{v}_0(n, b)$ and sample standard deviation $s_v(n, b)$ of n independent replications of the low estimator. We can then form a $1 - \delta$ confidence interval for $\mathsf{E}[\hat{v}_0]$,

$$\bar{v}_0(n, b) \pm z_{\delta/2} \frac{s_v(n, b)}{\sqrt{n}},$$

just as we did for the high estimator. Taking the lower limit for the low estimator and the upper limit for the high estimator, we get the interval

$$\left(\bar{v}_0(n,b) - z_{\delta/2} \frac{s_v(n,b)}{\sqrt{n}}, \bar{V}_0(n,b) + z_{\delta/2} \frac{s_V(n,b)}{\sqrt{n}} \right).$$

For this interval to fail to include the true option value $V_0(X_0)$, we must have

$$\mathsf{E}[\hat{v}_0] \le \bar{v}_0(n,b) - z_{\delta/2} \frac{s_v(n,b)}{\sqrt{n}}$$

or

$$\mathsf{E}[\hat{V}_0] \ge \bar{V}_0(n,b) + z_{\delta/2} \frac{s_V(n,b)}{\sqrt{n}}.$$

Each of these events has probability $\delta/2$ (for large n), so their union can have probability at most δ. We thus have a conservative confidence interval for $V_0(X_0)$. Moreover, because $\mathsf{E}[\hat{V}_0]$ and $\mathsf{E}[\hat{v}_0]$ both approach $V_0(X_0)$ as b increases, we can make the confidence interval as tight as we want by increasing n and b. This simple technique for error control is a convenient feature of the method.

8.3.3 Implementation

A naive implementation of the random tree method generates all m^b nodes (over m steps with branching parameter b) and then computes high and low estimators recursively as described in Sections 8.3.1 and 8.3.2. By noting that the high and low values at each node depend only on the subtree rooted at that node, we can dramatically reduce the storage requirements of the method. It is never necessary to store more than $mb + 1$ nodes at a time.

Depth-First Processing

The key to this reduction lies in depth-first generation and processing of the tree. Recall that we may label nodes in the tree through a string of indices $j_1 j_2 \cdots j_i$, each taking values in the set $\{1, \dots, b\}$. The string $j_1 j_2 \cdots j_i$ labels the node reached by following the j_1th branch out of the root node, then the j_2th branch out of the node reached at step 1, and so on. In the depth-first algorithm, we follow a single branch at a time rather than generating all branches simultaneously.

Consider the case of a four-step tree. We begin by generating the following nodes:

$$1, \ 11, \ 111, \ 1111.$$

At this point we have reached the terminal step and can go no deeper, so we generate nodes

$$1112, \ \dots, \ 111b.$$

From these values, we can calculate high and low estimators at node 111. We may now discard all b successors of node 111. Next we generate 112 and its successors

1121, 1122, ..., 112b.

We discard these after using them to calculate high and low estimators at 112. We repeat the process to calculate the estimators at nodes 113, ..., 11b. These in turn can be discarded after we use them to calculate high and low estimators at node 11. We repeat the process to compute estimators at nodes 12, ..., 1b to get estimators at node 1, and then to get estimators at nodes 2, ..., b, and finally at the root node.

Four stages of this procedure are illustrated in a tree with $m = 4$ and $b = 3$ in Figure 8.5. The dashed lines indicate branches previously generated, processed, and discarded. Detailed pseudo-code for implementing this method is provided in Appendix C of Broadie and Glasserman [65].

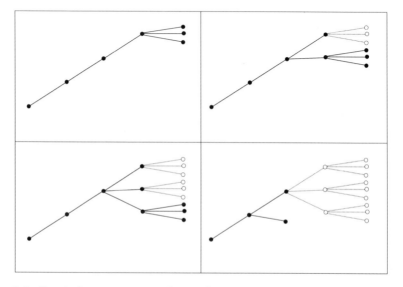

Fig. 8.5. Depth-first processing of tree. Solid circles indicate nodes currently in memory; hollow circles indicate nodes previously processed and no longer stored.

The maximal storage requirements of this method are attained in calculating the value at node b, at which point we are storing the values at nodes 1, ..., $b - 1$. Just before determining the value at b, we need to know the values at $b1$, ..., bb. But just before determining the value at bb we need to store $b1$, ..., $bb - 1$ and $bb1$, ..., bbb. And just before processing bbb we also need to store $bbb1$, ..., $bbbb$. We thus need to store up to b nodes at every time step plus the root node, leading to a total of $mb + 1$ nodes.

Pruning and Variance Reduction

Broadie et al. [69] investigate potential enhancements of the random tree method, including the use of variance reduction techniques in combination with a pruning technique for reducing the computational burden of the method. Their pruning technique is based on the observation that branching is needed only where the optimal exercise decision is unknown. If we know it is optimal not to exercise at a node, than it would suffice to generate a single branch out of that node. When we work backward through the tree, the value we assign to that node for both the high and low estimators is simply the (discounted) value of the corresponding estimator at the unique successor node. This is what we do implicitly in an ordinary simulation to price a European option, where we never have the choice to exercise early. By pruning branches, we reduce the time needed to calculate estimators from a tree.

But how can we determine that the optimal decision at a node is to continue? Broadie et al. [69] suggest the use of bounds. Suppose, as we have in Section 8.1, that the payoff functions h_i, $i = 1, \ldots, m$, are nonnegative. Then at any node at which the payoff from immediate exercise is 0, it is optimal to continue. This simple rule is often applicable at a large number of nodes.

European prices, when computable, can also be used as bounds. Consider an arbitrary node $X_i = X_i^{j_1 \cdots j_i}$ at the ith exercise date. Consider a European option expiring at t_m with discounted payoff function h_m. Suppose the value of this European option at node X_i is given by $g(X_i)$. This, then, is a lower bound on the value of the American option. If $g(X_i)$ exceeds $h_i(X_i)$, then the value of the American option must also exceed the immediate exercise value $h_i(X_i)$ and it is therefore optimal to continue. The same argument applies if the European option expires at one of the intermediate dates $t_{i+1}, \ldots, t_{m-1}$.

When European counterparts of American options can be priced quickly (using either a formula or a deterministic numerical method), the last step of the tree can be completely pruned. At the $(m - 1)$th exercise date, the value of the American option is the maximum of the immediate exercise value and the value of a European option expiring at t_m. If the European value is easily computable, there is no need to simulate the tree from t_{m-1} to t_m. This reduces the size of the tree by a factor of b.

Broadie et al. [69] also discuss the use of antithetic variates and Latin hypercube sampling in generating branches. In calculating the low estimator with antithetics, they apply the steps in (8.24)–(8.25) to averages over antithetic pairs. To apply Latin hypercube sampling, they generate two independent Latin hypercube samples at each node, compute low estimators from each set, and then combine the two by applying (8.24)–(8.25) with $b = 2$ to the two low estimators.

Figure 8.6, based on results reported in Broadie et al. [69], illustrates the performance of the method and the enhancements. The results displayed are for an American option on the maximum of two underlying assets. This allows comparison with a value computed from a two-dimensional binomial lattice. The assets follow a bivariate geometric Brownian motion with a 0.30 correlation parameter, 20% volatilities, 10% dividend yields, and initial values $(90, 90)$, $(100, 100)$, or $(110, 110)$. The risk-free rate is 5%. The option has a strike price of 100 and can be exercised at time 0, 1, 2, or 3, with the time between exercise opportunities equal to one year.

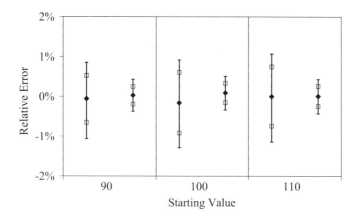

Fig. 8.6. Relative error in price estimates for an American option on the maximum of two underlying assets. Each interval shows high and low estimates and their midpoint. Shorter intervals use pruning and antithetic branching.

The results displayed in Figure 8.6 apply to the initial values 90, 100, and 110 of the initial assets, for which the option prices are 7.234, 12.412, and 19.059. The vertical scale is measured in percent deviations from these values. Within each of the three cases, the second confidence interval is computed using pruning, antithetic branching, and a European option as a control variate, whereas the first confidence interval uses only the control variate. Within each interval, the two squares indicate the high and low estimators and the diamond indicates their midpoint. The use of the midpoint as point estimator is somewhat arbitrary but effective. The confidence intervals are based on a nominal coverage of 90% ($z_{\delta/2} = 1.645$), but because these are conservative intervals numerical tests indicate that the actual coverage is much higher.

The results displayed are based on branching parameter $b = 50$. The number of replications used to estimate standard errors is set so that computing times are the same for all cases. Because pruning can substantially decrease the computing time required per tree, it allows a much larger number of replications. The first interval in each case uses about one hundred replications whereas the second interval in each case uses several thousand. The results indicate that for problems of this size the random tree method can readily produce estimates accurate to within about 1%, and that the proposed enhancements can be quite effective in increasing precision.

More extensive numerical tests are reported in [65] and [69], including tests on problems with five underlying assets for which no "true" value is available. The ability to compute a confidence interval is particularly valuable in such settings. These results support the view that the random tree method is an attractive technique for problems with a high-dimensional state vector and a small number of exercise opportunities. Its ability to produce a reliable interval estimate sets it apart from heuristic approximations.

8.4 State-Space Partitioning

State-space partitioning (called *stratification* by Barraquand and Martineau [38], *quantization* by Bally and Pagès [33]) uses a finite-state dynamic program to approximate the value of an American option. Whereas the dynamic program in the random tree method is based on randomly sampled states, in this method the states are defined in advance based on a partitioning of the state space of the underlying Markov chain $X_0, X_1, \ldots, X_m$.

We continue to use the notation of Section 8.1 and the dynamic programming formulation in (8.6)–(8.7). For each exercise date t_i, $i = 1, \ldots, m$, let $A_{i1}, \ldots, A_{ib_i}$ be a partition of the state space of X_i into b_i subsets. For the initial time 0, take $b_0 = 1$ and $A_{01} = \{X_0\}$. Define transition probabilities

$$p^i_{jk} = P(X_{i+1} \in A_{i+1,k} | X_i \in A_{ij}),$$

for all $j = 1, \ldots, b_i$, $k = 1, \ldots, b_{i+1}$, and $i = 0, \ldots, m-1$. (Take this to be zero if $P(X_i \in A_{ij}) = 0$.) These are not transition probabilities for X in the Markovian sense (the probability that X_{i+1} will fall in $A_{i+1,k}$ given $X_i \in A_{ij}$ will in general depend on past values of the chain), but we may nevertheless use them to define an approximating dynamic program.

For each $i = 1, \ldots, m$ and $j = 1, \ldots, b_i$, define

$$h_{ij} = \mathsf{E}[h_i(X_i) | X_i \in A_{ij}],$$

taking this to be zero if $P(X_i \in A_{ij}) = 0$. Now consider the backward induction

$$V_{ij} = \max\{h_{ij}, \sum_{k=1}^{b_{i+1}} p^i_{jk} V_{i+1,k}\}, \tag{8.26}$$

$i = 0, 1, \ldots, m-1$, $j = 1, \ldots, b_i$, initialized with $V_{mj} \equiv h_{mj}$. This method takes the value V_{01} calculated through (8.26) as an approximation to $V_0(X_0)$, the value in the original dynamic program.

Implementation of this method requires calculation of the transition probabilities p^i_{jk} and averaged payoffs h_{ij}, and this is where simulation is useful. We simulate a reasonably large number of replications of the Markov chain $X_0, X_1, \ldots, X_m$ and estimate these quantities from the simulation. In more detail, we record N^i_{jk}, the number of paths that move from A_{ij} to $A_{i+1,k}$, for all $i = 0, \ldots, m-1$, $j = 1, \ldots, b_i$, and $k = 1, \ldots, b_{i+1}$. We then calculate the estimates

$$\hat{p}^i_{jk} = N^i_{jk} / (N^i_{j1} + \cdots + N^i_{jb_i}),$$

taking the ratio to be zero whenever the denominator is zero. Similarly, we calculate $\hat{h}_{ij}$ as the average value of $h(X_i)$ over those replications in which $X_i \in A_{ij}$.

Using these estimates of the transition probabilities and average payoffs, we can calculate estimates $\hat{V}_{ij}$ of the approximating value function. We set $\hat{V}_{mj} = \hat{h}_{mj}$ for all $j = 1, \ldots, b_m$, and then recursively define

$$\hat{V}_{ij} = \max\{\hat{h}_{ij}, \sum_{k=1}^{b_{i+1}} \hat{p}_{jk}^i \hat{V}_{i+1,k}\},$$

for $j = 1, \ldots, b_i$, $i = 0, 1, \ldots, m - 1$. Our estimate of the approximate option value V_{01} is then $\hat{V}_{01}$.

By the strong law of large numbers, each $\hat{p}_{jk}^i$ and $\hat{h}_{ij}$ converges with probability 1 to the corresponding p_{jk}^i and h_{ij} as the number of replications increases. Moreover, the mapping from the transition probabilities and average payoffs to the value functions is continuous (because max, addition, and multiplication are continuous), so each $\hat{V}_{ij}$ converges to V_{ij} with probability 1 as well. The simulation procedure thus gives a strongly consistent estimator of V_{01}.

This, however, says nothing about the relation between the approximation V_{01} and the true option value $V_0(X_0)$. For any finite number of replications, the induction argument used to prove (8.21) in the random tree method shows that

$$\mathsf{E}[\hat{V}_{01}] \geq V_{01},$$

but the sign of the error $V_{01} - V_0(X_0)$ is unpredictable.

By adding a second simulation phase to the procedure, we can produce an estimate that is guaranteed to be biased low, relative to the true value $V_0(X_0)$. The idea — similar to one we used in Section 8.2 — is to turn the approximation V_{ij} into an implementable stopping policy. The option value thus produced is not just an ad hoc approximation; it is a value achievable by the option holder by following a well-specified exercise policy.

For the ith state X_i of the underlying Markov chain, define J_i to be the index of the subset containing X_i:

$$X_i \in A_{iJ_i}.$$

To each path $X_0, X_1, \ldots, X_m$, associate a stopping time

$$\tau = \min\{i : h_i(X_i) \geq V_{iJ_i}\},$$

defining τ to be m if the inequality is never satisfied. This is the stopping rule defined by the approximate value function V_{ij}. Because no stopping rule can be better than an optimal stopping rule, we have

$$\mathsf{E}[h_\tau(X_\tau)] \leq V_0(X_0),$$

which says that $h_\tau(X_\tau)$ is biased low.

If the V_{ij} were known, we could simulate replications of the Markov chain, record $h_\tau(X_\tau)$ on each path, and average over paths. This would produce a consistent estimator of $\mathsf{E}[h_\tau(X_\tau)]$. The quantity $\mathsf{E}[h_\tau(X_\tau)]$ does not entail any approximations not already present in the V_{ij}, and it has the advantage of being an achievable value: at each exercise date, the holder of the option could observe X_i and compare $h_i(X_i)$ with V_{iJ_i} to determine whether or not to stop.

Similar comments apply even if we rely on the estimates $\hat{V}_{ij}$ rather than true values V_{ij}. Define $\hat{\tau}$ by replacing V_{iJ_i} with $\hat{V}_{iJ_i}$ in the definition of τ. This, too, is an implementable and suboptimal stopping rule so

$$\mathsf{E}[h_{\hat{\tau}}(X_{\hat{\tau}})|\hat{V}_{ij}, j = 1, \ldots, b_i, i = 1, \ldots, m] \leq V_0(X_0),$$

and the inequality then also holds for the unconditional expectation. The conditional expectation on the left is the quantity to which the procedure converges as the number of second-phase replications increases.

The main challenge in using any variant of this approach lies in the selection of the state-space partitions. Bally and Pagès [33] discuss criteria for "optimal" partitions and propose simulation-based procedures for their construction. These appear to be computationally demanding. The effort might be justified if a partition, once constructed, could be applied to price many different American options. This, however, would not lend itself to tailoring the partition to the form of the payoff.

It is natural to expect that as the resolution of the partitions increases, the option value produced by this approach converges to the true value; see Theorem 1 of Bally and Pagès [33] for a precise statement. But methods that rely on refining *a priori* partitions are not well-suited to high-dimensional state spaces, just as deterministic numerical integration procedures are not well-suited to high-dimensional problems. In our discussion of stratified sampling we noted (cf. Example 4.3.4) that the number of strata required to maintain the same resolution along all dimensions grows exponentially with the number of dimensions. In the absence of problem-specific information, state-space partitioning exhibits similar dependence on dimension.

8.5 Stochastic Mesh Methods

8.5.1 General Framework

Like the random tree algorithm of Section 8.3, the stochastic mesh method solves a randomly sampled dynamic programming problem to approximate the price of an American option. The key distinction is that in valuing the option at a node at time step i, the mesh uses values from all nodes at time step $i + 1$, not just those that are successors of the current node. This in fact is why it produces a mesh rather than a tree. It keeps the number of nodes at each time step fixed, avoiding the exponential growth characteristic of a tree.

A general construction is illustrated in Figure 8.7. In the first phase, we simulate independent paths of the Markov chain $X_0, X_1, \ldots, X_m$; in the second phase, we "forget" which node at time i generated which node at $i+1$ and interconnect all nodes at consecutive time steps for the backward induction. We will consider other mechanisms for generating the nodes at each time step, but this is the most important one. We refer to it as the *independent-path construction*.

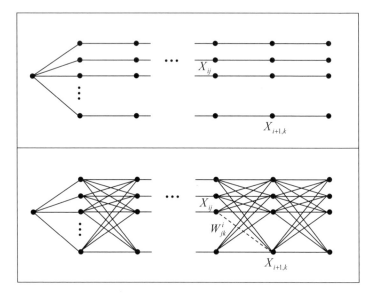

Fig. 8.7. Construction of stochastic mesh from independent paths. Nodes are generated by independent paths (top); weight W^i_{jk} is then assigned to a transition from the jth node at step i to the kth node at step $i+1$ (bottom).

For the pricing problem, we use the dynamic programming formulation and notation of (8.6)–(8.7). In the mesh, we use X_{ij} to denote the jth node at the ith exercise date, for $i = 1, \ldots, m$ and $j = 1, \ldots, b$. We use $\hat{V}_{ij}$ to denote the estimated value at this node, computed as follows. At the terminal nodes we set $\hat{V}_{mj} = h_m(X_{mj})$; we then work backward recursively by defining

$$\hat{V}_{ij} = \max \left\{ h_i(X_{ij}), \frac{1}{b} \sum_{k=1}^{b} W^i_{jk} \hat{V}_{i+1,k} \right\} \tag{8.27}$$

for some set of weights W^i_{jk}. At the root node, we set

$$\hat{V}_0 = \frac{1}{b} \sum_{k=1}^{b} \hat{V}_{1k}, \tag{8.28}$$

or the maximum of this and $h_0(X_0)$ if we want to allow exercise at time 0.

A fundamental distinction between (8.27) and the superficially similar recursion in (8.26) is that (8.27) evolves over randomly sampled nodes, whereas (8.26) evolves over fixed subsets of the state space.

As we have already seen several algorithms with a structure similar to (8.27), the main issue we need to address is the selection of the weights W_{jk}^i. A closely related issue is the sampling of the nodes X_{ij}. The independent-path construction in Figure 8.7 provides one mechanism but by no means the only one. We could, for example, generate b nodes at each time step i by drawing independent samples from the marginal distribution of X_i. Different rules for sampling the nodes and selecting the weights correspond to different versions of the method.

The formulation of the stochastic mesh method we give here is a bit more general than the one originally introduced by Broadie and Glasserman [66]. The advantage of a more general formulation is that it allows a unified treatment of related techniques, clarifying which features distinguish the methods and which features are shared. We postpone discussion of the detailed construction of weights — the central issue — to Sections 8.5.2 and 8.6.2 and proceed with the general formulation.

Conditions on the Mesh

We first impose two conditions on the mesh construction and weights that are unlikely to exclude any cases of practical interest, then add a third, more restrictive condition. To state these, let

$$\mathbf{X}_i = (X_{i1}, \ldots, X_{ib})$$

denote the "mesh state" at step i consisting of all nodes at step i, for $i = 1, \ldots, m$, and let $\mathbf{X}_0 = X_0$. We assume that the mesh construction is Markovian in the following sense:

(M1) $\{\mathbf{X}_0, \ldots, \mathbf{X}_{i-1}\}$ and $\{\mathbf{X}_{i+1}, \ldots, \mathbf{X}_m\}$ are independent given $\mathbf{X}_i$, for all $i = 1, \ldots, m-1$.

This condition is satisfied by the independent-path construction, because in that case $\{X_{1j}, \ldots, X_{mj}\}$, $j = 1, \ldots, b$, are independent copies of the Markov chain. It is also satisfied if nodes at different exercise dates are generated independently of each other. We also require

(M2) Each weight W_{jk}^i is a deterministic function of $\mathbf{X}_i$ and $\mathbf{X}_{i+1}$.

This includes as a special case the possibility that W_{jk}^i is a function only of X_{ij} and $X_{i+1,k}$.

Recall from (8.11) that $C_i(x)$ denotes the continuation value in state x at time i. Our next condition restricts the choice of weights to those that correctly estimate continuation values, on average:

(M3) For all $i = 1, \ldots, m - 1$ and all $j = 1, \ldots, b$,

$$\frac{1}{b} \sum_{k=1}^{b} \mathsf{E}\left[W_{jk}^i V_{i+1}(X_{i+1,k}) | \mathbf{X}_i\right] = C_i(X_{ij}).$$

This says that if we knew the true option values at time $i + 1$, the expected weighted average calculated at a node at time i would be the true continuation value.

High Bias

As a first implication of (M1)–(M3), we show that these conditions imply that the mesh estimator $\hat{V}_0$ defined by (8.27)–(8.28) is biased high. The argument is similar to the one we used for the random tree algorithm in Section 8.3.1 but, because of the mesh weights, sufficiently different to merit presentation.

We show that if, for some i,

$$\mathsf{E}[\hat{V}_{i+1,j} | \mathbf{X}_i] \geq V_{i+1}(X_{i+1,j}), \quad j = 1, \ldots, b, \tag{8.29}$$

then the same holds for all smaller i. Once this is established, noting that (8.29) holds (with equality) at $i = m - 1$ completes the induction argument.

Using (8.27) and Jensen's inequality, we get

$$\mathsf{E}[\hat{V}_{ij} | \mathbf{X}_i] \geq \max\left\{ h_i(X_{ij}), \frac{1}{b} \sum_{k=1}^{b} \mathsf{E}[W_{jk}^i \hat{V}_{i+1,k} | \mathbf{X}_i] \right\}. \tag{8.30}$$

We examine the conditional expectations on the right. By further conditioning on $\mathbf{X}_{i+1}$ and using (M2), we get

$$\mathsf{E}[W_{jk}^i \hat{V}_{i+1,k} | \mathbf{X}_i, \mathbf{X}_{i+1}] = W_{jk}^i \mathsf{E}[\hat{V}_{i+1,k} | \mathbf{X}_i, \mathbf{X}_{i+1}]$$
$$= W_{jk}^i \mathsf{E}[\hat{V}_{i+1,k} | \mathbf{X}_{i+1}], \tag{8.31}$$

the second equality following from (M1) and the definition of $\hat{V}_{i+1,k}$ through the backward induction in (8.27). Applying the induction hypothesis (8.29) to (8.31), we get

$$\mathsf{E}[W_{jk}^i \hat{V}_{i+1,k} | \mathbf{X}_i, \mathbf{X}_{i+1}] \geq W_{jk}^i V_{i+1}(X_{i+1,k}),$$

from which follows

$$\frac{1}{b} \sum_{k=1}^{b} \mathsf{E}[W_{jk}^i \hat{V}_{i+1,k} | \mathbf{X}_i] \geq \frac{1}{b} \sum_{k=1}^{b} \mathsf{E}[W_{jk}^i V_{i+1}(X_{i+1,k}) | \mathbf{X}_i]$$
$$= C_i(X_{ij})$$

using (M3). Applying this inequality to (8.30), we conclude that

$$\mathsf{E}[\hat{V}_{ij}|\mathbf{X}_i] \geq \max\{h_i(X_{ij}), C_i(X_{ij})\} = V_i(X_{ij}),$$

which is what we needed to show.

This induction argument offers some insight into the types of conditions needed to ensure convergence of the mesh values $\hat{V}_{ij}$ to the true values $V_i(X_i)$, given X_{ij}. Applying the contraction property (8.22) and (M3) to the mesh recursion (8.27) yields the bound

$$
\begin{aligned}
|\hat{V}_{ij} - V_i(X_{ij})| &\leq \left| \frac{1}{b} \sum_{k=1}^{b} W_{jk}^i \hat{V}_{i+1,k} - \mathsf{E}[W_{jk}^i V_{i+1}(X_{i+1,k})|\mathbf{X}_i] \right| \\
&\leq \left| \frac{1}{b} \sum_{k=1}^{b} W_{jk}^i V_{i+1}(X_{i+1,k}) - \mathsf{E}[W_{jk}^i V_{i+1}(X_{i+1,k})|\mathbf{X}_i] \right| \\
&\quad + \left| \frac{1}{b} \sum_{k=1}^{b} W_{jk}^i [\hat{V}_{i+1,k} - V_{i+1}(X_{i+1,k})] \right|.
\end{aligned}
$$

To require that the first term on the right side of the last inequality vanish is to require that the summands satisfy a law of large numbers. The second term requires a sufficiently strong induction hypothesis on the convergence of the mesh estimate at time $i+1$. Because $\hat{V}_{mj} = V_m(X_{mj})$ at all terminal nodes for all b, it is natural to proceed backward by induction.

Broadie and Glasserman [66] use these observations to prove convergence of the mesh estimator when $\mathbf{X}_1, \ldots, \mathbf{X}_m$ are independent of each other, $X_{i1}, X_{i2}, \ldots$ are i.i.d. for each i, and each weight W_{jk}^i is a function only of X_{ij} and $X_{i+1,k}$. The independence assumptions facilitate the application of laws of large numbers. In the more general cases encompassed by conditions (M1)–(M2), the problem is complicated by the dependence between nodes and the generality of the weights. Avramidis and Matzinger [29] derive a probabilistic upper bound on the error in the mesh estimator for a type of dependence structure that fits within conditions (M1)–(M2). They use this bound to prove convergence as $b \to \infty$.

Rust [314] proves convergence of a related method for a general class of dynamic programming problems, not specifically focused on American options or optimal stopping. In his method, the $\{X_{ij}, j = 1, \ldots, b\}$ are independent and uniformly distributed over a compact set, and the same nodes are used for all i.

Low Estimator

Broadie and Glasserman [66] supplement the high-biased mesh estimator with a low-biased estimator. Their low estimator uses a stopping rule defined by the mesh.

To define the stopping rule, we need to extend the weights W_{jk}^i from $X_{i1}, \ldots, X_{ib}$ to all points in the state space at time i. The details of how we do this will be evident once we consider specific choices of weights; for now, let us simply suppose that we have a function $W_k^i(\cdot)$ on the state space at time i. We interpret $W_k^i(x)$ as the weight from state x at time i to node $X_{i+1,k}$.

Through this weight function, the mesh defines a continuation value throughout the state space, and not just at the nodes in the mesh. The continuation value at state x at time i, $i = 1, \ldots, m-1$, is given by

$$\hat{C}_i(x) = \frac{1}{b} \sum_{k=1}^b W_k^i(x) \hat{V}_{i+1,k}. \tag{8.32}$$

If we impose the reasonable requirement that $W_k^i(X_{ij}) = W_{jk}^i$ (so that the weight function does in fact extend the original weights), then $\hat{C}_i(X_{ij})$ coincides with the continuation value estimated by the mesh at node X_{ij}, and $\hat{C}_i$ interpolates from the original nodes to the rest of the state space. Set $\hat{C}_m \equiv 0$.

With the mesh held fixed, we now simulate a path $X_0, X_1, \ldots, X_m$ of the underlying Markov chain, independent of the paths used to construct the mesh. Define a stopping time

$$\hat{\tau} = \min\{i : h_i(X_i) \geq \hat{C}_i(X_i)\}; \tag{8.33}$$

this is the first time the immediate exercise value is as great as the estimated continuation value. The low estimator for a single path is

$$\hat{v} = h_{\hat{\tau}}(X_{\hat{\tau}}), \tag{8.34}$$

the payoff from stopping at $\hat{\tau}$.

That this is indeed biased low follows from the observation that no policy can be better than an optimal policy. Conditioning on the mesh fixes the stopping rule and yields

$$\mathsf{E}[\hat{v}|\mathbf{X}_1, \ldots, \mathbf{X}_m] \leq V_0(X_0).$$

The same then applies to the unconditional expectation.

By simulating multiple paths independently, with each following the mesh-defined stopping rule, we can calculate an average low estimator conditional on the mesh. We can then generate independent copies of the mesh and calculate high and low estimators from each copy. From these independent replications of the high and low estimators we can calculate sample means and standard deviations to form a confidence interval for each estimator. The two intervals can then be combined as in Figure 8.2. Assuming independently generated nodes in the mesh, Broadie and Glasserman [66] give conditions under which the low estimator is asymptotically unbiased, meaning that $\mathsf{E}[\hat{v}] \to V_0(X_0)$ as $b \to \infty$ in the mesh. When this holds, the combined confidence interval shrinks to the point $V_0(X_0)$ if we let $b \to \infty$ and then $n \to \infty$.

An Interleaving Estimator

At several points in this chapter we have noted two sources of bias affecting simulation estimates of American option prices: high bias resulting (through Jensen's inequality) from applying backward induction over a finite set of paths, and low bias resulting from suboptimal exercise. The development of high and low estimators in this section keeps the two sources of bias separate, first applying backward induction and then applying a mesh-defined stopping rule. But blending the two techniques has the potential to produce a more accurate value by not compounding each source of bias. We present one such combination from the method of Longstaff and Schwartz [241], though they present it in a different setting.

Suppose value estimates $\hat{V}_{i+1,j}, \ldots, \hat{V}_{mj}$, $j = 1, \ldots, b$, have already been determined for steps $i + 1, \ldots, m$, starting with the initialization $\hat{V}_{mj} = h_m(X_{mj})$. These determine estimated continuation values:

$$\hat{C}_{\ell j} = \frac{1}{b} \sum_{k=1}^{b} W_{jk}^{\ell} \hat{V}_{\ell+1,k},$$

$\ell = i, \ldots, m - 1$, $\hat{C}_m \equiv 0$. Extending the weights throughout the state space as before then defines continuation values $\hat{C}_\ell(x)$ for every state x at time ℓ.

To assign a value to a node X_{ij} at step i, we consider two cases. If the immediate exercise value $h_i(X_{ij})$ is at least as great as the estimated continuation value $\hat{C}_{ij}$, then we exercise at node X_{ij} and thus set

$$V_{ij} = h_i(X_{ij});$$

this coincides with the high estimator (8.27). If, however, $h_i(X_{ij}) < \hat{C}_{ij}$, then rather than assign the high value $\hat{C}_{ij}$ to the current node (as (8.27) does), we simulate a path of the Markov chain $\tilde{X}_i, \tilde{X}_{i+1}, \ldots, \tilde{X}_m$ starting from the current node $\tilde{X}_i = X_{ij}$. To this path we apply the stopping rule (8.33) defined by the continuation values at $i+1, \ldots, m-1$. We record the payoff received at exercise and assign this as the value at V_{ij}, just as we would using the low estimator starting from node X_{ij}. (In the method of Longstaff and Schwartz [241], the path $\tilde{X}_i, \tilde{X}_{i+1}, \ldots, \tilde{X}_m$ is just the original path $X_{ij}, X_{i+1,j}, \ldots, X_{mj}$ passing through X_{ij}, but one could alternatively simulate an independent path from X_{ij}.)

To be more precise, we need to define a new stopping time for each i and apply it to a path starting from X_{ij}. Thus, for the path $\tilde{X}_i, \tilde{X}_{i+1}, \ldots, \tilde{X}_m$ starting at $\tilde{X}_i = X_{ij}$, define

$$\hat{\tau}_i = \min\{\ell \in \{i, i+1, \ldots, m\} : h_\ell(\tilde{X}_\ell) \geq \hat{C}_\ell(\tilde{X}_\ell)\}.$$

Then the two cases for assigning a value to V_{ij} can be combined into the rule

$$\hat{V}_{ij} = h_{\hat{\tau}_i}(\tilde{X}_{\hat{\tau}_i}). \tag{8.35}$$

This procedure interleaves elements of the high and low estimators, alternating between the two by applying backward induction to estimate a continuation value and then applying a suboptimal stopping rule starting from each node. This may partly offset the two sources of bias, but a precise comparison remains open for investigation. If every path $\tilde{X}_i, \tilde{X}_{i+1}, \ldots, \tilde{X}_m$ used in (8.35) is independent of the original set of paths (given $\tilde{X}_i = X_{ij}$), then because the $\hat{V}_{ij}$ in (8.35) result from stopping rules, these value estimates are biased low. In practice, using the original mesh path $X_{ij}, \ldots, X_{mj}$ rather than an independent path will usually also result in a low bias, because in this method the high bias resulting from Jensen's inequality tends to be less pronounced than the low bias resulting from a suboptimal stopping rule.

8.5.2 Likelihood Ratio Weights

In this section and in Section 8.6.2, we take up the question of defining weights W^i_{jk} that satisfy conditions (M2) and (M3) of Section 8.5.1. The alternatives we consider differ in their scope and computational requirements.

This section discusses weights defined through likelihood ratios, as proposed by Broadie and Glasserman [66]. Some of the ideas from our discussion of importance sampling in Section 4.6.1 are relevant here as well. But whereas the motivation for changing probability measure in importance sampling is variance reduction, here we use it to correct pricing as we move backward through the mesh.

Suppose that the state space of the Markov chain $X_0, X_1, \ldots, X_m$ is $\Re^d$ and that the chain admits transition densities $f_1, \ldots, f_m$, meaning that for $x \in \Re^d$ and $A \subseteq \Re^d$,

$$P(X_i \in A | X_{i-1} = x) = \int_A f_i(x, y) \, dy, \quad i = 1, \ldots, m.$$

With X_0 fixed, $g_1(\cdot) = f_1(X_0, \cdot)$ is the marginal density of X_1, and then by induction,

$$g_i(y) = \int g_{i-1}(x) f_i(x, y) \, dx$$

gives the marginal density of X_i, $i = 2, \ldots, m$. The optimal continuation value in state x at time i is

$$C_i(x) = \mathsf{E}[V_{i+1}(X_{i+1}) | X_i = x] = \int V_{i+1}(y) f_{i+1}(x, y) \, dy,$$

an integral with respect to a transition density. The main purpose of weights in the mesh is to estimate these continuation values.

Fix a node X_{ij} in the mesh and consider the estimation of the continuation value at that node. To motivate the introduction of likelihood ratios, we begin with some simplifying assumptions. Suppose that the nodes $X_{i+1,k}$, $k = 1, \ldots, b$, at the next step in the mesh were generated independently of each other and of all other nodes in the mesh from some density g. Suppose also that we know the true option values $V_{i+1}(X_{i+1,k})$ at these downstream nodes. Averaging over the b nodes at step $i+1$ and letting $b \to \infty$ yields

$$\frac{1}{b} \sum_{k=1}^{b} V_{i+1}(X_{i+1,k}) \to \int V_{i+1}(y)g(y)\,dy,$$

which will not in general equal the desired continuation value $C_i(X_{ij})$. In particular, if g is the marginal density g_{i+1} of the Markov chain at time $i+1$, then the limit

$$\frac{1}{b} \sum_{k=1}^{b} V_{i+1}(X_{i+1,k}) \to \mathsf{E}[V_{i+1}(X_{i+1})]$$

is the *unconditional* expected value at $i+1$, whereas what we want is a conditional expectation.

The purpose of the mesh weights is thus to correct for the fact that the downstream nodes were sampled from g rather than from the transition density $f_{i+1}(X_{ij}, \cdot)$. (In contrast, in the random tree method of Section 8.3 all successor nodes of a given node are generated from the same transition law and thus get equal weight.) Suppose we set each weight to be

$$W_{jk}^i = \frac{f_{i+1}(X_{ij}, X_{i+1,k})}{g(X_{i+1,k})}, \tag{8.36}$$

the likelihood ratio relating the transition density to the mesh density g. The pairs $(W_{jk}^i, V_{i+1}(X_{i+1,k}))$, $k = 1, \ldots, b$, are i.i.d. given X_{ij}, so we now get

$$\frac{1}{b} \sum_{k=1}^{b} W_{jk}^i V_{i+1}(X_{i+1,k}) \to \mathsf{E}_g[W_{jk}^i V_{i+1}(X_{i+1,k}) | X_{ij}]$$

$$= \int \frac{f_{i+1}(X_{ij}, y)}{g(y)} V_{i+1}(y)g(y)\,dy$$

$$= \int f_{i+1}(X_{ij}, y)V_{i+1}(y)\,dy = C_i(X_{ij}),$$

which is what we wanted. (We have subscripted the expectation by g to emphasize that $X_{i+1,k}$ has density g.) This is in fact stronger than the requirement in (M3), and the weights in (8.36) clearly satisfy (M2).

Likelihood ratios provide the only completely general weights, in the following sense. If W_{jk}^i is a function only of X_{ij} and $X_{i+1,k}$, and if

$$\mathsf{E}_g[W_{jk}^i h(X_{i+1,k}) | X_{ij}] = \int f_{i+1}(X_{ij}, y)h(y)\,dy \tag{8.37}$$

for all bounded $h : \Re^d \to \Re$, then the Radon-Nikodym Theorem (cf. Appendix B.4) implies that W_{jk}^i equals the likelihood ratio in (8.36) with probability

1. Interpret (8.37) as stating that the weights give the correct "price" at node X_{ij} for payoff $h(\cdot)$ at time $i+1$. Uniqueness holds with other sufficiently rich classes of functions h. This indicates that alternative strategies for selecting weights must in part rely on restricting the class of admissible functions h, though the restriction should not exclude the value function V_{i+1}.

Weights in a Markovian Mesh

We now drop the assumption that nodes in the mesh are generated independently and consider more general constructions consistent with the Markovian condition (M1). Even if we fix the mechanism used to generate the mesh, there is some flexibility in the choice of likelihood ratio weights that results from the flexibility in (M2) to allow the weights W_{jk}^i to depend on all nodes at times i and $i+1$. The alternatives we present below satisfy (M1)–(M3).

Consider the independent-path construction of Figure 8.7 based on independent paths $(X_{1j}, \ldots, X_{mj})$, $j = 1, \ldots, b$. For $k \neq j$, $X_{i+1,k}$ is independent of X_{ij}; its conditional distribution given X_{ij} is therefore just its unconditional marginal distribution, which has density g_{i+1}. For $k = j$, the conditional distribution is given by the transition density out of X_{ij}, so no weight is needed. We thus arrive at the weights

$$
W_{jk}^i = \begin{cases} f_{i+1}(X_{ij}, X_{i+1,k})/g_{i+1}(X_{i+1,k}), & k \neq j, \\ 1 & k = j. \end{cases} \tag{8.38}
$$

Alternatively, we could use the fact that the pairs $(X_{i\ell}, X_{i+1,\ell})$, $\ell = 1, \ldots, b$, are i.i.d. (in the independent-path construction) with joint density $g_i(x)f_{i+1}(x, y)$. The relevant likelihood ratios now take the form $W_{jj}^i = 1$ and

$$
W_{jk}^i = \frac{f_{i+1}(X_{ij}, X_{i+1,k})}{g_i(X_{ik})f_{i+1}(X_{ik}, X_{i+1,k})}, \quad k \neq j. \tag{8.39}
$$

In contrast to (8.38), these weights use information about the node X_{ik} from which $X_{i+1,k}$ was generated. A simple calculation shows that each weight in (8.38) is the conditional expectation given $X_{i+1,k}$ of the corresponding weight in (8.39).

Next consider yet another construction of the mesh in which the nodes at $i+1$ are generated from the nodes at i as follows. We pick a node $X_{i\ell}$ randomly and uniformly from the b nodes at time i and generate a successor by sampling from the transition density $f_{i+1}(X_{i\ell}, \cdot)$ out of $X_{i\ell}$. We repeat the procedure to generate a total of b nodes at time $i+1$, each time drawing randomly and uniformly from all b nodes at time i. (In other words, we sample b times "with replacement" from the nodes at time i and generate a successor from each selected node.) This construction is clearly consistent with the Markovian condition (M1). Given $\mathbf{X}_i$, the nodes at time $i+1$ are i.i.d. with density

$$
\frac{1}{b} \sum_{\ell=1}^{b} f_{i+1}(X_{i\ell}, \cdot), \tag{8.40}
$$

the average of the transition densities out of the nodes at time i. The corresponding likelihood ratios are

$$W_{jk}^i = \frac{f_{i+1}(X_{ij}, X_{i+1,k})}{\frac{1}{b}\sum_{\ell=1}^{b} f_{i+1}(X_{i\ell}, X_{i+1,k})}. \tag{8.41}$$

As $b \to \infty$, the average density (8.40) converges to the marginal density g_{i+1}, so the weights in (8.41) are close to those in (8.39). But the weights in (8.41) have a property that makes them appealing. If we fix k and sum over j, we get

$$\frac{1}{b}\sum_{j=1}^{b} W_{jk}^i = 1; \tag{8.42}$$

the average weight *into* a node is 1. This property may be surprising. If we were to interpret the ratios W_{jk}^i/b as transition probabilities, we would expect the average weight *out* of a node to sum to 1. Broadie and Glasserman [66] point out at an attractive feature of the less obvious condition (8.42), which we now explain.

Implications of the Weight Sum

Consider the pricing of a European option in a mesh. Suppose the option has (discounted) payoff h_m at t_m, and to price it in the mesh we use the recursion (8.27), but always taking the second term inside the max because early exercise is not permitted. The resulting estimate $\hat{V}_0$ at the root node can be written as

$$\frac{1}{b^m} \sum_{j_1,\dots,j_m} \prod_{i=2}^{m} W_{j_{i-1}j_i}^{i-1} h_m(X_{mj_m}),$$

the sum ranging over all $j_i = 1,\dots,b$, $i = 1,\dots,m$. In other words, $\hat{V}_0$ is the average over all b^m paths through the mesh of the payoff per path multiplied by the product of weights along the arcs of the path. By grouping paths that terminate at the same node, we can write this as

$$\frac{1}{b} \sum_{j_m=1}^{b} h_m(X_{mj_m}) \left(\frac{1}{b^{m-1}} \sum_{j_1,\dots,j_{m-1}} \prod_{i=2}^{m} W_{j_{i-1}j_i}^{i-1} \right).$$

If the weights are defined by likelihood ratios, then the factor in parentheses has expected value 1; this can be verified by induction using the fact that a likelihood ratio always has expected value 1. Rewriting this factor as

$$\frac{1}{b} \sum_{j_{m-1}=1}^{b} W_{j_{m-1}j_m}^{m-1} \cdot \frac{1}{b} \sum_{j_{m-2}=1}^{b} W_{j_{m-2}j_{m-1}}^{m-2} \cdots \frac{1}{b} \sum_{j_1=1}^{b} W_{j_1 j_2}^1$$

reveals that it is identically 1 if (8.42) holds. Thus, when (8.42) holds, the mesh estimate of the European option price is simply

$$\frac{1}{b} \sum_{j_m=1}^{b} h_m(X_{mj_m}),$$

the average of the terminal payoffs.

This simplification is important because multiplying weights along steps of a path through the mesh can produce exponentially growing variance. We encountered this phenomenon in our discussion of importance sampling over long time horizons; see the discussion surrounding (4.82). The property in (8.42) thus replaces factors that have exponentially growing variance with the constant 1. Of course, there is no reason to use a mesh to price a European option, so this should be taken as indirect evidence that (8.42) is beneficial in the more interesting case of pricing American options.

We arrived at the weights (8.41) through a construction that generates nodes at time $i + 1$ by drawing b independent samples from the average density (8.40), given $\mathbf{X}_i$. Now consider applying stratified sampling to this construction using as stratification variable the index j of the node X_{ij} selected at time i. This gives us b equiprobable strata, so in a sample of size b we draw exactly one value from each stratum. But this simply means that we draw one successor from each density $f_{i+1}(X_{ij}, \cdot)$, $j = 1, \ldots, b$, which is exactly what the independent-path construction does. In short, we may use the independent-path construction (which carries out the stratification implicitly) and then apply the weights (8.41). This is what Broadie and Glasserman [66] do in their numerical tests. Boyle, Kolkiewicz, and Tan [55] also implement this approach and combine it with low discrepancy sequences.

Weights for the Low Estimator

As explained in Section 8.5.1, a stochastic mesh defines an exercise policy throughout the state space (and not just at the nodes of the mesh) once the weights are extended to all points. This is the key to the low estimator defined through (8.32)–(8.33) and also to the interleaving estimator in (8.35). Both rely on the ability to estimate a continuation value $\hat{C}_i(x)$ at an arbitrary state x and time i.

As in Section 8.5.1, we use $W_k^i(x)$ to denote the weight on a hypothetical arc from state x at time i to mesh node $X_{i+1,k}$. In all of the likelihood ratio weights (8.38)–(8.41) discussed in this section, the current node X_{ij} appears as an explicit argument of a function (a transition density). The obvious way to extend (8.38)–(8.41) to arbitrary points x is thus to replace X_{ij} with x. This is the method used in Broadie and Glasserman [66]. To avoid computing additional values of transition densities, one might alternatively use interpolation.

Example 8.5.1 *American geometric average option on seven assets.* We consider a call option on the geometric average of multiple assets modeled by geometric Brownian motion. This provides a convenient setting for testing algorithms in high dimensions because the problem can be reduced to a single dimension in order to compute an accurate price for comparison. We used this idea in Section 5.5.1 to test quasi-Monte Carlo methods; see equation (5.33) and the surrounding discussion.

We consider a specific example from Broadie and Glasserman [66]. There are seven uncorrelated underlying assets. Each is modeled as $\mathrm{GBM}(r - \delta, \sigma^2)$ with interest rate $r = 3\%$, dividend yield $\delta = 5\%$, and volatility $\sigma = 0.40$. The assets have an initial price of 100 and the option's strike price is also 100. The option expires in 1 year and can be exercised at intervals of 0.1 years starting at 0 and ending at 1. A binomial lattice (applied to the one-dimensional geometric average) yields a price of 3.27. The corresponding European option has a price of 2.42, so the early-exercise feature has significant value in this example.

In pricing the option in the stochastic mesh, we treat it as a seven-dimensional problem and do not use the reduction to a single dimension. We use the weights in (8.41). Because the seven underlying assets evolve independently of each other, the transition density from one node to another factors as the product of one-dimensional densities. More explicitly, the transition density from one generic node $x = (x_1, \ldots, x_7)$ to another $y = (y_1, \ldots, y_7)$ over a time interval of length Δt is

$$f(x, y) = \prod_{i=1}^{7} \frac{1}{\sigma\sqrt{\Delta t} y_i} \phi\left(\frac{\log(y_i/x_i) - (r - \delta - \frac{1}{2}\sigma^2)\Delta t}{\sigma\sqrt{\Delta t}}\right),$$

with ϕ the standard normal density. This assumes that each x_i and y_i records the level S_i of the ith underlying asset. If instead we record $\log S_i$ in x_i and y_i, then the transition density simplifies to

$$f(x, y) = \prod_{i=1}^{7} \frac{1}{\sigma\sqrt{\Delta t}} \phi\left(\frac{y_i - x_i - (r - \delta - \frac{1}{2}\sigma^2)\Delta t}{\sigma\sqrt{\Delta t}}\right). \tag{8.43}$$

Figure 8.8 displays numerical results from Broadie and Glasserman [66] for this example. Computational effort increases as we move from left to right in the figure: the label (50,500) indicates a mesh constructed from $b = 50$ paths through which 500 additional paths are simulated to compute a low estimator, and the other labels should be similarly interpreted. The solid lines in the figure show the high and low estimators; the dashed lines show the upper and lower confidence limits (based on a nominal coverage of 90%) using standard errors estimated from 25 replications at each mesh size. The dotted horizontal line shows the true price.

The bias of the low estimator is much smaller than that of the high estimator in this example. Recall that the value computed by the low estimator

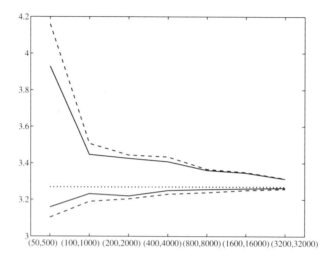

Fig. 8.8. Convergence of high and low mesh estimators for an American geometric average option on seven underlying assets.

reflects the exercise policy determined by the mesh. This means that even at relatively small values of b (100, for example), the mesh does a reasonable job of identifying an effective exercise policy, even though the high estimator shows a large bias. The bias in the high estimator results from the backward induction procedure, which overestimates the value implicit in the mesh stopping rule. □

Computational Costs and Limitations

Estimating a continuation value at a single node in the mesh requires calculating b weights and then a weighted average and is thus an $O(b)$ operation. Each step in the backward induction procedure (8.27) requires estimating continuation values at b nodes and is therefore an $O(b^2)$ operation. Applying the mesh to an m-step problem requires $O(mb^2)$ time. The implicit constant in this computing-time magnitude depends on the time required to generate a single transition of the underlying Markov chain (which in turn depends on the problem dimension d) and on the time to evaluate each weight. Each replication of the low estimator requires calculation of an additional weighted average at each step along a path and thus requires $O(mb)$ computing time. It is therefore practical to run many more low-estimator paths than the number of mesh paths b; but reducing the bias in the low estimator requires increasing b.

Based on numerical experiments, Broadie and Glasserman [66] tentatively suggest that the root mean square error of the mesh estimator is $O(b^{-1/2})$ in problems for which exact simulation of $X_1, \ldots, X_m$ is possible. This is the same convergence rate that would be obtained in estimating the price of a European option from b independent replications. But because the computational requirements of the mesh are quadratic in b, the convergence rate is a rather slow $O(s^{-1/4})$ when measured in units of computing time s. In contrast, European option pricing through independent replications retains square-root convergence in s as well as b. In Section 8.6.2 we will see that using regression the time required for each step in the backward induction is proportional to b rather than b^2, resulting in faster overall run times, usually at the cost of some approximation error from the choice of regressors.

Broadie and Yamamoto [70] use a fast Gauss transform to accelerate backward induction calculations. This method reduces the computational effort for each step in the mesh from $O(b^2)$ to $O(b)$, but entails an approximation in the evaluation of the weights. Broadie and Yamamoto [70] find experimentally that the method works well in up to three dimensions.

An important feature of likelihood ratio weights is that they do not depend on the payoff functions h_i. Once a mesh is constructed and all its weights computed, it can be used to price many different American options. This requires storing all weights, but significantly reduces computing times compared to generating a new mesh.

Based on numerical tests, Broadie and Glasserman [66] emphasize the importance of using control variates with the stochastic mesh. They use both inner and outer controls: inner controls apply at each node in the mesh, outer controls apply across independent replications of the mesh. Candidate control variates are moments of the underlying assets and the prices of European options, when available.

The main limitation on applying the mesh with likelihood ratio weights is the need for a transition density. Transition densities for the underlying Markov chain may be unknown or may fail to exist. We encountered similar issues in estimating sensitivities in Section 7.3; see especially Section 7.3.2.

Recall that the state X_i of the underlying Markov chain records information at the ith exercise date t_i. The intervals $t_{i+1} - t_i$ separating exercise dates could be large, in which case we may need to simulate X_{i+1} from X_i through a series of smaller steps. Even if we know the transition density over a single step, we may not be able to evaluate the transition density from t_i to t_{i+1}. We faced a similar problem in the setting of the LIBOR market model in Section 7.3.4. In such settings, it may be necessary to use a multivariate normal or lognormal approximation to the transition density in order to compute (approximate) likelihood ratios.

As discussed in Section 7.3.2, transition densities often fail to exist in singular models, meaning models in which the dimension of the state vector exceeds the dimension of the driving Brownian motion. This is often the case in interest rate models. It can also arise through the introduction of supplementary variables in the state vector X_i. Consider, for example, the pricing of an American option on the average of a single underlying asset S. One might take as state X_i the pair $(S(t_i), \bar{S}(t_i))$, in which $S(t_i)$ records the current level of the underlying asset and $\bar{S}(t_i)$ records the average over $t_1, \ldots, t_i$. This formulation eliminates path-dependence through the augmented state but results in a singular model: given $(S(t_i), \bar{S}(t_i))$ and $S(t_{i+1})$, the value of $\bar{S}(t_{i+1})$ is completely determined, so the Markov chain X_i does not admit a transition density on $\Re^2$. Weights based on likelihood ratios are infeasible in such settings.

Constrained Weights

To address the problem of unknown or nonexistent transition densities, Broadie et al. [68] propose a method that selects weights through a constrained optimization problem. This method relies on the availability of known conditional expectations, which are used to constrain the weights.

Suppose, then, that for some $\Re^M$-valued function G on the state space of the underlying Markov chain, the conditional expectation

$$g(x) = E[G(X_{i+1})|X_i = x]$$

is a known function of the state x. For example, moments of the underlying assets and simple European options often provide candidate functions. Fix a node X_{ij} and consider weights W^i_{jk}, $k = 1, \ldots, b$, satisfying

$$\frac{1}{b}\sum_{k=1}^{b} W^i_{jk}G(X_{i+1,k}) = g(X_{ij}); \tag{8.44}$$

these are weights that correctly "price" the payoff G, to be received at $i + 1$, from the perspective of node X_{ij}. If V_{i+1} is well-approximated by a linear combination of the components of G, then such weights should provide a good approximation to the continuation value at X_{ij} when used in the basic mesh recursion (8.27).

Taking one of the components of G to be a constant imposes the constraint

$$\frac{1}{b}\sum_{k=1}^{b} W^i_{jk} = 1. \tag{8.45}$$

We noted previously that the alternative condition (8.42) — which constrains the sum of the weights into a node rather than out of a node — has an appealing feature. But that condition links the choice of all b^2 weights from every node at step i to every node at step $i + 1$, whereas (8.44) and (8.45) apply separately to the b weights used for each node X_{ij}.

The number M of easily computed conditional expectations is likely to be much smaller than the number of nodes b, in which case the constraints (8.44) do not completely determine the weights. From all feasible weights, Broadie et al. [68] select those minimizing an objective of the form

$$\sum_{k=1}^{b} H(W_{jk}^{i}),$$

for some convex function H. The specific cases they consider are a quadratic objective $H(x) = x^2/2$ and the entropy objective $H(x) = x \log x$. The quadratic leads to a simpler optimization problem but could produce negative weights, whereas the entropy objective will provide a nonnegative feasible solution if one exists. Broadie et al. [68] report numerical results showing that the two objectives result in similar price estimates and that these estimates are close to those produced using likelihood ratio weights in cases in which a transition density is available.

This approach to selecting mesh weights is closely related to weighted Monte Carlo as presented in Section 4.5.2 and further analyzed in Glasserman and Yu [147]. As shown in Glasserman and Yu [148], it is also closely related to the regression-based methods we discuss next. A distinction between the methods is that the constrained optimization problem produces weights W_{jk}^{i} that depend on both X_{ij} and $X_{i+1,k}$, whereas weights defined through regression (in the next section) depend on X_{ij} and X_{ik}.

8.6 Regression-Based Methods and Weights

Several authors — especially Carrière [78], Longstaff and Schwartz [241], and Tsitsiklis and Van Roy [350, 351] — have proposed the use of regression to estimate continuation values from simulated paths and thus to price American options by simulation. Each continuation value $C_i(x)$ in (8.11) is the regression of the option value $V_{i+1}(X_{i+1})$ on the current state x, and this suggests an estimation procedure: approximate C_i by a linear combination of known functions of the current state and use regression (typically least-squares) to estimate the best coefficients in this approximation. This approach is relatively fast and broadly applicable; its accuracy depends on the choice of functions used in the regression. The flexibility to choose these functions provides a mechanism for exploiting knowledge or intuition about the pricing problem.

Though not originally presented this way, regression-based methods fit well within the stochastic mesh framework of Section 8.5: they start from the independent-path construction illustrated in Figure 8.7, and we will see that the use of regression at each step corresponds to an implicit choice of weights for the mesh. (We encountered a related observation in (4.19)–(4.20) where we observed that using control variates for variance reduction is equivalent to a particular way of assigning weights to replications.) This section therefore uses notation and ideas from Section 8.5.

8.6.1 Approximate Continuation Values

Regression-based methods posit an expression for the continuation value of the form

$$E[V_{i+1}(X_{i+1})|X_i = x] = \sum_{r=1}^{M} \beta_{ir}\psi_r(x), \qquad (8.46)$$

for some basis functions $\psi_r : \Re^d \mapsto \Re$ and constants β_{ir}, $r = 1,\dots,M$. We discussed this as an approximation strategy in Section 8.2. Using the notation C_i for the continuation value at time i, we may equivalently write (8.46) as

$$C_i(x) = \beta_i^\top \psi(x), \qquad (8.47)$$

with

$$\beta_i^\top = (\beta_{i1},\dots,\beta_{iM}), \quad \psi(x) = (\psi_1(x),\dots,\psi_M(x))^\top.$$

One could use different basis function at different exercise dates, but to simplify notation we suppress any dependence of ψ on i.

Assuming a relation of the form (8.46) holds, the vector β_i is given by

$$\beta_i = (E[\psi(X_i)\psi(X_i)^\top])^{-1}E[\psi(X_i)V_{i+1}(X_{i+1})] \equiv B_\psi^{-1}B_{\psi V}. \qquad (8.48)$$

Here, B_ψ is the indicated $M \times M$ matrix (assumed nonsingular) and $B_{\psi V}$ the indicated vector of length M. The variables (X_i, X_{i+1}) inside the expectations have the joint distribution of the state of the underlying Markov chain at dates i and $i+1$.

The coefficients β_{ir} could be estimated from observations of pairs $(X_{ij}, V_{i+1}(X_{i+1,j}))$, $j = 1,\dots,b$, each consisting of the state at time i and the corresponding option value at time $i+1$. Consider, in particular, independent paths $(X_{1j},\dots,X_{mj})$, $j = 1,\dots,b$, and suppose for a moment that the values $V_{i+i}(X_{i+1,j})$ are known. The least-squares estimate of β_i is then given by

$$\hat{\beta}_i = \hat{B}_\psi^{-1}\hat{B}_{\psi V},$$

where $\hat{B}_\psi$ and $\hat{B}_{\psi V}$ are the sample counterparts of B_ψ and $B_{\psi V}$. More explicitly, $\hat{B}_\psi$ is the $M \times M$ matrix with qr entry

$$\frac{1}{b}\sum_{j=1}^{b}\psi_q(X_{ij})\psi_r(X_{ij})$$

and $\hat{B}_{\psi V}$ is the M-vector with rth entry

$$\frac{1}{b}\sum_{k=1}^{b}\psi_r(X_{ik})V_{i+1}(X_{i+1,k}). \qquad (8.49)$$

All of these quantities can be calculated from function values at pairs of consecutive nodes $(X_{ij}, X_{i+1,j})$, $j = 1,\dots,b$. In practice, V_{i+1} is unknown and must be replaced by estimated values $\hat{V}_{i+1}$ at downstream nodes. The estimate $\hat{\beta}_i$ then defines an estimate

$$\hat{C}_i(x) = \hat{\beta}_i^\top \psi(x), \qquad (8.50)$$

of the continuation value at an arbitrary point x in the state space $\Re^d$. This in turn defines a procedure for estimating the option value:

Regression-Based Pricing Algorithm

(i)   Simulate b independent paths $\{X_{1j}, \ldots, X_{mj}\}$, $j = 1, \ldots, b$, of the Markov chain.

(ii) At terminal nodes, set $\hat{V}_{mj} = h_m(X_{mj})$, $j = 1, \ldots, b$.

(iii) Apply backward induction: for $i = m - 1, \ldots, 1$,
- given estimated values $\hat{V}_{i+1,j}$, $j = 1, \ldots, b$, use regression as above to calculate $\hat{\beta}_i = \hat{B}_{\psi}^{-1}\hat{B}_{\psi V}$;
- set
$$\hat{V}_{ij} = \max\left\{h_i(X_{ij}), \hat{C}_i(X_{ij})\right\}, \quad j = 1, \ldots, b, \qquad (8.51)$$

   with $\hat{C}_i$ as in (8.50).

(iv) Set $\hat{V}_0 = (\hat{V}_{11} + \cdots + \hat{V}_{1b})/b$.

This is the approach introduced by Tsitsiklis and Van Roy [350, 351]. They show in [351] that if the representation (8.46) holds at all $i = 1, \ldots, m - 1$, then the estimate $\hat{V}_0$ converges to the true value $V_0(X_0)$ as $b \to \infty$. Longstaff and Schwartz [241] combine continuation values estimated using (8.50) with their interleaving estimator discussed in Section 8.5.1. In other words, they replace (8.51) with

$$\hat{V}_{ij} = \begin{cases} h_i(X_{ij}), & h_i(X_{ij}) \geq \hat{C}_i(X_{ij}); \\ \hat{V}_{i+1,j} & h_i(X_{ij}) < \hat{C}_i(X_{ij}). \end{cases} \qquad (8.52)$$

They recommend omitting nodes X_{ij} with $h_i(X_{ij}) = 0$ in estimating β_i. Clément, Lamberton, and Protter [86] prove convergence of the Longstaff-Schwartz procedure as $b \to \infty$. The limit attained coincides with the true price $V_0(X_0)$ if the representation (8.46) holds exactly; otherwise, the limit coincides with the value under a suboptimal exercise policy and thus underestimates the true price. In practice, (8.52) therefore produces low-biased estimates.

The success of any regression-based approach clearly depends on the choice of basis functions. Polynomials (sometimes damped by functions vanishing at infinity) are a popular choice ([350, 241]). Through Taylor expansion, any sufficiently smooth value function can be approximated by polynomials. However, the number of monomials of a given degree grows exponentially in the number of variables, so without further restrictions on the value function the number of basis functions required could grow quickly with the dimension of the underlying state vector X_i. Longstaff and Schwartz [241] use 5–20 basis functions in the examples they test.

We have presented the regression-based pricing algorithm above in the discounted formulation (8.6)–(8.7) of the dynamic programming problem, as we have all methods in this chapter. This assumes that payoffs and value estimates are denominated in time-0 dollars. In practice, payoffs and value estimates at time t_i are usually denominated in time-t_i dollars, and this requires including explicit discounting in the algorithm. Consider, for example, the case of a constant continuously compounded interest rate r. In step (ii) of the algorithm, we would use $\tilde{h}_m$, as in (8.4); in step (iii) we would regress $e^{-r(t_{i+1}-t_i)}\hat{V}_{i+1,j}$ (rather than $\hat{V}_{i+1,j}$) against $\psi(X_i)$; in (8.49) we would replace $V_{i+1}(X_{i+1,k})$ with $e^{-r(t_{i+1}-t_i)}V_{i+1}(X_{i+1,k})$. With these modifications, $\hat{C}_i(x)$ in (8.50) would be interpreted as the present value of continuing (denominated in time-t_i dollars).

Low Estimator

The estimated vectors of coefficients $\hat{\beta}_i$ determine approximate continuation values $\hat{C}_i(x)$ for every step i and state x, through (8.50). These in turn define an exercise policy, just as in (8.33), and thus a low estimator as in (8.34). This is the low estimator of the stochastic mesh applied to continuation values estimated through regression. The method of Longstaff and Schwartz [241] in (8.52) usually produces a low-biased estimator as well, though as explained in Section 8.5.1, their interleaving estimator mixes elements of high and low bias.

Example 8.6.1 *American max option.* To illustrate regression-based pricing, we consider an American option on the maximum of underlying assets modeled by geometric Brownian motion. This example is used in Broadie and Glasserman [65, 66] and Andersen and Broadie [15] with up to five underlying assets; for simplicity, here we consider just two underlying assets, S_1 and S_2. Each is modeled as GBM$(r-\delta,\sigma^2)$ with interest rate $r=5\%$, dividend yield $\delta=10\%$, and volatility $\sigma=0.20$. The two assets are independent of each other. The (undiscounted) payoff upon exercise at time t is

$$\tilde{h}(S_1(t),S_2(t)) = (\max(S_1(t),S_2(t)) - K)^+.$$

We take $S_1(0)=S_2(0)$ and $K=100$. The option expires in $T=3$ years and can be exercised at nine equally spaced dates $t_i=i/3$, $i=1,\ldots,9$. Valuing the option in a two-dimensional binomial lattice, as in Boyle, Evnine, and Gibbs [54], yields option prices of 13.90, 8.08, and 21.34 for $S_i(0)=100$, 110, and 90, respectively. These are useful for comparison.

To price the option using regression and simulation, we need to choose basis functions. We compare various combinations of powers of the underlying assets, the immediate exercise value $\tilde{h}$, and related functions. For each case we apply the regression-based pricing algorithm (i)–(iv) with 4000 paths. The continuation values estimated by regression define an exercise policy so we then simulate a second set of 4000 paths that follow this exercise policy: this is the low estimator in (8.33)–(8.34) based on the regression estimate (8.47) of the continuation value. We also apply the method of Longstaff and Schwartz [241] in (8.52). We replicate all three of these estimators 100 times to calculate standard errors.

Tables 8.1 and 8.2 display numerical results for various sets of basis functions and initial values of the underlying assets. The basis functions appear in Table 8.1. We treat S_1 and S_2 symmetrically in each case; the abbreviation S_i^2, for example, indicates that both S_1^2 and S_2^2 are included. We also include a constant in each regression. The results in Table 8.1 are for $S_i(0) = 100$; Table 8.2 is based on the same basis functions but with $S_i(0) = 90$ (left half) and $S_i(0) = 110$ (right half). The estimates have standard errors of approximately 0.02 to 0.03.

The results in the tables show that the pure regression-based estimator can have significant high bias even with nine reasonable basis functions in a two-dimensional problem (as in the third row of each table). The choice of basis functions clearly affects the price estimate. In this example, including the interaction term $S_1 S_2$ and the exercise value $\tilde{h}(S_1, S_2)$ appears to be particularly important. (Andersen and Broadie [15] get nearly exact estimates for this example using twelve basis functions, including the value of a European max option.)

Basis Functions	Regression	Low	LSM
$1, S_i, S_i^2, S_i^3$	15.74	13.62	13.67
$1, S_i, S_i^2, S_i^3, S_1 S_2$	15.24	13.65	13.68
$1, S_i, S_i^2, S_i^3, S_1 S_2, \max(S_1, S_2)$	15.23	13.64	13.63
$1, S_i, S_i^2, S_i^3, S_1 S_2, S_1^2 S_2, S_1 S_2^2$	15.07	13.71	13.67
$1, S_i, S_i^2, S_i^3, S_1 S_2, S_1^2 S_2, S_1 S_2^2, \tilde{h}(S_1, S_2)$	14.06	13.77	13.79
$1, S_i, S_i^2, S_1 S_2, \tilde{h}(S_1, S_2)$	14.08	13.78	13.78

Table 8.1. Price estimates for an American option on the maximum of two assets. The true price is 13.90. Each estimate has a standard error of approximately 0.025.

Regression	Low	LSM	Regression	Low	LSM
9.49	7.93	7.92	24.52	20.79	21.14
9.39	7.97	7.87	23.18	21.02	21.15
9.44	7.98	7.87	22.76	20.98	21.02
9.25	7.95	7.87	22.49	21.08	21.15
8.24	8.01	7.95	21.42	21.25	21.20
8.27	7.99	7.99	21.38	21.26	21.16

Table 8.2. Price estimates for out-of-the-money (left) and in-the-money (right) American option on the maximum of two assets. True prices are 8.08 and 21.34. Each estimate has a standard error of approximately 0.02–0.03.

The low estimates appear to be less sensitive to the choice of basis functions. As the high bias in the regression estimate decreases, the low estimate generally increases; both properties result from a better fit to the continuation value. The ordinary low estimator (labeled "Low") and the Longstaff-Schwartz estimator (labeled "LSM") give nearly identical results. Longstaff and Schwartz [241] recommend including only in-the-money nodes X_{ij} in the regression used to estimate the continuation value C_i; this alternative gives inferior results in this example and is therefore omitted from the tables.

Though it is risky to extrapolate from limited numerical tests, this example suggests that using either of the low-biased estimators is preferable to relying on the pure regression-based estimator, and that neither of the low-biased estimators consistently outperforms the other. As expected, the choice of basis functions has a significant impact on the estimated prices.

Figure 8.9 displays exercise regions at $i = 4$, the fourth of the nine exercise opportunities. The dashed lines show the optimal exercise boundary computed from a binomial lattice: it is optimal to exercise the max option if the price of one — but not both — of the underlying assets is sufficiently high. (This and other properties of the exercise region are proved in Broadie and Detemple [63].) The shaded area shows the exercise region estimated through the last regression in Table 8.1. More specifically, the shaded area corresponds to points at which $\tilde{h}(S_1, S_2)$ is greater than or equal to the estimated continuation value. The regression estimate generally comes close to the optimal boundary but erroneously indicates that it is optimal to exercise in the lower-left corner where the regression estimate of the continuation value is negative. This can be corrected by replacing $\hat{C}_i(x)$ with $\max\{0, \hat{C}_i(x)\}$ and exercising only if $\tilde{h}(S_1, S_2)$ is strictly greater than the continuation value. The results in Table 8.1 use this correction.

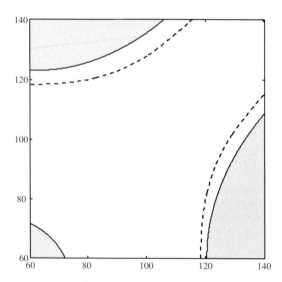

Fig. 8.9. Exercise region for American max option on two underlying assets. The shaded area is the exercise region determined by regression on seven basis functions; dashed lines shows optimal boundary.

8.6.2 Regression and Mesh Weights

We indicated at the beginning of this section that the regression-based algorithm (i)–(iv) corresponds to a stochastic mesh estimator with an implicit choice of mesh weights. We now make this explicit.

Using the regression representation (8.50) and then (8.49), we can write the estimated continuation value at node X_{ij} as

$$
\begin{aligned}
\hat{C}_i(X_{ij}) &= \psi(X_{ij})^\top \hat{\beta}_i \\
&= \psi(X_{ij})^\top \hat{B}_\psi^{-1} \hat{B}_{\psi V} \\
&= \frac{1}{b} \sum_{k=1}^{b} \left(\psi(X_{ij})^\top \hat{B}_\psi^{-1} \psi(X_{ik}) \right) \hat{V}_{i+1,k}.
\end{aligned}
\tag{8.53}
$$

Thus, the estimated continuation value at node X_{ij} is a weighted average of the estimated option values at step $i+1$, with weights

$$
W_{jk}^i = \psi(X_{ij})^\top \hat{B}_\psi^{-1} \psi(X_{ik}).
\tag{8.54}
$$

In other words, (8.53) is a special case of the general mesh approximation

$$
\hat{C}_i(X_{ij}) = \frac{1}{b} \sum_{k=1}^{b} W_{jk}^i \hat{V}_{i+1,k}.
\tag{8.55}
$$

This extends to arbitrary points x in the state space, as in (8.32), if we simply replace X_{ij} with x in (8.54) and (8.55).

We made similar observations in our discussion of the link between control variates and weighted Monte Carlo in (4.19)–(4.20) and Example 4.5.6, and we can put the weights in (8.54) in the form appearing in (4.20). In the representation (8.46) of the value function, one would often take one of the basis functions to be a constant. To make this explicit, let $\psi_0 \equiv 1$ and add a term $\beta_{i0}\psi_0(x)$ in (8.46). Let S_ψ denote the sample covariance matrix of the other basis functions: this is the $M \times M$ matrix with qr entry

$$
\frac{1}{b-1} \sum_{j=1}^{b} \left(\psi_q(X_{ij})\psi_r(X_{ij}) - b\bar{\psi}_q\bar{\psi}_r \right),
$$

where each $\bar{\psi}_\ell$ is the sample mean of ψ_ℓ values at time i,

$$
\bar{\psi}_\ell = \frac{1}{b} \sum_{j=1}^{b} \psi_\ell(X_{ij}).
$$

If S_ψ is nonsingular, then some matrix algebra shows that the regression weights (8.54) can also be written as

$$W^i_{jk} = 1 + \frac{b}{b-1}(\psi(X_{ij}) - \bar{\psi})^\top S_\psi^{-1}(\psi(X_{ik}) - \bar{\psi}), \qquad (8.56)$$

where $\bar{\psi}$ is the M-vector of sample means $\bar{\psi}_r$, $r = 1, \ldots, M$. (Had we not removed the constant ψ_0, the sample covariance matrix would have no chance of being invertible.) This expression has the same form as (4.20). Like (8.54), it extends to an arbitrary point x in the state space if we replace X_{ij} with x; this substitution defines the extended weights $W^i_k(x)$ needed for the low estimator and interleaving estimator discussed in Section 8.5.1.

We make the following observations regarding this formulation:

○ The regression-based weights (8.54) and (8.56) are symmetric in j and k. They sum to b if we either fix j and sum over k or fix k and sum over j. This is relevant to the discussion following (8.42).

○ The regression-based weights satisfy conditions (M1) and (M2) of Section 8.5.1. They satisfy condition (M3) if the regression equation (8.46) is valid.

○ From (8.56) we see that $\mathsf{E}[W^i_{jk}|X_{ij}]$ is nearly 1 for any $k \neq j$. In this respect the regression weights resemble likelihood ratios. But in contrast to likelihood ratios the regression weights can take negative values.

○ All of the likelihood ratio weights considered in Section 8.5.2 share the property that the weight W^i_{jk} depends on both the origin and destination nodes X_{ij} and $X_{i+1,k}$. In contrast, the weight assigned through regression depends on X_{ij} and X_{ik}, but not on any values at step $i+1$. This results in weights that are less variable but also less tuned to the observed outcomes. It also means, curiously, that the weights implicit in regression are insensitive to the spacing $t_{i+1} - t_i$ between exercise dates. Using least-squares and (8.44) produces weights W^i_{jk} that depend on X_{ij} and $X_{i+1,k}$.

○ The regression weight W^i_{jk} depends on a type of distance between X_{ij} and X_{ik}, or rather between $\psi(X_{ij})$ and $\psi(X_{ik})$. Consider, for example, what happens when $\hat{B}_\psi$ is the identity matrix. If $\psi(X_{ij})$ and $\psi(X_{ik})$ are orthogonal vectors, the weight W^i_{jk} is zero; the absolute value of the weight is greatest when these vectors are multiples of each other. Points $\psi(X_{ik})$ that are equidistant from $\psi(X_{ij})$ in the Euclidean sense can get very different weights W^i_{jk}, as illustrated (for $M = 2$ dimensions) in Figure 8.10.

○ As noted in the discussion surrounding (8.37), only likelihood ratio weights can correctly price all payoffs at time $i + 1$ from the perspective of node X_{ij}. The regression weights correctly price payoffs whose price at time i as a function of the current state is a linear combination of the basis functions ψ_r. In practice, the true continuation value C_i is unlikely to be exactly a linear combination of the basis functions, so the regression procedure produces (an estimate of) the projection of C_i onto the span of $\psi_1, \ldots, \psi_M$.

○ Calculating all estimated continuation values $\hat{C}_i(X_{ij})$, $j = 1, \ldots, b$, using the regression representation (8.50) requires a single $O(M^3)$ calculation to compute $\hat{\beta}_i$ and then $O(Mb)$ calculations to compute $\hat{\beta}_i^\top \psi(X_{ij})$ at all b nodes. In contrast, making explicit use of the weights as in (8.55) requires

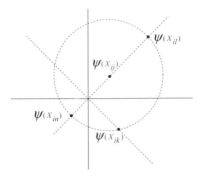

Fig. 8.10. The three points on the circle are the same distance from the center but are assigned different weights by node j: if $\hat{B}_\psi = I$, the lth node gets positive weight, the nth gets negative weight, and the kth gets zero weight.

$O(M^3)$ operations to calculate the weights and then $O(b^2)$ operations to estimate all b continuation values. Because one would typically set b much larger than M, this comparison favors taking advantage of the special structure of the weights, as in (8.50). If one prices multiple options using a single mesh, the coefficients $\hat{\beta}_i$ need to be recomputed for each option whereas the weights (8.56) do not. The computational effort required for (8.55) can be reduced to $O(M^3 + M^2 b)$ by rewriting (8.54) as

$$W_{jk}^i = \varepsilon_j^\top \varepsilon_k$$

with

$$\varepsilon_k = A^\top \psi(X_{ik}), \quad AA^\top = \hat{B}_\psi^{-1}.$$

Computing A is an $O(M^3)$ operation. Computing ε_k given A is $O(M^2)$ and repeating this for $k = 1, \ldots, b$ is $O(M^2 b)$. We can rearrange the calculation of the sum in (8.55) as

$$\sum_{k=1}^b W_{jk}^i \hat{V}_{i+1,k} = \sum_{k=1}^b \varepsilon_j^\top \varepsilon_k \hat{V}_{i+1,k} = \varepsilon_j^\top \left(\sum_{k=1}^b \varepsilon_k \hat{V}_{i+1,k} \right).$$

Calculating the term in parentheses is $O(Mb)$. This term is independent of j, so calculating its inner product with all ε_j, $j = 1, \ldots, b$, is also $O(Mb)$. This is advantageous if several options are to be priced using the same mesh and basis functions because the matrix inversion is executed just once, rather than separately for each option.

Example 8.6.2 *Weights for a Brownian mesh.* To illustrate the idea of regression-based weights, we consider a simple example in which the underlying Markov chain X_i is the state of a standard one-dimensional Brownian motion at dates $t_i = i$. We consider the estimation of a continuation value (or other conditional expectation) at $t_1 = 1$, at which time the Brownian motion has a standard normal distribution. For basis functions we use the constant 1 and powers x^n for $n = 1, \ldots, M$. With $M = 4$ and Z denoting a standard normal random variable, the matrix B_ψ becomes

$$B_\psi = \mathsf{E} \begin{pmatrix} 1 & Z & Z^2 & Z^3 & Z^4 \\ Z & Z^2 & Z^3 & Z^4 & Z^5 \\ Z^2 & Z^3 & Z^4 & Z^5 & Z^6 \\ Z^3 & Z^4 & Z^5 & Z^6 & Z^7 \\ Z^4 & Z^5 & Z^6 & Z^7 & Z^8 \end{pmatrix} = \begin{pmatrix} 1 & 0 & 1 & 0 & 3 \\ 0 & 1 & 0 & 3 & 0 \\ 1 & 0 & 3 & 0 & 15 \\ 0 & 3 & 0 & 15 & 0 \\ 3 & 0 & 15 & 0 & 105 \end{pmatrix}.$$

Its inverse is

$$B_\psi^{-1} = \begin{pmatrix} 15/8 & 0 & -5/4 & 0 & 1/8 \\ 0 & 5/2 & 0 & -1/2 & 0 \\ -5/4 & 0 & 2 & 0 & -1/4 \\ 0 & -1/2 & 0 & 1/6 & 0 \\ 1/8 & 0 & -1/4 & 0 & 1/24 \end{pmatrix}.$$

Consider two nodes $X_{1j} = x_j$ and $X_{1k} = x_k$; these correspond to two possible values of the Brownian motion at time 1. The corresponding weight $W_{jk} \equiv W_{jk}^1$ is given by

$$\begin{aligned}
W_{jk} &= \psi(x_j) B_\psi^{-1} \psi(x_k) \\
&= \frac{15}{8} - \frac{5}{4}(x_j^2 + x_k^2) + \frac{5}{2} x_j x_k + \frac{1}{8}(x_j^4 + x_k^4) - \frac{1}{2}(x_j x_k^3 + x_j^3 x_k) + 2x_j^2 x_k^2 - \\
&\quad \frac{1}{4}(x_j^2 x_k^4 + x_j^4 x_k^2) + \frac{1}{6} x_j^3 x_k^3 + \frac{1}{24} x_j^4 x_k^4.
\end{aligned}$$

More generally, regressing on powers up to order M makes each W_{jk} an M-th order polynomial in x_k for each value of x_j.

The solid line in Figure 8.11 plots W_{jk} as a function of x_k for $x_j = 1$. Taking $x_j = 1$ means we are computing a conditional expectation (or price) given that the Brownian motion is at 1 at time t_1. Regression computes this conditional expectation as a weighted average of downstream values that are successors of other nodes x_k at time t_1. The solid line in the figure shows the weight assigned to the successor of x_k as the value of x_k varies along the horizontal axis. The weights oscillate and then drop to $-\infty$ as x_k moves away from $x_j \equiv 1$; these properties result from the polynomial form of the weights.

The dashed line in Figure 8.11 shows the conditional expectation of likelihood ratio weights based on (8.36). A direct comparison between likelihood ratio weights and regression weights is not possible because (8.36) depends on X_{ij} and $X_{i+1,k}$, whereas the corresponding regression weight depends on X_{ij} and X_{ik}. By taking the conditional expectation of (8.36) given X_{ik} and X_{ij}, we make it a function of these nodes. In more detail, we consider the step from $t_i = 1$ to $t_{i+1} = 2$. The transition density is normal, $f(x,y) = \phi(y - x)$ with ϕ the standard normal density. The marginal distribution at $t_2 = 2$ is $N(0, 2)$. The likelihood ratio (8.36) is then

$$W_{jk} = \sqrt{2}\exp\left(\tfrac{1}{2}X_{ij}^2 - \tfrac{1}{4}(X_{i+1,k} - 2X_{ij})^2\right). \tag{8.57}$$

Using the fact that $X_{i+1,k} - X_{ik}$ has a standard normal distribution, we can calculate the conditional expectation of W_{jk} given X_{ij} and X_{ik} to get

$$\mathsf{E}[W_{jk}|X_{ij} = x_j, X_{ik} = x_k] = \frac{2}{\sqrt{3}}\exp\left(-(x_k^2 - 4x_jx_k + x_j^2)/6\right).$$

Viewed as a function of x_k with x_j held fixed, this is a scaled normal density centered at $2x_j$. It is plotted in Figure 8.11 with $x_j = 1$. The unconditional weights in (8.57) also traverse a scaled normal density, centered at $2X_{ij}$, if we fix X_{ij} and let $X_{i+1,k}$ vary. The dashed line in the figure is more consistent with intuition for how downstream nodes should be weighted than is the solid line. □

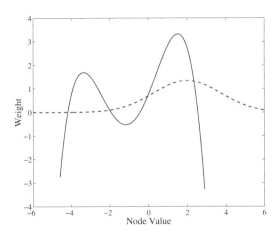

Fig. 8.11. Comparison of regression-based weight (solid) and conditional expectation of likelihood ratio weight (dashed). The curves show the weight assigned by node $x_j = 1$ to the successor of node x_k as a function of x_k.

Example 8.6.3 *Option on a single underlying asset.* Consider the pricing of an American call option on an asset modeled by geometric Brownian motion. The option expires in three years and can be exercised at any of 10 equally spaced exercise opportunities: $m = 10$ and $t_i = 0.3i$ for $i = 1,\ldots, m$. The payoff upon exercise at t_i is $(S(t_i) - K)^+$, with $K = 100$ and the underlying asset S described by GBM$(r - \delta, \sigma^2)$, with $S(0) = 100$, volatility $\sigma = 0.20$, interest rate $r = 5\%$, and dividend yield $\delta = 10\%$. (In the absence of dividends, early exercise would be suboptimal and the option would reduce to a European call.) This option is simple enough to be priced efficiently in a binomial lattice, which yields a price of 7.98.

Figure 8.12 shows the value of the option at t_1 as a function of the level $S(t_1)$ of the underlying asset. The solid line shows the value function computed using a 2000-step binomial lattice. The other two lines show estimates computed from an 800-path mesh generated using the independent-path construction. The dashed line shows the value estimated using regression with basis functions $1, x, x^2, x^3$, and x^4 and the dotted line is based on the likelihood ratio weights (8.41) with transition densities evaluated as in each factor of (8.43). In both cases, the value displayed is for time t_1 and therefore results from nine applications of the backward induction starting from the terminal time t_m. Both estimates come quite close to the exact value (as computed in the binomial lattice), though the estimates based on likelihood ratio weights are somewhat closer. As the number of paths increases, we should expect the likelihood ratio values to approach the exact values while errors in the regression values persist, because the exact value is not given by a fifth-order polynomial. The greater accuracy of the likelihood ratio weights suggested by the figure must be balanced against their much greater computational burden; see the discussion of computational considerations near the end of Section 8.5.2. □

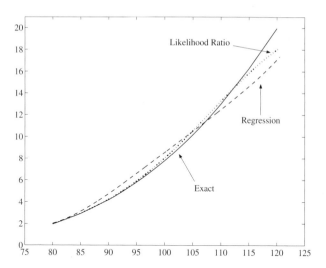

Fig. 8.12. Comparison of exact and estimated value functions for a call option on a single underlying asset as functions of the price of the underlying asset.

8.7 Duality

Throughout this chapter, we have formulated the American option pricing problem as one of maximizing over stopping times. Haugh and Kogan [172] and Rogers [308] have established dual formulations in which the price is represented through a minimization problem. The dual minimizes over a class of supermartingales or martingales. These duality results lead to useful approximations and upper bounds on prices, in addition to having theoretical interest.

We continue to work with discounted variables, as in (8.6)–(8.7), meaning that all payoffs and values are denominated in time-0 dollars. An immediate consequence of the dynamic programming recursion (8.7) is that

$$V_i(X_i) \geq \mathsf{E}[V_{i+1}(X_{i+1})|X_i],$$

for all $i = 0, 1, \ldots, m-1$, and this is the defining property of a supermartingale. Also, $V_i(X_i) \geq h_i(X_i)$ for all i. The value process $V_i(X_i)$, $i = 0, 1, \ldots, m$, is in fact the minimal supermartingale dominating $h_i(X_i)$. Haugh and Kogan [172] extend this characterization to formulate the pricing of American options as a minimization problem.

Rogers [308] proves a continuous-time duality result which we specialize to the case of m exercise dates. Let $M = \{M_i, i = 0, \ldots, m\}$ be a martingale with $M_0 = 0$. By the optional sampling property of martingales, for any stopping time τ taking values in $\{1, \ldots, m\}$ we have

$$\mathsf{E}[h_\tau(X_\tau)] = \mathsf{E}[h_\tau(X_\tau) - M_\tau] \leq \mathsf{E}[\max_{k=1,\ldots,m} \{h_k(X_k) - M_k\}],$$

and thus

$$\mathsf{E}[h_\tau(X_\tau)] \leq \inf_M \mathsf{E}[\max_{k=1,\ldots,m} \{h_k(X_k) - M_k\}],$$

the infimum taken over martingales with initial value 0. Because this inequality holds for every τ, it also holds for the supremum over τ, so

$$V_0(X_0) = \sup_\tau \mathsf{E}[h_\tau(X_\tau)] \leq \inf_M \mathsf{E}[\max_{k=1,\ldots,m} \{h_k(X_k) - M_k\}]. \tag{8.58}$$

The minimization problem on the right is the dual problem.

What makes (8.58) particularly interesting is that it holds with equality. We show this by constructing a martingale for which the expectation on the right equals $V_0(X_0)$. To this end, define

$$\Delta_i = V_i(X_i) - \mathsf{E}[V_i(X_i)|X_{i-1}], \quad i = 1, \ldots, m, \tag{8.59}$$

and set

$$M_i = \Delta_1 + \cdots + \Delta_i, \quad i = 1, \ldots, m, \tag{8.60}$$

with $M_0 = 0$. That this process is indeed a martingale follows from the property

$$\mathsf{E}[\Delta_i|X_{i-1}] = 0 \tag{8.61}$$

of the differences Δ_i.

We now use induction to show that

$$V_i(X_i) = \max\{h_i(X_i), h_{i+1}(X_{i+1}) - \Delta_{i+1}, \ldots,$$
$$h_m(X_m) - \Delta_m - \cdots - \Delta_{i+1}\}, \tag{8.62}$$

for all $i = 1, \ldots, m$. This holds at $i = m$ because $V_m(X_m) = h_m(X_m)$. Assuming it holds at i, then using

$$V_{i-1}(X_{i-1}) = \max\{h_{i-1}(X_{i-1}), \mathsf{E}[V_i(X_i)|X_{i-1}]\}$$
$$= \max\{h_{i-1}(X_{i-1}), V_i(X_i) - \Delta_i\}$$

we see that it extends to $i - 1$.

The option price at time 0 is

$$V_0(X_0) = \mathsf{E}[V_1(X_1)|X_0] = V_1(X_1) - \Delta_1.$$

By rewriting $V_1(X_1)$ using (8.62), we find that

$$V_0(X_0) = \max_{k=1,\ldots,m} (h_k(X_k) - M_k), \tag{8.63}$$

thus verifying that equality is indeed achieved in (8.58) and that optimality is attained by the martingale defined in (8.59)–(8.60). This optimal martingale is a special case of one specified more generally by the Doob-Meyer decomposition of a supermartingale; see Rogers [308].

The martingale differences (8.59) can alternatively be written as

$$\Delta_i = V_i(X_i) - C_{i-1}(X_{i-1}) \tag{8.64}$$

with C_{i-1} denoting the continuation value, as in (8.12). Because $C_0(X_0) = V_0(X_0)$, finding the optimal martingale appears to be as difficult as solving the original optimal stopping problem. But if we can find a martingale $\hat{M}$ that is close to the optimal martingale, then we can use

$$\max_{k=1,\ldots,m} \left(h_k(X_k) - \hat{M}_k \right) \tag{8.65}$$

to estimate (an upper bound for) the option price.

Where can we find nearly optimal martingales? A general strategy is to construct a martingale from an approximate value function or stopping rule. This idea is developed and shown to be effective in Andersen and Broadie [15] and Haugh and Kogan [172]. A suboptimal exercise policy provides a lower bound on the price of an American option; the dual value defined by extracting a martingale from the suboptimal policy complements the lower bound with an upper bound.

There are various characterizations of the martingale associated with the optimal value function and these suggest alternative strategies for constructing approximating martingales. We discuss two such approaches based on (8.64).

Martingales from Approximate Value Functions

Let $\hat{C}_i$ denote an approximation to the continuation value C_i, $i = 0, 1, \ldots, m - 1$. For example, $\hat{C}_i$ could result from a parametric approximation, as in Section 8.2, or through regression, as in (8.47). The associated approximate value function is $\hat{V}_i(x) = \max\{h_i(x), \hat{C}_i(x)\}$.

Along a simulated path $X_0, X_1, \ldots, X_m$, one could evaluate the differences

$$\hat{\Delta}_i = \hat{V}_i(X_i) - \hat{C}_{i-1}(X_{i-1})$$

approximating (8.64). However, these are not in general martingale differences because they may fail to satisfy (8.61); in particular, there is no guarantee that the approximate value function satisfies

$$\hat{C}_{i-1}(X_{i-1}) = \mathsf{E}[\hat{V}_i(X_i)|X_{i-1}].$$

If the $\hat{\Delta}_i$ do not satisfy (8.61), then using their sum $\hat{M}_k$ in (8.65) will not produce a valid upper bound.

To extract martingale differences from an approximate value function, we need to work a bit harder and compute (an estimate of)

$$\hat{\Delta}_i = \hat{V}_i(X_i) - \mathsf{E}[\hat{V}_i(X_i)|X_{i-1}]. \tag{8.66}$$

The first term on the right merely requires evaluating the approximation $\hat{V}_i$ along a simulated path of the Markov chain. For the second term, we can use a nested simulation. At each step X_{i-1} of the Markov chain, we generate n successors $X_i^{(1)}, \ldots, X_i^{(n)}$ and use

$$\frac{1}{n}\sum_{j=1}^{n}\hat{V}_i(X_i^{(j)}) \tag{8.67}$$

to estimate the conditional expectation of $V_i(X_i)$ given X_{i-1}. We discard these n successors and generate a new one to get the next step X_i of the path of the Markov chain. Thus, each nested simulation evolves over just a single step of the Markov chain. (As discussed in Glasserman and Yu [148], nested simulations become unnecessary if one selects basis functions for which the required conditional expectations can be evaluated in closed form.)

The simulated values

$$\hat{\Delta}_i = \hat{V}_i(X_i) - \frac{1}{n}\sum_{j=1}^{n}\hat{V}_i(X_i^{(j)})$$

are martingale differences, even though the second term on the right is only an estimate of the conditional expectation. This term is conditionally unbiased (given X_{i-1}), so the conditional expectation of $\hat{\Delta}_i$ given X_{i-1} is zero, as required by (8.61). It follows that using these $\hat{\Delta}_i$ in (8.65) is guaranteed to produce a high-biased estimator, even for finite n.

Haugh and Kogan [172] propose and test other methods for constructing upper bounds from approximate value functions. Their formulation of the dual problem takes an infimum over supermartingales rather than martingales, and they therefore use $\hat{V}_i$ to construct supermartingales.

Martingales from Stopping Rules

Consider an exercise policy defined by stopping times $\tau_1, \ldots, \tau_m$ with τ_i interpreted as the exercise time for an option issued at the ith exercise date — in particular, $\tau_i \geq i$. Suppose these stopping times are defined by approximate continuation values $\hat{C}_i$, $i = 1, \ldots, m$, with $\hat{C}_m \equiv 0$, through

$$\tau_i = \min\{k = i, \ldots, m : h_k(X_k) \geq \hat{C}_k(X_k)\}, \qquad (8.68)$$

for $i = 1, \ldots, m$, much as in (8.13). For example, $\hat{C}_i$ might be specified through the regression representation (8.50).

If these stopping times defined an optimal exercise policy, then the differences

$$\hat{\Delta}_i = \mathsf{E}[h_{\tau_i}(X_{\tau_i})|X_i] - \mathsf{E}[h_{\tau_i}(X_{\tau_i})|X_{i-1}] \qquad (8.69)$$

would define an optimal martingale: this difference is just another way of writing (8.64). Even under a suboptimal policy, these $\hat{\Delta}_i$ are martingale differences and thus lead to a valid upper bound. This therefore provides a general mechanism for defining martingales from stopping rules.

Evaluating these martingale differences requires computing (or estimating) conditional expectations and for these we again use nested simulations. The expression in (8.69) involves two conditional expectations which may appear to require separate treatment. Rewriting the first term on the right side of (8.69) as

$$\mathsf{E}[h_{\tau_i}(X_{\tau_i})|X_i] = \begin{cases} h_i(X_i), & \text{if } h_i(X_i) \geq \hat{C}_i(X_i); \\ \mathsf{E}[h_{\tau_{i+1}}(X_{\tau_{i+1}})|X_i], & \text{if } h_i(X_i) < \hat{C}_i(X_i), \end{cases} \qquad (8.70)$$

reveals that the only conditional expectations we need to estimate are

$$\mathsf{E}[h_{\tau_{k+1}}(X_{\tau_{k+1}})|X_k], \quad k = 0, 1, \ldots, m - 1.$$

These can then be used in (8.70) and (8.69) to compute the $\hat{\Delta}_i$.

The method proceeds by replicating the following steps:

○ Simulate a path $X_0, X_1, \ldots, X_m$ of the underlying Markov chain.
○ At each X_i, $i = 0, 1, \ldots, m - 1$,
  – evaluate $h_i(X_i)$ and $\hat{C}_i(X_i)$ and check which is larger, as in (8.70), taking $h_0 \equiv 0$;
  – simulate n subpaths starting from X_i and following the exercise policy τ_{i+1}; record the payoff $h_{\tau_{i+1}}(X_{\tau_{i+1}})$ from each subpath and use the average as an estimate of $\mathsf{E}[h_{\tau_{i+1}}(X_{\tau_{i+1}})|X_i]$.

○ Combine the estimates of the conditional expectations as in (8.69) and (8.70) to estimate the differences $\hat{\Delta}_j$.
○ Sum these differences to get $\hat{M}_k = \hat{\Delta}_1 + \cdots + \hat{\Delta}_k$, $k = 1, \ldots, m$.
○ Evaluate the maximum of $h_k(X_k) - \hat{M}_k$ over $k = 1, \ldots, m$ as in (8.65).

Observe that whereas each subpath used in (8.67) evolves for exactly one step of the Markov chain, here the subpaths evolve for a random number of steps determined by the stopping rule.

This is the method of Andersen and Broadie [15], but formulated in terms of the differences (8.69). They apply their method in combination with stopping rules defined as in Longstaff and Schwartz [241] and, for Bermudan swaptions, Andersen [12]. Their numerical results indicate excellent performance with reasonable computational effort. As they point out, in addition to providing a confidence interval, the spread between the estimated lower and upper bounds gives an indication of whether additional computational effort should be dedicated to improving the exercise policy — for example, by increasing the number of basis functions.

Numerical Example

Table 8.3 displays numerical results obtained by applying these dual estimates to the American max option of Example 8.6.1. The labels "Dual-$\hat{V}$" and "Dual-τ" refer, respectively, to dual estimates based on approximate value functions (as in (8.66)) and dual estimates based on stopping rules (as in (8.69)). Each of these methods requires simulating subpaths; the table shows results for $n = 10$ and $n = 100$ subpaths per replication in the last two pairs of columns. The regression and low estimates in the first pair of columns are repeated from Tables 8.1 and 8.2 for comparison. The dual results in Table 8.3 are based on 4000 initial paths used to estimate regression coefficients followed by 100 independent paths along which the duals are computed using $n = 10$ or $n = 100$ subpaths at each step; the entire procedure is then replicated 100 times to allow estimation of standard errors. We consider initial values $S(0) = 100, 110$, and 90, for which the correct prices are 13.90, 21.34, and 8.08. The first three rows in Table 8.3 use the first set of basis functions displayed in Table 8.1 (a poor set), and the second three rows use the last set of basis functions from Table 8.1 (a much better set). Standard errors for all estimates in the table range between 0.02 and 0.03.

The dual estimates are markedly better (giving tighter upper bounds) with $n = 100$ subpaths than $n = 10$. Recall that the subpaths are used to estimate the conditional expectations defining the $\hat{\Delta}_i$ so increasing n gives better estimates of these conditional expectations; the results in the table show the importance of estimating these accurately. The results also indicate that the two dual estimates give similar values in this example. When both methods use the same number of paths and subpaths, and when these sample sizes are large, we expect that Dual-τ will usually give better estimates than Dual-$\hat{V}$

because it is based directly on an implementable exercise rule rather than an approximate value function. However, Dual-$\hat{V}$ has the advantage of using only single-step subpaths and thus requires less computing time per path. A definitive comparison of the two methods would evaluate each under an optimal implementation, and this would require finding the optimal combination of sample sizes for the initial set of paths used for regression, the second-phase paths, and the subpaths.

| | | | $n = 10$ | | $n = 100$ | |
$S(0)$	Regression	Low	Dual-$\hat{V}$	Dual-τ	Dual-$\hat{V}$	Dual-τ
100	15.74	13.62	15.86	15.96	14.58	14.26
110	24.52	20.79	24.09	24.56	22.38	21.94
90	9.49	7.93	9.43	9.21	8.59	8.24
100	14.08	13.78	15.46	15.82	14.16	14.16
110	21.38	21.26	23.55	24.21	21.68	21.72
90	8.27	7.99	9.07	9.16	8.25	8.20

Table 8.3. Estimates for the American max option of Example 8.6.1. Correct values for $S(0) = 100$, 110, and 90 are 13.90, 21.34, and 8.08. Standard errors for all estimates in the table are approximately 0.02–0.03.

The dual estimates with $n = 100$ are noticeably better (lower) than the regression estimates in the first three rows, but roughly the same as the regression estimates in the last three rows. The dual estimates are thus most valuable when the chosen basis functions do not provide a good fit to the option value. This pattern is explained in part by a connection between the regression estimates (and any other estimate based on approximate dynamic programming) and the dual estimates, to which we now turn.

Connection with Regression

If the regression relation (8.47) holds — meaning that the optimal continuation value is linear in the basis functions — then the regression residuals are the optimal martingale differences. To see why, observe from the definition (8.59) of the Δ_i that

$$V_{i+1}(X_{i+1}) = E[V_{i+1}(X_{i+1})|X_i] + \Delta_{i+1},$$

and that the equivalent properties (8.46) and (8.47) then imply

$$V_{i+1}(X_{i+1}) = \hat{\beta}_i^\top \psi(X_i) + \Delta_{i+1}.$$

This shows that Δ_{i+1} is the residual in the regression of $V_{i+1}(X_{i+1})$ against $\psi(X_i)$. If we knew with certainty that the regression relation (8.47) held, then this would provide a simple mechanism for finding the Δ_i as a by-product of estimating the coefficients $\hat{\beta}_i$. However, in the more typical case that (8.47) holds only approximately, the regression residuals will not satisfy the martingale difference condition (8.59) and are therefore not guaranteed to provide a valid upper bound on the option price.

Even if the regression equation (8.47) does not hold exactly, this link sheds light on the relation between the regression-based estimator calculated through the dynamic programming recursion and the dual estimators. We end this section by developing this idea.

Given coefficients vectors β_i, $i = 1, \ldots, m-1$, estimated or exact, consider the sequence of value estimates

$$\hat{V}_m(X_m) = h_m(X_m) \tag{8.71}$$
$$\hat{V}_i(X_i) = \max\{h_i(X_i), \beta_i^\top \psi(X_i)\}, \quad i = 1, \ldots, m-1, \tag{8.72}$$

applied to a path $X_1, \ldots, X_m$ of the underlying Markov chain. Set $\hat{V}_0(X_0) = E[\hat{V}_1(X_1)]$. The limit as $b \to \infty$ of the regression-based estimator defined through (8.51) has this form; this follows from Theorem 2 of Tsitsiklis and Van Roy [351].

Define residuals ϵ_i, $i = 1, \ldots, m-1$, through the equation

$$\hat{V}_{i+1}(X_{i+1}) = \beta_i^\top \psi(X_i) + \epsilon_{i+1},$$

and

$$\hat{V}_1(X_1) = \hat{V}_0(X_0) + \epsilon_1.$$

Using exactly the same algebraic steps leading to (8.62) and (8.63), we find that

$$\hat{V}_0(X_0) = \max_{k=1,\ldots,m} \left(h_k(X_k) - \sum_{i=1}^{k} \epsilon_i \right). \tag{8.73}$$

Thus, the approximation $\hat{V}_0(X_0)$ defined through regression and dynamic programming admits a representation analogous to the dual formulation of the true value $V_0(X_0)$ in (8.63), except that the cumulative sum of the ϵ_i is not in general a martingale.

This explains the pattern of numerical results in Table 8.3. With a good choice of basis functions, $\beta_i^\top \psi(X_i)$ is nearly equal to the continuation value at X_i so the regression residual ϵ_{i+1} is nearly equal to the optimal martingale difference Δ_{i+1} and the regression approximation $\hat{V}_0(X_0)$ is nearly the same as the dual value. With a poor choice of basis functions, the residuals are farther from the optimal Δ_i and we see a greater difference between the regression estimate and the dual value.

These observations apply more generally to any approximation computed through backward induction (including a stochastic mesh with likelihood ratio weights), even without the use of regression. Suppose (8.71)–(8.72) hold with $\beta_i^\top \psi(X_i)$ replaced with an arbitrary approximation $\hat{C}_i$ to the continuation value at step i. Define residuals through

$$\hat{V}_{i+1}(X_{i+1}) = \hat{C}_i(X_i) + \epsilon_{i+1},$$

and then (8.73) holds. This shows that the high-biased estimates calculated through backward induction in Sections 8.3 and 8.5–8.6 have nearly the same form as high-biased estimates based on duality.

8.8 Concluding Remarks

The pricing of American options through Monte Carlo simulation is an active and evolving area of research. We have limited this chapter to techniques that address the problem generally, and not discussed the many ad hoc methods that have been developed for specific models. Some of the history of the topic and some early proposals are surveyed in Boyle et al. [53].

Because the field is still in flux, comparisons and conclusions are potentially premature. Based on the current state of knowledge and an admittedly subjective view of the field, we offer the following summary of the methods discussed in this chapter.

○ Parametric approximations provide a relatively simple way of computing rough estimates, particularly in problems for which good information is available about the main drivers of early exercise. A sound implementation of this approach uses a second pass (Step 3 in Section 8.2) to compute a low-biased estimator and also an estimate of the dual value. The gap between the two estimates reflects the quality of the approximation.

○ The random tree method is very simple to implement and relies on no more than the ability to generate paths of the underlying processes. It provides a conservative confidence interval that shrinks to the true value under minor regularity conditions. Because its complexity is exponential in the number of exercise dates, it is suitable only for options with few exercise opportunities — not more than about five.

○ State-space partitioning is relatively insensitive to the number of exercise dates but its complexity is exponential in the dimension of the state space, making it inapplicable to high-dimensional problems.

○ The stochastic mesh and regression-based methods provide the most powerful techniques for solving high-dimensional problems with many exercise opportunities. Mesh weights based on likelihood ratios are theoretically the most general but have two shortcomings: they require evaluating transition densities for the underlying processes (which may be unknown or may fail to exist), and their computational complexity is $O(b^2)$, with b the number of paths. Regression-based methods correspond to an implicit choice of mesh weights. They reduce the $O(b^2)$ complexity of general weights to $O(Mb)$, with M the number of basis functions. This makes it practical to simulate many more paths and thus achieve lower variability, though not necessarily greater accuracy. Accuracy depends on the choice of basis functions, which may require experimentation or good information about the structure of the problem. A sound implementation of either the stochastic mesh or regression-based estimator uses a second pass to estimate a low-biased estimator. This can be paired with an estimate of the dual value to produce an interval for the true price. The first-pass estimator based on dynamic programming is often biased high; this is essentially always true using likelihood ratio weights and typically true using regression. In the case of regression, the gap between high and low estimates provides a measure of the quality of fit achieved with the chosen basis functions and can thus alert the user to a need to change the set of basis functions.

9

Applications in Risk Management

This chapter discusses applications of Monte Carlo simulation to risk management. It addresses the problem of measuring the risk in a portfolio of assets, rather than computing the prices of individual securities. Simulation is useful in estimating the profit and loss distribution of a portfolio and thus in computing risk measures that summarize this distribution. We give particular attention to the problem of estimating the probability of large losses, which entails simulation of rare but significant events. We separate the problems of measuring market risk and credit risk because different types of models are used in the two domains.

There is less consensus in risk management around choices of models and computational methods than there is in derivatives pricing. And while simulation is widely used in the practice of risk management, research on ways of improving this application of simulation remains limited. This chapter emphasizes a small number of specific techniques for specific problems in the broad area of risk management.

9.1 Loss Probabilities and Value-at-Risk

9.1.1 Background

A prerequisite to managing market risk is measuring market risk, especially the risk of large losses. For the large and complex portfolios of assets held by large financial institutions, this presents a significant challenge. Some of the obstacles to risk measurement are administrative — creating an accurate, centralized database of a firm's positions spanning multiple markets and asset classes, for example — others are statistical and computational. Any method for measuring market risk must address two questions in particular:

o   What statistical model accurately yet conveniently describes the movements in the individual sources of risk and co-movements of multiple sources of risk affecting a portfolio?

o   How does the value of a portfolio change in response to changes in the underlying sources of risk?

The first of these questions asks for the joint distribution of changes in risk factors — the exchange rates, interest rates, equity, and commodity prices to which a portfolio may be exposed. The second asks for a mapping from risk factors to portfolio value. Once both elements are specified, the distribution of portfolio profit and loss is in principle determined, as is then any risk measure that summarizes this distribution.

Addressing these two questions inevitably involves balancing the complexity required by the first with the tractability required by the second. The multivariate normal, for example, has known deficiencies as a model of market prices but is widely used because of its many convenient properties. Our focus is more on the computational issues raised by the second question than the statistical issues raised by the first. It is nevertheless appropriate to mention two of the most salient features of the distribution of changes in market prices and rates: they are typically heavy-tailed, and their co-movements are at best imperfectly described by their correlations. The literature documenting evidence of heavy tails is too extensive to summarize — an early reference is Mandelbrot [246]; Campbell, Lo, and MacKinlay [74] and Embrechts, Klüppelberg, and Mikosch [111] provide more recent accounts. Shortcomings of correlation and merits of alternative measures of dependence in financial data are discussed by, among others, Embrechts, McNeil, and Straumann [112], Longin and Solnik [240], and Mashal and Zeevi [255]. We revisit these issues in Section 9.3, but mostly work with simpler models.

To describe in more detail the problems we consider, we introduce some notation:

$$S = \text{vector of } m \text{ market prices and rates;}$$
$$\Delta t = \text{risk-measurement horizon;}$$
$$\Delta S = \text{change in } S \text{ over interval } \Delta t;$$
$$V(S,t) = \text{portfolio value at time } t \text{ and market prices } S;$$
$$L = \text{loss over interval } \Delta t$$
$$= -\Delta V = V(S,t) - V(S + \Delta S, t + \Delta t);$$
$$F_L(x) = P(L < x), \text{ the distribution of } L.$$

The number m of relevant risk factors could be very large, potentially reaching the hundreds or thousands. In bank supervision the interval Δt is usually quite short, with regulatory agencies requiring measurement over a two-week horizon, and this is the setting we have in mind. The two-week horizon is often interpreted as the time that might be required to unwind complex positions in the case of an adverse market move. In other areas of market risk, such as asset-liability management for pension funds and insurance companies, the relevant time horizon is far longer and requires a richer framework.

The notation above reflects some implicit simplifying assumptions. We consider only the net loss over the horizon Δt, ignoring for example the maximum and minimum portfolio value within the horizon. We ignore the dynamics of the market prices, subsuming all details about the evolution of S in the vector of changes ΔS. And we assume that the composition of the portfolio remains fixed, though the value of its components may change in response to the market movement ΔS and the passage of time Δt, which may bring assets closer to maturity or expiry.

The portfolio's value-at-risk (VAR) is a percentile of its loss distribution over a fixed horizon Δt. For example, the 99% VAR is a point x_p satisfying

$$1 - F_L(x_p) \equiv P(L > x_p) = p$$

with $p = 0.01$. (For simplicity, we assume throughout that F_L is continuous so that such a point exists; ties can be broken using (2.14).) A quantile provides a simple way of summarizing information about the tail of a distribution, and this particular value is often interpreted as a reasonable worst-case loss level. VAR gained widespread acceptance as a measure of risk in the late 1990s, in large part because of international initiatives in bank supervision; see Jorion [203] for an account of this history. VAR might more accurately be called a measure of capital adequacy than simply a measure of risk. It is used primarily to determine if a bank has sufficient capital to sustain losses from its trading activities.

The widespread adoption of VAR has been accompanied by frequent criticism of VAR as a measure of risk or capital adequacy. Any attempt to summarize a distribution in a single number is open to criticism, but VAR has a particular deficiency stressed by Artzner, Delbaen, Eber, and Heath [19]: combining two portfolios into a single portfolio may result in a VAR that is larger than the sum of the VARs for the two original portfolios. This runs counter to the idea that diversification reduces risk. Many related measures are free of this shortcoming, including the conditional excess $\mathsf{E}[L|L > x]$, calling into question the appropriateness of VAR.

The significance of VAR (and related measures) lies in its focus on the tail of the loss distribution. It emphasizes a probabilistic view of risk, in contrast to the more formulaic accounting perspective traditionally used to gauge capital adequacy. And through this probabilistic view, it calls attention to the importance of co-movements of market risk factors in a portfolio-based approach to risk, in contrast to an earlier "building-block" approach that ignores correlation. (See, for example, Section 4.2 of Crouhy, Galai, and Mark [93].) We therefore focus on the more fundamental issue of measuring the tail of the loss distribution, particularly at large losses — i.e., on finding $P(L > x)$ for large thresholds x. Once these loss probabilities are determined, it is a comparatively simple matter to summarize them using VAR or some other measure.

The relevant loss distribution in risk management is the distribution under the objective probability measure describing observed events rather than the risk-neutral or other martingale measure used as a pricing device. Historical data is thus directly relevant in modeling the distribution of ΔS. One can imagine a nested simulation (alluded to in Example 1.1.3) in which one first generates price-change scenarios ΔS, and then in each scenario simulates paths of underlying assets to revalue the derivative securities in a portfolio. In such a procedure, the first step (sampling ΔS) takes place under the objective probability measure and the second step (sampling paths of underlying assets) ordinarily takes place under the risk-neutral or other risk-adjusted probability measure. There is no logical or theoretical inconsistency in this combined use of the two measures. It is useful to keep the roles of the different probability measures in mind, but we do not stress the distinction in this chapter. Over a short interval Δt, it would be difficult to distinguish the real-world and risk-neutral distributions of ΔS.

9.1.2 Calculating VAR

There are several approaches to calculating or approximating loss probabilities and VAR, each representing some compromise between realism and tractability. How best to make this compromise depends in part on the complexity of the portfolio and on the accuracy required. We discuss some of the principal methods because they are relevant to our treatment of variance reduction in Section 9.2 and because they are of independent interest.

Normal Market, Linear Portfolio

By far the simplest approach to VAR assumes that ΔS has a multivariate normal distribution and that the change in value ΔV (hence also the loss L) is linear in ΔS. This gives L a normal distribution and reduces the problem of calculating loss probabilities and VAR to the comparatively simple task of computing the mean and standard deviation of L.

It is customary to assume that ΔS has mean zero because over a short horizon the mean of each component ΔS_j is negligible compared to its standard deviation, and because mean returns are extremely difficult to estimate from historical data. Suppose then that ΔS has distribution $N(0, \Sigma_S)$ for some covariance matrix Σ_S. Estimation of this covariance matrix is itself a significant challenge; see, for example, the discussion in Alexander [10].

Further suppose that

$$\Delta V = \delta^\top \Delta S, \tag{9.1}$$

for some vector of sensitivities δ. Then $L \sim N(0, \sigma_L^2)$ with $\sigma_L^2 = \delta^\top \Sigma_S \delta$, and the 99% VAR is $2.33\sigma_L$ because $\Phi(2.33) = 0.99$.

One might object to the normal distribution as a model of market movements because it can theoretically produce negative prices and because it is inconsistent with, for example, a lognormal specification of price levels. But all we need to assume is that the change ΔS over the interval $(t, t + \Delta t)$ is *conditionally* normal given the price history up to time t. Given S_j, assuming that ΔS_j is normal is equivalent to assuming that the return $\Delta S_j / S_j$ is normal. For small Δt,

$$S_j(1 + \Delta S_j / S_j) \approx S_j \exp(\Delta S_j / S_j),$$

so the distinction between normal and lognormal turns out to be relatively minor in this setting.

It should also be noted that assuming that ΔS is conditionally normal imposes a much weaker condition than assuming that changes over disjoint intervals of length Δt are i.i.d. normal. In our calculation of the distribution of ΔV based on (9.1), Σ_S is the conditional covariance matrix for changes from t to $t + \Delta t$, given the price history to time t. At different times t, one would ordinarily estimate different covariance matrices. The unconditional distribution of the changes ΔS would then be a mixture of normals and could even be heavy-tailed. This occurs, for example, in GARCH models (see Section 8.4 of Embrechts et al. [111]). Similar ideas are implicit in the discretization methods of Chapter 6: the increments of the Euler scheme (6.2) are conditionally normal at each step, but the distribution of the state can be far from normal after multiple steps.

Delta-Gamma Approximation

The assumption that V is linear in S holds, for example, for a stock portfolio if S is the vector of underlying stock prices. But a portfolio with options has a nonlinear dependence on the prices of underlying assets, and fixed-income securities depend nonlinearly on interest rates. The model in (9.1) is thus not universally applicable.

A simple way to extend (9.1) to capture some nonlinearity is to add a quadratic term. The quadratic produced by Taylor expansion yields the *delta-gamma* approximation

$$\Delta V \approx \frac{\partial V}{\partial t} \Delta t + \delta^\top \Delta S + \tfrac{1}{2} \Delta S^\top \Gamma \Delta S, \tag{9.2}$$

where

$$\delta_i = \frac{\partial V}{\partial S_i}, \quad \Gamma_{ij} = \frac{\partial^2 V}{\partial S_i \partial S_j}$$

are first and second derivatives of V evaluated at $(S(t), t)$. This in turn yields a quadratic approximation to $L = -\Delta V$.

For this approximation to have practical value, the coefficients must be easy to evaluate and finding the distribution of the approximation must be substantially simpler than finding the distribution of L itself. As discussed in Chapter 7, calculating δ and Γ can be difficult; however, these sensitivities are routinely calculated for hedging purposes by individual trading desks and can be aggregated (at the end of the day, for example) for calculation of firmwide risk. This is a somewhat idealized description — for example, many off-diagonal gammas may not be readily available — but is sufficiently close to reality to provide a valid premise for analysis.

If ΔS has a multivariate normal distribution, then finding the distribution of the approximation in (9.2) requires finding the distribution of a quadratic function of normal random variables. This can be done numerically through transform inversion. We detail the derivation of the transform because it is relevant to the techniques we apply in Section 9.2.

Delta-Gamma: Diagonalization

The first step derives a convenient expression for the approximation. As in Section 2.3.3, we can replace the correlated normals $\Delta S \sim N(0, \Sigma_S)$ with independent normals $Z \sim N(0, 1)$ by setting

$$\Delta S = CZ \quad \text{with} \quad CC^\top = \Sigma_S.$$

In terms of Z, the quadratic approximation to $L = -\Delta V$ becomes

$$L \approx a - (C^\top \delta)^\top Z - \tfrac{1}{2} Z^\top (C^\top \Gamma C) Z \tag{9.3}$$

with $a = -(\Delta t)\partial V/\partial t$ deterministic.

It is convenient to choose the matrix C to diagonalize the quadratic term in (9.3), and this can be accomplished as follows. Let $\tilde{C}$ be any square matrix for which $\tilde{C}\tilde{C}^\top = \Sigma_S$, such as the one found by Cholesky factorization. The matrix $-\tfrac{1}{2}\tilde{C}^\top \Gamma \tilde{C}$ is symmetric and thus admits the representation

$$-\tfrac{1}{2}\tilde{C}^\top \Gamma \tilde{C} = U\Lambda U^\top$$

in which

$$\Lambda = \begin{pmatrix} \lambda_1 & & & \\ & \lambda_2 & & \\ & & \ddots & \\ & & & \lambda_m \end{pmatrix}$$

is a diagonal matrix and U is an orthogonal matrix ($UU^\top = I$) whose columns are eigenvectors of $-\tfrac{1}{2}\tilde{C}^\top \Gamma \tilde{C}$. The λ_j are eigenvalues of this matrix and also of $-\tfrac{1}{2}\Gamma\Sigma_S$. Now set $C = \tilde{C}U$ and observe that

$$CC^\top = \tilde{C}UU^\top \tilde{C}^\top = \Sigma_S$$

and

$$-\tfrac{1}{2}C^\top \Gamma C = -\tfrac{1}{2}U^\top (\tilde{C}^\top \Gamma \tilde{C})U = U^\top (U\Lambda U^\top)U = \Lambda.$$

Thus, by setting $b = -C^\top \delta$ we can rewrite (9.3) as

$$L \approx a + b^{\top} Z + Z^{\top} \Lambda Z$$

$$= a + \sum_{j=1}^{m} (b_j Z_j + \lambda_j Z_j^2) \equiv Q. \tag{9.4}$$

The delta-gamma approximation now becomes $P(L > x) \approx P(Q > x)$, so we need to find the distribution of Q.

Delta-Gamma: Moment Generating Function

We determine the distribution of Q by deriving its moment generating function and characteristic function. The moment generating function is finite in a neighborhood of the origin and, in light of the independence of the summands in (9.4), factors as

$$\mathsf{E}[e^{\theta Q}] = e^{a\theta} \prod_{j=1}^{m} \mathsf{E}[e^{\theta(b_j Z_j + \lambda_j Z_j^2)}] \equiv e^{a\theta} \prod_{j=1}^{m} e^{\psi_j(\theta)}.$$

If $\lambda_j = 0$, then (2.26) yields $\psi_j(\theta) = b_j^2 \theta^2 / 2$. Otherwise, we write

$$b_j Z_j + \lambda_j Z_j^2 = \lambda_j \left(Z_j + \frac{b_j}{2\lambda_j} \right)^2 - \frac{b_j^2}{4\lambda_j},$$

which is a linear transformation of a noncentral chi-square random variable; see (3.72). Using the identity (equation (29.6) of Johnson, Kotz, and Balakrishnan [202]),

$$\mathsf{E}[\exp(\theta(Z_j + c)^2)] = (1 - 2\theta)^{-1/2} \exp\left(\frac{\theta c^2}{1 - 2\theta} \right),$$

for $\theta < 1/2$, we arrive at the expression

$$\psi(\theta) \equiv a\theta + \sum_{j=1}^{m} \psi_j(\theta) = a\theta + \frac{1}{2} \sum_{j=1}^{m} \left(\frac{\theta^2 b_j^2}{1 - 2\theta \lambda_j} - \log(1 - 2\theta \lambda_j) \right) \tag{9.5}$$

for $\log \mathsf{E}[\exp(\theta Q)]$, the cumulant generating function of Q. This equation holds for all θ satisfying $\max_j \theta \lambda_j < 1/2$.

The characteristic function of Q is obtained by evaluating the moment generating function at a purely imaginary argument:

$$\hat{\phi}(u) = \mathsf{E}[e^{iuQ}] = e^{\psi(iu)}, \quad i = \sqrt{-1}.$$

The distribution of Q can now be computed through the inversion integral (Chung [85], p.153)

$$P(Q \le x) - P(Q \le x - y) = \frac{1}{\pi} \int_0^{\infty} \mathrm{Re}\left(\hat{\phi}(u) \left[\frac{e^{iuy} - 1}{iu} \right] e^{-iux} \right) du. \tag{9.6}$$

Abate, Choudhury, and Whitt [1] discuss algorithms for numerical evaluation of integrals of this type. To find $P(Q \leq x)$, choose y large so that $P(Q \leq x - y) \approx 0$.

Although the derivation of this method involves several steps, its implementation is relatively straightforward and fast. It is developed in this form by Rouvinez [310]. Studer [338] suggests replacing the ordinary Taylor expansion in (9.2) with an Itô-Taylor expansion (of the type in Section 6.3.1), leading to a slightly different quadratic. Britten-Jones and Schaefer [61] use approximations to the distribution of quadratic functions of normals in place of transform inversion. Extensions to non-normal risk factors are derived in Duffie and Pan [104] and (as discussed in Section 9.3.2) Glasserman, Heidelberger, and Shahabuddin [144].

Figure 9.1 illustrates the delta-gamma approximation for a simple portfolio. The portfolio consists of short positions of 10 calls and 5 puts on each of 10 underlying assets, with all options 0.10 years from expiration. We value the portfolio using the Black-Scholes formula for each option, and the quadratic approximation uses Black-Scholes deltas and gammas. The underlying assets are uncorrelated and each has a 0.40 volatility.

The scatter plot in Figure 9.1 shows the results of 1000 randomly generated scenarios ΔS. These scenarios are sampled from a multivariate normal distribution with independent components ΔS_j, $j = 1, \ldots, m$. Each ΔS_j has mean 0 and standard deviation $S_j(0)\sigma_j\sqrt{\Delta t}$, with $\Delta t = 0.04$ years or about two weeks. For each ΔS, we revalue the portfolio at time $t + \Delta t$ and underlying prices $S + \Delta S$ using the Black-Scholes formula; this gives the horizontal coordinate for each scenario. We also compute the quadratic approximation by substituting ΔS and Δt in the delta-gamma formula (9.2); this gives the vertical coordinate for each scenario. The scatter plot illustrates the strong relation between the exact and approximate losses in this example.

There is an evident inconsistency in this example between the model used to generate scenarios and the formula used to value the portfolio. This is representative of the standard practice of using fairly rough models to describe market risk while using more detailed models to price derivatives. It should also be noted that, in theory, ΔS should be sampled from the objective probability measure describing actual market movements — whereas pricing formulas ordinarily depend on the risk-neutral dynamics of underlying assets — so some inconsistency in how we model S for the two steps is appropriate. Most importantly, the use of the Black-Scholes formula in this example is convenient but by no means essential.

Our main interest in the delta-gamma approximation lies in accelerating Monte Carlo simulation. Even in cases in which the method may not by itself provide an accurate approximation to the loss distribution, it can provide a powerful tool for variance reduction.

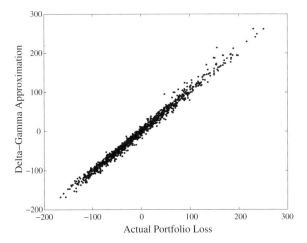

Fig. 9.1. Comparison of delta-gamma approximation and actual portfolio losses in 1000 randomly generated scenarios for a portfolio of 150 options.

Monte Carlo Simulation

Estimating loss probabilities and VAR by simulation is simple in concept, as illustrated by the following algorithm:

- For each of n independent replications
    - generate a vector of market moves ΔS;
    - revalue portfolio and compute loss $V(S, t) - V(S + \Delta S, t + \Delta t)$.
- Estimate $P(L > x)$ using

$$\frac{1}{n} \sum_{i=1}^{n} \mathbf{1}\{L_i > x\}$$

where L_i is the loss on the ith replication.

The bottleneck in this algorithm is the portfolio revaluation step. For a large portfolio of complex derivative securities, each revaluation may require running thousands of numerical pricing routines. Individual pricing routines may involve numerical integration, solving partial differential equations, or even running a separate simulation (as alluded to in Example 1.1.3). This makes variance reduction essential to achieving accurate estimates with reasonable computational effort. We return to this topic in Section 9.2.

"Historical simulation" is a special case of this algorithm in which the scenarios ΔS are drawn directly from historical data — daily price changes over the past year, for example. Past observations of changes ΔS in the underlying risk factors are applied to the current portfolio to produce a histogram of changes in portfolio value. This approach is sometimes defended on the grounds that it is easy to explain to a nontechnical audience; but the historical distribution of ΔS inevitably has gaps, especially in the tails. Fitting a theoretical distribution (or at least smoothing the tails) and then applying Monte Carlo gets around this artificial feature of the purely data-driven implementation.

Quantile Estimation

The simulation algorithm above estimates loss probabilities $P(L > x)$ rather than VAR and it is in this context that we discuss variance reduction techniques. Before doing so, we briefly discuss the estimation of VAR itself.

Let $\hat{F}_{L,n}$ denote the empirical distribution of portfolio losses based on n simulated replications,

$$\hat{F}_{L,n}(x) = \frac{1}{n} \sum_{i=1}^{n} \mathbf{1}\{L_i \leq x\}.$$

A simple estimate of the VAR at probability p (e.g., $p = 0.01$) is the empirical quantile

$$\hat{x}_p = \hat{F}_{L,n}^{-1}(1-p),$$

with the inverse of the piecewise constant function $\hat{F}_{L,n}$ defined as in (2.14). Applying piecewise linear interpolation to $\hat{F}_{L,n}$ before taking the inverse generally produces more accurate quantile estimates. (See Avramidis and Wilson [30] for a comparison of simulation-based quantile estimators, including several using antithetic and Latin hypercube sampling.)

Under minimal conditions (as in Serfling [326], p.75), the empirical quantile $\hat{x}_p$ converges to the true quantile x_p with probability 1 as $n \to \infty$. A central limit theorem provides additional information on the quality of convergence. For this we assume that L has a strictly positive density f in a neighborhood of x_p. Then

$$\sqrt{n}(\hat{x}_p - x_p) \Rightarrow \frac{\sqrt{p(1-p)}}{f(x_p)} N(0,1), \tag{9.7}$$

as shown, for example, in Serfling [326], p.77. The term $p(1-p)$ is the variance of the loss indicator $\mathbf{1}\{L > x_p\}$. We see from (9.7) that this variance is magnified by a factor of $1/f(x_p)^2$ when we estimate the quantile rather than the loss probability. This factor is potentially very large, especially if p is small, because the density is likely to be close to zero in this case. We will see that by reducing variance in estimates of $P(L > x)$ for x near x_p, we can reduce the variance in this central limit theorem for the VAR estimate.

Equation (9.7) provides the basis for a large-sample $1 - \alpha$ confidence interval for x_p of the form

$$\hat{x}_p \pm z_{\alpha/2} \frac{\sqrt{p(1-p)}}{f(x_p)\sqrt{n}},$$

with $1 - \Phi(z_{\alpha/2}) = \alpha/2$. The interval remains asymptotically valid with $f(x_p)$ replaced by $f(\hat{x}_p)$ if f is continuous at x_p. However, its reliance on evaluating the density f makes this interval estimate impractical. To avoid this step, one may divide the sample of n observations into batches, compute an estimate $\hat{x}_p$ from each batch, and form a confidence interval based on the sample standard deviation of the estimates across batches.

An alternative confidence interval not relying on a central limit theorem and valid for finite n uses the fact that the number of samples exceeding x_p has a binomial distribution with parameters n and p. Let

$$L_{(1)} \le L_{(2)} \le \cdots \le L_{(n)} \tag{9.8}$$

denote the order statistics of the L_i. An interval of the form $[L_{(r)}, L_{(s)}), r < s$, covers x_p with probability

$$P(L_{(r)} \le x_p < L_{(s)}) = \sum_{i=r}^{s-1} \binom{n}{i} (1-p)^i p^{n-i}.$$

The values r and s can be chosen to bring this probability close to the desired confidence level $1 - \alpha$. These values do not depend on the loss distribution F_L.

Approximate Simulation

We noted previously that the bottleneck in using Monte Carlo simulation for estimating loss probabilities and VAR lies in portfolio revaluation. It follows that there are two basic strategies for accelerating simulation:

○ reduce the number of scenarios required to achieve a target precision by applying a variance reduction technique; or
○ reduce the time required for each scenario through approximate portfolio revaluation.

Our focus in the next section is on a specific set of techniques for the first strategy. The second strategy is also promising but has received less systematic study to date.

A general approach to approximate revaluation undertakes exact valuation in a moderate number of scenarios (generated randomly or deterministically) and fits a function to the value surface using some form of interpolation or nonlinear regression. If the fitted approximation is easy to evaluate, then using it in place of $V(S + \Delta S, t + \Delta t)$ in the simulation algorithm makes it feasible to generate a much larger number of scenarios and compute (approximate) losses in each. The challenge in this approach lies in selecting an appropriate functional form and — especially for high-dimensional S — computing enough exact values to obtain a good fit.

The problem simplifies if the portfolio value is "separable" in the sense that each instrument in the portfolio depends on only a small number of risk factors, though the total number of risk factors may be large. In the extreme case of total separation, the portfolio value can be expressed as

$$V(S,t) = V_1(S_1,t) + \cdots + V_m(S_m,t)$$

where V_j gives the value of all instruments sensitive to S_j (and only to S_j).

Suppose evaluating the functions V_j is time-consuming. Evaluating each V_j at just two values of S_j yields the value of V at 2^m points if we take all vectors $(S_1, \ldots, S_m)$ formed by combinations of the pairs of S_j values. Similar though less dramatic savings apply if V decomposes as a sum of functions of, say, 2–5 arguments each if m is much larger than 5.

Variants of this general approach are described by Jamshidian and Zhu [198], Picoult [297], and Shaw [329]. Jamshidian and Zhu [198] propose a partitioning and weighting of the set of possible S, using principal components analysis (keeping only the most important components) to reduce the number of risk factors. Abken [2] tests the method on portfolios of multicurrency interest rate derivatives and reports mixed results.

9.2 Variance Reduction Using the Delta-Gamma Approximation

We stressed in Chapter 4 that effective variance reduction takes advantage of special features of a simulated model. In simulating portfolio losses we should therefore look for information available that could be used to improve precision. One source of information are the results of simulations of the portfolio's value in recent days under presumably similar market conditions; this potentially effective direction has not received systematic study to date. Another source of information for variance reduction is the quadratic delta-gamma approximation (9.2) to portfolio value (or the linear delta approximation (9.1)), and this is the case we treat in detail. This strategy for variance reduction leads to interesting theoretical results (especially in importance sampling) and effective variance reduction. The techniques we discuss are primarily from Glasserman, Heidelberger, and Shahabuddin (henceforth GHS) [143, 142, 144].

In order to use the delta-gamma approximation, we assume throughout this section that the changes in risk factors ΔS are multivariate normal $N(0, \Sigma_S)$. Section 9.3 develops extensions to a class of heavy-tailed distributions. The techniques discussed in this section are applicable with any quadratic approximation and thus could in principle be used with a hedged portfolio in which all deltas and gammas are zero, provided a less local quadratic approximation could be computed. As discussed in Section 9.1.2, the delta-gamma approximation has the advantage that it may be available with little or no additional computational effort.

9.2.1 Control Variate

The simplest way to take advantage of the delta-gamma approximation $L \approx Q$ applies it as a control variate. On each replication (meaning for each simulated market move ΔS), evaluate Q along with the portfolio loss L. Using the representation in (9.4) is convenient if we generate ΔS as CZ with $Z \sim N(0, I)$ and C the matrix constructed in the derivation of (9.4). Let (L_i, Q_i), $i = 1, \ldots, n$, be the values recorded on n independent replications. A control variate estimator of $P(L > x)$ is given by

$$1 - \hat{F}_L^{\mathrm{cv}}(x) = \frac{1}{n} \sum_{i=1}^{n} \mathbf{1}\{L_i > x\} - \hat{\beta} \left(\frac{1}{n} \sum_{i=1}^{n} \mathbf{1}\{Q_i > y\} - P(Q > y) \right). \quad (9.9)$$

An estimate $\hat{\beta}$ of the variance-minimizing coefficient can be computed from the (L_i, Q_i) as explained in Section 4.1.1.

As suggested by the notation in (9.9), the threshold y used for Q need not be the same as the one applied to L, though it often would be in practice. We could construct multiple controls by using multiple thresholds for Q. The exact probability $P(Q > y)$ needed for the control is evaluated using the inversion integral in (9.6).

The control variate estimator (9.9) is easy to implement and numerical experiments reported in Cardenas et al. [76] and GHS [142] indicate that it often reduces variance by a factor of 2–5. However, the estimator suffers from two related shortcomings:

○ Even if L and Q are highly correlated, the loss indicators $\mathbf{1}\{L > x\}$ and $\mathbf{1}\{Q > y\}$ may not be, especially for large x and y. This is relevant because the variance reduction achieved is (at best) ρ^2, with ρ the correlation between the indicators; see the discussion in Section 4.1.1.

○ If x and y are large, few simulations produce values of L or Q greater than these thresholds. This makes it difficult to estimate the optimal coefficient β and thus further erodes the variance reduction achieved.

Both of these points are illustrated in Figure 9.1. Although the correlation between exact and approximate losses is very high in this example, it is weakest in the upper-right corner, the area of greatest interest, and few observations fall in this area.

An indirect illustration of the first point above is provided by the normal distribution. If a pair of random variables (X, Y) has a bivariate normal distribution with correlation ρ, then the correlation of the indicators $\mathbf{1}\{X > u\}$ and $\mathbf{1}\{Y > u\}$ approaches zero as $u \to \infty$ if $|\rho| \neq 1$; see Embrechts et al. [112].

Quantile Estimation

Using the control variate to estimate a quantile x_p, (at which $P(L > x_p) = p$) requires inverting $\hat{F}_L^{\mathrm{cv}}$ in (9.9) to find a point $\hat{x}_p$ at which $\hat{F}_L^{\mathrm{cv}}(\hat{x}_p) \approx 1 - p$.

This appears to require recomputing the coefficient $\hat{\beta}$ for multiple values of x. But Hesterberg and Nelson [178] show that this can be avoided, as we now explain.

Their method uses the connection between control variate estimators and weighted Monte Carlo discussed in Section 4.1.2. Using (4.19), the control variate estimator (9.9) can be expressed as

$$1 - \hat{F}_L^{\mathrm{cv}}(x) = \sum_{i=1}^{m} W_i \mathbf{1}\{L_i > x\} = \sum_{i:L_i > x} W_i,$$

where the weights W_i depend on $Q_1, \ldots, Q_n$ and y but not on $L_1, \ldots, L_n$ or x. Thus, $\hat{F}_L^{\mathrm{cv}}(x)$ can be evaluated at multiple values of x simply by summing over the appropriate set of weights, without recalculation of $\hat{\beta}$.

Through this representation of the control variate estimator of the distribution of L, the quantile estimator

$$\hat{x}_p = \inf\{x : 1 - \hat{F}_L^{\mathrm{cv}}(x) \leq p\}$$

becomes

$$\hat{x}_p = \inf\{x : \sum_{i:L_i > x} W_i \leq p\}.$$

Let $L_{(i)}$, $i = 1, \ldots, n$, denote the order statistics of the L_i, as in (9.8) and let $W^{(i)}$ denote the weight corresponding to $L_{(i)}$. Then $\hat{x}_p = L_{(i_p)}$ where

$$i_p = \min\{k : \sum_{i=k+1}^{n} W^{(i)} \leq p\}.$$

To smooth the estimator using linear interpolation, find α for which

$$\alpha \sum_{i=i_p}^{n} W^{(i)} + (1 - \alpha) \sum_{i=i_p+1}^{n} W^{(i)} = p$$

and then estimate the quantile as

$$\hat{x}_p = \alpha L_{(i_p-1)} + (1 - \alpha) L_{(i_p)}.$$

This procedure is further simplified by the observation in Hesterberg and Nelson [178] that for (9.9), the weights reduce to

$$W_i = \begin{cases} nP(Q \leq y)/\sum_{j=1}^{n} \mathbf{1}\{Q_j \leq y\}, & \text{if } Q_i \leq y, \\ nP(Q > y)/\sum_{j=1}^{n} \mathbf{1}\{Q_j > y\}, & \text{if } Q_i > y. \end{cases}$$

This shows that the control variate estimator in (9.9) weights each indicator $\mathbf{1}\{L_i > x\}$ by the ratio of an exact and estimated probability for Q. As further observed by Hesterberg and Nelson [178], this makes it a poststratified estimator of $P(L > x)$ with stratification variable Q and strata $\{Q \leq y\}$ and $\{Q > y\}$. We will have more to say about using Q as a stratification variable in Section 9.2.3.

9.2.2 Importance Sampling

The control variate estimator (9.9) uses knowledge of the delta-gamma approximation Q after simulating replications of ΔS in order to adjust the average of the loss indicators $\mathbf{1}\{L_i > x\}$. But as already noted, if we observe few replications in which $Q_i > y$, we have little information on which to base the adjustment. This is evident in the connection with poststratification as well.

Through importance sampling, we can use the delta-gamma approximation to guide the sampling of scenarios before we compute losses, rather than to adjust our estimate after the scenarios are generated. In particular, we can try to use our knowledge of the distribution of Q to give greater probability to "important" scenarios in which L exceeds x. This idea is developed in GHS [142, 143, 144].

Before proceeding with the details of this approach, consider the qualitative information provided by the approximation

$$L \approx Q = a + \sum_{j=1}^{m} b_j Z_j + \sum_{j=1}^{m} \lambda_j Z_j^2,$$

with the Z_j independent standard normals, as in (9.4). The Z_j do not correspond directly to the original changes in market prices ΔS_j, but we can think of them as primitive sources of risk that drive the market changes through the relation $\Delta S = CZ$. Also, since $Z = C^{-1}\Delta S$, each Z_j could be interpreted as the change in value of a portfolio of linear positions in the elements of S.

What does this approximation to L say about how large losses occur? It suggests that large losses occur when

(i) Z_j is large and positive for some j with $b_j > 0$;
(ii) Z_j is large and negative for some j with $b_j < 0$; or
(iii) Z_j^2 is large and positive for some j with $\lambda_j > 0$.

This further suggests that to make large losses more likely, we should

(i')   give a positive mean to those Z_j for which $b_j > 0$;
(ii') give a negative mean to those Z_j for which $b_j < 0$; and
(iii') increase the variance of those Z_j for which $\lambda_j > 0$.

Each of the changes (i')–(iii') increases the probability of the corresponding event (i)–(iii). We develop an importance sampling procedure that makes these qualitative changes precise.

Exponential Twisting

In our discussion of importance sampling in Section 4.6, we saw examples in which an exponential change of measure leads to dramatic variance reduction. If our goal were to estimate $P(Q > x)$ rather than $P(L > x)$, this would lead us to consider importance sampling based on "exponentially twisting" Q. In more detail (cf. Example 4.6.2), this means defining a family of probability measures P_θ through the likelihood ratio

$$\frac{dP_\theta}{dP} = e^{\theta Q - \psi(\theta)}, \tag{9.10}$$

with ψ the cumulant generating function in (9.5) and θ any real number at which $\psi(\theta) < \infty$. Writing E_θ for expectation under the new measure, we have

$$P(Q > x) = \mathsf{E}_\theta\left[\left(\frac{dP}{dP_\theta}\right)\mathbf{1}\{Q > x\}\right] = \mathsf{E}_\theta\left[e^{-\theta Q + \psi(\theta)}\mathbf{1}\{Q > x\}\right].$$

To use this in simulation, we would need to generate Q from its distribution under P_θ and then use the average of independent replications of

$$e^{-\theta Q + \psi(\theta)}\mathbf{1}\{Q > x\}$$

to estimate $P(Q > x)$.

If θ is positive, then P_θ gives greater probability to large values of Q than $P = P_0$ does, thus increasing the probability of the event $\{Q > x\}$ for large x, which is intuitively appealing. More precisely, the second moment of this importance sampling estimator is

$$\mathsf{E}_\theta\left[e^{-2\theta Q + 2\psi(\theta)}\mathbf{1}\{Q > x\}\right] = \mathsf{E}\left[e^{-\theta Q + \psi(\theta)}\mathbf{1}\{Q > x\}\right] \le e^{-\theta x + \psi(\theta)}, \tag{9.11}$$

which decreases exponentially in x if θ is positive.

This idea extends to the more relevant problem of estimating $P(L > x)$. Using P_θ as defined in (9.10), we have

$$P(L > x) = \mathsf{E}_\theta\left[e^{-\theta Q + \psi(\theta)}\mathbf{1}\{L > x\}\right];$$

the expression inside the expectation on the right is an unbiased importance sampling estimator of the loss probability. Its second moment is

$$\mathsf{E}\left[e^{-\theta Q + \psi(\theta)}\mathbf{1}\{L > x\}\right],$$

which is small if $\theta > 0$ and Q is large on the event $\{L > x\}$.

Sampling from the Twisted Distribution

To use this estimator, we need to be able to generate independent replications of

$$e^{-\theta Q + \psi(\theta)}\mathbf{1}\{L > x\}$$

under the θ-twisted distribution. In other words, we need to be able to simulate the pair (Q, L) under P_θ. But recall from the generic simulation algorithm of Section 9.1.2 that, even in the absence of importance sampling, we do not generate (Q, L) directly; instead we generate Z and then evaluate Q and L from Z (through $\Delta S = CZ$). Thus, $(Q, L) = f(Z)$ for some deterministic function f. Sampling Z from the standard multivariate normal distribution gives (Q, L) its distribution under the original probability measure P; to simulate (Q, L) under P_θ, it suffices to sample Z from its distribution under the new measure.

What is the distribution of Z under P_θ? In other words, what is

$$P_\theta(Z_1 \leq z_1, \ldots, Z_m \leq z_m) = \mathsf{E}\left[e^{\theta Q - \psi(\theta)} \mathbf{1}\{Z_1 \leq z_1, \ldots, Z_m \leq z_m\}\right]?$$

It is shown in GHS [143] that, rather remarkably, the P_θ-distribution of Z is still multivariate normal, but with a new mean vector and covariance matrix determined by the delta-gamma approximation. In particular, $Z \sim N(\mu(\theta), \Sigma(\theta))$ where $\Sigma(\theta)$ is a diagonal matrix with diagonal entries $\sigma_j^2(\theta)$,

$$\mu_j(\theta) = \frac{\theta b_j}{1 - 2\lambda_j \theta}, \quad \sigma_j^2(\theta) = \frac{1}{1 - 2\lambda_j \theta}. \tag{9.12}$$

Recall that in (9.5) we required $2\lambda_j \theta < 1$ so that $\psi(\theta) < \infty$.

To show that this is indeed the P_θ-distribution of Z, it is easiest to argue in the opposite direction. For arbitrary μ and Σ, the likelihood ratio relating the density of $N(\mu, \Sigma)$ to $N(0, I)$ is given by

$$\frac{|\Sigma|^{-1/2} \exp\left(-\frac{1}{2}(Z - \mu)^\top \Sigma^{-1}(Z - \mu)\right)}{\exp\left(-\frac{1}{2}Z^\top Z\right)}, \tag{9.13}$$

the ratio of the two densities evaluated at Z. Substituting the specific parameters (9.12) and applying some simplifying algebra shows that this reduces to $\exp(\theta Q - \psi(\theta))$, with Q related to Z through (9.4). Because this is the likelihood ratio used to define P_θ, we may indeed conclude that under P_θ the Z_j are independent normals with the means and variances in (9.12). If μ and Σ are chosen arbitrarily, (9.13) depends on the entire vector Z. For the special case of (9.12), the dependence of the likelihood ratio on Z collapses to dependence on Q. This is a key feature of this importance sampling strategy.

The new means and variances defined by (9.12) have the qualitative features (i')–(iii') listed above for any $\theta > 0$, so (9.10) is consistent with these insights into how the delta-gamma approximation should guide the sampling of scenarios. If all the λ_j are close to zero, we might use just the linear term in Q (a delta-only approximation) for importance sampling. The parameters in (9.12) would then reduce to a change of mean for the vector Z in a manner consistent with features (i') and (ii').

Importance Sampling Algorithm

We can now summarize the ideas above in the following algorithm for estimating portfolio loss probabilities through importance sampling:

1. Choose a value of $\theta > 0$ at which $\psi(\theta) < \infty$. (More on this shortly.)
2. For each of n replications
   (a) generate Z from $N(\mu(\theta), \Sigma(\theta))$ with parameters given in (9.12);
   (b) evaluate Q from Z using (9.4);
   (c) set $\Delta S = CZ$;
   (d) calculate portfolio value $V(S + \Delta S, t + \Delta t)$ and loss L;
   (e) calculate

$$e^{-\theta Q + \psi(\theta)} \mathbf{1}\{L > x\}. \tag{9.14}$$

3. Calculate average of (9.14) over the n replications.

Rather little in this algorithm differs from the standard Monte Carlo algorithm given in Section 9.1.2. In particular, the core of the algorithm — invoking all the routines necessary to revalue the portfolio in step 2(d) — is unchanged. Step 2(a) is only slightly more complicated than generating Z from $N(0, I)$. Step 2(c) would be needed even without importance sampling, though here we require C to be the specific matrix constructed in deriving (9.4) rather than an arbitrary matrix for which $CC^\top = \Sigma_S$.

Choice of Twisting Parameter

Step 1 of the importance sampling algorithm requires us to choose θ, so we now discuss the selection of this parameter. In the absence of additional information, we choose a value of θ that would be effective in estimating $P(Q > x)$ and apply this value in estimating $P(L > x)$.

In (9.11) we have an upper bound on the second moment of the importance sampling estimator of $P(Q > x)$ for all x and all θ at which $\psi(\theta) < \infty$. We can minimize this upper bound for fixed x by choosing θ to minimize $\psi(\theta) - \theta x$. The fact that ψ is convex (because it is a cumulant generating function) implies that this expression is minimized at θ_x, the root of the equation

$$\psi'(\theta_x) = x. \tag{9.15}$$

This equation is illustrated in Figure 9.2.

In addition to minimizing an upper bound on the second moment of the estimator, the parameter θ_x has an interpretation that sheds light on this approach to importance sampling. Equation (9.10) defines an exponential family of probability measures, a concept we discussed in Section 4.6.1, especially in Example 4.6.2. As a consequence, we have

$$\psi'(\theta) = \mathsf{E}_\theta[Q] \tag{9.16}$$

for any θ at which $\psi(\theta)$ is finite. This follows from differentiating the definition $\psi(\theta) = \log \mathsf{E}[\exp(\theta Q)]$ to get

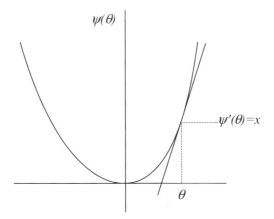

Fig. 9.2. Twisting parameter θ satisfying $\psi'(\theta) = x$.

$$\psi'(\theta) = \frac{\mathsf{E}[Qe^{\theta Q}]}{\mathsf{E}[e^{\theta Q}]} = \mathsf{E}[Qe^{\theta Q - \psi(\theta)}].$$

Choosing the parameter θ to use for importance sampling is thus equivalent to choosing $\mathsf{E}_\theta[Q]$, the expected value of the delta-gamma approximation under the new distribution. By choosing $\theta = \theta_x$ as in (9.15), we are sampling from a distribution under which

$$\mathsf{E}_{\theta_x}[Q] = x.$$

Whereas x was in the tail of the distribution under the original distribution, it is near the center of the distribution when we apply importance sampling.

Asymptotic Optimality

GHS [143] prove an asymptotic optimality result for this approach to importance sampling when applied to estimation of $P(Q > x)$. This should be interpreted as indirect evidence that the approach is also effective in estimating the actual loss probability $P(L > x)$. Numerical examples in GHS [142, 143] applying the approach to estimating loss probabilities in test portfolios lend further support to the method.

The asymptotic optimality result (from Theorems 1 and 2 of [143]) is as follows. Suppose $\lambda_{\max} \equiv \max_{1 \le i \le m} \lambda_i$, the largest of the eigenvalues of $-\Gamma \Sigma_S/2$ (as in (9.4)), is strictly positive; then the tail of Q satisfies

$$P(Q > x) = \exp\left(-\tfrac{1}{2}\lambda_{\max} x + o(x)\right),$$

and the second moment of the importance sampling estimator satisfies

$$\mathsf{E}_{\theta_x}\left[e^{-2\theta_x Q + 2\psi(\theta_x)}\mathbf{1}\{Q > x\}\right] = \exp\left(-\lambda_{\max} x + o(x)\right).$$

Thus, as x increases, the second moment decreases at twice the exponential rate of the probability itself. As explained in Example 4.6.3, this is the fastest possible rate of decrease of any unbiased estimator. In this sense, the importance sampling estimator is asymptotically optimal.

Theorem 3 of GHS [143] establishes an asymptotic optimality result when $\lambda_{\max} < 0$; this case is less interesting because it implies that Q is bounded above. Theorem 4 of GHS [143] analyzes the effect of letting the number of risk factors m increase and establishes a type of asymptotic optimality in this case as well.

To estimate the tail $P(Q > y)$ at many values of y — in order to estimate a quantile, for example — one would want to use estimators of the form

$$e^{-\theta_x Q + \psi(\theta_x)} \mathbf{1}\{Q > y\},$$

rather than use a separate value of θ for each point y. Theorem 5 of GHS [143] shows that this can be done without sacrificing asymptotic optimality. More fundamentally, the effectiveness of the estimator is not very sensitive to the value of θ used. In practice, to estimate loss probabilities over a range of large values of y, it is advisable to choose $\theta = \theta_x$ for x near the middle or left endpoint of the range. At loss thresholds that are not "large" there is no need to apply importance sampling.

Through a result of Glynn [154], variance reduction for tail probabilities using importance sampling translates to variance reduction for quantile estimates. In the central limit theorem (9.7) for quantile estimation, the factor $p(1-p)$ is the variance of the indicator function of the event that the quantile x_p is exceeded. With importance sampling, this factor gets replaced by the variance of the importance sampling estimator of the probability of this event. See Theorem 6 of GHS [144].

9.2.3 Stratified Sampling

To further reduce variance, we now apply stratified sampling to the estimator (9.14) by stratifying Q. If the true loss L were exactly equal to the Q, this would remove all variance, in the limit of infinitely fine stratification. (See the discussion following equation (4.46).) This makes the approach attractive even if Q is only an approximation to L.

Stratifying Q in (9.14) is a special case of a more general variance reduction strategy of applying importance sampling by exponentially twisting a random variable and then stratifying that random variable. This is then equivalent to stratifying the likelihood ratio and thus to removing variance that may have been introduced through the likelihood ratio. We encountered this strategy in Section 4.6.2 where we applied importance sampling to add a mean vector μ to a normal random vector Z and then stratified $\mu^\top Z$.

Recall from our general discussion in Section 4.3 that to apply stratified sampling to the pair (Q, L) with stratification variable Q, we need to address two issues:

o We need to find intervals A_k, $k = 1, \ldots, K$, of known (and perhaps of equal) probability for Q; these are the strata.
o We need a mechanism for sampling (Q, L) conditional on Q falling in a particular stratum.

Once we have strata partitioning the real line, we use the decomposition

$$E_\theta[e^{-\theta Q + \psi(\theta)} \mathbf{1}\{L > x\}] = \sum_{k=1}^{K} P_\theta(Q \in A_k) E_\theta[e^{-\theta Q + \psi(\theta)} \mathbf{1}\{L > x\} | Q \in A_k].$$

(9.17)

We estimate each conditional expectation by sampling (Q, L) conditional on $Q \in A_k$ and combine these estimates with the known stratum probabilities $P_\theta(Q \in A_k)$.

Defining Strata

As explained in Example 4.3.2, defining strata for a random variable is in principle straightforward given its cumulative distribution function. In the case of Q, the cumulative distribution is available through transform inversion. However, some modification of the inversion integral (9.6) is necessary because for (9.17) we need the distribution of Q under the importance sampling measure P_θ.

Conveniently, Q remains a quadratic function of normal random variables under P_θ because each Z_j remains normal under P_θ. Using the parameters in (9.12), we find that

$$\tilde{Z}_j = \frac{Z_j - \mu_j(\theta)}{\sigma_j(\theta)}, \quad j = 1, \ldots, m,$$

are independent $N(0, 1)$ random variables under P_θ. From (9.4), we get

$$Q = a + \sum_{j=1}^{m} (b_j(\mu_j + \sigma_j \tilde{Z}_j) + \lambda_j(\mu_j + \sigma_j \tilde{Z}_j)^2)$$

$$\equiv \tilde{a} + \sum_{j=1}^{m} (\tilde{b}_j \tilde{Z}_j + \tilde{\lambda}_j \tilde{Z}_j)^2,$$

the new coefficients defined by matching terms. The cumulant generating function of Q under P_θ now has the form in (9.5) but with the new coefficients.

A somewhat simpler derivation observes that the new cumulant generating function — call it ψ_θ — satisfies

$$\psi_\theta(u) = \log E_\theta[e^{uQ}] = \log E_\theta[e^{uQ} e^{\theta Q - \psi(\theta)}] = \psi(\theta + u) - \psi(\theta).$$

The characteristic function of Q under the new measure is thus given by

$$\hat{\phi}_\theta(u) = \mathsf{E}_\theta[e^{iuQ}] = e^{\psi(\theta+iu)-\psi(\theta)}, \quad i = \sqrt{-1}. \qquad (9.18)$$

Using this function in the inversion integral (9.6) yields the distribution

$$F_\theta(x) = P_\theta(Q \le x).$$

By computing (9.6) iteratively we can find points a_k at which $F_\theta(a_k) = p_1 + \cdots + p_k$ for desired stratum probabilities p_i and then set $A_k = (a_{k-1}, a_k)$. The endpoints of the intervals have zero probability and may be omitted or included.

Conditional Sampling

Given a mechanism for evaluating the distribution F_θ (in this case through transform inversion), we could generate Q conditional on $Q \in A_k$ using the inverse transform method as in Example 4.3.2. However, to use the stratified decomposition (9.17), we need a mechanism to generate the pair (Q, L) conditional on $Q \in A_k$ and this is less straightforward.

Whether or not we apply stratified sampling, we do not generate (Q, L) directly. Rather, we generate a normal random vector Z and then evaluate both Q and L as functions of Z; see the algorithm of Section 9.2.2. To sample (Q, L) conditional on $Q \in A_k$, we therefore need to sample Z conditional on $Q \in A_k$ and then evaluate L (and Q) from Z.

The analogous step in Section 4.6.2 required us to generate Z conditional on $\mu^\top Z$, and this proved convenient because the conditional distribution is itself normal. However, no similar simplification applies in generating Z given Q; see the discussion of radial stratification in Section 4.3.2.

In the absence of a direct way of generating Z conditional on the value of a quadratic function of Z, GHS [143] use a brute-force acceptance-rejection method along the lines of Example 2.2.8. The method generates independent replications of Z from $N(\mu(\theta), \Sigma(\theta))$, its unconditional distribution under P_θ, evaluates Q, and assigns Z to stratum k if $Q \in A_k$. If no more samples for this stratum are needed, Z is simply discarded.

This is illustrated in Figure 9.3 for an example with $Q = \lambda_1 Z_1^2 + \lambda_2 Z_2^2$ and positive λ_1, λ_2. The four strata of the distribution of Q in the lower part of the figure define the four elliptical strata for (Z_1, Z_2) in the upper part of the figure. The target sample size is eight. The labels on the points show the order in which they were generated and the labels under the density show which samples have been accepted for which strata. The points labeled 6, 9, and 10 have been rejected because their strata are full; the second stratum needs one more observation, so more candidates need to be generated.

To formulate this procedure more generally, suppose we have defined strata $A_1, \ldots, A_K$ for Q and want to generate n_k samples from stratum k, $k = 1, \ldots, K$. Consider the following algorithm:

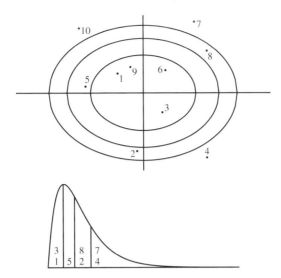

Fig. 9.3. Illustration of stratified sampling by acceptance-rejection. The strata of the density of Q define elliptical strata for (Z_1, Z_2). The labels under the density indicate the points accepted for each stratum in generating a sample of size eight.

- initialize stratum counters $N_k \leftarrow 0$, $k = 1, \dots, K$
- repeat
  - generate $Z \sim N(\mu(\theta), \Sigma(\theta))$
  - evaluate Q
  - find κ such that $Q \in A_\kappa$
  - if $N_\kappa < n_\kappa$
    $$N_\kappa \leftarrow N_\kappa + 1$$
    $$Z_{\kappa N_\kappa} \leftarrow Z$$
  until all $N_k = n_k$, $k = 1 \dots, K$

For each stratum $k = 1, \dots, K$, this algorithm produces samples Z_{kj}, $j = 1 \dots, n_k$, with the property that Q evaluated at Z_{kj} is in A_k. Also, each accepted Z_{kj} has the conditional distribution of Z given $Q \in A_k$.

This algorithm potentially generates many more candidate values of Z than the total number required (which is $n_1 + \cdots + n_K$) and in this respect may seem inefficient. But executing the steps in this algorithm can take far less time than revaluing the portfolio at each outcome of Z. In that case, expending the effort to generate stratified scenarios is justified. Appendix A of GHS [143] includes an analysis of the overhead due to this acceptance-rejection algorithm. It indicates, for example, that in generating 20 samples from each of 100 equiprobable strata, there is less than a 5% chance that the number of candidates generated will exceed 3800.

Write Q_{kj} and L_{kj} for the values of Q and L computed from Z_{kj}, $j = 1, \dots, n_k$, $k = 1, \dots, K$. Then the (Q_{kj}, L_{kj}) form a stratified sample from the distribution of (Q, L) with stratification variable Q. These can now be combined as in (4.32) to produce the estimator

$$\sum_{k=1}^{K} P_\theta(Q \in A_k) \frac{1}{n_k} \sum_{j=1}^{n_k} e^{-\theta Q_{kj} + \psi(\theta)} \mathbf{1}\{L_{kj} > x\}$$

of the loss probability $P(L > x)$.

Numerical results reported in GHS [143] show that this combination of importance sampling and stratification can substantially reduce variance, especially at values of x for which $P(L > x)$ is small. The results in [143] use strata that are equiprobable under the θ-twisted distribution, meaning that

$$P_\theta(Q \in A_1) = \cdots = P_\theta(Q \in A_K) = \frac{1}{K},$$

and a proportional allocation of samples to strata (in the sense of Section 4.3.1). GHS [141] propose and test other allocation rules. They find that some simple heuristics can further reduce variance by factors of 2–5.

If we use only the linear part of Q (a delta-only approximation) for importance sampling and stratification, then Q and Z are jointly normal and the conditional distribution of Z given Q is again normal. In this case, the conditional sampling can be implemented more efficiently by sampling directly from this conditional normal distribution; see Section 4.3.2.

Poststratification

An alternative to using acceptance-rejection for stratification uses unconditional samples of Z and weights the stratum averages by the stratum probabilities. This is poststratification, as discussed in Section 4.3.3. Let $Z_1, \ldots, Z_n$ be independent samples from $N(\mu(\theta), \Sigma(\theta))$. Let N_k be the random number of these falling in stratum k, $k = 1, \ldots, K$, and label these Z_{kj}, $j = 1, \ldots, N_k$. Let (Q_{kj}, L_{kj}) be the values computed from Z_{kj}. The poststratified estimator is

$$\sum_{k=1}^{K} P_\theta(Q \in A_k) \frac{1}{N_k} \sum_{j=1}^{N_k} e^{-\theta Q_{kj} + \psi(\theta)} \mathbf{1}\{L_{kj} > x\}.$$

Take the kth term to be zero if $N_k = 0$.

As shown in Section 4.3.3, the poststratified estimator achieves the same variance reduction as the genuinely stratified estimator in the limit as the same size increases. If L is much more time-consuming to evaluate than Q (as would often be the case in practice), then stratification through acceptance-rejection is preferable; otherwise, the poststratified estimator offers a faster alternative.

Numerical Examples

Table 9.1 compares variance reduction factors using the delta-gamma approximation as a control variate (CV), for importance sampling (IS), and for combined importance sampling and stratification (IS-S). These results are from GHS [142]; more extensive results are reported there and in [143].

The results in the table are for portfolios of options on 10 or 100 underlying assets. The assets are modeled as geometric Brownian motion, and option prices, deltas, and gammas use Black-Scholes formulas. Each underlying asset has a volatility of 0.30 and the assets are assumed uncorrelated except in the last case. We assume a risk-free interest rate of $r = 5\%$; for purposes of these examples, we do not distinguish between the real and risk-neutral rates of return on the assets, using r for both. We use the following simple test portfolios:

(A) Short positions in 10 calls and 5 puts on each of 10 underlying assets, all options expiring in 0.1 years.
(B) Same as (A), but with the number of puts increased to produce a net delta of zero.
(C) Short positions in 10 calls and 10 puts on each of 100 underlying assets, all options expiring in 0.1 years.
(D) Same as (C), but with each pair of underlying assets having correlation 0.2.

We use portfolios of short positions so that large moves in the prices of the underlying assets produce losses. In each case, we use a loss threshold x resulting in a loss probability $P(L > x)$ near 5%, 1%, or 0.5%. This threshold is specified in the table through x_{std}, the number of standard deviations above the mean in the distribution of Q:

$$x = \mathsf{E}[Q] + x_{\mathrm{std}}\sqrt{\mathsf{Var}[Q]}$$

$$= a + \sum_{j=1}^{m}\lambda_j + x_{\mathrm{std}}\sqrt{\sum_{j=1}^{m}(b_j^2 + 2\lambda_j^2)}.$$

The results displayed in the last three columns of Table 9.1 are variance reduction factors. Each is the ratio of variances using ordinary Monte Carlo and using a variance reduction technique. The results indicate that the delta-gamma control variate (column CV) yields modest variance reduction and that its effectiveness generally decreases at smaller loss probabilities. Importance sampling (IS) yields greater variance reduction, especially at the smallest loss probabilities. The combination of importance sampling and stratification (IS-S) yields very substantial variance reduction. The variance ratios in the table are estimated using 120,000 replications for each method, with 40 strata and 3000 replications per strata for the IS-S results.

Portfolio	x_{std}	$P(L > x)$	Variance Ratios		
			CV	IS	IS-S
(A)	1.8	5.0%	5	7	30
	2.6	1.1%	3	22	70
	3.3	0.3%	2	27	173
(B)	1.9	4.7%	3	6	14
	2.8	1.1%	2	18	30
	3.2	0.5%	2	28	48
(C)	2.5	1.0%	3	27	45
(D)	2.5	1.0%	2	10	23

Table 9.1. Variance reduction factors in estimating loss probabilities using the delta-gamma approximation as a control variate (CV), for importance sampling (IS), and for importance sampling with stratification (IS-S).

More extensive results are reported in GHS [143]. Some of the test cases included there are specifically designed to challenge the methods. These include portfolios of digital and barrier options combined in proportions that yield a net delta of zero. None of the variance reduction methods based on Q is effective in the most extreme cases, but the overall pattern is similar to the results in Table 9.1: importance sampling is most effective at small loss probabilities and stratification yields substantial additional variance reduction. Similar observations apply in estimating a conditional excess loss; the variance reduction achieved in estimating $\mathsf{E}[L|L > x]$ is usually about the same as that achieved in estimating $P(L > x)$.

9.3 A Heavy-Tailed Setting

9.3.1 Modeling Heavy Tails

We noted in Section 9.1.1 that the normal distribution has shortcomings as a model of changes in market prices: in virtually all markets, the distribution of observed price changes displays a higher peak and heavier tails than can be captured with a normal distribution. This is especially true over short time horizons; see, for example, the daily return statistics in Appendix F of Duffie and Pan [103]. High peaks and heavy tails are characteristic of a market with small price changes in most periods accompanied by occasional very large price changes.

While other theoretical distributions may do a better job of fitting market data, the normal distribution offers many convenient properties for modeling and computation. Choosing a description of market data thus entails a compromise between realism and tractability. This section shows how some of the methods of Section 9.2 can be extended beyond the normal distribution to capture features of market data.

The qualitative property of having a high peak and heavy tails is often measured through kurtosis. The kurtosis of a random variable X with mean μ is given by

$$\frac{\mathsf{E}[(X - \mu)^4]}{(\mathsf{E}[(X - \mu)^2])^2},$$

assuming X has a finite fourth moment. Every normal random variable has a kurtosis of 3; distributions are sometimes compared on the basis of *excess kurtosis*, the difference between the kurtosis and 3. A sample kurtosis can be calculated from data by replacing the expectations in the definition with sample averages.

Kurtosis normalizes the fourth central moment of a distribution by the square of its variance. If two distributions have the same standard deviation, the one with higher kurtosis will ordinarily have a higher peak and heavier tails. Such a distribution is called leptokurtotic.

The simplest extension of the normal distribution exhibiting higher kurtosis is a mixture of two normals. Consider a mixture

$$qN(0, \sigma_1^2) + (1 - q)N(0, \sigma_2^2)$$

with $q \in (0, 1)$. By this we mean the distribution of a random variable drawn from $N(0, \sigma_1^2)$ with probability q and drawn from $N(0, \sigma_2^2)$ with probability $1 - q$. Its variance is

$$\sigma^2 = q\sigma_1^2 + (1 - q)\sigma_2^2,$$

and its kurtosis is

$$\frac{3(q\sigma_1^4 + (1 - q)\sigma_2^4)}{(q\sigma_1^2 + (1 - q)\sigma_2^2)^2}.$$

Whereas the normal distribution $N(0, \sigma^2)$ with the same variance has kurtosis 3, the mixture can achieve arbitrarily high kurtosis if we let σ_1 increase and let q approach zero while keeping the overall variance σ^2 unchanged. Figure 9.4 compares the case $\sigma_1 = 1.4$, $\sigma_2 = 0.6$, and $q = 0.4$ with the standard normal density, both having a standard deviation of 1.

This mechanism describes a market in which a fraction q of periods (e.g., days) have high variability and a fraction $1 - q$ have low variability. It can be extended to a multivariate model by mixing multivariate normals $N(0, \Sigma_1)$ and $N(0, \Sigma_2)$. All of the variance reduction techniques discussed in Section 9.2 extend in a straightforward way to this class of models. The techniques apply to samples from each $N(0, \Sigma_i)$, and results from the two distributions can be combined using the weights q and $1 - q$.

Heavy Tails

Kurtosis provides some information about the tails of a distribution, but it is far from a complete measure of the heaviness of the tails. Further information

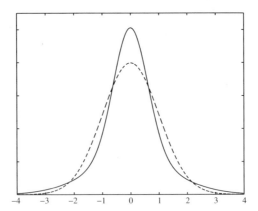

Fig. 9.4. Solid line is the mixed normal density with $\sigma_1 = 1.4$, $\sigma_2 = 0.6$, and $q = 0.4$. Dashed line is the standard normal density.

is provided by the rate of decay of a probability density or, equivalently, the number of finite moments or exponential moments.

To focus on just one tail, consider the case of a nonnegative random variable X, which could be the positive part of a random variable taking both negative and positive values. The following conditions define three important categories of distributions:

(i) $\mathsf{E}[\exp(\theta X)] < \infty$ for all $\theta \in \Re$;
(ii) $\mathsf{E}[\exp(\theta X)] < \infty$ for all $\theta < \theta^*$ and $\mathsf{E}[\exp(\theta X)] = \infty$ for all $\theta > \theta^*$, for some $\theta^* \in (0, \infty)$;
(iii) $\mathsf{E}[X^r] < \infty$ for all $r < \nu$ and $\mathsf{E}[X^r] = \infty$ for all $r > \nu$, for some $\nu \in (0, \infty)$.

The first category includes all normal random variables (or their positive parts) and all bounded random variables. The second category describes distributions with exponential tails and includes all gamma distributions and in particular the exponential density $\theta^* \exp(-\theta^* x)$. The third category (for which $\mathsf{E}[\exp(\theta X)] = \infty$ for all $\theta > 0$) describes heavy-tailed distributions. This category includes the stable Paretian distributions discussed in Section 3.5.2, for which $\nu \le 2$. This category includes distributions whose tails decay like $x^{-\nu}$ and, more generally, regularly varying tails, as defined in, e.g., Embrechts et al. [111]. These three categories are not exhaustive; the lognormal, for example, fits just between the second and third categories having $\theta^* = 0$ and $\nu = \infty$.

Empirical data is necessarily finite, making it impossible to draw definite conclusions about the extremes of a distribution. Many studies have found that the third category above provides the best description of market data, with ν somewhere in the range of 3–7, depending on the market and the time horizon over which returns are measured. Although a mixture of two normal distributions can produce an arbitrarily large kurtosis, it lies within the first category above (because the tail is ultimately determined by the larger of the two standard deviations), so it does not provide an entirely satisfactory model.

Student t Distribution

An alternative extension of the normal distribution that provides genuinely heavy tails is the Student t distribution with density

$$f_\nu(x) = \frac{\Gamma((\nu+1)/2)}{\sqrt{\nu\pi}\Gamma(\nu/2)}\left(1+\frac{x^2}{\nu}\right)^{-(\nu+1)/2}, \quad -\infty < x < \infty,$$

$\Gamma(\cdot)$ here denoting the gamma function. The parameter ν (the degrees of freedom) controls the heaviness of the tails. If X has this t_ν density, then

$$P(X > x) \sim \text{constant} \times x^{-\nu}$$

as $x \to \infty$, and ν determines the number of finite moments of $|X|$ in the sense of category (iii) above. If $\nu > 2$, then X has variance $\nu/(\nu-2)$. The standard normal density is the limit of f_ν as $\nu \to \infty$.

Figure 9.5 compares the t_5 density (solid line) with a normal density (dashed line) scaled to have the same variance. The higher peak of the t distribution is evident from the left panel of the figure, which plots the densities. The right panel shows the logarithms of the densities from which the heavier tails of the t distribution are evident.

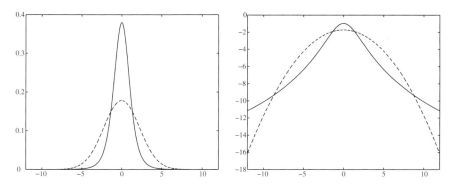

Fig. 9.5. Comparison of t_5 (solid) and normal distribution (dashed) with the same variance. Left panel shows the densities, right panel shows the log densities.

A t_ν random variable can be represented as a ratio $Z/\sqrt{Y/\nu}$ in which Z has the standard normal distribution and Y has the chi-square distribution χ^2_ν and is independent of Z. This representation shows that t_ν is a mixture of normals, but rather than mixing just two normals, the t_ν mixes infinitely many. If we interpret a mixture of two normals as a normal distribution with random variance taking values σ_1^2 and σ_2^2, then the t_ν can be thought of as a normal with random variance equal to ν/Y. This suggests a mechanism for generating an even richer class of distributions by replacing ν/Y with other random variables.

Multivariate t

Measuring the risk in a portfolio requires a multivariate model of changes in market prices, so we need a multivariate model with heavy-tailed marginals. A t density in $\Re^m$ (as defined in Anderson [17]) is given by

$$f_{\nu,\Sigma}(x) = \frac{\Gamma((\nu+m)/2)}{(\nu\pi)^{m/2}\Gamma(\nu/2)|\Sigma|^{1/2}} \left(1 + \frac{1}{\nu}x^\top \Sigma^{-1} x\right)^{-(\nu+m)/2}, \qquad x \in \Re^m.$$
(9.19)

Here, Σ is a symmetric, positive definite matrix, $|\Sigma|$ is its determinant, and ν is again the degrees of freedom parameter. If $\nu > 2$, then the distribution has covariance matrix $\nu\Sigma/(\nu-2)$. If all diagonal entries of Σ are equal to 1, then Σ is the correlation matrix of the distribution (assuming $\nu > 2$), and each marginal is a univariate t_ν distribution. Without this restriction on the diagonal of Σ, each coordinate has the distribution of a scaled t_ν random variable. In the limit as $\nu \to \infty$, (9.19) becomes the density of the multivariate normal distribution $N(0, \Sigma)$.

If $(X_1, \ldots, X_m)$ have (9.19) as their joint density, then they admit the representation

$$(X_1, \ldots, X_m) =_d \frac{(\xi_1, \ldots, \xi_m)}{\sqrt{Y/\nu}},$$
(9.20)

where $=_d$ denotes equality in distribution, $\xi = (\xi_1, \ldots, \xi_m)$ has distribution $N(0, \Sigma)$, and Y has distribution χ_ν^2 independent of ξ. A multivariate t random vector is therefore a multivariate normal vector with a randomly scaled covariance matrix. Also, it follows from (9.19) that the vector X with density $f_{\nu,\Sigma}$ can be represented as

$$X = \frac{AZ}{\sqrt{Y/\nu}} = A\tilde{X},$$
(9.21)

where $AA^\top = \Sigma$, $Z \sim N(0, I)$, and $\tilde{X}$ is a multivariate t random vector with density $f_{\nu,I}$. The components of $\tilde{X}$ are uncorrelated (their correlation matrix is the identity), but not independent. Dependence is introduced by the shared denominator.

The representation (9.20) and the factorization in (9.21) make (9.19) a particularly convenient multivariate distribution and provide a mechanism for simulation. Much of what we discuss in this section extends to other random vectors with the representation (9.20) for some other choice of denominator.

A shortcoming of (9.19) is that it requires all marginals to have the same parameter ν and thus the same degree of heaviness in their tails. Because empirical evidence suggests that for most markets ν should be in the range of 3–7, this is not a fatal flaw. It does, however, suggest considering more general multivariate t distributions.

The t-Copula

One mechanism for allowing different coordinates to have different parameters starts with a multivariate t vector of the type in (9.20) and then modifies the marginals. Let F_ν denote the cumulative distribution function of the univariate t_ν distribution. Let the vector X have the representation in (9.20) with Σ having all diagonal entries equal to 1. This implies that $X_i \sim t_\nu$ and then that $F_\nu(X_i)$ is uniformly distributed on the unit interval. Just as in the inverse transform method (Section 2.2.1), applying an inverse distribution $F_{\nu_i}^{-1}$ gives $F_{\nu_i}^{-1}(F_\nu(X_i))$ the t distribution with ν_i degrees of freedom. Applying such a transformation to each coordinate produces a vector

$$(\tilde{X}_1, \dots, \tilde{X}_m) = (F_{\nu_1}^{-1}(F_\nu(X_1)), \dots, F_{\nu_m}^{-1}(F_\nu(X_1))), \tag{9.22}$$

the components of which have t distributions with arbitrary parameters ν_1, ..., ν_m.

This is a special case of a more general mechanism (to which we return in Section 9.4.2) for constructing multivariate distributions through "copulas," in this case a t-copula. The transformation in (9.22) does not preserve correlations, making it difficult to give a concise description of the dependence structure of $(\tilde{X}_1, \dots, \tilde{X}_m)$; it does however preserve Spearman *rank* correlations. This and related properties are discussed in Embrechts et al. [112].

To simulate changes ΔS_i in market prices using (9.22), we would set

$$\Delta S_i = \tilde{\sigma}_i \sqrt{\frac{\nu_i - 2}{\nu_i}} \tilde{X}_i, \tag{9.23}$$

assuming $\nu_i > 2$. This makes $\tilde{\sigma}_i^2$ the variance of ΔS_i and gives ΔS_i a scaled t_{ν_i} distribution. The specification of the model would run in the opposite direction: using market data, we would first estimate $\tilde{\sigma}_i$ and ν_i for each marginal ΔS_i, $i = 1, \dots, m$; we would then apply the transformation $X_i = F_\nu^{-1}(F_{\nu_i}(\Delta S_i))$ to each coordinate of the data, and finally estimate the correlation matrix Σ of $(X_1, \dots, X_m)$. A similar procedure is developed and tested in Hosking, Bonti, and Siegel [188] based on the normal distribution (i.e., with $\nu = \infty$); GHS [144] recommend using a value of ν closer to $\nu_1, \dots, \nu_m$. Estimation of ν_i is discussed in Hosking et al. [188] and Johnson, Kotz, and Balakrishnan [202].

9.3.2 Delta-Gamma Approximation

The variance reduction techniques in Section 9.2 rely on the tractability of
the delta-gamma approximation (9.2) when ΔS has a multivariate normal
distribution. To extend these techniques to the multivariate t, we therefore
need to extend the analysis of the delta-gamma approximation. This analysis,
carried out in GHS [144], is of independent interest because it permits use
of the approximation as a rough but fast alternative to simulation. The extension applies to a large class of distributions admitting the representation
(9.20), in addition to the multivariate t. We comment on the application to
the transformed multivariate t in (9.22) after considering the simpler case of
(9.19).

Suppose, then, that ΔS has the multivariate t_ν density in (9.19) with
$\Sigma = \Sigma_S$. By following exactly the steps leading to (9.4), we arrive at the
representation

$$Q = a + \sum_{j=1}^{m}(b_j X_j + \lambda_j X_j^2) \tag{9.24}$$

for the quadratic approximation to the loss L, with

$$CX = \Delta S, \quad X = (X_1, \ldots, X_m)^\top,$$

$$b = -C^\top \delta, \quad b = (b_1, \ldots, b_m)^\top,$$

$$CC^\top = \Sigma_S, \quad -\tfrac{1}{2}C^\top \Gamma C = \begin{pmatrix} \lambda_1 & & & \\ & \lambda_2 & & \\ & & \ddots & \\ & & & \lambda_m \end{pmatrix},$$

where $\lambda_1, \ldots, \lambda_m$ are the eigenvalues of $-\Sigma_S\Gamma/2$ and X has the multivariate
t_ν density (9.19) with $\Sigma = I$ the identity matrix. The constant a in (9.24) is
$-(\Delta t)\partial V/\partial t$, just as in (9.3).

At this point, the analyses of the normal and t distributions diverge. Uncorrelated jointly normal random variables are independent of each other, so
(9.4) expresses Q as a sum of independent random variables. But as explained
following (9.21), the components $X_1, \ldots, X_m$ of the multivariate t random
vector are not independent even if they are uncorrelated. The independence
of the summands in (9.4) allowed us to derive the moment generating function
(and then characteristic function) of Q as the product of the moment generating functions of the summands. No such factorization applies to (9.24) because
of the loss of independence. Moreover, in the heavy-tailed setting, the X_i and
Q do not have moment generating functions; they belong to category (iii)
of Section 9.3.1. This further complicates the derivation of the characteristic
function of Q.

Indirect Delta-Gamma

These obstacles can be circumvented through an *indirect* delta-gamma approximation. Use (9.20) to write

$$Q = a + \sum_{j=1}^{m} \left(b_j \frac{Z_j}{\sqrt{Y/\nu}} + \lambda_j \frac{Z_j^2}{Y/\nu} \right)$$

with $Z_1, \ldots, Z_m$ independent $N(0,1)$ random variables. For fixed $x \in \mathfrak{R}$, define

$$Q_x = (Y/\nu)(Q - x) = (a - x)(Y/\nu) + \sum_{j=1}^{m} (b_j \sqrt{Y/\nu} Z_j + \lambda_j Z_j^2) \qquad (9.25)$$

and observe that

$$P(Q \le x) = P(Q_x \le 0). \qquad (9.26)$$

Thus, we can evaluate the distribution of Q at x indirectly by evaluating that of Q_x at zero. The random variable Q_x turns out to be much more convenient to work with.

Conditional on Y, Q_x is a quadratic function of independent standard normal random variables. We can therefore apply (9.5) to get (as in Theorem 3.1 of GHS [144])

$$\mathsf{E}[e^{\theta Q_x}|Y] = \exp\left((a - x)\theta Y/\nu + \tfrac{1}{2} \sum_{j=1}^{m} \frac{\theta^2 b_j^2 Y/\nu}{1 - 2\theta\lambda_j} \right) \prod_{j=1}^{m} \frac{1}{\sqrt{1 - 2\theta\lambda_j}}, \qquad (9.27)$$

provided $2\max_j \theta\lambda_j < 1$. The coefficient of Y in this expression is

$$\alpha(\theta) = (a - x)\theta/\nu + \frac{1}{2} \sum_{j=1}^{m} \frac{\theta^2 b_j^2 \nu}{1 - 2\theta\lambda_j}. \qquad (9.28)$$

Suppose the moment generating function of Y is finite at $\alpha(\theta)$:

$$\phi_Y(\alpha(\theta)) \equiv \mathsf{E}[e^{\alpha(\theta)Y}] < \infty.$$

Then (9.27) yields

$$\phi_x(\theta) \equiv \mathsf{E}[e^{\theta Q_x}] = \phi_Y(\alpha(\theta)) \prod_{j=1}^{m} \frac{1}{\sqrt{1 - 2\theta\lambda_j}}. \qquad (9.29)$$

This is the moment generating function of Q_x. Replacing the real argument θ with a purely imaginary argument $\sqrt{-1}u$, $u \in \mathfrak{R}$, yields the value of the characteristic function at u. This characteristic function can be inserted in the inversion integral (9.6) to compute $P(Q_x \le 0)$ and thus $P(Q \le x)$.

In the specific case that X has a multivariate t_ν distribution, Y is χ_ν^2 and

$$\phi_Y(\theta) = (1 - 2\theta)^{-\nu/2}, \quad \theta < 1/2,$$

which completes the calculation of (9.29). However, this derivation shows that Q_x has a tractable moment generating function and characteristic function in greater generality.

Calculation of the distribution of Q through inversion of the characteristic function of Q_x is not quite as efficient as the approach in the multivariate normal setting of Section 9.1.2 based on direct inversion of the characteristic function of Q. Using the Fast Fourier Transform (as discussed in, e.g., Press et al. [299]), the distribution of Q can be evaluated at N points using N terms in the inversion integral in $O(N \log N)$ time, given the characteristic function of Q. In the indirect approach based on (9.26), evaluating the distribution of Q at N points requires N separate inversion integrals, because Q_x depends on x. With N terms in each inversion integral, the total effort becomes $O(N^2)$. This is not a significant issue in our application of the approximation to variance reduction because the number of points at which we need the distribution of Q is small — often just one.

The derivation of (9.29) relies on the representation (9.20) of the multivariate t distribution (9.19) with common parameter ν for all marginals. To extend it to the more general case of (9.22) with arbitrary parameters $\nu_1, \ldots, \nu_m$, we make a linear approximation to the transformation in (9.22). This converts a quadratic approximation in the $\tilde{X}_i$ variables into a quadratic approximation in the X_i variables.

In more detail, write $\Delta S_i = K_i(X_i)$, $i = 1, \ldots, m$, for the mapping from $X_1, \ldots, X_m$ to ΔS defined by (9.22) and (9.23). The deltas and gammas of the portfolio value V with respect to the X_i are given by

$$\frac{\partial V}{\partial X_i} = \delta_i \frac{\partial K_i}{\partial X_i},$$

$$\frac{\partial^2 V}{\partial X_i \partial X_j} = \Gamma_{ij} \frac{\partial K_i}{\partial X_i} \frac{\partial K_j}{\partial X_j}, \ i \neq j, \quad \frac{\partial^2 V}{\partial X_i^2} = \Gamma_{ii} \left(\frac{\partial K_i}{\partial X_i}\right)^2 + \delta_i \frac{\partial^2 K_i}{\partial X_i^2},$$

with all derivatives of K evaluated at 0. Using these derivatives, we arrive at an approximation to the change in portfolio value (as in (9.2)) that is quadratic in $X_1, \ldots, X_m$ and to which the indirect approach through Q_x and (9.29) then applies.

9.3.3 Variance Reduction

We turn now to the use of the delta-gamma approximation for variance reduction in the heavy-tailed setting. With normally distributed ΔS, we applied importance sampling by defining an exponential change of measure through Q in (9.10). A direct application of this idea in the heavy-tailed setting is impossible, because Q no longer has a moment generating function (it belongs to category (iii) of Section 9.3.1), so there is no way to normalize the exponential twisting factor $\exp(\theta Q)$ to produce a probability distribution for positive values of θ. Indeed, importance sampling with heavy-tailed distributions is a notoriously difficult problem. Asmussen et al. [23] demonstrate the failure of several plausible importance sampling estimators in heavy-tailed problems. Beyond its relevance to risk management, the example we treat here is of interest because it is one of few examples of effective importance sampling with heavy-tailed distributions.

Importance Sampling

We circumvent the difficulty of working with Q by exponentially twisting the indirect approximation Q_x instead. We derived the moment generating function of Q_x in (9.29). Let $\psi_x(\theta)$ denote its cumulant generating function, the logarithm of (9.29). Define an exponential family of probability measures P_θ through the likelihood ratio

$$\frac{dP_\theta}{dP} = e^{\theta Q_x - \psi_x(\theta)} \tag{9.30}$$

for all θ at which $\psi_x(\theta) < \infty$. This allows us to write loss probabilities as

$$P(L > y) = \mathsf{E}_\theta\left[e^{-\theta Q_x + \psi_x(\theta)}\mathbf{1}\{L > y\}\right]$$

and this yields the importance sampling estimator

$$e^{-\theta Q_x + \psi_x(\theta)}\mathbf{1}\{L > y\}$$

with (Q_x, L) sampled from their joint distribution under P_θ.

Before proceeding with the mechanics of this estimator, we should point out what makes this approach appealing. We are free to choose x, so suppose we have $y = x$. The second moment of the estimator is then given by

$$\mathsf{E}_\theta\left[e^{-2\theta Q_x + 2\psi_x(\theta)}\mathbf{1}\{L > x\}\right] = \mathsf{E}_\theta\left[e^{-\theta Q_x + \psi_x(\theta)}\left(\frac{dP}{dP_\theta}\right)\mathbf{1}\{L > x\}\right]$$

$$= \mathsf{E}\left[e^{-\theta Q_x + \psi_x(\theta)}\mathbf{1}\{L > x\}\right],$$

which shows that to reduce variance we need the likelihood ratio to be small on the event $\{L > x\}$. If Q provides a reasonable approximation to L, then when $L > x$ we often have $Q > x$ and thus $Q_x > 0$, and this tends to produce a small likelihood ratio if θ is positive.

This tendency must be balanced against the magnitude of ψ_x. We strike a balance through the choice of parameter θ. Choosing θ equal to θ_x, the root of the equation

$$\psi'_x(\theta_x) = 0,$$

minimizes the convex function ψ_x. Finding θ_x is a one-dimensional minimization problem and can be solved very quickly. This choice of θ imposes the condition (see (9.16) and the surrounding discussion)

$$\mathsf{E}_{\theta_x}[Q_x] = 0$$

and thus centers the distribution of Q_x near 0. This centers the distribution of Q near x which in turn makes the event $\{L > x\}$ less rare.

To achieve these properties, we need to be able to sample the pair (Q_x, L) from their distribution under P_θ. As in Section 9.2.2, we do not sample these directly; rather, we sample Y and $Z_1, \ldots, Z_m$. From these variables we compute Q_x and $X_1, \ldots, X_m$ and from the X_i we compute L. Thus, it suffices to find the distribution of $(Y, Z_1, \ldots, Z_m)$ under P_θ.

This problem is solved by Theorem 4.1 of GHS [144] and the answer is in effect embedded in the derivation of the moment generating function (9.29). Let f_Y denote the density of Y under the original probability measure $P = P_0$; this is the χ^2_ν density in the case of the multivariate t model. Let $\alpha(\theta)$ be as in (9.28). Then under P_θ, Y has density

$$f_{Y,\theta}(y) = e^{\alpha(\theta)y} f_Y(y)/\phi_Y(\theta), \qquad (9.31)$$

with ϕ_Y the moment generating function of f_Y. Conditional on Y, the components of $(Z_1, \ldots, Z_m)$ are independent normal random variables $Z_j \sim N(\mu_j(\theta), \sigma_j^2(\theta))$, with

$$\mu_j(\theta) = \frac{\theta b_j \sqrt{Y/\nu}}{1 - 2\lambda_j \theta}, \quad \sigma_j^2(\theta) = \frac{1}{1 - 2\lambda_j \theta}. \qquad (9.32)$$

This is verified by writing the likelihood ratio for the change of distribution in (9.31) and multiplying it by the likelihood ratio corresponding to (9.32) given Y. (The second step coincides with the calculation in Section 9.2.2.) The product of these factors simplifies to (9.30), as detailed in GHS [144]. Notice that (9.32) also follows from (9.25) and (9.12) once we condition on Y.

This result says that the change of measure defined by exponentially twisting Q_x can be implemented by exponentially twisting Y and then changing the means and variances of the Z_j given Y. We have circumvented the difficulty of applying importance sampling directly to the t random variables X_j by instead applying exponential twisting to the denominator and numerator random variables.

In the specific case that $Y \sim \chi^2_\nu$, the change of measure (9.31) gives Y a gamma distribution with shape parameter $\nu/2$ and scale parameter $2/(1 - 2\alpha(\theta))$; i.e.,

$$f_{Y,\theta}(y) = \left(\frac{2}{1 - 2\alpha(\theta)}\right)^{-\nu/2} \frac{y^{(\nu-2)/2}}{\Gamma(\nu/2)} \exp\left(\frac{-y}{2/(1 - 2\alpha(\theta))}\right).$$

The χ^2_ν distribution is the gamma distribution with shape parameter $\nu/2$ and scale parameter 2, so a negative value of $\alpha(\theta)$ decreases the scale parameter. This produces smaller values of Y and thus larger values of the X_j and ΔS_j.

Importance Sampling Algorithm

The following algorithm summarizes the implementation of this importance sampling technique in estimating a loss probability $P(L > x)$:

1. Find θ_x solving $\psi'(\theta_x) = 0$
2. For each of n replications
   (a) generate Y from f_{Y,θ_x}
   (b) given Y, generate Z from $N(\mu(\theta), \Sigma(\theta))$ with parameters given in (9.32)
   (c) set $X = Z/\sqrt{Y/\nu}$
   (d) set $\Delta S = CX$
   (e) calculate portfolio loss L and approximation Q_x
   (f) evaluate
   $$e^{-\theta_x Q_x + \psi_x(\theta_x)} \mathbf{1}\{L > x\} \qquad (9.33)$$
3. Calculate average of (9.33) over the n replications.

It is not essential (or necessarily optimal) to take the x in the definition of Q_x to be the same as the loss threshold in the probability $P(L > x)$. In estimating $P(L > y)$ at several values of y from a single set of replications, we would need to use a single value of x. This value should be chosen near the middle of the range of y under consideration, or a bit smaller.

Asymptotic Optimality

GHS [144] establish an asymptotic optimality result for this approach applied to estimating $P(Q > x)$. This should be viewed as indirect support for use of the method in the real problem of estimating $P(L > x)$. The notion of asymptotic optimality in this setting is appropriate to a heavy-tailed distribution and thus refers to a polynomial rather than exponential rate of decay. The precise statement of the result is a bit involved (see Theorem 5.1 of [144]), but it says roughly that $P(Q > x)$ is $O(x^{-\nu/2})$ and that the second moment of the estimator (9.33) (with L replaced by Q) is $O(x^{-\nu})$. Thus, the second moment decreases at twice the rate of the probability itself. This result is relatively insensitive to the choice of parameter θ, as explained in GHS [144].

Stratified Sampling

To further reduce the variance of the estimator (9.33), we stratify Q_x. The procedure is nearly the same as the one in the light-tailed setting of Section 9.2.3.

To define strata, we need the distribution of Q_x (for fixed x) under P_θ. We get this distribution through numerical inversion of the characteristic function of Q_x under P_θ. The moment generating function of Q_x under P_θ is given by

$u \mapsto \phi(\theta+u)/\phi(\theta)$ and evaluating this function along the imaginary axis yields the characteristic function.

Once we have defined strata for Q_x, we can use the acceptance-rejection procedure of Section 9.2.3 to generate samples of X conditional on the stratum containing Q_x. From these we generate price changes ΔS and portfolios losses L. Through these steps, we produce pairs (Q_x, L) with Q_x stratified.

Numerical Examples

Table 9.2 shows variance reduction factors in estimating loss probabilities for four of the test cases reported in GHS [144]. Portfolios (A) and (B) are as described in Section 9.2.3, except that each ΔS_j is now a scaled t random variable,

$$\Delta S_j = \tilde{\sigma}_j \sqrt{\frac{\nu_j - 2}{\nu_j}} t_{\nu_j}.$$

For cases (A) and (B) in the table, all marginals have $\nu_j = 5$. Cases (A') and (B') use the same portfolios of options but of the ten underlying assets, five have $\nu_j = 3$ and five have $\nu_j = 7$. These are generated using the t-copula in (9.22) starting from a reference value of $\nu = 5$.

In these examples, as in those of Section 9.2.3, we assume the portfolio value is given by applying the Black-Scholes formula to each option. We use an implied volatility of 0.30 in evaluating the Black-Scholes formula. The variance of ΔS_j is $\tilde{\sigma}_j^2$, which corresponds to an annual volatility of $\tilde{\sigma}_j/S_j\sqrt{\Delta t}$, given S_j. To equate this to the implied volatility, we set $\tilde{\sigma}_j = 0.30 S_j \sqrt{\Delta t}$. There is still an evident inconsistency in applying the Black-Scholes formula with t-distributed price changes, but option pricing formulas are commonly used this way in practice. Also, the variance reduction techniques do not rely on special features of the Black-Scholes formula, so the results should be indicative of what would be obtained using more complex portfolio revaluation.

The stratified results use 40 approximately equiprobable strata. The variance ratios in the table are estimated from 40,000 replications. With stratified sampling, this allows 1000 replications per stratum.

			Variance Ratios	
Portfolio	x	$P(L > x)$	IS	IS-S
(A)	469	1%	46	134
(B)	617	1%	42	112
(A')	475	1%	38	55
(B')	671	1%	39	57

Table 9.2. Variance reduction factors in estimating loss probabilities using the delta-gamma approximation for importance sampling (IS) and for importance sampling with stratification (IS-S).

The results in Table 9.2 show even greater variance reduction in the heavy-tailed setting than in the normal setting of Table 9.1. The results also suggest that the methods are less effective when applied with the t-copula (9.22) than when all marginals have the same parameter ν. Both of these observations are further supported by the more extensive numerical tests reported in GHS [144]. The worst cases found in [144], resulting in little or no variance reduction, combine the t-copula model of underlying assets with portfolios of digital and barrier options. Results in [144] also indicate that the variance reduction achieved in estimating a conditional excess $\mathsf{E}[L|L > x]$ is usually about the same as that achieved in estimating the loss probability $P(L > x)$.

Figure 9.6 (from [144]) gives a graphical illustration of the variance reduction achieved with portfolio (A). The figure plots estimated loss probabilities $P(L > x)$ for multiple values of x. The two outermost lines (solid) show 99% confidence intervals using standard simulation; these lines are formed by connecting confidence limits computed at each value of x. The two dotted lines show the tight intervals achieved by combining importance sampling and stratified sampling. The results using the variance reduction techniques were all estimated simultaneously: all use the same twisting parameter θ_x, the same strata, and the same values of (Q_x, L), with $x \approx 400$. This is important because in practice one would not want to generate separate scenarios for different thresholds x.

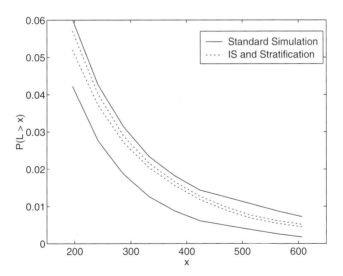

Fig. 9.6. Confidence intervals for loss probabilities for portfolio (A) using standard simulation and importance sampling combined with stratification, from GHS [144].

The figure also illustrates how reducing variance in estimates of loss probabilities results in more precise quantile estimates. To estimate the 99% VAR, for example, we would read across from 0.01 on the vertical axis until we hit the estimated curve $P(L > x)$, and then read down to find the corresponding value of x. A rough indication of the uncertainty in the VAR estimate is the space between the confidence interval curves at a height of 0.01. The outer confidence bands (based on standard simulation) produce a wide interval ranging roughly from $x = 375$ to 525, whereas the inner confidence bands produce a tight interval for the quantile.

9.4 Credit Risk

Sections 9.1–9.3 of this chapter deal with market risk — the risk of losses resulting from changes in market prices. *Credit risk* refers to losses resulting from the failure of an obligor (a party under a legal obligation) to make a contractual payment. Credit risk includes, for example, the possibility that a debtor will fail to repay a loan, a bond issuer will miss a coupon payment, or a counterparty to a swap will fail to make an interest payment. Market risk and credit risk interact, so the distinction between the two is not always sharp; but the models and methods used in the two settings differ in the features they emphasize.

This section introduces some of the methods used to model credit risk, with emphasis on their implementation through simulation. The development of credit risk models remains an active area of research, and the field is currently spread wide over modeling variations too numerous to survey. Our goal is to illustrate some of the key features of models used in this area.

Section 9.4.1 discusses models of the time to default for a single obligor. Such models are used in valuing corporate bonds and credit derivatives tied to a single source of credit risk. Section 9.4.2 discusses mechanisms for capturing dependence between the default times of multiple obligors, loosely referred to as "default correlation." This is a central issue in measuring the credit risk in a portfolio and in valuing credit derivatives tied to the creditworthiness of multiple obligors. Section 9.4.3 deals with a simulation method for measuring portfolio credit risk.

9.4.1 Default Times and Valuation

The simplest setting in which to discuss credit risk is the valuation of a zero-coupon bond subject to possible default. The issuer of the bond is scheduled to make a fixed payment of 1 at a fixed time T. If the issuer goes into default prior to time T, no payment is made; otherwise the payment of 1 is made as scheduled. By letting τ denote the time of default, we can combine the two cases by writing the obligor's payment as $\mathbf{1}\{\tau > T\}$.

At the heart of most models of credit risk is a mechanism describing the occurrence of default and thus the distribution of τ. In referring to the distribution of τ, we should distinguish the distribution under a risk-neutral probability measure from the distribution under the objective probability measure describing the observed time to default. The risk-neutral distribution is appropriate for valuation and the objective distribution is appropriate for measuring risk. Some models of default are used in both settings; we will not explicitly distinguish between the two and instead let context determine which is relevant.

Given the (risk-neutral) distribution of τ and a short rate process $r(t)$, the value of the bond payment $\mathbf{1}\{\tau > T\}$ is in principle determined as

$$\mathsf{E}\left[e^{-\int_0^T r(t)\,dt}\mathbf{1}\{\tau > T\}\right]. \tag{9.34}$$

More generally, if default at time $\tau < T$ yields a cashflow of $X(\tau)$, then the value of the bond is given by

$$\mathsf{E}\left[e^{-\int_0^\tau r(t)\,dt}X(\tau) + e^{-\int_0^T r(t)\,dt}\mathbf{1}\{\tau > T\}\right],$$

assuming the event $\{\tau = T\}$ has zero probability. Expressions of this type are sometimes used in reverse to find the distribution of the time to default implied by the market prices of corporate bonds.

Default and Capital Structure

A line of research that includes Merton [262], Black and Scholes [50], Leland [230], and much subsequent work values corporate bonds by starting from fundamental principles of corporate finance. This leads to a characterization of the default time τ as a first-passage time for the value of the issuing firm.

This approach posits a stochastic process for the value of a firm's assets. The value of the firm's assets equals the value of its equity plus the value of its debt. The limited liability feature of equity makes it an option on the firm's assets: if the value of the assets becomes insufficient to cover the debt, the equity holders may walk away and surrender the firm to the bond holders.

Valuing equity and debt thus reduces to valuing a type of barrier option in which the firm's assets act as the underlying state variable and the barrier crossing triggers default. Different models make different assumptions about the dynamics of the firm's value and about how the level of the barrier is determined; see Chapter 11 of Duffie [98] for examples and extensive references.

In a sufficiently complex model, one might consider using simulation to determine the distribution of the default time. Importance sampling would then be potentially useful, especially if default is rare. Approximations of the type discussed in Section 6.4 can be used to reduce discretization error associated with the barrier crossing; see Caramellino and Iovino [75] for work on this topic. But looking at default from the perspective of capital structure and corporate finance has proved more successful as a conceptual framework than as a valuation tool. Simpler models are usually used in practice. We therefore omit further discussion of simulation in this setting.

Default Intensity

A stochastic intensity for a default time τ is a process $\lambda(t)$ for which the process

$$1\{\tau \leq t\} - \int_0^t \lambda(u)\, du \tag{9.35}$$

is a martingale. For small Δt, interpret $\lambda(t)\Delta t$ as the conditional probability of default in $(t, t + \Delta t)$ given information available at time t. This generalizes the intensity of a Poisson process, for which λ is deterministic.

To make precise the notion of available information and the martingale property of (9.35) we would need to specify a filtration. A default time τ may admit different intensities with respect to different filtrations and may admit no intensity with respect to some filtrations. This is not just a technical issue — it is significant in modeling as well. For example, if τ is the time of a barrier crossing for Brownian motion, then τ has no intensity with respect to the history of the Brownian motion. This follows from the fact that (9.35) does not have the form required by the martingale representation theorem (Theorem B.3.2 in Appendix B.3). This also makes precise the idea that an observer of the Brownian path could anticipate the barrier crossing. The same τ could, however, admit an intensity with respect to a different filtration that records imperfect information about the Brownian path; see Duffie and Lando [102].

A consequence of the existence of an intensity is the identity

$$P(\tau > T) = \mathsf{E}\left[\exp\left(-\int_0^T \lambda(u)\, du\right)\right]. \tag{9.36}$$

The bond-valuation formula (9.34) becomes

$$\mathsf{E}\left[\exp\left(-\int_0^T [r(u) + \lambda(u)]\, du\right)\right]. \tag{9.37}$$

This is evident if r and λ are independent processes, but it holds more generally if r is adapted to the filtration with respect to which λ is an intensity; see Section 11.J of Duffie [98] and the references given there. As formulated in [98] the result requires boundedness of r and λ, but this condition can be relaxed and is routinely ignored in applications.

An appealing feature of (9.37) is that it values a defaultable bond the same way one would value a default-free bond, but with the discount rate increased from r to $r + \lambda$. This is consistent with the market practice of pricing credit risk by discounting at a higher rate. It is also consistent with the practice of interpreting yield spreads on corporate bonds as measures of their likelihood of default.

Because (9.37) is a valuation result discounting at a risk-free interest rate, it should be understood as an expectation under a risk-neutral measure. The relevant processes r and λ thus describe the risk-neutral dynamics of the short rate and the default intensity. A counterpart of the Girsanov Theorem (see Section VI.2 of Brémaud [60]) shows that changing probability measures has a multiplicative effect on the intensity. The multiplicative factor may be interpreted as a risk premium for default risk.

The valuation formula (9.37) assumes that in the event of default prior to T the bond becomes worthless. In practice, the holder of a bond in default usually recovers some fraction of the bond's promised payments as a result of a complex legal process. Incorporating partial recovery in a model requires making simplifying assumptions about the amount received in the event of default. Duffie and Singleton [106] take the recovery to be a fraction of the value of the bond just prior to default; Jarrow and Turnbull [200] take it to be a fraction of the value of an otherwise identical default-free bond; Madan and Unal [244] use a random recovery rate. These and other assumptions are discussed in Bielecki and Rutkowski [46].

The approach of Duffie and Singleton [106] leads to a simple extension of (9.37). Suppose that the value of the bond just after default at time τ equals a fraction $1 - L(\tau)$ of its value just before default. This time-dependent fraction may be stochastic. Subject to regularity conditions, (9.37) generalizes to

$$
\mathsf{E}\left[\exp\left(-\int_0^T [r(u) + \lambda(u)L(u)]\,du\right)\right].
$$

In this case, the spread between a defaultable and default-free bond reflects the loss given default as well as the probability of default.

Intensity-Based Modeling

Intensity-based modeling of default uses a stochastic intensity to model the time to default (rather than deriving the intensity from τ). A key property for this construction is the fact that the cumulative intensity to default,

$$
\int_0^\tau \lambda(t)\,dt,
$$

is exponentially distributed with mean 1. (This follows from Theorem II.16 of Brémaud [60].) Given an arbitrary nonnegative process λ and an exponentially distributed random variable ξ independent of λ, we may define

$$
\tau = \inf\left\{t \geq 0 : \int_0^t \lambda(u)\,du = \xi\right\} \tag{9.38}
$$

(much as in (3.83)), and then τ has intensity λ. This construction is also useful in simulating a default time from its intensity.

A natural candidate for the process λ is the square-root diffusion of Section 3.4, because it is positive and mean-reverting. With this choice of intensity, the tail of the distribution of τ in (9.36) has exactly the same form as the bond pricing formula of Cox, Ingersoll, and Ross [91]; see Section 3.4.3. Thus,

$$F_\tau(t) \equiv P(\tau \leq t) = 1 - e^{A(t)+C(t)\lambda(0)}, \qquad (9.39)$$

with A and C as defined in Section 3.4.3. To simulate τ it therefore suffices to apply the inverse transform method of Section 2.2.1 to this distribution — simulation of λ itself is unnecessary. The same method applies, more generally, when λ belongs to the class of affine jump-diffusions defined in Duffie and Kan [101]. For example, Duffie and Gârleanu [99] consider in detail the case of a square-root diffusion modified to take exponentially distributed jumps at the epochs of a Poisson process.

Given an expression like (9.39) for the distribution of a default time τ, there would be little reason to model the dynamics of the intensity if we were interested only in τ. But in valuing credit derivatives or measuring credit risk we are often interested in capturing dependence between default times and dependence between defaults and interest rates. One approach to this specifies correlated processes for interest rates and intensities. A multifactor affine process, as in Duffie and Singleton [106], provides a convenient framework in which to model these correlated processes.

Ratings Transitions

Corporate bonds and other securities with significant credit risk can lose value through credit events less severe than outright default. A change in credit rating, for example, can produce a change in the market price of a bond.

A simple model of changes in credit ratings uses a Markov process to describe ratings transition. Consider, for example, a finite state space $\{0, 1, \ldots, N\}$ in which each state represents a level of credit quality. States with higher indices correspond to higher credit quality; state 0 is an absorbing state representing default. Let $\{X(t), t \geq 0\}$ be a Markov process on this state space with transition rate $q(i, j)$ from state i to state j, for $i, j \in \{0, 1, \ldots, N\}$. If default has not occurred by time t, then the default intensity at time t (with respect to the history of the process X) is $q(X(t), 0)$, the transition rate from the current state $X(t)$ to the default state 0. A model of this type is developed by Jarrow, Lando, and Turnbull [199].

One can simulate paths of X by simulating the sojourns in each state and the transitions between states. At the entry of X to state $i \neq 0$, generate a random variable ξ exponentially distributed with mean $1/q(i)$, where

$$q(i) = \sum_{j \neq i} q(i, j);$$

this is the length of the sojourn in state i. After advancing the simulation clock by ξ time units, generate a transition out of state i by choosing state j with probability $q(i, j)/q(i)$, $j \neq i$.

9.4.2 Dependent Defaults

The models described in the previous section focus on a single default time. The credit risk in a portfolio and the value of some credit derivatives depend on the joint distribution of the default times of multiple obligors; capturing this dependence requires a mechanism for linking the marginal default times of the individual obligors. The likelihood of default for an individual obligor can often be at least partly determined from the credit rating and yield spreads on bonds it has issued, so capturing the dependence between obligors is the primary challenge in modeling credit risk.

Using Equity Correlations

Direct estimation of dependence between default times using historical data is difficult because defaults are rare. Whereas daily fluctuations in market prices provide a nearly continuous flow of information about market risk, the time to default for an investment-grade issue is measured in years.

The link between default and capital structure discussed in Section 9.4.1 provides a framework for translating information in equity prices to information about credit risk. Taken literally, this approach would require building a multivariate model of the dynamics of firm value for multiple obligors and capturing the dependence in their default times through dependence in first-passage times to multiple boundaries, each boundary determined by the capital structure of one of the obligors. In practice, this idea is used in a much simpler way to build models of correlated defaults.

As a first example of this, we discuss the method of Gupton, Finger, and Bhatia [163]. They model the occurrence or non-occurrence of default over a fixed horizon (a year, for example) rather than the time to default. To model m obligors they use a normally distributed random vector $(X_1, \ldots, X_m)$. Each X_i has a standard normal distribution, but the components are correlated. The components are (loosely) interpreted as centered and scaled firm values and the correlations are taken from the correlations in equity prices.

The ith obligor defaults within the fixed horizon if $X_i < x_i$, with x_i a constant. In analogy with Merton's [262] setting, the threshold x_i measures the firm's debt burden. Gupton et al. [163] assume a known default probability p_i for each obligor and set $x_i = \Phi^{-1}(p_i)$ so that

$$P(X_i < x_i) = p_i.$$

This calibrates the default indicator $Y_i = \mathbf{1}\{X_i < x_i\}$ for each obligor. The joint distribution of the default indicators $Y_1, \ldots, Y_m$ is implicitly determined by the joint distribution of $X_1, \ldots, X_m$. The transformation from the X_i to the Y_i does not preserve linear correlations but, because it is monotone, it does preserve rank correlations.

The setting considered by Gupton et al. [163] is richer than what we have described in that it considers ratings transitions as well as default. For each obligor they partition the real line into intervals, each interval corresponding to a credit rating (or default). Each interval is chosen so that its probability (under the standard normal distribution) matches the probability that the obligor will be in that ratings category at the end of the period. (These probabilities are assumed known.) The intervals are defined using the same steps used to define strata in Example 4.3.2. The correlations in the X_i thus introduce dependence in the ratings transitions of the various obligors.

The use of equity price correlations for the X_i is not essential. As in Gupton et al. [163], one could also model the X_i using a representation of the form

$$X_i = a_{i1}Z_1 + \cdots + a_{ik}Z_k + b_i\epsilon_i, \quad i = 1,\ldots,m, \tag{9.40}$$

with $Z_1,\ldots,Z_k$ and ϵ_i independent $N(0,1)$ random variables and $a_{i1}^2 + \cdots + a_{ik}^2 + b_i^2 = 1$. The Z_j represent factors affecting multiple obligors whereas ϵ_i affects only the ith obligor. The common factors could, for example, represent risks specific to an industry or geographical region or a market-wide factor affecting all obligors.

Correlation and Intensities

As mentioned in Section 9.4.1 and demonstrated in Duffie [98], affine jump-diffusion processes provide a convenient framework for modeling stochastic intensities. Through a multivariate process of this form, one can specify the joint dynamics of the intensities of multiple obligors together with the dynamics of interest rates for various maturities while retaining some tractability,

To simulate default times in this framework, we can extend (9.38) and set

$$\tau_i = \inf\{t \geq 0 : \int_0^t \lambda_i(u)\,du = \xi_i\}, \quad i = 1,\ldots,m,$$

with λ_i the default intensity for the ith obligor and $\xi_1,\ldots,\xi_m$ are independent unit-mean exponential random variables. Except in special cases, this would require simulation of the paths of the intensities because the comparatively simple expression in (9.39) for the marginal distribution of a default time does not easily extend to sampling from the joint distribution of $\tau_1,\ldots,\tau_m$.

There are several ways of making the τ_i dependent in this framework. The simplest mechanism uses correlated Brownian motions to drive the intensity processes λ_i. This, however, introduces rather weak dependence among the default times: if the intensities for two obligors are close, then their instantaneous default probabilities are close but the default times themselves can be far apart because of the independence of the ξ_i.

An alternative defines the default intensities to be overlapping sums of state variables. Consider a model driven by a d-dimensional state vector $X(t)$ with nonnegative components and suppose the default intensity for obligor i is given by

$$\lambda_i(t) = \sum_{j \in A_i} X_j(t),$$

with A_i a subset of $\{1, \ldots, d\}$. Interpret each X_j as an intensity process and using the mechanism in (9.38), let T_j be the first arrival time generated by this intensity. At T_j, obligor i defaults if it has not previously defaulted and if $j \in A_i$. In this formulation, the X_j act like factors driving default and affecting one or more obligors. A shortcoming of this approach is that it introduces dependence by making defaults occur simultaneously.

A third approach introduces dependence through the exponential random variables $\xi_1, \ldots, \xi_m$ rather than through the intensities. This can provide rather strong dependence between default times without requiring simultaneous defaults. Moreover, if the intensities of the obligors are independent and if the marginal distributions of the τ_i admit tractable expressions F_i, as in (9.39), then we may avoid simulating the intensities by setting

$$\tau_i = F_i^{-1}(U_i), \quad U_i = 1 - \exp(-\xi_i). \tag{9.41}$$

Each U_i is uniformly distributed on the unit interval, but $U_1, \ldots, U_m$ inherit dependence from $\xi_1, \ldots, \xi_m$. This reduces the problem of specifying dependence among the default times to one of specifying a multivariate exponential or multivariate uniform distribution, which we turn to next.

Normal Copula

A simple way to specify a multivariate uniform distribution starts from a vector $(X_1, \ldots, X_m)$ of correlated $N(0, 1)$ random variables and sets

$$U_i = \Phi^{-1}(X_i), \quad i = 1, \ldots, m. \tag{9.42}$$

The uniform random variables $U_1, \ldots, U_m$ can then be used to generate other dependent random variables. For example, setting $\xi_i = -\log(1 - U_i)$, $i = 1, \ldots, m$, makes $\xi_1, \ldots, \xi_m$ dependent exponential random variables.

This mechanism for introducing dependence is called a *normal copula*. It is implicit in the method of Gupton et al. [163] discussed above (set $Y_i = 1\{U_i < p_i\}$) and similar to the *t*-copula used in Section 9.3. This construction is convenient because it captures dependence through the correlation matrix of a normal random vector, though as we noted above, the transformation from the X_i to other distributions does not in general preserve correlation. For more on the correlation of the transformed variables, see Cario and Nelson [77], Embrechts et al. [112], Ghosh and Henderson [137], and Li [236]. Shaw [329], applying a method of Stein [337], combines a normal copula with Latin hypercube sampling in estimating value-at-risk.

Other Copulas

We have thus far referred to normal copulas and *t*-copulas; we now define copulas more generally. A copula is a function describing a multivariate distribution in terms of its marginals. A copula function with m arguments is a distribution function C on $[0,1]^m$ with uniform marginals. The requirement of uniform marginals means that $C(u_1,\ldots,u_m) = u_i$ if $u_j = 1$ for all $j \neq i$; the requirement that C be a distribution on $[0,1]^m$ cannot be formulated so succinctly. It implies, of course, that C is increasing in each argument for all values of the other arguments and that $C(0,\ldots,0) = 0$, $C(1,\ldots,1) = 1$, though these properties do not suffice. See Nelsen [276] for a detailed discussion.

Given univariate distributions $F_1,\ldots,F_m$, a copula function C determines a multivariate distribution on $\Re^m$ with the F_i as marginals through the definition

$$F(x_1,\ldots,x_m) = C(F_1(x_1),\ldots,F_m(x_m)). \tag{9.43}$$

Conversely, given a multivariate distribution F with marginals $F_1,\ldots,F_m$, setting

$$C_F(u_1,\ldots,u_m) = F(F_1^{-1}(u_1),\ldots,F_m^{-1}(u_m)) \tag{9.44}$$

shows that every multivariate distribution admits the representation in (9.43). Copulas therefore provide a natural framework for modeling dependence between default times: the marginal distribution of the time to default for each obligor is at least in part determined by information specific to that obligor; the goal of much credit risk modeling is to link these marginal distributions into a multivariate distribution.

The specific constructions based on the t distribution in (9.22) and based on the normal distribution in (9.42) are special cases of the general copula representation in (9.43). If F is a multivariate normal distribution, C_F in (9.44) is a normal copula and if F is a multivariate t distribution, C_F is a t-copula. These copula functions can then be applied to other marginal distributions to link those marginals using the dependence structure of the multivariate normal or t.

For example, the generalized random t vector in (9.22) has as its joint distribution the function

$$\begin{aligned}
P(\tilde{X}_1 &\leq x_1,\ldots,\tilde{X}_m \leq x_m) \\
&= P(F_\nu(X_1) \leq F_{\nu_1}(x_1),\ldots,F_\nu(X_m) \leq F_{\nu_m}(x_m)) \\
&= C_{t_\nu}(F_{\nu_1}(x_1),\ldots,F_{\nu_m}(x_m)),
\end{aligned}$$

where we have written C_{t_ν} for the copula function of the multivariate t distribution with ν degrees of freedom. Here and in (9.22) we apply a t-copula to marginals that happen to be t distributions themselves, but this is not essential. We could replace C_{t_ν} with a normal copula (as Hosking et al. [188]) do), and we could apply the t-copula to other marginal distributions. Mashal and Zeevi [255], for example, apply a t-copula to empirical distributions of asset returns.

A simple application of a normal copula can be used to generate dependent default times with marginal distributions $F_1,\ldots,F_m$. Let $X_1,\ldots,X_m$ be correlated $N(0,1)$ random variables and set

$$\tau_i = F_i^{-1}(\Phi(X_i)), \quad i = 1, \dots, m. \tag{9.45}$$

This combines (9.41) and (9.42) and is among the methods discussed in Li [236]. One might look to the correlations in equity returns to determine the correlations between the X_i, though it is less clear how to make this translation here than in the fixed-horizon setting of Gupton et al. [163]. Finger [120] compares the dependence between default times achieved through four constructions (including (9.45)) calibrated to the same parameters and finds that different mechanisms can give significantly different results.

Various types of copulas are surveyed with a view towards risk management applications in Embrechts, McNeil, and Straumann [112], Hamilton, James, and Webber [167], and Li [236]. Scönbucher and Schubert [321] analyze the combination of copulas with intensity-based modeling of marginals defaults. Hamilton et al. [167] use historical default data to estimate a copula empirically. In simulation, it is convenient to work with a copula associated with a distribution from which it is easy to draw samples (like the normal and t distributions). This facilitates the implementation of the type of mechanism in (9.45). Marshall and Olkin [252] define an interesting class of copula functions that also lend themselves to simulation. Schönbucher [320] compares credit-loss distributions under various members of this class.

9.4.3 Portfolio Credit Risk

The most basic problem in measuring portfolio credit risk is determining the distribution of losses from default over a fixed horizon. This is the credit risk counterpart of the market risk problem considered in Sections 9.1–9.3. For credit risk, one usually considers a longer horizon — a year, for example.

Consider a fixed portfolio exposed to the credit risk of m obligors. Let

Y_i = indicator that ith obligor defaults within the fixed horizon;

$p_i = P(Y_i = 1)$ = marginal probability of default of the ith obligor;

c_i = loss resulting from default of ith obligor;

$$L = \sum_{i=1}^{m} Y_i c_i = \text{portfolio loss.}$$

Estimating the distribution of the loss L often requires simulation, particularly in a model that captures dependence among the default indicators Y_i. In estimating $P(L > x)$ at large values of x, it is natural to apply importance sampling. This section describes some initial steps in this direction based on joint work with Jingyi Li.

Independent Obligors

We begin by considering the simple case of independent obligors. The loss c_i given default of the ith obligor could be modeled as a random variable but to further simplify, we take it to be a constant. These simplifications make L a sum of independent random variables with moment generating function

$$\mathsf{E}[e^{\theta L}] = \prod_{i=1}^{m} \mathsf{E}[e^{\theta Y_i c_i}] = \prod_{i=1}^{m} \left(p_i e^{c_i \theta} + (1 - p_i) \right),$$

finite for all $\theta \in \Re$. Replacing θ with a purely imaginary argument yields the characteristic function of L, which can be inverted numerically to find the distribution of L. An alternative to exact inversion is a saddlepoint approximation, as applied in Martin, Thompson, and Browne [253].

Although simulation is unnecessary in this setting, it provides a convenient starting point for developing variance reduction techniques. To simplify the problem even further, suppose all obligors have the same default probabilities $p_i \equiv p$ and exposures $c_i \equiv 1$. Consider the effect of exponentially twisting Y_i, in the sense we encountered in Example 4.6.2 and Section 9.2.2. In other words, define a change of probability measure through the likelihood ratio

$$\exp(\theta Y_i - \psi(\theta))$$

with

$$\psi(\theta) = \log \left(pe^{\theta} + (1 - p) \right)$$

and $\theta \in \Re$ a parameter. The default probability under the new distribution is the probability that $Y_i = 1$ and this is also the mean of Y_i under the new distribution. Using a standard property of exponential twisting, also used in Examples 4.6.2–4.6.4 and (9.16), this mean is given by

$$p(\theta) \equiv \psi'(\theta) = \frac{pe^{\theta}}{pe^{\theta} + (1 - p)}.$$

By choosing $\theta > 0$, we thus increase the probability of default.

Now apply this exponential twist to all the Y_i. Because the Y_i are independent, the resulting likelihood ratio is the product of the individual likelihood ratios and is thus given by

$$\prod_{i=1}^{m} \exp(\theta Y_i - \psi(\theta)) = \exp(\theta L - m\psi(\theta)).$$

In other words, in this simple setting, exponentially twisting every Y_i defines the same change of measure as exponentially twisting L itself. Write P_θ for the new probability measure so defined and E_θ for expectation under this measure. This provides the representation

$$P(L > x) = \mathsf{E}_\theta \left[e^{-\theta L + m\psi(\theta)} \mathbf{1}\{L > x\} \right] \tag{9.46}$$

for the loss probability. The expression inside the expectation on the right provides an unbiased estimator of the loss probability if sampled under the new probability measure. Sampling under P_θ is easy: it simply involves replacing the original default probability p with $p(\theta)$.

It remains to choose the parameter θ. The argument leading to (9.15), based on minimizing an upper bound on the second moment, similarly leads here to the value of θ_x solving

$$\psi'(\theta_x) = x/m,$$

which then also satisfies

$$\mathsf{E}_{\theta_x}[L] = \mathsf{E}_{\theta_x}\left[\sum_{i=1}^m Y_i\right] = m\psi'(\theta_x) = x. \tag{9.47}$$

Thus, to estimate $P(L > x)$ for large x, we increase the individual default probabilities to make x the expected loss. It follows from more general results on importance sampling (as in Bucklew, Ney, and Sadowsky [72] and Sadowsky [315]) that this method is asymptotically optimal as x and m increase in a fixed proportion.

This approach extends easily to the case of unequal p_i and c_i. Set

$$\psi_i(\theta) = \log\left(p_i e^{\theta c_i} + (1 - p_i)\right),$$

and $\psi_L = \psi_1 + \cdots + \psi_m$. Then ψ_L is the cumulant generating function of L and exponentially twisting L defines the change of measure

$$\frac{dP_\theta}{dP} = \exp(\theta L - \psi_L(\theta)).$$

This has the same effect as exponentially twisting every $Y_i c_i$. Under P_θ, each term $Y_i c_i$ has mean $\psi_i'(\theta)$, which is to say that the ith obligor defaults with probability $p_i(\theta) = \psi_i'(\theta)/c_i$. This can also be written as

$$\frac{p_i(\theta)}{1 - p_i(\theta)} = \left(\frac{p_i}{1 - p_i}\right) e^{\theta c_i},$$

which shows that taking $\theta > 0$ increases the odds ratio for every obligor, with larger increases for obligors with larger exposures c_i. Choosing θ as the root of the equation $\psi_L'(\theta) = x$ again satisfies (9.47) and minimizes an upper bound on the second moment of the estimator.

Dependent Defaults

We now turn to the more interesting case in which the Y_i are dependent. Different models of dependence entail different approaches to importance sampling; we consider dependence introduced through a normal copula as discussed in Section 9.4.2.

For each obligor i, let

$$x_i = \Phi^{-1}(1 - p_i) \tag{9.48}$$

so that the probability of default p_i equals the probability that a standard normal random variable exceeds x_i. Construct default indicators by setting $Y_i = \mathbf{1}\{X_i > x_i\}$ with $X_i \sim N(0, 1)$. (We are free to choose the threshold x_i in the lower tail of the normal as in Section 9.4.2 or the upper tail as we do here.) Link the random variables $X_1, \ldots, X_m$ through a specification of the form (9.40); in vector notation, this becomes

$$X = AZ + B\epsilon, \tag{9.49}$$

in which ϵ has the $N(0, I)$ distribution in $\Re^m$, Z has the $N(0, I)$ distribution in $\Re^k$, and Z is independent of ϵ. The $m \times k$ matrix A determines the correlation matrix of X, whose off-diagonal entries are given by those of $AA^\top$. The $m \times m$ diagonal matrix B is chosen so that all diagonal entries of the covariance matrix $AA^\top + B^2$ equal 1, thus making the components of X standard normal random variables. We think of the dimension k of the common factors as substantially smaller than the number of obligors; for example, m could be in the thousands and k as small as 1–10.

We apply importance sampling conditional on the common factors. Given Z, the vector X is normally distributed with mean AZ and diagonal covariance matrix B^2. The conditional probability of default of the ith obligor is therefore

$$\tilde{p}_i = P(Y_i = 1|Z) = P(X_i > x_i|Z) = 1 - \Phi\left(\frac{x_i - a_i Z}{b_i}\right), \tag{9.50}$$

where a_i is the ith row of A and b_i is the ith element on the diagonal of B. Given Z, the portfolio loss L becomes a sum of m independent random variables, the ith summand taking the values c_i and 0 with probabilities $\tilde{p}_i$ and $1 - \tilde{p}_i$.

Conditioning on Z thus reduces the problem to the independent case with which we began this section. Define

$$\psi_{L|Z}(\theta) = \log \mathsf{E}[e^{\theta L}|Z] = \sum_{i=1}^{m} \log\left(\tilde{p}_i e^{\theta c_i} + (1 - \tilde{p}_i)\right); \tag{9.51}$$

this is the cumulant generating function of the distribution of L given Z. Let $\tilde{\theta}_x$ solve

$$\psi'_{L|Z}(\tilde{\theta}_x) = x.$$

Define the conditional default probabilities

$$\tilde{p}_i(\tilde{\theta}_x) = \frac{\tilde{p}_i e^{\tilde{\theta}_x c_i}}{\tilde{p}_i e^{\tilde{\theta}_x c_i} + 1 - \tilde{p}_i}, \quad i = 1, \ldots, m. \tag{9.52}$$

Given Z, the default indicators $Y_1, \ldots, Y_m$ are independent and, under the $\tilde{\theta}_x$-twisted distribution, Y_i takes the value 1 with probability $\tilde{p}_i(\tilde{\theta}_x)$; these are therefore easy to generate. Setting L equal to the sum of the $Y_i c_i$ yields the estimator

$$e^{-\tilde{\theta}_x L + \psi_{L|Z}(\tilde{\theta}_x)} \mathbf{1}\{L > x\}; \qquad (9.53)$$

this is the conditional counterpart of the expression in (9.46). Its conditional expectation is $P(L > x|Z)$ and its unconditional expectation is therefore $P(L > x)$.

Factor Twisting

We can apply further importance sampling to the normally distributed factors Z. To see why this might be useful, decompose the variance of the estimator (9.53) as

$$\mathsf{E}[\mathsf{Var}_{\tilde{\theta}_x}[e^{-\tilde{\theta}_x L + \psi_{L|Z}(\tilde{\theta}_x)} \mathbf{1}\{L > x\}|Z]] + \mathsf{Var}[P(L > x|Z)]. \qquad (9.54)$$

The second term is the variance of the conditional expectation of the estimator given Z. Twisting the default indicators conditional on Z, as in (9.53), makes the first term in this decomposition small but does nothing for the second term. Because $P(L > x|Z)$ is a function of Z, we can apply importance sampling to Z to try to reduce the contribution of the second term to the total variance.

The simplest application of importance sampling to an $N(0, I)$ random vector introduces a mean vector μ, as in Example 4.6.1 and Section 4.6.2. The likelihood ratio for this change of measure is

$$\exp(-\mu^\top Z + \tfrac{1}{2}\mu^\top \mu).$$

When multiplied by (9.53), this yields the estimator

$$\exp(-\mu^\top Z + \tfrac{1}{2}\mu^\top \mu - \tilde{\theta}_x L + \psi_{L|Z}(\tilde{\theta}_x)) \mathbf{1}\{L > x\}, \qquad (9.55)$$

in which Z is sampled from $N(\mu, I)$ and then L is sampled from the $\tilde{\theta}_x$-twisted distribution conditional on Z.

Importance Sampling Algorithm for Credit Losses

We now summarize the two-step importance sampling procedure in a single algorithm. We assume that the matrices A and B in the factor representation (9.49) and the thresholds x_i in (9.48) have already been defined. We also assume that a mean vector μ for the common factors has been selected; we return to this point below.

1. Generate $Z \sim N(\mu, I)$ in $\Re^k$
2. Compute conditional default probabilities $\tilde{p}_i$, $i = 1, \ldots, m$, as in (9.50)
3. Define $\psi_{L|Z}$ as in (9.51) and find θ solving $\psi'_{L|Z}(\theta) = x$;

set $\tilde{\theta}_x = \max\{0, \theta\}$

4. Compute twisted conditional default probabilities $\tilde{p}_i(\tilde{\theta}_x)$, $i = 1, \ldots, m$, using (9.52)

5. For $i = 1, \ldots, m$, let $Y_i = 1$ with probability $\tilde{p}_i(\tilde{\theta}_x)$ and $Y_i = 0$ otherwise. Calculate loss $L = c_1 Y_1 + \cdots + c_m Y_m$

6. Return estimator (9.55).

These steps compute a single replication of the estimator (9.55), which is an unbiased estimator of $P(L > x)$. The steps can be repeated and the results averaged over independent replications.

At some values of Z (large values, if all the coefficients in (9.40) are positive), the conditional default probabilities $\tilde{p}_i$ may be sufficiently large that the expected loss given Z exceeds x; i.e., $\psi'_{L|Z}(0) > x$. In this case, the θ calculated in Step 3 of the algorithm would be negative, and twisting by this parameter would *decrease* the default probabilities. To avoid this, Step 3 sets $\tilde{\theta}_x = 0$ if the root θ is negative.

Choice of Factor Mean

It remains to specify the new mean μ for the common factors Z. As noted in the discussion surrounding (9.54), the conditional twist (9.53) reduces variance in estimating $P(L > x|Z)$ and the purpose of applying importance sampling to Z is to reduce variance in estimating the expectation of $P(L > x|Z)$, viewed as a function of Z. We should therefore choose μ to minimize this variance or, equivalently, the second moment

$$\mathsf{E}_\mu \left[e^{-2\mu^\top Z + \mu^\top \mu} P(L > x|Z)^2 \right]. \tag{9.56}$$

We have subscripted the expectation by μ to emphasize that it is computed with Z having distribution $N(\mu, I)$.

At this point, we restrict attention to a very special case. We take all the default probabilities p_i to be equal and we take all $c_i = 1$. This makes L the number of defaults. We take Z to scalar and all X_i in (9.49) of the form

$$X_i = \rho Z + \sqrt{1 - \rho^2} \epsilon_i,$$

with the same scalar ρ for all $i = 1, \ldots, m$.

At $\rho = 0$, the default indicators $Y_i = \mathbf{1}\{X_i > x_i\}$ become independent and $P(L > x|Z)$ equals $P(L > x)$. Because Z drops out of the problem, the optimal choice of μ in this case is simply $\mu = 0$. At $\rho = 1$, all the default indicators become identical and L takes only the values 0 and m. Assuming $0 < x < m$, we thus have

$$P(L > x|Z) = \mathbf{1}\{L > x\} = \mathbf{1}\{Y_i = 1\} = \mathbf{1}\{Z > x_i\},$$

for any $i = 1, \ldots, m$. If we make this substitution in (9.56) and follow the same steps leading to (9.15)–(9.16), we arrive at the parameter value satisfying

$$\mathsf{E}_\mu[Z] = x_i; \tag{9.57}$$

that is, $\mu = x_i$. The default threshold x_i equals $\Phi^{-1}(1 - p)$ where p is the common value of the default probabilities p_i. This argument therefore suggests that μ should increase from 0 to $\Phi^{-1}(1 - p)$ as ρ increases from 0 to 1.

The numerical results in Figure 9.7 support this idea. The results apply to a portfolio with $m = 1000$ obligors, $c_i \equiv 1$ and $p_i \equiv p = 0.002$. The figure plots variance reduction factors (relative to ordinary Monte Carlo) for the estimator (9.55) as a function μ for $\rho = 0.05, 0.10, 0.25, 0.50$, and 0.80. Each curve corresponds to a value of ρ. For each ρ, we choose x to make $P(L > x)$ close to 1%, though this is not always possible because L is integer-valued. The values of x and the loss probabilities are as follows:

$$\begin{array}{rccccc}
\rho : & 0.05 & 0.10 & 0.25 & 0.50 & 0.80 \\
x : & 6 & 6 & 10 & 26 & 43 \\
P(L > x) : & 0.6\% & 1.0\% & 0.9\% & 0.9\% & 1.0\%
\end{array}$$

The curves in Figure 9.7 indeed show that the optimal μ (the point at which the variance reduction is greatest) tends to increase with ρ. Moreover, a default probability of $p = 0.002$ corresponds to a threshold of $x_i = \Phi^{-1}(1 - 0.002) = 2.88$, and the argument leading to (9.57) asserts that this should be close to the optimal μ for ρ close to 1. (Equation (9.57) defines the optimal μ for $\rho = 1$ as $x_i \to \infty$, but for finite x_i the optimal μ may not be exactly equal to x_i, even at $\rho = 1$.) The curves in the figure indicate that the limiting value of 2.88 is close to optimal for $\rho = 0.50$ and $\rho = 0.80$, and is very effective even at $\rho = 0.25$. We know that at $\rho = 0$ it would be optimal to take $\mu = 0$, but the figure shows that with a small increase in ρ to 0.05 there is already substantial benefit to taking $\mu > 0$. For the examples in the figure, we find variance reduction factors in the range of 30–50 with μ chosen optimally for loss probabilities near 1%. Greater variance reduction would be achieved at smaller loss probabilities.

This example is simple enough that it does not require simulation — L has a binomial distribution conditional on Z, so calculation of $P(L > x)$ reduces to a one-dimensional integral. It nevertheless shows the potential effectiveness of importance sampling in estimating the tail of the loss distribution. The dependence mechanisms used in credit risk models in turn pose interesting new challenges for research on variance reduction.

9.5 Concluding Remarks

In this chapter, we have presented some applications of Monte Carlo simulation to risk management. In our discussion of market risk, we have focused on the problem of estimating loss probabilities and value-at-risk and detailed the use of the delta-gamma approximation as a basis for variance reduction.

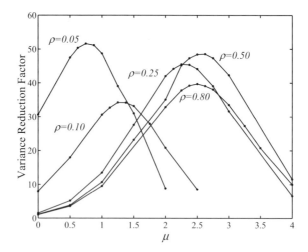

Fig. 9.7. Variance reduction achieved through importance sampling as a function of factor mean μ for various levels of ρ. The optimal μ increases with ρ and is close to $\Phi^{-1}(1 - p) = 0.288$ at larger values of ρ.

In discussing credit risk, we have described some of the main modeling approaches and simulation issues, and described some initial steps in research on efficient simulation for portfolio credit risk.

We have not attempted a comprehensive treatment of the use of simulation in risk management — the topic is too broad to permit that here. Simulation is widely used in areas of risk management not even touched on in this chapter — pension planning and insurance, for example. This section provides some additional references to relevant methods and applications.

In our discussion of market risk, we have focused on portfolios with assets whose value must be computed rather than simply observed. This is appropriate for a portfolio of options but overly complicated for, e.g., a portfolio of stocks. When time series of asset values are available, extreme value theory is useful for quantile estimation. See Bassi, Embrechts, and Kafetzaki [39] for an introduction to risk management applications and Embrechts, Klüppelberg, and Mikosch [111] for the underlying theory.

Quasi-Monte Carlo is a natural tool to consider in calculating risk measures. Methods for improving uniformity are not, however, specifically suited to estimating small probabilities or extreme quantiles.

Talay and Zheng [344] analyze the discretization error in using an Euler approximation to estimate quantiles of the law of a diffusion, with applications to value-at-risk. They show that the discretization error (like the variance in the central limit theorem (9.7)) involves the reciprocal of the density at the quantile and can therefore be very large at extreme quantiles.

Importance sampling for heavy-tailed distributions is an active area of current research, motivated by applications in telecommunications and in insurance risk theory. Work on this topic includes Asmussen and Binswanger [22], Asmussen, Binswanger, and Højgaard [23], and Juneja and Shahabuddin [206]. These papers address difficulties in extending importance sampling for the classical ruin problem of Example 4.6.3 to the case of heavy-tailed claims.

At the interface of market and credit risk lies the problem of calculating the evolution of credit exposures. The exposure is the amount that would be lost if a counterparty defaulted and is thus the positive part of the net market value of all contracts with that counterparty, irrespective of the probability of default. The exposure in an interest rate swap, for example, is zero at inception and at termination and reaches a maximum around a third or half of the way into the life of the swap. Simulation is useful in estimating the path of the mean exposure and a quantile of the exposure; see Wakeman [354], for example.

Dynamic financial analysis refers to simulation-based techniques for risk management of growing use in the insurance industry. These are primarily long-term simulations of interest rates and other financial variables coupled with insurance losses. See Kaufmann, Gadmer, and Klett [208] for an introduction.

As noted in Section 9.4.1, interest rates and default intensities play formally similar roles in the calculation of bond prices and survival probabilities. This analogy leads to HJM-like models of intensities, as presented in Chapters 13 and 14 of Bielecki and Rutkowski [46]. The simulation methods of Sections 3.6 and 3.7 are potentially relevant to these models.

For more on credit risk modeling and the valuation of credit derivatives, see the books by Bielecki and Rutkowski [46] and Duffie [98].

A

Appendix: Convergence and Confidence Intervals

This appendix summarizes basic convergence concepts and the application of the central limit theorem to the construction of confidence intervals. The results and definitions reviewed in this appendix are covered in greater detail in many textbooks on probability and statistics.

A.1 Convergence Concepts

Random variables $\{X_n, n = 1, 2, \ldots\}$ on a probability space $(\Omega, \mathcal{F}, P)$ converge almost surely (i.e., with probability 1) to a random variable X if

$$P\left(\lim_{n\to\infty} X_n = X\right) = 1,$$

meaning, more explicitly that the set

$$\left\{\omega \in \Omega : \lim_{n\to\infty} X_n(\omega) = X(\omega)\right\}$$

has P-probability 1. The convergence holds in probability if, for all $\epsilon > 0$,

$$P(|X_n - X| > \epsilon) \to 0$$

as $n \to \infty$. It holds in p-norm, $0 < p < \infty$, if all X_n and X have finite pth moment and

$$\mathsf{E}[|X_n - X|^p] \to 0.$$

Almost sure convergence implies convergence in probability. Convergence in probability implies the existence of a deterministic subsequence $\{n_k, k = 1, 2, \ldots\}$ through which $X_{n_k} \to X$ almost surely (Chung [85], p.73). Convergence in p-norm implies convergence in probability. Neither almost sure convergence nor convergence in p-norm implies the other.

For random vectors, convergence in probability, in norm, or almost surely is equivalent to the corresponding convergence of each component of the vector.

A sequence of estimators $\{\hat{\theta}_n, n \geq 1\}$ is consistent for a parameter θ if $\hat{\theta}_n$ converges to θ in probability. The sequence is strongly consistent if convergence holds with probability 1.

Convergence in Distribution

Random variables $\{X_n, n = 1, 2, \ldots\}$ with distribution functions F_n converge *in distribution* to a random variable X with distribution F if

$$F_n(x) \to F(x) \text{ for every } x \in \Re \text{ at which } F \text{ is continuous.} \qquad (A.1)$$

The random variables $X, X_1, X_2, \ldots$ need not be defined on a common probability space. Convergence in distribution is denoted by the symbol "$\Rightarrow$", as in

$$X_n \Rightarrow X,$$

and is also called *weak* convergence. It is equivalent to the convergence of $\mathsf{E}[f(X_n)]$ to $\mathsf{E}[f(X)]$ for all bounded continuous functions $f : \Re \to \Re$. This characterization is convenient in extending the definition of weak convergence to random elements of more general spaces; in particular, we take it as the definition of convergence in distribution for random vectors.

Convergence in distribution of random variables X_n to X is also equivalent to pointwise convergence of their characteristic functions:

$$\lim_{n \to \infty} \mathsf{E}\left[e^{itX_n}\right] = \mathsf{E}\left[e^{itX}\right]$$

for all $t \in \Re$, with $i = \sqrt{-1}$. More precisely, $X_n \Rightarrow X$ implies convergence of the characteristic functions; and if the characteristic functions of the X_n converge pointwise to a function continuous at zero, then this limit is the characteristic function of a random variable to which the X_n then converge in distribution.

Convergence in distribution is implied by convergence in probability, hence also by almost sure convergence and by convergence in norm. If $X_n \Rightarrow c$ with c a constant, then the X_n converge to c in probability; this follows from (A.1) when $F(x) = \mathbf{1}\{x \geq c\}$.

Suppose random variables $\{X_n, n \geq 1\}$ and $\{Y_n, n \geq 1\}$ are all defined on a common probability space and that $X_n \Rightarrow X$ and $Y_n \Rightarrow c$, with c a constant. Then

$$X_n + Y_n \Rightarrow X + c \quad \text{and} \quad X_n Y_n \Rightarrow Xc. \qquad (A.2)$$

The first assertion in (A.2) is sometimes called Slutsky's Theorem.

The requirement that the limit of the Y_n be constant is important. If $X_n \Rightarrow X$ and $Y_n \Rightarrow Y$, it is not in general the case that $X_n + Y_n \Rightarrow X + Y$ or $X_n Y_n \Rightarrow XY$. These limits do hold under the stronger hypothesis that $(X_n, Y_n) \Rightarrow (X, Y)$. Indeed, any bounded continuous function of $x+y$ or xy can be written as a bounded continuous function of (x, y), and weak convergence of vectors is defined by convergence of expectations of all bounded continuous functions.

Suppose $\{N_n, n = 1, 2, \ldots\}$ is a nondecreasing process taking positive integer values and increasing to infinity with probability 1. Suppose $X_n \Rightarrow X$. If $\{N_n, n = 1, 2, \ldots\}$ and $\{X_n, n = 1, 2, \ldots\}$ are independent sequences, then $X_{N_n} \Rightarrow X$. This holds even without independence if $N_n/n \Rightarrow c$ for some constant $c > 0$, a result sometimes called Anscombe's Theorem. We need this result in (1.12). See also Theorem 7.3.2 of Chung [85].

Convergence of Moments

Because the mapping $x \mapsto x^r$, $r > 0$, is unbounded, convergence in distribution does not imply convergence of moments. Suppose $X_n \Rightarrow X$. Then a necessary and sufficient condition for the convergence of $\mathsf{E}[|X_n|^r]$ to $\mathsf{E}[|X|^r]$ is uniform integrability:

$$\lim_{c \to \infty} \sup_{n \geq 1} \mathsf{E}\left[|X|^r \mathbf{1}\{|X|^r > c\}\right] = 0. \tag{A.3}$$

By the dominated convergence theorem, a simple sufficient condition is the existence of an integrable random variable Y such that $|X_n|^r < Y$ for all n. Another sufficient condition is

$$\sup_{n \geq 1} \mathsf{E}\left[|X_n|^{r+\epsilon}\right] < \infty,$$

for some $\epsilon > 0$.

A.2 Central Limit Theorem and Confidence Intervals

The elementary central limit theorem states the following: If $X_1, X_2, \ldots$ are independent and identically distributed with expectation μ and variance σ^2, $0 < \sigma < \infty$, then the sample mean

$$\bar{X}_n = \frac{1}{n} \sum_{i=1}^{n} X_i$$

satisfies

$$\frac{\bar{X}_n - \mu}{\sigma/\sqrt{n}} \Rightarrow N(0, 1), \tag{A.4}$$

with $N(0, 1)$ denoting the standard normal distribution. This is proved by showing that the characteristic function of the expression on the left converges to the characteristic function of the standard normal distribution.

If $X_1, X_2, \ldots$ are i.i.d. random vectors with mean vector μ and covariance matrix Σ, then

$$\sqrt{n}[\bar{X}_n - \mu] \Rightarrow N(0, \Sigma),$$

where $N(0, \Sigma)$ denotes the multivariate normal distribution with mean 0 and covariance matrix Σ. This result can be deduced from (A.4) by considering all linear combinations of the components of the vector $\bar{X}_n$.

By the definition of convergence in distribution, (A.4) means that for all $x \in \Re$,

$$P\left(\frac{\bar{X}_n - \mu}{\sigma/\sqrt{n}} \le x\right) \to \Phi(x),$$

with Φ the standard cumulative normal distribution. From this it follows that the probability that an interval of the form

$$\left(\bar{X}_n - a\frac{\sigma}{\sqrt{n}}, \bar{X}_n + b\frac{\sigma}{\sqrt{n}}\right),$$

$0 \le a, b < \infty$, covers μ approaches $\Phi(b) - \Phi(-a)$ as $n \to \infty$. We can choose a and b so that this limiting probability is $1 - \delta$, for any $\delta > 0$. Among all choices of a and b for which $\Phi(b) - \Phi(-a) = 1 - \delta$, the values minimizing the length of the interval $(-a, b)$ are given by $a = b = z_{\delta/2}$, where $1 - \Phi(z_{\delta/2}) = \delta/2$. The interval

$$\bar{X}_n \pm z_{\delta/2}\frac{\sigma}{\sqrt{n}} \tag{A.5}$$

covers μ with probability approaching $1 - \delta$ as $n \to \infty$ and is in this sense an asymptotically valid confidence interval for μ.

Now let s_n be any consistent estimator of σ, meaning that $s_n \Rightarrow \sigma$. Because $\sigma > 0$, we may modify s_n so that it is always positive without affecting consistency, and then we have $\sigma/s_n \Rightarrow 1$. From (A.2) and (A.4) it follows that

$$\frac{\bar{X}_n - \mu}{s_n/\sqrt{n}} \Rightarrow N(0, 1),$$

and thus that

$$\bar{X}_n \pm z_{\delta/2}\frac{s_n}{\sqrt{n}} \tag{A.6}$$

is also an asymptotically valid $1 - \delta$ confidence interval. Because σ is typically unknown, this interval is of more practical use than (A.5); often,

$$s_n = \sqrt{\frac{1}{n-1}\sum_{i=1}^{n}(X_i - \bar{X}_n)^2},$$

the sample standard deviation of $X_1, \ldots, X_n$.

If the X_i are normally distributed, then the ratio in (A.4) has the standard normal distribution for all $n \ge 1$. It follows that in this case the interval (A.5) covers μ with probability $1 - \delta$ for all $n \ge 1$. With s_n the sample standard deviation,

$$\frac{\bar{X}_n - \mu}{s_n/\sqrt{n}} \sim t_{n-1}$$

for all $n \geq 2$; i.e., the ratio on the left has the t distribution with $n-1$ degrees of freedom. Accordingly, if we replace $z_{\delta/2}$ with $t_{n-1,\delta/2}$, the $1 - \delta/2$ quantile of the t_{n-1} distribution, the interval

$$\bar{X}_n \pm t_{n-1,\delta/2}\frac{s_n}{\sqrt{n}}$$

covers μ with probability $1-\delta$. Even if the X_i are not normally distributed, this produces a more conservative confidence interval because the t multiplier is always larger than the corresponding z multiplier. But for even modest values of n, the multipliers are nearly equal so we do not stress this distinction.

In addition to providing information about the precision of an estimator, a confidence interval is useful in sample-size determination. From (A.5) we find that the number of replications required to achieve a confidence interval halfwidth of ϵ is

$$n_\epsilon = \frac{z_{\delta/2}^2 \sigma^2}{\epsilon^2}. \tag{A.7}$$

If σ is unknown, a two-stage procedure uses an initial set of replications to estimate it and then uses this estimate in (A.7) to estimate the total sample size required.

Similar error bounds and procedures can be derived from Chebyshev's inequality, which states that

$$P\left(|\bar{X} - \mu| \leq \frac{\sigma}{\sqrt{\delta n}}\right) \geq 1 - \delta,$$

for all $\delta > 0$. This is valid for all $n \geq 1$, but is more conservative than the normal approximation.

In (A.4) we have presented only the most elementary form of the central limit theorem. The sample mean and other estimators are asymptotically normal under more general conditions. The elementary result suffices whenever we simulate independent and identically distributed replications. But more general results are needed to handle other settings that arise in simulation; we mention two.

○ Variance reduction techniques often introduce dependence across replications. In some cases (e.g., control variates), the dependence becomes negligible as the number of replications increases, but in others (e.g., stratified sampling, Latin hypercube sampling) it does not. Simulating batches and allowing dependence within batches but not across batches reduces the problem to one of independent replications; but even without this limitation a more general central limit theorem often applies. See, for example, the discussion of output analysis in Section 4.4.

○ We are often interested in how the error in an estimator changes as some parameter of a simulation changes along with an increase in the number of replications. This applies, for example, when we discretize a model with time step h. Rather than fixing h and letting the number of replications n increase, we may want to analyze the convergence of an estimator as both $h \to 0$ and $n \to \infty$. Because changing h changes the distribution from which the replications are sampled, this setting requires a central limit theorem for an array (as opposed to a sequence) of random variables. The Lindeberg-Feller central limit theorem (as in, e.g., Chung [85], Section 7.2) is the key result of this type.

B

Appendix: Results from Stochastic Calculus

This appendix records some background results on stochastic integrals, stochastic differential equations, martingales, and measure transformations. For more comprehensive treatments of these topics see, e.g., Hunt and Kennedy [191], Karatzas and Shreve [207], Øksendal [284], Protter [300], and Revuz and Yor [306].

B.1 Itô's Formula

Our starting point is a probability space $(\Omega, \mathcal{F}, P)$ with a filtration $\{\mathcal{F}_t, t \geq 0\}$, meaning a family of sub-σ-algebras of $\mathcal{F}$ with $\mathcal{F}_s \subseteq \mathcal{F}_t$ whenever $s \leq t$. Depending on the context, t may range over all of $[0, \infty)$ or be restricted to an interval $[0, T]$ for some fixed finite T. Some results require that the filtration satisfy the "usual conditions," so we assume these hold: $\mathcal{F}_0$ contains all subsets of sets in $\mathcal{F}$ having P-probability 0, and each $\mathcal{F}_s$ is the intersection of all $\mathcal{F}_t$ with $t > s$. A stochastic process $\{X(t), t \geq 0\}$ is adapted to the filtration if $X(t) \in \mathcal{F}_t$ for all $t \geq 0$, meaning that $X(t)$ is $\mathcal{F}_t$-measurable. We assume that on this filtered probability space is defined a k-dimensional Brownian motion $W = (W_1, \ldots, W_k)^\top$ with respect to the filtration. In particular, W is adapted and if $t > s$ then $W(t) - W(s)$ is independent of $\mathcal{F}_s$. We denote by $\{\mathcal{F}_t^W, t \geq 0\}$ the filtration generated by the Brownian motion with $\mathcal{F}_0^W$ augmented so that the usual conditions are satisfied.

For any vector or matrix a, let $\|a\|$ denote the square root of the sum of squared entries of a.

If $\{\gamma(t), 0 \leq t \leq T\}$ is an $\Re^k$-valued adapted process for which

$$P\left(\int_0^T \|\gamma(t)\|^2 \, dt < \infty\right) = 1,$$

for some fixed T, then the Itô integral

$$\int_0^t \gamma(u)^\top \, dW(u) \tag{B.1}$$

is well-defined for all t in $[0, T]$. This is not a routine extension of the ordinary integral because the paths of W have infinite variation. See, e.g., Karatzas and Shreve [207] or Øksendal [284] for a development of the Itô integral. We can replace the integrand γ with an $\Re^{d \times k}$-valued process b by applying (B.1) to each row of b; this produces a d-dimensional vector, each component of which is a stochastic integral.

An $\Re^d$-valued process $\{X(t), 0 \le t \le T\}$ is an *Itô process* if it can be represented as

$$X(t) = X(0) + \int_0^t a(u) \, du + \int_0^t b(u) \, dW(u), \quad 0 \le t \le T, \tag{B.2}$$

where $X(0)$ is $\mathcal{F}_0$-measurable, a is an $\Re^d$-valued adapted process satisfying

$$P\left(\int_0^T |a_i(t)| \, dt < \infty\right) = 1, \quad i = 1, \ldots, d, \tag{B.3}$$

and b is an $\Re^{d \times k}$-valued adapted process satisfying

$$P\left(\int_0^T \|b(u)\|^2 \, dt < \infty\right) = 1. \tag{B.4}$$

The notation

$$dX(t) = a(t) \, dt + b(t) \, dW(t) \tag{B.5}$$

is shorthand for (B.2).

Theorem B.1.1 *(Itô's Formula). Let X be an $\Re^d$-valued Itô process as in (B.2) and let $f : [0, T] \times \Re^d \to \Re$ be continuously differentiable in its first argument and twice continuously differentiable in its second argument. Let $\Sigma(t) = b(t)b(t)^\top$. Then $Y(t) = f(t, X(t))$ is an Itô process with*

$dY(t)$

$$= \frac{\partial f}{\partial t}(t, X(t)) \, dt + \sum_{i=1}^d \frac{\partial f}{\partial x_i}(t, X(t)) \, dX_i(t) + \frac{1}{2} \sum_{i,j=1}^d \frac{\partial^2 f}{\partial x_i \partial x_j}(t, X(t)) \Sigma_{ij}(t) \, dt$$

$$= \left(\frac{\partial f}{\partial t}(t, X(t)) + \sum_{i=1}^d \frac{\partial f}{\partial x_i}(t, X(t)) a_i(t) + \frac{1}{2} \sum_{i,j=1}^d \frac{\partial^2 f}{\partial x_i \partial x_j}(t, X(t)) \Sigma_{ij}(t)\right) dt$$

$$+ \sum_{i=1}^d \frac{\partial f}{\partial x_i}(t, X(t)) b_{i.}(t) \, dW(t) \tag{B.6}$$

with $b_{i.}$ the ith row of b; i.e.,

$$Y(t) = f(0, X(0)) + \int_0^t \left(\frac{\partial f}{\partial t}(u, X(u)) + \sum_{i=1}^d \frac{\partial f}{\partial x_i}(u, X(u))a_i(u) \right.$$

$$+ \frac{1}{2} \sum_{i,j=1}^d \frac{\partial^2 f}{\partial x_i \partial x_j}(u, X(u))\Sigma_{ij}(u) \Bigg) \, du$$

$$+ \int_0^t \sum_{i=1}^d \frac{\partial f}{\partial x_i}(u, X(u))b_{i\cdot}(u) \, dW(u). \tag{B.7}$$

This is the chain rule of stochastic calculus. It differs from the corresponding result in ordinary calculus through the appearance of second derivatives of f in the dt term.

If f has no explicit dependence on t (so that $Y(t) = f(X(t))$), equation (B.6) simplifies to

$$dY(t) = \left(\sum_{i=1}^d \frac{\partial f}{\partial x_i}(X(t))a_i(t) + \frac{1}{2} \sum_{i,j=1}^d \frac{\partial^2 f}{\partial x_i \partial x_j}(X(t))\Sigma_{ij}(t) \right) dt$$

$$+ \sum_{i=1}^d \frac{\partial f}{\partial x_i}(X(t))b_{i\cdot}(t) \, dW(t). \tag{B.8}$$

If X, a, and b are scalar processes, it becomes

$$dY(t) = \left(f'(X(t))a(t) + \tfrac{1}{2}f''(X(t))b^2(t) \right) dt + f'(X(t))b(t) \, dW(t). \tag{B.9}$$

By applying Theorem B.1.1 to the mapping $(x, y) \mapsto xy$ we obtain the following useful special case:

Corollary B.1.1 *(Product Rule). Let (X, Y) be an Itô process on $\Re^2$,*

$$d\begin{pmatrix} X(t) \\ Y(t) \end{pmatrix} = \begin{pmatrix} a_X(t) \\ a_Y(t) \end{pmatrix} dt + \begin{pmatrix} b_X(t)^\top \\ b_Y(t)^\top \end{pmatrix} dW(t),$$

with a_X, a_Y scalar valued, b_X, b_Y taking values in $\Re^k$, and W a k-dimensional Brownian motion. Then

$$d(X(t)Y(t)) = X(t) \, dY(t) + Y(t) \, dX(t) + b_X(t)^\top b_Y(t) \, dt \tag{B.10}$$
$$= [X(t)a_Y(t) + Y(t)a_X(t) + b_X(t)^\top b_Y(t)] \, dt$$
$$+ [X(t)b_Y(t) + Y(t)b_X(t)]^\top \, dW(t).$$

This result can be interpreted as an integration-by-parts formula for Itô calculus because (after rearranging terms) it relates $X \, dY$ to $Y \, dX$.

B.2 Stochastic Differential Equations

Most models used in financial engineering can be described through a stochastic differential equation (SDE) of the form

$$dX(t) = a(X(t), t) \, dt + b(X(t), t) \, dW(t), \quad X(0) = X_0, \tag{B.11}$$

with W a k-dimensional Brownian motion, a mapping $\Re^d \times [0, \infty)$ into $\Re^d$, b mapping $\Re^d \times [0, \infty)$ into $\Re^{d \times k}$, and X_0 a random d-vector independent of W. In what sense do the functions a and b and the initial condition X_0 determine a process X?

A *strong solution* to (B.11) on an interval $[0, T]$ is an Itô process $\{X(t), 0 \le t \le T\}$ for which $P(X(0) = X_0) = 1$ and

$$X(t) = X(0) + \int_0^t a(X(u), u) \, du + \int_0^t b(X(u), u) \, dW(u), \quad 0 \le t \le T.$$

The requirement that X be an Itô process imposes conditions (B.3) and (B.4) on the processes $\{a(X(t), t), 0 \le t \le T\}$ and $\{b(X(t), t), 0 \le t \le T\}$. We now state the main result on strong solutions to SDEs.

Theorem B.2.1 *(Existence and Uniqueness of Solutions). Suppose* $\mathsf{E}[\|X_0\|^2]$ *is finite and that there is a constant K for which the following conditions are satisfied:*

(i) $\|a(x, t) - a(y, t)\| + \|b(x, t) - b(y, t)\| \le K\|x - y\|$ *(Lipschitz condition)*
(ii) $\|a(x, t)\| + \|b(x, t)\| \le K(1 + \|x\|)$ *(Linear growth condition)*

for all $t \in [0, T]$ and all $x, y \in \Re^d$. Then the SDE (B.11) admits a strong solution X. This solution is unique in the sense that if $\tilde{X}$ is also a solution, then $P(X(t) = \tilde{X}(t) \, \forall t \in [0, T]) = 1$. For all $t \in [0, T]$, the solution satisfies $\mathsf{E}[\|X(t)\|^2] < \infty$.

Proofs of this result can be found, e.g., Hunt and Kennedy [191], Karatzas and Shreve [207], and Øksendal [284].

In the definition of a strong solution, the probability space and driving Brownian motion W are specified as part of the SDE together with the functions a and b. If we ask for just a *weak* solution, we are free to define a different probability space supporting its own Brownian motion on which (B.11) holds. For modeling purposes we are generally only concerned about the law of a process, so there is little reason to insist on a particular probability space. The most relevant issue is whether a, b, and the distribution of X_0 uniquely determine the law of any weak solution to (B.11). The strong uniqueness implied by Theorem B.2.1 is more than enough to ensure this, but the simplicity of the conditions on a and b make this a particularly convenient result.

The square-root function is not Lipschitz continuous, so Theorem B.2.1 does not apply to the square-root diffusion of Section 3.4. A result covering that model appears in Karatzas and Shreve [207], p.291. See also Krylov [216] for existence and uniqueness results under conditions weaker than those in Theorem B.2.1.

Markov Property

The form of the SDE (B.11) suggests that $X(t)$ provides a complete description of the state of the system modeled by X: the dynamics of X in (B.11) do not depend on the past evolution of X except through its current value $X(t)$. This is the intuitive content of the Markov property, which is made precise through the requirement that

$$\mathsf{E}[f(X_t)|\mathcal{F}_s] = \mathsf{E}[f(X_t)|X_s] \qquad (B.12)$$

for all $0 \le s \le t \le T$ and all bounded Borel functions $f : \Re^d \to \Re$. The process X is a *strong* Markov process if (B.12) continues to hold with s replaced by any stopping time (with respect to $\{\mathcal{F}_t\}$) taking values in $[0, T]$. This property is confirmed by the following result, proved in, for example, Hunt and Kennedy [191] and Øksendal [284]:

Theorem B.2.2 *(Markov property). Under the conditions of Theorem B.2.1, the solution X of (B.11) is a strong Markov process.*

The Gaussian Case

An $\Re^d$-valued process $\{\xi(t), t \in [0, T]\}$ is called *Gaussian* if for all $n = 1, 2, \dots$ and all $t_1, \dots, t_n \in [0, T]$, the vector formed by concatenating $\xi(t_1), \dots, \xi(t_n)$ has a multivariate normal distribution. Brownian motion is an example of a Gaussian process. A Gaussian process need not be Markovian and the solution to an SDE need not be Gaussian. But the next result tells us that in the case of a linear SDE the solution is indeed Gaussian:

Theorem B.2.3 *(Linear SDE). Let A, c, and D be bounded measurable functions on $[0, T]$ taking values in $\Re^{d \times d}$, $\Re^d$, and $\Re^{d \times k}$, respectively. Let X_0 be normally distributed on $\Re^d$ independent of the k-dimensional Brownian motion W. Then the solution of the SDE*

$$dX(t) = (A(t)X(t) + c(t))\, dt + D(t)\, dW(t), \quad X(0) = X_0 \qquad (B.13)$$

is a Gaussian process.

For a proof see Karatzas and Shreve [207], Problem 5.6.2 (solution included).

The law of a Gaussian process is completely specified by its first- and second-order moments. These can be given fairly explicitly in the case of (B.13). We consider only the case of constant $A(t) \equiv A$; for the general case see, e.g., Karatzas and Shreve [207].

Proposition B.2.1 *If $A(t) \equiv A$ in Theorem B.2.3, then*

$$X(t) = e^{At}X(0) + \int_0^t e^{A(t-u)}c(u)\, du + \int_0^t e^{A(t-u)}D(u)\, dW(u). \qquad (B.14)$$

The mean $m(t) = \mathsf{E}[X(t)]$ is given by

$$m(t) = e^{At}m(0) + \int_0^t e^{A(t-u)}c(u)\,du$$

and the covariance by

$$\mathsf{E}[(X(t) - m(t))(X(s) - m(s))^\top] = \int_0^{\min(s,t)} e^{A(t-u)}D(u)D(u)^\top e^{A^\top(t-u)}\,du.$$

That the process in (B.14) satisfies (B.13) can be verified using Itô's formula. The expressions for the moments of X then follow from simple rules for calculating means and covariances of stochastic integrals with deterministic integrands, which we now make more explicit. If $\sigma : [0,T] \to \Re^k$ satisfies

$$\int_0^T \|\sigma(u)\|^2\,du < \infty, \tag{B.15}$$

then

$$\mathsf{E}\left[\int_0^T \sigma(u)^\top\,dW(u)\right] = 0, \tag{B.16}$$

and

$$\mathsf{Var}\left[\int_0^T \sigma(u)^\top\,dW(u)\right] = \int_0^T \|\sigma(u)\|^2\,du. \tag{B.17}$$

If σ_1, σ_2 both satisfy (B.15), then for any $s, t \in [0,T]$

$$\mathsf{Cov}\left[\int_0^t \sigma_1(u)^\top\,dW(u), \int_0^s \sigma_2(u)^\top\,dW(u)\right] = \int_0^{\min(s,t)} \sigma_1(u)^\top\sigma_2(u)\,du.$$

B.3 Martingales

This section summarizes some results relating stochastic integrals and martingales.

A real-valued adapted process $\{X(t), t \geq 0\}$ is a martingale if

(i) $\mathsf{E}[|X_t|] < \infty$ for all $t \geq 0$;
(ii) $\mathsf{E}[X_t|\mathcal{F}_s] = X_s$ for all $0 \leq s < t < \infty$.

Define a martingale on $[0,T]$ by restricting t and s to this interval. Throughout most of this book, we implicitly assume the integrability property (i) in calling a process a martingale.

The process X is a *local martingale* if there exists a sequence of stopping times $\{\tau_n, n = 1, 2, \ldots\}$ with $\tau_n \uparrow \infty$ for which each process $X_n(t) \equiv X(t \wedge \tau_n)$

is a martingale. All martingales are local martingales, but the converse does not hold. If, however, X is a local martingale and

$$\mathsf{E}\left[\sup_{0\leq t\leq T} |X(t)|\right] < \infty,$$

then X is in fact a martingale on $[0, T]$. This follows from the dominated convergence theorem, as demonstrated in Protter [300], Theorem I.47.

It is common in applied work to assume that the solution to an SDE with no drift term (a in (B.11)) is a martingale and, more generally, that any process of the form

$$X(t) = X(0) + \int_0^t \gamma(u)^\top dW(u), \quad X(0) \in \mathcal{F}_0, \tag{B.18}$$

is a martingale. We make this assumption in several places throughout this book, usually to derive implications for asset price dynamics from the absence of arbitrage. But the process in (B.18) is not automatically a martingale without additional hypotheses. The following result gives some indication of what we are leaving out in assuming the martingale property holds.

Theorem B.3.1 *(Stochastic integrals as martingales). Let X be as in (B.18) with*

$$\int_0^t \|\gamma(u)\|^2 \, du < \infty, \quad a.s., \text{ for all } t, \tag{B.19}$$

then (i) X is a local martingale. (ii) If $X(t) \geq 0$, a.s., for all t, then

$$\mathsf{E}[X_t|\mathcal{F}_s] \leq X_s, \quad 0 \leq s \leq t;$$

i.e., X is a supermartingale. If $\mathsf{E}[X_t]$ is constant then X is a martingale. (iii) If $\mathsf{E}[X(0)^2] < \infty$ and if for all $t > 0$,

$$\mathsf{E}\left[\int_0^t \|\gamma(u)\|^2 \, du\right] < \infty, \tag{B.20}$$

then X is a martingale and

$$\mathsf{E}[X(t)^2] = \mathsf{E}[X(0)^2] + \mathsf{E}\left[\int_0^t \|\gamma(u)\|^2 \, du\right].$$

From (i) we see that a "driftless" process is a local martingale but not necessarily a martingale. Property (ii) follows from the fact (Revuz and Yor [306], p.123) that every nonnegative local martingale is a supermartingale and any supermartingale with constant expectation is a martingale. Property (iii) states that as long as we restrict ourselves to integrands satisfying (B.20), stochastic integrals are indeed martingales and in fact square-integrable martingales. This is a special case of general results on stochastic integration with respect to martingales; for example, Proposition 3.2.10 of Karatzas and Shreve [207].

We frequently work with processes of the form

$$Y(t) = Y(0) \exp\left(-\frac{1}{2}\int_0^t \|\gamma(u)\|^2 \, du + \int_0^t \gamma(u)^\top \, dW(u)\right). \qquad \text{(B.21)}$$

If (B.19) holds, then

$$X(t) = -\frac{1}{2}\int_0^t \|\gamma(u)\|^2 \, du + \int_0^t \gamma(u)^\top \, dW(u) \qquad \text{(B.22)}$$

is an Itô process and hence (by Theorem B.1.1) Y is too. An application of Itô's formula shows that Y satisfies

$$dY(t) = Y(t)\gamma(u)^\top \, dW(u),$$

so Y is at least a local martingale. If $Y(0)$ is nonnegative and if $\mathsf{E}[Y(t)]$ is constant, then we see from part (ii) of Theorem B.3.1 that Y is in fact a martingale—an *exponential* martingale. In the special case that γ is deterministic and bounded on finite intervals, X is a Gaussian process; with $Y(0) \equiv 1$ we have

$$
\begin{aligned}
\mathsf{E}[Y(t)] &= \mathsf{E}[\exp(X(t))] \\
&= \exp\left(\mathsf{E}[X(t)] + \tfrac{1}{2}\mathsf{Var}[X(t)]\right) \\
&= \exp\left(-\frac{1}{2}\int_0^t \|\gamma(u)\|^2 \, du + \frac{1}{2}\int_0^t \|\gamma(u)\|^2 \, du\right) = 1,
\end{aligned}
$$

using (B.16) and (B.17). This verifies that Y is a martingale.

Theorem B.3.1 states that under appropriate additional conditions, stochastic integrals are martingales. The next result may be paraphrased as stating that if Brownian motion is the only source of uncertainty, then all martingales are stochastic integrals. To make this precise, let $\mathcal{F}_t^W$ be the σ-algebra generated by $\{W(u), 0 \leq u \leq t\}$ augmented to include all subsets of null sets. We now specialize to the filtration $\{\mathcal{F}_t^W\}$.

Theorem B.3.2 (*Martingale representation theorem*). *If X is a local martingale with respect to $\{\mathcal{F}_t^W\}$, then there exists a process γ such that (B.18) holds. If X is a square-integrable martingale, then γ satisfies (B.20).*

The second part is proved on pp.182-184 of Karatzas and Shreve [207]; a proof of the first part is provided in Hunt and Kennedy [191], p.113.

A simple consequence of this result is that any integrable random variable $\xi \in \mathcal{F}_T^W$ has a representation of the form

$$\xi = \mathsf{E}[\xi] + \int_0^T \gamma(u)^\top \, dW(u). \qquad \text{(B.23)}$$

This follows by applying Theorem B.3.2 to the martingale $X(t) = \mathsf{E}[\xi|\mathcal{F}_t]$. Equation (B.23) says that we can synthesize the "payoff" $\xi - \mathsf{E}[\xi]$ by "trading" in the underlying Brownian motion W.

A further consequence of Theorem B.3.2 is that any strictly positive local martingale (with respect to $\{\mathcal{F}_t^W\}$) has a representation of the form (B.21). More precisely, suppose Y is strictly positive and $\tilde{Y}(t) = Y(t)/Y(0)$ is a local martingale. From Theorem B.3.2, we get a representation of the form

$$\tilde{Y}(t) = \int_0^t \tilde{\gamma}(u)^\top \, dW(u).$$

Because $\tilde{Y}$ is strictly positive, we can define $X(t) = \log \tilde{Y}$ and Itô's formula shows that X satisfies (B.22) with $\gamma = \tilde{\gamma}/\tilde{Y}$. But $\tilde{Y}(t) = \exp(X(t))$ and $Y(t) = Y(0) \exp(X(t))$, so Y has the representation in (B.21).

B.4 Change of Measure

Let X be a nonnegative random variable on $(\Omega, \mathcal{F}, P)$ with $\mathsf{E}[X] = 1$. Define $Q : \mathcal{F} \to [0, 1]$ by setting

$$Q(A) = \mathsf{E}[\mathbf{1}_A X] = \int_A X(\omega) \, dP(\omega), \quad A \in \mathcal{F}, \tag{B.24}$$

with $\mathbf{1}_A$ the indicator function of the set A. It is easy to verify that the set function Q is a probability measure on $(\Omega, \mathcal{F})$. It is absolutely continuous with respect to P, meaning that

$$Q(A) > 0 \Rightarrow P(A) > 0$$

for every $A \in \mathcal{F}$. The Radon-Nikodym Theorem states that all such measures arise in this way: if P and Q are probability measures on $(\Omega, \mathcal{F})$ and if Q is absolutely continuous with respect to P, then there exists a random variable X such that (B.24) holds. Moreover, X is unique in the sense that if (B.24) holds for all $A \in \mathcal{F}$ for some other random variable X', then $P(X = X') = 1$. Because Q is a probability, we must have $P(X \geq 0) = 1$ and

$$\int_\Omega X(\omega) \, dP(\omega) = 1.$$

The random variable X is commonly written as dQ/dP and called the *Radon-Nikodym derivative* or *likelihood ratio* of Q with respect to P.

If P is also absolutely continuous with respect to Q, then P and Q are *equivalent*. Equivalent measures agree about which events have probability zero. If P and Q are equivalent, then

$$\frac{dP}{dQ} = \left(\frac{dQ}{dP}\right)^{-1}.$$

To illustrate these ideas, suppose the random variable Z on $(\Omega, \mathcal{F})$ has a density g under probability measure P; i.e.,

$$P(Z \leq z) = \int_{-\infty}^{z} g(x)\, dx$$

for all $z \in \mathfrak{R}$. Let f be another probability density on $\mathfrak{R}$ with the property that $g(z) = 0 \Rightarrow f(z) = 0$ and define a new probability measure Q on $(\Omega, \mathcal{F})$ by setting

$$Q(A) = \mathsf{E}_P\left[1_A \frac{f(Z)}{g(Z)} \right];$$

we have subscripted the expectation to emphasize that it is taken with respect to P. Interpret $f(z)/g(z)$ to be 1 whenever both numerator and denominator are 0. Clearly, $P(f(Z)/g(Z) \geq 0) = 1$ and

$$\int_{\Omega} \frac{f(Z(\omega))}{g(Z(\omega))}\, dP(\omega) = \int_{-\infty}^{\infty} \frac{f(x)}{g(x)} g(x)\, dx = 1.$$

Under the new measure, the event $\{Z \leq z\}$ has probability

$$\begin{aligned} Q(Z \leq z) &= \int_{\{\omega : Z(\omega) \leq z\}} \frac{f(Z(\omega))}{g(Z(\omega))}\, dP(\omega) \\ &= \int_{-\infty}^{z} \frac{f(x)}{g(x)} g(x)\, dx \\ &= \int_{-\infty}^{z} f(x)\, dx. \end{aligned}$$

Thus, under Q, the random variable Z has density f.

As a special case, let g be the standard normal density and let f be the normal density with mean μ and variance 1. Then $f(x)/g(x) = \exp(\mu x - \mu^2/2)$. If Z has the standard normal distribution under P and if we define Q by setting

$$\frac{dQ}{dP} = e^{-\frac{1}{2}\mu^2 + \mu Z},$$

then Z has mean μ under Q and $Z - \mu$ has the standard normal distribution under Q.

A similar calculation shows that if, under some probability measure P_n, the random variables $Z_1, \ldots, Z_n$ are independent with densities $g_1, \ldots, g_n$, and if we define Q_n by setting

$$\frac{dQ_n}{dP_n} = \prod_{i=1}^{n} \frac{f_i(Z_i)}{g_i(Z_i)},$$

then under Q_n, the Z_i are independent with densities f_i. In particular, if the Z_i are independent standard normals under P_n and we set

$$\frac{dQ_n}{dP_n} = \exp\left(-\frac{1}{2}\sum_{i=1}^{n}\mu_i^2 + \sum_{i=1}^{n}\mu_i Z_i\right), \tag{B.25}$$

then

$$Z_1 - \mu_1, Z_2 - \mu_2, \ldots, Z_n - \mu_n \tag{B.26}$$

are independent standard normals under Q_n.

Consider, now, what happens in this example as n becomes large. For each n, the support of the multivariate normal vector $(Z_1, \ldots, Z_n)$ is all of $\Re^n$ regardless of its mean and the measures P_n and Q_n are equivalent by construction. Suppose that under P the random variables $Z_1, Z_2, \ldots$ are independent standard normals, and suppose there is another measure Q on $(\Omega, \mathcal{F})$ such that the Z_i are independent normals with mean μ and variance 1 under Q. Then

$$P\left(\lim_{n\to\infty}\frac{1}{n}\sum_{i=1}^{n}Z_i = 0\right) = 1,$$

whereas if $\mu \neq 0$ this event has Q-probability zero. Not only do P and Q fail to be equivalent, they live on entirely different sets. The mutual absolute continuity that holds for each n through (B.25) breaks down in the limit as $n \to \infty$.

Our main interest in measure transformations lies in changes of measure that have the effect of adding a drift to a Brownian motion. This may be viewed as an extension of (B.25) and (B.26). In discussing measure transformations for continuous-time processes we restrict ourselves to finite time intervals, just as the transformation in (B.25) is feasible only for finite n.

Girsanov's Theorem

We now generalize the basic transformation in (B.24). For the filtration $\{\mathcal{F}_t\}$ of $(\Omega, \mathcal{F}, P)$, let P_t denote the restriction of P to $\mathcal{F}_t$. Let $\{X(t), t \in [0, T]\}$ be a nonnegative martingale with respect to $\{\mathcal{F}_t\}$ and suppose $\mathsf{E}[X(T)] = 1$. Define a probability measure Q_t on $\mathcal{F}_t$ by setting

$$Q_t(A) = \mathsf{E}_P[\mathbf{1}_A X(t)] = \mathsf{E}_{P_t}[\mathbf{1}_A X(t)], \quad A \in \mathcal{F}_t;$$

i.e., for each $t \in [0, T]$,

$$\frac{dQ_t}{dP_t} = X(t).$$

Then Q_t is the restriction of Q_T to $\mathcal{F}_t$ because for any $A \in \mathcal{F}_t$,

$$Q_T(A) = \mathsf{E}_P[\mathbf{1}_A X(T)] = \mathsf{E}_P[\mathbf{1}_A \mathsf{E}_P[X(T)|\mathcal{F}_t]] = \mathsf{E}_P[\mathbf{1}_A X(t)] = Q_t(A).$$

In this sense, the measures $\{Q_t, t \in [0, T]\}$ are consistent. If X is strictly positive, then Q_t and P_t are equivalent for all t. To replace $[0, T]$ with $[0, \infty)$

and still have all Q_t consistent, we would need to add the requirement that the martingale X be uniformly integrable (see (A.3)).

Suppose, now, that P and Q are equivalent probability measures on $(\Omega, \mathcal{F})$ with Radon-Nikodym derivative dQ/dP. Define

$$\left(\frac{dQ}{dP}\right)_t = \mathsf{E}_P\left[\frac{dQ}{dP}|\mathcal{F}_t\right], \; t \in [0, T].$$

We claim that $(dQ/dP)_t$ is a martingale and equals the Radon-Nikodym derivative of the restriction of Q to $\mathcal{F}_t$ with respect to the restriction of P to $\mathcal{F}_t$. The martingale property is immediate from the definition. For the second claim, observe that for any $A \in \mathcal{F}_t$,

$$\mathsf{E}_P\left[\mathbf{1}_A\left(\frac{dQ}{dP}\right)_t\right] = \mathsf{E}_P\left[\mathsf{E}_P\left[\mathbf{1}_A\frac{dQ}{dP}|\mathcal{F}_t\right]\right] = \mathsf{E}_P\left[\mathbf{1}_A\frac{dQ}{dP}\right] = Q(A).$$

We may summarize this discussion by saying that a nonnegative, unit-mean martingale defines a consistent (with respect to $\{\mathcal{F}_t, t \in [0, T]\}$) family of probability measures, and, conversely, the Radon-Nikodym derivatives for such a family define a unit-mean, nonnegative martingale.

The following simple rule for applying a change of measure to conditional expectations arises frequently in mathematical finance, most notably in applying a change of numeraire:

Proposition B.4.1 *If P and Q are equivalent and $\mathsf{E}_Q[|X|] < \infty$, then*

$$\mathsf{E}_Q[X|\mathcal{F}_t] = \left(\frac{dQ}{dP}\right)_t^{-1} \mathsf{E}_P\left[X\frac{dQ}{dP}|\mathcal{F}_t\right].$$

This follows immediately from the definition of $(dQ/dP)_t$; for an explicit proof, see Musiela and Rutkowski [275], p.458.

We can now state Girsanov's Theorem. In the following, let $\{W(t), t \in [0, T]\}$ denote a standard k-dimensional Brownian motion on $(\Omega, \mathcal{F}, P)$ and let $\{\mathcal{F}_t^W, t \in [0, T]\}$ denote the filtration generated by W augmented to include all subsets of sets having P-probability 0.

Theorem B.4.1 *(Girsanov Theorem).* (i) *Let γ be an $\Re^k$-valued process adapted to $\{\mathcal{F}_t^W\}$ satisfying (B.19) for $t \in [0, T]$ and let*

$$X(t) = \exp\left(-\frac{1}{2}\int_0^t \|\gamma(u)\|^2\, du + \int_0^t \gamma(u)\, dW(u)\right). \qquad (B.27)$$

If $\mathsf{E}_P[X(T)] = 1$, then $\{X(t), t \in [0, T]\}$ is a martingale and the measure Q on $(\Omega, \mathcal{F}_T^W)$ defined by

$$\frac{dQ}{dP} = X(T)$$

is equivalent to P. Under Q, the process

$$W^Q(t) \stackrel{\triangle}{=} W(t) - \int_0^t \gamma(u)\, du, \quad t \in [0, T]$$

is a standard Brownian motion with respect to $\{\mathcal{F}_t^W\}$. (ii) Conversely, if Q is a probability measure on $(\Omega, \mathcal{F}_T^W)$ equivalent to the restriction of P to $\mathcal{F}_T^W$, then $(dQ/dP)_t$ admits the representation in (B.27) for some γ and W^Q is a standard Brownian motion under Q.

Thus, the change of measure associated with a change of "drift" in a Brownian motion over a finite horizon is an absolutely continuous change of measure, and every absolutely continuous change of measure for Brownian motion is of this type.

There are several different results, applicable at different levels of generality, known as Girsanov's Theorem. The formulation given here is more precisely Girsanov's Theorem for Brownian motion and is closest to the one proved in Section 5.2.2 of Hunt and Kennedy [191]. See Revuz and Yor [306] for historical remarks as well as a more general formulation. Usually, only part (i) of this result is called Girsanov's Theorem. Part (ii) is a consequence of the Martingale Representation Theorem: as a positive unit-mean martingale, $(dQ/dP)_t$ must be of the form (B.27) and then Q and P must be related as in part (i). We have included the converse because of its importance in the theory of derivative pricing. It assures us that when we change from the objective probability measure to an equivalent martingale measure, the drifts in asset prices may change but their volatilities may not.

The requirement that $\mathsf{E}_P[X(T)] = 1$ in part (i) of Theorem B.4.1 is needed to ensure that X is a martingale and not merely a local martingale; see Theorem B.3.1. The Novikov condition is a widely cited sufficient condition for this requirement (see, e.g., Section 3.5 of Karatzas and Shreve [207] for a proof):

Proposition B.4.2 *(Novikov condition). If*

$$\mathsf{E}\left[\exp\left(\tfrac{1}{2}\int_0^T \|\gamma(u)\|^2\, du\right)\right] < \infty,$$

then X in (B.27) is a martingale on $[0, T]$.

C

Appendix: The Term Structure of Interest Rates

The *term structure* of interest rates refers to the dependence of interest rates on maturity. There are several equivalent ways of recording this relationship. This appendix reviews terminology used for this purpose and describes some of the most important interest rate derivative securities.

C.1 Term Structure Terminology

A unit of account (e.g., a dollar) invested at a continuously compounded rate R grows to a value of e^{RT} over the interval from time 0 to T. If instead the investment earns *simple* interest over $[0, T]$, the value grows to $1 + RT$ over this interval. The discount factors associated with continuous compounding and simple interest are thus e^{-RT} and $1/(1 + RT)$, respectively.

Many fixed income securities (including US Treasury bonds) follow an intermediate convention in determining how interest accrues: an interest rate is quoted on an annual basis with semi-annual compounding. In this case, an initial investment of 1 grows to a value of $1 + (R/2)$ at the end of half a year, to a value of $(1 + (R/2))^2$ at the end of one year, and so on. A bit more generally, if we let δ denote the fraction of a year over which interest is compounded (with $\delta = 1/2$ and $\delta = 1/4$ the most important cases), the interest accrued over $n\delta$ years is $(1 + \delta R)^n - 1$. Depending on what day-count convention is used, the exact lengths of nominally equal six-month or three-month intervals may vary; if we therefore generalize to allow unequal fractions $\delta_1, \delta_2, \ldots$, then at the end of n periods, an initial investment of 1 grows to a value of

$$\prod_{i=1}^{n} (1 + \delta_i R).$$

The associated discount factor is the reciprocal of this expression.

Bonds, Yields, and Forward Rates

Let $B(t,T)$ denote the price at time t of a security making a single payment of 1 at time T, $T \geq t$. This is a zero-coupon (or pure-discount) bond with maturity T. A coupon bond makes payments at multiple dates and may be viewed as a portfolio of zero-coupon bonds. A coupon bond paying c_i at T_i, $i = 1, \ldots, n$, and a principal payment of 1 at T_n has a value prior to maturity of

$$B_c(t) = B(t, T_n) + \sum_{i=\ell(t)}^{n} c_i B(t, T_i), \qquad \text{(C.1)}$$

with $\ell(t)$ the index of the next coupon date, defined by

$$T_{\ell(t)-1} < t \leq T_{\ell(t)}.$$

In a world with a constant continuously compounded interest rate, an investor could replicate a zero-coupon bond with maturity T by investing e^{-RT} in an interest bearing account at time 0 and letting it grow to a value of 1 at time T. It follows that $B(0,T) = e^{-RT}$ in this setting.

More generally, if the continuously compounded rate at time t (the short rate) is given by a stochastic process $r(t)$, an investment of 1 at time 0 grows to a value of

$$\beta(t) = \exp\left(\int_0^t r(u)\, du \right)$$

at time t. As explained in Chapter 1, the price of a bond is given by

$$B(0,T) = \mathsf{E}\left[\exp\left(-\int_0^T r(t)\, dt \right) \right], \qquad \text{(C.2)}$$

the expectation taken under the risk-neutral measure. This is the only identity in this section that involves probability.

The *yield* of a bond may be interpreted as the interest rate implied by the price of the bond; the yield therefore depends on the compounding convention assumed for the implied interest rate. The continuously compounded yield $Y(t,T)$ for a zero-coupon bond maturing at T is defined by

$$B(t,T) = e^{-Y(t,T)(T-t)} \quad \text{or} \quad Y(t,T) = -\frac{1}{T-t} \log B(t,T). \qquad \text{(C.3)}$$

The continuously compounded yield $Y_c(t)$ for the bond in (C.1) is defined by the condition

$$B_c(t) = e^{-Y_c(t)(T_n-t)} + \sum_{i=\ell(t)}^{n} c_i e^{-Y_c(t)(T_i-t)}.$$

Yields are more commonly quoted on a semi-annual basis. The yield $Y_\delta(t,T)$ associated with a compounding interval δ solves the equation

$$B(t,T) = \frac{1}{(1 + \delta Y_\delta(t,T))^n},$$

when $n = (T-t)/\delta$ is an integer. This extends to arbitrary $t \in (0,T)$ through the convention that interest accrues linearly between compounding dates. The yield of a coupon bond is similarly defined by discounting the coupons as well as the principal payment.

A *forward rate* is an interest rate set today for borrowing or lending at some date in the future. Consider, first, the case of a forward rate based on simple interest, and let $F(t,T_1,T_2)$ denote the forward rate fixed at time t for the interval $[T_1,T_2]$, with $t < T_1 < T_2$. An investor entering into an agreement at time t to borrow 1 at time T_1 and repay the loan at time T_2 pays interest at rate $F(t,T_1,T_2)$. More explicitly, the investor receives 1 at T_1 and pays $1 + F(t,T_1,T_2)(T_2 - T_1)$ at T_2; no other payments are exchanged.

An arbitrage argument shows that forward rates are determined by bond prices. At time t, an investor could buy a zero-coupon bond maturing at T_1, funding the purchase by issuing bonds maturing at T_2. If the number of bonds k is chosen to satisfy

$$kB(t,T_2) = B(t,T_1),$$

there is no net cashflow at time t. The investor will receive 1 at T_1 and pay k at T_2. To preclude arbitrage, the amount paid at T_2 in this transaction must be the same as the amount paid in the forward rate transaction, so

$$k = 1 + F(t,T_1,T_2)(T_2 - T_1).$$

But $k = B(t,T_1)/B(t,T_2)$, so we conclude that

$$F(t,T_1,T_2) = \frac{1}{T_2 - T_1}\left(\frac{B(t,T_1) - B(t,T_2)}{B(t,T_2)}\right). \tag{C.4}$$

For much of the financial industry, the most important benchmark interest rates are the London Inter-Bank Offered Rates or LIBOR. LIBOR is calculated daily through an average of rates offered by select banks in London. Separate rates are quoted for different maturities (e.g., three months and six months) and different currencies. LIBOR is quoted as a simple annualized interest rate. A *forward* LIBOR rate is a special case of (C.4) with a fixed length $\delta = T_2 - T_1$ for the accrual period, typically with $\delta = 1/2$ or $\delta = 1/4$. Thus, the δ-year forward LIBOR rate at time t with maturity T is

$$L(t,T) = F(t,T,T+\delta) = \frac{1}{\delta}\left(\frac{B(t,T) - B(t,T+\delta)}{B(t,T+\delta)}\right). \tag{C.5}$$

In taking forward LIBOR rates as a special case of (C.4), we are ignoring credit risk. The discussion leading to (C.4) assumes that bonds always make their scheduled payments, that issuers never default. But the banks whose rates set LIBOR may indeed default and this risk is presumably reflected in the rates they offer. Equation (C.5) ignores this feature. It is a convenient simplification often used in practice.

In deriving (C.4) we assumed a simple forward rate, but we could just as well have used a continuously compounded forward rate $f(t, T_1, T_2)$. The interest paid must be the same regardless of the compounding convention, so we must have

$$\exp\left(f(t, T_1, T_2)(T_2 - T_1)\right) - 1 = F(t, T_1, T_2)(T_2 - T_1).$$

With (C.4), this implies

$$f(t, T_1, T_2) = \frac{\log B(t, T_1) - \log B(t, T_2)}{T_2 - T_1}$$

for the continuously compounded forward rate for the accrual interval $[T_1, T_2]$.

Now define $f(t, T)$ to be the continuously compounded forward rate fixed at t for the instant T. This is the limit (assuming it exists) of $f(t, T, T + h)$ as h approaches 0, and is thus given by

$$f(t, T) = -\frac{\partial}{\partial T} \log B(t, T).$$

Inverting this relationship and using $B(T, T) = 1$, we get

$$B(t, T) = \exp\left(-\int_t^T f(t, u)\, du\right).$$

Thus, the forward curve $f(t, \cdot)$ is characterized by the property that discounting along this curve reproduces time-t bond prices.

Comparison with (C.3) reveals that

$$Y(t, T) = \frac{1}{T - t} \int_t^T f(t, u)\, du;$$

i.e., that yields are averages over forward rates. This suggests that forward rates are more fundamental quantities than yields and thus potentially a more attractive starting point in building models of term structure *dynamics*.

Swaps and Swap Rates

In a standard interest rate swap, two parties agree to exchange payments tied to a notional principal, one party paying interest at a floating rate, the other at a fixed rate. The principal is notional in the sense that it is never paid by either party; it is merely used to determine the magnitudes of the payments.

Fix a period δ (e.g., half a year) and a set of dates $T_n = n\delta$, $n = 0, 1, \ldots, M + 1$. Consider a swap with payment dates $T_1, \ldots, T_{M+1}$ on a notional principal of 100. At each T_n, the fixed-rate payer pays $100 R \delta$: this is the simple interest accrued on a principal of 100 over an interval of length δ at an annual rate of R. Denote by $L_{n-1}(T_{n-1})$ the simple annualized interest rate fixed at T_{n-1} for the interval $[T_{n-1}, T_n]$. (Thus, $L_{n-1}(T_{n-1}) = L(T_{n-1}, T_{n-1})$ in (C.5), the δ-year LIBOR rate fixed at T_{n-1}.) The floating-rate payer pays $100 L_{n-1}(T_{n-1})\delta$ at each T_n. The exchange of payments terminates at T_{M+1}.

Consider the value of the swap from the perspective of the party paying floating to receive fixed; the value to the other party has the same magnitude but opposite sign. Although no principal is ever exchanged in the swap, valuation is simplified if we pretend that each party pays the other 100 at T_{M+1}. These two fictitious payments cancel each other and thus have no effect on the value of the swap. With this modification, each party's payments look like those of a bond with a face value of 100, one bond having a fixed coupon rate of R, the other having a floating coupon. The value of the swap is the difference between the values of the two bonds.

At time $T_0 = 0$, the value of the fixed rate bond is (see (C.1))

$$100 R \delta \sum_{i=1}^{M+1} B(0, T_i) + 100 B(0, T_{M+1}).$$

To value the floating rate bond, we argue that it can be replicated with an initial investment of 100. Over the interval $[0, T_1]$, the initial investment earns $100 \delta L_0(0)$ in interest, precisely enough to pay the first coupon of the floating rate bond. The remaining 100 can then be invested at rate $L_1(T_1)$ until T_2 to fund the next coupon while preserving the original 100. This reinvestment process can be repeated until T_{M+1} when the 100 is used to pay the bond's principal. Because the cashflows of the floating rate bond can be replicated with an initial investment of 100, we conclude that the value of the floating rate bond must itself be 100. From the perspective of the floating-for-fixed party, the value of the swap is the difference

$$100 R \delta \sum_{i=1}^{M+1} B(0, T_i) + 100 B(0, T_{M+1}) - 100 \qquad (C.6)$$

between the fixed and floating rate bonds.

By definition, the swap rate at time 0 (for payments at $T_1, \ldots, T_{M+1}$) is the fixed rate R that makes the value of the swap (C.6) equal to zero. Both parties would willingly enter into a swap at this rate without either party having to make an additional payment to the other—the swap is costless. From (C.6), we find that the swap rate is

$$S_0(0) = \frac{1 - B(0, T_{M+1})}{\delta \sum_{i=1}^{M+1} B(0, T_i)}.$$

We have subscripted the swap rate by 0 to indicate that this is the rate for a swap beginning at time 0 with payments at $T_1, \ldots, T_{M+1}$. The same derivation shows that the rate at T_n for a swap with payments at $T_{n+1}, \ldots, T_{M+1}$ is

$$S_n(T_n) = \frac{1 - B(T_n, T_{M+1})}{\delta \sum_{i=n+1}^{M+1} B(T_n, T_i)}.$$

The *forward* swap rate at time $t < T_n$ for a swap with payment dates $T_{n+1}, \ldots, T_{M+1}$ is

$$S_n(t) = \frac{B(t, T_n) - B(t, T_{M+1})}{\delta \sum_{i=n+1}^{M+1} B(t, T_i)}. \tag{C.7}$$

An extension of the argument leading to (C.6) shows that this is the rate that makes a swap with payment dates $T_{n+1}, \ldots, T_{M+1}$ costless at time t. The forward swap rate for a single period ($M = n$) coincides with the forward LIBOR rate $L(t, T_n)$; see (C.5).

C.2 Interest Rate Derivatives

The relationships among forward rates, swap rates, and bonds in Section C.1 are purely algebraic and independent of any stochastic assumptions about interest rates. These relationships follow from static no-arbitrage arguments that must hold at each valuation date t irrespective of the dynamics of the term structure. This is not generally true of the value of interest rate futures and options. The relationship between these interest rate derivatives and the underlying term structure variables ordinarily depends on how one models term structure dynamics.

Futures

We describe a slightly simplified version of Eurodollar futures, which are among the most actively traded contracts in any market. The futures contract has a settlement value of

$$100 \cdot (1 - L(T, T))$$

at the expiration date T, where $L(T, T)$ is the δ-year LIBOR rate in (C.5). Through an argument detailed in Section 8.D of Duffie [98], the time-t *futures price* associated with a futures contract is the risk-neutral conditional expectation of the settlement value. Define

$$\hat{L}_T(t) = \mathsf{E}[L(T, T)|\mathcal{F}_t] = \mathsf{E}\left[\frac{1}{\delta}\left(\frac{1}{B(T, T+\delta)} - 1\right)\Big|\mathcal{F}_t\right], \tag{C.8}$$

with $\mathcal{F}_t$ the history of market prices up to time t, and the expectation taken under the risk-neutral measure. The Eurodollar futures price at time t (for settlement at T) is then $100(1 - \hat{L}_T(t))$. The futures contract commits the holder to making or receiving payments as the futures price fluctuates through a process called *resettlement* or *marking to market*. In the idealized case of continuous resettlement, each increment $d\hat{L}_T$ triggers a payment of $100 d\hat{L}_T$ by the holder of the contract, with a negative payment interpreted as a dividend.

The relationship (C.8) between the futures rate $\hat{L}_T$ and bond prices depends on stochastic elements of a model of the dynamics of the term structure, whereas the relationship between forward rates and bond prices is essentially algebraic and independent of the choice of model. The futures rate $\hat{L}_T(t)$ is a martingale under the risk-neutral measure; the forward rate $L(t,T)$ is a martingale under the *forward* measure for date T. The two processes coincide at $t = T$.

Caps and Floors

An interest rate cap is a portfolio of options that serve to limit the interest paid on a floating rate liability over a set of consecutive periods. Each individual option in the cap applies to a single period and is called a *caplet*. Because the value of a cap is simply the sum of the values of its component caplets, it suffices to discuss valuation of caplets.

Consider, then, a caplet for the interval $[T, T+\delta]$. A party with a floating rate liability over this interval would pay interest equal to $\delta L(T,T)$ times the principal at the end of the interval, with $L(T,T)$ the δ-year rate in (C.5). A security designed to limit the interest rate paid to some fixed level K refunds the difference $\delta(L(T,T)-K)$ (per unit of principal) if this difference is positive and pays nothing otherwise. Thus, the payoff of a caplet is

$$\delta(L(T,T) - K)^+, \qquad (C.9)$$

and this payment is made at $T + \delta$. This can also be written as

$$\left(\left[\exp\left(\int_0^\delta f(T, T+u)\right) - 1\right] - \delta K\right)^+,$$

using the curve $f(T, \cdot)$ of instantaneous forward rates at time T.

A floor similarly sets a lower limit on interest payments. A single-period floor (a *floorlet*) for the interval $[T, T + \delta]$ pays $\delta(K - L(T,T))^+$ at $T + \delta$.

The caplet payoff (C.9) is received at time $T+\delta$ but fixed at time T; there is no uncertainty in the payoff over the interval $[T, T + \delta]$. Hence, a security paying (C.9) at time $T + \delta$ is equivalent to one paying

$$\frac{\delta}{1 + \delta L(T,T)}(L(T,T) - K)^+ = \delta B(T, T+\delta)(L(T,T) - K)^+ \qquad (C.10)$$

at time T.

The payoff of a caplet (or floorlet) can be replicated by trading in two underlying assets, the bonds maturing at T and $T + \delta$. Valuing a caplet entails determining the initial cost of this trading strategy or, more directly, computing the expected present value of the caplet's payoff. This requires specifying a model of the dynamics of the term structure.

By market convention, caplet prices are quoted through *Black's formula*, after Black [49]. This formula equates the time-t price of the caplet to

$$\delta B(t, T + \delta) \left(L(t, T) \Phi \left(\frac{\log(L(t,T)/K) + \sigma^2(T - t)/2}{\sigma\sqrt{T - t}} \right) \right.$$
$$\left. - K\Phi \left(\frac{\log(L(t,T)/K) - \sigma^2(T - t)/2}{\sigma\sqrt{T - t}} \right) \right),$$

with Φ the standard cumulative normal distribution. This expression is what one would obtain for the expectation of $\delta B(t, T + \delta)(L(T, T) - K)^+$ if $L(\cdot, T)$ satisfied

$$\frac{dL(t, T)}{L(t, T)} = \sigma\, dW(t),$$

though this does not necessarily correspond to a price in the sense of the theory of derivatives valuation. In practice, the Black formula is typically used in reverse, to extract the "implied volatility" parameter σ from the market prices of caps. An obvious modification of the formula above produces the Black formula for a floor.

Swaptions

A swaption is an option to enter into a swap. A "2 × 5" or "2-into-5" swaption is a two-year option to enter into a five-year swap. A bit more generically, consider an option expiring at T_n to enter into a swap with payment dates $T_{n+1}, \ldots, T_{M+1}$. Suppose the option grants the holder the right to pay fixed and receive floating on a notional principal of 1. Denote by R the fixed rate specified in the underlying swap. At the expiration date T_n, the value of the underlying swap is then

$$V(T_n) = 1 - R\delta \sum_{i=n+1}^{M+1} B(T_n, T_i) - B(T_n, T_{M+1}),$$

by the argument used to derive (C.6). The holder of the option exercises if the swap has positive value and otherwise lets the option expire worthless. We may therefore think of the swaption as an instrument that pays $[V(T_n)]^+$ at T_n.

Using (C.7), we find that

$$[V(T_n)]^+ = \delta \sum_{i=n+1}^{M+1} B(T_n, T_i)(S_n(T_n) - R)^+. \tag{C.11}$$

Hence, the swaption looks like a call option on a swap *rate*. This formulation is convenient because modeling the dynamics of forward swap rates is more natural than modeling the dynamics of swap values, just as modeling the dynamics of forward interest rates is more natural than modeling bond prices. Furthermore, this representation makes it evident that a caplet is a special case of a swaption by taking $M = n$ and comparing with (C.10).

A swaption can be replicated by trading in bonds maturing at T_n, T_{n+1}, ..., T_{M+1}. Valuing a swaption entails determining the initial cost of this trading strategy or, more directly, computing the expected present value of the swaption payoff. This requires specifying a model of the dynamics of the term structure.

By market convention, swaption prices are quoted through a version of Black's formula. This formula equates the time-t price of the swaption to

$$\delta \sum_{i=n+1}^{M+1} B(t, T_i) \left(S_n(t) \Phi \left(\frac{\log(S_n(t)/K) + \sigma^2(T_n - t)/2}{\sigma \sqrt{T_n - t}} \right) \right.$$
$$\left. - K \Phi \left(\frac{\log(S_n(t)/K) - \sigma^2(T_n - t)/2}{\sigma \sqrt{T_n - t}} \right) \right).$$

This expression is what one would obtain for the expectation of

$$\delta \sum_{i=n+1}^{M+1} B(t, T_i)(S_n(T_n) - K)^+$$

if the forward swap rate satisfied

$$\frac{dS_n(t)}{S_n(t)} = \sigma \, dW(t).$$

As with the Black formula for caplets, this does not necessarily correspond to a price in the sense of the theory of derivatives valuation (but see Jamshidian [197] for a setting in which it does). In practice, the Black formula is used to extract an implied volatility for swap rates.

References

1. Abate, J., Choudhury, G., and Whitt, W. (1999) An introduction to numerical transform inversion and its application to probability models, pp.257–323 in *Computational Proability*, W. Grassman, ed., Kluwer Publishers, Boston.
2. Abken, P.A. (2000) An empirical evaluation of value at risk by scenario simulation, *Journal of Derivatives* 7(Summer):12–30.
3. Abramowitz, M., and Stegun, I.A. (1964) *Handbook of Mathematical Functions*, National Bureau of Standards, Washington D.C. Reprinted by Dover, New York.
4. Acworth, P., Broadie, M., and Glasserman, P. (1998) A comparison of some Monte Carlo and quasi Monte Carlo methods for option pricing, pp.1–18 in *Monte Carlo and Quasi-Monte Carlo Methods 1996*, P. Hellekalek, G. Larcher, H. Niederreiter, and P. Zinterhof, eds., Springer-Verlag, Berlin.
5. Adler, R.J. (1990) *An Introduction to Continuity, Extrema, and Related Topics for General Gaussian Processes*, Institute of Mathematical Statistics, Hayward, California.
6. Ahrens, J.H., and Dieter, U. (1974) Computer methods for sampling from the gamma, beta, Poisson, and binomial distributions, *Computing* 12:223–246.
7. Aït-Sahalia, Y. (2001) Closed-form likelihood expansions for multivariate diffusions, Working Paper 8956, National Bureau of Economic Research, Cambridge, Mass.
8. Åkesson, F., and Lehoczky, J. (1998) Discrete eigenfunction expansion of multidimensional Brownian motion and the Ornstein-Uhlenbeck process, working paper, Department of Statistics, Carnegie-Mellon University, Pittsburgh, PA.
9. Åkesson, F., and Lehoczky, J. (2000) Path generation for quasi-Monte Carlo simulation of mortgage-backed securities, *Management Science* 46:1171–1187.
10. Alexander, C. (1999) Volatility and correlation: measurement, models, and applications, pp.125–171 in *Risk Management and Analysis, Volume 1*, C. Alexander, ed., Wiley, Chichester, England.
11. Andersen, L. (1995) Simulation and calibration of the HJM model, working paper, General Re Financial Products, New York.
12. Andersen, L. (2000) A simple approach to the pricing of Bermudan swaptions in the multi-factor Libor Market Model, *Journal of Computational Finance* 3:5–32.

13. Andersen, L., and Andreasen, J. (2000) Volatility skews and extensions of the Libor market model, *Applied Mathematical Finance* 7:1–32.

14. Andersen, L., and Andreasen, J. (2001) Factor dependence of Bermudan swaption prices: fact or fiction?, *Journal of Financial Economics* 62:3–37.

15. Andersen, L., and Broadie, M. (2001) A primal-dual simulation algorithm for pricing multi-dimensional American options, working paper, Columbia Business School, New York.

16. Andersen, L., and Brotherton-Ratcliffe, R. (1996) Exact exotics, *Risk* 9(October):85–89.

17. Anderson, T.W. (1984) *An Introduction to Multivariate Statistical Analysis,* Second Edition, Wiley, New York.

18. Antanov, I.A., and Saleev, V.M. (1979) An economic method of computing LP_τ sequences, *USSR Journal of Computational Mathematics and Mathematical Physics* (English translation) 19:252–256.

19. Artzner, P., Delbaen, F., Eber, J.-M., and Heath, D. (1999) Coherent measures of risk, *Mathematical Finance* 9:203–228.

20. Asmussen, S. (1987) *Applied Probability and Queues,* Wiley, Chichester, England.

21. Asmussen, S. (1998) *Stochastic Simulation with a View Towards Stochastic Processes,* MaPhySto Lecture Notes No. 2, University of Aarhus, Aarhus, Denmark.

22. Asmussen, S., and Binswanger, K. (1997) Simulation of ruin probabilities for subexponential claims, *ASTIN Bulletin* 27:297–318.

23. Asmussen, S., Binswanger, K., and Højgaard, B. (2000) Rare events simulation for heavy tailed distributions, *Bernoulli* 6:303–322.

24. Asmussen, S., Glynn, P., and Pitman, J. (1995) Discretization error in simulation of one-dimensional reflecting Brownian motion, *Annals of Applied Probability* 5:875–896.

25. Asmussen, S., and Rosiński, J. (2001) Approximations of small jumps of Lévy processes with a view towards simulation, *Journal of Applied Probability* 38:482–493.

26. Avellaneda, M. (1998) Minimum-relative-entropy calibration of asset-pricing models, *International Journal of Theoretical and Applied Finance* 1:447–472.

27. Avellaneda, M., Buff, R., Friedman, C., Grandchamp, N., Kruk, L., and Newman, J. (2001) Weighted Monte Carlo: a new technique for calibrating asset-pricing models, *International Journal of Theoretical and Applied Finance* 4:1–29.

28. Avellandeda, M., and Gamba, R. (2000) Conquering the Greeks in Monte Carlo: efficient calculation of the market sensitivities and hedge-ratios of financial assets by direct numerical simulation, pp.336–356 in *Quantitative Analysis in Financial Markets, Vol. III*, M. Avellaneda, ed., World Scientific, Singapore.

29. Avramidis, A.N., and Matzinger, H. (2002) Convergence of the stochastic mesh estimator for pricing American options, pp.1560–1567 in *Proceedings of the Winter Simulation Conference*, IEEE Press, New York.

30. Avramidis, A.N., and Wilson, J.R. (1998) Correlation-induction techniques for estimating quantiles in simulation experiments, *Operations Research* 46:574–591.

31. Baldi, P. (1995) Exact asymptotics for the probability of exit from a domain and applications to simulation, *Annals of Probability* 23:1644–1670.

32. Baldi P., Caramellino L., and Iovino M.G. (1999) Pricing single and double barrier options via sharp large deviations techniques, *Mathematical Finance* 9:293–321.

33. Bally, V., and Pagès, G. (2000) A quantization algorithm for solving multidimensional optimal stopping problems, Preprint 628, Laboratoire de Probabilités et Modèles Aléatoires, Université de Paris VI.

34. Bally, V., and Talay, D. (1995) The law of the Euler scheme for stochastic differential equations (I): convergence rate of the distribution function, *Probability Theory and Related Fields* 102:43–60.

35. Barndorff-Nielsen, O.E. (1978) *Information and Exponential Families in Statistical Theory*, Wiley, New York.

36. Barndorff-Nielsen, O.E. (1998) Processes of normal inverse Gaussian type, *Finance and Stochastics* 2:41–68.

37. Barraquand, J. (1995) Numerical valuation of high dimensional multivariate European securities, *Management Science* 41:1882–1891.

38. Barraquand, J., and Martineau, D. (1995) Numerical valuation of high dimensional multivariate American securities, *Journal of Financial and Quantitative Analysis* 30:383–405.

39. Bassi, F., Embrechts, P., and Kafetzaki, M. (1998) Risk management and quantile estimation, pp.111-130 in *A Practical Guide to Heavy Tails*, R.J. Adler, R. Friedman, and M. Taqqu, eds., Birkhäuser, Boston.

40. Bauer, K.W., Venkatraman, S., and Wilson, J.R. (1987) Estimation procedures based on control variates with known covariance matrix, pp.334–341 in *Proceedings of the Winter Simulation Conference*, IEEE Press, New York.

41. Beaglehole, D.R., Dybvig, P.H., and Zhou, G. (1997) Going to extremes: correcting simulation bias in exotic option valuation, *Financial Analysts Journal* 53(1):62–68.

42. Beaglehole, D.R. and Tenney, M.S. (1991) General solutions of some interest-rate contingent claims pricing equations, *Journal of Fixed Income* 1(September):69-83.

43. Beasley, J.D., and Springer, S.G. (1977) The percentage points of the normal distribution, *Applied Statistics* 26:118–121.

44. Benhamou, E. (2000) An application of Malliavin calculus to continuous time Asian options Greeks, working paper, London School of Economics, London.

45. Berman, L. (1997) Accelerating Monte Carlo: quasirandom sequences and variance reduction, *Journal of Computational Finance* 1(Winter):79–95.

46. Bielecki, T.R., and Rutkowski, M. (2002) *Credit Risk: Modeling, Valuation and Hedging*, Springer-Verlag, Berlin.

47. Birge, J.R. (1994) Quasi-Monte Carlo approaches to option pricing, Technical Report 94-19, Department of Industrial and Operations Engineering, University of Michigan, Ann Arbor, MI.

48. Björk, T. (1998) *Arbitrage Pricing in Continuous Time*, Oxford University Press, Oxford.

49. Black, F. (1976) The pricing of commodity contracts, *Journal of Financial Economics*, 3:167-179.

50. Black, F., and Scholes, M. (1973) The pricing of options and corporate liabilities, *Journal of Political Economy* 81:637–654.

51. Box, G.E.P, and Muller, M.E. (1958) A note on the generation of random normal deviates, *Annals of Mathematical Statistics* 29:610–611.

52. Boyle, P.P. (1977) Options: a Monte Carlo approach, *Journal of Financial Economics* 4:323–338.

53. Boyle, P., Broadie, M., and Glasserman, P. (1997) Monte Carlo methods for security pricing, *Journal of Economic Dynamics and Control* 21:1267–1321.

54. Boyle, P., Evnine, J., and Gibbs, S. (1989) Numerical evaluation of multivariate contingent claims, *Review of Financial Studies* 2:241–250.

55. Boyle, P., Kolkiewicz, A., and Tan, K.-S. (2002) Pricing American derivatives using simulation: a biased low approach, pp.181–200 in *Monte Carlo and Quasi-Monte Carlo Methods 2000*, K.-T. Fang, F.J. Hickernell, and H. Niederreiter, eds., Springer-Verlag, Berlin.

56. Brace, A., Gatarek, D., and Musiela, M. (1997) The market model of interest rate dynamics, *Mathematical Finance* 7:127–155.

57. Bratley, P., and Fox, B.L. (1988) Algorithm 659: Implementing Sobol's quasirandom sequence generator, *ACM Transactions on Mathematical Software* 14:88–100.

58. Bratley, P., Fox, B.L., and Niederreiter, H. (1992) Implementation and tests of low-discrepancy sequences, *ACM Transactions on Modeling and Computer Simulation* 2:195–213.

59. Bratley, P., Fox, B.L., and Schrage, L. (1987) *A Guide to Simulation*, Second Edition, Springer-Verlag, New York.

60. Brémaud, P. (1981) *Point Processes and Queues*, Springer-Verlag, New York.

61. Britten-Jones, M., and Schaefer, S.M. (1999) Non-linear value-at-risk, *European Finance Review* 2:161–187.

62. Briys, E., Bellalah, M., Mai, H.M., and De Varenne, F. (1998) *Options, Futures, and Exotic Derivatives*, Wiley, Chichester, England.

63. Broadie, M., and Detemple, J. (1997) Valuation of American options on multiple assets, *Review of Financial Studies* 7:241–286.

64. Broadie, M., and Glasserman, P. (1996) Estimating security price derivatives using simulation, *Management Science* 42:269–285.

65. Broadie, M., and Glasserman, P. (1997) Pricing American-style securities by simulation, *Journal of Economic Dynamics and Control* 21:1323–1352.

66. Broadie, M., and Glasserman, P. (1997) A stochastic mesh method for pricing high-dimensional American options PaineWebber Working Papers in Money, Economics and Finance #PW9804, Columbia Business School, New York.

67. Broadie, M., and Glasserman, P. (1999) Simulation in option pricing and risk management, pp.173–207 in *Handbook of Risk Management and Analysis*, C. Alexander, ed., Wiley, Chichester, England.

68. Broadie, M., Glasserman, P., and Ha, Z. (2000) Pricing American options by simulation using a stochastic mesh with optimized weights, pp.32-50, in *Probabilistic Constrained Optimization: Methodology and Applications*, S. Uryasev, ed., Kluwer Publishers, Norwell, Mass.

69. Broadie, M., Glasserman, P., and Jain, G. (1997) Enhanced Monte Carlo estimates of American option prices, *Journal of Derivatives* 4(Fall):25–44.

70. Broadie, M., and Yamamoto, Y. (2002) Application of the fast Gauss transform to option pricing, working paper, Columbia Business School, New York.

71. Brown, R.H., and Schaefer, S.M. (1994) Interest rate volatility and the shape of the term structure, *Philosophical Transactions of the Royal Society of London, Series A* 347:563–576.

72. Bucklew, J.A., Ney, P., and Sadowsky, J.S. (1990) Monte Carlo simulation and large deviations theory for uniformly recurrent Markov chains, *Journal of Applied Probability* 27:44–59.

73. Caflisch, R.E., Morokoff, W., and Owen, A. (1997) Valuation of mortgage-backed securities using Brownian bridges to reduce effective dimension, *Journal of Computational Finance* 1:27–46.

74. Campbell, J.Y., Lo, A.W., and MacKinlay, A.C. (1997) *The Econometrics of Financial Assets,* Princeton University Press, Princeton, New Jersey.

75. Caramellino, L., and Iovino, M.G. (2002) An exit-probability-based approach for the valuation of defaultable securities, *Journal of Computational Finance* 6:1–25.

76. Cardenas, J., Fruchard, E., Picron, J.-F., Reyes, C., Walters, K., and Yang, W. (1999) Monte Carlo within a day, *Risk* 12(February):55–59.

77. Cario, M.C., and Nelson, B.L. (1997) Modeling and generating random vectors with arbitrary marginal distributions and correlation matrix, working paper, IEMS Department, Northwestern University, Evanston, IL.

78. Carriére, J. (1996) Valuation of early-exercise price of options using simulations and nonparametric regression, *Insurance: Mathematics and Economics* 19:19–30.

79. Chambers, J.M., Mallows, C.L., and Stuck, B.W. (1976) A method for simulating stable random variables, *Journal of the American Statistical Association* 71:340–344.

80. Chen, J., and Fu, M.C. (2002) Efficient sensitivity analysis of mortgage backed securities, *12th Annual Derivative Securities Conference*, New York.

81. Cheng, R.C.H. (1985) Generation of multivariate normal samples with given sample mean and covariance matrix, *Journal of Statistical Computation and Simulation* 21:39–49.

82. Cheng, R.C.H. (1985) Generation of inverse Gaussian variates with given sample mean and dispersion, *Applied Statistics — Journal of the Royal Statistical Society Series C* 33:309–316.

83. Cheng, R.C.H., and Feast, G.M. (1980) Gamma variate generators with increased shape parameter range, *Communications of the ACM* 23:389–394.

84. Cheng, R.C.H., and Feast, G.M. (1980) Control variables with known mean and variance *Journal of the Operational Research Society* 31:51–56.

85. Chung, K.L. (1974) *A Course in Probability Theory,* Second Edition, Academic Press, New York.

86. Clément, E., Lamberton, D., and Protter, P. (2002) An analysis of a least squares regression algorithm for American option pricing, *Finance and Stochastics* 6:449–471.

87. Clewlow, L., and Carverhill, A. (1994) On the simulation of contingent claims, *Journal of Derivatives* 2:66–74.

88. Coveyou, R.R., and MacPherson, R.D. (1967) Fourier analysis of uniform random number generators, *Journal of the ACM* 14:100-119.

89. Cox, J.C., and Ross, S.A. (1976) The valuation of options for alternative stochastic processes, *Journal of Financial Economics* 3:145–166.

90. Cox, J.C., Ingersoll, J.E., and Ross, S.A. (1981) The relation between forward prices and futures prices, *Journal of Financial Economics* 9:321–346.

91. Cox, J.C., Ingersoll, J.E., and Ross, S.A. (1985) A theory of the term structure of interest rates, *Econometrica* 53:129–151.

92. Cranley, R., and Patterson, T.N.L. (1976) Randomization of number theoretic methods for numerical integration, *SIAM Journal on Numerical Analysis* 13:904–914.

93. Crouhy, M., Galai, D., and Mark, R. (2001) *Risk Management*, McGraw-Hill, New York.

94. Cvitanić, J., Ma, J., and Zheng, J. (2003) Efficient computation of hedging portfolios for options with discontinuous payoffs, *Mathematical Finance* 13:135–151.

95. Devroye, L. (1986) *Non-Uniform Random Variate Generation*, Springer-Verlag, New York.

96. Duan, J.C., and Simonato, J.G. (1998) Empirical martingale simulation for asset prices, *Management Science* 44:1218–1233.

97. Duan, J.C., Gauthier, G., and Simonato, J.G. (2001) Asymptotic distribution of the EMS option price estimator, *Management Science* 47:1122–1132.

98. Duffie, D. (2001) *Dynamic Asset Pricing Theory*, Third Edition, Princeton University Press, Princeton, New Jersey.

99. Duffie, D., and Gârleanu, N. (2001) Risk and valuation of collateralized debt obligations, *Financial Analysts Journal* 57(Jan/Feb):41–59.

100. Duffie, D., and Glynn, P. (1995) Efficient Monte Carlo simulation of security prices, *Annals of Applied Probability* 5:897–905.

101. Duffie, D., and Kan, R. (1996) A yield-factor model of interest rates, *Mathematical Finance* 6:379–406.

102. Duffie, D., and Lando, D. (2001) Term structures of credit spreads with incomplete accounting information, *Econometrica* 69:633–664.

103. Duffie, D., and Pan, J. (1997) An overview of value at risk, *Journal of Derivatives* 4(Spring):7–49.

104. Duffie, D., and Pan, J. (2001) Analytical value-at-risk with jumps and credit risk, *Finance and Stochastics* 2:155–180.

105. Duffie, D., Pan, J., and Singleton, K. (2000) Transform analysis and option pricing for affine jump-diffusions, *Econometrica* 68:1343–1376.

106. Duffie, D., and Singleton, K. (1999) Modeling term structures of defaultable bonds, *Review of Financial Studies* 12:687–720.

107. Dupire, B. (1994) Pricing with a smile, *Risk* 7(January):18–20.

108. Dupuis, P., and Wang, H. (2002) Importance sampling, large deviations, and differential games, Technical Report LCDS 02-16, Division of Applied Mathematics, Brown University, Providence, Rhode Island.

109. Eberlein, E. (2001) Application of generalized hyperbolic Lévy motion to finance, pp.319–337 in *Lévy Processes: Theory and Applications*, O.E. Barndorff-Nielsen, T. Mikosch, and S. Resnick, eds., Birkhaüser, Boston.

110. Eichenauer-Herrmann, J., Herrmann, E., and Wegenkittl, S. (1998) A survey of quadratic and inversive congruential pseudorandom numbers, pp.66–97 in *Monte Carlo and Quasi-Monte Carlo Methods in Scientific Computing 1996*, P. Hellekalek, G. Larcher, H. Niederreiter, and P. Zinterhof, eds., Springer-Verlag, Berlin.

111. Embrechts, P., Klüppelberg, C., and Mikosch, T. (1997) *Modelling Extremal Events for Insurance and Finance*, Springer-Verlag, Berlin.

112. Embrechts, P., McNeil, A., and Straumann, D. (2000) Correlation and dependence properties in risk management: properties and pitfalls, pp.71-76 in *Extremes and Integrated Risk Management*, P. Embrechts, ed., Risk Books, London.

113. Esary, J.D., Proschan, F., and Walkup, D.W. (1967) Association of random variables, with applications, *Annals of Mathematical Statistics* 38:1466-1474.

114. Fang, K.-T., Kotz, S., and Ng, K.W. (1987) *Symmetric Multivariate and Related Distributions*, Chapman & Hall, London.

115. Fang, K.-T., and Wang, Y. (1994) *Number-Theoretic Methods in Statistics*, Chapman & Hall, London.

116. Faure, H. (1982) Discrépence de suites associées à un système de numération (en dimension s), *Acta Arithmetica* 41:337–351.

117. Faure, H. (2001) Variations on $(0, s)$-sequences, *Journal of Complexity* 17:1–13.

118. Feller, W. (1951) Two singular diffusion problems, *Annals of Mathematics* 54:173–182.

119. Feller, W. (1971) *An Introduction to Probability Theory and Its Applications, Volume II*, Second Edition, Wiley, New York.

120. Finger, C.C. (2000) A comparison of stochastic default rate models, Working Paper Number 00-02, The RiskMetrics Group, New York.

121. Fishman, G.S. (1996) *Monte Carlo: Concepts, Algorithms, and Applications*, Springer-Verlag, New York.

122. Fishman, G.S., and Huang, B.D. (1983) Antithetic variates revisited, *Communications of the ACM* 26:964–971.

123. Fishman, G.S., and Moore, L.R. (1986) An exhaustive analysis of multiplicative congruential random number generators with modulus $2^{31} - 1$, *SIAM Journal on Scientific and Statistical Computing* 7:24–45.

124. Fournié, E., Lasry, J.-M., Lebuchoux, J., Lions, P.-L., and Touzi, N. (1999) Applications of Malliavin calculus to Monte Carlo methods in finance, *Finance and Stochastics* 3:391–412.

125. Fournié, E., Lasry, J.M., and Touzi, N. (1997) Monte Carlo methods for stochastic volatility models, pp.146–164 in *Numerical Methods in Finance*, L.C.G. Rogers and D. Talay, eds., Cambridge University Press, Cambridge, UK.

126. Fox, B.L. (1986) Algorithm 647: Implementation and relative efficiency of quasirandom sequence generators, *ACM Transactions on Mathematical Software* 12:362–376.

127. Fox, B.L. (1999) *Strategies for Quasi-Monte Carlo*, Kluwer Academic Publishers, Boston, Mass.

128. Fox, B.L., and Glynn, P.W. (1989) Replication schemes for limiting expectations, *Probability in the Engineering and Information Sciences* 3:299–318.

129. Fox, B.L., and Glynn, P.W. (1989) Simulating discounted costs, *Management Science* 35:1297–1315.

130. Frolov, A.S., and Chentsov, N.N. (1963) On the calculation of definite integrals dependent on a parameter by the Monte Carlo method, *USSR Journal of Computational Mathematics and Mathematical Physics* (English translation) 4:802–808.

131. Fu, M.C., and Hu, J.-Q. (1995) Sensitivity analysis for Monte Carlo simulation of option pricing, *Probability in the Engineering and Information Sciences* 9:417–446.

132. Gaines, J.G., and Lyons, T.J. (1994) Random generation of stochastic area integrals, *SIAM Journal on Applied Mathematics* 54:1132–1146.

133. Gaines, J.G., and Lyons, T.J. (1997) Variable step size control in the numerical solution of stochastic differential equations, *SIAM Journal on Applied Mathematics* 57:1455–1484.

134. Garcia, D. (2003) Convergence and biases of Monte Carlo estimates of American option prices using a parametric exercise rule, *Journal of Economic Dynamics and Control* 27:1855–1879.

135. Geman, H., and Yor, M. (1993) Bessel processes, Asian options and perpetuities, *Mathematical Finance* 3:349–375.

136. Gentle, J.E. (1998) *Random Number Generation and Monte Carlo Methods*, Springer-Verlag, New York.

137. Ghosh, S., and Henderson, S.G. (2002) Properties of the NORTA method in higher dimensions, pp.263–269 in *Proceedings of the Winter Simulation Conference*, IEEE Press, New York.

138. Glasserman, P. (1991) *Gradient Estimation via Perturbation Analysis*, Kluwer Academic Publishers, Norwell, Mass.

139. Glasserman, P., Heidelberger, P., and Shahabuddin, P. (1999) Asymptotically optimal importance sampling and stratification for path-dependent options, *Mathematical Finance* 9:117–152.

140. Glasserman, P., Heidelberger, P., and Shahabuddin, P. (1999) Importance sampling in the Heath-Jarrow-Morton framework, *Journal of Derivatives* 6:(Fall)32–50.

141. Glasserman, P., Heidelberger, P., and Shahabuddin, P. (1999) Stratification issues in estimating value-at-risk, pp.351–358 in *Proceedings of the Winter Simulation Conference*, IEEE Press, New York.

142. Glasserman, P., Heidelberger, P., and Shahabuddin, P. (2000) Importance sampling and stratification for value-at-risk, pp.7–24 in *Computational Finance 1999 (Proceedings of the Sixth International Conference on Computational Finance)*, Y.S. Abu-Mostafa, B. LeBaron, A.W. Lo, and A.S. Weigend, eds., MIT Press, Cambridge, Mass.

143. Glasserman, P., Heidelberger, P., and Shahabuddin, P. (2000) Variance reduction techniques for estimating value-at-risk, *Management Science* 46:1349–1364.

144. Glasserman, P., Heidelberger, P., and Shahabuddin, P. (2002) Portfolio value-at-risk with heavy-tailed risk factors, *Mathematical Finance* 12:239–269.

145. Glasserman, P., and Merener, N. (2003) Numerical solution of jump-diffusion LIBOR market models, *Finance and Stochastics* 7:1–27.

146. Glasserman, P., and Staum, J. (2001) Conditioning on one-step survival in barrier option simulations, *Operations Research* 49:923–937.

147. Glasserman, P., and Yu, B. (2002) Large sample properties of weighted Monte Carlo, Working Paper DRO-2002-07, Columbia Business School, New York.

148. Glasserman, P., and Yu, B. (2003) Pricing American options by simulation: regression now or regression later?, to appear in *Monte Carlo and Quasi-Monte Carlo Methods 2002*, H. Niederreiter, ed., Springer-Verlag, Berlin.

149. Glasserman, P., and Wang, H. (2000) Discretization of deflated bond prices, *Advances in Applied Probability* 32:540–563.

150. Glasserman, P., and Zhao, X. (1999) Fast Greeks by simulation in forward LIBOR models, *Journal of Computational Finance* 3:5–39.

151. Glasserman, P., and Zhao, X. (2000) Arbitrage-free discretization of lognormal forward LIBOR and swap rate models, *Finance and Stochastics* 4:35–68.

152. Glynn, P.W. (1987) Likelihood ratio gradient estimation: an overview, pp.366–374 in *Proceedings of the Winter Simulation Conference*, IEEE Press, New York.

153. Glynn, P.W. (1989) Optimization of stochastic systems via simulation, pp.90–105 in *Proceedings of the Winter Simulation Conference,* IEEE Press, New York.

154. Glynn, P.W. (1996) Importance sampling for Monte Carlo estimation of quantiles, pp.180–185 in *Mathematical Methods in Stochastic Simulation and Experimental Design: Proceedings of the Second St. Petersburg Workshop on Simulation,* St. Petersburg University Press, St. Petersburg, Russia.

155. Glynn, P.W., and Heidelberger, P. (1990) Bias properties of budget constrained simulations, *Operations Research* 38:801–814.

156. Glynn, P.W., and Iglehart, D.L. (1988) Simulation methods for queues: an overview, *Queueing Systems: Theory and Applications* 3:221–256.

157. Glynn, P.W., and Iglehart, D.L. (1989) Importance sampling for stochastic simulations, *Management Science* 35:1367–1392.

158. Glynn, P.W., and L'Ecuyer, P. (1995) Likelihood ratio gradient estimation for regenerative stochastic recursions, *Advances in Applied Probability* 27:1019–1053.

159. Glynn, P.W., and Whitt, W. (1989) Indirect estimation via $L = \lambda W$, *Operations Research* 37:82–103.

160. Glynn, P.W., and Whitt, W. (1992) The efficiency of simulation estimators, *Operations Research* 40:505–520.

161. Gobet, E., and Munos, R. (2002) Sensitivity analysis using Itô-Malliavin calculus and martingales. Application to stochastic optimal control, Report 498, Centre de Mathématique Appliquées, Ecole Polytechnique, Palaiseau, France.

162. Golub, G., and Van Loan, C.F. (1996) *Matrix Computations,* Third Edition, Johns Hopkins University Press, Baltimore.

163. Gupton, G.M., Finger, C.C., and Bhatia, M. (1997) *CreditMetrics Technical Document,* The RiskMetrics Group, New York.

164. Hall, P. (1992) *The Bootstrap and Edgeworth Expansion,* Springer-Verlag, New York.

165. Halton, J.H. (1960) On the efficiency of certain quasi-random sequences of points in evaluating multi-dimensional integrals, *Numerische Mathematik* 2:84–90.

166. Halton, J.H., and Smith, G.B. (1964) Algorithm 247: Radical-inverse quasi-random point sequence, *Communications of the ACM* 7:701–702.

167. Hamilton, D., James, J., and Webber, N. (2001) Copula methods and the analysis of credit risk, working paper, Warwick Business School, Coventry, England.

168. Hammersley, J.M. (1960) Monte Carlo methods for solving multivariable problems, *Annals of the New York Academy of Sciences* 86:844–874.

169. Hammersley, J.M., and Handscomb, D.C. (1964) *Monte Carlo Methods,* Methuen, London.

170. Harrison, J.M., and Kreps, D. (1979) Martingales and arbitrage in multiperiod securities markets, *Journal of Economic Theory* 20:381–408.

171. Hastings, C., Jr. (1955) *Approximations for Digital Computers,* Princeton University Press, Princeton, New Jersey.

172. Haugh, M., and Kogan, L. (2001) Pricing American options: a duality approach, *Operations Research,* to appear.

173. Heath, D., Jarrow, R., and Morton, A. (1990) Bond pricing and the term structure of interest rates: a discrete time approximation, *Journal of Financial and Quantitative Analysis* 25:419–440.

174. Heath, D., Jarrow, R., and Morton, A. (1992) Bond pricing and the term structure of interest rates: a new methodology for contingent claims valuation, *Econometrica* 60:77–105.
175. Heidelberger, P. (1995) Fast simulation of rare events in queueing and reliability models, *ACM Transactions on Modeling and Computer Simulation* 5:43–85.
176. Hellekalek, P. (1998) On the assessment of random and quasi-random point sets, pp.49–108 in *Random and Quasi-Random Point Sets*, P. Hellekalek and G. Larcher, eds., Springer-Verlag, Berlin.
177. Hesterberg, T.C. (1995) Weighted average importance sampling and defensive mixture distributions, *Technometrics* 37:185–194.
178. Hesterberg, T.C., and Nelson, B.L. (1998) Control variates for probability and quantile estimation, *Management Science* 44:1295–1312.
179. Heston, S.I. (1993) A closed-form solution for options with stochastic volatility with applications to bond and currency options, *Review of Financial Studies* 6:327–343.
180. Hickernell, F.J. (1996) The mean square discrepancy of randomized nets, *ACM Transactions on Modeling and Computer Simulation* 6:274–296.
181. Hickernell, F.J. (1998) A generalized discrepancy and quadrature error bound, *Mathematics of Computation* 67:299-322.
182. Hickernell, F.J. (1998) Lattice rules: how well do they measure up? pp.109–166 in *Random and Quasi-Random Point Sets*, P. Hellekalek and G. Larcher, eds., Springer-Verlag, Berlin.
183. Hickernell, F.J., and Hong, H.S. (1999) The asymptotic efficiency of randomized nets for quadrature, *Mathematics of Computation* 68:767–791.
184. Hickernell, F.J., Hong, H.S., L'Ecuyer, P., and Lemieux, C. (2000) Extensible lattice sequences for quasi-Monte Carlo quadrature, *SIAM Journal on Scientific Computing* 22:1117–1138.
185. Ho, T.S.Y., and Lee, S.-B. (1986) Term structure movements and pricing interest rate contingent claims, *Journal of Finance* 41:1011–1029.
186. Ho, Y.C., and Cao, X.-R. (1983) Optimization and perturbation analysis of queueing networks, *Journal of Optimization Theory and Applications* 40:559–582.
187. Hong, H.S., and Hickernell, F.J. (2003) Implementing scrambled digital sequences, *ACM Transactions on Mathematical Software*, to appear.
188. Hosking, J.R.M., Bonti, G., and Siegel, D. (2000) Beyond the lognormal, *Risk* 13(May):59–62.
189. Hull, J. (2000) *Options, Futures, and Other Derivative Securities*, Fourth Edition, Prentice-Hall, Upper Saddle River, New Jersey.
190. Hull, J., and White, A. (1990) Pricing interest-rate-derivative securities, *Review of Financial Studies* 3:573–592.
191. Hunt, P.J., and Kennedy, J.E. (2000) *Financial Derivatives in Theory and Practice*, Wiley, Chichester, England.
192. Hunter, C.J., Jackel, P., and Joshi, M. (2001) Getting the drift, *Risk* 14(July):81–84.
193. Jacod, J., and Protter, P. (1998) Asymptotic error distributions for the Euler method for stochastic differential equations, *Annals of Probability* 26:267–307.
194. James, J., and Webber, N. (2000) *Interest Rate Modelling*, Wiley, Chichester, England.
195. Jamshidian, F. (1995) A simple class of square-root interest rate models, *Applied Mathematical Finance* 2:61–72.

196. Jamshidian, F. (1996) Bond, futures and options evaluation in the quadratic interest rate model, *Applied Mathematical Finance* 3:93–115.

197. Jamshidian, F. (1997) LIBOR and swap market models and measures, *Finance and Stochastics* 1:293–330.

198. Jamshidian, F., and Zhu, Y. (1997) Scenario simulation: theory and methodology, *Finance and Stochastics* 1:43–67.

199. Jarrow, R.A., Lando, D., and Turnbull, S.M. (1997) A Markov model for the term structure of credit risk spreads, *Review of Financial Studies* 10:481–523.

200. Jarrow, R.A., and Turnbull, S.M. (1995) Pricing derivatives on financial securities subject to credit risk, *Journal of Finance* 50:53–85.

201. Johnson, N.L., Kotz, S., and Balakrishnan, N. (1994) *Continuous Univariate Distributions, Volume 1*, Second Edition, Wiley, New York.

202. Johnson, N.L., Kotz, S., and Balakrishnan, N. (1995) *Continuous Univariate Distributions, Volume 2*, Second Edition, Wiley, New York.

203. Jorion, P. (2001) *Value at Risk*, Second Edition, McGraw-Hill, New York.

204. Joy, C., Boyle, P.P., and Tan, K.S. (1996) Quasi-Monte Carlo methods in numerical finance, *Management Science* 42:926–938.

205. Ju, N. (1998) Pricing an American option by approximating its early exercise boundary as a multipiece exponential function, *Review of Financial Studies* 11:627–646.

206. Juneja, S., and Shahabuddin, P. (2002): Simulating heavy tailed processes using delayed hazard rate twisting, *ACM Transactions on Modeling and Computer Simulation* 12:94–118.

207. Karatzas, I., and Shreve, S. (1991) *Brownian Motion and Stochastic Calculus*, Springer-Verlag, New York.

208. Kaufmann R., Gadmer, A., and Klett, R. (2001) Introduction to dynamic financial analysis, *ASTIN Bulletin* 31:213–249.

209. Kemna, A.G.Z., and Vorst, A.C.F. (1990) A pricing method for options based on average asset values, *Journal of Banking and Finance*, 14:113–129.

210. Khuri, A.I., and Cornell, J.A. (1996) *Response Surfaces: Designs and Analyses*, Second Edition, Marcel Dekker, New York.

211. Kloeden, P.E., and Platen, E. (1992) *Numerical Solution of Stochastic Differential Equations*, Springer-Verlag, Berlin.

212. Knuth, D.E. (1998) *The Art of Computer Programming, Volume II: Seminumerical Algorithms*, Third Edition, Addison Wesley Longman, Reading, Mass.

213. Kocis, L., and Whiten, W.J. (1997) Computational investigations of low-discrepancy sequences, *ACM Transactions on Mathematical Software* 23:266–294.

214. Kollman, C., Baggerly, K.A., Cox, D.D., and Picard, R.R. (1999) Adaptive importance sampling on discrete Markov chains, *Annals of Applied Probability* 9:391–412.

215. Kou, S.G. (2002) A jump diffusion model for option pricing, *Management Science* 48:1086–1101.

216. Krylov, N. V. (1995) *Introduction to the Theory of Diffusion Processes*, Translations of Mathematical Monographs 142. American Mathematical Society, Providence, Rhode Island.

217. Kunita, H. (1990) *Stochastic Flows and Stochastic Differential Equations*, Cambridge University Press, Cambridge, UK.

218. Lamberton, D., and Lapeyre, B. (1996) *Introduction to Stochastic Calculus Applied to Finance*, Chapman & Hall, London.

219. Larcher, G. (1998) Digital point sets: analysis and application, pp.167–222 in *Random and Quasi-Random Point Sets*, P. Hellekalek and G. Larcher, eds., Springer-Verlag, Berlin.

220. Larcher, G., Leobacher, G., and Scheicher, K. (2002) On the tractability of the Brownian bridge algorithm, working paper, Department of Financial Mathematics, Universtiy of Linz, Linz, Austria.

221. Lavenberg, S.S., Moeller, T.L., and Welch, P.D. (1982) Statistical results on control variables with application to queueing network simulation, *Operations Research* 30:182–202.

222. L'Ecuyer, P. (1988) Efficient and portable combined random number generators, *Communications of the ACM* 31:742–749, 774. *Correspondence* 32:1019–1024.

223. L'Ecuyer, P. (1994) Uniform random number generation, *Annals of Operations Research* 53:77–120.

224. L'Ecuyer, P. (1996) Combined multiple recursive random number generators, *Operations Research* 44:816–822.

225. L'Ecuyer, P. (1999) Good parameters and implementations for combined multiple recursive random number generators, *Operations Research* 47:159–164.

226. L'Ecuyer, P., and Lemieux, C. (2000) Variance reduction via lattice rules, *Management Science* 46:1214–1235.

227. L'Ecuyer, P., Simard, R., Chen, E.J., and Kelton, W.D. (2002) An object-oriented random-number package with many long streams and substreams, *Operations Research* 50:1073–1075.

228. L'Ecuyer, P., Simard, R., and Wegenkittl, S. (2002) Sparse serial tests of uniformity for random number generators, *SIAM Journal on Scientific Computing* 24:652–668.

229. Lehmer, D.H. (1951) Mathematical methods in large-scale computing units, pp.141–146 in *Proceedings of the Second Symposium on Large Scale Digital Computing Machinery*, Harvard University Press, Cambridge, Mass.

230. Leland, H.E. (1994) Corporate debt value, bond covenants, and optimal capital structure, *Journal of Finance* 49:1213–1252.

231. Lemieux, C., Cieslak, M., and Luttmer, K. (2002) *RandQMC user's guide: a package for randomized quasi-Monte Carlo methods in C*, Technical Report 2002-712-15, Department of Computer Science, University of Calgary, Calgary, Canada.

232. Lemieux, C., and L'Ecuyer, P. (2001) On selection criteria for lattice rules and other quasi-Monte Carlo point sets, *Mathematics and Computers in Simulation* 55:139–148.

233. Lévy, P. (1948) *Processus stochastique et mouvement brownien*, Gauthier-Villars, Paris.

234. Lewis, P.A.W., Goodman, A.S., and Miller, J.M. (1969) A pseudo-random number generator for the System/360, *IBM Systems Journal* 8:136–146.

235. Lewis, P.A.W., and Shedler, G.S. (1979) Simulation of nonhomogeneous Poisson processes by thinning, *Naval Logistics Quarterly* 26:403–413.

236. Li, D. (2000) On default correlation: a copula function approach, *Journal of Fixed Income* 9:43–54.

237. Linetsky, V. (2001) Exact Pricing of Asian Options: An Application of Spectral Theory, working paper, IEMS Department, Northwestern University, Evanston, Illinois.

238. Loh, W.-L. (1996) On Latin hypercube sampling, *The Annals of Statistics* 24:2058–2080.

239. Loh, W.W. (1997) On the method of control variates, Doctoral Dissertation, Department of Operations Research, Stanford University, Stanford, California.

240. Longin, F., and Solnik, B. (2001) Correlation structure of international equity markets during extremely volatile periods, *Journal of Finance* 46:649–676.

241. Longstaff, F.A., and Schwartz, E.S. (2001) Valuing American options by simulation: a simple least-squares approach, *Review of Financial Studies* 14:113–147.

242. Madan, D.B., Carr, P., and Chang, E.C. (1998) The variance gamma process and option pricing, *European Finance Review* 2:79–105.

243. Madan, D.B., and Seneta, E. (1990) The variance gamma (v.g.) model for share market returns, *Journal of Business* 63:511–524.

244. Madan, D.B., and Unal, H. (1998) Pricing the risk of default, *Review of Derivatives Research* 2:121–160.

245. Maghsoodi, Y. (1998) Exact solution and doubly efficient approximations of jump-diffusion Ito equations, *Stochastic Analysis and Applications* 16:1049–1072.

246. Mandelbrot, B. (1963) The variation of certain speculative prices, *Journal of Business* 36:394–419.

247. Margrabe, W. (1978) The value of an option to exchange one asset for another, *Journal of Finance* 33:177–186.

248. Marsaglia, G. (1968) Random numbers fall mainly in the planes, *Proceedings of the National Academy of Sciences* 61:25–28.

249. Marsaglia, G. (1972) The structure of linear congruential generators, in *Applications of Number Theory to Numerical Analysis*, S.K. Zaremba, ed., 249–286, Academic Press, New York.

250. Marsaglia, G., and Bray, T.A. (1964) A convenient method for generating normal variables, *SIAM Review* 6:260–264.

251. Marsaglia, G., Zaman, A., and Marsaglia, J.C.W. (1994) Rapid evaluation of the inverse of the normal distribution function, *Statistics and Probability Letters* 19:259–266.

252. Marshall, A.W., and Olkin, I. (1988) Families of multivariate distributions, *Journal of the American Statistical Association* 83:834–841.

253. Martin, R., Thompson, K., and Browne, C. (2001) Taking to the saddle, *Risk* 14(June):91–94.

254. Maruyama, G. (1955) Continuous Markov processes and stochastic equations, *Rendiconti del Circolo Matematico di Palermo, Serie II* 4:48–90.

255. Mashal, R., and Zeevi, A. (2002) Beyond correlation: extreme co-movements between financial assests, working paper, Columbia Business School, New York.

256. Matoušek, J. (1999) *Geometric Discrepancy: An Illustrated Guide*, Springer-Verlag, Berlin.

257. Matoušek, J. (1998) On the L_2-discrepancy for anchored boxes, *Journal of Complexity* 14:527–556.

258. Matsumoto, M., and Nishimura, T. (1998) Mersenne twister: a 623-dimensionally equidistributed uniform pseudo-random number generator, *ACM Transactions on Modeling and Computer Simulation* 8:3–30.

259. McKay, M.D., Conover, W.J., and Beckman, R.J. (1979) A comparison of three methods for selecting input variables in the analysis of output from a computer code, *Technometrics* 21:239–245.

260. McKean, H.P. (1969) *Stochastic Integrals*, Academic Press, New York.
261. Merton, R.C. (1973) Theory of rational option pricing, *Bell Journal of Economics and Management Science* 4:141-183.
262. Merton, R.C. (1974) On the pricing of corporate debt: the risk structure of interest rates, *Journal of Finance* 29:449–470.
263. Merton, R.C. (1976) Option pricing when underlying stock returns are discontinuous, *Journal of Financial Economics* 3:125-144.
264. Michael, J.R., Schucany, W.R., and Haas, R.W. (1976) Generating random variates using transformations with multiple roots, *American Statistician* 30:88–90.
265. Mikulevicius, R., and Platen, E. (1988) Time discrete Taylor approximations for Ito processes with jump component, *Mathematische Nachrichten* 138:93–104.
266. Milstein, G.N. (1974) Approximate integration of stochastic differential equations, *Theory of Probability and its Application* 19:557–562.
267. Milstein, G.N. (1978) A method of second-order accuracy integration of stochastic differential equations, *Theory of Probability and its Application* 19:557–562.
268. Miltersen, K.R., Sandmann, K., and Sondermann, D. (1997) Closed-form solutions for term structure derivatives with lognormal interest rates, *Journal of Finance* 52:409–430.
269. Mintz, D. (1998) Less is more, pp.69–74 in *Hedging with Trees*, M. Broadie and P. Glasserman, eds., Risk Publications, London.
270. Morland, W. (2001) *Valuation of Consumer Demand Deposits with High-Dimensional Sobol Sequences*, Masters thesis, Computer Science Department, University of Waterloo, Ontario, Canada.
271. Moro, B. (1995) The full monte, *Risk* 8(Feb):57–58.
272. Morokoff, W.J., and Caflisch, R.E. (1995) Quasi-Monte Carlo integration, *Journal of Computational Physics* 122:218–230.
273. Moskowitz, B., and Caflisch, R.E. (1996) Smoothness and dimension reduction in quasi-Monte Carlo methods, *Mathematical and Computer Modelling* 23:37–54.
274. Musiela, M., and Rutkowski, M. (1997) Continuous-time term structure models: forward measure approach, *Finance and Stochastics* 1:261–292.
275. Musiela, M., and Rutkowski, M. (1997) *Martingale Methods in Financial Modeling*, Springer-Verlag, New York.
276. Nelsen, R.B. (1999) *An Introduction to Copulas*, Springer-Verlag, New York.
277. Nelson, B.L. (1990) Control variate remedies, *Operations Research* 38:974–992.
278. Newton, N.J. (1994) Variance reduction for simulated diffusions, *SIAM Journal on Applied Mathematics* 51:542–567.
279. Newton, N.J. (1997) Continuous-time Monte Carlo methods and variance reduction, in *Numerical Methods in Finance*, L.C.G. Rogers and D. Talay, eds., Cambridge University Press, 22–42.
280. Niederreiter, H. (1987) Point sets and sequences with low discrepancy, *Monatshefte für Mathematik* 104:273–337.
281. Niederreiter, H. (1992) *Random Number Generation and Quasi-Monte Carlo Methods*, Society for Industrial and Applied Mathematics, Philadelphia.
282. Niederreiter, H., and Xing, C. (1998) Nets, (t, s)-sequences, and algebraic geometry, pp.267–302 in *Random and Quasi-Random Point Sets*, P. Hellekalek and G. Larcher, eds., Springer-Verlag, Berlin.

283. Ninomiya, S., and Tezuka, S. (1996) Toward real-time pricing of complex financial derivatives, *Applied Mathematical Finance* 3:1–20.

284. Øksendal, B. (1998) *Stochastic Differential Equations: An Introduction with Applications*, Fifth Edition, Springer-Verlag, Berlin.

285. Owen, A.B. (1992) A central limit theorem for Latin hypercube sampling, *Journal of the Royal Statistical Society, Series B* 54:541–551.

286. Owen, A.B. (1994) Lattice sampling revisted; Monte Carlo variance of means over randomized orthogonal arrays, *Annals of Statistics* 22:930–945.

287. Owen, A.B. (1995) Randomly permuted (t, m, s)-nets and (t, s)-sequences, pp.299–317 in *Monte Carlo and Quasi-Monte Carlo Methods in Scientific Computing*, H. Niederreiter and J.-S. Shiue, eds., Springer-Verlag, New York.

288. Owen, A.B. (1997) Monte Carlo variance of scrambled net quadrature, *SIAM Journal on Numerical Analysis* 34:1884–1910.

289. Owen, A.B. (1997) Scrambled net variance for integrals of smooth functions, *Annals of Statistics* 25:1541–1562.

290. Owen, A.B. (1998) Scrambling Sobol' and Niederreiter-Xing points, *Journal of Complexity* 14:466–489.

291. Owen, A.B. (2002) Necessity of low effective dimension, working paper, Statistics Department, Stanford University, Stanford, California.

292. Owen, A.B., and Zhou, Y. (2000) Advances in Importance Sampling, pp.53–66 in *Computational Finance 1999 (Proceedings of the Sixth International Conference on Computational Finance)*, Y.S. Abu-Mostafa, B. LeBaron, A.W. Lo, and A.S. Weigend, eds., MIT Press, Cambridge, Mass.

293. Papageorgiou, A., and Traub, J. (1996) Beating Monte Carlo, *Risk* 9(June):63–65.

294. Park, S.K., and Miller, K.W. (1988) Random number generators: good ones are hard to find, *Communications of the ACM* 31:1192–1201.

295. Paskov, S. (1997) New methodologies for valuing derivatives, pp.545–582 in *Mathematics of Derivative Securities*, S. Pliska and M. Dempster, eds., Cambridge University Press, Cambridge, UK.

296. Paskov, S., and Traub, J. (1995) Faster valuation of financial derivatives, *Journal of Portfolio Management* 22:113–120.

297. Picoult, E. (1999) Calculating value-at-risk with Monte Carlo simulation, pp.209–229 in *Monte Carlo: Methodologies and Applications for Pricing and Risk Management*, B. Dupire, ed., Risk Publications, London.

298. Pirsic, G. (2002) A software implementation of Niederreiter-Xing sequences, pp.434–445 in *Monte Carlo and Quasi-Monte Carlo Methods 2000*, K.-T. Fang, F.J. Hickernell, and H. Niederreiter, eds., Springer-Verlag, Berlin.

299. Press, W.H., Teukolsky, S.A., Vetterling, W.T., and Flannery, B.P. (1992) *Numerical Recipes in C*, Second Edition, Cambridge University Press, Cambridge, UK.

300. Protter, P. (1990) *Stochastic Integration and Differential Equations*, Springer-Verlag, Berlin.

301. Protter, P., and Talay, D. (1997) The Euler scheme for Lévy driven stochastic differential equations, *Annals of Probability* 25:393–423.

302. Rachev, S., and Mittnik, S. (2000) *Stable Paretian Models in Finance*, Wiley, Chichester, England.

303. Rebonato, R. (2000) *Volatility and Correlation In the Pricing of Equity, FX and Interest-Rate Options*, Wiley, Chichester, England.

304. Rebonato, R. (2002) *Modern Pricing of Interest-Rate Derivatives: The LIBOR Market Model and Beyond*, Princeton University Press, Princeton, New Jersey.

305. Reiman, M., and Weiss, A. (1989) Sensitivity analysis for simulations via likelihood ratios, *Operations Research* 37:830–844.

306. Revuz, D., and Yor, M. (1999) *Continuous Martingales and Brownian Motion*, Third Edition, Springer-Verlag, Berlin.

307. Rogers, L.C.G. (1995) Which model for the term structure of interest rates should one use?, pp.93–116 in *Mathematical Finance*, M.H.A. Davis, D. Duffie, and I. Karatzas, eds., IMA Vol. 65, Springer-Verlag, New York.

308. Rogers, L.C.G. (2002) Monte Carlo valuation of American options, *Mathematical Finance* 12:271–286.

309. Ross, R. (1998) Good point methods for computing prices and sensitivities of multi-asset European style options, *Applied Mathematical Finance* 5:83–106.

310. Rouvinez, C. (1997) Going Greek with VAR, *Risk* 10(February):57–65.

311. Rubinstein, R. (1989) Sensitivity analysis and performance extrapolation for computer simulation models, *Operations Research* 37:72–81.

312. Rubinstein, R.Y., Samorodnitsky, G., and Shaked, M. (1985) Antithetic variates, multivariate dependence and simulation of stochastic systems, *Management Science* 31:66–77.

313. Rubinstein, R.Y., and Shapiro, A. (1993) *Discrete Event Systems: Sensitivity Analysis and Stochastic Optimization*, Wiley, New York.

314. Rust, J. (1997) Using randomization to break the curse of dimensionality, *Econometrica* 65:487–516.

315. Sadowsky, J.S. (1993) On the optimality and stability of exponential twisting in Monte Carlo estimation, *IEEE Transactions on Information Theory* 39:119–128.

316. Samorodnitsky, G., and Taqqu, M.S. (1994) *Stable Non-Gaussian Random Processes*, Chapman & Hall, London.

317. Sato, K.-I. (1999) *Lévy Processes and Infinitely Divisible Distributions*, Cambridge University Press, Cambridge, England.

318. Schmeiser, B.W., Taaffe, M.R., and Wang, J. (2001) Biased control-variate estimation, *IIE Transactions* 33:219–228.

319. Schoenmakers, J.G., and Heemink, A.W. (1997) Fast valuation of financial derivatives, *Journal of Computational Finance* 1:47–62.

320. Schönbucher, P.J. (2002) Taken to the limit: simple and not-so-simple loan loss distributions, working paper, Department of Statistics, University of Bonn, Bonn, Germany.

321. Schönbucher, P.J., and Schubert, D. (2001) Copula-dependent default risk in intensity models, working paper, Department of Statistics, University of Bonn, Bonn, Germany.

322. Schroder, M. (1989) Computing the constant elasticity of variance option pricing formula, *Journal of Finance* 44:211–219.

323. Schruben, L.W., and Margolin, B.H. (1978) Pseudorandom number assignment in statistically designed simulation and distribution sampling experiments, *Journal of the American Statistical Association* 73:504–520.

324. Scott, L.O. (1996) Simulating a multi-factor term structure model over relatively long discrete time periods, in *Proceedings of the IAFE First Annual Computational Finance Conference*, Graduate School of Business, Stanford University.

325. Seber, G.A.F. (1984) *Multivariate Observations*, Wiley, New York.

326. Serfling, R.J. (1980) *Approximation Theorems of Mathematical Statistics*, Wiley, New York.

327. Shahabuddin, P. (1994) Importance sampling for the simulation of highly reliable Markovian systems, *Management Science* 40:333–352.

328. Shahabuddin, P. (2001) Rare event simulation, pp.163–178 in *Performability Modeling: Techniques and Tools*, B.R. Haverkort, R. Marie, K. Trivedi and G. Rubino, eds., Wiley, New York.

329. Shaw, J. (1999) Beyond VAR and stress testing, pp.231-244 in *Monte Carlo: Methodologies and Applications for Pricing and Risk Management*, B. Dupire, ed., Risk Publications, London.

330. Sidenius, J. (2000) LIBOR market models in practice, *Journal of Computational Finance* 3(Spring):5–26.

331. Siegmund, D. (1976) Importance sampling in the Monte Carlo study of sequential tests, *Annals of Statistics* 4:673–684.

332. Sloan, I.H. (2002) QMC integration — beating intractability by weighting the coordinate directions, pp.103–123 in *Monte Carlo and Quasi-Monte Carlo Methods 2000*, K.-T. Fang, F.J. Hickernell, and H. Niederreiter, eds., Springer-Verlag, Berlin.

333. Sloan, I.H., and Joe, S. (1994) *Lattice Methods for Multiple Integration*, Oxford University Press, Oxford, UK.

334. Sloan, I.H., and Wózniakowski, H. (1998) When are quasi-Monte Carlo algorithms efficient for high dimensional integrals? *Journal of Complexity* 14:1–33.

335. Sobol', I.M. (1967) On the distribution of points in a cube and the approximate evaluation of integrals, *USSR Journal of Computational Mathematics and Mathematical Physics* (English translation) 7:784–802.

336. Sobol', I.M. (1976) Uniformly distributed sequences with an additional uniform property, *USSR Journal of Computational Mathematics and Mathematical Physics* (English translation) 16:1332–1337.

337. Stein, M. (1987) Large sample properties of simulations using Latin hypercube sampling, *Technometrics* 29:143–151 (correction 32:367).

338. Studer, M. (2001) *Stochastic Taylor Expansions and Saddlepoint Approximations for Risk Management*, Dissertation 14242, Mathematics Department, ETH Zürich.

339. Suri, R., and Zazanis, M. (1988) Perturbation analysis gives strongly consistent sensitivity estimates for the M/G/1 queue, *Management Science* 34:39–64.

340. Talay, D. (1982) How to discretize stochastic differential equations, pp.276–292 in *Lecture Notes in Mathematics, Vol. 972*, Springer-Verlag, Berlin.

341. Talay, D. (1984) Efficient numerical schemes for the approximation of expectations of functionals of the solutions of a s.d.e., and applications, pp.294–313 in *Lecture Notes in Control and Information Sciences, Vol. 61*, Springer-Verlag, Berlin.

342. Talay, D. (1995) Simulation and numerical analysis of stochastic differential systems: a review, pp.63–106 in *Probabilistic Methods in Applied Physics*, P. Kree and W. Wedig, eds., Springer-Verlag, Berlin.

343. Talay, D., and Tubaro, L. (1990) Expansion of the global error for numerical schemes solving stochastic differential equations, *Stochastic Analysis and Applications* 8:483–509.

344. Talay, D., and Zheng, Z. (2003) Quantiles of the Euler scheme for diffusion processes and financial applications, *Mathematical Finance* 13:187–199.

345. Tan, K.S., and Boyle, P.P. (2000) Application of randomized low discrepancy sequences to the valuation of complex securities, *Journal of Economic Dynamics and Control* 24:1747–1782.

346. Tausworthe, R.C. (1965) Random numbers generated by linear recurrence modulo two, *Mathematics of Computation* 19:201–209.

347. Tezuka, S. (1993) Polynomial arithmetic analogue of Halton sequences, *ACM Transactions on Modeling and Computer Simulation* 3:99–107.

348. Tezuka, S. (1994) A generalization of Faure sequences and its efficient implementation, Research Report RT0105, IBM Tokyo Research Laboratory, Japan.

349. Tezuka, S., and Tokuyama, T. (1994) A note on polynomial arithmetic analogue of Halton sequences, *ACM Transactions on Modeling and Computer Simulation* 4:279–284.

350. Tsitsiklis, J., and Van Roy, B. (1999) Optimal stopping of Markov processes: Hilbert space theory, approximation algorithms, and an application to pricing high-dimensional financial derivatives, *IEEE Transactions on Automatic Control* 44:1840–1851.

351. Tsitsiklis, J., and Van Roy, B. (2001) Regression methods for pricing complex American-style options, *IEEE Transactions on Neural Networks* 12:694–703.

352. Vasicek, O.A. (1977) An equilibrium characterization of the term structure, *Journal of Financial Economics* 5:177-188.

353. Von Neumann, J. (1951) Various techniques used in connection with random digits, *Applied Mathematics Series*, 12, National Bureau of Standards, Washington, D.C.

354. Wakeman, L. (1999) Credit enhancement, pp.255–275 in *Risk Management and Analysis, Volume 1*, C. Alexander, ed., Wiley, Chichester, England.

355. Wichmann, B.A, and Hill, I.D. (1982) An efficient and portable pseudorandom number generator, *Applied Statistics* 31:188–190 (correction 33:123 (1984)).

356. Wiktorsson, M. (2001) Joint characteristic function and simultaneous simulation of iterated Ito integrals for multiple independent Brownian motions, *Annals of Applied Probability* 11:470-487.

357. Yan, L. (2002) The Euler scheme with irregular coefficients, *Annals of Probability* 30:1172-1222.

358. Zazanis, M.A. (1987) *Statistical Properties of Perturbation Analysis Estimates for Discrete Event Systems*, doctoral dissertation, Division of Applied Sciences, Harvard University, Cambridge, Mass.

Index

absolute continuity, 259, 407–408, 553
acceptance-rejection method, 58–63, 66, 125
  and quasi-Monte Carlo, 283
  for beta distribution, 59
  for conditional sampling, 61
  for gamma distribution, 126
  for normal distribution, 60
  in stratified sampling, 214, 502
adapted process, 545
affine model, 114, 118, 128, 524, 526
American option, 4, 421–479
  and nested conditional expectations, 15
  dual formulation, 471
  high bias, 427, 433, 446–447, 472
  in binomial lattice, 231, 462
  interleaving estimator, 449, 461
  low bias, 427, 428, 435–436, 443, 447–448, 450, 462
  parametric approximations, 426–430
  parametric value function, 430
  quantization, 441
  random tree methods, 430–441, 478
  regression-based methods, 430, 459–470, 478
  state-space partitioning, 441–443
  stochastic mesh methods, 443–459, 465–470, 478
  supermartingale property, 471
antithetics, 205–209, 275, 277, 361
  condition for variance reduction, 207
  confidence interval, 206
  in derivative estimation, 386
  in quantile estimation, 490
  in random tree method, 439
  variance decomposition, 207
  with acceptance-rejection, 63
  with the normal distribution, 205
arbitrage
  and discretization error, 159, 374
  and matching underlying assets, 245
  definition, 26
  from parallel shift, 153
  restrictions and control variates, 188, 191
  restrictions in HJM framework, 151
  restrictions in LIBOR market model, 169
arcsine law, 56, 59
Asian option, 8
  continuously monitored, 99
  control variates for, 189
  delta, 389
  discretely monitored, 99
  likelihood ratio method, 404
  on geometric average, 99, 189, 324
  using Latin hypercube sampling, 242
  vega, 390, 409
  with importance sampling, 270
associated random variables, 207
asymptotically optimal importance sampling, 264, 270, 499–500, 517, 531

b-ary expansion, 285
  algorithm, 296

barrier option, 100
  and conditional Monte Carlo, 267
  and default, 521
  and Koksma-Hlawka bound, 290
  discretization error, 368–370, 521
  likelihood ratio method, 405
  on multiple assets, 105
  pathwise method, 393, 395, 400
  random computing time, 11
  using Latin hypercube sampling, 242
  using quasi-Monte Carlo, 334
  with importance sampling, 264
base-b expansion, 285
batching, 543
  and control variates, 202
  and moment matching, 246
  and stratified sampling, 216
Beasley-Springer-Moro approximation,
  67
Bermudan option, 423
Bessel process, 132, 373
beta distribution, 59, 229
bias, 5
  from discretization, 339–376
  in American option pricing, 427–428,
    433, 435–436, 443, 446–448, 450,
    462, 472
  in control variates, 191, 200, 278
  in finite-difference estimators,
    378–379
  in likelihood ratio method, 407
  in moment matching, 245
  in pathwise method, 393
  in ratio estimators, 234
  sources of, 12–16
bias-variance tradeoff, 16–19, 365–366,
    381–383, 456
binomial lattice, 230
  Latin hypercube sampling, 238
  simulating paths, 231
  terminal stratification, 231–232
Black formula, 566, 567
  for cap, 173
Black-Scholes formula, 5
  as risk-neutral expectation, 31
  as solution to PDE, 25
  for geometric average option, 100
  in jump-diffusion option pricing, 137
  with dividends, 32

Black-Scholes PDE, 25
bonds, 560
  as control variates, 190
  as numeraire, 34, 116, 131, 154, 171
  in CIR model, 128
  in Gaussian short rate models,
    111–118
  in HJM framework, 151, 163
  in LIBOR market model, 166
  in LIBOR market model simulation,
    177
  in Vasicek model, 113–114
  second-order discretization, 358
  subject to default, 520
Box-Muller method, 65
  and radial stratification, 227
Brownian bridge, 83
  maximum of, 56, 367
Brownian bridge construction, 82–86
  algorithm, 84
  and quasi-Monte Carlo, 333–334, 337
  and stratified sampling, 221
  continuous limit, 89
  in multiple dimensions, 91
Brownian motion
  covariance matrix, 82
  definition, 79
  Latin hypercube sampling of, 238
  maximum of, 56, 360, 367
  multivariate, 90
  stratified sampling of, 221

cap, 274, 416, 565–566
  and calibration, 180
  delta, 399
  discretization error, 371
  in HJM framework, 163
  in LIBOR market model, 172, 180
  with importance sampling, 275
  with stratified sampling, 225
caplet, 565
central limit theorem, 9, 541–544
  contrasted with Koksma-Hlawka
    bound, 289
  for finite-difference estimators, 382
  for poststratified estimators, 235
  for quantile estimator, 490
  for ratio estimators, 234
  using delta method, 204

with antithetic sampling, 206
with control variates, 196
with discretization error, 366
with Latin hypercube sampling, 241
with nonlinear controls, 204
with random number of replications, 12, 541
with stratified sampling, 216
central-difference estimator, 378
change of measure, 255, 553–557
    and conditional expectation, 556
    and risk measurement, 483, 521
    as change of drift, 557
    as change of intensity, 523
    for risk-neutral pricing, 28
    in American option pricing, 450
    in heavy-tailed model, 279, 515–516, 537
    in HJM framework, 154
    in LIBOR market model, 171
    in Vasicek model, 117
    through change of numeraire, 33
    through delta-gamma approximation, 495
    through geometric Brownian motion, 106
    through weighted Monte Carlo, 253
characteristic function
    delta-gamma approximation, 487
    indirect delta-gamma approximation, 513
    inversion integral, 488
    stable laws, 148
Chebyshev's inequality, 289, 543
chi-square χ_ν^2, 122, 125, 227, 392, 509
    moment generating function, 514
Cholesky factorization, 72, 486
    for Brownian motion, 82
CIR model, 120, 128, 524
    delta, 392
combined random number generators, 50
commutativity condition, 353
conditional excess, 483, 506, 519
conditional Monte Carlo, 279, 369, 399
    for barrier options, 267
conditional sampling
    in stratified sampling, 211, 214, 502–504

using acceptance-rejection, 61
using inverse transform, 57
confidence interval, 6, 541–544
    combining high and low estimators, 431, 434, 437
    for American option, 431, 434, 437
    through batching, 216, 242, 246, 543
    with antithetic sampling, 206
    with control variates, 196, 198
    with Latin hypercube sampling, 241
    with moment matching, 246
    with stratified sampling, 216
constant diffusion transformation, 373
constant elasticity of variance (CEV), 133
continuation values, 426
control variates, 185–205, 277, 420, 440
    and moment matching, 245, 246, 249
    and price sensitivities, 418
    and weighted Monte Carlo, 199, 254
    biased, 278
    compared with stratified sampling, 220
    confidence intervals, 195–196, 198
    delta estimation, 417
    delta-gamma approximation, 493
    for quantile estimation, 493–494
    in stochastic mesh, 457
    loss factor, 201
    optimal coefficient, 186, 197
    score function, 411
    variance decomposition, 198
convergence
    modes of, 539–541
convergence in distribution, 540
convergence order, 344–348
copula, 511, 527, 529
credit rating, 524
credit risk, 520–535
    variance reduction techniques, 529–535
cumulant generating function
    after change of measure, 501
    conditional, 532
    definition, 260
    figure, 264, 265, 498
    of delta-gamma approximation, 487
    of indirect delta-gamma approximation, 515

of portfolio credit losses, 531

default, 121, 520
  and LIBOR, 166
default intensity, 522–524
deflator, 26
delta, 23, 377, 485
  Black-Scholes, 388, 403
  cap, 399
  finite-difference estimators, 383
  likelihood ratio method, 403, 404,
    406, 414
  mixed estimator, 417
  numerical comparison, 412
  path-dependent options, 389, 404
  pathwise, 388, 389, 391
  square-root diffusion, 391
  stochastic volatility, 414
delta hedging, 23, 377
  and control variates, 194
delta method, 203, 234
delta-gamma approximation, 485–488
  and importance sampling, 495–500,
    515–517
  as a control variate, 493–494
  cumulant generating function, 487
  in heavy-tailed model, 512–514
  indirect, 513–515
  with stratified sampling, 500–504,
    517–518
depth-first processing, 437
digital nets, 314
dimension (of integration problem), 3,
    62, 282, 285, 324, 327, 337
discounting, 4, 559
  by stochastic discount factor, 26
  in LIBOR market model, 168
  in stochastic short rate model, 108
discrepancy, 283–285
  extreme, 284
  isotropic, 284, 290
  L_2, 290
  star, 284
discretization error, 8, 110, 115, 159,
    339–376
  and change of variables, 371–375
  and control variates, 191
  and likelihood ratio method, 414
  and path adjustment, 250

and path-dependence, 357–360,
    366–370
  and pathwise method, 396
  as example of bias, 13
  bias-variance tradeoff, 365–366
  in barrier options, 368–370
  in cap deltas, 399
  in caplet pricing, 180
  in discounted bonds, 157
  in forward curve, 155
  in jump-diffusion processes, 363–364
  in LIBOR market model, 174
  in option payoff, 14
  in quantile estimation, 536
  second-order methods, 348–357
dividends, 31, 96
  and American options, 423, 469
dual formulation of American option
    pricing, 470–478
  connection with regression, 476–478
  optimal martingale, 472
dynamic financial analysis, 537
dynamic programming, 424, 426

empirical martingale method, 245
equivalent martingale measure, 28
error function (Erf), 70
Euler approximation, 7, 81, 121,
    339–340
  and likelihood ratio method, 414
  and pathwise method, 396
  convergence order, 345–347
  deterministic volatility function, 103
  in LIBOR market model, 175
  in square-root diffusion, 124
  in Vasicek model, 110
exercise region, 422, 425
  for max option, 464
  parametric approximation, 426
experimental design, 385
exponential family, exponential twisting,
    exponential change of measure,
    260, 262, 264, 278, 407
  and default indicators, 530
  and delta-gamma approximation,
    495, 498
  and indirect delta-gamma, 515
exponential tail, 508
extreme value theory, 536

factor loading, 75
Faure generator matrices, 298
  algorithm, 301
Faure net, 300
  cyclic property, 300, 331
Faure sequence, 297–303
  algorithm, 302
  as (t, d)-sequence, 300
  implementation, 300–303
  numerical examples, 325–330
  plateaus, 328
  starting point, 302, 325, 327, 329
filtration, 545
finite-difference estimators, 378–386
  bias-variance tradeoff, 381–383
forward measure
  defined, 34
  in CIR model, 131
  in HJM framework, 154, 159
  in LIBOR market model, 171–172
  in Vasicek model, 117–118
forward price, 98
  as input to simulation, 102
forward rate, 560–562
  continuous compounding, 149
  factors, 273
  simple compounding, 165
forward-difference estimator, 378
Fundamental Theorem of Asset Pricing,
    27
futures price, 97, 564

gamma, 378, 485
  central-difference estimator, 384
  likelihood ratio method, 411–413
  pathwise method, 392
gamma distribution, 125–127, 143, 508,
    516
  exponential family, 261
  sampling algorithm, 126, 127
gamma process, 143–144
Gaussian short rate models, 108–120
  multifactor, 118
generalized Faure sequences, 316, 323
generalized feedback shift register
    methods, 52
generalized Niederreiter sequences, 316
geometric Brownian motion, 93–107
  as numeraire, 106

derivation of SDE, 93
Girsanov theorem, 29, 35, 37, 107, 117,
    131, 155, 171, 409, 523, 555–557
Gray code, 307, 313
  in base b, 308
Greeks, 377

Halton sequence, 293–297
  discrepancy, 294
  implementation, 296–297
  in high dimensions, 295
  leaped, 295
Hammersley points, 294
Hardy-Krause variation, 287
  of an indicator function, 290
Heath-Jarrow-Morton (HJM) frame-
    work, 150–155
  discretized, 374
  simulation algorithm, 160–162
  with importance sampling, 273–276
  with path adjustment, 250
heavy-tailed distributions, 148, 279,
    506–511
Heston model, 121
  likelihood ratio method, 414
  pathwise method, 392
  second-order discretization, 356
high estimator, 432–434, 446–447, 472
  and duality, 478
HJM drift
  and control variates, 191
  discretized, 158, 160
  forward measure, 155
  risk-neutral measure, 152
Ho-Lee model, 109, 111
  in HJM framework, 154
hyperbolic model, 146

importance sampling, 255–276, 278
  and likelihood ratio method, 407
  and stochastic mesh methods, 450
  asymptotic optimality, 264, 270,
    499–500, 517
  combined with stratified sampling,
    271–273, 500–504, 517–518
  deterministic change of drift, 267
  for credit risk, 529–535
  for knock-in option, 264
  for path-dependent options, 267–271

optimal drift, 269, 274
optimal path, 269–271
  through log-linear approximation,
    268–269
  using delta-gamma approximation,
    495–500, 515–517
  with heavy-tailed distributions, 279,
    515–516, 537
  zero-variance estimator, 256
independent-path construction, 444
infinitely divisible distribution, 143–145
  stable distribution, 148
insurance risk, 262, 279, 537
intensity, 140, 141, 364, 522, 523
interleaving estimator, 449, 461
inverse Gaussian distribution, 144
  sampling algorithm, 146
inverse transform method, 54–58, 67,
    127, 128, 231, 367, 490, 502, 511,
    524
  and antithetics, 205
  and Latin hypercube sampling, 237
  and stratified sampling, 212
  avoiding zero, 294
  for conditional sampling, 57
  for Poisson distribution, 128
  in quasi-Monte Carlo, 331
inversive congruential generator, 52
Itô's formula, 545–547
  in operator notation, 348
Itô-Taylor expansions, 362–363

jump-diffusion process, 134, 524
  discretization, 363–364

Karhounen-Loève expansion, 89
Koksma-Hlawka inequality, 288
  generalizations, 289
Korobov rules, 317, 329
kurtosis, 134, 507

Lévy area, 344
Lévy construction of Brownian motion,
    90
Lévy process, 142–149, 364
Latin hypercube sampling, 236–243, 278
  and normal copula, 527
  confidence intervals for, 241
  in a binomial lattice, 238

in quantile estimation, 490
in random tree method, 439
of Brownian paths, 238
variance reduction, 240
lattice rules, 316–320, 325
  extensible, 319
  integration error, 318–319
  Korobov, 317
  rank, 316
LIBOR, 166, 561
LIBOR market model, 166–174
  Bermudan swaption, 429
  commutativity condition, 354
  control variate for, 192
  discretization error, 364, 371
  likelihood ratio method, 415–416
  pathwise method, 398–399
  stratified sampling, 224
  transition density, 415, 457
likelihood ratio, 256–259, 553
  for change of mean and covariance in
    normal family, 497
  for change of mean in normal family,
    260, 533
  for change of numeraire, 33
  for conditional process, 369
  for delta-gamma approximation, 496
  for increased default probability, 530
  for indirect delta-gamma, 515
  geometric Brownian motion, 106
  in Vasicek model, 117
  over long horizon, 259, 409
  over random horizon, 258
  relating objective and risk-neutral
    measures, 28
  skewness, 260
  weights in stochastic mesh, 450–456,
    466, 468
likelihood ratio method, 401–418
  and regression, 418
  combined with pathwise method, 412,
    416–417
  for barrier option, 405
  for general diffusion processes,
    413–415
  for LIBOR market model, 415–416
  for square-root diffusion, 406
  gamma, 411–413
  limitations, 407

variance, 407–411
vega, 405
with stochastic volatility, 414
linear congruential generator, 40, 43–49, 317
local martingale, 550
lognormal distribution, 94, 508
  moments, 95
lookback option, 101
  discretization error, 360, 367
loss factor (for control variates), 201
low discrepancy, 285
low estimator, 428, 434–437, 443, 447–448, 450, 462

Marsaglia-Bray method, 66
martingale, 550–553
  and duality, 471
  and moment matching, 245
  and stochastic intensity, 522
  control variate, 188
  deflated bond, 169
  discounted asset price, 26, 29
  discounted bond price, 151
  discrete discounted bond prices, 156
  exponential, 117, 151, 552
  from approximate value function, 473
  from Poisson process, 137
  from stochastic integral, 551
  from stopping rule, 474
  futures price, 98
  geometric Brownian motion, 96
  local, 550
  optimal, 472
  score, 410
martingale discretization, 158, 160, 374–375
  in LIBOR market model, 176–180
martingale representation theorem, 36, 522, 552
Milstein schemes, 343, 347, 351
mixed (iterated) Brownian integrals, 344
mod operation, 41
moment generating function
  chi-square distribution, 514
  delta-gamma approximation, 487
  failure with heavy tails, 512
  normal distribution, 65, 95

portfolio loss, 530
moment matching, 244–254, 277
  compared with control variates, 246
multifactor Gaussian models, 118
multiple recursive generator (MRG), 50
  in C, 51

$N(\mu, \sigma^2)$, 63
$N(\mu, \Sigma)$, 64
net, 291
Niederreiter sequences, 314
Niederreiter-Xing sequences, 316
noncentral chi-square $\chi_\nu'^2(\lambda)$, 122, 123, 130, 133, 392, 406, 414, 487
nonlinear control variates, 202–205
normal distribution, 63–77
  acceptance-rejection method, 60
  conditioning formula, 65
  copula, 527
  for change of variables, 372
  inverse of, 67
  mixture, 507, 509
  multivariate, 64, 71
  numerical approximation, 69
  sampling methods, 60, 65–69
normal inverse Gaussian process, 144–147
numeraire, 19, 32, 255
  and importance sampling, 267
  bond in HJM framework, 154, 160
  bond in LIBOR market model, 171
  for spot measure, 169
  geometric Brownian motion, 106
  in CIR model, 131
  in Gaussian short rate model, 116

obligor, 520
optimal allocation of samples, 217–218
Ornstein-Uhlenbeck process, 108, 132

parametric value function, 430
pathwise method, 386–401
  combined with likelihood ratio method, 412, 416–417
  conditions for unbiasedness, 393–396
  for general diffusion processes, 396–397
  for LIBOR market model, 398–399
  limitations, 392, 396

smoothing, 399–401
Poisson distribution, 123, 127–128
  exponential family, 261
  inverse transform method, 128
Poisson process, 56, 136, 138, 363, 524
  and stratified sampling, 228–229
  inhomogeneous, 140–141
  of insurance claims, 262
poststratification, 232–235
  and control variates, 494
  and quantile estimation, 494
  asymptotic variance, 235
  compared with stratified sampling,
    235
  of delta-gamma approximation, 504
predictor-corrector method, 372, 373
pricing kernel, 26
primitive polynomials
  defined, 304
  tables of, 305
principal components, 74
  and value-at-risk, 492
  of discretely observed Brownian
    motion, 88
  of forward rates, 184
  optimality of, 75
principal components construction,
    86–89
  and quasi-Monte Carlo, 334, 337
  in multiple dimensions, 92
put-call parity, 245

quantile (fractile, percentile), 55, 347,
    483
  control variates for, 493–494
  discretization error, 536
  estimation, 279, 490–491
  estimation with importance sampling,
    500, 519
  in stratified sampling, 212
quanto options, 105
quasi-Monte Carlo, 281–337
  and Brownian bridge construction,
    333
  and principal components, 334
  and stratified sampling, 333
  integration, 282
  numerical comparisons, 323–330

radial stratification, 227
radical inverse, 286, 315
Radon-Nikodym derivative, 28, 33, 256,
    553
Radon-Nikodym theorem, 553
random permutation, 237
random tree methods, 430–441, 478
  depth-first processing, 437
  pruning, 438
  variance reduction, 439
random walk construction of Brownian
    motion, 81, 91
randomized quasi-Monte Carlo, 320–323
ratio estimator
  as example of bias, 12
  central limit theorem for, 234
ratio-of-uniforms method, 125
Rayleigh distribution, 56, 367
regression
  and control variates, 187, 197, 200
  and likelihood ratio method, 418
  and price sensitivities, 418
  and weighted Monte Carlo, 200, 254
  in American option pricing, 459–470
  residuals and duality, 476
Richardson extrapolation, 360, 364, 375
risk management, 481–492, 520–535
risk-neutral measure, 4
  defined, 28
risk-neutral pricing, 27
Romberg extrapolation, 360
Runge-Kutta methods, 351

score function, 403–406, 408, 419
  as control variate, 411
  generalized for second derivatives,
    411
  variance increase, 410
scrambled nets, 322
second-order discretization, 348–357
  multidimensional, 351–357
  of Heston model, 356
  simplified, 355–357
seed, random number, 41, 50
self-financing trading strategy, 22
simple interest, 165, 559
Sobol' generator matrices, 304
  algorithm, 313
Sobol' sequence, 303–314

algorithm, 314
as (t,d)-sequence, 312
discrepancy, 312
implementation, 313–314
initialization, 309–312
numerical examples, 325–330
quality of coordinates, 312, 331
starting point, 325, 327, 329
Sobol's Property A, 309
spectral test, 48, 319
spot measure, 168–171
spread option, 105
square-root diffusion, 120–134
    algorithm, 124
    and CEV process, 133
    as model of default intensity, 523
    degrees of freedom, 122, 133
    delta, 391, 406
    discretization, 121, 356, 373
    existence and uniqueness, 548
    likelihood ratio method, 406
    noncentrality parameter, 122
    pathwise method, 391
    stationary distribution, 122
squared Gaussian models, 132
stable laws, 147, 508
stable processes, 147–149
star discrepancy, 284
stochastic differential equation (SDE),
        339, 548–550
    existence and uniqueness, 548
    for general system of assets, 21
    for geometric Brownian motion, 4
stochastic mesh methods, 443–459,
        465–470, 478
    computational costs, 456, 466, 478
    high estimator, 446–447
    independent-path construction, 444
    likelihood ratio weights, 450–456
    low estimator, 447–448
    with regression weights, 459, 465
stochastic volatility, 121, 278, 356, 392,
        400
    and control variate, 192
stopping rules (for American options),
        425, 428, 443, 447–448, 450, 474
stratified sampling, 209–235, 237, 277
    and American option pricing, 443

and delta-gamma approximation,
        500–504, 517–518
    and quasi-Monte Carlo, 333, 335
    compared with control variates, 220
    confidence intervals for, 216
    in stochastic mesh, 454
    of Brownian motion, 221
    optimal allocation, 217–218
    optimal directions, 226, 271
    variance reduction, 218–220
strong error criterion, 344
subordinator, 143
swap, 164, 562
    in LIBOR market model, 173
swaption, 566–567
    and calibration, 180
    Bermudan, 429
    in HJM framework, 164
    in LIBOR market model, 173
    with stratified sampling, 225
systematic sampling, 208, 321

t distribution, 509–511
    copula, 511
    in confidence interval, 6, 543
    multivariate, 510
(t,d)-sequence, 291
terminal stratification, 220–222,
        230–232
thinning, 141, 364
(t,m,d)-net, 291
transition density, 415
    and score function, 410
    in stochastic mesh, 450–453, 457
    of geometric Brownian motion, 404,
        455
    of square-root diffusion, 121

uniform distribution, Unif[0,1], 54
uniform integrability, 16, 19, 393, 395
    definition, 541
unit hypercube, 214
    open versus closed, 282

value-at-risk (VAR), 483–492, 520
Van der Corput sequences, 285–287
variance gamma process, 144
Vasicek model, 108, 113
    in HJM framework, 154
    stationary distribution, 110

vega, 378
    Asian option, 409
    Black-Scholes, 389, 404
    likelihood ratio method, 404, 405
    path-dependent options, 390
    pathwise, 389, 390

weak convergence, 540
weak error criterion, 344

weighted Monte Carlo, 251–254, 277
    and American option pricing, 253,
      459
    and calibration, 252
    and control variates, 199–200, 254
    and price sensitivities, 418

yield, 560

Applications of Mathematics

(continued from page ii)

35 Kushner/Yin, **Stochastic Approximation and Recursive Algorithms and Applications,** Second Ed. (2003)

36 Musiela/Rutkowski, **Martingale Methods in Financial Modeling: Theory and Application** (1997)

37 Yin/Zhang, **Continuous-Time Markov Chains and Applications** (1998)

38 Dembo/Zeitouni, **Large Deviations Techniques and Applications,** Second Ed. (1998)

39 Karatzas/Shreve, **Methods of Mathematical Finance** (1998)

40 Fayolle/Iasnogorodski/Malyshev, **Random Walks in the Quarter Plane** (1999)

41 Aven/Jensen, **Stochastic Models in Reliability** (1999)

42 Hernández-Lerma/Lasserre, **Further Topics on Discrete-Time Markov Control Processes** (1999)

43 Yong/Zhou, **Stochastic Controls: Hamiltonian Systems and HJB Equations** (1999)

44 Serfozo, **Introduction to Stochastic Networks** (1999)

45 Steele, **Stochastic Calculus and Financial Applications** (2000)

46 Chen/Yao, **Fundamentals of Queueing Networks: Performance, Asymptotics, and Optimization** (2001)

47 Kushner, **Heavy Traffic Analysis of Controlled Queueing and Communication Networks** (2001)

48 Fernholz, **Stochastic Portfolio Theory** (2002)

49 Kabanov/Pergamenshchikov, **Two-Scale Stochastic Systems** (2003)

50 Han, **Information-Spectrum Methods in Information Theory** (2003)

51 Asmussen, **Applied Probability and Queues** (2003)

52 Robert, **Stochastic Networks and Queues** (2003)

53 Glasserman, **Monte Carlo Methods in Financial Engineering** (2003)

54 Sethi/Zhang/Zhang, **Average-Cost Control of Stochastic Manufacturing Systems** (2004)

55 Yin/Zhang, **Discrete-Time Markov Chains: Two-Time-Scale Methods and Applications** (2004)